Bill Piltzecker

BRADY

Chemistry of Hazardous Materials

Fourth Edition

Eugene Meyer

PEARSON

Prentice Hall

Upper Saddle River, New Jersey 07458

Library of Congress Cataloging-in-Publication Data

Meyer, Eugene
   Chemistry of hazardous materials / Eugene Meyer. — 4th ed.
       p. cm.
Includes index.
   ISBN 0-13-112760-8
   1. Hazardous substances—Fires and fire prevention.   2. Hazardous
substances. I. Title.
   TH9446.H38M48 2004
   628.9′2—dc22   2004000484

Publisher: *Julie Levin Alexander*
Publisher's Assistant: *Regina Bruno*
Senior Acquisitions Editor: *Katrin Beacom*
Assistant Editor: *Kierra Kashickey*
Senior Marketing Manager: *Katrin Beacom*
Channel Marketing Manager: *Rachele Strober*
Director of Production and Manufacturing: *Bruce Johnson*
Managing Editor for Production: *Patrick Walsh*
Manufacturing Manager: *Ilene Sanford*
Manufacturing Buyer: *Pat Brown*
Production Liaison: *Julie Li*
Production Editor: *Penny Walker, The GTS Companies/York, PA Campus*
Creative Director: *Cheryl Asherman*
Senior Design Coordinator: *Christopher Weigand*
Cover Designer: *Christopher Weigand*
Cover Photo: *Joaquim Pol Martinez/911 Pictures*
Composition: *The GTS Companies/York, PA Campus*
Printing and Binding: *Courier Westford*
Cover Printer: *Coral Graphics*

Pearson Prentice Hall™ is a trademark of Pearson Education, Inc.
Pearson® is a registered trademark of Pearson plc
Prentice Hall® is a registered trademark of Pearson Education, Inc.

Pearson Education LTD.
Pearson Education Singapore, Pte. Ltd
Pearson Education, Canada, Ltd
Pearson Education–Japan

Pearson Education Australia PTY, Limited
Pearson Education North Asia Ltd
Pearson Educación de Mexico, S.A. de C.V.
Pearson Education Malaysia, Pte. Ltd

10 9 8 7 6 5 4 3 2 1
ISBN 0-13-112760-8

Contents

Preface

Unforeseen events occurred in the world since the publication of the third edition of *Chemistry of Hazardous Materials*. They include the intentional use by terrorists of hazardous materials capable of killing or severely harming large segments of the civilized population. These traumatic incidents have caused emergency responders to address special ways of effectively reducing the impact of a terrorist act. For this reason, in this fourth edition, I introduce the hazardous materials likely to be encountered when terrorists use destructive materials. I identify these materials and the properties that cause them to be hazardous and suggest ways of effectively responding when they are encountered. I also exercise a certain degree of care when discussing them. For obvious reasons, I intentionally avoid reporting on the manners by which they can be produced.

As in earlier editions of this book, I continue to emphasize the hazardous materials regulations promulgated by the Occupational Safety and Health Administration, the U.S. Department of Transportation, and the Environmental Protection Agency. In this edition, I have updated the regulations to reflect changes that have occurred since publication of the third edition.

I have worked to make this fourth edition more comprehensive and easier for nonscientists to learn and understand. To do so, I crafted performance goals so students are apprised up front of what they should learn in each section. I have also listed the names of chemical substances under *each* formula in *every* equation so students can more readily comprehend the relevant chemical change. I also constructed new Solved Exercises and Review Exercises, and I expanded the glossary to include the definitions of new technical terms and phrases in use by emergency responders.

During the preparation of this book, I have considered the advice of several individuals. For the combination of their comments, I am extremely grateful. Sincerest thanks are due to the following individuals who, despite their heavy responsibilities and workloads, found the time to provide careful reviews and critiques of the entire manuscript or selected chapters thereof: John M. Eversole, Chicago Fire Department (retired), Chicago, Illinois; Gerald LaFlamme, Quinsigamond Community College, Worcester, Massachusetts, and Chief, Shrewsbury Fire Department, Shrewsbury, Massachusetts; Jeffrey T. Lindsey, Estero Fire Rescue, Estero, Florida; Chris Hawley, Baltimore County Fire Department, Baltimore, Maryland; Gary Kistner, San Antonio College, San Antonio, Texas; James F. Ross, Mercer County Community College, Trenton, New Jersey; and Donald L. Walsh, Chicago Fire Department, Chicago, Illinois. Special thanks are also due to Ms. Katrin Beacom, Senior Editor, and Ms. Kierra Kashickey, Editorial Assistant, Prentice Hall/Brady, for their assistance and input during preparation of the manuscript. A big thank you to the copy editor,

Ms. Kristin Landon, and the project manager, Ms. Penny Walker, whose tireless efforts converted the manuscript into this book.

Finally, as with the preceding editions, I extend an extra special thank you to my wife, Phyllis, for her critical review of the manuscript and her support throughout the hours needed to complete this project. Her constant love, never-ending encouragement, and patience have always influenced my writing. To her, I dedicate this fourth edition.

Eugene Meyer

Introduction

Emergency incidents caused by natural phenomena have repeatedly been mitigated by firefighters, police, and other emergency responders. Wild fires, flash floods, earthquakes, active volcanoes, droughts, and storms are examples of such incidents. These natural events are not entirely predictable, and they cannot yet be controlled or prevented.

In the 1940s, new types of emergency incidents began to surface. They were linked with the behavior of certain chemical, petroleum, and nuclear products collectively called *hazardous materials*. These types of emergency incidents most often occur when hazardous materials are misused or involved in unintended fires and mishaps.

Hazardous materials are at the heart of our technology-based society. They are used in connection with the production and manufacture of a host of products. Among the most common of these products are plastics, rubber, leather, paper, textiles, paints, fertilizers, pesticides, solvents, detergents, fuels, medicines, automobiles, televisions, building materials, electronics, and sporting equipment. Their use makes it possible for us to experience the trappings of the "good life." To eliminate hazardous materials from our society would not only be impractical, it would be undesirable. Instead, to reduce the loss of lives and property and protect the environment, *we* must learn to live safely in their presence and effectively control the emergency-related incidents *they* produce.

On September 11, 2001, the United States fell victim to a third type of emergency incident. On "9/11," as it is now called, foreign terrorist groups hijacked and deliberately flew two commercial planes as missiles into the towers of the World Trade Center. They flew a third plane into the Pentagon. A fourth hijacked plane did not reach its intended target but crash-landed in rural Pennsylvania. In connection with these incidents, 2976 innocent people died with 19 hijackers and at least 857 were injured. To execute them, a hazardous material—aviation fuel—was used as a weapon of mass destruction. Although the 9/11 events were not the first terrorist actions undertaken against the United States, it was during their aftermath that emergency-response personnel acknowledged the evolution of a new component of their workload. Now, they must also respond to incidents in which hazardous materials are employed *en masse* for destructive purposes. Following a terrorist act, emergency responders are concerned not only with saving lives, property, and the environment, but with helping to secure and preserve our homeland.

The nature of hazardous materials did not change on 9/11, nor did the manner by which we respond to incidents in which they are involved. But the 9/11 terrorist acts did provide us with a heightened sense of the unorthodox ways in which individuals or groups can use hazardous materials to intentionally kill massive numbers of people, cause substantial property damage, and affect economic stability. The events of 9/11 also alerted emergency-response personnel to anticipate the presence of biological and chemical warfare agents within the communities in which we live.

We will begin our study in this first chapter by broadly examining the general features of all hazardous materials. We will also observe how several federal statutes aim to reduce the risks associated with the usage, storage, and transportation of hazardous materials. Finally, we will learn that certain interest groups stand ready to help responders when hazardous materials are involved in emergencies.

Performance Goals for Section 1.1
- ◆ Relate the field of chemistry to the profession of the emergency responder.
- ◆ Identify the two main branches of chemistry.

◆ 1.1 WHY MUST EMERGENCY RESPONDERS STUDY CHEMISTRY?

Odds are you have never considered what life would be like without chemistry. If you had, it would soon be apparent that chemistry affects everything we do. There is not a single point in time during which we are not affected by a chemical substance or a chemical process.

Chemistry is regarded as a *natural science;* that is, it is a subject concerned with studying natural phenomena. Specifically, the science of chemistry is the study of substances, their composition, properties, and the changes they undergo. Chemistry concerns itself not only with substances that occur naturally, but also with synthetic substances.

By understanding the interplay of various substances, chemistry has successfully contributed to combating the scourges of hunger, disease, and human deprivation throughout the world. By using established methods, chemistry aspires to further improve the quality of our lives in the future. For instance, during the 21st century, we anticipate that chemistry will radically alter the methods used to treat sicknesses and disease.

Chemistry is broadly divided into two main branches: *organic chemistry* and *inorganic chemistry*. At one time, chemists had presumed that certain substances could only exist in the world if they had originated in living things—plants and animals. They included sugars, alcohols, waxes, fats, and oils, all of which were called organic substances since their origin was believed to require the vital force of life itself.

In 1828, this hypothesis was disproved when Friedrich Wöhler prepared urea, a known constituent of urine. Wöhler successfully prepared urea from substances having no apparent connection to plants or animals. This caused chemists to abandon the idea that a vital force was required for the production of certain substances. Despite abandonment of the hypothesis, however, the name of this branch of chemistry has been historically retained.

Organic chemistry is now recognized as the study of substances that contain carbon in their chemical composition. While some organic substances are actually found in living systems, many others have been synthesized that have no known natural counterpart. The study of organic substances that affect the life process is now regarded as a subdivision of organic chemistry called *biochemistry*. By contrast, the study of those substances that do not contain carbon in their composition is known as *inorganic chemistry*. Inorganic substances include aluminum, iron, sulfur, oxygen, table salt, and many others.

To be an effective emergency responder, is it necessary to study organic and inorganic chemistry? It would be misleading to give the impression that the academic pursuit of chemistry is essential for a successful career as a firefighter, policeman, or other emergency responder. Nonetheless, the study of hazardous materials utilizes the fundamental concepts common to organic and inorganic chemistry courses. An understanding of this basic chemistry can equip the modern emergency responder with the technical background essential for the effective performance of duty.

Performance Goal for Section 1.2
  ◆ Name the general types of a "hazardous material" as this term is used within the emergency-response profession.

◆ 1.2 GENERAL CHARACTERISTICS OF HAZARDOUS MATERIALS

From the perspective of fire science, a hazardous material is often regarded as any substance or combination of substances that is potentially damaging to health, well-being, or the environment. Based on this viewpoint, there are seven general classes of hazardous materials. We shall note here the broad features of these classes, and in Chapter 6, review the manner by which the U.S. Department of Transportation specifically classifies hazardous materials. Although the latter system is much more inclusive than the traditional approach followed by emergency-response professionals, the general approach is useful here as an introduction to the characteristics that are likely to cause a substance or mixture of substances to be regarded as a hazardous material.

The following are the seven classes of hazardous materials traditionally noted in the field of fire science:

  ◆ *Flammable materials.* These are solid, liquid, vapor, or gaseous materials that ignite easily and burn rapidly when exposed to an ignition source. Examples of such flammable materials include the commercial solvents toluene and ethanol, flour dust, finely dispersed powders of aluminum and certain other metals, gasoline, and natural gas.
  ◆ *Spontaneously ignitable materials.* These are solid or liquid materials that ignite spontaneously without the need of an ignition source. Examples of spontaneously ignitable materials are white phosphorus (Section 7.5) and certain alkyl aluminum compounds (Section 9.4). Spontaneous ignition can be initiated by the concentration of heat within a material resulting from oxidation or microbiological action. In other instances, the cause can be traced to the production of a flammable gas or vapor at the surface of a material when it reacts with atmospheric water vapor.
  ◆ *Explosives.* These chemical substances detonate. Examples of explosives are dynamite and TNT. The cause of explosive detonations is typically an initiating mechanism such as shock or the localized concentration of heat.
  ◆ *Oxidizers.* These substances evolve or generate oxygen at ambient temperatures when exposed to heat. An example of an oxidizer is ammonium nitrate.
  ◆ *Corrosive materials.* These are solid or liquid materials that "burn" or otherwise damage skin tissue at the site of contact. An example of a liquid corrosive material is battery acid.
  ◆ *Toxic materials.* Broadly, these are substances that cause adverse health effects or death in individuals exposed to relatively small doses. An example of a toxic material of primary concern to firefighters is carbon monoxide (Section 10.8).
  ◆ *Radioactive materials.* These materials emit radiation, exposure to which could cause adverse health effects or death. An example of a radioactive material is uranium.

Performance Goal for Section 1.3
 ◆ Name the fuel types that comprise class A, class B, class C, and class D fires.

◆ 1.3 CLASSES OF FIRE

Based on emergency-response actions, four general types of fire have been devised: class A, class B, class C, and class D. We shall briefly examine their nature.

CLASS A FIRES

Class A fires result from the burning of ordinary cellulosic materials, such as wood and paper and similar natural and synthetic materials like rubber and plastics. Examples of materials involved in class A fires are the combustible items in Figure 1.1. The solid residue of a class A fire consists of ashes or embers. Class A fires are typically extinguished through the effective use of water.

CLASS B FIRES

Class B fires result when flammable gases and flammable and combustible liquids burn. Several examples are illustrated in Figure 1.2. Firefighters generally use carbon dioxide, dry chemical extinguishers, or foam to extinguish class B fires. The use of water is typically unsuitable for extinguishing large class B fires.

CLASS C FIRES

Class C fires result from the combustion of materials occurring in or originating from energized electrical circuits and equipment like those shown in Figure 1.3. The use of carbon dioxide or dry chemical extinguishers is recommended for extinguishing class C fires. The use of water is not recommended for extinguishing class C fires.

CLASS D FIRES

Class D fires result from the combustion of certain metals, some examples of which are shown in Figure 1.4. They include titanium, magnesium, aluminum, and sodium.

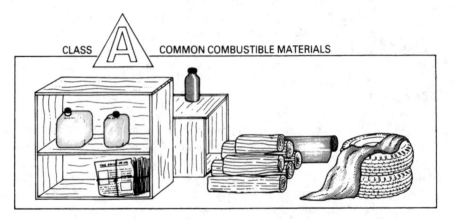

FIGURE 1.1 ◆ The combustion of wood, cotton, plastics, or rubber results in a class A fire.

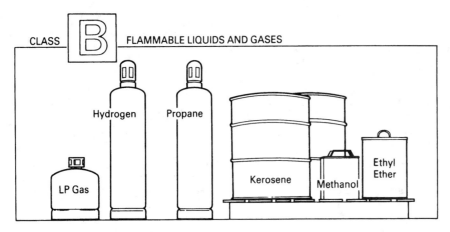

FIGURE 1.2 ◆ The combustion of a flammable gas or a flammable liquid results in a class B fire.

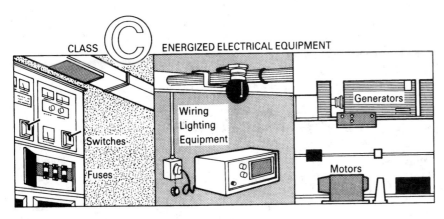

FIGURE 1.3 ◆ The combustion processes that occur in energized electrical equipment are denoted as class C fires.

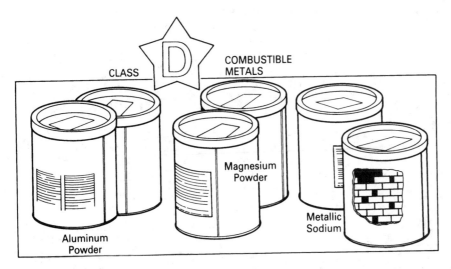

FIGURE 1.4 ◆ The combustion processes associated with certain chemically reactive metals are denoted as class D fires.

Class D fires are often extinguished with special fire extinguishers, such as graphite-based dry powder or sodium chloride. The use of water is suitable for extinguishing some class D fires, but only when it can be applied in a deluging volume.

Performance Goals for Section 1.4

- Identify the information the Consumer Product Safety Commission (CPSC) requires manufacturers to list on labels affixed to containers of consumer products.
- Identify the specific hazardous substances the CPSC bans in all consumer products.
- Identify the information the Food and Drug Administration (FDA) requires on labels affixed to containers of foods, drugs, and cosmetics.
- Identify the information the U.S. Environmental Protection Agency (EPA) requires on containers of approved pesticides.
- Describe the manner by which EPA regulates the manufacture, distribution, and use of chemical substances.
- Identify the specific hazardous substances EPA bans in all products.
- Name the federal laws that provide the CPC, FDA, and EPA with the legal authority to require manufacturers to identify the potential hazards of their products for American consumers.

◆ 1.4 HAZARDOUS SUBSTANCES WITHIN HOUSEHOLD PRODUCTS

The federal statutes listed in Table 1.1 affect the manner by which manufacturers identify the contents of products sold to consumers. We shall briefly examine some features of these statutes.

FEDERAL HAZARDOUS SUBSTANCES ACT

When Congress passed the *Federal Hazardous Substances Act,* it directed the Consumer Product Safety Commission (CPSC) to protect the public against unsuspecting exposure to hazardous substances contained within household products. One way by which the Consumer Product Safety Commission achieves this mandate is to require appropriate labeling on the containers of household products that contain hazardous substances.

When commercial products contain hazardous substances, the CPSC requires manufacturers to affix labels that provide the following information on the containers of their products:

- Initial advisory information, such as DANGER, WARNING, CAUTION, or POISON
- A description of the principal hazards involved in the use of the product, such as "FLAMMABLE" or "VAPOR HARMFUL"
- Instructions the user should follow to reduce or eliminate the risk of a hazard such as "USE ONLY IN WELL-VENTILATED AREA"
- The common names of the hazardous ingredients
- First-aid instructions
- The name and location of the manufacturer, distributor, or repacker
- Warning statements like "KEEP OUT OF REACH OF CHILDREN"

TABLE 1.1 ◆ Federal Laws That Affect the Contents and Labeling of Household and Other Products

Federal Statute	Responsible Agency	Authorizations
Federal Hazardous Substances Act	Consumer Product Safety Commission	Establishes requirements for identifying certain hazardous substances found in household products
Federal Food, Drug, and Cosmetic Act	U.S. Food and Drug Administration	Ensures that foods are safe to eat, drugs are safe and effective, and packaging and labeling of foods, drugs, and cosmetics are truthful and informative
Federal Insecticide, Fungicide, and Rodenticide Act	U.S. Environmental Protection Agency	Regulates the manufacture, use, and disposal of pesticides
Toxic Substances Control Act	U.S. Environmental Protection Agency	Requires pre-market evaluation of all new chemical products other than food additives, drugs, pesticides, alcohol, and tobacco; regulates existing hazards not covered by other federal laws dealing with toxic substances
Federal Alcohol Administration Act	Bureau of Alcohol, Tobacco, and Firearms	Includes ensuring that distilled spirits (beverage alcohols), wines, and malt beverages intended for nonindustrial use are safe for consumption, and their packaging and labeling are truthful and informative

An example of proper labeling on a household product is illustrated in Figure 1.5. The label on a container of aerosol oven cleaner provides the user with several advisory warnings. In addition, it lists first-aid actions to be implemented when the cleaner is swallowed or splashed into the eyes.

Another way by which the CPSC achieves its congressional mandate is to ban manufacturers from including a hazardous substance in household products under the following conditions:

"when the substance possesses such a degree or nature of hazard that adequate cautionary labeling cannot be written and the public health and safety can be served only by keeping such articles out of intrastate commerce."

Using the authority of the Federal Hazardous Substances Act, the CPSC banned manufacturers from including the following hazardous substances in household products:

- Carbon tetrachloride (formerly used for dry-cleaning)
- Extremely flammable water repellants
- Certain fireworks
- Liquid drain cleaners containing 10% or more potassium hydroxide
- Products containing soluble metallic cyanides

FIGURE 1.5 ◆ The Federal Hazardous Substances Act gives the Consumer Product Safety Commission the authority to compel manufacturers to label their products with adequate warning statements when warranted. The label on this aerosol oven cleaner advises the consumer of its flammable and health hazards.

- Paint having a lead content above 0.06% by dry weight (other than artist's paint)
- Aerosols containing vinyl chloride
- General-use garments, tape joint compounds, and artificial fireplace embers containing asbestos

FEDERAL FOOD, DRUG, AND COSMETIC ACT

This federal statute directs the U.S. Food and Drug Administration (FDA) to ensure that foods are safe to eat, drugs are safe and effective, and the labeling of foods and drugs is truthful and informative. FDA limits the concentration of certain chemical substances in food, assesses the safety of food additives, and prohibits the addition to food of any substance known to induce cancer in humans or animals. FDA is also responsible for monitoring and assessing the health risks potentially caused by exposure to all types of chemical substances in foods, drugs, and cosmetic products.

Finally, FDA requires those who produce and package foods to label their containers with ingredient and nutritional information. FDA requires the ingredient in the greatest amount to be listed first. For instance, look at the information posted on an aluminum beverage can of a nondiet soda. The first and second ingredients are carbonated water and high-fructose corn syrup and/or sugar. The listed information

Solved Exercise 1.1

A portion of the labeling on bottles of a concentrated chrome and tile cleaner follows:

CHROME AND TILE CLEANER

Dissolves rust, lime, and scale on glass shower doors, frames and tracks, fiberglass and tile showers, porcelain tubs, and sinks

Warning

May Be Injurious to Eyes and Mucous Membranes
Contains Phosphoric Acid
Wear Rubber Gloves to Avoid
Direct Skin Contact
Use Only in Well-Ventilated Area

Antidote

EXTERNAL: In case of skin or eye contact, flush skin or eyes with plenty of water for at least 15 minutes; for eyes, obtain medical attention

INTERNAL: Do not induce vomiting. Drink 1 oz of Milk of Magnesia or large quantities of water or egg whites. Obtain medical attention immediately.

Why has the manufacturer included these label statements on the bottles of this chrome and tile cleaner?

Solution: To warn the consumer of the presence of hazardous substances, the Consumer Product Safety Commission is authorized to *compel* manufacturers to label their products with the following minimum information: advisory warnings; initial advisory information; a description of the principal hazards involved in using the product; measures the user should take to reduce or eliminate the risk of a hazard; the common names of the hazardous ingredients; first-aid instructions; and other relevant information. By providing this information on bottles of the product, the manufacturer has complied with the obligations specified by the Consumer Product Safety Commission.

Solved Exercise 1.2

Which federal statute provides the legal authority for requiring battery manufacturers and distributors to post the following warnings on the packaging of batteries sold in the United States:

CAUTION

May explode or leak, and cause burn injury, if recharged, disposed of in fire, mixed with a different battery type, inserted backwards, or disassembled.

Replace all used batteries at the same time.

Do not carry batteries loose in your pocket or purse.

Do not remove the battery label.

Solution: The *Federal Hazardous Substances Act* provides the Consumer Product Safety Commission the legal authority to establish labeling requirements for hazardous substances likely to be encountered in household products. When ingredients of commercial products are hazardous substances, their containers must bear labels that provide initial advisory information (CAUTION), a description of the principal hazards involved in the use of the product, and a statement of measures the user should take to reduce or eliminate the risk of a hazard.

---

also identifies the number of calories, total fat, total carbohydrate, total sodium, and total protein per serving. FDA requires the manufacturers of prepared foods to post this combination of information so we can make judicious decisions concerning the nature of what we drink and eat.

FEDERAL INSECTICIDE, FUNGICIDE, AND RODENTICIDE ACT (FIFRA)

When Congress passed the *Federal Insecticide, Fungicide, and Rodenticide Act*, it directed the U.S. Environmental Protection Agency (EPA) to regulate the manufacture, use, and disposal of pesticides for agricultural, forestry, household, and other activities. A pesticide is defined as "any substance or mixture of substances intended for preventing, destroying, repelling, or mitigating insects, rodents, fungi, weeds, and other forms of plant and animal life."

FIFRA provides EPA with the authority to require manufacturers to register their pesticides and to classify them for general or restricted use. It also provides EPA with the authority to require informative and accurate labeling on pesticide containers. When the use of a pesticide is shown to pose an "imminent hazard" or "unreasonable adverse effects on the environment," FIFRA provides EPA the authority to stop its production and sale.

Using the authority of FIFRA, EPA requires manufacturers and distributors to label pesticide containers with the following information, prominently placed with such conspicuousness and in such terms as to render it likely to be read and understood by the ordinary individual:

- The name, brand, or trademark
- Directions for use that are "necessary for effecting the purpose for which the product is intended" and adequate (when complied with) to protect public health and the environment
- An ingredient statement, giving the name and percentage of each active ingredient and the percentage of all inert ingredients
- A statement of the use classification under which it is registered
- The name and address of the manufacturer or distributor
- Appropriate warning or caution statements

Solved Exercise 1.3

A portion of the labeling on aerosol cans of Insect Zappo, an insecticide, follows:

INSECT ZAPPO

Kills Fast—Keeps on Killing up to Four Months
Keep Out of Reach of Children

Active Ingredients:

Tralomethrin (1R,4S)-3[(1′-RS)(1′,2′,2′,2′-tetrabromomethyl)]-2,2-dimethyl-cyclopropanecarboxylic acid(S)-alpha-cyano-3-phenoxybenzyl ester	0.03%
d-*trans*-Allethrin	0.05%
Inert ingredients	99.92%

How to Use:

1. Shake well before each use.
2. Hold can upright. Aim opening away from people. Spray in short bursts to prevent excessive wetting and waste of product.
3. Do not allow the spray to contact food. Thoroughly wash dishes, utensils, and countertops with soap and water if they are sprayed with this product. Do not allow people or pets to contact treated surfaces until spray has dried. Remove pets and cover aquariums before spraying.

Indoors:

1. Contact as many insects as possible with the spray in addition to thoroughly treating all parts of the room where these pests may hide.
2. Pay special attention to cracks, dark corners, hidden surfaces under sinks, behind stoves and refrigerators, food storage areas and wherever these pests are suspected. The treatment continues to work for up to four months.
3. To kill insects and prevent them from entering your home, spray around door and window frames, inside and outside. Spray upward to treat active insect nests. Do not remove insect nests until spray has dried. Repeat as necessary.

Outdoors:

1. Spray outside surfaces of screens, doors, window frames, or wherever these pests may enter the home.
2. Also spray surfaces around light fixtures on porches, in garages, and other places where they alight or congregate. Repeat as necessary.

Precautionary Statements: Hazards to Humans and Domestic Animals. Harmful if absorbed through the skin. Avoid contact with skin, eyes, or clothing. Wash thoroughly with soap and water after handling. Avoid contamination of food and drinking water. All food preparation surfaces and utensils should be covered during treatment or thoroughly washed before use. Remove pets and cover aquariums before spraying.

Statement of Practical Treatment: If on skin, wash with plenty of soap and water. Get medical attention if irritation persists. If swallowed, do not induce vomiting because of aspiration hazard. Get medical attention. If in eyes, flush with plenty of water. Get medical attention.

Environmental Hazards: This product is toxic to fish. Do not apply directly to water.

Physical and Chemical Hazards: Contents under pressure. Do not use or store near heat or open flame. Do not puncture or incinerate container. Exposure to temperatures above 130°F may cause bursting.

Why has the pesticide manufacturer included these label statements on the cans of this product?

Solution: Using the authority of FIFRA, EPA *compels* manufacturers and distributors to label their pesticide products with the following minimum information: the name, brand, or trademark; directions for use that are "necessary for effecting the purpose for which the product is intended" and adequate to protect public health and the environment; an ingredient statement, giving the name and percentage of each active ingredient and the percentage of all inert ingredients; a statement of the use classification under which it is registered; the name and address of the manufacturer; and appropriate warning or caution statements. By providing this information on the pesticide containers, the manufacturer or distributor has complied with the obligations specified by EPA.

TOXIC SUBSTANCES CONTROL ACT (TSCA)

When Congress passed this law in 1976, it provided EPA with the authority for regulating the manufacture, distribution, and use of chemical substances. To accomplish the mandate of this law, EPA uses its authority in two ways. First, it obtains production and test data from industry on commercial chemical products whose manufacture, processing, distribution, use, or disposal could pose an unreasonable risk of injury to public health and the environment. Then, it evaluates these data and, when warranted, imposes limitations or prohibitions on their manufacture and use.

Eight product categories are exempt from TSCA's authority: pesticides, tobacco, nuclear materials, firearms and ammunition, food, food additives, and cosmetics. These product categories are regulated under other federal laws. For example, health and environmental issues related to pesticides and drugs are covered by the *Federal Insecticide, Fungicide, and Rodenticide Act* and the *Federal Food, Drug, and Cosmetic Act,* respectively.

Using the authority of TSCA, EPA selects a control action to prevent unreasonable risks to public health and the environment from chemical products. The nature of the control action varies with the severity of the risk. In some instances, EPA has imposed a hazard warning on product labels. In other instances, it has banned outright the manufacture or severely restricted the use of a product.

In future chapters, we will encounter hazardous materials whose manufacture or use has been banned or restricted by EPA using the authority of TSCA. These hazardous materials include polychlorinated biphenyls (PCBs), chlorofluorocarbons (CFCs), 2,3,7,8-tetrachlorodibenzo-*p*-dioxin (TCDD), asbestos, and hexavalent chromium.

Solved Exercise 1.4

Polychlorinated biphenyls (PCBs) are chemical substances that were formerly used as electrical transformer and capacitor fluids. When EPA evaluated PCB production and test data obtained from industry, it concluded that exposure to PCBs could cause cancer, birth defects, liver damage, acne, impotence, and death. EPA then banned their manufacture and restricted the future use of PCBs. Which federal statute provides EPA with the legal authority to ban the manufacture of PCBs and restrict their use in the United States?

Solution: Using the authority of the *Toxic Substances Control Act*, EPA evaluates commercial chemical products to determine whether their manufacture, processing, distribution, use, or disposal could present an unreasonable risk of injury or damage to public health and the environment. When these data inform EPA that these actions pose a substantial risk to public health and the environment, EPA is obligated to prohibit the manufacture of the product or restrict its future use.

Upon determining that ill effects could result from the manufacture, processing, distribution, use, or disposal of PCBs, EPA used the authority of the Toxic Substances Control Act to prohibit their manufacture and restrict their future use.

Performance Goals for Section 1.5

- ◆ Name the six criteria air pollutants and distinguish between the nature of the information provided by their "primary" standard and "secondary" standard.
- ◆ Describe the general nature of the hazardous air pollutants identified by EPA whose airborne emissions are regulated by EPA through national emission standards called NESHAPs.
- ◆ Describe the general nature of the toxic pollutants identified by EPA whose discharge into waterways is regulated by EPA.
- ◆ Describe the general nature of the hazardous wastes whose treatment, storage, and disposal are regulated by EPA.
- ◆ Describe the general nature of the hazardous substances identified by EPA whose improper treatment, storage, and disposal at facilities mandates EPA to clean up and restore the natural resources at the sites.
- ◆ Name the federal laws that provide EPA with the legal authority to regulate emissions and other discharges of criteria air pollutants, hazardous air pollutants, toxic pollutants, and hazardous wastes.

◆ 1.5 HAZARDOUS CONSTITUENTS OF POLLUTANTS AND WASTES

Using the authority of the federal statutes in Table 1.2, EPA has been charged with the responsibility of assuring protection of human health and the environment from exposure to the hazardous substances contained in polluted air and water and chemical wastes. We shall note the important components of each statute separately.

CLEAN AIR ACT

The modern approach to the regulation of air pollution was initiated with passage in 1970 of the *Clean Air Act*. To assure Americans of their right to breathe clean air, Congress empowered EPA to establish national ambient air quality standards for a number of air contaminants called *criteria air pollutants*. To date, EPA has established standards for six criteria air pollutants: sulfur dioxide, particulate matter, carbon monoxide, ozone, nitrogen dioxide, and lead.

Each national ambient air quality standard is composed of the following parts:

- *Primary standard.* This standard identifies a contaminant concentration in the ambient air that is protective of human health. When EPA sets the primary standard, it considers not only healthy individuals, but the health of "sensitive" populations such as children, asthmatics, and the elderly.
- *Secondary standard.* This standard identifies a contaminant concentration that is protective of public welfare. In setting the standard, EPA accounts for such subjects as reduced visibility within the air and damage to animals and vegetation.

TABLE 1.2 ◆ Some Federal Environmental Statutes

Statute	U.S. EPA's Responsibility
Clean Air Act	Establishes national ambient air quality standards for specified air contaminants; devises and promulgates technology-based emission reduction standards
Federal Water Pollution Prevention and Control Act	Restores and maintains the chemical, physical, and biological integrity of the nation's waters
Resource Conservation and Recovery Act	Regulates the generation, transportation, treatment, storage, and disposal of hazardous wastes within a cradle-to-grave framework
Comprehensive Environmental Response, Compensation, and Liability Act	Leads efforts to cleanup sites at which hazardous substances were improperly disposed of in the past
Superfund Amendments and Reauthorization Act	Considers the standards and requirements found in other environmental laws and regulations and includes citizen participation when considering a remedial-response action

The contaminant concentrations provided in primary and secondary standards are generally annual arithmetic means. When they are exceeded, EPA is compelled to take action so public health and welfare are protected. For example, much of the nitrogen dioxide in the ambient air is linked with its presence in vehicular exhaust. Concerned with reducing its concentration, EPA approached automobile manufacturers to determine whether they could construct and operate automobiles in a fashion that would reduce the amount of nitrogen dioxide within the exhaust. In response, the manufacturers built new automobiles in which they installed catalytic converters. The converters facilitate the change of nitrogen dioxide into nitrogen, thereby reducing the amount of nitrogen dioxide discharged into the air.

The Clean Air Act also provides EPA with the authority to establish national emission standards for airborne substances that cause or are suspected to cause serious health problems. These substances are called *hazardous air pollutants* and the standards are called *National Emission Standards for Hazardous Air Pollutants* (NESHAPs). To date, EPA has published NESHAPs for 189 hazardous air pollutants.

The NESHAP regarding airborne asbestos (Section 10.17) is familiar to emergency-response personnel providing standby services during demolition work involving asbestos-containing materials. This NESHAP requires persons intending to conduct asbestos-related activities to notify EPA of the proposed action, to follow certain procedures to protect public health and the environment, and to adopt specific work procedures to prevent the release of asbestos fibers into the air.

The Clean Air Act also provides EPA with the authority to regulate the prevention of accidental, potentially catastrophic releases of airborne hazardous substances. This program obligates the owners and operators of facilities using large quantities of toxic or flammable substances to identify potential sources from which they could be released into the air. It also places responsibility on them to prevent and minimize the risks associated with these potential catastrophes. Companies that make or use large quantities of toxic or flammable substances are obligated to compile *risk management plans,* which provide key information to emergency planners, fire departments, and local residents for responding to emergencies involving them.

Finally, the Clean Air Act provided for the establishment of the U.S. Chemical Safety and Hazard Investigation Board. Modeled after the National Transportation Safety Board, the Chemical and Hazard Investigation Board conducts investigations and reports on findings regarding the causes of major chemical accidents. These findings are then used to recommend actions to improve the safety of operations involved in the production, transportation, industrial handling, use, and disposal of chemical substances.

FEDERAL WATER POLLUTION PREVENTION AND CONTROL ACT

The federal water pollution control effort began seriously with the passage of this law in 1948, but a comprehensive regulatory program for controlling water pollution discharges was first set by EPA with passage of the Federal Water Pollution Prevention and Control Act amendments of 1972. This combination of amendments is commonly called the *Clean Water Act*. The primary regulatory approach of the act now consists of the issuance of industry by industry, technology-based effluent limitations, that apply to the discharge of toxic pollutants in all waters within the United States and its territories. Thus, EPA may permit the discharge of an aqueous waste into a nearby river, but only when the toxic pollutant concentrations in the waste are equal to or less than established values.

A *toxic pollutant* is defined in the statute as a substance that —

"after discharge and upon exposure, ingestion, inhalation or assimilation into any organism . . . , will, on the basis of information . . . , cause death, disease, behavioral abnormalities, cancer, genetic mutations, physiological malfunctions (including malfunctions in reproduction) or physical deformations, in organisms or their offspring."

To date, EPA uses the authority of the Clean Water Act to regulate the discharge of approximately 150 toxic pollutants. In addition, it regulates the conventional water pollutants designated as *biochemical oxygen demand,* total suspended solids, pH, fecal coliform, and oil and grease. Biochemical oxygen demand (BOD) is a standardized means of estimating the degree of contamination of a water supply, particularly one contaminated by sewage or industrial wastes.

Using the authority of the Clean Water Act, EPA regulates not only the direct discharges of toxic pollutants into waters of the United States, but also indirect discharges, such as those directed into storm drains or sewers that have the potential for ultimate entrance into navigable waters. The U.S. Coast Guard has jurisdiction over the discharges into marine waters, while EPA has jurisdiction over discharges into fresh waters.

When these agencies receive notification of a discharge, they respond by locating the point from which the toxic pollutant is entering the waterway. Appropriate measures are then taken to stop the discharge and correct the damage to natural resources.

RESOURCE CONSERVATION AND RECOVERY ACT (RCRA)

In 1976, Congress enacted the *Resource Conservation and Recovery Act,* and in 1979, EPA promulgated regulations that aim to ensure protection of public health and the environment by facilities that generate, transport, treat, store, or dispose of hazardous wastes. The statute specifically defines the term *hazardous waste* as a material that, "because of its quantity, concentration, or physical, chemical, or infectious characteristics may:

- cause, or significantly contribute to an increase in mortality or an increase in serious irreversible, or incapacitating, reversible illness; *or*
- pose a substantial present or potential hazard to human health or the environment when improperly treated, stored, or disposed of, or otherwise managed."

EPA identifies hazardous wastes in two ways. First, EPA *lists* a series of wastes, each with an EPA identification number. Second, EPA provides criteria for *characterizing* a hazardous waste as ignitable, corrosive, reactive, or toxic. Each RCRA characteristic shall be discussed in detail at appropriate points in later chapters.

COMPREHENSIVE ENVIRONMENTAL RESPONSE, COMPENSATION, AND LIABILITY ACT

In 1980, Congress enacted the *Comprehensive Environmental Response, Compensation, and Liability Act* to correct for the past mistakes resulting from the improper disposal of hazardous substances. The law is commonly known by its acronym *CERCLA,* and even more commonly, as the *Superfund law.*

CERCLA addresses the prevention and cleanup of releases of hazardous substances and the restoration and replacement of the natural resources that were damaged or lost because of such releases. As defined in the statute, a hazardous substance is any of the following:

- Any substance designated as a hazardous substance under the Federal Water Pollution Control Act

- Any substance specifically designated as a hazardous substance pursuant to the act
- Any hazardous waste listed in the RCRA regulations or possessing the characteristic of ignitability, corrosivity, reactivity, or toxicity
- Any substance designated as a toxic pollutant pursuant to the Federal Water Pollution Prevention and Control Act
- Any substance designated as a hazardous air pollutant pursuant to the Clean Air Act
- Any imminently hazardous substance designated pursuant to the Toxic Substances Control Act

EPA has used the legal authority of CERCLA to correct the problems of abandoned or uncontrolled sites that threaten public health or the environment. This is accomplished through use of a tax imposed on the chemical and petroleum industries. The tax monies were initially deposited into a trust fund called the *superfund*. When those responsible cannot be located, or when responsible parties are unwilling or unable to pay for the damages they caused, EPA uses superfund monies to investigate and remedy the problems. CERCLA also gives EPA the authority to require those responsible for contaminating sites to investigate and conduct remedial activities or to reimburse EPA for doing the work. To date, EPA has used the authority of CERCLA to clean up hundreds of sites at which hazardous substances were improperly treated, stored, or disposed, and to recover the cleanup costs from the parties responsible for the contamination.

In 1986, Congress enacted the *Superfund Amendments and Reauthorization Act,* commonly called *SARA*. This law requires EPA, when selecting a remedial action, to consider the standards and requirements found in other environmental laws and regulations and to include citizen participation in the decision-making process. SARA also establishes an assistance program for training and educating workers engaged in activities such as emergency response.

Performance Goals for Section 1.6

- Describe the manner by which the Occupational Safety and Health Administration (OSHA) aspires to reduce or eliminate the incidence of occupational illness and injury caused by exposure to hazardous substances in the workplace.
- Identify the 16 categories of information OSHA requires manufacturers to provide on a Material Safety Data Sheet (MSDS).

◆ 1.6 HAZARDOUS SUBSTANCES WITHIN THE WORKPLACE

The *Occupational Safety and Health Act* is a federal statute that aims to protect employees in the workplace from occupational illness and injuries caused by exposure to hazardous substances. When Congress passed this act, it empowered the Occupational Safety and Health Administration (OSHA) to regulate certain aspects of the workplace, thereby reducing or eliminating the incidence of chemically induced occupational illnesses and injuries.

When hazardous substances are present in the workplace, OSHA requires employers to limit employee exposure to certain permissible limits averaged over the 8-hour workday. OSHA establishes these limits, but on occasion, the National Institute for Occupational Safety and Health (NIOSH) recommends a revision of the limits based upon its own research efforts. It is the responsibility of OSHA and NIOSH to ensure that employers provide every working individual with a safe and healthful working environment.

FIGURE 1.6 ◆ The label affixed to containers of 50% hydrogen peroxide provides a hazard statement, storage and handling information, and the name and address of the manufacturer. The United States Department of Transportation requires shippers and carriers to affix labels to containers of hydrogen peroxide designated for shipment. The nature of these labels is described in Chapter 6. *Courtesy of FMC Corporation, Hydrogen Peroxide Division, Philadelphia, Pennsylvania.*

In 1983, OSHA first enacted a standard that sets minimum requirements to which employers must adhere for communicating information to workers; it is often referred to as the *right-to-know law*. Briefly, its intent is to assure workers of their right to know about the hazards associated with substances to which they are being exposed in their places of employment.

The right-to-know law requires chemical manufacturers and importers to assess the hazards of the chemical substances and products they produce or import and to transmit the hazard information to their users. This information must be sufficiently comprehensive to allow user–employers to develop appropriate employee protection programs, and to give employees the information that they require to protect themselves against the potential risks associated with the substances and products. This regulation applies to all chemical substances known to be present within the workplace, substances to which employees could be exposed under normal working conditions, and substances to which employees could be exposed in a foreseeable emergency.

WARNING LABELS

OSHA requires every manufacturer and importer of a chemical substance to ensure that each container of a hazardous substance is labeled, tagged, and marked with the identity of the product, appropriate hazard warnings, storage and handling information, and the name and address of the manufacturer, importer, or other responsible party. Symbols, pictures, and/or words are used on the labels to present the hazard-warning messages.

An example of a warning label for a chemical product is shown in Figure 1.6. This label is affixed to containers of 50% hydrogen peroxide distributed by FMC Corporation.

MATERIAL SAFETY DATA SHEETS (MSDSs)

An MSDS is a technical bulletin containing detailed information about a hazardous substance. OSHA requires manufacturers to prepare an MSDS for each hazardous substance it sells in the marketplace. OSHA requires manufacturers and distributors to provide a copy of the relevant MSDS when each sample or order of a hazardous substance is shipped by the manufacturer or other carrier to a location for the first time.

To ensure that the MSDS serves as a comprehensive source of written information on a product, OSHA requires that it address the following 16 categories of information:

Chemical product and company identification	Physical and chemical properties
Chemical composition and information on ingredients	Stability and reactivity
Hazards identification	Toxicological properties
First aid measures	Ecological information
Firefighting measures	Disposal considerations
Accidental release measures	Transport information
Handling and storage information	Regulatory information
Exposure controls/personal protection	Other information deemed important

An example of an MSDS is provided in Figure 1.7. It consists of nine pages of information on the chemical product hydrogen peroxide. Note that it contains information relating to each of the 16 categories previously cited.

MATERIAL SAFETY DATA SHEET

Hydrogen Peroxide (40 to 60%)

MSDS Ref. No: 7722-84-1-4
Version: US/Canada
Date Approved: 06/10/2002
Revision No: 6

1. PRODUCT AND COMPANY IDENTIFICATION

PRODUCT NAME: Hydrogen Peroxide (40 to 60%)
ALTERNATE TRADE NAME(S): Durox® Reg. & LR 50%, Hybrite® 50%, Oxypure® 50%, Semiconductor Reg & Seg 50%, Standard 50%, Technical 50%, Chlorate Grade 50%, Super D® 50%
GENERAL USE: Durox® 50% Reg. and LR - meets the Food Chemical Codex requirements for aseptic packaging and other food related applications.

Oxypure® 50% - certified by NSF to meet ANSI/NSF Std 60 requirements for drinking water treatment.

Standard 50% - most suitable for industrial bleaching, processing, pollution abatement and general oxidation reactions.

Semiconductor Reg. & Seg. 50% - conforms to ACS and Semi Specs., for water etching and cleaning, and applications requiring low residues.

Super D® 50% - meets US Pharmacopoeia specifications for 3% topical solutions when diluted with proper quality water. While manufactured to th eUSP standards or purity and to FMC's demanding ISO 9002 quality standards, FMC does not claim that its Hydrogen Peroxide is manufactured in accordance with all pharmaceutical cGMP conditions.

Technical 50% - essentially free of inorganic metals, suitable for chemical synthesis.

Chlorate Grade 50% - specially formulated for use in chlorate manufacture or processing.

MANUFACTURER	**Emergency Telephone Numbers:**
FMC of Canada Ltd.	**CHEMTREC (U.S.):** (800) 424-9300
Hydrogen Peroxide Division	**Emergency Phone** 613-996-6666 (Canutec)
PG Pulp Mill Road	
Prince George, BC V2N2S6	
General Information: 604-561-4200	
FMC Corporation	**Emergency Phone** (303) 595-9048 (Medical)
Hydrogen Peroxide Division	Call Collect
1735 Market Street	**Emergency Phone** (609) 924-6677 (Plant)
Philadelphia, PA 19103	Call Collect
General Information: (215) 299-6000	

FIGURE 1.7 ◆ The Material Safety Data Sheet for solutions containing 40–60% hydrogen peroxide. The various points of technical information regarding this product are noted throughout the textbook. *Courtesy of FMC Corporation, Hydrogen Peroxide Division, Philadelphia, Pennsylvania.*

2. COMPOSITION / INFORMATION ON INGREDIENTS

Chemical Name	CAS#	Wt.%
Hydrogen Peroxide	7722-84-1	40 - 60
Water	7732-18-5	40 - 60

3. HAZARDS IDENTIFICATION

EMERGENCY OVERVIEW

IMMEDIATE CONCERNS: Oxidizer. Contact with combustibles may cause fire. Decomposes yielding oxygen that supports combustion of organic matters and can cause overpressure if confined.

POTENTIAL HEALTH EFFECTS: Corrosive to eyes, skin, nose, throat and lungs. May cause irreversible tissue damage to the eyes including blindness.

4. FIRST AID MEASURES

EYES: Immediately flush with water for at least 15 minutes, lifting the upper and lower eyelids intermittently. See a medical doctor or ophthalmologist immediately.

SKIN: Immediately flush with plenty of water while removing contaminated clothing and/or shoes, and thoroughly wash with soap and water. See a medical doctor immediately.

INGESTION: Rinse mouth with water. Dilute by giving 1 or 2 glasses of water. Do not induce vomiting. Never give anything by mouth to an unconscious person. See a medical doctor immediately.

INHALATION: Remove to fresh air. If breathing difficulty or discomfort occurs and persists, contact a medical doctor.

NOTES TO MEDICAL DOCTOR: Hydrogen peroxide at these concentrations is a strong oxidant. Direct contact with the eye is likely to cause corneal damage especially if not washed immediately. Careful ophthalmologic evaluation is recommended and the possibility of local corticosteroid therapy should be considered. Because of the likelihood of corrosive effects on the gastrointestinal tract after ingestion, and the unlikelihood of systemic effects, attempts at evacuating the stomach via emesis induction or gastric lavage should be avoided. There is a remote possibility, however, that a nasogastric or orogastric tube may be required for the reduction of severe distension due to gas formation.

FIGURE 1.7 ◆ *Continued*

5. FIRE FIGHTING MEASURES

FLASH POINT AND METHOD: Non-combustible

FLAMMABLE LIMITS: Non-combustible

AUTOIGNITION TEMPERATURE: Non-combustible

EXTINGUISHING MEDIA: Flood with water.

FIRE / EXPLOSION HAZARDS: Product is non-combustible. On decomposition releases oxygen which may intensify fire.

FIRE FIGHTING PROCEDURES: Any tank or container surrounded by fire should be flooded with water for cooling. Wear full protective clothing and self-contained breathing apparatus.

SENSITIVITY TO STATIC DISCHARGE: No data available

SENSITIVITY TO IMPACT: No data available

HAZARDOUS DECOMPOSITION PRODUCTS: Oxygen which supports combustion.

6. ACCIDENTAL RELEASE MEASURES

RELEASE NOTES: Dilute with a large volume of water and hold in a pond or diked area until hydrogen peroxide decomposes. Hydrogen peroxide may be decomposed by adding sodium metabisulfite or sodium sulfite after diluting to about 5%. Dispose according to methods outlined for waste disposal.

Combustible materials exposed to hydrogen peroxide should be immediately submerged in or rinsed with large amounts of water to ensure that all hydrogen peroxide is removed. Residual hydrogen peroxide that is allowed to dry (upon evaporation hydrogen peroxide can concentrate) on organic materials such as paper, fabrics, cotton, leather, wood or other combustibles can cause the material to ignite and result in a fire.

7. HANDLING AND STORAGE

HANDLING: Wear cup type chemical safety goggles and full-face shield, impervious clothing, such as rubber, PVC, etc., and rubber or neoprene gloves and shoes. Avoid cotton, wool and leather. Avoid excessive heat and contamination. Contamination may cause decomposition and generation of oxygen gas which could result in high pressures and possible container rupture. Hydrogen peroxide should be stored only in vented containers and transferred only in a prescribed manner (see FMC Technical Bulletins). Never return unused hydrogen peroxide to original container, empty drums should be triple rinsed with water before discarding. Utensils used for handling hydrogen peroxide should only be made of glass, stainless steel, aluminum or plastic.

FIGURE 1.7 ◆ *Continued*

STORAGE: Store drums in cool areas out of direct sunlight and away from combustibles. For bulk storage refer to FMC Technical Bulletins.

COMMENTS: VENTILATION:
Provide mechanical general and/or local exhaust ventilation to prevent release of vapor or mist into the work environment.

---

8. EXPOSURE CONTROLS / PERSONAL PROTECTION

EXPOSURE LIMITS

Chemical Name	TWA (ACGIH)	STEL/Ceiling (ACGIH)	PEL (OSHA)	STEL/Ceiling (OSHA)
Hydrogen Peroxide	1 ppm		1 ppm	

ENGINEERING CONTROLS: Ventilation should be provided to minimize the release of hydrogen peroxide vapors and mists into the work environment. Spills should be minimized or confined immediately to prevent release into the work area. Remove contaminated clothing immediately and wash before reuse.

PERSONAL PROTECTIVE EQUIPMENT

EYES AND FACE: Use cup type chemical goggles. Full face shield may be used.

RESPIRATORY: If concentrations in excess of 10 ppm are expected use approved self-contained breathing apparatus. Do not use oxidizable sorbants such as activated carbon.

PROTECTIVE CLOTHING: Liquid proof rubber or neoprene gloves. Rubber or neoprene footwear (avoid leather). Impervious clothing materials such as rubber, neoprene, nitrile or polyvinyl chloride (avoid cotton, wool and leather). Completely submerge hydrogen peroxide contaminated clothing or other materials in water prior to drying. Residual hydrogen peroxide, if allowed to dry on materials such as paper, fabrics, cotton, leather, wood or other combustibles can cause the material to ignite and result in a fire.

---

9. PHYSICAL AND CHEMICAL PROPERTIES

ODOR: Odorless

APPEARANCE: Clear, colorless liquid

pH: (as is) 1.0 to 3.0

PERCENT VOLATILE: 100%

VAPOR PRESSURE: 22 mmHg @ 30°C (40%); 18.3 mmHg @ 30°C (50%)

VAPOR DENSITY: (Air = 1): Not available

FIGURE 1.7 ◆ *Continued*

BOILING POINT: 110°C (229°F) (40%); 114°C (237°F) (50%)

FREEZING POINT: -41.4°C (-42.5°F) (40%); -52°C (-62°F) (50%)

SOLUBILITY IN WATER: (in H2O % by wt) 100%

EVAPORATION RATE: (Butyl Acetate = 1) Above 1

DENSITY: Not available

SPECIFIC GRAVITY: (H20 = 1) 1.15 @ 20°C/4°C (40%); 1.19 @ 20°C/4°C (50%)

COEFF. OIL/WATER: Not available

ODOR THRESHOLD: Not available

OXIDIZING PROPERTIES: Strong oxidizer

COMMENTS: pH (1% solution) : 5.0 - 6.0

10. STABILITY AND REACTIVITY

CONDITIONS TO AVOID: Excessive heat or contamination could cause product to become unstable.

STABILITY: Stable (heat and contamination could cause decomposition)

POLYMERIZATION: Will not occur

HAZARDOUS DECOMPOSITION PRODUCTS: Oxygen which supports combustion.

INCOMPATIBLE MATERIALS: Reducing agents, wood, paper and other combustibles, iron and other heavy metals, copper alloys and caustic.

COMMENTS: Materials to Avoid : Dirt, organics, cyanides and combustibles such as wood, paper, oils, etc.

11. TOXICOLOGICAL INFORMATION

EYE EFFECTS: Severe irritant (corrosive), (rabbit), (70% hydrogen peroxide) [FMC Study Number: ICG/T-79.027]

SKIN EFFECTS: Severe irritant (corrosive), (rabbit), (50% hydrogen peroxide) [FMC Study Number: I89-1079]

FIGURE 1.7 ◆ *Continued*

DERMAL LD$_{50}$: >6.5 g/kg (rabbit) (70% hydrogen peroxide) [FMC Study Number: ICG/T-79.027]

ORAL LD$_{50}$: >225 mg/kg (rat) (50% hydrogen peroxide) [FMC Study Number: I86-914]

INHALATION LC$_{50}$: >0.17 mg/L (rat) (50% hydrogen peroxide) [FMC Study Number: I89-1080]

TARGET ORGANS: Eye, skin, nose, throat, lungs

ACUTE EFFECTS FROM OVEREXPOSURE: Severe irritant/corrosive to eyes, skin and gastrointestinal tract. May cause irreversible tissue damage to the eyes including blindness. Inhalation of mist or vapors may be severely irritating to nose, throat and lungs.

CHRONIC EFFECTS FROM OVEREXPOSURE: There are reports of limited evidence of carcinogenicity of hydrogen peroxide to mice administered high concentrations in their drinking water (IARC Monograph 36, 1985). However, the International Agency For Research on Cancer concluded that hydrogen peroxide could not be classified as to its carcinogenicity to humans (Group III carcinogen).

CARCINOGENICITY

Chemical Name	NTP Status	IARC Status	OSHA Status	Other
Hydrogen Peroxide	Not listed	Not listed	Not listed	(ACGIH) Listed (A3, Animal Carcinogen)

12. ECOLOGICAL INFORMATION

ECOTOXICOLOGICAL INFORMATION: Channel catfish 96 hour LC50 = 37.4 mg/L
Fathead minnow 96 hour LC50 = 16.4 mg/L
Daphnia magna 24 hour EC50 = 7.7 mg/L
Daphnia pulex 48 hour LC50 = 2.4 mg/L
Freshwater snail 96 hour LC50 = 17.7 mg/L

For more information refer to ECETOC "Joint Assessment of Commodity Chemicals No. 22, Hydrogen Peroxide." ISSN-0773-6339, January 1993

CHEMICAL FATE INFORMATION: Hydrogen peroxide in the aquatic environment is subject to various reduction or oxidation processes and decomposes into water and oxygen. Hydrogen peroxide half-life in freshwater ranged from 8 hours to 20 days, in air from 10-20 hrs. and in soils from minutes to hours depending upon microbiological activity and metal contaminants.

FIGURE 1.7 ◆ *Continued*

13. DISPOSAL CONSIDERATIONS

DISPOSAL METHOD: An acceptable method of disposal is to dilute with a large amount of water and allow the hydrogen peroxide to decompose followed by discharge into a suitable treatment system in accordance with all regulatory agencies. The appropriate regulatory agencies should be contacted prior to disposal.

14. TRANSPORT INFORMATION

U.S. DEPARTMENT OF TRANSPORTATION (DOT)

PROPER SHIPPING NAME: Hydrogen peroxide, aqueous solutions with more than 40% but not more than 60% hydrogen peroxide.

PRIMARY HAZARD CLASS/DIVISION: 5.1 (Oxidizer)

UN/NA NUMBER: UN 2014

PACKING GROUP: II

PLACARDS: 5.1 (Oxidizer)

LABEL: Oxidizer Corrosive

OTHER SHIPPING INFORMATION:
DOT Marking: Hydrogen Peroxide, aqueous solution with more than 40%, but not more than 60% Hydrogen Peroxide, UN 2014
Hazardous Substance/RQ: Not applicable
49 STCC Number : 4918776

Aluminum tanks, drum/DOT 42D

SPECIAL SHIPPING NOTES: IMDG: Hydrogen Peroxide, aqueous solutions with more than 40%, but not more than 60% hydrogen peroxide.

IATA: Hydrogen Peroxide (40 - 60%) is forbidden on Passenger and Cargo Aircraft, as well as Cargo Only Aircraft.

Protect from physical damage. Keep drums in upright position. Drums should not be stacked in transit. Do not store drum on wooden pallets.

15. REGULATORY INFORMATION

UNITED STATES

SARA TITLE III (SUPERFUND AMENDMENTS AND REAUTHORIZATION ACT)

SARA TITLE III SECTION 302 EXTREMELY HAZARDOUS SUBSTANCES

FIGURE 1.7 ◆ *Continued*

(40 CFR 355):
Hydrogen Peroxide > 52%
RQ: 1000 lbs.
Planning Threshold: 1000 lbs.

SECTION 311 HAZARD CATEGORY (40 CFR 370):
Fire Hazard
Immediate (Acute) Health Hazard

SECTION 312 THRESHOLD PLANNING QUANTITY (40 CFR 370): 1000 lbs.
(conc. >52%); 10000 lbs. (conc. <52%)

SECTION 313 REPORTABLE INGREDIENTS (40 CFR 372): Not listed

CERCLA (COMPREHENSIVE ENVIRONMENTAL RESPONSE COMPENSATION AND LIABILITY ACT)

CERCLA REGULATORY (40 CFR 302.4): Unlisted (Hydrogen Peroxide); RQ = 100 lbs.;
Ignitability, Corrosivity

TSCA (TOXIC SUBSTANCE CONTROL ACT)

TSCA STATUS (40 CFR 710): Listed

RCRA STATUS: Waste No. D001 Waste No. D002

CANADA

WHMIS (WORKPLACE HAZARDOUS MATERIALS INFORMATION SYSTEM):
Hazard Classification: Class C (Oxidizer), Class D, Div. 2 Subdiv. B, Class E (Corrosive)
Product Identification No. : 2014
Ingredient Disclosure List: Listed

---

16. OTHER INFORMATION

REVISION SUMMARY
This MSDS replaces Revision #5, dated September 29, 2000. Changes in information are as follows:

Section 16 (Other Information): HMIS Headings

HMIS RATING	
HEALTH:	3
FLAMMABILITY	0
PHYSICAL HAZARD:	1
PERSONAL PROTECTION (PPE):	H

NFPA RATING	
HEALTH:	3
FLAMMABILITY	0
REACTIVITY:	1
SPECIAL:	OX

FIGURE 1.7 ◆ *Continued*

Key
4 = Severe
3 = Serious
2 = Moderate
1 = Slight
0 = Minimal

HMIS RATINGS NOTES: Protection = H (Safety goggles, gloves, apron, the use of a supplied air or SCBA respirator is required in lieu of a vapor cartridge respirator)

GENERAL STATEMENTS: Note: NFPA - Reactivity is 3, when greater than 52%

The contents and format of this MSDS are in accordance with OSHA Hazard Communication Standard and Canada's Workplace Hazardous Information System (WHMIS).

National Fire Protection Association (NFPA)

SPECIAL = OX (Oxidizer)

Hazardous Materials Identification System (HMIS)

FIGURE 1.7 ◆ *Continued*

Performance Goal for Section 1.7
- Discuss the intent of the Federal Hazardous Materials Transportation Law, especially as it bears upon the emergency-response profession.

◆ 1.7 HAZARDOUS MATERIALS IN TRANSIT

Using the authority of the *Federal Hazardous Materials Transportation Law*, the U.S. Department of Transportation regulates shippers and carriers who offer or accept hazardous materials for intrastate, interstate, or international transportation through certain marking, labeling, placarding, and packaging requirements. The information conveyed by these warning labels, placards, and shipping papers is especially critical for emergency responders. We will discuss the details of these requirements in Chapter 6.

Performance Goal for Section 1.8
- Discuss the intent of the Emergency Planning and Community Right-to-Know Act, especially as it bears upon the emergency-response profession.

◆ 1.8 HAZARDOUS SUBSTANCES WITHIN COMMUNITIES

To better prepare communities with information concerning the presence of chemical substances within their boundaries, Congress enacted the *Emergency Planning and Community Right-to-Know Act* (EPCRA) in 1986. The statute requires any facility handling chemical substances either to submit copies of MSDSs for certain chemical substances present at the facility in amounts greater than specified threshold quantities or to provide a list of such substances along with each substance's acute health hazard, chronic health hazard, fire hazard, sudden-release-of-pressure hazard, and reactivity hazard. The relevant substances include 77 toxic substances, 63 flammable substances, and certain explosive substances. Threshold quantities, that is, minimum amounts required for reporting, have been established for toxic substances ranging from 500 to 20,000 pounds. For all listed flammable substances, the threshold quantity is 10,000 pounds, and for the explosive substances, the threshold quantity is established at 5000 pounds.

EPCRA provides EPA with the authority to require submission of the list of hazardous substances or copies of relevant MSDSs to the local emergency-planning committee, the state emergency-response commission, and the local fire department having jurisdiction over the facility. From a review of this information, the three groups can readily determine the substances that are present within any community above threshold quantities.

When utilized effectively, this information serves to improve the quality of accident-planning exercises, preparedness, mitigation, response, and recovery capabilities of emergency-response teams. This information also ensures citizens that sufficient information is available beforehand to prepare for an emergency involving chemical substances within their local communities.

Following 9/11, EPA acknowledged that it would be imprudent to publish data concerning amounts of hazardous substances present at individual chemical companies. Although this information is useful to communities preparing for an emergency, radical organizations could also find the information useful for preparing terrorist actions. Whereas the information was once available on EPA's Web site, this information is now provided to the public only at EPA's reading rooms.

Performance Goals for Section 1.9

- Describe the procedure used by the National Fire Protection Association (NFPA) for conveying to emergency responders the potential hazards associated with exposure to a given hazardous substance.
- Identify the colors and the significance of each NFPA number denoting a relative rating for health, flammability, and chemical reactivity.

◆ 1.9 NFPA SYSTEM OF IDENTIFYING POTENTIAL HAZARDS

At the scene of an emergency, how may we rapidly identify the potential hazards associated with a given hazardous material? The answer to this question is based on recognizing certain markings that are posted on bulk tanks, exterior building walls, pipelines, and other relevant locations at which hazardous materials are stored or used. This procedure for identifying the potential hazards associated with a hazardous material is known as the *NFPA system*, where the acronym refers to the National Fire Protection Association.*

---

*National Fire Protection Association, 1 Batterymarch Park, Quincy, Massachusetts 02269-9101. As we study the properties of individual hazardous materials beginning in Chapter 7, the appropriate NFPA diamond will be displayed in the page margin near the point at which a discussion of each material first begins.

TABLE 1.3 ◆ The NFPA Codes for Health, Flammability, and Reactivity

Identification of Health		Identification of Flammability		Identification of Reactivity (Stability)	
Hazard Color Code: BLUE		Hazard Color Code: RED		Hazard Color Code: YELLOW	
Signal	Type of Possible Injury	Signal	Susceptibility of Materials to Burning	Signal	Susceptibility to Release of Energy
4	Materials that on very short exposure could cause death or major residual injury even though prompt medical treatment was given	4	Materials that will rapidly or completely vaporize at atmospheric pressure and normal ambient temperature, or that are readily dispersed in air, and will burn readily	4	Materials that in themselves are readily capable of detonation or of explosive decomposition or reaction at normal temperatures and pressures
3	Materials that on short exposure could cause serious temporary or residual injury even though prompt medical treatment was given	3	Liquids and solids that can be ignited under almost all ambient temperature conditions	3	Materials that in themselves are capable of detonation or explosive reaction but require a strong initiating source or that must be heated under confinement before initiation or that react explosively with water
2	Materials that on intense or continued exposure could cause temporary incapacitation or possible residual injury unless prompt medical treatment was given	2	Materials that must be moderately heated or exposed to relatively high ambient temperatures before ignition can occur	2	Materials that in themselves are normally unstable and readily undergo violent chemical change but do not detonate. Also, materials that may react violently with water or may form potentially explosive mixtures with water
1	Materials that on exposure would cause irritation but only minor residual injury even if no treatment was given	1	Materials that must be preheated before ignition can occur	1	Materials that in themselves are normally stable, but which can become unstable at elevated temperatures and pressures or which may react with water with some release of energy but not violently
0	Materials that on exposure under fire conditions would offer no hazard beyond that of ordinary combustible material	0	Materials that will not burn	0	Materials that in themselves are normally stable, even under fire exposure conditions, and which are not reactive with water

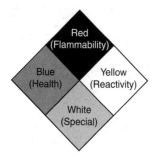

FIGURE 1.8 ◆ The potential hazards of a substance may be identified by use of a color-coded numeral system on this diamond-shaped diagram. The relative degree of severity is expressed by numbers ranging from 0 (no hazard) to 4 (maximum hazard). The substance illustrated by the number 4 (left) indicates that exposure may rapidly cause death or major residual illness; the 4 (top) indicates the material readily vaporizes at ambient conditions and burns if ignited; and the 2 (right) indicates the material is normally unstable, but does not detonate.

The NFPA system provides a procedure for rapidly identifying the relative degree of three hazards associated with a given hazardous material: its health, flammability, and chemical reactivity hazards. NFPA implements this procedure by assigning one of five numbers, zero through four, to each property for a given hazardous material. The numbers in Table 1.3 identify the relative degree of hazard that corresponds to the relevant property. The numbers 1 to 4 communicate the severity of a given hazard. The number 0 signifies that the hazardous material at issue does not possess the relevant hazard, whereas the number 4 denotes that it possesses the highest degree of that hazard.

Solved Exercise 1.5

To aid firefighters in an emergency-response action, identify the markings that NFPA recommends on the exterior walls of 15 storage tanks containing flammable liquids within a tank farm.

Solution: Firefighters responding to an emergency at a tank farm need readily accessible and accurate information concerning the contents of the tanks. Accordingly, the information marked on the storage tanks should be easily visible and legible and should provide the names of the liquids stored within them, their general hazards, and special firefighting precautions of which firefighters should be aware. Appropriate NFPA diamond symbols provide this general information by denoting the relative degrees of health, flammability, and reactivity hazards of the substances.

Each of the appropriate numbers is then displayed in the top three quadrants of the diamond-shaped diagram in Figure 1.8. Each quadrant of the diamond is often color-coded, beginning with the left-hand quadrant and proceeding clockwise: blue for the health hazard; red for the fire hazard; and yellow for the chemical reactivity hazard.

Four symbols are also employed in the NFPA system, each of which is displayed, when appropriate, in the bottom quadrant of the diamond:

- A radiation hazard symbol resembling a three-bladed propeller
- The letter "W" with a line drawn through its center to caution against the application of water
- The letters "OXY" to indicate that the material is an oxidizer
- The letter "P" to indicate that the material is prone to autopolymerize (Section 13.3)

◆ 1.10 CHEMTREC

CHEMTREC stands for <u>Chem</u>ical <u>T</u>ransportation <u>E</u>mergency <u>C</u>enter, a public service of the American Chemistry Council, Arlington, Virginia. By design, CHEMTREC is confined to dealing with transportation emergencies, to reinforcing the effectiveness of specialized emergency-response groups, and to enhancing hazardous materials transportation security.

CHEMTREC is a state-of-the-art communications center that provides immediate advice for those at the scene of a transportation emergency involving the release of a hazardous material. It operates around the clock, 24 hours a day, 7 days a week, to receive direct-dial, toll-free calls from any point in the United States through a wide-area service (WATS) telephone number.* During an average year, CHEMTREC receives over ten thousand calls relating to transportation emergencies involving the release of hazardous materials.

An emergency reported to CHEMTREC is received by the communicator on duty, who records the details in writing and tape-records the conversation. During the phone conversation, the communicator attempts to determine the following:

- The nature of the emergency and where and when it occurred
- The hazardous material involved and the type and condition of its tanks or containers
- The identity of the shipper and the shipping point
- The identity of the carrier, consignee, and destination
- The general nature and extent of the injuries, if any, to people, property, and the environment
- The prevailing weather conditions
- The composition of the surrounding area
- The name of the caller and a means by which the caller can be located
- A means by which telephone contact can be reestablished with the caller or another responsible person at the scene

With the caller remaining on the line, the communicator draws upon the best available information concerning the hazardous materials that are involved in the emergency-response action. As immediate first steps in controlling the emergency, specific hazard information and directions on what to do and what not to do are provided to the caller. CHEMTREC obtains this information by searching among the nearly 2.8 million MSDSs in its data base. To preclude unfounded personal speculation regarding the specific features of an emergency, the communicator is under instructions to abide strictly by information previously prepared for use.

*Within the United States, telephone (800) 424-9300 or (908) 859-2151. For calls originating outside the United States, telephone collect (703) 527-3887. Within Canada, telephone CANUTEC at (631) 996-6666. These numbers have been widely circulated in the professional literature distributed to emergency-service personnel, carriers, and the chemical industry, and has been further circulated in bulletins of governmental agencies, trade associations, and similar groups.

Having advised the caller, the communicator proceeds immediately to notify the shipper by telephone. The known particulars of the emergency incident thus relayed, the responsibility for further guidance to the emergency-response crew, including dispatching personnel to the scene, passes to the shipper.

Although proceeding to the second stage of assistance becomes more difficult when the shipper is unknown, communicators are armed with other resources upon which they can rely. For instance, CHEMTREC can contact the U.S. Department of Energy in emergencies involving the release of radioactive materials, or it can contact the Chlorine Institute to implement the Chlorine Emergency Plan (CHLOREP) in emergencies involving the release of chlorine (Section 7.4).

The notification of a transportation emergency involving hazardous materials to CHEMTREC does not constitute compliance with the RCRA, CERCLA, or U.S. Department of Transportation (DOT) reporting requirements noted in the following section and elsewhere in this book.

Performance Goal for Section 1.11
- ◆ Identify the general function of the National Response Center.

◆ 1.11 NATIONAL RESPONSE CENTER

The National Response Center (NRC) first began operating when EPA required the services of a federal group to whom responsible parties could report spills, discharges, fires, explosions, and other incidents in which oil and hazardous substances were released into waters of the United States. The existence of this group was first mandated by the Federal Water Pollution Prevention and Control Act and has been operated by the U.S. Coast Guard since 1974.

The responsibilities of the NRC were broadened subsequent to enactment of federal transportation and environmental regulations. Today, the NRC serves as the sole federal point of contact for reporting oil, hazardous materials, hazardous substances, biological, etiological (infectious), and radiological spills throughout the United States and its territories. The NRC serves as the link to the National Response Team (NRT), a group of individuals from 16 federal agencies having varying interest and expertise in emergency-response matters. When the situation warrants, the NRT is dispatched to the scene of a release incident to provide assistance.

Reporting a release incident initially involves telephoning NRC at (800) 424-8802 or (202) 426-2675. The notification should minimally include such information as the date, time, nature of the release incident, the quantity and type of material involved in the release, and the extent of injuries, if any. A follow-up written report to the relevant federal agency is also required.

The NRC maintains a communication link with CHEMTREC personnel, who provide the caller with emergency-response information. The NRC also informs the Chemical Safety and Hazard Investigation Board, which subsequently investigates the nature of the physical evidence found at the accident scene to determine the cause of the accident. EPA dispatches an on-scene coordinator who ensures that the subsequent cleanup follows the relevant environmental laws.

■ ■

REVIEW EXERCISES

Classification of Hazardous Materials

1.1 Assign each of the following to one or more of the seven classes of hazardous materials denoted in Section 1.2:

(a) Blasting agents
(b) Carbon monoxide
(c) Bottled gas
(d) Fissile material
(e) Home heating oil
(f) Garden insecticide
(g) Fireworks
(h) Gasoline
(i) Battery acid
(j) Lye
(k) Freshly mowed hay
(l) Safety matches
(m) Hand grenades
(n) Mustard gas
(o) Naphtha

Classes of Fire

1.2 Which class of fire is involved when each of the following ignites?

(a) Freshly mowed hay
(b) Bottled gas
(c) Cooking oil
(d) Magnesium engine parts
(e) Naphtha
(f) Acetylene
(g) Solvent-based paint
(h) Cardboard
(i) Rubber tires
(j) Liquefied petroleum gas
(k) Newspaper
(l) Energized transformer
(m) Aluminum dust
(n) Plastic container

Hazardous Substances Within Household Products

1.3 Why has a pesticide manufacturer included the following statements on the labels affixed to aerosol cans of a wasp and hornet killer?

Keep Out of Reach of Children

Active Ingredients:

Chloropyrifos[*O,O*-diethyl *O*-3,5,6-trichloro-2-pyridinyl phosphorothioate]	0.25%
d-*trans*-Allethrin	0.05%
Inert ingredients	99.70%

Hazards to Humans and Domestic Animals: Avoid contact with eyes, skin, and clothing. Avoid breathing vapors or spray mist. Keep away from food, feed-stuffs, and domestic water supplies. Wash thoroughly after handling.

Statement of Practical Treatment: If swallowed, call a physician or Poison Control Center. Drink 1 or 2 glasses of water and induce vomiting by touching back of throat with finger. Do not induce vomiting or give anything by mouth to an unconscious person. In case of skin contact, remove contaminated clothing and immediately wash skin with plenty of water. If in eyes, flush eyes with plenty of water. Note to physician: Chloropyrifos is a cholinesterase inhibitor. Treat symptomatically. Atropine only by injection is an antidote.

Environmental Hazards: This pesticide is toxic to fish, birds, and other wildlife. Do not apply directly to water. This product is highly toxic to bees exposed to direct treatment or residues on plants.

Physical or Chemical Hazards: Contents under pressure. Do not use or store near heat or open flame. Do not puncture or incinerate. Exposure to temperatures above 150°F may cause bursting.

1.4 Which federal statute provides the legal authority for requiring manufacturers, producers, and distributors to post the following warnings on the bottles of all alcoholic beverages sold in the United States?

Government Warning

(1) According to the Surgeon General, women should not drink alcoholic beverages during pregnancy because of the risk of birth defects.

(4) Consumption of alcoholic beverages impairs your ability to drive a car or operate machinery, and may cause health problems.

Hazardous Constituents of Pollutants and Wastes

1.5 EPA takes appropriate measures to ensure that the ozone concentration in the air does not exceed 0.12 parts per million (ppm) within a single hour. Which federal statute provides EPA with the legal authority to take such measures?

1.6 A chrome-plating company generates washwaters containing chromium. Which federal statute provides EPA with the legal authority to issue the company a permit granting it approval to discharge washwaters containing no more than a specified chromium concentration into a nearby river?

1.7 Which federal statute provides EPA with the legal authority to clean up a facility in the United States at which used oil was processed during the 1960s, and to file a legal action against the used oil generators for the cleanup expense?

1.8 Which federal statute provides EPA with the legal authority to require manufacturing and process industries to manage their chemical wastes in a fashion that protects public health and the environment?

Hazardous Substances in the Workplace

1.9 An adhesive caulk used by carpenters is characterized as a thick white or tan paste with a mild sweet odor. Canisters of the caulk bear the following warning statements:

Warning

May cause eye, skin, nose, throat and respiratory tract irritation. Harmful if swallowed or absorbed through the skin. Presents little or no hazard (if spilled) and/or no unusual hazard if involved in a fire.

Potential Health Effects
Eye contact: May cause eye irritation.
Skin contact: May cause allergic skin reaction or sensitization. Prolonged or repeated contact with skin may cause irritation.
Inhalation: Harmful if swallowed.
Chronic Hazards: Repeated or prolonged exposure may cause skin, respiratory, kidney, cardiovascular, and liver damage.

How can the manufacturer use this warning information when designing the Material Safety Data Sheet for this product?

1.10 A supplier of aluminum carbide provides its customers with the following information concerning the product:

(a) **Personal safety precautions**   Wear protective equipment. Keep unprotected persons away. Ensure adequate ventilation. Keep away from ignition sources

(b) **Measures for environmental protection**   Do not allow material to be released to the environment without proper governmental permits.

(c) **Measures for cleaning/collecting**   Ensure adequate ventilation. Do not flush with water or aqueous cleaning agents. Keep away from ignition sources.

Is the supplier required by OSHA to provide this information on the MSDS for this chemical product?

Hazardous Substances Within Communities

1.11 Why do many fire departments retain a compilation of the MSDSs for the chemical products known to be stored or used within their areas of jurisdiction?

NFPA Diamonds

1.12 The first-on-the-scene responders observe the numbers "1," "0," and "3" encoded within the three quadrants, proceeding clockwise from the far left, respectively, of an NFPA diamond affixed to the front of a burning storage shed. How is the observation of these numbers used by the emergency responders?

1.13 The first-on-the-scene responders observe the numbers "3," "3," and "3" encoded within the blue, red, and yellow quadrants, respectively, of an NFPA diamond affixed to the front of a burning storage shed. How is the observation of these numbers used by the emergency responders?

1.14 Acetyl chloride is a colorless liquid. When unconfined, the liquid emits a vapor having a pungent and choking odor. The vapor forms a mixture with air that readily ignites when exposed to an ignition source. Acetyl chloride reacts rapidly with many other chemical substances, forming toxic fumes. With water, it reacts violently, producing a vapor that stings the eyes and is corrosive to the skin. Based solely upon this characterization, assign appropriate numbers to each quadrant of the NFPA diamond that describes the hazards of this chemical product.

1.15 The labels affixed to 55-gallon drums of carbon disulfide provide the following hazard information about this chemical product:

> **Extremely Flammable Liquid Harmful if Inhaled or Swallowed**
>
> Extremely flammable liquid. Immediately evacuate area. Firefighters should use pressure-demand, self-contained breathing apparatus to avoid possible exposure to sulfur dioxide. Use water fog or spray and cool area after extinguishing to avoid possible reignition. The autoignition temperature of carbon disulfide is 212°F. Use carbon dioxide or dry chemicals for small fires.
>
> *Inhalation:*   Remove to fresh air. If the person is not breathing, give artificial respiration, preferably mouth-to-mouth. If breathing is difficult, administer oxygen. Call a physician.
>
> *Ingestion:*   *If conscious*, drink a full quart of water. Then, induce vomiting by placing a finger far back in the throat. Call a physician. If vomiting cannot be induced, take immediately to a physician or hospital. *If unconscious, or in convulsions*, take immediately to a physician or hospital. Do not induce vomiting or give anything by mouth to an unconscious person.

Based solely upon this characterization, assign appropriate numbers to each quadrant of the NFPA diamond that describes the hazards of carbon disulfide.

1.16 For each of the following pairs of NFPA diamonds, indicate which conveys the greater degree of the following hazards:

(a) Fire

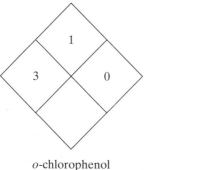

o-chlorophenol

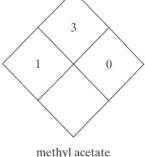

methyl acetate

(b) Health

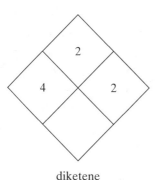

diketene

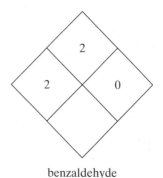

benzaldehyde

(c) Chemical reactivity

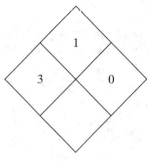

m-nitrotoluene

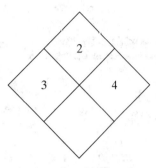

peracetic acid diluted
with 60% acetic acid

(d) Combined fire and health

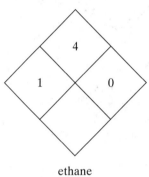

ethane

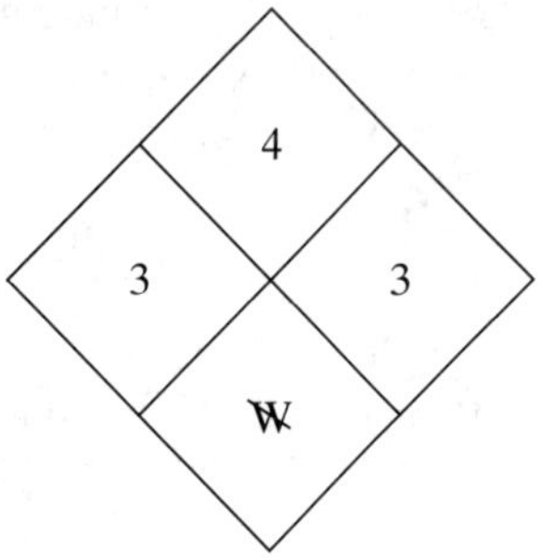

diethylzinc

Some Features of Matter and Energy  2 CHAPTER

At first glance, the mere number of hazardous materials is likely to overwhelm the average nonscientist. How is it possible to learn the individual properties of so many substances and recall them under the disordered conditions that often prevail when lives and property are in jeopardy?

Fortunately, the hazardous properties of many substances can be associated with their state of matter. For instance, all gases possess certain common properties; upon studying the chemistry of gases, we learn to identify these common properties and then turn our attention later to the features that cause individual gases to be regarded as unique substances.

In this chapter, we shall learn about some of the general properties of matter and energy and how they influence certain phenomena such as the spread of fire. Also, as we review the properties of matter and energy, we will learn how they relate to the issues in fire science. Specifically, in this chapter we will learn how the modes of heat transfer contribute to the propagation of fire, the reason liquid water often effectively extinguishes a fire, and why a gas confined within a cylinder is likely to rupture the cylinder when excessively heated.

Performance Goals for Section 2.1
- ◆ Describe the general features of matter.
- ◆ Distinguish between mass and weight.
- ◆ Identify the properties that characterize the three physical states of matter.

◆ 2.1 MATTER DEFINED

Each day, we encounter countless types of matter, such as air, water, metals, stone, dirt, animals, and plants. These materials of which the world is made are different kinds of *matter*. When we search for common features among its forms, we note that all matter possesses mass and occupies space. In other words, matter is distinguishable from empty space by its presence in it.

The concept of *mass* is closely related to the concept of *weight*. In our universe, every form of matter is attracted to all other forms by the force we call *gravity*. On Earth, the weight of matter is a measure of the force with which gravity pulls it toward Earth's center. As we leave Earth's surface, the gravitational pull decreases until it becomes virtually insignificant. The *weight* of matter accordingly reduces to zero. Yet the matter still possesses the same *mass* as it did on the surface of Earth.

As usually experienced,* matter exists in three different forms or states of aggregation: solid, liquid, and gas. These are called the *physical states of matter*.

Since matter occupies space, a given form of matter is also associated with a definite volume or capacity. Space should not be confused with air, since air is itself a form of matter. *Volume* refers to the actual amount of space that a given form of matter occupies.

SOLIDS

A *solid* is the form of matter existing in a rigid state independent of the size and shape of its container. Consider this book. It retains its shape regardless of its position in space and does not need to be placed into a container to retain that shape. Left to itself, it will never spontaneously assume a shape different from what it has now.

Solids also occupy a definite volume at a given temperature and pressure. We can squeeze most solids with all our might or heat or cool them, but their total volume change is relatively insignificant.

Certain solid plastics can behave unlike most solids. Because of their inherent elasticity, they could assume different shapes and volumes when squeezed, stretched, or otherwise manipulated. Solid foams can also behave uncharacteristically, since they contain encapsulated air.

LIQUIDS

A *liquid* is a form of matter that does not possess a characteristic shape; rather, its shape depends upon the shape of the container it occupies. Consider water within a glass. The liquid water takes the shape of the glass up to the level that it occupies. If we pour the water into a cup, the water takes the shape of the cup; or, if we pour it into a bowl, the water takes the shape of the bowl. Of course, sufficient space must always be available for the water within the container; otherwise, the liquid overflows. But, assuming that space is available, any liquid assumes whatever shape its container possesses.

Like solids, liquids occupy a specific volume at a given temperature and pressure. They tend to maintain a relatively fixed volume when they are exposed to a change in either of these conditions. Liquids and solids are often considered incompressible, since the application of pressure barely changes their volumes. When heated, liquids do expand, more than solids do, but not nearly to the degree that gases do. We shall revisit the expansion of heated liquids in Section 2.11.

GASES

A *gas* is a form of matter that does not possess a characteristic shape and will take the shape of its container. If a gas or mixture of gases, such as air, is put into a balloon, it assumes the shape of the balloon; or, if it is transferred into a tire, it assumes the shape of the tire.

---

*Other physical states of matter aside from the solid, liquid, and gas are known. A fourth state, called *plasma*, exists only at very high temperatures, while a fifth state, called the *Bose–Einstein condensate*, exists only at very low temperatures. Although we will not encounter materials in the fourth and fifth states during the study of hazardous materials, it is interesting to note that the sun, other stars, and other forms of intergalactic matter exist in the plasma state. It is the predominant state of matter in the universe.

Gases also lack a characteristic volume. When confined to a container with non-rigid, flexible walls, the volume that a confined gas occupies depends upon its temperature and pressure. When confined to a balloon, for instance, the gas's volume expands and contracts depending on the prevailing temperature and pressure. When confined to a container with rigid walls, however, the volume of the gas is forced to remain constant. This property of gases could cause rigid containers to explode, a topic we shall note later in Section 2.12.

These properties of solids, liquids, and gases can now be used to formally define the three states of matter:

- Matter in the solid state possesses a definite volume and a definite shape.
- Matter in the liquid state possesses a definite volume, but lacks a definite shape.
- Matter in the gaseous state possesses neither a definite volume nor a definite shape.

Performance Goals for Section 2.2
- Identify the common English and metric units used to measure length, mass, and volume.
- Learn the equivalent English and metric units of measurement for length, mass, and volume.

◆ 2.2 UNITS OF MEASUREMENT

The necessity to measure, and to measure with accuracy, is essential to any kind of scientific or technological endeavor. To *measure* means to find the number of units in a sample of something. For instance, when we measure the distance from one point to another along a wall, we generally determine how many feet, yards, or meters are between the two points. The foot, yard, and meter are examples of common units of length.

Two systems of units have survived the test of time: the English system and the metric system. In the United States (and Liberia and Myanmar), the *English system* is still used. Even the English no longer use it. This system comprises an array of units that have no obvious interrelationship, such as inches (in.), feet (ft), yards (yd), miles (mi), ounces (oz), pounds (lb), tons (tn), pints (pt), quarts (qt), and gallons (gal), the combination of which are still called the *English units of measurement*.

The majority of the world's population and the worldwide scientific community use a system of measurement called the *metric system*. The metric system has itself been modified so that today, we actually use the *SI system*, a name derived from its official French name, *Le Système International d'Unités*. The SI system encourages the use of certain fundamental units from which all other measurements are constructed. These fundamental units are called the *SI base units*. Examples of SI base units are the meter and kilogram, for length and mass, respectively.

Certain prefixes are used in the SI system to denote multiples and fractions of the units of measurement. Each prefix is a fraction or multiple of the number 10. For example, when we wish to refer to 1000 meters, we use the word *kilometer*. The prefix *kilo* means 1000 times the meter, the SI base unit. Only the four prefixes listed in Table 2.1 are commonly used in studying the chemistry of hazardous materials. These particular prefixes should be committed to memory. They are used to measure certain

TABLE 2.1 ◆ Common Prefixes Used in the SI (Metric) System	
Prefix	*Meaning*
kilo-	One thousand times the SI base unit*
centi-	One-hundredth of the SI base unit
milli-	One-thousandth of the SI base unit
micro-	One-millionth of the SI base unit

*See text for an exception in the case of the SI base unit of mass.

properties of matter, particularly its length, mass, and volume. It is appropriate to discuss each type of measurement separately.

LENGTH

Today, the *meter* is defined as the distance light travels in 1/299,792,458 of a second. In practice, we measure length in the metric system with a metric ruler. By so doing, we discover that 1 meter (m) is slightly longer than a yard; specifically, 1 meter equals 39.37 inches (in.).

$$1 \text{ m} = 39.37 \text{ in.}$$

One meter is equivalent to 100 centimeters (cm) and to 1000 millimeters (mm).

$$1 \text{ m} = 100 \text{ cm} = 1000 \text{ mm}$$

For very large lengths, we use the kilometer (km); once again, 1000 meters are equivalent to 1 kilometer.

$$1 \text{ km} = 1000 \text{ m}$$

For very small lengths, we use the micron (μm). One micron is one-millionth of a meter.

$$1 \text{ μm} = 0.000001 \text{ m}$$

The interrelationship between the inch and centimeter is shown in Figure 2.1. One inch equals 2.54 centimeters.

$$1 \text{ in.} = 2.54 \text{ cm}$$

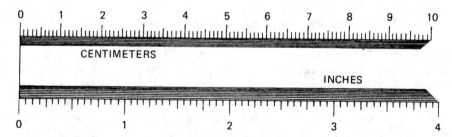

FIGURE 2.1 ◆ The relationship between the inch and centimeter. Note that 1 inch (in.) equals 2.54 centimeters (cm).

MASS

The SI unit of mass is the *kilogram*. A bit of attention needs to be given to constructing the multiples and fractions of mass measurements, since this unit of mass is the only one that already contains a prefix (kilo-). The names of the various multiples and fractions of the unit of mass are constructed by attaching the appropriate prefix to the word *gram*, not kilogram. One kilogram is equivalent to 2.2 pounds. In other words, the gram is used as though it is the SI unit of mass, even though it actually is not. One gram is approximately the mass of a peanut.

Three common metric units of mass are the milligram, microgram, and kilogram. One one-thousandth of a gram is called a *milligram* (mg); one one-millionth of a gram is called a *microgram* (μg); and one thousand grams is the *kilogram* (kg).

$$1 \text{ mg} = 0.001 \text{ g}$$
$$1 \text{ μg} = 0.000001 \text{ g}$$
$$1 \text{ kg} = 1000 \text{ g}$$

One milligram is approximately the mass of a grain of sand, while one microgram is approximately the mass of a fleck of dust.

For relatively large mass measurements, use is made of the metric ton, or *tonne*.

$$1 \text{ tonne} = 1000 \text{ kg}$$

VOLUME

The approved SI unit of volume is the *cubic meter* (m^3). This unit is derived directly from the SI unit of length and is not a SI base unit itself. We can easily derive the cubic meter by considering the cube in Figure 2.2, which measures 1 meter on each edge. The volume of this cube is determined by taking the product of its length, width, and height.

$$\text{Volume} = 1 \text{ m} \times 1 \text{ m} \times 1 \text{ m} = 1 \text{ m}^3$$

Using simple arithmetic, the volume of this cube is determined to be 1 cubic meter. Since it is possible to derive the cubic meter directly from a previously defined unit, it is unnecessary to define some other unit as the SI unit of volume.

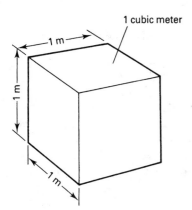

1 cubic meter

1 m

1 m

1 m

FIGURE 2.2 ◆ A cube that measures 1 meter to each edge. The volume occupied by this cube is 1 *cubic meter* (m^3), the approved SI unit of volume.

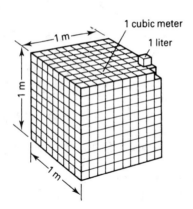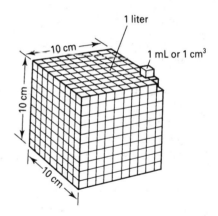

FIGURE 2.3 ◆ The cube on the left measures 1 meter to each edge and has been divided into 1000 equally sized cubes. The volume occupied by each of the smaller cubes is 1 liter (L). The cube on the right measures 10 cm to each edge and has been divided into 1000 equally sized cubes. The volume of the larger cube is 1 liter and the volume of each smaller cube is 1 milliliter (mL) (or 1 cubic centimeter).

Although the cubic meter is the approved unit of volume in the SI system, another unit has been used for many years to measure volume. This unit is the *liter* (L). One liter is slightly less than one quart.

$$946 \text{ mL} = 1 \text{ qt}$$

Chemists continue to use the liter for measuring volume because its size is so convenient for laboratory-scale measurements. By comparison, the cubic meter is generally too large.

One liter is equivalent to one one-thousandth of a cubic meter.

$$1 \text{ L} = 0.001 \text{ m}^3$$

If we construct a cube measuring 1 meter to each edge, and then divide the resulting volume into 1000 equally sized cubes, the volume of each small cube is 1 liter. Imagine further dividing each liter cube into another 1000 equally sized cubes. These subdivisions are illustrated in Figure 2.3. Since the prefix milli- means one one-thousandth of a unit, the volume of each of the new cubes is 1 *milliliter* (mL).

$$1 \text{ L} = 1000 \text{ mL}$$

Formerly, the milliliter was known as a cubic centimeter (cm^3). The cubic centimeter is still occasionally used in chemistry.

GENERAL METRIC USE IN THE UNITED STATES

In the international community, the United States stands alone as the major country that has not officially adopted the use of the metric system. To address the concern that the United States' nonuse of the metric system was contributing to unsuccessful competition in some international markets, Congress passed the *Metric Conversion Act* in 1975. Notwithstanding its passage, the act called only for voluntary compliance and accomplished little change in the manner by which Americans measure the properties of matter.

Then, in 1988, Congress passed the *Omnibus Trade and Competitiveness Act,* which aimed to accomplish the following:

- Identification of the metric system as the preferred system for United States' trade and commerce
- Required use of the metric system by all U.S. government agencies
- Required use of metric units for all federally funded construction projects costing over $10 million
- Required use of metric units for all federally assisted highway construction projects that began on or after October 1, 1996

This act obligated building and highway construction companies to begin using the metric system. In this fashion, the use of the metric system has slowly crept into the United States and taken us one step further toward common worldwide usage.

Performance Goal for Section 2.3
- Use the factor-unit method for converting between "English" and metric units of measurement.

◆ 2.3 CONVERTING BETWEEN UNITS OF THE SAME KIND

Suppose we wish to convert between grams and kilograms, pounds and grams, or liters and gallons. How would we accomplish this task? Problems like these are best solved by using a simple procedure called the *factor-unit method*. Briefly, it consists of the following key steps:

- Identify the desired unit.
- Choose the proper conversion factor(s).
- Multiply the given measurement by the conversion factor(s), being certain to multiply, divide, and cancel equal units.

Choosing the proper conversion factor is a crucial step to obtaining the correct answer to a problem. The conversion factor is a fraction numerically equal to 1 that equates the quantity in the numerator to the quantity in the denominator. For example, we know there are 1000 m in 1 kilometer. Either of the following fractions serves as a correct conversion factor:

$$\frac{1000 \text{ m}}{1 \text{ km}} \quad \text{or} \quad \frac{1 \text{ km}}{1000 \text{ m}}$$

The factors are respectively read as follows: 1000 meters per kilometer; one kilometer per 1000 meters.

Suppose we know that the height of the Sears Tower in Chicago is 443.2 m, measured from ground level to the top. What is its height in kilometers? Using the factor-unit method, we multiply the given measurement, 443.2 m, by a conversion factor that relates meters to kilometers.

$$443.2 \text{ m} \times \frac{1 \text{ km}}{1000 \text{ m}} = 0.4432 \text{ km}$$

TABLE 2.2 ◆ Common SI (Metric) Units and Their English Equivalents
1 m = 39.37 in.
2.54 cm = 1 in.
1 kg = 2.2 lb
454 g = 1 lb
946 mL = 1 qt

Note that the "m" in 443.2 m cancels with the "m" in 1000 m. Then, dividing 443.2 km by 1000, which gives 0.4432 km.

Individuals who are practiced in using the metric system simply move the decimal point from left to right, or vice versa, as the need arises.

Some of the more frequently used metric units and their equivalent English counterparts are given in Table 2.2. These interrelationships permit us to write conversion factors like the following:

$$\frac{1\ m}{39.37\ in.} \qquad \frac{2.54\ cm}{1\ in.} \qquad \frac{1\ kg}{2.2\ lb} \qquad \frac{1\ qt}{946\ mL}$$

Solved Exercise 2.1

The standard length of fire hose in the United States is 50 ft. What is the equivalent length when expressed in meters?

Solution: We first need one or more conversion factors to convert 50 ft into its equivalent length in meters. Since one meter is 39.37 in., and one foot is 12 in., we can construct the following conversion factors:

$$\frac{1\ m}{39.37\ in.} \qquad \frac{12\ in.}{1\ ft}$$

Through use of the factor-unit method, we then convert 50 ft into meters as follows:

$$50\ ft \times \frac{12\ in.}{1\ ft} \times \frac{1\ m}{39.37\ in.} = 15\ m$$

Solved Exercise 2.2

A firefighter must possess the physical agility to easily carry a 60-lb "bundle" from the base to the top of a 148-ft ladder. Express the bundle mass in kilograms and the length of the ladder in meters.

Solution: According to the information in Table 2.2, 1 kg = 2.2 lb and 1 m = 39.37 in. Using the factor-unit method, a 60-lb bundle is equivalent to a 27-kg bundle.

$$60 \text{ lb} \times \frac{1 \text{ kg}}{2.2 \text{ lb}} = 27 \text{ kg}$$

Furthermore, a 148-ft ladder is equivalent to a 45-m ladder.

$$148 \text{ ft} \times \frac{12 \text{ in.}}{1 \text{ ft}} \times \frac{1 \text{ m}}{39.37 \text{ in.}} = 45 \text{ m}$$

Performance Goals for Section 2.4
- Describe the concepts of percentage, parts per million, and parts per billion.
- Express the concentration of a substance in percentage, parts per million, and parts per billion.

◆ **2.4 CONCENTRATION**

Scientists often measure the amount of a substance as a stated unit in a given mass or volume of a mixture, such as the number of micrograms of carbon monoxide per cubic meter of air. This is called the substance's *concentration*.

The percentage is a common method of expressing a concentration. The word *percentage* means parts per hundred. To calculate the percentage of an item in a total, we divide the number of items by the total number and multiply the product by 100. The percentage is expressed as a number followed by the percent sign (%).

In chemistry, we often express the concentration of a substance in a mixture as either the percentage by mass or percentage by volume. In the former instance, mass units of the same type are used to determine the concentration, and in the latter case, volume units of the same type are used. For example, suppose we have 100 g of a solution in which 5 g of table salt is dissolved. The concentration of table salt in the solution is expressed as 5% by mass. If we have 100 mL of a solution in which 5 mL of a substance is dissolved, the concentration of table salt in the solution is expressed as 5% by volume.

In this textbook, we shall also encounter hazardous materials in the air or water whose concentrations are expressed in parts per million (ppm) or parts per billion (ppb). A concentration of 1 ppm means 1 part of a substance measured as an arbitrary unit in a million parts of the same unit. For example, depending on the units used, 1 ppm could be one molecule in a million molecules, one gram in a million grams, or one ounce in a million ounces.

Similarly, one part per billion means 1 part of a substance measured as a unit in a billion parts of the same unit. Parts per million are related to parts per billion as follows:

$$1 \text{ ppm} = 1000 \text{ ppb}$$

To appreciate the magnitude of a part per million, imagine drawing one million circles on a large piece of paper. Now darken one of them. This represents a concentration of one part per million. We can experience ill effects, injury, or death from breathing air or drinking water containing certain hazardous substances having concentrations of one part per million or less.

Performance Goals for Section 2.5
- ◆ Calculate the density of a substance from its mass and volume.
- ◆ Relate the density of a substance to its specific gravity or vapor density.
- ◆ Use the specific gravity of a specific liquid to evaluate its behavior when mixed with water.
- ◆ Use the vapor density of a specific gas or vapor to evaluate its behavior when released from confinement.

◆ 2.5 DENSITY OF MATTER

If the volume of a given mass of a substance is known, we can readily use division to compute its mass per unit volume. This ratio is a property of substances known as the *density*.

$$\text{Density} = \frac{\text{mass}}{\text{volume}}$$

For instance, suppose we weigh exactly 50 pounds of water on a scale.

$$\text{Mass} = 50.00 \text{ lb}$$

Then, we transfer this water into containers of known capacity. When this simple exercise is performed, we discover that 50.00 lb of water completely fills six 1-gallon containers to their brims at 20° (68°F)

$$\text{Volume} = 6.00 \text{ gal}$$

The density of water is then computed to be 8.33 lb/gal, as follows:

$$\text{Density of water} = \frac{50.00 \text{ lb}}{6.00 \text{ gal}} = 8.33 \text{ lb/gal}$$

In the English system of units, the densities of solids and liquids are usually determined in pounds per gallon (lb/gal) or pounds per cubic foot (lb/ft^3), while the densities of gases or vapors are determined in pounds per cubic foot. In the metric system, densities of solids and liquids are normally determined in grams per milliliter (g/mL), while the densities of gases and vapors are determined in grams per milliliter and grams or kilograms per cubic meter (g/m^3 or kg/m^3, respectively).

The densities of some common substances are provided in Table 2.3. When the density of water is determined using metric units, the obvious result at 20°C (68°F) is 1.00 g/mL. This result is obtained because water was originally selected as the reference standard for establishing the units of mass and volume in the metric system. One gram was once defined as the mass of water that occupied a volume of 1 milliliter at 20°C (68°F).

TABLE 2.3 ◆ Densities of Some Common Liquids and Solids

Substance	Grams/Milliliter (g/mL) at 20°C	Pounds/Gallon (lb/gal) at 68°F
Acetone	0.792	6.6
Aluminum	2.7	22.5
Benzene	0.879	7.3
Carbon disulfide	1.274	10.6
Carbon tetrachloride	1.595	13.3
Chloroform	1.489	12.4
Diethyl ether	0.730	6.1
Ethyl alcohol	0.791	6.6
Gasoline	0.66–0.69	5.5–5.7
Kerosene	0.82	6.8
Lead	11.34	94.4
Mercury	13.6	113
Silver	10.5	87.4
Sulfur	2.07	17.2
Turpentine	0.87	7.2
Water, 39°F (4°C)	1.00	8.3

Solved Exercise 2.3

The density of dry air in kilograms per cubic meter (kg/m^3) varies with temperature as follows: 1.29 (0°C); 1.25 (10°C); 1.21 (20°C); and 1.16 (30°C). What do these data illustrate about the nature of dry air during building fires?

Solution: These data denote that the density of dry air decreases as the temperature of the air increases. In other words, a given volume of dry air gets lighter in mass as it becomes hotter. This lighter, hotter air is buoyant, that is, it rises upward. During building fires, the lighter, hotter air concentrates near the undersides of ceilings and roofs, whereas the denser, colder air concentrates near the floors.

All substances possess densities that vary with the prevailing temperature, and some possess a maximum density at a unique temperature. Figure 2.4 shows that the density of melted ice increases as the temperature of the water increases until the temperature becomes 39.1°F (3.97°C). Then, the density of the water decreases as the temperature continues to increase.

SPECIFIC GRAVITY

Often, the mass of a liquid and solid substance is compared to the mass of an equal volume of water. This comparison yields a dimensionless number called its *specific gravity*. For example, if we weigh 1 gallon of sulfuric acid, we find that it weighs 15.33 lb.

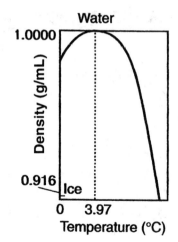

FIGURE 2.4 ◆ The density of water near its freezing point as a function of decreasing temperature.

This means that the density of sulfuric acid is 15.33 lb/gal. We can compare the density of sulfuric acid to the density of water to compute the specific gravity of sulfuric acid, as follows:

$$\text{Specific gravity} = \frac{15.33 \text{ lb/gal}}{8.33 \text{ lb/gal}} = 1.84$$

This computation informs us that any volume of sulfuric acid is 1.84 times heavier than an equal volume of water.

The specific gravities of liquids are directly determined through the use of a device called a *hydrometer*, a cylindrical glass stem containing a bulb weighted with shot so that it floats upright in a liquid. A graduated scale is printed on the hydrometer's stem. A sample of the liquid is generally poured into a tall jar, as in Figure 2.5, and the hydrometer is lowered into the sample until it floats. To determine the specific gravity of the liquid, read the graduated scale at the point where the surface of the liquid touches the stem. The specific gravity of water is 1.0. The specific gravities of some other common liquids are provided in Table 2.4.

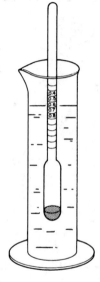

FIGURE 2.5 ◆ A hydrometer is a device that is often employed to determine the specific gravity of a liquid. The specific gravity of this liquid is 0.68. The hydrometer operates on the principle that an object floats "high" in a liquid of relatively high specific gravity (greater than 1.0) and "low" in a liquid of relatively low specific gravity (less than 1.0).

TABLE 2.4 ◆ Specific Gravities of Some Liquids[a]

Substance	Specific Gravity at 20°C (68°F)
Acetic acid	1.05
Allyl chloride[a]	0.94
Chlorobenzene	1.11
Cyclopentanone[a]	0.95
2-Ethylhexanol[a]	0.83
Heptane[a]	0.68
Hydrochloric acid	1.19
Methyl acetate	0.97
Nitric acid	1.50
Sulfuric acid	1.84
Tetraethyllead[a]	1.66
Trichlorofluoromethane[a]	1.49

[a]Immiscible with water and flammable, except that trichlorofluoromethane is nonflammable.

While many substances are soluble in water to some extent, there are also other substances that are relatively insoluble in water. When a liquid substance does not appreciably dissolve in water, we say that they are mutually *immiscible*. Oil and water are examples of two mutually immiscible liquids. When combined, mutually immiscible liquids coexist as two separate and distinct phases, one on top of the other.

When a liquid is immiscible with water, its specific gravity tells us how the liquid will behave when it is mixed with water as follows:

- An immiscible liquid *floats* on water when it possesses a specific gravity less than 1.0.
- An immiscible liquid *sinks* below water when it possesses a specific gravity greater than 1.0.

All grades of fuel oils possess specific gravities less than 1.0. Consequently, when an oil tanker ruptures at sea, the oil that spills from the tanker floats upon the water, forming an oil slick.

Knowing the specific gravity of a flammable liquid can be useful in certain instances when ascertaining whether water will function effectively as a fire extinguisher. The usefulness of this information is demonstrated in Figure 2.6 by the following situations:

- In the first situation, a burning liquid immiscible with water and having a specific gravity less than 1.0 is confined within a drum or tank. When water is added, it sinks below the surface of the liquid where it is incapable of extinguishing the fire. If excess water is added, the burning liquid overflows the container or tank before the water. Consequently, the application of water could potentially worsen the situation by spreading the fire to adjoining areas.
- In the second situation, a burning liquid immiscible with water and possessing a specific gravity greater than 1.0 is confined within a drum or tank. When water is added, it floats upon the heavier liquid and smothers it. In this instance, the fire is effectively extinguished.

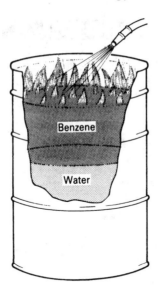 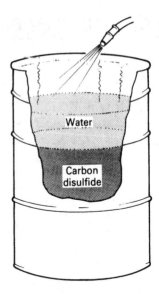

FIGURE 2.6 ◆ On the left, water is added to a drum containing burning benzene (specific gravity = 0.879). Water and benzene are immiscible liquids. The water settles below the benzene and does not ordinarily extinguish the fire. On the right, water is added to a drum containing burning carbon disulfide (specific gravity = 1.274). Water and carbon disulfide are also immiscible liquids. The water floats on the carbon disulfide, which prevents further contact of the fuel with atmospheric oxygen. In this case, the addition of water effectively extinguishes the fire.

VAPOR DENSITY

Air is a mixture of several gases that possesses a density of 0.08 lb/ft^3, or 1.29 g/L, at 32°F (0°C) and 1 atm of pressure (Section 2.8). Often, the mass of a given volume of a substance is compared to the mass of an equal volume of air. The comparison yields the *vapor density* of that substance. The vapor density is a dimensionless property of all gases.

Suppose we compare the masses of equal volumes of oxygen and air. From Table 2.3, we see that a liter of oxygen weighs 1.43 g. This information can be used to compute the vapor density of oxygen relative to air, as follows:

$$\text{Vapor density of oxygen} = \frac{1.43 \text{ g/L}}{1.29 \text{ g/L}} = 1.11$$

The value tells us that an arbitrary volume of oxygen is 1.11 times heavier in mass than an equal volume of air. The vapor densities of some common gases are provided in Table 2.5.

An awareness of the vapor density of a gas or vapor is useful for identifying the initial location of these substances when they are released into the air. The information can be summarized as follows:

- When the vapor density of a gas or vapor is greater than 1.0, an arbitrary volume of the gas or vapor is heavier than the same volume of air. When the gas or vapor is released from its container, it tends to concentrate in low spots.
- When the vapor density of a gas or vapor is less than 1.0, an arbitrary volume of the gas or vapor is lighter than the same volume of air. When it is released, the gas or vapor naturally rises into the air.

TABLE 2.5 ◆ Vapor Densities of Some Common Gases

Substance	Vapor Density (air = 1)
Acetylene	0.899
Ammonia	0.589
Carbon dioxide	1.52
Carbon monoxide	0.969
Chlorine	2.46
Fluorine	1.7
Hydrogen	0.07
Hydrogen chloride	1.26
Hydrogen cyanide	0.938
Hydrogen sulfide	1.18
Methane	0.553
Nitrogen	0.969
Oxygen	1.11
Ozone	1.66
Propane	1.52
Sulfur dioxide	2.22

A leak of a gaseous substance whose vapor density is greater than 1.0 constitutes a potentially serious risk, since the gas tends to displace the air and concentrate in low spots before ultimately dissipating into the air. When the substance is poisonous, its presence poses the risk of inhalation toxicity. When the substance is flammable, its presence poses a serious risk of fire and explosion. For example, a given volume of gasoline vapor is heavier than an equal volume of air. When gasoline evaporates, its vapor accumulates in low spots such as the bilges in boats, where exposure to an ignition source could cause the vapor to ignite.

On the other hand, a gaseous substance whose vapor density is less than 1.0 tends to rise in the atmosphere before it ultimately dissipates. In the open air, a gaseous substance whose vapor density is less than 1.0 dissipates rapidly. On the other hand, when this substance is released within a room, it first accumulates near its ceiling. If it is flammable, the localized concentration of this substance may pose the risk of fire and explosion.

All gases and vapors are totally miscible with air. This means that given adequate time, they become completely mixed with the components of the air. We say that gas and vapors are *infinitely miscible* with air.

Solved Exercise 2.4

Carbon monoxide is a poisonous gas emitted to the atmosphere from the exhaust pipes of operating motor vehicles. When first emitted from an exhaust pipe, does carbon monoxide tend to concentrate in the lower or upper regions of a confined garage?

Solution: Table 2.5 reveals that carbon monoxide possesses a vapor density of 0.969, indicating that an arbitrary volume of carbon monoxide is only slightly lighter than an equal volume of air. In addition, the gases emitted from exhaust pipes of operating motor vehicles are hotter than the air into which they are released. From these two facts, we can conclude that the carbon monoxide emitted from a vehicle's exhaust pipe is likely to rise toward the ceiling of the garage. Since the exhaust is hot and the densities of carbon monoxide and air are so similar, the carbon monoxide will not concentrate at the ceiling for long. Instead, it is likely to disperse rapidly within the confined space of the garage and mix with the air.

Performance Goals for Section 2.6
- Define energy and identify the different forms in which it is manifested.
- Identify the common units in the English and metric systems used to measure energy.
- Describe how mass and energy are interrelated.

◆ 2.6 ENERGY

Energy is defined as the capacity to do work; thus, energy is proportional to work. In a broad sense, when we expend an effort to accomplish some act, the activity is referred to as work. Scientists describe work in terms of a force applied over a distance, like a push or pull of matter that results in moving it.

Energy exists in a variety of forms, although these forms are often abstract: radiant energy (light, heat and X rays), thermal energy (heat), acoustical energy (sound), mechanical energy, chemical energy, electrical energy, and atomic energy. All forms of matter possess *chemical energy*, which is the energy stored in its chemical makeup. Each hazardous material possesses chemical energy, which could constitute the basis for its hazardous nature. Dynamite, for example, possesses a large amount of chemical energy. When dynamite detonates, some of its chemical energy is released to the environment as light, heat, and sound.

Energy is measured in various units. In the United States, we commonly encounter energy units like the British thermal unit (Btu) and the calorie. The SI unit of energy is called the *joule*. One joule (J) is the energy possessed by a 2-kilogram mass moving at a velocity of 1 meter per second. We will revisit these units in Section 2.9.

Mass and energy are closely interrelated. Neither mass nor energy can be created or destroyed. Certain phenomena require the conversion of matter to energy, and vice versa, but regardless of the nature of the transformation, the total amount of mass and energy in the universe remains constant. This observation is an important law of nature known as the *law of conservation of mass and energy*.

Performance Goals for Section 2.7
- Describe the concept of temperature.
- Describe the differences between temperature readings on the Fahrenheit, Celsius, Kelvin, and Rankine scales.
- When provided with a reading on one temperature scale, convert it to its equivalent on the other temperature scales.

◆ 2.7 TEMPERATURE AND ITS MEASUREMENT

Temperature is the property of matter associated with its degree of hotness or coldness. Hot matter is associated with high temperatures, and cold matter is associated with low temperatures. To say that something is hot or cold merely points out a relative condition, whereas a temperature reading tells us just how hot or cold matter is. The temperature is an indication of a condition of matter, just as the measurement of its size is an indication of how large or small it is.

The temperature of many substances is most commonly determined using the simple mercury-in-glass column thermometer. This is a sealed glass capillary tube that has been partially filled with mercury and then calibrated according to a prescribed procedure. When inserted into a substance, the temperature of the glass and mercury rises or falls—causing the mercury to expand or contract, respectively—until it reaches the temperature of the substance itself. By observing the height of a calibrated column of mercury, we then measure the temperature of this substance.

Temperature is measured on a number of scales, four of which are illustrated in Figure 2.7 and discussed independently below.

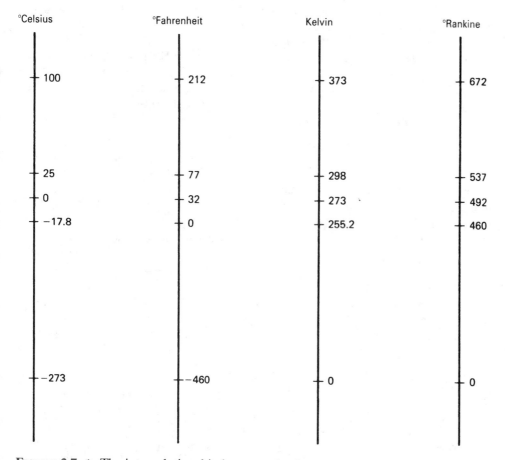

FIGURE 2.7 ◆ The interrelationship between the four temperature scales: Celsius (°C), Fahrenheit (°F), Kelvin (0 K), and Rankine (°R). Note that 32°F = 0°C and 212°F = 100°C. Absolute zero is −273.15°C (0 K), or −459.67°F. Also note that 25°C = 77°F = 298 K = 5537°R.

FAHRENHEIT AND CELSIUS TEMPERATURE SCALES

Suppose we take an unmarked capillary tube containing mercury and place it in the steam that evolves above boiling water at sea level. The mercury level rises to a certain point and then remains stationary. This height can now be etched on the glass and serves as the first reference point, called the *steam point*. If we place the same tube in ice water, the mercury level drops and again comes to a position where it remains stationary. This height of the mercury column serves as the second reference point, called the *ice point*. These calibrations are assigned temperature units in at least two common ways:

- On the *Fahrenheit scale*, the ice point and steam point are assigned the values of 32 degrees and 212 degrees, respectively. There are 180 equally spaced divisions between these two reference points, each division called one degree Fahrenheit (°F). A thermometer that has been calibrated in this fashion is called a *Fahrenheit thermometer*.
- On the *Celsius scale*, the ice point and steam point are assigned the values of 0 degrees and 100 degrees, respectively. There are 100 equally spaced divisions between these two reference points; each division is called one degree Celsius (°C). This is called a *Celsius thermometer*.

Thus, the Fahrenheit and Celsius thermometers are alike insofar as the freezing and boiling points of water are used to define their reference points, but they differ by the numbers that are assigned to these points.

Since the same temperature interval is divided into 180 degrees on the Fahrenheit scale and 100 degrees on the Celsius scale, the temperature range corresponding to 1 Celsius degree is 180/100, or 9/5, as great as the temperature range corresponding to 1 Fahrenheit degree. This factor can be used to relate these two temperature scales to one another by formulas like the following*:

$$t(°F) = 9/5 \ t(°C) + 32$$
$$t(°C) = 5/9 \ [t(°F) - 32]$$

When using either formula, it is essential to perform the arithmetic exactly as the formula is written. To determine a Fahrenheit reading, nine-fifths of a Celsius reading is taken first and 32 degrees are added to the product. Suppose we desire to convert 46°C to its equivalent temperature reading on the Fahrenheit scale. We take nine-fifths of 46, which equals 83, and add 32:

$$t(°F) = (9/5 × 46) + 32 = 83 + 32 = 115°F$$

To determine a Celsius reading, 32 is subtracted from the Fahrenheit reading, and then five-ninths of this difference is taken. Suppose we desire to convert 115°F to its equivalent temperature on the Celsius scale. We subtract 32 from 115, which equals 83, and then take five-ninths of this number:

$$t(°C) = 5/9 × (115 - 32) = 5/9 × 83 = 46°C$$

Subdivision of the Fahrenheit and Celsius scales can be continued above and below the two reference points. When a thermometer is read below the zero mark, the divisions are read as minus degrees, or as degrees Fahrenheit, or Celsius, *below zero*. When a temperature reading below zero on either scale is converted into

---

*In this textbook, the lowercase "*t*" is used to denote a temperature on the Fahrenheit and Celsius scales, whereas the capital "*T*" is used to denote a temperature on the Kelvin and Rankine scales.

a reading on the other scale, it is important to account for the negative signs algebraically.

Solved Exercise 2.5

When exposed to an ignition source, polystyrene begins to burn in the temperature range from 345°C to 360°C. It self-ignites in air (without exposure to an ignition source) at 490°C.

(a) Will coaxial television cable made of polystyrene ignite at a fire scene whose maximum temperature is 555°F?

(b) Will the cable ignite without exposure to an ignition source when the temperature is further elevated to 950°F?

Solution: (a) The minimum temperature at which polystyrene begins to burn when exposed to an ignition source is 345°C, or 653°F.

$$(9/5 \times 345) + 32 = 653°F$$

Consequently, when exposed to an ignition source at 555°F, the polystyrene will not burn.

(b) Polystyrene autoignites in air at 490°C, or 914°F.

$$(9/5 \times 490) + 32 = 914°F$$

Consequently, when the temperature of the fire scene has been elevated to at least 914°F, polystyrene self-ignites.

KELVIN AND RANKINE TEMPERATURE SCALES

No upper temperature limit appears to exist. The highest known value is 3.6 million degrees, the average temperature of the sun's corona, or outer atmosphere.

On the other hand, scientific theory and experiment establish the limit below which it is impossible to cool matter. This coldest temperature is −273.15°C, or −459.67°F, each of which is called *absolute zero*. By referring to this lowest attainable temperature as zero, two additional temperature scales have been defined: the Kelvin and Rankine temperatures scales.

The Kelvin temperature is defined as follows:

$$T(K) = t(°C) + 273.15°C$$

Thus, a Kelvin temperature is attained by merely adding 273.15 to any Celsius temperature reading. The unit of temperature on the Kelvin scale is called the *kelvin*, without a degree sign.

The Rankine temperature scale is defined as follows:

$$T(°R) = t(°F) + 459.69°F$$

A Rankine temperature is attained by adding 459.69 degrees to the Fahrenheit temperature reading. The unit of temperature on the Rankine scale is the degree Rankine (°R).

Since 0 K, or 0°R, represent the lowest attainable temperature, the use of negative numbers is not encountered on either the Kelvin or Rankine temperature scales.

Solved Exercise 2.6

Metallic mercury loses all resistance to the flow of electricity when it has been cooled to 9 K. Under such conditions, it is said to be a *superconductor*. At which Fahrenheit temperature does mercury become a superconductor?

Solution: A temperature in kelvins is obtained by adding 273.15°C to a Celsius temperature.

$$K = °C + 273.15°C$$

Consequently, 9 K is equivalent to −264.15°C.

$$9\ K = −264.15°C + 273.15°C$$

A temperature on the Fahrenheit scale is obtained by adding 32 to 9/5ths of a Celsius temperature. Consequently, −264.15°C is equivalent to −443.47°F.

$$(9/5 × −264.15) + 32 = −443.47°F$$

SOME WAYS IN WHICH TEMPERATURE VALUES ARE USED

Temperature values are often used in chemistry to characterize certain features of substances like their melting points and boiling points. These temperature values can also be used in the fire service. Arson investigators, for instance, often estimate the temperatures achieved during fires of questionable origin. This can be accomplished by examining the condition of the objects remaining after the fires.

The melting points of metals, plastics, and other building materials assist investigators in establishing the approximate temperature of fires. Some materials likely to be encountered following a fire in a typical home are listed in Table 2.6. Suppose a fire

TABLE 2.6 ◆ Approximate Melting Points of Some Materials Likely Encountered After a Fire

Material	Melting Point	
	°F	°C
Aluminum	1220	660
Chromium	3407	1875
Copper	1981	1083
Gold	1945	1063
Iron	2781	1527
Lead	618	326
Magnesium	1202	650
Silver	1761	961
Solder[a]	361	183
Tin	449	232
Zinc	878	471

[a]"Solder" refers to alloys of tin and lead.

investigator identifies a necklace that melted during the fire and then solidified. If the jewelry is gold, the investigator can conclude from the information in Table 2.6 that the fire minimally achieved a temperature of 1945°F (1063°C) at the location where the necklace was found.

Performance Goals for Section 2.8

- ◆ Describe the concept of pressure.
- ◆ Distinguish between the gauge pressure and absolute pressure of a confined gas or vapor.
- ◆ Identify the common English and metric units used to measure pressure.
- ◆ Identify the equivalent units in which the standard atmospheric pressure is measured.
- ◆ Convert a pressure reading in one unit into its equivalent in another unit.
- ◆ Describe the concept of vapor pressure.
- ◆ Describe the general impact that a liquid's vapor pressure may have upon its fire and health hazards.
- ◆ Describe the concept of blood pressure.
- ◆ Describe generally how a blood pressure reading is used to identify the severity of certain health problems.

◆ 2.8 PRESSURE AND ITS MEASUREMENT

Pressure is the force exerted over a specified area; hence, pressure is often expressed as a unit of force per unit of area, such as pounds per square inch (psi). In the SI system, the unit of force is the *newton* (N); it is the force that accelerates a 1-kilogram body 1 meter per second for each second. The SI unit of pressure is the newton per square meter (N/m^2), which is called a *pascal* (Pa). A unit of pressure commonly encountered is the *kilopascal* (kPa).

The force resulting from the mass of the overlying air produces *atmospheric pressure*. Earth's gravity gives the atmosphere an average downward force of 14.7 psi at sea level; that is, on the average, 14.7 lb of air bears down on each square inch of Earth's surface. The average pressure exerted by the atmosphere at sea level supports a column of mercury 760 mm high (760 mm Hg); it is also called 1 *atmosphere* (atm). For the purpose of converting a pressure expressed in one unit to another unit, each of the following is equivalent:

$$1 \text{ atm} = 760 \text{ mm Hg} = 760 \text{ torr} = 101.3 \text{ kPa} = 29.9 \text{ in. Hg} = 14.7 \text{ psi}$$

Each pressure is referred to as *standard atmospheric pressure*.

Atmospheric pressure readings are normally recorded on an instrument called a *barometer*. A simple barometer can be easily constructed by filling a glass tube, longer than 760 mm and closed at one end, with mercury. When the tube is inverted, open-side down, in a dish of mercury, the liquid flows out of the submerged open bottom until the level in the tube reaches an average height of 760 mm, or 1 atm. When the atmospheric pressure decreases below the average value of 1 atm, the mercury level on the barometer is said to "fall"; when it increases above 1 atm, the mercury level is said to "rise." Such changes in the atmospheric pressure are closely monitored and used by meteorologists to forewarn of inclement weather conditions.

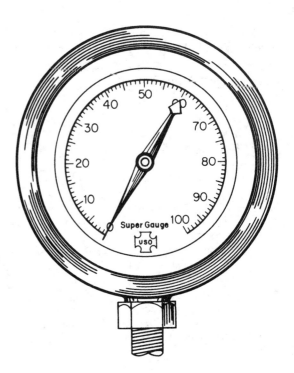

FIGURE 2.8 ◆ A pressure gauge. The difference between the actual pressure that a substance exerts and the pressure of the atmosphere is called the *gauge pressure.* When the needle reads zero, as illustrated here, the gauge pressure is zero. This corresponds to an absolute pressure on the average of 14.7 psi (101.3 kPa).

A pressure reading is also used for monitoring the amount of a gas or vapor confined within a cylinder or tank. For such purposes, the affixed pressure gauge in Figure 2.8 is read. This reading provides the pressure exerted by the contents of the vessel. As portions are removed, the amount remaining in the vessel becomes less and less and the gauge readings approach zero. Since an atmospheric pressure is continually applied to all forms of matter upon Earth, a pressure gauge measures the amount of pressure by which the gas or vapor exceeds atmospheric pressure. The reading is called the *gauge pressure.* The units of gauge pressure are expressed with the inclusion of a subscript "g," such as psi_g.

The true pressure of the contents within a containment vessel, called its *absolute pressure,* is the sum of the gauge pressure and the atmospheric pressure. The absolute pressure is measured with respect to a value of zero, whereas the gauge pressure is measured with respect to the prevailing pressure of the atmosphere; thus, in the English system of units, we express the absolute pressure on the average as follows:

$$P(psi_a) = P(psi_g) + 14.7$$

The units of absolute pressure are sometimes noted by inclusion of the subscript "a" with the unit as shown in the previous expression. Thus, a gauge pressure of 19.4 psi_g corresponds to an average absolute pressure of 34.1 psi_a.

$$19.4\ psi_g + 14.7\ psi = 34.1\ psi_a$$

We sometimes refer to the temperature and pressure at *room conditions.* For our purposes, *room conditions* means a temperature of 70°F (21°C) and a pressure of 14.7 psi (101.3 kPa).

VAPOR PRESSURE

When a liquid is confined within a closed container at a given temperature, some of it evaporates into the space above the liquid until equilibrium between the liquid and its vapor has been attained. This equilibrium is characterized by two opposing changes occurring simultaneously: the rate at which the liquid evaporates and the rate at which the vapor condenses. At equilibrium, these rates are equal. The pressure exerted by the vapor in equilibrium with its liquid is called the liquid's *vapor pressure*.

While the vapor pressure of a liquid could be expressed in any unit of pressure, chemists normally measure it in millimeters of mercury (mm Hg) at the prevailing temperature. When expressed in this fashion, the vapor pressure is the height of a column of mercury (measured on a meter stick) that the liquid's vapor supports at a given temperature.

The vapor pressure of all liquids (as well as some solids) is a characteristic property of the substance. As its temperature increases, more and more of the substance vaporizes into the headspace of the container in which it is confined. This causes the substance's vapor pressure to increase accordingly as shown by the information in Table 2.7 and Figure 2.9.

Liquids with low boiling points possess relatively high vapor pressures. They evaporate readily. On the other hand, liquids with high boiling points possess relatively low vapor pressures. They evaporate more slowly.

Generally speaking, a gas or the vapor of a substance is its most hazardous physical form. It is the vapor of a flammable liquid that burns, and it is a liquid's vapor that can elicit adverse health effects when inhaled. Clearly, the vapor pressure of a flammable or toxic substance has a direct impact upon its potential fire and health hazards.

When a flammable or toxic liquid possesses a relatively low vapor pressure, little vapor evolves at the prevailing temperature. The likelihood that the substance can ignite or be inhaled is accordingly low. On the other hand, when a liquid possesses a relatively high vapor pressure, a substantially greater volume of vapor evolves at the given temperature. It poses an appreciably higher risk of flammability and inhalation toxicity compared to the former substance. As a general guideline, 10 mm Hg is often used to differentiate "high" from "low" when noting the degree of these hazards.

TABLE 2.7 ◆ Vapor Pressures of Some Common Liquids

Temperature		Water	Ethyl Alcohol	Benzene
°F	°C	(mm Hg)	(mm Hg)	(mm Hg)
14	−10	2.1	5.6	15
32	0	4.6	12.2	27
50	10	9.2	23.6	45
68	20	17.5	43.9	74
86	30	31.8	78.8	118
122	50	92.5	222.2	271
167	75	289.1	666.1	643
212	100	760.0	1693.3	1360

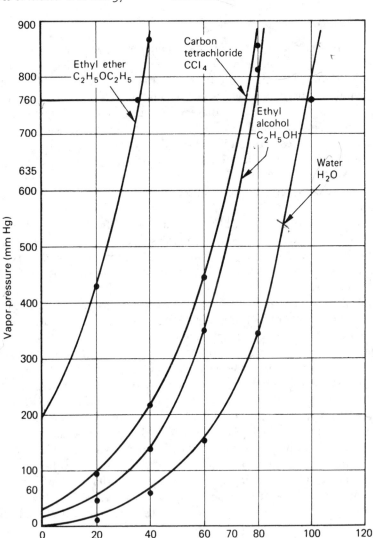

FIGURE 2.9 ◆ The variance of the vapor pressure of some common liquids with temperature. When the vapor pressure equals atmospheric pressure (760 mm Hg), the liquid boils; the corresponding temperature is called the *boiling point* of the liquid. Note that the boiling point of water is 100°C at atmospheric pressure.

A flammable substance possessing a relatively high vapor pressure poses unique transportation problems. During the course of its transportation, a liquid confined within a tank or other vessel could absorb heat from the surroundings, thereby increasing in temperature. As the temperature of the liquid increases, a considerable volume of vapor is produced within the tank. This vapor enters the headspace above the liquid and exerts pressure upon the walls of the confining vessel. To retain its integrity, the vessel must be constructed to withstand this internal pressure.

Even when a container or transport tank containing a flammable or toxic liquid has been essentially emptied of its contents, the vessel may still constitute health, fire, or explosion hazards. These hazards are caused by the presence of the residual vapor within the vessel, which can ignite or explode when exposed to an ignition source or pose a health-related problem when inhaled.

Solved Exercise 2.7

At room conditions, nerve "gases" are actually highly volatile liquids. When inhaled, they cause their victims to suffocate by paralyzing the muscles around the lungs. At least five nerve gases could be used by terrorists to tyrannize the general population and coerce governments toward their position. Their code names are GA, GB, GD, GF, and VX, and their vapor pressures at 68°F (20°C) are 0.037, 2.10, 0.4, 0.044, and 0.0007 mm Hg, respectively. Which of these nerve gases possesses the greatest rate of vaporization at the indicated temperature?

Solution: GB, also called *Sarin*, possesses the highest vapor pressure of these five nerve gases. Therefore, it vaporizes at a faster rate compared to the other four.

Solved Exercise 2.8

Benzene vapor is both flammable and toxic. Use the data in Table 2.7 to determine the degree of caution that should be exercised when encountering a spill of liquid benzene at 86°F.

Solution: Table 2.7 reveals that the vapor pressure of benzene at 86°F (30°C) is 118 mm Hg. Since this value is an order of magnitude greater than 10 mm Hg, liquid benzene is very volatile at 86°F. Since its vapor is both flammable and toxic, prudence dictates that considerable caution should be exercised to prevent its ignition and inhalation.

BLOOD PRESSURE

When dispatched to an emergency scene, paramedics often measure an individual's blood pressure. This is the constant force per unit area that is exerted on the walls of our arteries as blood is pumped to the tissues of the body. Blood pressure is measured using a cuff-like device called a *sphygmomanometer*. It can also be measured electronically through use of a device that provides digital readings, as in Figure 2.10. When measuring an individual's blood pressure, two numbers are obtained: the *systolic*, which records the pressure as the heart beats; and the *diastolic*, which indicates the pressure as the heart relaxes between beats. A typical "normal" blood pressure reading is 120 mm Hg/80 mm Hg, referred to as "120 over 80." When a paramedic relays an

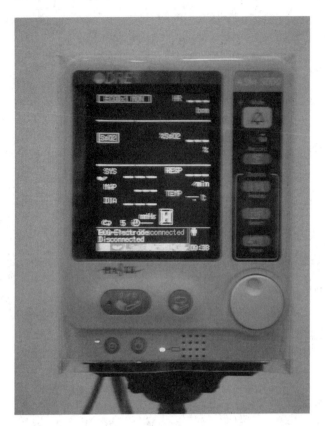

FIGURE 2.10 ◆ Today, blood pressure is generally measured by means of an electronic sphygmomanometer. A flexible cuff is placed around a patient's upper arm over the brachial artery and inflated to about 200 mm Hg. The pressure squeezes the artery shut and prevents blood flow. When the pressure is relieved, blood spurts through the artery at the systolic pressure, creating a turbulent tapping sound that is detected by the device. The pressure registered at the moment the first sound is detected is the systolic blood pressure. As the pressure in the cuff is further lowered, the sounds in the artery continue to be detected until the pressure becomes low enough to allow diastolic blood flow. At this point, the blood flow becomes smooth, no sounds are heard, and the device records a diastolic blood pressure. *Courtesy of Frank S. Drongowski, DDS, Red Rock Oral and Maxillofacial Surgery, Las Vegas, Nevada.*

individual's blood pressure reading to a physician, the critical nature of an illness can be estimated by noting the deviation from the norm.

Blood pressure readings of approximately 140 over 90 are considered abnormal. Individuals having such readings are victims of the disease known as *hypertension,* or *high blood pressure.* These elevated readings signify that the blood vessels are being subjected to more strain than they normally experience. Although the use of medication can lower an individual's blood pressure, regular exercise, weight control, and lifestyle changes are equally important. Hypertension can damage not only the blood vessels, but also the organs that supply blood. In more severe instances, it causes a traumatic health problem, such as a stroke.

Performance Goals for Section 2.9

- ◆ Define the concept of heat.
- ◆ Describe the mechanisms by which heat is transmitted from one material to another or from spot to spot.
- ◆ Identify the common English and metric units used to measure heat.
- ◆ Convert a heat measurement in one unit into its equivalent in another unit.
- ◆ Describe the manner by which convection and radiation impact the spread of a freely burning fire.
- ◆ Describe the manner by which the human body responds to the presence of excessive heat.
- ◆ Describe the reactions likely to be experienced by individuals exposed to excessive heat.
- ◆ Describe the nature of a first-, second-, and third-degree burn.

◆ 2.9 HEAT, ITS TRANSMISSION, AND ITS IMPACT ON THE BODY

Heat is the form of energy associated with the motion of molecules or atoms (Sections 4.4 and 4.6), small particles of which all matter is composed. Heat is manifested when any of the following changes occur:

- ◆ A material's temperature
- ◆ The physical state of a substance
- ◆ The chemical identity of a substance

Heat is either absorbed or emitted as these processes occur. A process that results in the absorption of heat is called an *endothermic process,* while a process that results in the release of heat to the surroundings is called an *exothermic process*.

When chemical reactions occur, substances change their chemical identities. One substance transforms into another substance. The thermal energy accompanying the chemical reaction is called its *heat of reaction;* when a substance burns, the evolved energy is called the *heat of combustion*. Heat is evolved from combustion processes such as the burning of coal, gasoline, wood, natural gas, and other materials. This energy is an important source of heat, power, and light.

The proper control of the heat emitted during combustion is also extremely important in fire control. Fires continue as self-sustained phenomena only when sufficient heat has been released during combustion to substitute for the input energy initially provided from a burning match or similar ignition source. Conversely, many fires cannot be extinguished until some means is undertaken to reduce or eliminate this heat.

As noted in Section 2.6, energy is measured in British thermal units and calories. One *British thermal unit* represents the heat that must be supplied to raise 1 pound of water 1 degree Fahrenheit, specified at the temperature of water's maximum density, 39°F (4°C). One *calorie* is defined as the amount of heat required to raise the temperature of 1 gram of water 1 degree Celsius from 14.5 to 15.5°C. Two hundred fifty-two (252) calories are equivalent to 1 British thermal unit, and 1 calorie is equivalent to 4.184 joules.

$$1 \text{ Btu} = 252 \text{ cal}$$
$$1 \text{ cal} = 4.184 \text{ J}$$

Heat is always transferred from warm materials to cooler ones. If several materials near one another have different temperatures, those that are warm become cooler and those that are cool become warmer, until they have achieved a common temperature. Heat is transmitted from one material to another or from one spot to another spot by three independent modes: conduction, convection, and radiation. The nature of these modes of heat transmission is an important factor associated with understanding how fire spreads.

CONDUCTION

The handle of a metal spoon in hot coffee gets hot itself, even though it is not placed directly in the hot liquid. This transfer of heat between two or more stationary materials in contact—in this case, from the coffee to the spoon—is called *conduction*.

Every material conducts heat to some extent. Metals, such as silver, copper, iron, and aluminum, conduct heat most efficiently; nonmetals, such as glass and air, are not good conductors of heat. Materials that are not good conductors are good insulators, because they delay the transfer of heat.

CONVECTION

Heat can also be transferred from spot to spot within a given substance or even between two or more substances by the natural mixing of their component parts. This happens when cold milk or cream disperses throughout hot coffee without the use of a spoon, or when the air in a room gets warmer when it is hot outside the room. This transmission of heat within a substance or between substances by means of natural mixing is known as *convection*.

The circulation of warm air from a heat vent throughout a room is caused by convection. The popular expression "heat rises" actually means "hot air rises." As warm, less-dense air issues from a heat vent into a room, it naturally rises toward the ceiling. The surrounding cool air descends to replace the warm air that rose, is heated at the heat vent, and then rises toward the ceiling. In this fashion, warm and cool air exchange to generate air currents.

The origin of the convection phenomenon is associated with the impact that gravity has on matter. In zero gravity, convection can only occur if it is artificially induced. When a substance burns, convection moves its combustion products away from the flame to dissipate into the surrounding atmosphere; but where zero gravity exists, as in a space ship orbiting the Earth, the combustion products remain in the immediate area of the combustion zone, where they quickly extinguish the flame.

RADIATION

Imagine a 200-watt light bulb hanging from a ceiling. When the light is turned on, heat can be felt when we hold our hands *under* the bulb. This transfer of heat from the bulb to our hands cannot be caused by conduction, since the air between the bulb and our hands is a poor conductor of heat. It also cannot be caused by convection, since hot air currents rise upward. The heat from the bulb is transmitted by a third means called *radiation*.

Unlike conduction and convection, radiation occurs even when there is no material contact between two objects. Heat from the sun radiates through space and warms the Earth and other celestial bodies.

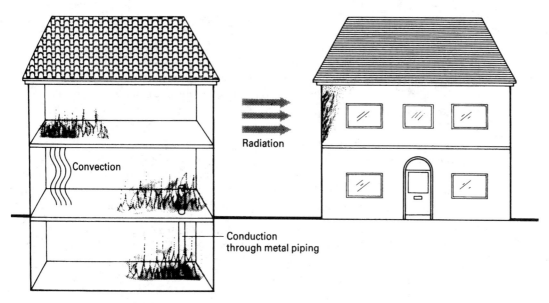

FIGURE 2.11 ◆ Conduction, convection, and radiation contribute to the spread of fire. In the building on the left, heat is first conducted through metal piping from a basement fire to the first floor. Heat is then transmitted to the second floor by convection and to the nearby adjacent building on the right by radiation.

All matter radiates heat at elevated temperatures. When the temperature of a heated object exceeds approximately 932°F (500°C), the radiation becomes visible. Burning flames, glowing coals, and molten metal are examples of matter hot enough to radiate visible light. Hot objects also radiate energy at lesser temperatures, but it is generally emitted in the infrared region of the electromagnetic spectrum, which cannot be observed with the naked eye.

SPREAD OF FIRE

The conduction process does not significantly contribute to the sustenance and spread of a normal fire. As Figure 2.11 illustrates, convection and radiation are the modes of heat transmission primarily responsible for sustaining and spreading freely burning fires in the following ways:

- ◆ Convection affects the spread of fire by means of the natural movement of hot and cold air. As hot air rises, heat is transmitted to adjoining materials and initiates their ignition. Cool, fresh air simultaneously flows inward into the fire to replace the hot air, thereby providing the fire with a supply of oxygen.
- ◆ Radiation affects the spread of fire by transmitting heat, often laterally, from the immediate fire scene to nearby materials and initiating their combustion.

ADVERSE HEALTH EFFECTS RESULTING FROM EXPOSURE TO HEAT

The routine work activities of a firefighter require exposure to high-temperature environments. Heat is transmitted from these environments directly to the body. The human organism reacts to the presence of heat and attempts to reduce the body's temperature by the following mechanisms, which typically function in combination:

- *The body circulates blood to the skin.* The body releases excess heat to the environment through the skin. When the muscles are being used for physical labor, as during firefighting activities, less blood flows to the skin, which can hinder the body's capacity to release its excess heat.
- *The body perspires.* Our bodies attempt to maintain a stable internal temperature by producing perspiration; but, perspiring is an effective means of cooling only when the prevailing atmospheric humidity is low enough to permit evaporation and if the fluids and salts lost during the process are adequately replaced. Evaporative cooling is limited in the case of individuals—such as firefighters—who are obligated to wear protective gear incapable of adequately releasing perspiration to the atmosphere.

Individuals store the heat they are incapable of eliminating. They first sense a rise in body temperature and heart rate. As additional heat is stored, they experience a lethargic or sick feeling, irritability, an inability to effectively concentrate, and difficulty with focusing on a task. To overcome these sensations, the afflicted persons often respond by fainting.

Overheated individuals may also experience one or more of the following reactions:

- *Heat rash.* When perspiration is not readily removed from the surface of the skin by evaporation, small itchy, red spots appear on the surface of the skin. In severe cases, this heat rash is so uncomfortable to the victim that it impedes performance.
- *Heat cramps.* These painful, involuntary spasms of the muscles are often caused by the failure of individuals to replace the minerals lost in perspiration, despite consuming water after exposure to a heated environment. Heat cramps frequently develop in individuals who have exerted muscular activity for an extended period of time.
- *Heat exhaustion.* This condition generally manifests certain common symptoms such as weakness, fatigue, giddiness, nausea, and headache. The skin becomes clammy and moist, the complexion pale or flushed, and the body temperature increases slightly higher than normal. Heat exhaustion results when an individual fails to drink adequate volumes of fluids or to replace the salts lost in perspiration following exposure to a heated environment.
- *Heatstroke, or hyperthermia.* This condition is caused by the failure of the body to satisfactorily regulate its internal temperature upon exposure to excessive heat. The body temperature of this individual could rapidly rise to 106°F (41°C), or higher. The common symptoms associated with heatstroke are mental confusion, delirium, loss of consciousness, convulsions, and coma. The skin generally becomes red or bluish. Victims of heatstroke must be promptly treated or they will die.

Care can be provided to individuals who have been exposed to excessive heat by quickly providing them with an adequate volume of fluids, especially commercially available products known to be rich in mineral salts. Minerals, including sodium and potassium, are essential for proper cardiac and other bodily functions. As individuals perspire, these minerals are lost from the body. For this reason, individuals who have been exposed to excessive heat should also be assessed for cardiac dysfunction. While awaiting medical assistance, victims of heatstroke should be cooled by soaking their clothing with cool water.

Solved Exercise 2.9

Gatorade is a drink that is rich in minerals, especially sodium. Why are over-heated athletes and firefighters encouraged by physicians to drink *Gatorade?*

   Solution: Overheated athletes and firefighters are encouraged to drink *Gatorade* to replace the minerals lost during perspiration. By replacing the minerals required by the body for muscular activities, including proper cardiac function, athletes and firefighters mitigate the risk of experiencing muscular cramps.

The skin and its surrounding areas may be damaged or destroyed when a person contacts very hot materials, including superheated air. The extent of the damage and destruction is designated as a first-, second-, or third-degree burn, or a superficial, partial-thickness, or full-thickness burn, respectively. These burn types are associated with the features summarized below:

- During a *first-degree burn,* the surface of an inner layer of the skin, called the *dermis,* is superficially damaged. The afflicted area visually appears red, resembling a bad sunburn, but it is entirely intact. The first-degree burn is typically accompanied with minor swelling. The victim of a first-degree burn experiences relatively little pain.
- During a *second-degree burn,* the dermis is gravely damaged to a limited depth, such that the surrounding skin swells and blisters. A second-degree burn is intensely red in appearance. The afflicted person can experience severe pain.
- During a *third-degree burn,* the damage or destruction of the skin typically involves the subsurface muscular structures, fat, and nerves. When the skin receives a third-degree burn, it may be charred. If the nerves remain intact, a third-degree burn is associated with considerable pain; but, when the nerves have been destroyed, the afflicted person experiences little or no pain. Firefighters are at risk of experiencing a third-degree burn whenever superheated air has been inhaled. In the most egregious instances, the lungs are seared, which can seriously hamper the individual's ability to breathe.

Care should be provided to burn victims by running cool tap water on the afflicted skin for a period of 5 minutes to an hour. The cool water absorbs the residual heat, which reduces the possibility of further damage to the skin. Cool water also acts as an anesthetic to reduce pain. Ice should *never* be applied to burns, since the extreme cold can damage the skin.

First- and second-degree burns should be loosely covered with a gauze pad until medical attention is received. Petroleum jelly and bandages should *not* ordinarily be applied, since this practice traps heat and restricts air circulation. To reduce the likelihood of infection, the blisters should *not* be punctured. The healing period for first- and second-degree burns varies, but it is typically no more than 3 weeks.

A person who has received a third-degree burn must be provided with specialized medical assistance as quickly as feasible. Ultimately, the corrective measures associated with a full-thickness burn may include grafting and reconstructive surgery. The healing period for a third-degree burn is lengthy, generally no less than 1 year. An individual with a serious third-degree burn could also be at risk of death. When 70% or more of the skin on the body has been burned, the person has only a 1-in-2 chance of remaining alive.

Performance Goals for Section 2.10
- ◆ Describe the concepts of heat capacity and specific heat.
- ◆ Calculate the amount of heat evolved or absorbed by a substance when provided with its heat capacity and the magnitude of a temperature change.
- ◆ Identify the latent heat of vaporization and the latent heat of fusion of water.
- ◆ Describe the interrelationship between the vapor pressure of a liquid and its boiling point.
- ◆ Describe the interrelationship between melting and freezing and between boiling and condensation.
- ◆ Calculate the amount of heat manifested during each step of a process involving the change of one pound of water into one pound of steam over the temperature range from −20°F to 300°F (−29°C to 149°C), and relate the significance of these calculations to the use of water as a fire extinguisher.

◆ 2.10 CALCULATION OF HEAT

Two or more substances differ from one another by the quantity of heat required to produce a given elevation of temperature in the same mass. This quantity of heat is called the *heat capacity*. The heat capacities of some common liquids are provided in Table 2.8. For water, the magnitude of the heat capacity varies as a function of its temperature, as indicated. The units of heat capacity in the English and metric systems are the British thermal unit per pound per degree Fahrenheit (Btu/lb °F) and the calorie per gram per degree Celsius (cal/g °C), respectively. The magnitude of these units is equal.

$$1 \text{ Btu/lb °F} = 1 \text{ cal/g °C}$$

TABLE 2.8 ◆ Heat Capacities of Some Common Liquids

Liquid	Heat Capacity (cal/g °C or Btu/lb °F)
Acetone	0.506
Benzene	0.406
Carbon tetrachloride	0.201
Chloroform	0.234
Diethyl ether	0.547
Ethyl alcohol	0.581
Methyl alcohol	0.600
Turpentine	0.411
Water (from 0° to 100°C)	1.00
Water (below 0° and above 100°C)	0.5

The ratio of the heat capacity of a substance to the heat capacity of water at the same temperature is a dimensionless number called the *specific heat* of the substance. The specific heats of most substances are less than 1.

When heat is transferred to a substance, it can cause a change in its physical state or its chemical identity, or it can cause the substance to assume a different temperature. The heat that is transferred to a substance causing a change in its temperature, but no change in its physical state or chemical identity, is calculated by the following formula, in which Q is the heat, m is the mass of the substance, C is its heat capacity, and Δt (delta *t*) is the difference between the final and initial temperatures of the substance:

$$Q = m \times C \times \Delta t$$

For example, given that copper has a heat capacity of 0.093 cal/g °C, what amount of heat is transferred to 100 g of copper when it is heated from 30°C to 100°C? First, the difference in temperature, Δt, is 100°C − 30°C, or 70°C. Then, the heat transferred is calculated to be 651 cal by multiplication, as follows:

$$Q = 100 \text{ g} \times 0.093 \text{ cal/g °C} \times 70°C = 651 \text{ cal}$$

As noted, a change in temperature can also result in a change in the physical state of matter. If liquid water is exposed to the cold, its temperature falls until 32°F (0°C) has been attained. Then, the temperature remains fixed until the entire liquid mass freezes to solid ice. This latter temperature is called the *freezing point* of water. The quantity of heat per unit mass required to change a liquid substance to a solid at its freezing point is called the *latent heat of fusion* of the substance. For example, the conversion of liquid water to solid ice liberates 144 Btu/lb, or 80 cal/g. This is the latent heat of fusion of water.

We can similarly heat a mass of liquid water to 212°F (100°C). Then, the temperature remains fixed while the liquid water changes into steam. This is the temperature at which the vapor pressure of water equals the atmospheric pressure of 1 atm; it is called the *boiling point* of water. The amount of heat that must be supplied to a material at its boiling point to convert it from a liquid to a vapor is called its *latent heat of vaporization*. The heat of vaporization of water is 970 Btu/lb, or 540 cal/g. This is a relatively high value compared to the heats of vaporization of other substances.

Two phenomena are closely associated with freezing and boiling:

+ *Melting* is the reverse of freezing. As ice melts, 144 Btu/lb, or 80 cal/g, is absorbed from the surroundings.
+ *Condensation* is the reverse of boiling. When steam condenses into liquid water, 970 Btu/lb, or 540 cal/g, is liberated to the surroundings. This is the reason that exposure of the skin to steam at 212°F (100°C) causes a more severe burn than exposure to liquid water at the same temperature.

The amount of heat associated with the conversion of 1 pound of water into 1 pound of steam over the temperature range from −20°F to 300°F (−29°C to 149°C) is illustrated in Figure 2.12. The relevant quantities of heat in Btu can be calculated at each juncture as follows:

+ Ice undergoes a change in temperature from −20°F to 32°F (−29°C to 0°C)

$$Q = 1 \text{ lb} \times 0.5 \text{ Btu/lb °F} \times 52°F = 26 \text{ Btu}$$

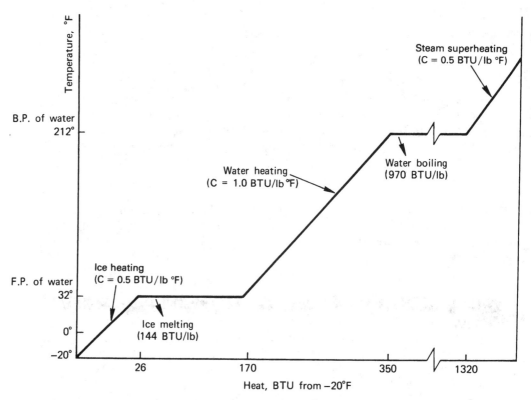

FIGURE 2.12 ◆ The heat evolved in Btu for one pound of pure water at atmospheric pressure (14.7 psi$_a$ or 101.3 kPa) as a function of Fahrenheit temperature (not to scale). B.P. and F.P. refer to the boiling and freezing points of water, respectively.

◆ Ice melts at 32°F (0°C)

$$Q = 1 \text{ lb} \times 144 \text{ Btu/lb} = 144 \text{ Btu}$$

◆ Liquid water undergoes a change in temperature from 32°F to 212°F (0°C to 100°C)

$$Q = 1 \text{ lb} \times 1.0 \text{ Btu/lb °F} \times 180°\text{F} = 180 \text{ Btu}$$

◆ Liquid water evaporates into steam at 212°F (100°C)

$$Q = 1 \text{ lb} \times 970 \text{ Btu/lb} = 970 \text{ Btu}$$

◆ Steam is superheated from 212°F to 300°F (100°C to 149°C)

$$Q = 1 \text{ lb} \times 0.5 \text{ Btu/lb °F} \times 88°\text{F} = 44 \text{ Btu}$$

These calculations clearly demonstrate that the largest quantity of heat is liberated to the surroundings when liquid water vaporizes.

Water is commonly employed as a fire extinguisher since it is nonflammable, plentiful, and inexpensive; but the primary reason water effectively extinguishes fire is associated with its ability to remove heat from the burning material. As water is applied to a fire, it absorbs heat and ultimately vaporizes, each pound carrying 970 Btu of heat into the atmosphere and away from the burning material. Simultaneously, the temperature of the burning material decreases until it ultimately drops to the ambient temperature of the surroundings. A temperature reduction is normally sufficient to extinguish most fires.

Solved Exercise 2.10

When wood and high-density polyethylene burn, the average heat released into the environment is 8500 Btu/lb and 20,050 Btu/lb, respectively. Compare the potential impact that the burning of products made from equal amounts of these materials has on an ongoing fire.

   Solution: The heat of combustion of high-density polyethylene is approximately 2.4 times greater than the heat of combustion of wood. When these two materials burn, energy in the form of heat and light is released into the immediate environment. Since substantially more heat is available from burning polyethylene compared to burning an equivalent amount of wood, the presence of polyethylene products at a fire scene increases the risk of spreading fire in the immediate vicinity. Flammable wooden materials in the same setting may not even ignite, since a lesser amount of heat is released when the same amount of wood burns.

Performance Goals for Section 2.11
   - ◆ Calculate the volume assumed by a liquid when provided with its coefficient of volume expansion and the magnitude of its temperature change.
   - ◆ Describe the nature of the potential hazard associated with the expansion of a heated liquid confined within its storage vessel.

◆ 2.11 EXPANSION OF LIQUIDS

With few exceptions, the dimensions of a material expand when it is heated and contract when it is cooled. As we will note in more detail in Section 2.12, gases and vapors expand more than liquids and solids; liquids expand less; and solids even less. Water is the common exception to this general rule: Water begins to expand when it is cooled below 39.1°F (3.97°C).

   The volume to which liquids and solids expand when heated is determined from use of the following formula:

$$V_2 = V_1 + (V_1 \times \alpha \times \Delta t)$$

When using it, we first determine the product of V_1, α, and Δt, and then add the product to V_1. V_1 and V_2 are the initial and final volumes of an arbitrary liquid, respectively; Δt is the difference between its initial and final temperature; and the Greek letter alpha (α) is a measure of the change in volume per unit of original volume per degree change in temperature. The last term is called the *coefficient of*

TABLE 2.9 ◆ Coefficient of Volume Expansion for Some Common Liquids

Liquid	$\alpha(°F^{-1})$	$\alpha(°C^{-1})$
Acetic acid	0.00059	0.00107
Acetone	0.00085	0.00153
Benzene	0.00071	0.00128
Carbon tetrachloride	0.00069	0.00124
Diethyl ether	0.00098	0.00176
Ethyl alcohol	0.00062	0.00112
Gasoline	0.00080[a]	0.00144
Glycerine	0.00028	0.00051
Methyl alcohol	0.00072	0.00130
Pentane	0.00093	0.00168
Toluene	0.00063	0.00113
Water	0.00012	0.00022

[a]The coefficient of volume expansion for gasoline varies from $0.00055°F^{-1}$ to $0.00090°F^{-1}$. The coefficient provided in this table is a representative value.

volume expansion, the units of which are reciprocal degrees Fahrenheit ($°F^{-1}$) and reciprocal degrees Celsius ($°C^{-1}$). Numerical values of this coefficient are given in Table 2.9 for some common liquids.

Imagine a tank or metal drum filled to its brim with a liquid. When the liquid expands, it must overtop the brim. If the tank or container is sealed so that the contents cannot overflow, it must either expand to compensate for the increase in liquid volume or rupture. To prevent its rupture, a tank or container should *never* be completely filled with a liquid product. The amount by which a tank or container of liquid falls short of being completely filled is called its *outage,* or *ullage*. It is generally expressed in percentage by volume.

The manufacturers of liquid chemical products normally acknowledge that an outage must be provided in containers. Nonetheless, the allowance could be inadequate to compensate for the volume increase experienced by liquid materials during a fire. Consider a 55-gallon steel drum filled to the brim with flammable benzene at 68°F (20°C) that is subsequently heated during a fire to 170°F (77°C). As the liquid benzene is heated, it expands to occupy a volume of 59 gallons at the new temperature.

$$V_2 = 55 \text{ gal} + \left(55 \text{ gal} \times \frac{0.000686}{°F} \times 102°F\right) = 59 \text{ gal}$$

This increase in volume is 4 gallons, or 7.3% by volume. To relieve the internal pressure accompanying this volume increase, the drum ruptures and releases the benzene into the environment. If the benzene is exposed to an ignition source, its vapor will burst into flame.

When liquids are transferred into containers, railroad tankcars, and other cargo tanks for transportation, DOT requires the presence of a headspace above the liquid so the vessels are not completely filled to the brim. DOT requires the space to prevent the leakage from, or distortion of, the transport vessel by accounting for the expansion of its contents due to a temperature rise. In containers of 110 gallons or less, sufficient outage must be provided so that they are not completely filled with liquid at 130°F (55°C). In railroad tankcars, an adequate headspace must be provided within the shell when the dome of the tankcar does not itself provide sufficient outage.

Solved Exercise 2.11

A manufacturer stores 54 gal of liquid methyl alcohol within each of six 55-gallon steel drums. The six drums are assembled next to one another along an interior wall within a refrigerated vault whose average temperature is routinely maintained at 20°F (7°C). However, the cooling system inadvertently fails for a period of 6 days during summer weather, causing the temperature within the vault to rise to 90°F (32°C). Are these drums likely to rupture at this increased temperature?

Solution: We need to determine the volume of the liquid at 90°F (32°C). To do so, we use the following formula:

$$V_2 = V_1 + V_1 \alpha \Delta t = V_1(1 + \alpha \Delta t)$$

$V_1 = 54$ gal, $\Delta t = 90°F - 20°F = 70°F$, and $\alpha = 0.00072°F^{-1}$ (from Table 2.9).

$$V_2 = 54 \text{ gal } [1 + (0.00072°F^{-1} \times 70°F)]$$
$$= 54 \text{ gal} \times 1.05 = 57 \text{ gal}$$

Although the manufacturer provided a 1-gallon headspace in each drum, 57 gal of liquid cannot be accommodated within a 55-gallon drum at a temperature of 90°F (32°C). Consequently, the sealed drums are likely to rupture.

Performance Goals for Section 2.12

- State Boyle's law, Charles's law, and the combined gas law.
- Use Boyle's law, Charles's law, and the combined gas law to calculate the volume occupied by a confined gas when it is exposed to different pressure and temperature conditions.
- Describe the nature of the potential hazard associated with the expansion of a heated gas or vapor that is confined within a storage vessel.

◆ 2.12 GENERAL PROPERTIES OF THE GASEOUS STATE

Of the three states of matter, only the gaseous state is capable of being described in comparatively simple terms. This description relates the volume of a gas to its temperature and pressure. We shall independently examine the laws of nature that apply to the gaseous state of matter: Boyle's law, Charles's law, and the combined gas law.

BOYLE'S LAW

Robert Boyle, an Irish physicist, demonstrated experimentally that the volume of a confined gas varies inversely with its absolute pressure when the temperature remains fixed. The constant-temperature experiments performed by Boyle demonstrated that the mathematical product of the volume and absolute pressure of a gas is always constant. This observation is commonly called *Boyle's law*. It is expressed mathematically as follows:

$$V_1 \times P_1 = V_2 \times P_2$$

Here, V_1 and P_1 are the initial volume and initial absolute pressure, respectively; V_2 and P_2 are the final volume and final absolute pressure, respectively.

Boyle's law is used to determine the volume that a gas occupies when it remains at a fixed temperature but undergoes a change in pressure. Suppose the pressure of 40 cubic feet (40 ft^3) of an arbitrary gas is changed from 14.7 psi to 450 psi at a fixed temperature; what is the new volume assumed by the gas? Common sense tells us that the new volume must be less than 40 cubic feet since the application of any pressure squeezes the gas into a smaller volume. Using Boyle's law, we can readily compute the new volume as follows:

$$V_2 = 40 \text{ ft}^3 \times \frac{14.7 \text{ psi}}{450 \text{ psi}} = 1.3 \text{ ft}^3$$

CHARLES'S LAW

Jacques Charles and Joseph Gay-Lussac, two French scientists, independently demonstrated experimentally that the volume of a confined gas increases proportionately to the increase in its absolute temperature when the pressure remains fixed. This statement is now known as *Charles's law*. It is expressed mathematically as follows:

$$V_1 \times T_2 = V_2 \times T_1$$

Here, V_1 and T_1 are the initial volume and initial absolute temperature, respectively; V_2 and T_2 are the final volume and final absolute temperature, respectively.

Suppose an arbitrary gas at atmospheric pressure occupies a volume of 300 mL at 0°C; what volume does it occupy at 100°C and the same pressure? Common sense tells us that a heated gas always occupies a larger volume. We must first convert the temperatures in degrees Celsius into absolute temperatures in kelvins. In Section 2.7, we learned that the conversion to absolute temperature is accomplished by adding 273.15°C to the Celsius temperature readings. On the Kelvin scale, the temperature readings are as follows:

$$0°C = 273.15 \text{ K}$$
$$100°C = 373.15 \text{ K}$$

Now, we can use Charles's law to compute the new volume.

$$V_2 = 300 \text{ mL} \times \frac{373.15 \text{ K}}{273.15 \text{ K}} = 410 \text{ mL}$$

COMBINED GAS LAW

Boyle's and Charles's laws can also be combined into the following mathematical expression:

$$V_1 \times P_1 \times T_2 = V_2 \times P_2 \times T_1$$

This is called the *combined gas law*. It is used to calculate the volume of a gas at a new temperature and pressure.

The volume of a gas can be forced to remain fixed, as when a gas is confined within a steel cylinder or other storage vessel. Under this circumstance, $V_1 = V_2$, and the combined gas law assumes the following form:

$$P_1 \times T_2 = P_2 \times T_1$$

When a gas enclosed in a sealed steel tube is heated, its pressure increases accordingly. Figure 2.13 dramatizes this pressure increase as follows:

- The first tube contains a gas that has been confined at 1000 psi at normal room temperature, 70°F (21°C)
- The second tube contains the same gas that has been heated to 570°F (299°C)
- The third tube contains the same gas that has been heated to 1070°F (577°C)

When heated to 570°F, the pressure of the gas nearly doubles; when heated to 1070°F, the pressure of the gas nearly triples. These elevated temperatures noted here are akin to those routinely encountered during fires. To relieve the strain on the walls caused by these excessive internal pressures, the second and third tubes are likely to rupture.

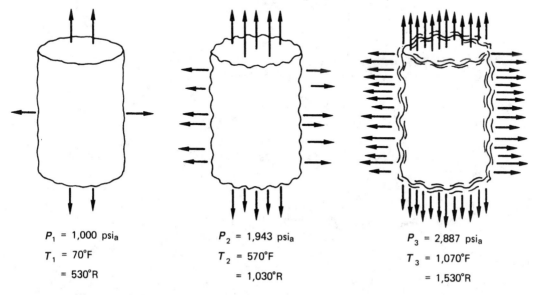

P_1 = 1,000 psi$_a$ P_2 = 1,943 psi$_a$ P_3 = 2,887 psi$_a$

T_1 = 70°F T_2 = 570°F T_3 = 1,070°F

    = 530°R   = 1,030°R   = 1,530°R

FIGURE 2.13 ◆ The effect of applied temperature on a gas confined to a constant-volume container, such as a sealed metal drum or barrel. An increase in the temperature of the gas by 500°F causes the internal pressure to nearly double, which often causes the container to rupture.

Solved Exercise 2.12

A welder purchases a gas cylinder of flammable hydrogen gas and chains it to a wall within a workshop. After periodic usage, the gauge pressure reads 235 psi_g when the temperature is 65°F. What is the gauge pressure reading when the temperature of the cylinder contents become 350°F during a fire?

Solution: Since a steel cylinder is a constant-volume container, $P_1 \times T_2 = P_2 \times T_1$, where these symbols refer to the *absolute* initial and final pressures and temperatures, respectively.

$$P_1 = 235 \; psi_g + 14.7 \; psi = 250 \; psi_a$$
$$T_1 = 65°F + 459.69°F = 525°R$$
$$T_2 = 350°F + 459.69°F = 810°R$$
$$P_2 = 250 \; psi_a \times \frac{810°F}{525°F} = 386 \; psi_a$$

Then, the gauge pressure is determined by subtracting 14.7 psi from the absolute pressure to obtain 371 psi_g.

$$P_2 = 386 \; psi_a - 14.7 \; psi = 371 \; psi_g$$

The absorption of heat causes the internal pressure of the cylinder contents to increase from 235 psi_g to 371 psi_g.

Performance Goals for Section 2.13
- ◆ Describe the general nature of a cryogen.
- ◆ Describe the general hazards associated with exposure to cryogens.

◆ 2.13 GENERAL HAZARDS RESULTING FROM EXPOSURE TO CRYOGENS

The study of matter at temperatures less than approximately 90 K is called *cryogenics*. When substances have been cooled to these very low temperatures, they are called *cryogens*, or *cryogenic liquids*, from the Greek word *kryos*, meaning icy cold.

A gas or vapor can be reduced eventually to a liquid by lowering its temperature and/or increasing its pressure. This liquid can then be reduced to a solid by lowering its temperature and/or increasing its pressure further. Nonetheless, there is a temperature above which an increased pressure alone cannot cause a gas or vapor to condense; this is called the *critical temperature* of that substance.

Consider gaseous carbon dioxide. When it is cooled to a minimum temperature of 88°F (31°C) and compressed by sufficient pressure, carbon dioxide liquefies. But when its temperature is above 88°F (31°C), applied pressure is unable to liquefy carbon dioxide. The pressure required to liquefy a gas or vapor maintained at its critical temperature is called the *critical pressure*. The critical pressure of carbon dioxide is 73 atm (7400 kPa). Consequently, when carbon dioxide gas is confined within a vessel and maintained at a temperature equal to or less than 88°F (31°C), it liquefies when compressed by a pressure at least equal to 73 atm (7400 kPa).

TABLE 2.10 ◆ **Critical Temperature and Critical Pressure of Some Selected Substances**

Substance	Critical Temperature		Critical Pressure		
	°C	°F	atm	psi$_a$	kPa
Ammonia	130	266	115	1690	11,650
Butane	152	306	38	559	3849
Carbon dioxide	31.1	88	73	1073	7395
Diethyl ether	197	387	35.8	526	3627
Hydrogen	−234.5	−390	20	294	2026
Nitrogen	−146	−231	33	485	3343
Oxygen	−118	−180	50	735	5065
Propane	96.7	206	42	617	4255
Sulfur dioxide	155.4	311.7	78.9	1160	7993

The critical temperatures and critical pressures of some common substances are provided in Table 2.10.

To retain a cryogenic liquid, it must be stored in a specially designed and insulated vessel fitted with pressure-regulating valves to control its internal maximum pressure. DOT requires shippers and carriers to transport cryogenic liquids within a specialized delivery unit or vessel like those in Figures 2.14 and 2.15. DOT regulates the design

FIGURE 2.14 ◆ A liquid carbon dioxide delivery unit mounted upon a tilt cab. When in actual use, DOT regulations stipulate a marking on the unit indicating "CARBON DIOXIDE—REFRIGERATED LIQUID" (not illustrated here). *Courtesy of TOMCO$_2$ Equipment Company, Loganville, Georgia.*

FIGURE 2.15 ◆ A tank truck used to transport cryogenic nitrogen by public highway. When in use, DOT regulations stipulate a marking on the vessel indicating "NITROGEN—REFRIGERATED LIQUID." *Courtesy of Air Liquide America Corporation, Houston, Texas.*

specifications for these transport vehicles, including their temperature and pressure control systems, venting mechanisms, valves, and piping.

Tanks approved by DOT for the transportation of cryogenic liquids sometimes utilize a liquid-nitrogen refrigeration system. The cold liquefied nitrogen is dispersed through internal coils around the transport tank to cool the cryogen. These specialized tanks used for the transportation of cryogenic liquids are designated as DOT Specification 113 and 204 tankcars.

EXPANSION OF CRYOGENS

When it becomes impossible to adequately maintain the temperature and pressure conditions at which a cryogen is confined, the substance must ultimately vaporize. This event is likely to occur catastrophically, since the difference in volume between the liquid and its gas or vapor is always tremendous.

Consider cryogenic methane confined within a storage vessel whose pressure-reducing equipment malfunctions. When the entire bulk of the liquid vaporizes, it assumes a volume approximately 630 times its initial volume. This results in a buildup of internal pressure within the storage vessel, which most likely would rupture from the force exerted upon its walls. As the tank or container bursts, the mass of methane is released into the environment. Methane is a flammable substance. Consequently, this situation poses a significant risk of fire and explosion.

LIQUEFACTION AND SOLIDIFICATION OF OTHER GASES AND VAPORS

Cryogens are so cold they can cause other substances to liquefy or solidify upon contact. Even some components of the air solidify when cryogens vaporize into the atmosphere. When transferring cryogens from one closed vessel into another, the solidification of an air component constitutes a major hazard, since the "ice" could block the passageways within tubing. The same mechanism can cause venting valves to become blocked. This prevents the release of internal pressure.

HEALTH RISKS ASSOCIATED WITH EXPOSURE TO CRYOGENS

When skin is exposed to a cryogenic liquid, serious burns may develop, especially when the specific heat of the liquid is relatively high as in the case of liquid oxygen. Depending on the depth to which the skin tissue has been impacted, these skin burns resemble first-, second-, and third-degree thermal burns in physical appearance. To avoid the emergence of burns, individuals should exercise precaution when handling cryogenic liquids. To protect the eyes, a face shield or visor should be used, and to protect the hands, loosely fitted gloves should be worn. If a mishap occurs during which a cryogen mistakenly flows inside gloves or boots, these items should be removed immediately to minimize the length of time during which the skin and liquid remain in contact.

When living tissue is exposed to cryogenic liquids for an extended period, the tissue solidifies. In physical appearance, the tissue resembles a case of severe frostbite. The solidification may cause a local arrest in the circulation of blood. Widespread cellular damage can occur within solidified tissue, causing it to become vulnerable to bacterial infection. When the tissue has been solidified for many hours or days, physicians may decide to surgically isolate the affected tissue.

The extreme coldness of a cryogenic liquid or solid can also cause bonding between the substance and the uninsulated vessels or pipes containing it. The bond may be so firm that the flesh rips or tears when a separation attempt is made. Bonding also occurs when a piece of dry ice (solid carbon dioxide) is inserted into the mouth. The dry ice bonds to the tongue. To prevent damage to the mouth and its organs, such acts should be avoided.

Normal body temperature should be restored to tissues that have been exposed to cryogens as soon as practical. The rapid restoration of normal body temperature minimizes the potential for further tissue damage. This is best accomplished by flushing the affected area with large volumes of tepid—not hot—water.

Individuals generally experience discomfort when they breathe the cold gas or vapor of a cryogen. Prolonged breathing of a very cold gas or vapor could cause serious lung disease.

Finally, we must exercise precautions as a cryogenic liquid vaporizes, since its gas or vapor displaces a corresponding amount of air. In a confined or unventilated area, the resulting atmosphere may contain a reduced concentration of oxygen that is incapable of supporting life. Individuals who breathe this atmosphere could suffocate or experience other ill effects.

Solved Exercise 2.13

Why do experts encourage the use of a water fog at emergency scenes where flammable cryogens have spilled but discourage the use of streams of water?

Solution: The discharge of a water fog at an emergency scene involving a cryogen spill dissipates the gas or vapor produced as the cryogen evaporates. When the cryogen is flammable, the use of a water fog accomplishes each of the following:

- ◆ A water fog dilutes the flammable vapor within the air, thereby rendering the atmosphere less flammable.
- ◆ A water fog quenches or prevents the propagation of an incipient or developed flame front.

Since these results are not as readily attained by discharging streams of water, experts discourage their use.

■■

REVIEW EXERCISES

States of Matter

2.1 In which state of matter does each of the following materials exist at room conditions [70°F (21°C) and 101.3 kPa]:

(a) Iron

(b) Air

(c) Nitrogen

(d) Household ammonia

(e) Propane

(f) Lye

(g) Mercury

(h) Oxygen

(i) Steel

(j) Carbon dioxide

(k) Kerosene

(l) Blasting powder

2.2 Which state(s) of matter exhibit each of the following properties?

(a) A characteristic shape

(b) A characteristic volume

(c) A characteristic shape *and* volume

(d) A characteristic volume, but not a characteristic shape

Conversions Between Units of Measurement

2.3 A "small hose system" is generally defined as a fire hose having a diameter ranging from 5/8 in. to 1½ in. The small hose system is often used to control and extinguish relatively small fires in their initial stages. Express the diameter range for a small hose system in centimeters.

2.4 A 5000-gallon steel storage tank is used at a manufacturing facility to hold acetone, a flammable liquid. When the volume is expressed in liters, what is the maximum volume of acetone that the manufacturer can store in this tank?

2.5 The volume of an average aluminum can of soft drink is 12 fl oz. When expressed in milliliters, what is the volume of the soft drink contained in an average aluminum can? [One fluid ounce (fl oz) is equivalent to 29.6 milliliters.]

2.6 Portable tanks used to haul water to regions unequipped with fire hydrants normally carry from 1000 to 3000 gal of water.

(a) What is the minimum volume expressed in liters that these tanks carry?

(b) What is the maximum volume expressed in cubic meters that these tanks carry?

2.7 To isolate and preserve potential arson evidence, a fire inspector wishes to cordon off an area that measures 15.0 ft × 12.0 ft. What size tarpaulin measured in square meters would completely cover this area?

2.8 A fire line measuring 400 ft × 120 ft is engulfed in flames within a forest. During a single pass over the fire, a water-bearing helicopter drops sufficient water to extinguish a burning area measuring 20 m × 20 m.

(a) Assuming the fire remains confined along the original fire line, how many passes must four helicopters make to extinguish the fire?

(b) An average time of 20 minutes is required for one helicopter to travel one way to reach the fire scene from the water source. What period of time is required for four helicopters to extinguish the fire?

2.9 When a fire truck travels at 60 mi/hr, what is its equivalent speed when expressed in meters per minute?

2.10 The average thickness of an oil slick on water is 0.1 μm. What volume of oil in liters is contained within a 10-km-long, 100-m-wide oil slick?

2.11 Throughout the world, unintentional fires scorch approximately 71 million hectares of forest and grassland annually.

    (a) How many acres are scorched by these unintentional fires annually? [One hectare (ha) is equivalent to 2.471 acres (A.).]

    (b) How many square miles are scored by these unintentional fires annually? [One acre is equivalent to 0.0015625 square miles (mi^2).]

Density of Matter

2.12 At 68°F (20°C), 5 gallons of a flammable liquid weighs 53 lb.

    (a) Use Table 2.3 to identify the name of this liquid.

    (b) What is the specific gravity of the liquid?

2.13 A piece of metal weighs 13.21 g. When inserted into a cylinder containing exactly 50.00 mL of water, the metal and water occupy a volume of 54.89 mL at 68°F (20°C). Use Table 2.3 to identify the metal.

2.14 Drums of a nonflammable, water-insoluble liquid are identified at a fire scene. A sample of the liquid collected from a drum weighs 1.661 lb and occupies a volume of 0.125 gal.

    (a) Using the information in Table 2.3, identify the name of this liquid.

    (b) What is the specific gravity of the liquid?

    (c) If water is added to the contents of the drum, will it float on or sink beneath the liquid?

2.15 Five partially filled drums of an unidentified liquid are discovered by a fire investigator in an abandoned warehouse. Using a small volume of the liquid from one of the drums, the investigator ascertains that the liquid is flammable and immiscible with water. Using a hydrometer of the type in Figure 2.5, the investigator measures the specific gravity of the liquid as 0.68.

    (a) Using Table 2.4, identify the name of this liquid.

    (b) Would water serve as an effective fire extinguisher if the liquid ignited?

    (c) If water was added to a drum of the liquid when it is burning, which would overflow first: water or the burning liquid?

2.16 Chlorine is a poisonous gas that was first used as a chemical warfare agent on April 22, 1915, when Germany released 600,000 lb against an unprotected enemy. Use the information in Table 2.5 to determine whether the chlorine, when first discharged, concentrated in the trenches or rose into the air.

2.17 Two nonflammable liquids readily vaporize at room temperature [70°F (21°C)]. They possess vapor densities of 0.699 and 1.91, respectively. Based solely on this information, which liquid is more likely to be effective as a fire extinguisher?

2.18 Methyl isobutyl ketone (MIBK) is a flammable, water-soluble liquid substance that boils at 246°F (118.9°C) and possesses a specific gravity of 0.802 at 68°F (20°C). MIBK vapor possesses a density of 3.5 when measured relative to air. Demonstrate that this combination of information is useful to firefighters responding to an emergency at a burning warehouse in which seventy-five 55-gallon steel drums containing MIBK are stored.

Conversion of Temperature Readings

2.19 The surface of an iron burns the skin at approximately 65°C, a temperature at which most individuals experience pain. This temperature is called the *threshold for pain*. Will an average person sense pain when touching a steel radiator whose surface temperature is 175°F?

2.20 The approximate temperature range of an incandescent metal surface is 800–1000°C when the surface is a bright, cherry red in color. What is the minimum temperature of this incandescent metal surface when expressed in degrees Fahrenheit?

2.21 Carbon dioxide is a colorless gas at ambient conditions, but when the gas is compressed and cooled to 31.1°C, it liquefies. At which temperature does carbon dioxide liquefy when expressed in the following units:
(a) Fahrenheit degrees
(b) Rankine degrees
(c) kelvins

2.22 The four reaction zones of color and temperature of a candle flame are illustrated in Figure 2.16. Incandescent soot particles burning in the "yellow zone," sometimes called the "luminous zone," provide most of the light from a burning candle.
(a) Identify the coolest and hottest zones of the flame.
(b) What is the temperature in degrees Fahrenheit of the blue zone?
(c) What is the temperature in degrees Rankine of the yellow zone?

Pressure

2.23 Industrial diamonds are produced when graphite is compressed at 1.3 million psi and heated to approximately 3000°F. At which Celsius temperature and pressure in kilopascals can industrial diamonds be produced?

2.24 The OSHA regulation at 29 CFR §1910.158(a)(3)(iii) stipulates that employers must provide the personnel in fire brigades, industrial fire departments, and private or contractually arranged fire departments with fire hose of such a length that friction loss resulting from the flow of water through the hose will still provide a dynamic pressure at the nozzle within the approximate range of 210 kPa

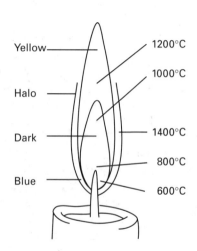

FIGURE 2.16 ◆ The zones of color and temperature of a candle flame.

to 860 kPa. What pressure range at the nozzle does OSHA require in the fire hose when the pressure is expressed in pounds per square inch (psi)?

2.25 Ethylamine and ethyl mercaptan are flammable liquids whose vapor pressures at 68°F (20°C) are 1.18 atm and 442 mm Hg, respectively. Which liquid is most likely to pose the greater risk of fire and explosion?

2.26 Nonflammable, toxic vapor evolves from liquid nitric acid. At which temperature is the degree of inhalation toxicity from exposure to this vapor greater: 0°C or 30°C?

2.27 The OSHA regulation at 29 CFR. §1910.157(f)(6) stipulates that employers must hydrostatically test carbon dioxide hose assemblies having a shutoff nozzle at 300 psi. If an employer hydrostatically tests such a hose assembly at 2100 kPa, is the company in compliance with this OSHA regulation?

Transmission of Heat

2.28 When fire begins on the fourth floor of a high-rise building, why are firefighters more concerned about its immediate spread to upper floors than to lower floors?

2.29 Why do firefighters often ventilate a burning building by opening a large hole at the highest point on its roof?

2.30 Which mechanism of heat transfer is involved in each of the following situations:
   (a) A tile floor feels cold to bare feet, while a wool carpet at the same temperature feels warm.
   (b) While pain is not experienced by placing your hand *next* to a candle flame, it is barely tolerable when the hand is placed *above* the flame.
   (c) Most of the heat generated by burning logs within a fireplace is lost up the chimney.
   (d) Warmth is experienced by an individual sitting in front of an operating fireplace.
   (e) A ceiling heat lamp warms the air through an entire bathroom.

Calculation of Heat

2.31 The human body is approximately 60% water by mass. Why is it capable of maintaining a relatively steady internal temperature of 98.6°F (37°C), regardless of the surrounding temperature?

2.32 Sweating cools the body through evaporation. Approximately 540 calories of heat are removed when 1 gram of sweat evaporates. How many joules of heat are removed when 1 gram of sweat evaporates?

2.33 Taking a bath could use approximately 450 lb of water. How much heat measured in British thermal units is needed to heat the water from a relatively cold 60°F to a warm 104°F?

2.34 How many Celsius degrees will the temperature of 1 kg of liquid water be raised by 4500 cal of heat?

2.35 Use Table 2.8 to determine the number of calories required to raise the temperature of 50.0 g of each of the following liquids from 20.0°C to 50.0°C?
   (a) Acetone
   (b) Carbon tetrachloride
   (c) Methyl alcohol
   (d) Water

Expansion of Liquids

2.36 The DOT regulation at 49 CFR §173.116(b) requires shippers and carriers to confine liquids within containers having a capacity of 110 gal or less so they are not completely full at a temperature of 130°F (55°C). To comply with this regulation, what is the maximum volume of liquid pentane that may be transported in 55-gallon steel drums at 70°F?

General Properties of the Gaseous State

2.37 Natural gas is stored in a gas cylinder measuring 5.0 ft long and having an internal diameter of 8.0 in. What volume in cubic feet does the natural gas occupy within the cylinder? (The volume of a cylinder is determined by using the geometrical formula $V = \pi r^2 h$, where V is the volume, $\pi = 3.1416$, r is the radius, and h is the height of the cylinder.)

2.38 When exactly 4.00 L of oxygen is compressed into a small evacuated cylinder at room temperature, the affixed gauge reads 1600 psi_g. What volume in liters will the oxygen occupy when it is transferred from the cylinder into a balloon at standard atmospheric pressure and the same temperature?

2.39 Exactly 4.00 L of air is inserted into a balloon indoors at 75.2°F (24°C). If the balloon is taken outside where the temperature is −9.4°F (−23°C), what will its volume in liters become? (Assume the atmospheric pressure remains constant.)

2.40 Ten cubic meters ($10\,\text{m}^3$) of argon is squeezed into a balloon at 73.4°F (23.0°C) and a pressure of 740 torr. What will the volume of the argon become when the temperature and pressure are reduced to 14°F (−10°C) and 370 torr, respectively?

2.41 A paint manufacturer affixes the following hazard warning statements to aerosol cans whose contents have been pressurized.

> **CONTENTS UNDER PRESSURE**
> **DO NOT INCINERATE CONTAINER**
> **DO NOT EXPOSE TO HEAT**
> **OR STORE AT TEMPERATURES**
> **ABOVE 120°F**

    (a) Why has the manufacturer provided this information to the consumer?
    (b) What is the technical basis for the hazard warning statement?

2.42 The air within a tire on a fire truck is compressed at 32.0 psi_g at 68°F (20°C). Following use of the fire truck, the new air pressure in the tire is measured as 36.0 psi_g. What is the new temperature in degrees Celsius of the air within the tire?

Cryogenic Liquids

2.43 Use Table 2.10 to determine whether each of the following substances exists as a gas, liquid, or solid at the specified temperature and pressure:
    (a) Sulfur dioxide at −51°F (−46°C) and 500 psi_a
    (b) Propane at 68°F (20°C) and 200 psi_a

FIGURE 2.17 ◆ Over 6500 gallons (25 m^3) of liquid nitrogen was released to the atmosphere when this tank truck overturned. *Courtesy of Los Angeles Times, Los Angeles, California; Los Angeles Times photo by Steve Osman.*

2.44 When adding a cryogenic liquid to a cylinder or other pressure vessel, why is it generally advisable to fill the vessel to no more than 80% of its total capacity?

2.45 Nitrogen gas is neither flammable nor toxic. When responding to an incident involving a leak or spill of cryogenic nitrogen, what is the most likely reason emergency-response crews are encouraged to use self-contained breathing apparatus and wear total-encapsulating suits as in Figure 2.17?

Flammable Gases and Flammable Liquids

CHAPTER 3

In industrial settings, large inventories of flammable gases and flammable liquids are often stored for eventual use in connection with the manufacturing and processing of commercial products. Some flammable gases and flammable liquids are also found within residential environments, albeit in lesser amounts. Given the commonplace nature of these hazardous materials, thousands of emergency incidents involving them occur annually.

While we will note the hazardous properties of individual flammable gases and flammable liquids elsewhere within this book, we shall note their common features in this chapter. We shall also review the safety practices that OSHA mandates when employers store flammable gases and flammable liquids in the workplace. Finally, we will examine the procedures implemented by emergency responders when an incident involves the release of flammable gases and flammable liquids into the environment.

Performance Goals for Section 3.1

- For flammable liquids, discuss the concept of a lower and upper explosive limit, flammable range, flashpoint, fire point, kindling point, and autoignition point.
- Describe a flammable liquid as the term is used by OSHA and the NFPA.
- When provided with a liquid's flashpoint and boiling point, demonstrate how Figure 3.3 may be used to appropriately classify it as a flammable liquid or combustible liquid.

◆ 3.1 LIQUID FLAMMABILITY

Even when exposed to an ignition source, there are unique minimum and maximum vapor concentrations below and above which a flammable liquid does not burn. These minimum and maximum concentrations are called the *lower explosive limit* and *upper explosive limit*, respectively. For example, xylene is a flammable liquid. Its lower explosive limit is 1.1% by volume, whereas its upper explosive limit is 7.0% by volume in air. These limits signify that a mixture of xylene vapor having a concentration less than 1.1% by volume in air is too *lean* in xylene vapor to burn; similarly, a mixture containing more than 7.0% by volume is too *rich* in xylene vapor to burn.

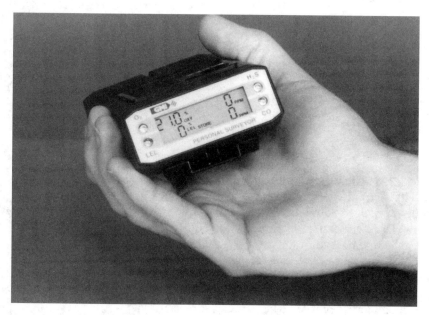

FIGURE 3.1 ◆ This personal multi-gas confined-space entry monitor simultaneously monitors, displays, and alerts personnel to the presence of one to four gases. *Courtesy of CEA Instruments, Inc., Emerson, New Jersey.*

The *flammable range,* sometimes called the *explosive range,* of flammable gases and the vapors of flammable liquids is the numerical difference between their upper and lower explosive limits. Thus, the flammable range of xylene vapor is 5.9% by volume.

A flammable gas or vapor burns only when the concentration of the gas or vapor is exposed to an ignition source *and* its concentration in the air lies within the flammable range for the given substance. Thus, all concentrations of xylene vapor between 1.1% and 7.0% by volume ignite when exposed to an ignition source.

A gas or vapor often escapes from its confinement vessel at hatches, apertures, vents, and other openings. If it is a flammable gas or vapor, it either ignites or disperses upon exposure to the air. Immediate corrective action must be taken when a leaking gas or vapor is discovered. When such a leak is identified during an emergency-response action, personnel are often obligated to stop or seal the leak.

When responding to an emergency involving the release of a flammable gas or vapor, the first-on-the-scene responder often determines the concentration of the substance in the air through use of a portable combustible gas monitor or sensor like the one in Figure 3.1. Using this device, emergency-response crews readily estimate the likelihood as to whether a spark, flame, or other ignition source will ignite a flammable gas or vapor.

When a flammable liquid is stored within a container or tank, its vapor evolves into the headspace above it, where the vapor mixes with the confined air. When the liquid is removed, the vapor remains within the container or storage tank. This ostensibly emptied tank or container could pose substantial risks for the following reasons:

◆ An "emptied" vessel poses the risk of fire and explosion. It is especially hazardous to conduct cutting or welding operations on a tank that had been used to store a flammable liquid, unless specific actions to completely discharge the vapor from the tank have been implemented.

FIGURE 3.2 ◆ The flashpoint of a flammable or combustible liquid is experimentally determined by using specialized equipment, like this Pensky-Martens closed-cup tester. A sample of the liquid is contained in the test cup and slowly heated. An ignition source is periodically applied in the vapor space above the liquid. When a flash of fire is momentarily observed, the temperature on the thermometer is recorded. This is the flash point of the liquid. *Courtesy of Fisher Scientific Company, Pittsburgh, Pennsylvania.*

- An "emptied" vessel also poses the risk of inhalation toxicity. It is especially hazardous to enter an "emptied" tank that had been used to store a flammable liquid. While the inhalation of a nontoxic gas or vapor causes dizziness, illness, and suffocation, the inhalation of a toxic gas could be fatal.

For this combination of reasons, precautions should always be exercised when handling or working in or around a tank or container in which a flammable liquid was previously stored.

Three temperatures are used to describe the ability with which a flammable liquid burns: the flashpoint, fire point, and autoignition point.

FLASHPOINT

The *flashpoint* of a flammable liquid is the minimum temperature at which it gives off sufficient vapor to form an ignitable mixture with air near the surface of the liquid or within a test vessel. The flashpoint of the substance is determined using the specialized apparatus in Figure 3.2. Using such an apparatus, the flashpoint of xylene is determined to be 84°F (29°C).

FIRE POINT

Above the flashpoint of a substance is a temperature at which self-sustained combustion occurs. This temperature is called the *fire point*. At its fire point, a flammable liquid gives off sufficient vapor so that continued combustion is maintained. The fire points of many liquids are approximately 30 to 50°F (17 to 27°C) higher than their flashpoints; for example, the fire point of xylene vapor has been measured to be 111°F (44°C), almost 30°F (17°C) higher than its flashpoint.

AUTOIGNITION POINT

As the temperature of the confined vapor of a flammable liquid further increases, a minimum temperature is attained at which self-sustained combustion occurs even in the absence of an ignition source. This is called the *autoignition point*. For example, the autoignition point of xylene is 924°F (496°C).

Solved Exercise 3.1

The flashpoints of *tert*-butyl mercaptan and phenyl acetate are −15°F and 80°C, respectively. Other factors being equal, which of the liquids poses the greater fire and explosion hazard?

Solution: When two liquids possess flashpoints that are widely different, the liquid having the lower flashpoint generally poses the greater fire and explosion hazard. To compare them, their flashpoints must be expressed on the same temperature scale. We calculate that a temperature of −15°F is equivalent to −26°C:

$$t(°C) = [9/5 \times (-15)] + 32 = -26°C$$

Since the flashpoint of *tert*-butyl mercaptan [−15°F(−26°C)] is substantially less than the flashpoint of phenyl acetate [176°F(80°C)], *tert*-butyl mercaptan poses the greater fire and explosion hazard.

THE OSHA/NFPA DEFINITION OF A FLAMMABLE LIQUID AND COMBUSTIBLE LIQUID

We can now define the term "flammable liquid" following the system used by OSHA and NFPA. A liquid is called a *flammable liquid* as well as a *class I liquid* if it has a flashpoint below 100°F (37.8°C), unless it is a liquid mixture having 99% or more of the volume of its components with flashpoints of 100°F (37.8°C) or greater. OSHA and NFPA recognize the following three classes of flammable liquids:

- A *class IA flammable liquid* is a liquid with a flashpoint below 73°F (22.8°C) and a boiling point below 100°F (37.8°C). An example of a class IA liquid is *n*-pentane, since its flashpoint and boiling point are −56°F (−49°C) and 97°F (36°C), respectively.
- A *class IB flammable liquid* is a liquid with a flashpoint below 73°F (22.8°C) and a boiling point at or above 100°F (37.8°C). An example of a class IB liquid is acetone, since its flashpoint and boiling point are 0°F (−18°C) and 133°F (56°C), respectively.
- A *class IC flammable liquid* is a liquid with a flashpoint at or above 73°F (22.8°C) and below 100°F (37.8°C). An example of a class IC liquid is turpentine, since its flashpoint lies in the range from 95 to 102°F (35 to 39°C).

OSHA and NFPA also recognize a *combustible liquid* as any liquid having a flashpoint at or above 100°F (37.8°C). There are three types of combustible liquids:

- A *class II combustible liquid* is a liquid with a flashpoint at or above 100°F (37.8°C) but below 140°F (60°C), other than a liquid mixture having 99% or more of the volume of its components with flashpoints equal to or greater than 200°F (93.3°C). An example of a class II combustible liquid is acetic acid, since its flashpoint is 109°F (43°C).

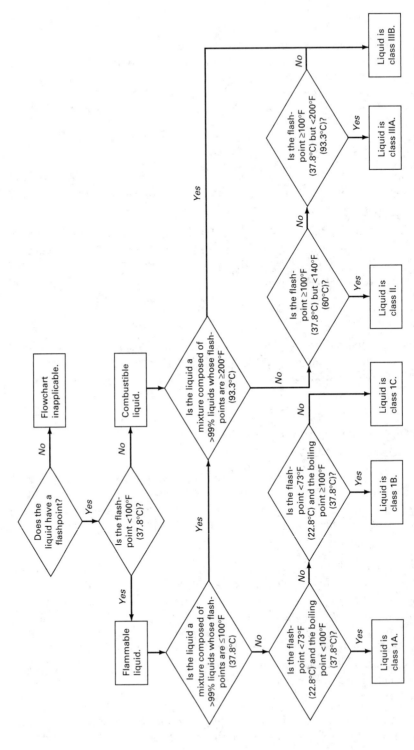

FIGURE 3.3 ◆ A flowchart from which the OSHA/NFPA classes of flammable and combustible liquids may be readily established from the flashpoints and boiling points of unique substances and mixtures.

- A *class IIIA combustible liquid* is a liquid with a flashpoint at or above 140°F (60°C) but below 200°F (93.3°C). An example of a class IIIA liquid is creosote oil, since its flashpoint lies in the range from 165 to 185°F (74 to 85°C).
- A *class IIIB combustible liquid* is a liquid having a flashpoint at or above 200°F (93.3°C). An example of a class IIIB liquid is ethylene glycol, the major constituent of many antifreeze agents, since its flashpoint is 232°F (111°C).

The flowchart in Figure 3.3 is useful for classifying flammable and combustible liquids.

Solved Exercise 3.2

Vinyl ether is a liquid that boils at 102°F (39°C) when the atmospheric pressure is 101.3 kPa. It also possesses a flashpoint of −22°F (−30°C). To which NFPA class of flammable liquid does vinyl ether belong?

Solution: By definition, a liquid having a flashpoint less than 73°F (22.8°C) and a boiling point below 100°F (37.8°C) is a class IA liquid. For this reason, vinyl ether is a NFPA class IA liquid.

Performance Goals for Section 3.2
- Describe the concept of ignitability as it is used by EPA in RCRA regulations.
- Determine whether a chemical waste exhibits the RCRA characteristic of ignitability.

◆ 3.2 RCRA CHARACTERISTIC OF IGNITABILITY

As we previously noted in Section 1.5, the *Resource Conservation and Recovery Act* (RCRA) gives EPA the authority to regulate the treatment, storage, and disposal of certain materials called *hazardous wastes*. Whereas EPA denotes some hazardous wastes by their chemical names, the nature of other hazardous wastes is established by determining whether they exhibit certain characteristics, one of which is relevant here.

A waste exhibits the RCRA characteristic of *ignitability* if it is any one of the following:

- A *flammable gas*, as this term is defined in the DOT regulations (Section 6.7)
- A liquid that possesses a flashpoint equal to or less than 140°F (60°C), other than an aqueous solution (with water as the solvent) containing less than 24% alcohol by volume, when tested through use of the Pensky-Martens closed-cup tester in Figure 3.2
- A material other than a liquid that is capable "of causing fire by friction, adsorption of moisture, or spontaneous chemical changes and, when ignited, burns so vigorously and persistently as to create a hazard"
- A *flammable solid*, as this term is defined in the DOT regulations (Section 6.7)
- An *oxidizer*, as this term is defined in the DOT regulations (Section 6.7)

A waste exhibiting the RCRA characteristic of ignitability is a hazardous waste and its treatment, storage, and disposal are subject to EPA's regulations. A waste exhibiting the characteristic of ignitability is assigned the hazardous waste number "D001."

Solved Exercise 3.3

Three terms are routinely used to describe the potential by which a liquid burns: flammable, combustible, and ignitable. What differences are denoted by each of them?

Solution: Although all three terms are commonly used to denote the relative ease by which a liquid burns, these terms have specific meanings in regulations enacted pursuant to federal laws, as follows:

- OSHA and NFPA define a *flammable liquid* as a liquid having a flashpoint below 100°F (37.8°C), unless it is a liquid mixture having 99% or more of the volume of its components with flashpoints of 100°F (37.8°C) or greater.
- OSHA defines a *combustible liquid* as any liquid having a flashpoint at or above 100°F (37.8°C).
- In the RCRA regulations, EPA characterizes an *ignitable liquid* as a liquid waste, other than an aqueous alcoholic solution containing less than 24% alcohol by volume, that possesses a flashpoint equal to or less than 140°F (60°C).

Performance Goals for Section 3.3
- Distinguish between a compressed gas and a liquefied compressed gas.
- Describe generally the design features of a gas cylinder.
- Describe the significance of the "service pressure" that is marked on a gas cylinder.
- Describe the test used to determine whether a gas cylinder is suitable for use.
- Identify the nature of the markings DOT requires the shipper or carrier to place on cylinders and railroad tankcars used to transport compressed gases and liquefied compressed gases.
- Describe the general practices for safely handling, storing, and transporting compressed gases, including flammable compressed gases.

◆ 3.3 STORING AND TRANSPORTING COMPRESSED GASES

During storage and shipment, a gas is confined under pressure within a steel cylinder or tank. It is called a *compressed gas.* Examples of compressed gases are hydrogen, oxygen, and nitrogen. When a confined gas liquefies under the application of moderate pressure, it is called a *liquefied compressed gas.* Examples of liquefied compressed gases are methane and propane. In this section, we shall use these terms interchangeably.

A compressed gas and a liquefied compressed gas may be either nonflammable or flammable. Oxygen and nitrogen are nonflammable, whereas hydrogen, methane, and propane are flammable. The following discussion is pertinent to both types.

COMPRESSED GASES IN STORAGE

Bulk volumes of compressed gases are ordinarily stored at facilities within stationary spherical tanks, whereas nonbulk volumes are stored within the steel cylinders that were used to transport them from their place of purchase.

FIGURE 3.4 ◆ To ensure stability, cylinders of compressed gases should be securely strapped or chained to a wall or other permanent structure. When not in use, a good practice is to keep them capped. Care should always be taken to keep compressed gas cylinders away from sources of heat or ignition. *Courtesy of Compressed Gas Association, Arlington, Virginia.*

It is in this latter fashion that a compressed gas is most frequently encountered. During storage and use, precautions should be exercised to avoid damaging its cylinder. Experts recommend maintaining the cylinder in an upright position and strapping or chaining it to a fixed position, as in Figure 3.4. The valve protection cap should be kept in place and only removed after the cylinder has been secured.

When it must be moved from place to place, a gas cylinder should again be secured or restrained by straps or chains to a cylinder cart having wheels. A cylinder should never be rolled or dragged, as such actions could seriously damage it. Dropping a cylinder could result in the catastrophic release of its contents.

COMPRESSED GASES DURING TRANSPORT

To ensure that a gas cylinder can be used to safely transport a compressed gas, DOT regulates the manner by which each gas cylinder is designed and constructed. Some DOT design features of a typical gas cylinder are provided in Figure 3.5. Foremost among DOT's concerns is the ability of the cylinder to maintain its integrity when its contents are under pressure and exposed to the range of temperatures likely to be encountered during transportation.

Each type of gas cylinder is assigned a unique pressure at 70°F (21°C) called the cylinder's *service pressure*. DOT requires the service pressure to be marked on the exterior surface of the cylinder in units of psi_g. For instance, when the service pressure of a gas cylinder is 1800 psi_g (1792 kPa), it is marked on the cylinder as "DOT-3E1800"; the "DOT-3E" specifies the type of cylinder.

Each type of gas cylinder is constructed in a specific fashion and tested to determine whether its integrity is maintained when the temperature of the contents is increased from 70°F (21°C) to 130°F (54°C). DOT approves the cylinder for use only when the test reveals that upon heating, the pressure of its contents did not exceed

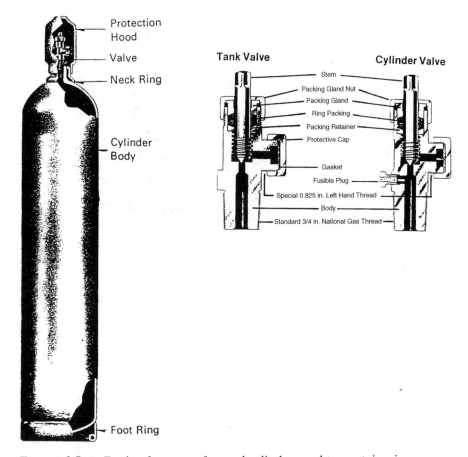

FIGURE 3.5 ◆ Design features of a steel cylinder used to containerize a compressed gas. Located at the top of the cylinder is a single discharge valve that has a fusible plug in the valve body just below the seat. The plug is constructed from a special forged bronze alloy that melts at a temperature ranging from 158° to 163°F (70° to 73°C).

five-fourths of the cylinder's service pressure. When the cylinder is subsequently loaded aboard a truck or railroad car, DOT requires it to be securely strapped or chained in a fixed position, marked, and labeled for transportation.

Solved Exercise 3.4

"DOT-3A2000" is marked on a steel cylinder whose gauge pressure reads 1750 psi at 70°F (21°C). Is this cylinder likely to rupture during a fire if the temperature of its contents elevates to 300°F (149°C)?

Solution: To ensure the integrity of a gas cylinder at elevated temperatures, DOT stipulates that the actual pressure of its contents cannot exceed five-fourths of the service pressure when the cylinder is subjected to a temperature of 130°F

(54°C). The service pressure of a cylinder marked "DOT-3A2000" is 2000 psi. If the gas manufacturer complied with the relevant DOT regulation, the integrity of the cylinder should be maintained as long as the pressure of the contents does not exceed 5/4 × 2000 psi, or 2500 psi, at a temperature of 130°F (54°C).

On the other hand, an internal pressure greater than 2500 psi is generated when the temperature of the gas cylinder exceeds 130°F (54°C). For this reason, a cylinder marked "DOT-3A2000" is likely to rupture when heated to 300°F (149°C).

Bulk volumes of a compressed gas are often transported within a cargo tank or a railroad tankcar, each of which is distinctively characterized by the hemispherical shape of its head and end. The shipment of bulk volumes of certain compressed gases is also authorized by DOT in the jumbo-sized tubes in Figure 3.6. DOT authorizes these bulk shipments only when the transport vessel has been constructed according to specified requirements, and the shipper or carrier has complied with allowable maximum filling limits, testing, maintenance, marking, placarding, and other relevant requirements.

The type of railroad tankcar in Figure 3.7 is commonly used for transporting bulk volumes of a compressed gas over long distances. DOT requires the tankcar to be marked with certain information, including its specification number. An example of such a specification number is DOT-105A500W. The "DOT-105A" denotes a type of tankcar constructed to specifications for transporting certain contents; the "500" denotes the maximum pressure in psi_g under which the contents can be safely confined; and the "W" denotes that the tank was built using fusion-welded construction methods.

FIGURE 3.6 ◆ The side view of a 60-tube semitrailer used to transport compressed hydrogen gas by public highway. Each tube shown here contains 77.65 ft^3 (2.2 m^3) of hydrogen under a pressure of 2400 psi (17,000 kPa). *Courtesy of MG Industries, Malvern, Pennsylvania.*

Figure 3.7 ◆ A 23,800-gal (90 m³) DOT-105 pressure railroad tankcar used to transport a liquefied flammable gas, such as vinyl chloride. *Courtesy of Union Tank Car Company, Chicago, Illinois.*

Performance Goals for Section 3.4
- ◆ Identify the common hazards associated with the presence of compressed gases in cylinders, stationary storage tanks, and transport vessels.
- ◆ Discuss the nature of the vessel types in which compressed gases are stored and transported.

◆ 3.4 GENERAL HAZARDS OF COMPRESSED GASES

We shall examine the general hazards of compressed gases when their cylinders, stationary storage tanks, and transport vessels are involved in emergency-response actions.

CYLINDERS AND STATIONARY STORAGE TANKS

When a gas cylinder has been constructed and tested according to DOT specifications, it will not ordinarily rupture under normal conditions of transport. Even when it is heated slowly to temperatures above ambient conditions, a small quantity of the contents slowly escapes into the environment through pressure-relief devices or fusible plugs, each of which serves to prevent the cylinder from bursting.

On the other hand, when it is heated rapidly to temperatures exceeding approximately 130°F (54°C), a gas cylinder could readily rupture. At such temperatures, the steel cylinder has clearly exceeded its authorized service pressure. The heated contents cannot discharge fast enough through the pressure-relief valves or fusible plugs to adequately reduce the internal pressure. Then, the cylinder is likely to rupture. If unsecured, it could behave as an active airborne missile and jettison from spot to spot, perhaps for hundreds of yards.

When a cylinder or tank containing a flammable gas initially ruptures, the concentration of the released gas is usually within its flammable range. As it mixes with the air, the gas ignites explosively, typically engulfing the entire area in flames. Even when the contents are nonflammable, the environment in which the rupture occurs is an extremely fearsome one. An immense force is exerted upon the surroundings by the recoil, which causes substantial physical destruction within the immediate area.

A stationary storage tank can also be used for confining a compressed gas. When it is situated inside a building, sensing devices electronically linked to exhaust fans are often installed to provide maximum protection for workers. Once activated, they evacuate escaping gases from the enclosure into the outside atmosphere at very high speeds, thereby minimizing the possibility of concentrating the gases indoors. For those compressed gases that possess vapor densities greater than 1.0, the sensing devices and fans are positioned near the floors of the buildings in which their storage tanks are located; for gases possessing vapor densities less than 1.0, they are positioned near the ceilings.

Solved Exercise 3.5

A gas cylinder contains compressed nitrogen, a nonflammable gas, at 2000 psi_g and 65°F. During a fire, the cylinder is engulfed in flames and the temperature of the cylinder and its contents is rapidly elevated to 350°F.

(a) What is the gauge pressure on the cylinder at 350°F?

(b) Is the cylinder likely to withstand this elevated temperature?

Solution: **(a)** The pressure of the nitrogen confined within the cylinder can be calculated using the combined gas law previously noted in Section 2.12. Since a steel cylinder is essentially a constant-volume container, $V_1 = V_2$.

First, the initial absolute pressure of the nitrogen at 65°F in the cylinder is calculated as follows:

$$2000 \text{ psi} + 14.7 \text{ psi} = 2015 \text{ psi}_a.$$

This is P_1.

Then, the initial and final Fahrenheit temperatures are converted into their equivalent readings on the Rankine scale by adding 460°F to each temperature:

$$T(°R) = 65°F + 460°F = 525°R$$
$$T(°R) = 350°F + 460°F = 810°R$$

These temperatures are T_1 and T_2, respectively.

The absolute pressure of nitrogen at 350°F, P_2, is then calculated as follows:

$$2015 \text{ psi}_a \times \frac{810°F}{525°F} = 3109 \text{ psi}_a$$

The gauge pressure is then calculated by subtracting 14.7 psi from the absolute pressure:

$$3109 \text{ psi} - 14.7 \text{ psi} = 3094 \text{ psi}_g$$

(b) The internal strain on the walls of the steel cylinder has increased from 2000 psi at 65°F to 3094 psi at 350°F. To relieve this excessive strain, the cylinder is likely to rupture and expel the contents into the immediate environment. Although nitrogen does not burn, its release from a ruptured cylinder generates an immense force upon the surroundings that causes considerable damage.

TRANSPORT VESSELS

Compressed gases and liquefied compressed gases are routinely transported within cargo tanks and railroad tankcars. The liquefied compressed gases exist within these transport vessels as two phases: a heavier liquid that settles to the bottom of the vessel and a gas that coexists in the headspace above the liquid. It is the presence of this gaseous phase that gives rise to special concern, not only when these transport vessels are involved in transportation mishaps, but whenever they are engulfed in fire.

Performance Goal for Section 3.5

- Use Figures 3.8, 3.9, 3.10, and 3.11 to describe the general practices recommended for emergency responders when they encounter the release of a compressed gas.

◆ 3.5 RESPONDING TO INCIDENTS INVOLVING THE RELEASE OF COMPRESSED GASES

There have been countless occasions involving the release of compressed gases from cylinders, storage tanks, and transport vessels. For the first-on-the-scene responders, it is always prudent to acknowledge up front that these vessels are likely to rupture when they are exposed to intense heat.

Procedures to be implemented during emergency-response actions involving compressed gases in cylinders and tanks are provided in Figures 3.8, 3.9, 3.10, and 3.11. We shall give special attention to emergency-response actions involving compressed gases that are toxic in Section 10.14.

In general, before emergency responders can exercise any action, the answers to the following questions should be sought:

- Are the cylinders or tanks releasing their contents? If so, where are they leaking and what is the relative size of the leak?
- Is an ongoing fire near the cylinders or tanks?
- Are the cylinder or tank contents flammable, nonflammable, or toxic, or do they support combustion? For compressed gases in cylinders, the answer to this question may be obtained by noting the cylinder labels from a distance.
- What is the vapor density of the cylinder or tank contents? The answer to this question reveals whether the compressed gas is lighter or heavier than air and whether a leak will settle in low spots or rise into the atmosphere.
- What is the approximate volume of gas that could be released to the environment?
- What are the prevailing weather conditions?
- What is the population density of the immediate area?
- Is sufficient water available for cooling the cylinders or tanks and for extinguishing fires?

As the nature of the scene is being assessed, emergency responders should proceed to undertake the following actions as deemed necessary and appropriate:

GENERAL PROCEDURES

- If the emergency-response incident involves a transportation mishap, CHEMTREC (Section 1.10) should be contacted as soon as practical and provided with relevant information about the unique substances at issue. Minimal information can be obtained by speaking with the driver or conductor, reviewing the transportation manifest, and noting the labels, markings, and placards used by the shipper and carrier.
- If the response action involves the storage of a gas in a stationary tank, personnel should use the information posted on an NFPA diamond to ascertain the degree of hazard relating to fire, health, and chemical reactivity.
- When there is a fire in an area where cylinders or tanks containing compressed gases are located, unmanned monitors should be situated to cool them with direct streams of water.
- When a small leak from a cylinder is discovered, a water fog should be used to disperse the gas or vapor.
- When personnel must decide whether to move a leaking cylinder containing a flammable, nontoxic gas, a portable combustible gas monitor or similar device can be used to ascertain the concentration of the gas within the emergency-response area. No attempt to move a leaking cylinder should be made if the concentration of an escaping gas or vapor is in the flammable range.

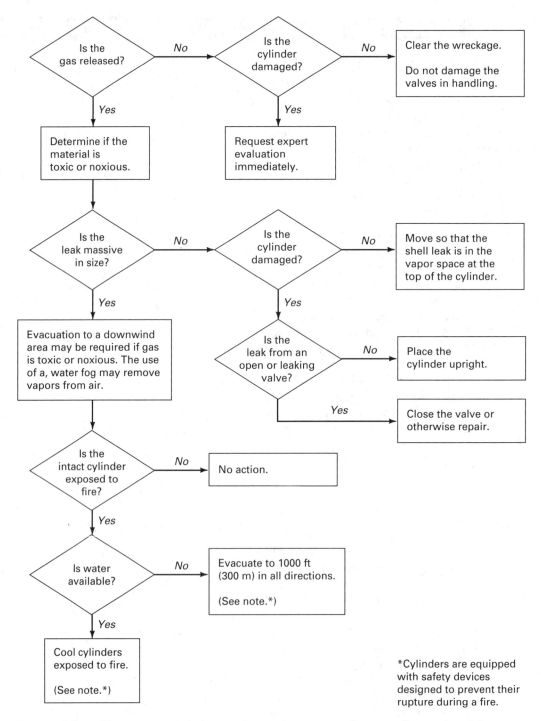

FIGURE 3.8 ◆ The recommended procedures when responding to a disaster involving a nonflammable compressed gas contained within a steel cylinder. *Adapted with permission of the American Society for Testing and Materials, from a figure in ASTM STP 825, A Guide to the Safe Handling of Hazardous Materials Accidents. Copyright © 1983, American Society for Testing and Materials.*

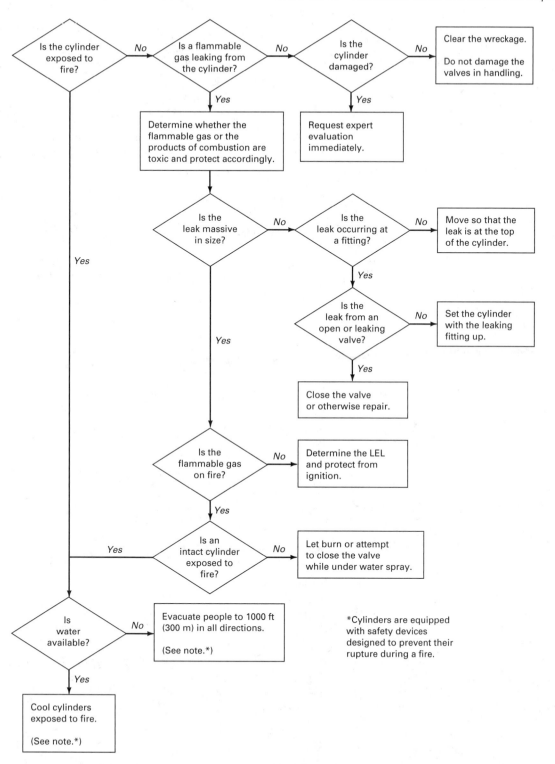

FIGURE 3.9 ◆ The recommended procedures when responding to a disaster involving a flammable compressed gas contained within a steel cylinder. *Adapted with permission of the American Society for Testing and Materials, from a figure in ASTM STP 825, A Guide to the Safe Handling of Hazardous Materials Accidents. Copyright © 1983, American Society for Testing and Materials.*

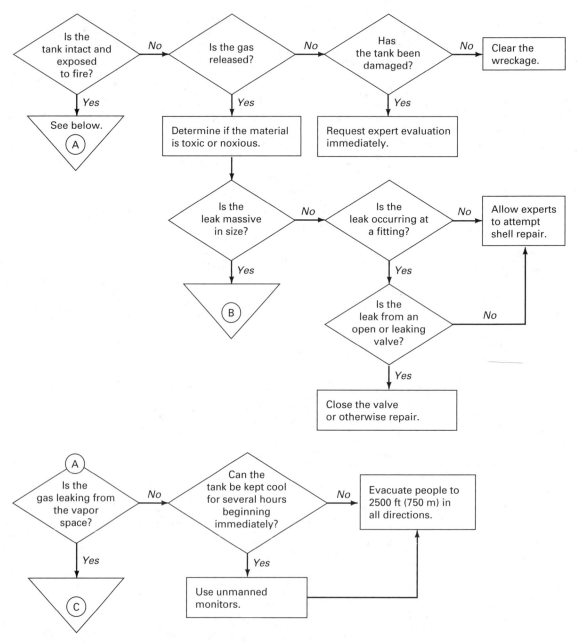

FIGURE 3.10 ◆ The recommended procedures when responding to a disaster involving a nonflammable compressed gas contained within bulk packaging. *Adapted with permission of the American Society for Testing and Materials, from a figure in ASTM STP 825, A Guide to the Safe Handling of Hazardous Materials Accidents. Copyright © 1983, American Society for Testing and Materials.*

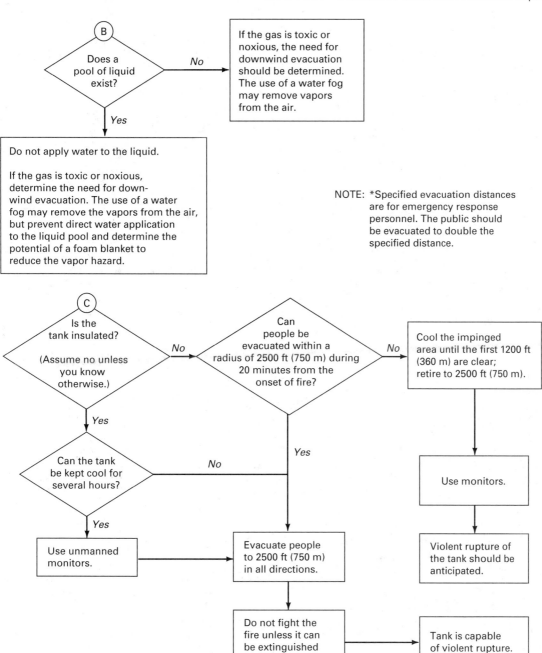

FIGURE 3.10 ◆ *(continued)*

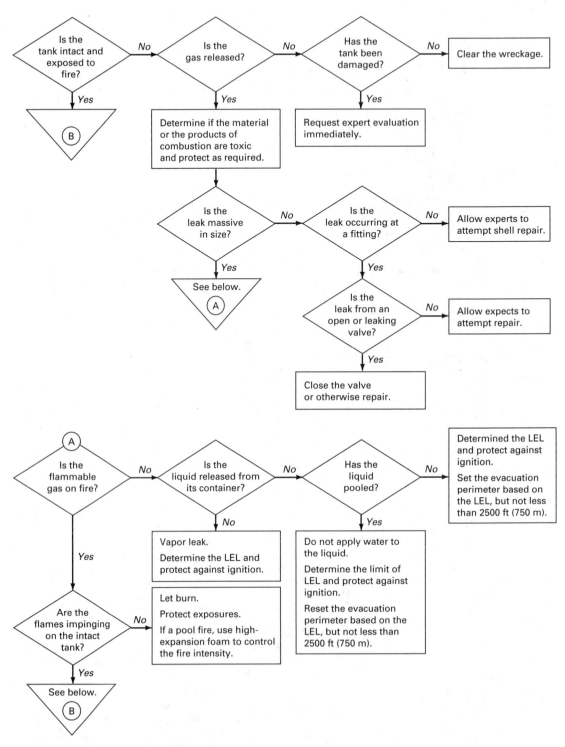

FIGURE 3.11 ◆ The recommended procedures when responding to a disaster involving a flammable compressed gas contained within bulk packaging. *Adapted with permission of the American Society for Testing and Materials, from a figure in ASTM STP 825, A Guide to the Safe Handling of Hazardous Materials Accidents. Copyright © 1983, American Society for Testing and Materials.*

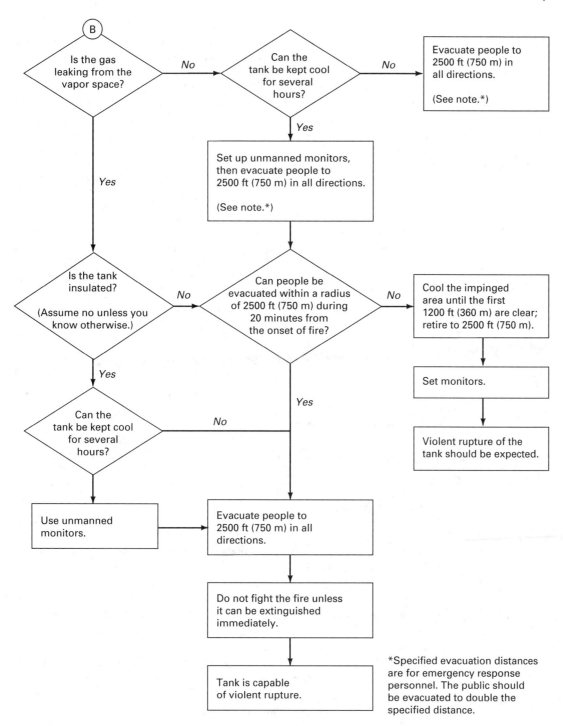

FIGURE 3.11 ◆ (*continued*)

- A slowly leaking cylinder containing a flammable, nontoxic gas can be moved into an open isolated area, but *only* when the gas has not ignited and when the operation can be safely conducted. When moving a leaking cylinder, it is prudent to wear a total-encapsulating suit and use self-contained breathing apparatus. The latter action is warranted since the release of many compressed gases causes dizziness, illness, or suffocation.
- When a cylinder is moved into an open area, it should be rotated so that the point of leakage becomes the uppermost part of the cylinder. Although emergency responders may elect to allow the gas to slowly escape into the air, the leak can often be stopped by closing the valve or tightening the packing gland nut.
- Experience tells us that vessels generally rupture at a tank's head or end, near the seam where the head or end was welded to the tank's body. For this reason, emergency responders should *never* approach the head or end of a cylinder or tank containing a flammable compressed gas during an ongoing fire.
- Since a compressed gas exerts pressure equally on all internal points of a cylinder or tank to which it is confined, an overturned, nonleaking railroad tankcar can be safely righted, when practical, without first unloading its contents. This action should *only* be undertaken by experienced personnel.
- Sometimes, to lighten a load, a decision is made to first transfer a portion of the tank contents before righting a railroad tankcar. When a flammable gas is transferred from one tank to another, the transfer line must be electrically grounded.
- When emergency-response personnel must be physically present in an area where a cylinder or tank containing a flammable and *toxic* gas has been located, they should wear total-encapsulating suits, use self-contained breathing apparatus, and work downwind from the tanks.
- Based on the combined experience of fire personnel, combating a major fire involving a flammable compressed gas should not be undertaken *unless* the escape of the flammable gas from its cylinder or tank can be stopped. Sealing a gas leak should only be attempted by experienced personnel.

EVACUATION DISTANCES

- When a *cylinder* containing a flammable or nonflammable gas is exposed to fire, and when insufficient water is available for cooling the cylinder, all unauthorized personnel should be evacuated to a distance at least 1000 ft (300 m) in all directions from the fire.* The public should be evacuated to double the distance.
- When a *tank* containing a flammable or nonflammable gas is exposed to fire, and when insufficient water is available for cooling the tank, all unauthorized personnel should be evacuated to a distance at least 2500 ft (750 m) in all directions from the fire. The public should be evacuated to double the distance.
- When a tank containing a flammable or nonflammable gas is exposed to fire, all unauthorized personnel should be evacuated to a distance of at least 2500 ft (750 m) in all directions from the fire *even while unmanned monitors are used to cool the tank*. The public should be evacuated to double the distance.

---

*Recommended isolation and evacuation distances vary with the emergency-response situation. A distance can generally be established by considering the size of the leak, the population density within the emergency area, and the specific characteristics of the contents at issue. Recommended initial isolation and protective action distances are provided in the green-bordered section of the DOT publication *Emergency Response Guidebook* (Section 6.8).

◆ 3.6 STORING FLAMMABLE LIQUIDS

At manufacturing and production facilities, flammable liquids are routinely stored for future use in safety cans, bottles, metal cans, metal drums, barrels, portable tanks, and stationary tanks. Whichever way they are stored, the following general guidelines should be followed:

- When not in use, containers and tanks should always be closed by means of a lid or other device that prevents the liquids and their vapors from escaping at ordinary temperatures.
- To protect against fires ignited by static electricity, class I flammable liquids should *only* be dispensed from containers or tanks that have been electrically grounded and bonded. The flowing of flammable liquids within pipes generates electrical charges capable of producing a spark that may ignite a combustible mixture of vapor and air.

To minimize or eliminate employee exposure to flammable liquids, OSHA regulates the manner by which employers store flammable liquids within the workplace. We shall review some aspects of these storage regulations most relevant to fire-service personnel.*

When employers choose to store flammable liquids within an inside room, they must comply with size and construction conditions and allowances on the volume of the flammable liquid that can be stored. These conditions are provided in Table 3.1. The storage room must be constructed in compliance with relevant NFPA standards and use liquid-tight sills or ramps at least 4 in. (10 cm) in height, approved self-closing doors, and a liquid-tight seal between the walls and floor.

When employers choose to store flammable liquids in the workplace, OSHA requires that they provide either an inside room used solely for storage purposes or an approved storage cabinet similar to those in Figure 3.12. OSHA requires employers to store no more than 60 gal (230 L) of class I or class II flammable liquids, and no more than 120 gal (450 L) of class III flammable liquids within a storage

---

*These regulations do not apply to the following circumstances: liquids stored in bulk plants, service stations, processing plants, refineries, or distilleries; a liquid used in vehicle tanks, portable engines, or stationary engines; when the liquid is a paint, oil, or varnish used for maintenance; a liquid beverage stored in containers having a capacity of less than 1 gallon (3.8 L); and liquid medicines, beverages, foodstuffs, cosmetics, and other common consumer items.

TABLE 3.1 ◆ Storage of Flammable Liquids in Inside Rooms[a]

Fire Protection System Needed[b]	Fire Resistance[c]	Maximum Size		Total Allowable Volumes	
		ft²	m²	gal/ft²	L/m²
Yes	2 hours	500	46.5	10	407
No	2 hours	500	46.5	5	200
Yes	1 hour	150	13.9	4	160
No	1 hour	150	13.9	2	80

[a]29 CFR §1910.106(d), Table H.13
[b]The fire protection system must be sprinkler, water spray, carbon dioxide, or a similarly approved system.
[c]The fire resistance rating of construction materials is determined by exposing them to standardized gas burners within a furnace. The rating is expressed in units of time. Materials rated fire resistances of "1 hour" and "2 hours" have met the furnace test without failure for 1 hour and 2 hours, respectively.

FIGURE 3.12 ◆ Storage cabinets approved by OSHA for the indoor storage of limited quantities of flammable liquids. *Courtesy of SE-Cur-All® Cabinets, Laporte, Indiana.*

TABLE 3.2 ◆ Maximum Allowable Sizes of Containers and Portable Tanks[a]

| | Flammable Liquids | | | | | | Combustible Liquids | | | |
| | Class IA | | Class IB | | Class IC | | Class II | | Class III | |
Type	*gal*	*L*	*gal*	*L*	*gal*	*L*	*gal*	*L*	*gal*	*L*
Glass or approved plastic	0.12	0.5	0.25	1	1	4	1	4	1	4
Metal (not DOT approved)	1	4	5	20	5	20	5	20	5	20
Safety cans[b]	2	7.5	5	20	5	20	5	20	5	20
Metal drums (DOT approved)	60	230	60	230	60	230	60	230	60	230
Portable tanks	660	2500	660	2500	660	2500	660	2500	660	2500

[a]29 CFR §1910.106(d), Table H.12
[b]A "safety can" is an approved container when it holds no more than 5 gallons (19 L) and has a spring-closing lid and spout cover designed to relieve internal pressure safely when exposed to fire.

cabinet at any time. The volume of flammable liquid that may be located *outside* of an inside storage room or storage cabinet located within a building cannot exceed the following:

- ◆ 25 gal (95 L) of Class IA liquids in containers
- ◆ 120 gal (455 L) of Class IB, IC, II, or III liquids in containers
- ◆ 660 gal (2500 L) of Class IB, IC, II, or III liquids in a single portable tank

OSHA also restricts the volumes of flammable liquids that employers store within different types of containers. These restrictions are provided in Table 3.2. The maximum volume of any flammable liquid that OSHA permits to be stored in a container is 600 gal (2500 L), but only when it is contained within a portable tank.

STORAGE WITHIN CONTAINERS

OSHA restricts the volume of flammable liquids that employers store within the workplace in containers. This allowable maximum volume varies with the flashpoint of the liquid, the manner in which the liquid is stored, and whether the storage occurs indoors or outdoors.

When employers choose to store flammable liquids in containers indoors, they must comply with the restrictions in Table 3.3. When two or more classes of flammable or combustible materials are stored in containers in a single pile or group, the maximum volume permitted in the pile is the smallest of their separate maximum volumes. OSHA also requires that no container be situated more than 12 ft (4 m) from an aisle. The main aisles must be at least 3 ft (0.9 m) wide, and the side aisles must be at least 4 ft (1.2 m) wide.

When employers choose to store flammable liquids in containers outdoors, they must comply with the restrictions in Table 3.4. When two or more classes of flammable or combustible materials are stored in containers in a single pile or group, the maximum volume permitted in the pile is the smallest of their separate maximum volumes.

TABLE 3.3 ◆ Indoor Container Storage[a]

Class of Liquid	Storage Level	Protected Storage, Maximum per Pile		Unprotected Storage, Maximum per Pile	
		gal	L	gal	L
IA	Ground and upper floors	2750 (50)[b]	10,400	660 (12)	2300
	Basement	Not permitted		Not permitted	
IB	Ground and upper floors	5500 (100)	21,000	1375 (25)	5200
	Basement	Not permitted		Not permitted	
IC	Ground and upper floors	16,500 (300)	62,400	4125 (75)	15,600
	Basement	Not permitted		Not permitted	
II	Ground and upper floors	16,500 (300)	62,400	4125 (75)	15,600
	Basement	5500 (100)	21,000	4125 (75)	15,600
III	Ground and upper floors	55,000 (1000)	210,000	13,750 (250)	52,000
	Basement	8250 (450)	31,000	Not permitted	

[a]29 CFR §1910.106(d), Table H.14
[b]The numbers in parentheses indicate the number of 55-gallon drums equivalent to the indicated volumes.

OSHA also requires the availability of a 12-foot-wide (4-m-wide) access way for the approach of fire control apparatus.

The distances in Table 3.4 apply to a facility that is "protected for exposure," meaning that all adjacent structures on the facility have been adequately provided with fire protection systems. When these structures have not been protected for exposure, the

TABLE 3.4 ◆ Outdoor Container Storage[a]

Class of Liquid	Maximum per Pile		Distance between Piles		Distance to the Property Line that can be Built Upon		Distance to the Street, Alley, or Other Public Way	
	gal	L	ft	m	ft	m	ft	m
IA	1100	4100	5	1.5	20	6	10	3
IB	2200	8300	5	1.5	20	6	10	3
IC	4400	16,600	5	1.5	20	6	10	3
II	8800	33,200	5	1.5	10	3	5	1.5
III	22,000	83,000	5	1.5	10	3	5	1.5

[a]29 CFR §1910.106(d), Table H.16

distances in the fourth column of Table 3.4 must be doubled. When the total volume of flammable liquids does not exceed 50% of the maximum per pile, the distances in the fourth and fifth columns can be reduced by one-half, but in any event, the distance must never be less than 3 ft (0.9 m).

Solved Exercise 3.6

Methyl formate is a liquid that boils at 90°F (32°C) and possesses a flashpoint of −2°F (−19°C). Since it possesses a pleasant odor, methyl formate is often a component of commercial air fumigants. A company that manufactures air fumigants has constructed an indoor storage room for flammable liquids from fire-resistant materials and equipped it with a sprinkling system. What is the maximum number of 55-gallon drums containing methyl formate that can be stored per pile in the room in compliance with the OSHA regulations?

Solution: From the boiling point and flashpoint data, we determine that methyl formate is an OSHA class IA liquid. From Table 3.3, we determine that, when the storage is protected, the maximum number of 55-gallon drums of methyl formate that OSHA allows to be stored per pile in an indoor storage room on either the ground or upper floors is 50. (When the storage is unprotected, the maximum number of 55-gallon drums of methyl formate that OSHA permits to be stored per pile in an indoor storage room is only 12.)

Solved Exercise 3.7

Ethanol and acetone boil at 174°F (79°C) and 133°F (56°C) and possess flashpoints of 54°F (12°C) and 0°F (−18°C), respectively. What maximum volume of these liquids does OSHA allow to be stored at any time within a storage cabinet situated in an inside room?

Solution: From the boiling point and flashpoint data, we determine that ethanol and acetone are OSHA class IB liquids. To be in compliance with OSHA regulations, no more than a total of 60 gal (230 L) of class I flammable liquids can be stored within a storage cabinet at any time.

Solved Exercise 3.8

A lubricating oil boils at 680°F (360°C) and possesses a flashpoint greater than 300°F (>149°C). What is the maximum volume in gallons of this lubricating oil that OSHA allows to be stored outdoors within containers?

Solution: From the flashpoint data, we determine that the lubricating oil is an OSHA class IIIC liquid. From Table 3.4, a maximum of 22,000 gallons of a class IIIC liquid can be stored outdoors within containers, as long as the adjacent structures to the storage area are provided with fire protection systems.

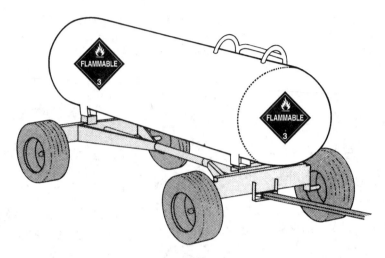

FIGURE 3.13 ◆ A portable tank is used to store regulated volumes of flammable liquids within certain workplaces. A portable tank is defined by OSHA as a closed container that is not fixed in position and possesses a liquid capacity of 60 gallons (227 L) or more.

STORAGE WITHIN PORTABLE TANKS

For the purpose of this section, a portable tank is a closed container that is not intended for fixed installation and possesses a liquid capacity of 60 gallons (227 L) or more. The portable tank in Figure 3.13 has been designed with mountings to facilitate handling by mechanical means.

Portable tanks approved for the storage of flammable liquids have one or more devices installed in their tops with sufficient emergency-venting capability to limit the internal pressure under fire conditions to 10 psi_g, or 30% of the bursting pressure, whichever is greater.

When employers choose to store flammable liquids indoors in portable tanks, they must comply with the restrictions in Table 3.5. When one or more classes of flammable or combustible materials are stored in a single pile or group of portable tanks, the maximum volume permitted is the smallest of the separate maximum volumes. All piles are required to be separated by at least 4 ft (1.2 m).

When employers choose to store flammable liquids outdoors in portable tanks, they must comply with the restrictions in Table 3.6. When one or more classes of flammable or combustible materials are stored in a single pile or group of portable tanks, the maximum volume permitted is the smallest of the separate maximum volumes.

The distances in Table 3.6 apply to a property that has been protected for exposure. When the property has not been protected for exposure, the distances in the fourth column of Table 3.6 must be doubled. When the total volume of flammable liquids does not exceed 50% of the maximum per pile, the distances in the fourth and fifth columns can be reduced by one-half, but in any event, the distance must never be less than 3 ft (0.9 m).

TABLE 3.5 ◆ Indoor Portable Tank Storage[a]

Class of Liquid	Storage Level	Protected Storage, Maximum per Pile		Unprotected Storage, Maximum per Pile	
		gal	L	gal	L
IA	Ground and upper floors	Not permitted		Not permitted	
	Basement	Not permitted		Not permitted	
IB	Ground and upper floors	20,000	75,700	2000	7600
	Basement	Not permitted		Not permitted	
IC	Ground and upper floors	40,000	151,000	5500	20,800
	Basement	Not permitted		Not permitted	
II	Ground and upper floors	40,000	151,000	5500	20,800
	Basement	20,000	75,700	Not permitted	
III	Ground and upper floors	60,000	227,000	22,000	83,000
	Basement	20,000	75,700	Not permitted	

[a]29 CFR §1910.106(d), Table H.15

STORAGE WITHIN STATIONARY TANKS

Flammable liquids are also stored in stationary tanks, of which there are three common types:

- An *atmospheric tank*. This is a storage tank designed to operate at pressures ranging from atmospheric pressure through 0.5 psi_g (3 kPa).
- A *low-pressure tank*. This is a storage tank designed to operate at pressures above 0.5 psi_g (3 kPa), but not more than 15 psi_g (100 kPa).
- A *pressure vessel*. This is a storage tank or other vessel designed to operate at pressures above 15 psi_g (100 kPa).

TABLE 3.6 ◆ Outdoor Portable Tank Storage[a]

Class of Liquid	Maximum per Pile		Distance between Piles		Distance to the Property Line that can be Built Upon		Distance to the Street, Alley, or Other Public Way	
	gal	L	ft	m	ft	m	ft	m
IA	2200	8300	5	1.5	20	6	10	3
IB	4400	16,600	5	1.5	20	6	10	3
IC	8800	33,200	5	1.5	20	6	10	3
II	17,600	66,400	5	1.5	10	3	5	1.5
III	44,000	166,000	5	1.5	10	3	5	1.5

[a]29 CFR §1910.106(d), Table H.17

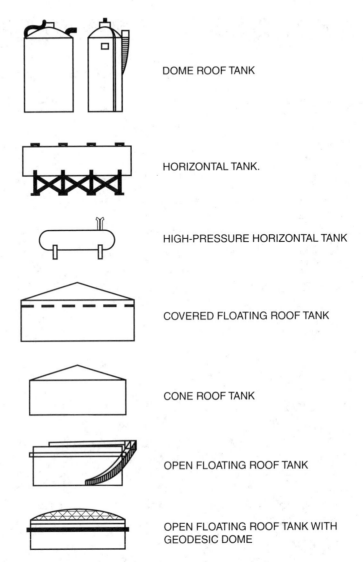

DOME ROOF TANK

HORIZONTAL TANK.

HIGH-PRESSURE HORIZONTAL TANK

COVERED FLOATING ROOF TANK

CONE ROOF TANK

OPEN FLOATING ROOF TANK

OPEN FLOATING ROOF TANK WITH GEODESIC DOME

FIGURE 3.14 ◆ Silhouettes of the commonly encountered types of stationary tanks used for the storage of flammable liquids.

The silhouettes of the most common types of stationary tanks are shown in Figure 3.14.

Multiple stationary tanks of varying sizes and shapes are often situated on the same piece of property. This collection of storage tanks is called a *tank farm*. Water cannons and other cooling devices are generally available in a tank farm to rapidly extinguish fires and cool the surrounding tanks.

When encountered, stationary tanks are routinely situated within secondary containment areas designed to provide a degree of protection in the event of tank failure. The liquid generated from overfilling, maintenance activities, and piping failures is collected within this containment area.

NFPA has provided guidelines for situating stationary tanks used for the storage of flammable liquids. These guidelines include restrictions on the distances from property lines, public ways, and important buildings on the same property.* NFPA also maintains that the storage tanks be situated within a secondary containment area having a capacity equal to at least the capacity of the largest tank within the area. In certain instances, diversion curbs or grading must be provided to protect adjoining property. Released liquids can then drain from the area in which the tank is located into a remote impounding area. (See Review Exercise 3.24.)

NFPA and OSHA also address the design, construction, fabrication and installation conditions relevant to these tanks. Some relevant information follows:

- An atmospheric tank should not be used to store a flammable liquid at a temperature equal to or above its boiling point.
- The operating pressure of a low-pressure or pressure vessel should never exceed the design pressure of the vessel.
- The distance between any two *adjacent* outside aboveground tanks storing flammable liquids cannot be less than one-sixth the sum of their diameters. When the diameter of one tank is less than one-half the diameter of the adjacent tank, the distance between the two tanks cannot be less than one-half the diameter of the smaller tank.
- The distance between any two outside aboveground tanks storing flammable liquids cannot be less than 3 ft (0.9 m).
- When tanks are connected in three or more rows or in an irregular pattern, greater spacing or other means must be provided so that the innermost tanks are accessible to firefighting crews.
- Where crude petroleum tanks are located at production facilities in isolated areas and have capacities not exceeding 126,000 gal (477,000 L or 3000 42-gal barrels), the distance between such tanks cannot be less than 3 ft (0.9 m).
- The minimum separation between a liquefied petroleum gas container and an aboveground atmospheric tank storing flammable liquids must be 20 ft (6 m).

Performance Goals for Section 3.7
- Identify the general ways in which flammable liquids are transported.
- Distinguish between the nature of nonpressure and pressure railroad tankcars.
- Distinguish between the nature of thermally insulated and thermally protected railroad tankcars.

◆ 3.7 TRANSPORTING FLAMMABLE LIQUIDS

When transported, flammable liquids are often confined within bulk transport vessels such as portable tanks, cargo tanks, railroad tankcars, or multiunit tankcars. DOT authorizes their shipment when the shipper and carrier have loaded them into transport vessels constructed according to specified requirements and complied with certain allowable maximum filling limits, testing, maintenance, marking, placarding, and other relevant requirements.

*NFPA No. 30, *Flammable and Combustible Liquids Code* (Quincy, Massachusetts: National Fire Protection Association).

FIGURE 3.15 ◆ Flammable liquids are frequently transported by public highway in semitrailers. The top of the figure shows the side view of a DOT-412 semitrailer; the lower portion shows a back view (left) and a side view (right) of another DOT-412 tank trailer. The capacity of each tank is 6300 gallons (24,000 liters). *Courtesy of Marsh Industrial, DRM Inc., Kalkasha, Michigan.*

TRANSPORT BY MOTOR CARRIER

Two types of motor trucks are commonly used to transport bulk volumes of hazardous materials: cargo vans and the cargo tank/semitrailer combination in Figure 3.15. Cargo vans are used to transport flammable liquids contained within steel drums and other approved containers.

TRANSPORT BY RAILROAD TANKCAR

The railroad tankcars in Figure 3.16 vary significantly in size and capacity. They are of two types:

- *Nonpressure tanks.* These are DOT Specification 103, 104, 111, and 115 tankcars. Prior to their use, the integrity of the tanks is tested by applying internal pressures ranging

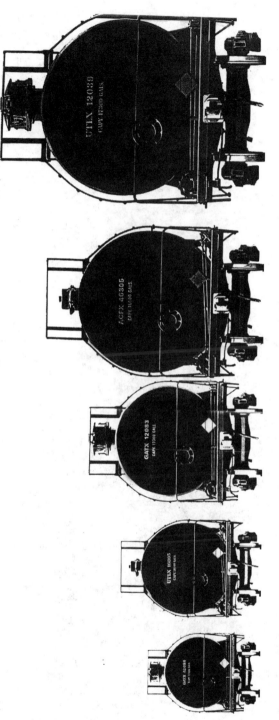

FIGURE 3.16 ◆ Railroad tankcars are used to transport flammable liquids in sizable volumes.

FIGURE 3.17 ◆ A 25,000-gal (95 m³) DOT-111 nonpressure, all-purpose railroad tankcar used to transport liquids, including flammable liquids. *Courtesy of Union Tank Car Company, Chicago, Illinois.*

from 60 to 100 psi (410 to 690 kPa). While the outlets and external heater lines are located on the bottom of these tankcars, all other fittings, valves, and pressure-relief devices are located topside and externally within the dome. Approximately 75% of the railroad tankcars used to transport flammable liquids are non-pressure tanks like the type in Figure 3.17.

◆ *Pressure tanks.* These are DOT Specification 105, 109, 112, 114, and 120 tankcars. The integrity of these tanks is tested by applying internal pressures ranging from 100 to 600 psi (690 to 4100 kPa). Whereas the DOT 114 tankcar has a bottom unloading valve, the fittings, valves, and pressure-relief devices on the DOT 105, 109, 112, and 120 tankcars are located topside and externally within the dome. Approximately 23% of the railroad tankcars used to transport flammable liquids resemble the pressure tankcar in Figure 3.7.

The railroad tankcars in use today have typically been constructed so that their contents are thermally insulated or thermally protected during transportation.

◆ Thermally "insulated" tankcars have been constructed so that external heat is transferred to the contents very slowly under the *normal conditions* of transport.

◆ Thermally "protected" tankcars have been constructed so that external heat cannot be transferred to the contents by either conduction or radiation under *abnormal conditions*, such as a transportation mishap.

Today, DOT requires tankcars used for the shipment of a flammable gas or flammable liquid to be equipped with thermal protection, either by spraying an insulating material directly upon the tank shell or by enclosing it within a jacket that surrounds the shell.

Performance Goals for Section 3.8

◆ Using Figures 3.19 and 3.20, describe the general practices recommended for emergency responders when they encounter the release of a flammable liquid.

◆ Describe the cause of a BLEVE.

◆ Identify the combination of conditions necessary for the occurrence of a BLEVE.

◆ Describe the actions emergency responders take to prevent the occurrence of a BLEVE.

◆ 3.8 RESPONDING TO INCIDENTS INVOLVING THE RELEASE OF A FLAMMABLE LIQUID

Mishaps involving the release of a flammable liquid occur all too frequently. They can occur when the liquid is transported, as in Figure 3.18, and they can also occur when the liquid is in storage and in use at manufacturing and processing industries. Most commonly, these mishaps originate when the flammable liquid leaks or spills from its containment vessel. The liquid then vaporizes, giving rise to the risk of fire once the flammable range has been achieved.

Procedures recommended for implementation during emergency-response actions involving the release of a flammable liquid are provided in Figures 3.19 and 3.20, respectively. Many procedures are analogous to those associated with the release of a flammable gas.

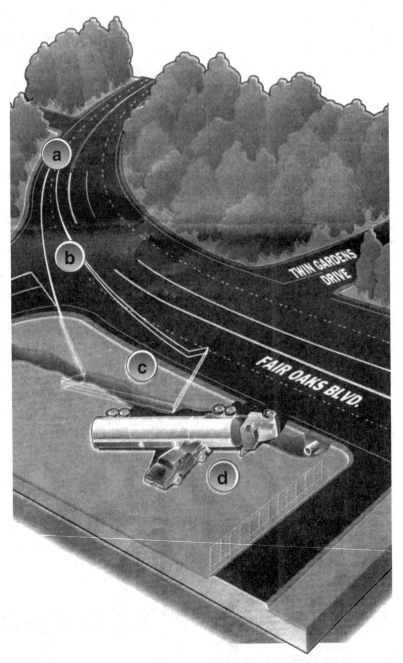

FIGURE 3.18 ◆ An example of a transportation mishap: In (a), the driver of a motor tank truck exceeds the speed limit designated for local road conditions; in (b), the driver proceeds to speed downhill and around a curve, struggling to maintain the vehicle upright; in (c), the vehicle overturns and skids before halting on the roadside; and in (d), a liquid hazardous material gushes from the damaged truck.

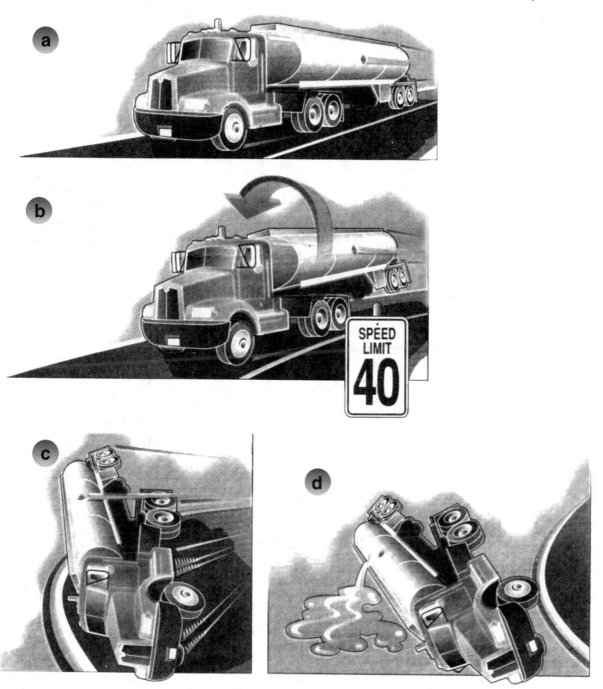

FIGURE 3.18 ◆ *(continued)*

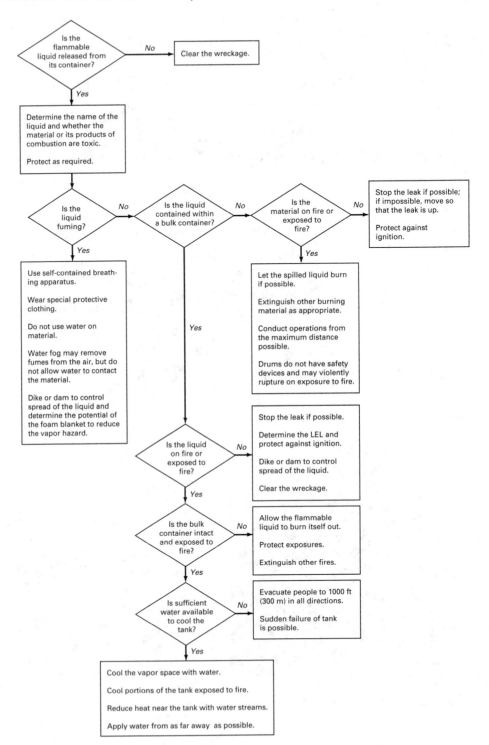

FIGURE 3.19 ◆ The recommended procedures when responding to a disaster involving a flammable liquid. *Adapted with permission of the American Society for Testing and Materials from a figure in ASTM STP 825, A Guide to the Safe Handling of Hazardous Materials Accidents. Copyright © 1983, American Society for Testing and Materials.*

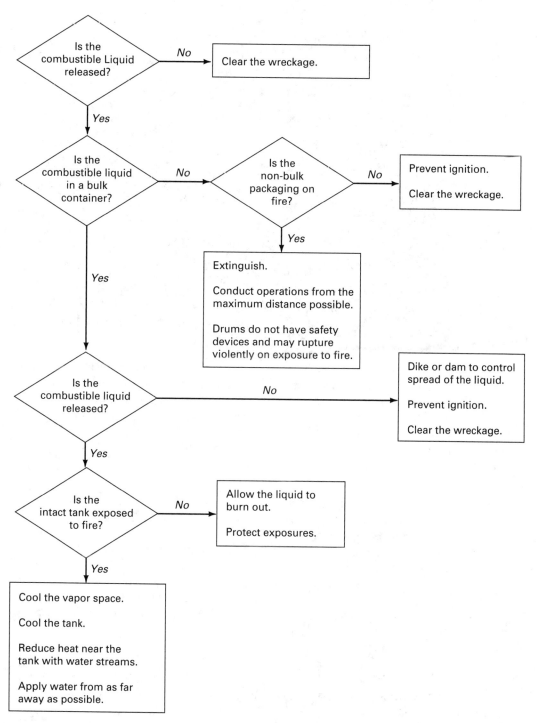

FIGURE 3.20 ◆ The recommended procedures when responding to a disaster involving a combustible liquid. *Adapted with permission of the American Society for Testing and Materials from a figure in ASTM STP 825, A Guide to the Safe Handling of Hazardous Materials Accidents. Copyright © 1983, American Society for Testing and Materials.*

FIGURE 3.21 ◆ A BLEVE results when a flammable liquid is rapidly heated to relatively high temperatures above its boiling point. The venting mechanisms of cargo tanks are generally incapable of releasing the huge buildup of internal pressure caused by the expansion of the vapor. Structural failure results, causing a massive explosion with an accompanying fireball.

The most serious incident involving the release of a flammable liquid occurs when an ongoing fire impinges upon a vessel in which a flammable liquid is stored. During a rail tankcar mishap, for example, the fire associated with a burning tankcar can impinge upon the shell of a nearby tankcar containing a flammable liquid. The heat from the fire causes the shell of the second tankcar to weaken, but it also causes the liquid to vaporize. This gives rise to the existence of two phases within the vessel: a liquid and its vapor. As more and more vapor is produced, the internal pressure increases within the vessel, ultimately causing it to rupture. The liquid discharges into the atmosphere as a colossal mass of vapor that ignites in a fireball.

This phenomenon is called a *BLEVE* (*blev-ey* or blé-vée); the acronym for a *b*oiling *l*iquid *e*xpanding *v*apor *e*xplosion. A BLEVE can produce a fireball having a radius as sizable as 900 ft (270 m). The relative size of the fireball in Figure 3.21 can be estimated by visually comparing its dimensions with those of the standing water tower in the foreground.

When first-on-the-scene responders arrive at a mishap involving the release of a flammable liquid, it is critical that immediate action be taken to prevent the occurrence of a BLEVE. This can be accomplished by situating unmanned monitors so cooling water is directed upon the vessel impacted by the fire. The water should be directed at the vessel's main vulnerable points, the uppermost area where the vapor phase is present, and the actual site of flame contact upon the shell of the vessel. Experts recommend discharging the cooling water at a rate of approximately 500 gal/min (1.9 m³/min). Unless it is effectively cooled, failure in the metal structure of the vessel is likely to occur within 10 to 30 minutes after the initial contact of the fire.

The danger to property, firefighters, and bystanders from the occurrence of a BLEVE cannot be exaggerated. The fragments of the rupturing vessel often travel like missiles for hundreds of feet in all directions, potentially causing death or injury to anyone in the way. Nearby structures are destroyed by the secondary fires that originate from radiation and the rocketing shells. Following the occurrence of a BLEVE, the main assistance firefighters can provide is to extinguish the secondary fires.

The nature of an emergency-response action involving flammable liquids does not always approach the magnitude of an operation concerned with preventing a BLEVE. As the nature of the emergency scene is being assessed, the first-on-the-scene responders should proceed to undertake the following actions as deemed necessary and appropriate:

Solved Exercise 3.9

A major fire at a manufacturing plant within a densely populated industrial park threatens to engulf a stationary tank that is located on adjacent property. The captain of the firefighting team is informed that the tank contains Cellosolve acetate, an NFPA/OSHA class II liquid whose lower explosive limit is 1.75% by volume and whose flashpoint is 131°F (55°C). Aside from extinguishing the major fire, what action should the responding team undertake?

Solution: When emergency-response personnel arrive at an emergency incident involving the potential release of a flammable liquid, it is critical that they undertake immediate action to prevent the occurrence of a BLEVE. This can be accomplished by discharging cooling water from unmanned monitors upon the storage tank containing Cellosolve acetate. The water should be directed at the uppermost area of the tank and the actual sites where there is flame contact with the shell of the tank.

GENERAL PROCEDURES

- As quickly as possible, the chemical or product name of the flammable liquid should be determined, either by reviewing transportation manifests, labels, markings, and placards, or by speaking with plant or transportation personnel. When the mishap involves a bulk volume of a flammable liquid, attention should also be focused upon other hazardous materials that are in the immediate vicinity.
- When the emergency incident is associated with a transportation mishap, CHEMTREC and, if warranted, the National Response Center should be contacted.
- If the response action involves the storage of a flammable liquid in a stationary tank, personnel should use the information posted on an NFPA diamond to ascertain the degree of hazard relating to fire, health, and chemical reactivity.
- Once the specific flammable liquid is known by either its chemical or product name, its specific gravity, water solubility, and vapor density should be determined. CHEMTREC can usually provide this information. The specific gravity of the liquid tells us whether the liquid is lighter or heavier than water and whether water can be used to extinguish a fire confined within a container or tank. The solubility of the liquid in water tells us

whether the use of water would effectively extinguish the fire. As we noted in Section 2.5, immiscible liquids float on water if they possess specific gravities less than 1.0, and they sink below water if they possess specific gravities greater than 1.0. The vapor density tells us whether the liquid's vapor is lighter or heavier than air, from which a determination can be made as to whether the leaking vapor will first settle in low spots or rise into the atmosphere.

- Once the chemical nature and approximate volume of the flammable liquid and availability of extinguishers are known, an appropriate fire extinguisher can be selected. The use of either water or carbon dioxide is typically effective when extinguishing a fire involving a relatively small volume of a flammable liquid, whereas the use of an aqueous film-forming foam (Section 5.11) is warranted when extinguishing a class B fire involving a bulk volume of a flammable liquid.

- When a small spill or leak of flammable liquid from a container or tank is discovered, a water fog should be used to disperse the vapor.

- When a container or tank containing a flammable liquid is situated in an area near an ongoing fire, or when such containers or tanks are exposed directly to flames, water, fog, foam, or carbon dioxide should be applied for extinguishing or cooling purposes. Quick attention should be given to cooling steel drums used to store flammable liquids, as when heated, they are particularly susceptible to rupturing. While they may have tiny pinholes through which the accumulated vapor slowly releases, they are never outfitted with the safety-relief devices or rupture disks typically found on compressed gas cylinders and railroad tankcars. To prevent the drums from rupturing, they should be cooled with direct streams of water.

- When a liquid's vapor has not yet ignited, or when its fire has been momentarily extinguished, the concentration of the flammable vapor in the air should be measured. If the reading establishes that the vapor concentration is within the flammable range, emergency responders should vacate the area until the use of a water fog within the area reduces the concentration to less than the lower explosive limit. Only when the vapor has not ignited and the vapor concentration is outside the flammable range should emergency-response personnel cautiously attempt to seal or otherwise stop a leak on a stationary tank or transport vessel. Sometimes, this task is nothing more than tightening a valve. If it is determined that the leak cannot be stopped, experienced personnel may elect to transfer the liquid into a new storage vessel. When this is impractical, the tank or transport vessel should be cooled with unmanned water monitors, but no further attempt to extinguish the fire should be made.

- The use of water for cooling a leaking storage tank often generates a volume of water so large that the secondary containment system is unable to hold it. This water is contaminated with the flammable liquid leaking from the tank. When the water is likely to overtop the secondary containment system, action should be undertaken to preclude its entrance into nearby sewers or bodies of water or flow to areas where other flammable or combustible materials are stored. This can be accomplished by pumping the mixture into a portable tank or another temporary storage system.

- When emergency responders encounter a massive leak of a flammable liquid, a foam blanket should be applied to confine the flammable vapor. To the extent feasible, the liquid should be dammed or diked to prevent widespread exposure and contamination.

- During mishaps associated with the transportation of a flammable liquid, upright cabs and railroad cars should be disconnected and moved from overturned tankcars whenever practical.

- When the vapor of a flammable liquid has been released into the atmosphere, it is prudent to wear total-encapsulating suits, use self-contained breathing apparatus, and perform as many response actions as practical from a downwind location. Although some flammable liquids are nontoxic, the use of respiratory protection equipment is nonetheless warranted since inhalation of the liquid's vapor can cause dizziness, illness, or suffocation.

◆ An overturned *pressure* transport tank can be safely righted without first unloading its liquid contents; however, an overturned *nonpressure* transport tank should not be righted until the bulk of the tank's liquid contents has first been transferred into a different tankcar. This transference should only be undertaken by experienced personnel.

EVACUATION DISTANCES

◆ When a container or tank containing a flammable liquid is exposed to fire, and when insufficient water is available for cooling purposes, all unauthorized personnel should be evacuated to a distance at least 2500 ft (750 m) in all directions from the fire. The public should be evacuated to double the distance.

◆ When a tank containing a flammable liquid is exposed to fire, all unauthorized personnel should be evacuated to a distance of at least 1000 ft (300 m) in all directions from the fire *even while unmanned monitors are used to cool the tank*. The public should be evacuated to double the distance.

■ ■

REVIEW EXERCISES

Flammability and Ignitability

3.1 The flashpoints of acetone and carbon disulfide are 0°F and −30°C, respectively. Other factors being equal, can these flashpoint data discern which liquid poses the greater fire and explosion hazard?

3.2 Given the following data, identify the NFPA/OSHA class of flammable or combustible liquid to which each substance belongs:

Substance	*Boiling point*	*Flashpoint*
(a) cyclohexanone	313°F (156°C)	146°F (63°C)
(b) methyl mercaptan	45°F (7.6°C)	0°F (−18°C)
(c) mesityl oxide	266°F (130°C)	88°F (31°C)
(d) *n*-butyl acetate	257°F (125°C)	81°F (27°C)
(e) tetrahydrofuran	151°F (65°C)	6°F (−14°C)
(f) toluene	231°F (111°C)	40°F (4.4°C)
(g) cyclopentadiene	108°F (42°C)	77°F (405°C)

3.3 When measured through the use of the Pensky-Martens closed-cup tester in Figure 3.2, a representative sample of a liquid chemical waste is found to possess a flashpoint of 138°F. Does the waste exhibit the RCRA characteristic of ignitability?

3.4 The lower and upper explosive limits of liquid chloroprene are 4% and 20% by volume, respectively. When exposed to an ignition source, will chloroprene vapor ignite at each of the following concentrations in air by volume:
(a) 1%;
(b) 10%;
(c) 30%?

3.5 Methyl isobutyl ketone, or *MIBK*, is a liquid constituent of the solvent mixture in many solvent-based paints. Some of its properties are provided below:

Boiling point	243°F (117°C)
Freezing point	−121°F (−85°C)
Vapor density (air = 1)	3.5
Specific gravity at 68°F (20°C)	0.80
Flashpoint	73°F (23°C)
Vapor pressure at 68°F (20°C)	15.7 mm Hg
Lower explosive limit	1.4% by volume
Upper explosive limit	7.5% by volume
Solubility in water	Slightly soluble

(a) In which range of temperature does MIBK exist as a liquid?
(b) Identify the NFPA class of flammable liquid to which MIBK belongs.
(c) If steel drums containing MIBK are exposed to fire, are they likely to rupture when the surrounding temperature reaches 350°F (177°C)?
(d) If steel drums containing MIBK rupture and produce a vapor concentration ranging from 20% to 45% by volume in air, is the vapor likely to ignite if exposed to an ignition source?
(e) If steel drums containing MIBK rupture, will its vapor initially accumulate in low areas or diffuse upward?

Storing and Transporting Compressed Gases

3.6 Why do compressed gas suppliers recommend that users securely strap or chain gas cylinders to stationary items?

3.7 "DOT-3E1800" is marked on a steel cylinder when its gauge pressure is 1610 psi at 70°F (21°C). Is this cylinder likely to rupture if the temperature of its contents rises during a fire to 300°F (149°C)?

3.8 When first purchased from a compressed gas supplier, the gauge reading on a DOT-3AA480 cylinder is 480 psi at 70°F (21°C). If the temperature of the surroundings rises to 130°F (54°C), at which pressure expressed in psi will this cylinder be safe against rupturing?

General Hazards of Compressed Gases

3.9 Why is the rupture of a steel cylinder containing compressed carbon dioxide gas a gravely hazardous incident, despite the fact that carbon dioxide is nonflammable?

3.10 Unlike steel cylinders, aerosol cans are not outfitted with safety relief valves or rupture disks. Determine which of the following is more likely to rupture first at 200°F (93°C): five aerosol cans containing a room deodorant dissolved in a compressed inert gas; or steel cylinders marked DOT-3A1800 containing compressed nitrogen?

3.11 A gas cylinder contains hydrogen, a flammable gas, at 235 psi_g and 65°F. During a fire, the rupture disks on the cylinder fail to release the contents, and the temperature of the cylinder and its contents is elevated to 350°F.
(a) What is the gauge pressure on the cylinder at 350°F?
(b) Is the cylinder likely to withstand this elevated temperature?

Responding to Incidents Involving the Release of Compressed Gases

3.12 When a leaking cylinder is moved into an open area, why do experts recommend that it be rotated so that the point of leakage becomes the uppermost part of the cylinder?

3.13 When moving a leaking cylinder containing any compressed gas, why should emergency responders wear total-encapsulating suits and use self-contained breathing apparatus?

3.14 Why do experts insist that the only acceptable way for emergency responders to approach a tank containing a flammable gas during an ongoing fire is from the direction of one of its elongated sides?

3.15 Why do experts recommend that emergency-response personnel work as far away as feasible when a transportation mishap occurs involving multiple railroad tankcars that are leaking gases?

3.16 Three cylinders containing chlorine gas are strapped to a wall within a burning building located near a public swimming pool. Although chlorine is nonflammable, it is highly toxic.

  (a) Why is it advisable to move the cylinders to a remote location before they become engulfed in the fire, but only if they can be removed safely and promptly?

  (b) When moving the cylinders from the building, what procedures should be implemented?

3.17 Upon arrival at a transportation mishap, the first-on-the-scene responders observe an ongoing fire and learn from the driver of an overturned cargo tank that his consignment is compressed liquefied butane. They also observe a ruptured discharge line from which releasing vapors have ignited.

  (a) What immediate action should the team take upon arrival at this scene?

  (b) What is the approximate distance from this scene to which unauthorized personnel should be evacuated?

Storing Flammable Liquids

3.18 Methyl acetate is a liquid that boils at 135°F (57°C) and possesses a flashpoint of 15°F (−9°C). It is a constituent of paint removers. A company that manufactures paint removers has constructed an indoor storage room for flammable liquids from fire-resistant materials and equipped it with a sprinkling system. What is the maximum number of 55-gallon drums containing methyl acetate that can be stored per pile in the room in compliance with the OSHA regulations?

3.19 A solvent-based paint possesses a boiling point of 171°F (77°C) and a flashpoint less than 80°F (<27°C). To comply with the OSHA regulations, what is the maximum volume in gallons that can be stored outdoors within steel drums?

3.20 The boiling point and flashpoint of furfural are 324°F (162°C) and 140°F (60°C), respectively. To comply with the OSHA regulations, what is the maximum number of gallons of furfural that can be stored within a safety can for future use?

3.21 An aerospace research laboratory stores 2-methylhydrazine, a flammable liquid, in an inside room measuring 20 ft × 20 ft (6 m × 6 m) and equipped with a carbon dioxide fire protection system. The walls and ceiling of the room have been constructed according to the specifications in NFPA No. 220, *Standard Types of Building Construction*, using construction materials having a 2-hour fire resistance rating. To comply with OSHA's storage regulations, what is the maximum volume in liters of 2-methylhydrazine that can be stored within this room at any one time?

3.22 A manufacturing company stores twenty-one 55-gallon drums containing 2-nitropropane outdoors in a building located 20 ft (6 m) from the property line. The company stores these drums in groups of seven drums per pile and separates each of the three piles from the others by a distance of five feet. The flashpoint and boiling point of 2-nitropropane are 103°F (39°C) and 248°F (120°C), respectively.

    (a) If all buildings on the property have been equipped with approved fire protection systems, is the company in compliance with OSHA's flammable liquids storage regulations?

    (b) If the building is not protected from exposure, is the company in compliance with OSHA's flammable liquids storage regulations?

    (c) Which number should be included within the red zone of an NFPA label affixed to the outside of the building?

3.23 The boiling point and flashpoint of methyl isobutyrate are 198°F (92°C) and 55°F (13°C), respectively. What maximum volume in liters of methyl isobutyrate does

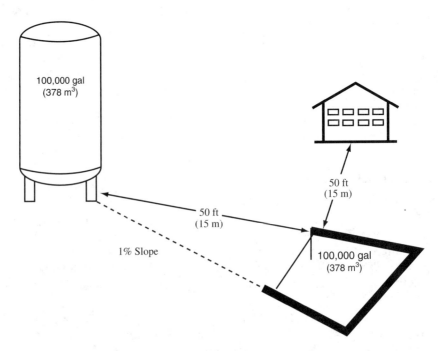

FIGURE 3.22 ◆ When essential, NFPA recommends the use of a remote impounding area into which a spilled or leaked flammable liquid may be drained from a stationary tank in accordance with each condition referenced herein. *Reprinted with permission from NFPA 30,* Flammable and Combustible Liquids Code. *Copyright © 1996, National Fire Protection Association, Quincy, MA 02269. This reprinted material is not the complete and official position of the National Fire Protection Association on the referenced subject, which is represented only by the standard in its entirety.*

OSHA permit to be stored within portable tanks on the second floor of an exposure-protected building in which 125 people are employed?

3.24 Adjoining property and waterways can be protected from the hazards posed by the presence of a flammable or combustible liquid in a stationary storage tank by constructing a means to drain a spilled or leaked liquid from the tank into a remote impounding area. For such purposes, NFPA No. 30, *Flammable and Combustible Liquids Code*, advocates each of the following for the 100,000-gal (378 m³) tank in Figure 3.22:

- ◆ A slope of not less than 1% away from the tank for at least 50 ft (15 m) toward the impounding area
- ◆ A capacity of the impounding area not less than the capacity of the tank
- ◆ A route of drainage so that, if ignited, the burning liquid within the drainage system will not seriously expose tanks or adjoining property
- ◆ The location of the impounding area not closer than 50 ft (15 m) from any property line that is or can be built upon, or from any tank

If a processing company stores 50,000 gal of *n*-amyl alcohol, a class IC liquid, in a tank located between two buildings on the same property, each separated from the tank's remote 45,000-gal impounding area by a distance of 50 m, has the company complied with this NFPA recommendation?

3.25 The EPA regulation at 40 C. F. R. §264.175(b)(3) requires that the capacity of a secondary containment system for a hazardous waste storage area must be capable of containing 10% of the volume of the containers within the area, or the entire volume of the largest container, whichever is greater. If a manufacturing facility stores four 55-gallon drums of flammable acetone waste in its hazardous waste storage area, what capacity must be provided in the secondary containment system to comply with this regulation?

Responding to Incidents Involving the Release of Flammable Liquids

3.26 Arriving at the scene of a railroad transportation mishap, the first-on-the-scene responders note that a fire associated with one tankcar is impinging upon the outer surface of an adjacent tankcar. No information is immediately available concerning the chemical nature of the contents in either tankcar. To protect public health and the environment, what immediate action should these responders take?

3.27 Emergency responders arriving at the scene of a transportation mishap along a frequently trafficked route in a densely populated municipality observe smoke from the interior of an overturned truck van placarded FLAMMABLE. The shipper's manifest indicates that the consignment consists of 45 drums of *n*-heptane, a class IB liquid, in addition to other nonhazardous but combustible materials. What actions should the emergency-response personnel undertake to ensure the safety of other motorists and individuals in nearby business buildings?

Chemical Forms of Matter

CHAPTER 4

Commercial chemical products like paint, pesticide formulations, ammunition, gasoline, and other petroleum fuels are hazardous materials that consist of mixtures of distinctly different components, each separable from the others. Paint, for example, is a mixture of some resin or other film-forming compound, a solvent or thinner, and organic or inorganic pigments. Each chemical product is characterized in a similar fashion as a uniform or nonuniform blend of different components, but capable of separation from one another.

When any mixture has been separated into its components, the individual components are unique forms of matter. In this chapter, we will identify them as either elements or compounds possessing a characteristic set of physical and chemical properties. We will also observe how chemists describe the structural nature of these substances in terms of atoms, molecules, and ions. Finally, we will learn how chemists name these substances and write their chemical formulas.

```
Performance Goals for Section 4.1
    ◆ Distinguish between the nature of elements and compounds.
    ◆ Learn the symbols of the elements listed in Table 4.2.
```

◆ 4.1 ELEMENTS AND COMPOUNDS

A material that has been separated from all other materials is called a *pure substance*, or more simply, a *substance*. Examples include aluminum metal, copper metal, oxygen, distilled water, and table sugar. No matter what procedures are used to purify them, or what their origin is, samples of the same substance are indistinguishable from each other. All samples of distilled water are alike and indistinguishable from all other samples; that is, water is the same substance whether it has been distilled from rainwater, well water, seawater, or any other source.

A substance is characterized as a material having a fixed composition, usually expressed in terms of percentage by mass. Distilled water is a pure substance consisting of 11.2% hydrogen and 88.8% oxygen by mass. By contrast, coal is not a pure substance;

its carbon content alone may vary anywhere from 35 to 90% by mass. Such materials that are not pure substances are *mixtures*.

Suppose we take a pure substance like limestone (calcium carbonate) and heat it, as shown in Figure 4.1. The limestone ultimately crumbles into a white powder. Careful examination shows that carbon dioxide evolved from the limestone while it was heated. Substances like limestone that can be broken down into two or more simpler substances are called *compound substances,* or simply *compounds*. There are more than 18 million different compounds.

The substances that cannot be decomposed into simpler forms of matter are called *elements*. The elements are the fundamental substances of which all matter is composed. There are only 114 known elements, but there are billions of known compounds. Of the 114 elements, only 88 are present in detectable amounts on Earth, and many are very rare. Table 4.1 shows that only 10 elements make up approximately 99% by mass of the Earth, including its crust, the atmosphere, the surface layer, and the bodies of water. Oxygen is the most abundant element on Earth, which we find in the free state in the atmosphere, as well as in combined form with other elements in numerous minerals and ores.

Each element has a specific name and symbol. The names and symbols of some common elements are listed in Table 4.2. The symbols of the elements consist of either one or two letters with the first letter capitalized. The selection of the symbols of many elements was originally based upon their Latin names: Pb for *plumbum* (lead), Ag for *argentum* (silver), and Sn for *stannum* (tin). It is best to memorize these symbols of the elements listed in Table 4.2, since they are frequently encountered in the study of hazardous materials.

All elements beyond uranium (atomic number 92) have been synthesized in nuclear laboratories by scientists. The International Union of Pure and Applied Chemistry

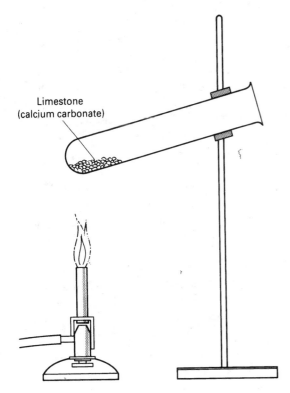

Limestone
(calcium carbonate)

FIGURE 4.1 ◆ The heating of limestone (calcium carbonate) in a test tube. The limestone chemically decomposes, evolving carbon dioxide. This illustrates that calcium carbonate is a compound and not an element.

TABLE 4.1 ◆ Natural Abundance of the Elements (Earth's Crust, Oceans, and the Atmosphere)

Element	%	Element	%
Oxygen	49.5%	Chlorine	0.19%
Silicon	25.7%	Phosphorus	0.12%
Aluminum	7.5%	Manganese	0.09%
Iron	4.7%	Carbon	0.08%
Calcium	3.4%	Sulfur	0.06%
Sodium	2.6%	Barium	0.04%
Potassium	2.4%	Chromium	0.033%
Magnesium	1.9%	Nitrogen	0.030%
Hydrogen	1.9%	Fluorine	0.027%
Titanium	0.58%	Zirconium	0.023%
		All others	<0.1%

TABLE 4.2 ◆ Symbols of the Most Common Elements

Element	Symbol	Element	Symbol
Aluminum	Al	Lithium	Li
Antimony	Sb	Magnesium	Mg
Argon	Ar	Manganese	Mn
Arsenic	As	Mercury	Hg
Barium	Ba	Neon	Ne
Bismuth	Bi	Nickel	Ni
Boron	B	Nitrogen	N
Bromine	Br	Oxygen	O
Cadmium	Cd	Phosphorus	P
Calcium	Ca	Platinum	Pt
Carbon	C	Potassium	K
Chlorine	Cl	Radium	Pt
Chromium	Cr	Silicon	Si
Cobalt	Co	Silver	Ag
Copper	Cu	Sodium	Na
Fluorine	F	Strontium	Sr
Gold	Au	Sulfur	S
Helium	He	Tin	Sn
Hydrogen	H	Titanium	Ti
Iodine	I	Tungsten	W
Iron	Fe	Uranium	U
Lead	Pb	Zinc	Zn

(IUPAC) has accepted the following names for these elements: rutherfordium (atomic number 104), symbol Rf; dubnium (atomic number 105), symbol Db; seaborgium (atomic number 106), symbol Sg; bohrium (atomic number 107), symbol Bh; hassium (atomic number 108), symbol Hs; meitnerium (atomic number 109), symbol Mt; and darmstadtium (atomic number 110), symbol Ds. The elements having atomic numbers 111 and 112 have not yet been named.*

A complete listing of all known elements is provided on the inside back cover of this book.

Performance Goal for Section 4.2

♦ Identify the general properties of metals, nonmetals, and metalloids.

◆ 4.2 METALS, NONMETALS, AND METALLOIDS

Each element can be classified as a metal, nonmetal, or metalloid. *Metals* are elements that usually possess the following properties:

- ♦ Conduct heat and electricity well
- ♦ Melt and boil at relatively high temperatures
- ♦ Possess relatively high densities
- ♦ Are normally malleable (able to be hammered into sheets) and ductile (able to be drawn into a wire)
- ♦ Display a brilliant luster

Examples of metals are iron, platinum, and silver. Almost all metals are solids at room temperature [68°F (20°C)]. Mercury is the only common metal that is liquid at room temperature. No metal is gaseous at room temperature.

Elements that do not possess the physical properties of metals are called *nonmetals*. These elements have the following general properties:

- ♦ Melt and boil at relatively low temperatures
- ♦ Do not possess a luster
- ♦ Are less dense than metals
- ♦ Poorly conduct heat and electricity

At room temperature, most nonmetals generally exist as either solids or gases. Bromine is the only liquid nonmetal at room temperature. Oxygen, fluorine, and nitrogen are examples of gaseous nonmetals, while carbon, sulfur, and phosphorus are examples of solid nonmetals.

Not all metals and nonmetals exhibit the general properties noted here. There are exceptions; for instance, carbon conducts heat and electricity very well, and one form of carbon melts at about 6656°F (~3680°C), a relatively high temperature.

Several elements have properties resembling both metals and nonmetals. They are called *metalloids*. The metalloids are boron, silicon, germanium, arsenic, antimony, tellurium, and polonium. For example, silicon and germanium have the luster associated with metals, but they do not conduct heat and electricity.

*In 1999, researchers at Lawrence Berkeley National Laboratory announced the discovery of elements having atomic numbers 116 and 118. However, in 2001, they retracted their claim when they were unable to duplicate their experimental results.

◆ 4.3 CHEMICAL AND PHYSICAL CHANGES

The constant composition associated with a given substance is maintained by internal linkages among its units. These linkages are called *chemical bonds*. When a particular process occurs that makes or breaks these bonds, we say that a *chemical reaction,* or a *chemical change,* has occurred. Combustion and corrosion are common examples of chemical changes associated with some hazardous materials.

Let's briefly consider the nature of the combustion process. When something burns, it combines with oxygen. The resulting products of combustion are compounds containing oxygen, called *oxides*. For instance, many commercial gasolines are mixtures of several substances, including one called octane. When octane burns completely, it becomes carbon dioxide gas and water vapor. Carbon dioxide is an oxide of carbon as its name implies; while water is an oxide of hydrogen. Carbon dioxide and water vapor are unlike octane or other gasoline components. They have different properties and different compositions. This conversion of octane to carbon dioxide and water is typical of a chemical change.

By contrast, substances can undergo changes in which its composition remains the same. Such changes are called *physical changes*. Let's consider octane again. Some of its physical and chemical changes are illustrated in Figure 4.2. When

FIGURE 4.2 ◆ An example of a physical and chemical change in octane, a component of gasoline. On the left the substance evaporates (that is, it changes its physical state from the liquid to vapor); this is a physical change. On the right a spill of octane ignites and burns, thereby changing to carbon dioxide and water vapor; this is a chemical change.

TABLE 4.3 ◆ Characteristics of Some Substances		
Substance	*Physical Properties*	*Chemical Properties*
Oxygen, an element	Odorless, colorless gas; does not conduct heat or electricity; density = 1.43 g/L; becomes liquid at −297°F (−183°C)	Combines readily with many elements (a chemical reaction called oxidation)
Phosphorus, an element	White or red solid; does not conduct heat or electricity; density = 1.82 g/mL (white) and 2.34 g/mL (red)	Readily combines with oxygen, chlorine, and fluorine; white form spontaneously ignites in dry air
Carbon dioxide, a compound	Odorless and colorless gas; solidifies at −83°F (−67°C) under pressure, forming dry ice; soluble in water under pressure	Does not burn; reacts with water-soluble metal compounds, forming metallic carbonates
Hydrogen chloride, a compound	Strong-smelling, colorless gas; density = 1.20 g/mL; soluble in water, forming hydrochloric acid	Reacts with many minerals, forming water-soluble products; reacts with ammonia, forming ammonium chloride

exposed to the ambient environment, liquid octane evaporates, but its chemical composition remains unchanged. Such alterations in the physical state of a substance, such as from a liquid to a vapor, are considered physical changes. Other examples of physical changes are melting, freezing, boiling, crushing, and pulverizing.

The types of behavior that a substance exhibits when undergoing chemical changes are called its *chemical properties*. The characteristics that do not involve changes in the chemical identity of a substance are called its *physical properties*. All substances can be distinguished from one another by these properties, in much the same way as certain features—fingerprints or DNA, for example—distinguish one human being from another. The study of hazardous materials is concerned to a great extent with learning the chemical and physical properties of appropriate substances, some examples of which are listed in Table 4.3.

Performance Goals for Section 4.4

- ◆ Identify the names of the common particles present in an atom.
- ◆ Identify the common properties of the common particles present in an atom.
- ◆ Determine the number of protons and electrons in an atom from an element's atomic number.
- ◆ Describe the general structure of an atom.
- ◆ Describe the nature of atomic orbitals.
- ◆ Distinguish between the atomic number and atomic weight of an element.
- ◆ Given the atomic mass of each stable isotope of an element and its natural abundance, calculate the atomic weight of the element.

◆ 4.4 SOME BASIC FEATURES OF ATOMS

If a small piece of an element, say aluminum, could be hypothetically divided and subdivided into smaller and smaller pieces, until subdivision is no longer possible, the result would be one particle of aluminum. This smallest particle of the element, which is still representative of the element, is called an *atom,* from the Greek word *atomos,* meaning indivisible.

An atom is infinitesimally small. Yet it is also composed of even smaller particles known as electrons, protons, and neutrons. *Electrons* are negatively charged particles that are responsible for the chemical reactivity of a given element. *Protons* are positively charged particles. *Neutrons* are neutral particles. Electrons and protons bear the same magnitude of charge, but of opposite signs. For convenience, the electron has a charge of -1, the proton of $+1$, and the neutron of 0.

Protons are relatively heavy particles; they are 1836 times more massive than electrons. Neutrons are slightly more massive than protons. The fundamental characteristics of electrons, protons, and neutrons are summarized in Table 4.4.

The protons and neutrons of an atom reside in a central area called the *nucleus*. Electrons reside primarily in designated regions of space surrounding the nucleus, called *atomic orbitals*. There are several types of atomic orbitals; some are close to the nucleus, while others are relatively remote from it. Scientists have learned that only a prescribed number of electrons reside in a given type of atomic orbital. Two electrons are always close to the nucleus, in an atom's innermost atomic orbital (with the exception of a hydrogen atom, which possesses only one electron). Most atoms have additional electrons in further atomic orbitals.

The number of protons in an atom is called the *atomic number*. We often use atomic numbers in the study of chemistry to determine the number of electrons possessed by neutral atoms of an element. An atom of hydrogen has one electron, helium has two, lithium has three, and so forth. Carbon is an element composed only of carbon atoms, and all carbon atoms exhibit nearly the same physical and chemical properties. Some atoms may have slightly different masses due to different numbers of neutrons in their nuclei, but these carbon atoms all act the same when they undergo chemical changes. Similarly, oxygen is an element composed of oxygen atoms, and all oxygen atoms possess nearly the same properties. But carbon and oxygen atoms are not alike, since their atoms possess different numbers of electrons, protons, and neutrons.

While neutral atoms of the same element have an identical number of protons and electrons, they may differ by the number of neutrons in their nuclei. Atoms of the same element having different numbers of neutrons are called the *isotopes* of that element. We shall note later in Chapter 15 that some isotopes are radioactive.

TABLE 4.4 ◆ Some Basic Atomic Particles

Particle	Proton	Electron	Neutron
Symbol	P^+	e^-	n
Relative charge	$+1$	-1	0
Relative mass	1	about 0[a]	1

[a]The mass of an electron is 1/1836 the mass of the proton.

Over the past two centuries, scientists have determined the relative masses of many atoms of the known elements. These relative masses are called *atomic weights*. The atomic weights of the elements are not absolute masses, but rather masses measured relative to the mass of another atom that has been selected as a reference standard. Since atomic weights are relative parameters, they are unitless numbers. A specific isotope of carbon, called carbon-12, has been selected as the atom whose mass serves as the reference standard. The atoms of this carbon isotope consists of six electrons, six protons, and six neutrons. It is assigned a mass of exactly 12.

The atomic weight of a given element is a number that tells us how the mass of an *average* atom of that element compares with the mass of the carbon-12 atom. Scientists establish the atomic weight of an element by first determining the atomic mass and natural abundance of each of its stable isotopes. Suppose an element has three stable isotopes, *X-a*, *X-b*, and *X-c*, whose atomic masses and natural abundances are known. The atomic weight of the element is then determined as follows:

$$\text{Atomic weight} = \frac{(\text{Mass}_{X\text{-}a} \times \%_{X\text{-}a}) + (\text{Mass}_{X\text{-}b} \times \%_{X\text{-}b}) + (\text{Mass}_{X\text{-}c} \times \%_{X\text{-}c})}{100}$$

The atomic weight of any element is obtained from the relative mass of its naturally occurring isotopes, weighted according to their natural abundances.

Use of the carbon-12 standard results in an atomic weight for natural carbon of 12.011. This slight difference results from averaging the masses of its natural isotopic forms. The atomic numbers and atomic weights of the elements have been provided with the listing on the inside back cover of this book.

Performance Goals for Section 4.5

- Explain the basis for the arrangement of the elements on the periodic table.
- Associate the chemical reactivity of an element with the electrons in the outermost atomic orbitals of its atoms.
- Use the periodic table to distinguish between a family and a period.
- Note the location of the principal families on the periodic table: the alkali metals, alkaline earth metals, chalcogens, halogens, and noble gases.

◆ 4.5 THE PERIODIC CLASSIFICATION OF THE ELEMENTS

During the last half of the 19th century, several scientists first noted that the chemical properties of any given element were similar to the chemical properties of certain other elements. For example, they noted that sodium metal reacts explosively with water and burns spontaneously in the air. When these two chemical properties of sodium were compared with the properties of other elements, they found that potassium also explodes on contact with water and burns spontaneously in air. These scientists summarized this observation in the *periodic law*: The properties of the elements vary periodically with their atomic numbers. The term "periodic" reflects this repetition of chemical properties.

Suppose we list each element in a square, and then arrange the squares by order of increasing atomic number. This means that the total number of electrons possessed by each element increases in this arrangement, one at a time, as we move from one square to the next. Then, let's further arrange them into columns of elements that possess similar properties. Of course, one would need to know a great deal of chemistry to accomplish this feat. A similar exercise was first performed more than 130 years ago, when many elements known today had not yet been discovered.

Such an arrangement of the chemical elements into a chart designed to represent the periodic law is called a *periodic table*, a modern version of which is illustrated in Figure 4.3. The elements positioned within the same column of the periodic table are called a *family* of elements. Each family is identified by a Roman numeral and a capital letter at the top of the column, such as IA, IIA, and so on. Thus, for example, helium, neon, argon, krypton, xenon, and radon belong to the same family, identified by VIIIA. Elements in the same row of a periodic table are said to belong to the same *period*. The periods are numbered on the far left of the table from 1 to 7. There is one period of 2 elements, two periods of 8 elements each, two more of 18 elements each, one period of 32 elements, and a final period that presently has 28 elements.

The periodic table of the elements is one of the most powerful icons in science: a single table that consolidates an array of valuable information. Some version of the table hangs on the wall of nearly every chemistry laboratory throughout the world. Simply by glancing at the periodic table, we can observe immediately the atomic number and atomic weight of any element. We can also readily distinguish between those elements that are metals, nonmetals, and metalloids: the bold, zigzag line separates metals from nonmetals, while those elements lying to each immediate side of the line are the metalloids. Generally, metals fall to the left of the line, and nonmetals fall to the right.

The usefulness of a periodic table consists in the manner by which it displays the periodicity, or repetition, in the properties of the elements at regular intervals. In particular, when we observe the elements as members of the same family, we know they possess similar chemical properties. Five families deserve special recognition in this regard. They are frequently identified by unique names:

- Group IA is called the *alkali metal family;* its members are lithium, sodium, potassium, rubidium, cesium, and francium. As noted earlier, each of these metals reacts explosively with water and ignites on exposure to the air.
- Group IIA is called the *alkaline earth family;* its members are beryllium, magnesium, calcium, strontium, barium, and radium. These elements are also chemically reactive, but not nearly as reactive as the alkali metals. They cause water to decompose, but the rate of decomposition is slow at ambient temperatures. They ignite in the air, but only after they have been heated or exposed to an ignition source.
- Group VIA is called the *chalcogen family;* its members are oxygen, sulfur, selenium, tellurium, and polonium. These five elements are moderately reactive substances.
- Group VIIA is called the *halogen family;* its members are fluorine, chlorine, bromine, iodine, and astatine. These elements are nonmetals, each of which is especially reactive. Fluorine ranks as the most chemically reactive of all elements.
- Group VIIIA is called the *noble gas family;* its members are helium, argon, krypton, xenon, and radon. Chemists originally thought that these gases were all inert to chemical combination and called them *inert gases*. Although some of them, such as krypton, are now known to form compounds, the noble gases uniquely stand out as a group of elements lacking the chemical reactivity observed for all other elements.

Hydrogen is a unique element in that it occupies two positions on the periodic table, once in family IA, and then again in VIIA. This is because hydrogen has some properties of the alkali metals and some properties of the halogens. Although positioned in each family, hydrogen is considered neither an alkali metal nor a halogen; rather, chemists regard hydrogen as an element by itself, thereby illustrating its uniqueness and individuality.

Periodic table (Figure 4.3):

IA	IIA	IIIB	IVB	VB	VIB	VIIB	VIIIB	VIIIB	VIIIB	IB	IIB	IIIA	IVA	VA	VIA	VIIA	VIIIA
1 H 1.0079																	2 He 4.0026
3 Li 6.941	4 Be 9.0122											5 B 10.811	6 C 12.011	7 N 14.007	8 O 15.999	9 F 18.998	10 Ne 20.190
11 Na 22.990	12 Mg 24.305											13 Al 26.982	14 Si 28.094	15 P 30.974	16 S 32.086	17 Cl 35.453	18 Ar 38.948
19 K 39.095	20 Ca 40.075	21 Sc 44.956	22 Ti 47.58	23 V 50.492	24 Cr 51.996	25 Mn 54.938	26 Fe 55.547	27 Co 55.933	28 Ni 58.89	29 Cu 63.546	30 Zn 65.39	31 Ga 69.723	32 Ge 72.61	33 As 74.922	34 Se 75.96	35 Br 79.904	36 Kr 83.80
37 Rb 85.468	38 Sr 87.62	39 Y 88.906	40 Zr 91.224	41 Nb 92.906	42 Mo 95.94	43 Tc 96.906	44 Ru 101.07	45 Rh 102.91	46 Pd 106.42	47 Ag 107.57	48 Cd 112.41	49 In 114.82	50 Sn 118.71	51 Sb 121.75	52 Te 127.60	53 I 126.90	54 Xe 131.29
55 Cs 132.90	56 Ba 137.33	57-71 La- Lu	72 Hf 178.49	73 Ta 180.96	74 W 183.96	75 Re 186.21	76 Os 190.2	77 Ir 192.22	78 Pt 195.08	79 Au 196.97	80 Hg 200.59	81 Tl 204.36	82 Pb 207.2	83 Bi 208.96	84 Po 208.98	85 At 209.99	86 Rn 222.02
87 Fr 223.02	88 Ra 226.03	89-103 Ac- Lr	104 Rf	105 Db	106 Sg	107 Bh	108 Hs	109 Mt	110 Ds	111	112						

57 La 139.91	58 Ce 140.12	59 Pr 140.91	60 Nd 144.24	61 Pm 146.92	62 Sm 150.36	63 Eu 151.97	64 Gd 157.25	65 Tb 158.83	66 Dy 162.30	67 Ho 164.93	68 Er 167.26	69 Tm 168.93	70 Yb 173.04	71 Lu 174.97
89 Ac 227.03	90 Th 232.04	91 Pa 231.04	92 U 236.03	93 Np 237.05	94 Pu 244.06	95 Am 243.06	96 Cm 247.07	97 Bk 247.07	98 Cf 251.08	99 Es 252.06	100 Fm 257.09	101 Md 258.10	102 No 259.10	103 Lr 260.10

FIGURE 4.3 ◆ A version of the periodic table of the elements. The zigzag solid line separates metals (on the left) from nonmetals (on the right). The numbers above and below the symbols of the elements are their atomic numbers and atomic weights, respectively.

Solved Exercise 4.1

Since compounds of thallium are highly toxic, they are sometimes commercially available in rodenticides, products that kill rodents.* Using the periodic table, answer the following questions:

(a) What is the symbol for thallium?

(b) What are the symbols for the elements immediately adjacent to thallium on the periodic table?

(c) Is thallium a metal, nonmetal, or metalloid?

(d) Provide the symbols of all the elements in the family of which thallium is a member.

(e) Identify the group number of the family of which thallium is a member.

Solution: (a) By scanning the periodic table, thallium is located in the sixth period. Its chemical symbol is Tl.

(b) The elements immediately adjacent to thallium on the periodic table are mercury and lead, whose symbols are Hg and Pb, respectively.

(c) Thallium is located to the left of the zigzag line on the periodic table. For this reason, it is a metal.

(d) Boron, aluminum, gallium, indium, and thallium are members of the same family of elements.

(e) Thallium is a member of the family of elements denoted as IIIA.

*It is believed that thallium sulfate, a colorless, tasteless salt, was administered by Iraqi security forces to poison individuals opposed to the regime of Saddam Hussein. In 1992, two members of the Iraqi opposition were told by Qusay Hussein, Saddam's son, to leave Iraq. One day, during an occasion in which they drank tea, the two began to experience the symptoms of having been poisoned. Both escaped from Iraq into Damascus and then to London, where physicians diagnosed thallium poisoning. They were successfully treated. [Human Rights Watch, *Endless Torment: The 1991 Uprising in Iraq and Its Aftermath* (June 1992).] Qusay and his brother Uday were killed in Iraq by the coalition forces on July 22, 2003. Saddam Hussein was captured on December 14, 2003 by American military forces in Baghdad.

Performance Goals for Section 4.6
- Describe the nature of the molecule as a basic unit of matter.
- Demonstrate the manner by which ions are formed from atoms.

◆ 4.6 MOLECULES AND IONS

Although the smallest representative particle of an element is the atom, not all uncombined elements exist as single atoms. In fact, only six elements actually exist as single atoms. These are the noble gases. We say that these elements are *monatomic*.

Other gases or liquids at room conditions consist of units containing pairs of like atoms. These units are called *molecules*. For example, hydrogen, oxygen, nitrogen, and

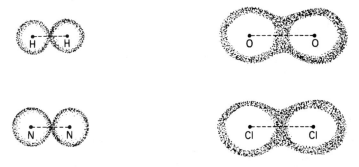

FIGURE 4.4 ◆ Some simple diatomic molecules (not to scale).

chlorine are gaseous elements as we generally encounter them, each of which is composed of a molecule having two atoms. These molecules are said to be *diatomic* and are symbolized by the notations H_2, O_2, N_2, and Cl_2, respectively. They are illustrated in Figure 4.4.

The smallest particle of many compounds is the molecule. Molecules of compounds contain atoms of two or more elements. For example, the water molecule consists of two hydrogen atoms and one oxygen atom; a molecule of methane consists of one carbon atom and four hydrogen atoms. Chemists denote these molecules as H_2O and CH_4, respectively.

Not all compounds occur naturally as molecules. Many occur as aggregates of oppositely charged atoms or groups of atoms called *ions*. Atoms become charged by gaining or losing some of their electrons. In general, metal atoms *lose* electrons, while nonmetal atoms *gain* electrons. Atoms of metals that lose their electrons become positively charged; atoms of nonmetals that gain electrons become negatively charged.

Let's consider the difference between the sodium atom and the sodium ion. By examination of Figure 4.3 or the listing on the inside back cover, we learn that the atomic number of sodium is 11. Thus, the neutral sodium atom has 11 electrons and 11 protons. If a sodium atom loses one electron, it then has only 10 left, although it still retains its 11 protons. By losing the electron, the sodium atom becomes a sodium ion. Its net charge is +1; that is, $+11 + (-10) = +1$.

Consider magnesium as a second example. The atomic number of magnesium is 12. Thus, the neutral magnesium atom has 12 electrons and 12 protons. When a magnesium atom loses two electrons, it becomes a magnesium ion. The magnesium ion possesses 10 electrons and 12 protons, and its net charge is +2.

These two ionization processes for metals can be represented as follows:

$$Na \longrightarrow Na^+ + e^-$$
$$Mg \longrightarrow Mg^{2+} + 2e^-$$

Here, e^- represents an electron; writing e^- to the right of the arrow means an electron is lost from the atom of the metal whose symbol appears to the left of the arrow. Na^+ and Mg^{2+} denote the sodium ion and the magnesium ion, respectively.

As noted earlier, the atoms of a nonmetal tend to *gain* electrons. Let's consider fluorine. The atomic number of fluorine is 9. The neutral fluorine atom possesses nine electrons and nine protons. If the fluorine atom somehow attains another electron, it will then have 10 electrons, but still only 9 protons. Its net charge is −1, that is, $+9 + (-10) = -1$. Since fluorine exists in the form of diatomic molecules, we represent the ionization of fluorine using F_2 to symbolize the fluorine molecule on the left of the arrow. This process can be designated as follows:

$$F_2 + 2e^- \longrightarrow 2F^-$$

Here, a 2 is written in front of e^- and F^- so that an equal number of electrons exists on each side of the arrow.

Metal atoms that have lost one or more electrons to become positively charged ions retain the name of the metal. As noted earlier, Na^+ and Mg^{2+} are called the sodium ion and the magnesium ion, respectively. Atoms of nonmetals that have gained one or more electrons to become negatively charged ions are named by modifying the name of the nonmetal so it ends in *-ide*; F^- is thus named the fluoride ion.

Frequently, two or more nonmetal atoms unite to form a *polyatomic ion*. For instance, one sulfur atom and four oxygen atoms unite to form an ion with a net charge of -2; it is called the *sulfate ion* and is symbolized as SO_4^{2-}. We shall observe other examples of polyatomic ions in Section 4.10.

Performance Goals for Section 4.7
- ◆ Describe the nature of a chemical bond.
- ◆ Associate the octet rule with the formation of chemical bonds.

◆ 4.7 THE NATURE OF CHEMICAL BONDING

As we noted earlier, when compounds form, the atoms of one element become attached to, or associated with, atoms of other elements by forces called *chemical bonds*. But, how do these chemical bonds form? Chemists have pondered the answer to this question for centuries.

A clue to understanding the manner by which chemical bonds form can be deduced from the observation that the noble gases are relatively inert substances. This implies that the electrons in the atoms of the noble gases are specially arranged. Today, we know that the electronic stability of the noble gases is associated with the number of electrons in their outmost atomic orbitals, as follows:

- ◆ An atom of helium possesses two electrons, both located in the innermost atomic orbital.
- ◆ The atoms of the noble gases other than helium possess eight electrons in their outermost atomic orbitals, those orbitals farthest from their nuclei.

Scientists discovered long ago that molecules and ions often form when their constituent atoms acquire the electronic stability of the noble gases. Since there are only a prescribed number of electrons in an atom, they must somehow interact with the electrons from other atoms to acquire the electronic structure of the noble gases. For the elements in the first and second periods, this occurs as follows:

- ◆ Hydrogen, lithium, and beryllium atoms attain the electronic structure of the helium atom. This means that electronic stability is achieved when these atoms have two electrons.
- ◆ Atoms of the elements other than hydrogen, helium, lithium, and beryllium attain the electronic structure of the noble gas nearest in atomic number. For the atoms of these

elements, electronic stability is achieved by attaining a total of eight electrons in their outermost atomic orbitals. The observation that atoms other than hydrogen, helium, lithium, and beryllium achieve electronic stability by attaining a total of eight electrons is called the *octet rule*.

There are two mechanisms by which atoms are capable of achieving these electronic structures, called *ionic* and *covalent* bonding. Each represents an extreme situation; in actuality, some degree of each description exists in all substances. The mechanisms are discussed independently in Sections 4.9 and 4.10.

Performance Goals for Section 4.8
- ◆ Distinguish between the bonding (valence) and nonbonding electrons in an atom.
- ◆ Describe the use of a Lewis symbol.
- ◆ Illustrate how the Lewis symbols can be readily determined using the periodic table and Table 4.5.

◆ **4.8 LEWIS SYMBOLS**

While an atom of any element possesses a definite number of electrons, only some of them are involved in chemical bonding. These electrons are called the atom's *valence electrons*. The valence electrons are those electrons in an atom's outermost atomic orbital. Electrons that do not participate in bonding are called *nonbonding* electrons.

We display valence electrons by means of a *Lewis symbol,* named after an American chemist, Gilbert N. Lewis. A Lewis symbol consists of the symbol of an element together with a certain number of dots that represent the atom's valence electrons. For example, Na· is the Lewis symbol for the sodium atom. The sodium atom has only one electron that participates in chemical bonding.

Table 4.5 lists the Lewis symbols of some representative elements important to the study of hazardous materials. Note that a simple way exists for determining the number of valence electrons for any element of an "A" family: the number that identifies the "A" family on the periodic table is also the number of valence electrons possessed by elements in that family. For instance, the halogens are located in the family identified VIIA; note from Table 4.5 that each halogen has seven valence electrons.

TABLE 4.5 ◆ Lewis Symbols of Some Representative Elements

Family	IA	IIA	IIIA	IVA	VA	VIA	VIIA	
	Li·	·Be·	·B·	·C·	·N·	·O·	:F·	H·
	Na·	·Mg·	·Al·	·Si·	·P·	·S·	:Cl·	
	K·	·Ca·	·Ga·	·Ge·	·As·	·Se·	:Br·	
	Rb·	·Sr·				·Te·	:I·	

◆ 4.9 IONIC BONDING

Electrons can be *transferred* from an atom of one element to an atom of another element resulting in the formation of positive and negative ions. This phenomenon generally occurs between the atoms of metals and nonmetals. The ions that form are attracted to each other by virtue of their opposite charges. Chemists call this electrostatic force of attraction an *ionic bond*. The formation of the ionic bond is based upon a fundamental law of nature by which forms of matter with like charges ($+/+$ or $-/-$) repel each other, whereas forms of matter with unlike charges ($+/-$) attract.

Many atoms of the elements transfer or accept just the number of electrons that gives them eight electrons in their outermost atomic orbitals. By transferring only this number of electrons, these atoms attain the electronic stability of the nearest noble gas to them in atomic number.

For illustration, consider the ionic bonding in sodium fluoride. From Table 4.5, we learn that the Lewis symbols of sodium and fluorine are Na· and :F·, respectively. When these two elements combine at the atomic level, an atom of sodium transfers its single electron to a fluorine atom. By transferring one electron, the sodium atom electronically resembles neon. By accepting it, the fluorine atom electronically resembles neon. The atoms become charged; that is, they become ions. The process can be written schematically as follows:

$$\mathrm{Na\cdot} + \mathrm{:\ddot{F}\cdot} \longrightarrow \mathrm{Na^+} \ \mathrm{:\ddot{F}:^-}$$

The attraction between these oppositely charged ions constitutes the ionic bond that binds the two ions together.

◆ 4.10 COVALENT BONDING

Electrons can also be *shared* by atoms of identical or different elements to form molecules of elements or compounds. This sharing of electrons is usually between nonmetal atoms. Atoms of the same nonmetal bond to one another forming molecules of the element; atoms of different nonmetals bond to one another forming molecules of a compound.

The atoms of nonmetals acquire their electronic stability by sharing electrons in a manner that permits them to resemble atoms of the noble gases nearest to them in atomic number. For the atoms in the first and second periods other than hydrogen, helium, lithium, and beryllium, this means sharing a total of eight electrons in their outermost atomic orbital. This includes the atom's valence electrons plus those it shares with another atom. Hydrogen atoms share only two electrons. One shared pair of electrons between any two atoms is called a *covalent bond*.

Let's observe how two atoms of hydrogen combine to form a hydrogen molecule. H· is the Lewis symbol of hydrogen. To achieve the electronic structure of helium, the nearest noble gas, one hydrogen atom shares its only electron with the single electron from a second hydrogen atom. We represent this process as follows:

$$H· + H· \longrightarrow H:H$$

The pair of electrons shared between the two hydrogen atoms is a covalent bond. This manner of representing the hydrogen molecule (H:H) is called its *Lewis structure*.

Two chlorine atoms combine to form a chlorine molecule. :Cl· is the Lewis symbol of chlorine. To achieve the electronic structure of argon, the nearest noble gas, each chlorine atom shares its unpaired electron. This formation of the chlorine molecule from two chlorine atoms can be represented as follows:

$$:Cl· + :\ddot{C}l· \longrightarrow :\ddot{C}l:\ddot{C}l:$$

Let's consider next the combination of hydrogen and chlorine atoms. A hydrogen atom and a chlorine atom can share an electron pair to form a molecule of the substance called *hydrogen chloride*. The formation of a hydrogen chloride molecule from hydrogen and chlorine atoms can be represented as follows:

$$H· + :\ddot{C}l· \longrightarrow H:\ddot{C}l:$$

The hydrogen atom shares an electron pair so its electronic structure resembles that of helium, while the chlorine atom shares an electron pair so that its electronic structure resembles that of argon.

An atom can also form several covalent bonds with other atoms by simply sharing more than one pair of electrons. For instance, consider the formation of the methane molecule. This molecule consists of one carbon atom and four hydrogen atoms. The Lewis symbols for carbon and hydrogen are ·C· and H·, respectively. In the methane molecule, the carbon atom shares each of its four electrons with the electrons from four hydrogen atoms, as represented below:

$$·C· + 4H· \longrightarrow \begin{matrix} H \\ H:\ddot{C}:H \\ H \end{matrix}$$

The sharing of electrons in the methane molecule results in an electronic arrangement like the neon atom for the carbon atom and an electronic arrangement like the helium atom for each of the four hydrogen atoms.

Sometimes, two nonmetallic atoms share more than one pair of electrons. This behavior is particularly characteristic of carbon atoms. It results in the formation of *multiple bonds*. Two types of multiple bonds exist: double and triple bonds. A *double bond* consists of the sharing of two pairs of electrons (::), while a *triple bond* consists of the sharing of three pairs of electrons (:::).

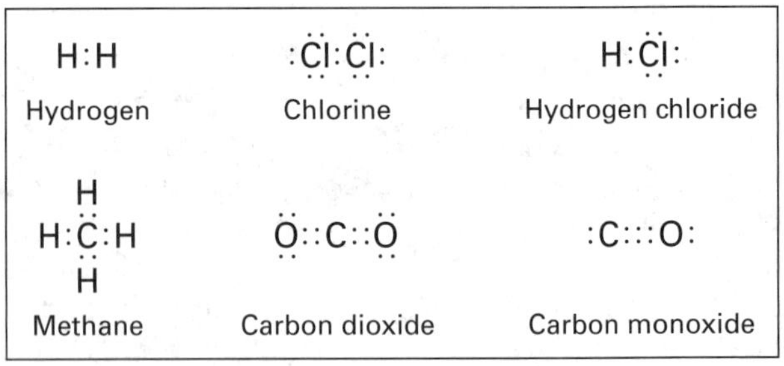

$$H:H \qquad\qquad :\ddot{C}l:\ddot{C}l: \qquad\qquad H:\ddot{C}l:$$
Hydrogen   Chlorine   Hydrogen chloride

$$\begin{array}{c} H \\ H:\overset{\displaystyle .}{\underset{\displaystyle .}{C}}:H \\ H \end{array} \qquad\qquad \ddot{O}::C::\ddot{O} \qquad\qquad :C:::O:$$
Methane   Carbon dioxide   Carbon monoxide

FIGURE 4.5 ◆ The Lewis structures of some simple molecules.

The carbon dioxide molecule consists of one atom of carbon and two atoms of oxygen. The Lewis symbols for carbon and oxygen are $\cdot\overset{.}{C}\cdot$ and $\cdot\ddot{O}\cdot$, respectively. The formation of a carbon dioxide molecule from carbon and oxygen atoms can be represented as follows:

$$\cdot\ddot{O}\cdot + \cdot\overset{.}{C}\cdot + \cdot\ddot{O}\cdot \longrightarrow \ddot{O}::C::\ddot{O}$$

Notice the existence of two pairs of electrons on each side of the carbon atom in the carbon dioxide molecule. Each of these shared pairs of electrons is a double bond. By sharing electrons in this fashion, the carbon atom and the two oxygen atoms each achieve the electronic arrangement of neon.

The formation of the carbon monoxide molecule from carbon and oxygen atoms can be represented as follows:

$$\cdot\overset{.}{C}\cdot + \cdot\ddot{O}\cdot \longrightarrow :C:::O:$$

The three shared pairs of electrons between the carbon and oxygen atoms constitutes a triple bond. Once again, the carbon and oxygen atoms achieve the electronic stability of the neon atom by sharing eight electrons between them. The Lewis structures of several other molecules are illustrated in Figure 4.5.

For the sake of simplicity, Lewis structures are usually written by drawing a long dash to represent the shared pair of electrons. The dots representing all other electrons are omitted. By using this notation, the compounds previously discussed can be represented as follows:

$$\text{H}-\text{H} \qquad \text{Cl}-\text{Cl} \qquad \text{H}-\text{Cl} \qquad \begin{array}{c} \text{H} \\ | \\ \text{H}-\text{C}-\text{H} \\ | \\ \text{H} \end{array} \qquad \text{O}=\text{C}=\text{O} \qquad \text{C}\equiv\text{O}$$

hydrogen   chlorine   hydrogen chloride   methane   carbon dioxide   carbon monoxide

---

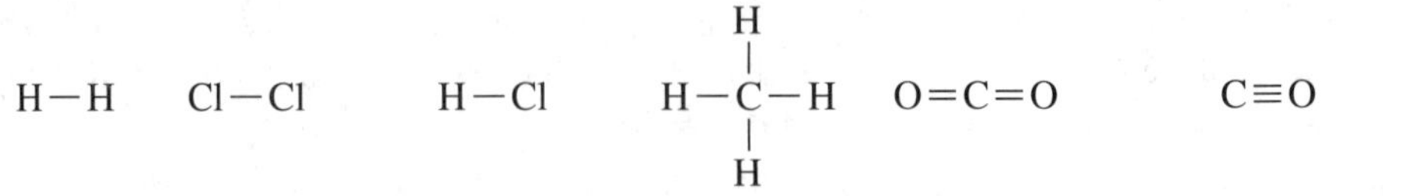

Solved Exercise 4.2

During World War I, the chemical warfare agent (Section 12.16) called *phosgene* was responsible for approximately 80% of the casualties associated with exposures to chemical substances. It was called a *choking agent*, since the troops who inhaled phosgene subsequently experienced bronchial constriction and were unable to properly breathe. The chemical formula of phosgene is $COCl_2$. What is its Lewis structure?

Solution: The Lewis symbols of the carbon, oxygen, and chlorine atoms are provided in Table 4.5. The Lewis structure of a phosgene molecule can be represented by examining its formation from one carbon atom, one oxygen atom, and two chlorine atoms as follows:

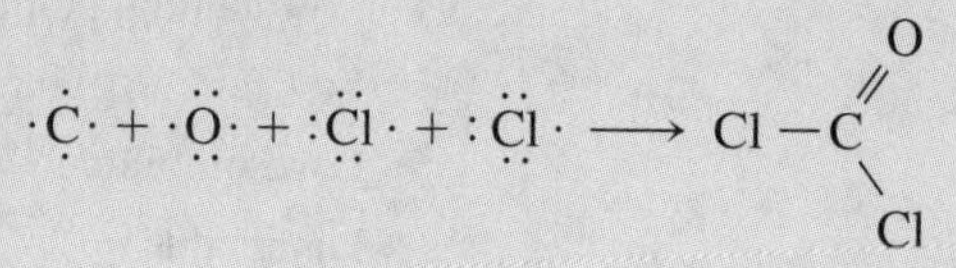

$$\dot{\underset{.}{C}}\cdot + \cdot \ddot{\underset{..}{O}}\cdot + :\ddot{\underset{..}{C}}l\cdot + :\ddot{\underset{..}{C}}l\cdot \longrightarrow Cl-C\underset{Cl}{\overset{O}{\diagup}}$$

Solved Exercise 4.3

The chemical warfare agent (Section 12.16) called *diphosgene* is denoted as a *delay-action casualty agent*, since approximately 3 hours pass following its inhalation before individuals experience ill effects. The chemical formula of diphosgene is $C_2O_2Cl_4$. What is its Lewis structure?

Solution: The Lewis symbols of the carbon, oxygen, and chlorine atoms are provided in Table 4.5. From them, the Lewis structure of a diphosgene molecule can be represented by examining its formation from two carbon atoms, two oxygen atoms, and four chlorine atoms. It is helpful to first recognize that an oxygen atom bonds between the two carbon atoms. Then, the process can be denoted as follows:

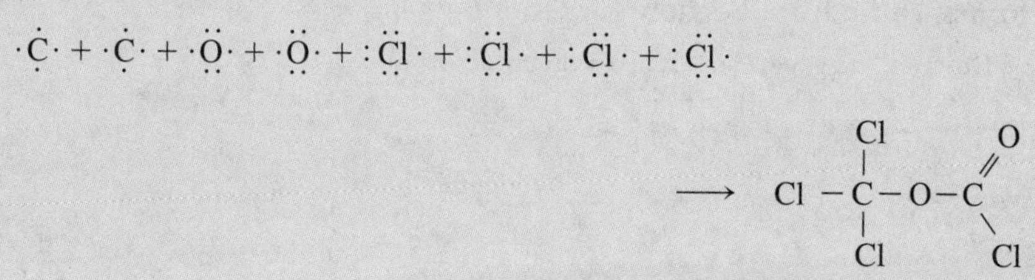

$$\dot{\underset{.}{C}}\cdot + \cdot\dot{\underset{.}{C}}\cdot + \cdot \ddot{\underset{..}{O}}\cdot + \cdot\ddot{\underset{..}{O}}\cdot + :\ddot{\underset{..}{C}}l\cdot + :\ddot{\underset{..}{C}}l\cdot + :\ddot{\underset{..}{C}}l\cdot + :\ddot{\underset{..}{C}}l\cdot$$

$$\longrightarrow \quad Cl-\underset{Cl}{\overset{Cl}{\underset{|}{\overset{|}{C}}}}-O-C\underset{Cl}{\overset{O}{\diagup}}$$

(Chemists refer to diphosgene as trichloromethyl chloroformate.)

Performance Goals for Section 4.11
  ◆ Distinguish between the nature of ionic compounds and covalent compounds.
  ◆ Contrast the general properties of ionic compounds and covalent compounds.

◆ **4.11 IONIC AND COVALENT COMPOUNDS**

Chemical compounds are classified into two groups based on the predominating nature of the bonding between their atoms. Chemical compounds with mainly ionic bonds are called *ionic compounds*. Compounds whose atoms are mainly bonded together by covalent bonds are called *covalent compounds*.

Table 4.6 ◆ Contrasting Properties Between Ionic and Covalent Compounds

Ionic Compounds	Covalent Compounds
High melting points	Low melting points
High boiling points	Low boiling points
High solubility in water	Low solubility in water
Nonflammable	Flammable
Molten substances and their water solutions conduct an electric current	Molten substances do not conduct an electric current
Exist predominantly as solids at room temperature	Exist as gases, liquids, and solids at room temperature

Ionic and covalent compounds are associated with a number of general properties summarized in Table 4.6. Note from this table that ionic compounds usually boil and melt at much higher temperatures than covalent compounds. Most ionic compounds dissolve in water and will conduct electricity when melted or dissolved in solution, while the opposite is true of covalent compounds. These properties are generalizations that only reflect the characteristics of the majority of ionic and covalent compounds.

Performance Goal for Section 4.12

- ◆ Describe the general nature of a chemical formula.

◆ 4.12 THE CHEMICAL FORMULA

Chemists condense information regarding the chemical composition of substances by writing a *chemical formula* for each of them. For substances consisting of molecules, the chemical formula indicates the kinds and numbers of atoms present in each molecule. For instance, we observed earlier in this chapter that hydrogen and oxygen are symbolized as H_2 and O_2, respectively; these expressions are the chemical formulas of elemental hydrogen and oxygen. The subscript 2 means that each molecule contains two atoms. These chemical formulas are read verbally as "H-two" and "O-two," respectively.

We have also observed that the chemical formulas for hydrogen chloride and methane are HCl and CH_4, respectively. These formulas represent the composition of the individual molecules. One molecule of hydrogen chloride consists of one hydrogen atom and one chlorine atom, while a molecule of methane consists of one carbon atom and four hydrogen atoms. We verbally read these chemical formulas as "HCl" and "CH-four," respectively. Figure 4.6 further illustrates the chemical formulas of two substances based upon their composition.

On the other hand, substances are not always composed of molecules. For instance, sodium fluoride, an ionic compound, is composed of ions held together by an

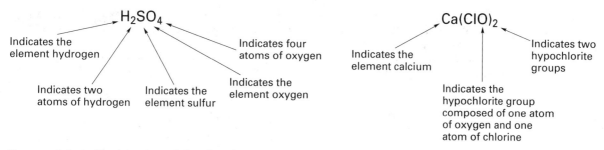

FIGURE 4.6 ◆ Explanation of the chemical formulas H_2SO_4 and $Ca(ClO)_2$. Sulfuric acid is denoted by the formula H_2SO_4; it is a corrosive liquid used to produce rayon, certain explosives, dyes, other acids and detergents. Calcium hypochlorite is denoted by $Ca(ClO)_2$; it is a constituent of some solid bleaching agents and disinfecting agents.

ionic bond. Experimental evidence indicates that the components of ionic compounds are not molecules, but rather aggregates of positive and negative ions. When we analyze a sample of sodium fluoride, we learn that it is composed of an equal number of sodium ions (Na^+) and fluoride ions (F^-). Chemists write its chemical formula as Na^+ F^-, representing the lowest number of formula units. In common practice, the ionic charges are omitted and we simply write NaF.

When responding to a disaster involving a hazardous material, knowing the names or chemical formulas of its constituents could be vital. When the constituents have been identified, a prompt and appropriate response can be directed towards handling the hazardous material while minimizing the risk to personnel. During emergency-response actions, the names and formulas of commercial chemical commodities can be determined in several ways. For instance, when the disaster occurs within a manufacturing plant or other workplace, the formulas are provided on MSDSs (Section 1.6). During a transportation mishap, the specific names of the hazardous material being transported can be readily obtained from a review of the accompanying shipping papers. Sometimes, when a commodity is transported in bulk, its name is stenciled on the side of the transport vehicle. In Chapter 7, we will review other ways to rapidly determine the nature of the hazardous materials involved in transportation mishaps.

Performance Goals for Section 4.13
- ◆ Use Table 4.7 to learn the names and symbols of the common ions.
- ◆ Write the chemical formula of an ionic compound when provided with its name.
- ◆ Provide the name of an ionic compound when provided with its chemical formula.

◆ 4.13 WRITING CHEMICAL FORMULAS AND NAMING IONIC COMPOUNDS

The chemical formulas of ionic compounds are obtained from the symbols of the ions that make up a given substance. The most common ions are listed in Table 4.7. The names and symbols of these ions should be memorized. There are several simple rules you can follow.

TABLE 4.7 ◆ Some Common Ions

Positive Ions	Negative Ions
Ammonium, NH_4^+	Acetate, $C_2H_3O_2^-$
Copper(I) or cuprous, Cu^+	Bromide, Br^-
Hydrogen, H^+	Chloride, Cl^-
Silver, Ag^+	Chlorate, ClO_3^-
Sodium, Na^+	Chlorite, ClO_2^-
Potassium, K^+	Cyanide, CN^-
	Fluoride, F^-
Barium, Ba^{2+}	Hydrogen carbonate or bicarbonate, HCO_3^-
Cadmium, Cd^{2+}	Hydrogen sulfate or bisulfate, HSO_4^-
Calcium, Ca^{2+}	Hydrogen sulfite or bisulfite, HSO_3^-
Cobalt(II) or cobaltous, Co^{2+}	Hydroxide, OH^-
Copper(II) or cupric, Cu^{2+}	Hypochlorite, ClO^-
Iron(II) or ferrous, Fe^{2+}	Iodide, I^-
Lead(II) or plumbous, Pb^{2+}	Nitrate, NO_3^-
Magnesium, Mg^{2+}	Nitrite, NO_2^-
Manganese(II) or manganous, Mn^{2+}	Perchlorate, ClO_4^-
Mercury(I) or mercurous, Hg_2^{2+}	Permanganate, MnO_4^-
Nickel, Ni^{2+}	
Strontium, Sr^{2+}	
Tin(II) or stannous, Sn^{2+}	Carbonate, CO_3^{2-}
Zinc, Zn^{2+}	Oxide, O^{2-}
	Peroxide, O_2^{2-}
Chromium(III) or chromic, Cr^{3+}	Sulfate, SO_4^{2-}
Iron(III) or ferric, Fe^{3+}	Sulfide, S^{2-}
	Sulfite, SO_3^{2-}
Tin(IV) or stannic, Sn^{4+}	
	Phosphate, PO_4^{3-}

POSITIVE IONS

◆ Alkali metals, alkaline earth metals, aluminum, zinc, and hydrogen form monatomic positive ions, each taking the name of the element from which it was derived. For the hydrogen ion and each alkali metal ion, the net charge is +1, and for each alkaline earth metal ion, the net charge is +2.

Na^+ sodium ion Ba^{2+} barium ion

Al^{3+} aluminum ion H^+ hydrogen ion

- If a metal forms more than one positive ion, the ion is named in either of two ways:
  - **a.** The ion takes the English name of the metal from which it is derived, immediately followed by a Roman numeral written in parentheses that indicates the net charge on the ion. This system of naming ionic compounds is called the *Stock system*.

$$Cu^+ \text{ copper(I) ion} \qquad Sn^{2+} \text{ tin(II) ion}$$
$$Cu^{2+} \text{ copper(II) ion} \qquad Sn^{4+} \text{ tin(IV) ion}$$

  - **b.** The ion takes the Latin name of the metal from which it was derived, together with one of the suffixes, -ous or -ic, representing the lower and higher net ionic charge, respectively. We call this system of naming ionic compounds the *older system*.

$$Cu^+ \text{ cuprous ion} \qquad Sn^{2+} \text{ stannous ion}$$
$$Cu^{2+} \text{ cupric ion} \qquad Sn^{4+} \text{ stannic ion}$$

- There are only two common polyatomic positive ions, noted below:

$$NH_4^+ \text{ ammonium ion} \qquad Hg_2^{2+} \text{ mercury(I) or}$$
$$\text{mercurous ion}$$

NEGATIVE IONS

- Monatomic negative ions are named by adding the suffix -*ide* to the stem of their names.

$$F^- \text{ fluoride ion} \qquad Cl^- \text{ chloride ion}$$
$$O^{2-} \text{ oxide ion} \qquad S^{2-} \text{ sulfide ion}$$

- Polyatomic negative ions are named as follows.
  - Two polyatomic negative ions are named with the suffix -ide:

$$CN^- \text{ cyanide ion} \qquad OH^- \text{ hydroxide ion}$$

- The polyatomic negative ions containing carbon are named uniquely, two of which, the carbonate and acetate ions, are commonly encountered.

$$CO_3^{2-} \text{ carbonate ion} \qquad C_2H_3O_2^- \text{ acetate ion}$$

- When a nonmetal forms two different negative ions containing oxygen, the suffixes -*ite* and -*ate* are used to name the ion with the lesser and higher number of oxygen atoms, respectively.

$$NO_2^- \text{ nitrite ion} \qquad SO_3^{2-} \text{ sulfite ion}$$
$$NO_3^- \text{ nitrate ion} \qquad SO_4^{2-} \text{ sulfate ion}$$

- When a nonmetal forms more than two different negative ions containing oxygen, the prefixes *hypo-* (meaning lower than usual) and *per-* (meaning higher than usual) are used together with the suffixes, -*ite* and -*ate*, in order of increasing number of oxygen atoms, as follows: *hypo-* _____ -*ite*, _____ -*ite*, _____ -*ate*, and *per-* _____ -*ate*, where the dashed line represents the stem of the identifying nonmetal.

$$ClO^- \qquad \text{hypochlorite ion}$$
$$ClO_2^- \qquad \text{chlorite ion}$$
$$ClO_3^- \qquad \text{chlorate ion}$$
$$ClO_4^- \qquad \text{perchlorate ion}$$

• Negative ions containing hydrogen and oxygen atoms (except hydroxide and acetate ions) are named hydrogen _____, or alternatively, bi _____, where the line represents the stem of the identifying nonmetal.

HCO_3^- hydrogen carbonate, or bicarbonate

HSO_4^- hydrogen sulfate, or bisulfate

When we know the chemical formulas of substances, we name the associated compounds by simply listing the names of the positive and negative ions in that order. For instance, NaCl is sodium chloride, MgS is magnesium sulfide, $BaCO_3$ is barium carbonate, and so forth.

Frequently, we encounter a chemical formula containing parentheses, such as $Fe(ClO_4)_3$. To name such a formula, it is sometimes best to rewrite it and indicate the net charges on the constituent ions. In this case, we have $Fe^{3+}(ClO_4^-)_3$. This shows us that the +3 charge on the iron(III) ion balances the −3 charge on three perchlorate ions. The total positive and negative charges of the ions are equal (+3 and −3), since the compound itself is uncharged, or neutral. The compound is named iron(III) perchlorate by the Stock system and ferric perchlorate by the older system.

Before proceeding further, note the names of the following chemical formulas:

$Al_2(SO_3)_3$	aluminum sulfite
$(NH_4)_2CO_3$	ammonium carbonate
$Ba(CN)_2$	barium cyanide
$Cd_3(PO_4)_2$	cadmium phosphate
$Ca(MnO_4)_2$	calcium permanganate

Suppose next that we are interested in writing a chemical formula when we know the name of a substance. It is again best to follow two simple rules, as follows:

• If the net charge of the positive and negative ions is equal (but opposite in sign), the formula of a substance is obtained by simply listing the symbol of the positive ion first, then the symbol of the negative ion. For instance, NaF, CaS, ZnO, KOH, $NaClO_3$, $BaSO_4$, and $AlPO_4$ all have component ions with numerically equal charges. Their names are sodium fluoride, calcium sulfide, zinc oxide, potassium hydroxide, sodium chlorate, barium sulfate and aluminum phosphate, respectively.

• If the net charges of the positive and negative ions are not numerically equal, first write the symbol of the positive and negative ions together with their respective charges. Then, simply cross the numbers representing the charges (but not the signs) to the opposite ion, as shown in the following examples:

Potassium sulfate	$K_2^+ \, SO_4^{2-}$,	or K_2SO_4 (1 is not written)
Aluminum oxide	$Al^{3+}_2 \, O^{2-}_3$,	or Al_2O_3
Barium iodide	$Ba^{2+} \, I^-_2$,	or BaI_2 (1 is not written)
Ammonium sulfate	$NH_4^+{}_2 \, SO_4^{2-}$,	or $(NH_4)_2SO_4$

When using this rule, one special case should be noted. When the charges are multiples of one another, such as +4 and −2, the lowest multiple is used. For instance, the chemical formula of tin(IV) oxide is SnO_2, not Sn_2O_4. In the study of hazardous materials, this exception arises only for compounds containing the tin(IV) ion.

Solved Exercise 4.4

Solutions of monoammonium hydrogen phosphate and diammonium hydrogen phosphate have been used as fire retardants, especially to control the spread of forest fires. What are the chemical formulas of these two substances?

 Solution: By consulting Table 4.7, we determine that the charges of the ammonium, hydrogen, and phosphate ions are $+1$, $+1$, and -3, respectively. Using this information, the chemical formulas of monoammonium hydrogen phosphate and diammonium hydrogen phosphate are written as $NH_4H_2PO_4$ and $(NH_4)_2HPO_4$, respectively.

Solved Exercise 4.5

Provide acceptable names for the compounds whose chemical formulas are **(a)** $CaCO_3$, **(b)** $FeSO_4$, and **(c)** $Ni_3(PO_4)_2$.

 Solution: **(a)** Since Ca^{2+} and CO_3^{2-} are the symbols for the calcium and carbonate ions, respectively, $CaCO_3$ is the formula for the compound whose name is calcium carbonate.

 (b) Iron has two ions: ferrous, or iron(II) (Fe^{2+}); and ferric, or iron(II) (Fe^{3+}). The listing in Table 4.7 indicates that SO_4^{2-} is the symbol for the sulfate ion. Since a subscript is absent adjacent to the "Fe" in $FeSO_4$, the symbol represents the ferrous ion. Acceptable names for the compound whose formula is $FeSO_4$ are ferrous sulfate and iron(II) sulfate.

 (c) Since Ni^{2+} and PO_4^{3-} are the symbols for the nickel and phosphate ions, $Ni_3(PO_4)_2$ is the formula for the compound whose name is nickel phosphate.

Performance Goals for Section 4.14
- ◆ Learn the common names and formulas of the substances in Table 4.8.
- ◆ Using Table 4.9, learn the names of the prefixes that identify the number of atoms of a given type in a covalent compound.
- ◆ Write the chemical formula of a simple covalent compound when provided with its name.
- ◆ Provide the name of a simple covalent compound when provided with its chemical formula.

◆ 4.14 SOME CHEMICAL FORMULAS AND NAMES OF COVALENT COMPOUNDS

It is generally difficult to write the chemical formula of a covalent compound. In this book, we will learn many formulas of covalent compounds as the occasions present themselves. Nevertheless, a great many covalent names have acquired common names that are still used as part of the chemical language. Some examples are listed in Table 4.8. These names and formulas should be memorized.

TABLE 4.8 ◆ Common Names for Some Simple Covalent Compounds

Formula of Compound	Common Name
C_2H_2	Acetylene
NH_3	Ammonia
C_6H_6	Benzene
C_2H_6	Ethane
N_2H_4	Hydrazine
H_2S	Hydrogen sulfide
CH_4	Methane
PH_3	Phosphine
C_3H_8	Propane
H_2O	Water

A number of covalent compounds contain only two elements, and these substances can be named in a simple fashion. We name the element that is written first in a chemical formula, proceeding it with a Greek prefix indicating the number of atoms of the element in the compound; then, we name the other element, preceding it also with its relevant prefix to indicate the number of its atoms and modifying its ending to -*ide*. Table 4.9 lists the Greek prefixes corresponding to the numbers used to name simple covalent compounds.

TABLE 4.9 ◆ Greek Prefixes Used in Naming Simple Covalent Compounds

Prefix		Compound
Mono-	1	Carbon monoxide, CO[a]
Di-	2	Carbon dioxide, CO_2
Tri-	3	Phosphorus trichloride, PCl_3
Tetra-	4	Carbon tetrachloride, CCl_4
Penta-	5	Phosphorus pentachloride, PCl_5
Hexa-	6	Sulfur hexafluoride, SF_6
Hepta-	7	Dichlorine heptoxide, Cl_2O_7[a]
Octa-	8	Dichlorine octoxide, Cl_2O_8
Ennea-[b]	9	Tetraiodine enneaoxide, I_4O_9
Deca-	10	Tetraphosphorus decoxide, P_4O_{10}[a]

[a] When two vowels appear next to one another, as *oo* in monoxide and *ao* in heptaoxide, the vowel from the Greek prefix is dropped for the sake of euphony.
[b] The Latin prefix *nona-* is also used to denote nine atoms. Thus, I_4O_9 may also be correctly named tetraiodine nonoxide.

We have encountered the chemical formulas of several simple covalent compounds containing only two elements: HCl, CO, and CO_2. The compounds represented by these formulas are named hydrogen chloride, carbon *mono*xide, and carbon *di*oxide, respectively. Other examples of naming covalent compounds are provided in Table 4.9.

Solved Exercise 4.6

Dinitrogen tetroxide has been used to oxidize rocket fuels. What is its chemical formula?

Solution: From Table 4.9, we see that the Greek prefixes *di-* and *tetra-* refer to two and four, respectively. This means that the dinitrogen tetroxide molecule contains two nitrogen atoms and four oxygen atoms. Accordingly, the chemical formula of dinitrogen tetroxide is N_2O_4.

Solved Exercise 4.7

In 2002, to disinfect anthrax-laden mail in postal areas and in the Hart Senate Office Building, EPA contractors used a substance having the chemical formula ClO_2. What is the name of this substance?

Solution: From Table 4.9, we determine that each molecule of ClO_2 is composed of one atom of chlorine and two atoms of oxygen. Consequently, the name of the substance used to disinfect the anthrax-laden mail was chlorine dioxide.

Performance Goals for Section 4.15
- ◆ Write the names and chemical formulas of the common binary acids.
- ◆ Learn the names and chemical formulas of the oxyacids listed in Table 4.10.
- ◆ Relate the names of an oxyacid with the name of its corresponding polyatomic ion.

◆ **4.15 NAMING ACIDS**

Acids are a class of chemical compounds that we shall study more extensively in Chapter 8. For now, we shall consider acids as compounds that produce hydrogen ions when dissolved in water. This means that all acids contain hydrogen in their chemical structures.

Binary acids contain hydrogen and one other nonmetal. The chemical formula of a binary acid is HX or H_2X, where X is the symbol of the nonmetal other than hydrogen. These acids are named by placing the prefix *hydro-* before the stem of the name of an identifying nonmetal, attaching the suffix *-ic* to this stem, and adding the word

TABLE 4.10 ◆ Names and Chemical Formulas of Some Oxyacids

Name	Chemical Formula
Sulfurous acid	H_2SO_3
Sulfuric acid	H_2SO_4
Nitrous acid	HNO_2
Nitric acid	HNO_3
Phosphorous acid	H_3PO_3
Phosphoric acid	H_3PO_4
Hypochlorous acid	$HClO$
Chlorous acid	$HClO_2$
Chloric acid	$HClO_3$
Perchloric acid	$HClO_4$

"acid" to the name. There are two important binary acids whose chemical formulas should be noted:

$$HF \qquad \text{hydrofluoric acid}$$
$$HCl \qquad \text{hydrochloric acid}$$

The names of these binary acids are valid only when the acids are dissolved in water. As single substances, they are named as compounds of hydrogen: hydrogen fluoride and hydrogen chloride.

Another important group of acids are the *oxyacids*. In addition to hydrogen and the identifying nonmetal, these acids contain one or more oxygen atoms. The chemical formula of the oxyacids is H_mXO_n, where X is the symbol of the identifying nonmetal, and m and n are numbers. There are only 10 oxyacids important to the study of hazardous materials. Their names and chemical formulas are listed in Table 4.10. They should be memorized.

The names of oxyacids correlate directly with the names of their associated polyatomic ions:

- Polyatomic ions whose names end in *-ate* are derived from acids whose names end in *-ic*; for example, the nit*rate* ion is derived from nit*ric* acid, and the perchlo*rate* ion is derived from perchlo*ric* acid.
- Polyatomic ions whose names end in *-ite* are derived from acids whose names end in *-ous*; for example, the nit*rite* ion is derived from nit*rous* acid, and the hypochlo*rite* ion is derived from hypochlo*rous* acid.

Performance Goals for Section 4.16
- Describe the manner by which the molecular weight or formula weight of a substance is determined.
- Describe the concept of a mole.
- Calculate the number of moles of a substance when its amount is provided.
- Identify the Avogadro number.
- Use the Avogadro number to calculate how many atoms, molecules, or formula units are present in a given amount.

◆ 4.16 MOLECULAR WEIGHTS, FORMULA WEIGHTS, AND THE MOLE

The relative weight of a compound that occurs as molecules is called the *molecular weight*. It is the sum of the atomic weights of each atom that are part of the molecule. Consider the water molecule. Its molecular weight is determined as follows (to five significant figures):

$$2 \text{ hydrogen atoms} = 2 \times 1.008 = 2.016$$
$$1 \text{ oxygen atom} = 1 \times 15.999 = \underline{15.999}$$
$$\text{Molecular weight of } H_2O = 18.015$$

The relative weight of a compound that occurs as formula units is called the *formula weight*. It is the sum of the atomic weights of all atoms that make up one formula unit. Consider sodium fluoride. Its formula weight is determined as follows (to five significant figures):

$$1 \text{ sodium ion} = 22.990$$
$$1 \text{ fluoride ion} = \underline{18.998}$$
$$\text{Formula weight of NaF} = 41.988$$

Chemists also frequently make use of a unit quantity called the *mole*. The mole is the approved SI unit for the amount of any substance. As a unit, it is symbolized as *mol* (without the *-e*). One mole of atoms, molecules, or formula units represents the amount of substance that has a mass in grams equal to its atomic weight, molecular weight, or formula weight, respectively. Thus, 1 mol of carbon is an amount of carbon that has a mass of 12.011 g. Also, 1 mol of water is an amount of water that has a mass of 18.105 g.

Since the concept of the mole applies to any type of particle, it is important to identify just what the particle is. For instance, 1 mol of hydrogen *atoms* has a mass of 1.0079 g, but 1 mol of hydrogen *molecules* has a mass of 2.0158 g.

Finally, 1 mol of atoms, molecules, or formula units contains a definite number of these units: 6.022×10^{23}. This number is called the *Avogadro number*, in honor of the scientist who first suggested its existence. The exponential notation refers to a number in which the decimal point has been moved 23 places to the right. We can also write this number without using an exponent as 602,200,000,000,000,000,000,000. The mole is analogous to units like the dozen or ream, which mean 12 and 500 items, respectively. One mol of hydrogen atoms is 6.022×10^{23} H atoms; 1 mol of hydrogen molecules is 6.022×10^{23} H_2 molecules, and 1 mol of sodium fluoride units is 6.022×10^{23} NaF units.

The Avogadro number is not just huge—it is so enormous that it is difficult to appreciate how large it is. Imagine that you are counting jelly beans at the rate of 3 beans per second. At this rate, it would take 6×10^{15} years to count 1 mol of jelly beans! This is a million times older than the age of the Earth. The enormous magnitude of this number reflects the minute dimensions of atoms and molecules on the scale of "ordinary" measurements.

Solved Exercise 4.8

When hydrazine burns, 148.6 kcal/mol of heat evolves to the environment. How many kcal/g of heat evolves?

Solution: From Table 4.8, we see that the chemical formula of hydrazine is N_2H_4. Using this formula, the molecular weight of hydrazine is calculated to be 32.045:

$$2 \text{ nitrogen atoms, } 2 \times 14.0067 = 28.013$$
$$4 \text{ hydrogen atoms, } 4 \times 1.008 = \underline{4.032}$$
$$\text{Molecular weight } = 32.045$$

One mole of hydrazine is an amount equal to 32.045 g. When hydrazine burns, 4.5 kcal/g of heat evolves:

$$\frac{148.6 \text{ kcal/mol}}{32.045 \text{ g/mol}} = 4.5 \text{ kcal/g}$$

REVIEW EXERCISES

Elements and Compounds

4.1 Classify each of the following forms of matter as an element, a compound, or a mixture:

(a) water vapor
(b) hydrogen
(c) an apple
(d) air
(e) dry ice

(f) nitrogen
(g) chromium
(h) sugar
(i) carbon monoxide

4.2 Name the elements having these symbols: H, He, Ne, Al, S, Cl, O, F, Mg, Mn, Pb, Br, Cr, Ag, Si, Na, K, Cu, Fe, Ca, Ba, Zn, and Co.

4.3 Identify the symbol that represents each of the following elements:

(a) carbon
(b) fluorine
(c) sodium
(d) lead
(e) aluminum
(f) barium

(g) chlorine
(h) lithium
(i) zinc
(j) magnesium
(k) nickel
(l) iron

4.4 What are the four most abundant elements in Earth's crust, oceans, and the atmosphere?

Physical and Chemical Changes

4.5 Classify each of the following phenomena as a physical or chemical change:

(a) Calcium hydride reacts with water to produce hydrogen and calcium hydroxide.
(b) Benzene vapor ignites at 1044°F (562°C) without an ignition source.
(c) The acid in vinegar is neutralized when mixed with sodium bicarbonate.
(d) Ethyl acrylate, a raw material used to manufacture acrylic paints, is dissolved in water.

(e) Water freezes at 32°F (0°C).

(f) Tungsten metal melts at 6165°F (3407°C).

4.6 Isoamyl nitrite is a liquid that is sometimes mixed with the fuel used in racing cars. Some characteristics of isoamyl nitrite are noted below. Classify each characteristic as either a physical or chemical property:

(a) Isoamyl nitrite is a yellowish liquid having an ethereal odor.

(b) At 77°F (25°C), the specific gravity of isoamyl nitrite is in the range from 0.865 to 0.875.

(c) Isoamyl nitrite is nearly insoluble in water.

(d) Isoamyl nitrite boils when heated to a temperature within the range from 205 to 210°F (96 to 99°C).

(e) When exposed to light, isoamyl nitrite decomposes.

(f) When it ignites, isoamyl nitrite forms carbon monoxide, carbon dioxide, nitric oxide, nitrogen dioxide, and water vapor.

4.7 Iodine has only one naturally occurring isotope, iodine-127. Its atomic mass is 126.9044. What is the atomic weight of iodine?

4.8 Strontium has four naturally occurring isotopes: strontium-84, strontium-86, strontium-87, and strontium-88. Their natural abundances and atomic masses are noted below:

Isotope	Mass	Natural abundance (%)
Strontium-84	83.9134	0.50
Strontium-86	85.9094	9.90
Strontium-87	86.9089	7.00
Strontium-88	87.9056	82.60

Show that the atomic weight of strontium is 87.62.

The Periodic Table

4.9 Which elements in the following list possess chemical properties similar to those of nitrogen: magnesium, phosphorus, arsenic, lead, or oxygen?

4.10 Use the periodic table to determine the symbol of each of the following elements:

(a) Radon, exposure to which is regarded as the *second* leading cause of lung cancer. (The leading cause is exposure to tobacco smoke.)

(b) Selenium, a nutrient needed in the human diet.

(c) Zirconium, found in the gemstones called zirconia.

(d) Erbium, terbium, yttrium, and ytterbium, four metals named after Ytterby, a village in Sweden where a mineral containing yttrium and the lanthanide elements was first discovered.

4.11 Which two of the following elements possess chemical properties similar to those of barium: chromium, tungsten, selenium, strontium, cesium, iodine, magnesium, iron, or silver?

4.12 An isotope of americium is used in household and commercial smoke detectors. Using the periodic table, answer the following questions:

(a) What is the symbol for americium?

(b) What are the symbols of the elements immediately adjacent to americium on the periodic table?

(c) Is americium a metal, nonmetal, or metalloid?

(d) Explain why americium acts chemically more like scandium than like calcium.

4.13 Consider the element that possesses an atomic number of 114.
  (a) Identify the family of which this element is a member.
  (b) Identify the period of which this element is a member.
  (c) Identify the element that occupies the position immediately above it on the period table.

4.14 Radioactive isotopes of cesium and strontium were identified in the atmospheric fallout from atomic bomb explosions.
  (a) Locate these two elements on the periodic table and identify their chemical families.
  (b) As common components of the human diet, certain compounds are routinely found in the body. For example, sodium and potassium are constituents of all the body's tissues, and calcium is found in our teeth and bones. Although cesium and strontium are *not* components of the human diet, they can be inadvertently absorbed into the human body as contaminants in milk and other dairy products. Based solely on their positions on the periodic table, in what parts of the body would you expect cesium and strontium to concentrate?

Atoms, Moles, and Ions

4.15 How many valence electrons do atoms of each of the following elements possess:
  (a) sodium   **(d)** aluminum
  (b) carbon   **(e)** bromine
  (c) silicon   **(f)** oxygen

4.16 Write the Lewis symbol of the following atoms:
  (a) nitrogen   **(d)** phosphorus
  (b) calcium   **(e)** potassium
  (c) boron   **(f)** sulfur

4.17 Determine the net charge of each of the following ions:
  (a) sodium   **(d)** chloride
  (b) oxide   **(e)** magnesium
  (c) hydrogen   **(f)** sulfide

Chemical Bonding

4.18 Heat sufficient to weld metallic sheets together is commonly derived by the burning of acetylene. Draw the Lewis structure of the acetylene molecule.

4.19 Anaerobic bacteria initiate the decomposition of some forms of matter in the absence of oxygen. The phenomenon is often associated with the generation of hydrogen sulfide, a gas possessing the offensive odor of rotten eggs. Draw the Lewis structure of the hydrogen sulfide molecule.

4.20 Hydrazine is a substance sometimes used as a rocket propellant. Its chemical formula is N_2H_4. Write the Lewis structure of the hydrazine molecule.

4.21 In the military, cyanogen chloride is known as a blood agent, since inhalation prevents the blood from utilizing oxygen in the normal fashion. The chemical formula of cyanogen chloride is CNCl. What is the Lewis structure of the cyanogen chloride molecule?

4.22 Methyl bromide is a poisonous liquid that boils at 40°F (4°C) and does not possess a flashpoint. Although it was once a popular fumigant applied to agricultural crops, its use in the United States has now been banned (Section 12.7). The chemical formula of methyl bromide is CH_3Br. What is the Lewis structure of the methyl bromide molecule?

4.23 Dentists frequently use a 2% sodium fluoride solution in water to prevent tooth decay. Illustrate the manner by which a unit of sodium fluoride forms from atoms of the elements sodium and fluorine.

4.24 Fluorite is a mineral that is commonly used in making opalescent glass, as well as enameling cooking utensils. The principal chemical constituent of fluorite is calcium fluoride. Describe the nature of the chemical bonds in a unit of calcium fluoride and thereby show that the chemical formula of calcium fluoride is CaF_2.

4.25 Diazomethane is considered one of the most severely shock-sensitive substances known. Its chemical formula is H_2CN_2. In molecules of diazomethane, one of the nitrogen atoms is bonded directly to the carbon atom and the second nitrogen atom is bonded to the first nitrogen atom. Draw a Lewis structure of a molecule of diazomethane in which the following are true:
  (a) The two nitrogen atoms are bonded to one another by means of a triple bond.
  (b) The central nitrogen atom forms double bonds to the carbon atom and to the second nitrogen atom.

Chemical Formulas

4.26 Magnesium hydroxide and aluminum hydroxide are incorporated into textiles to serve as flame retardants. Write their chemical formulas.

4.27 Saltpeter, known chemically as potassium nitrate, is used to make gunpowder, fireworks, and matches. It is also used when smelting ores, preserving meats, and dyeing fabrics. What is the chemical formula of potassium nitrate?

4.28 In 1995, 169 people died when a mixture of oil and ammonium nitrate was intentionally detonated in a federal office building in Oklahoma City, OK. Write the chemical formula for ammonium nitrate.

4.29 Formaldehyde is produced during the combustion of methane. The chemical formula of formaldehyde is HCHO. Illustrate the Lewis formula of a molecule of formaldehyde.

4.30 The tips of "strike-anywhere" matches contain the substance whose chemical formula is P_4S_3. This substance possesses such a low ignition temperature that it ignites by friction when struck against a hard surface. What is the name of this flammable solid?

4.31 Write the chemical formula of the ionic compound that contains the following ions:
  (a) Na^+ and ClO_2^- **(h)** Cu^{2+} and S^{2-}
  (b) NH_4^+ and SO_4^{2-} **(i)** H^+ and F^-
  (c) K^+ and $C_2H_3O_2^-$ **(j)** Ba^{2+} and O^{2-}
  (d) Zn^{2+} and OH^- **(k)** Sn^{2+} and PO_4^{3-}
  (e) Pb^{2+} and NO_3^{2-} **(l)** Fe^{3+} and SO_4^{2-}
  (f) Mg^{2+} and ClO_3^- **(m)** Ca^{2+} and F^-
  (g) Fe^{2+} and CN^- **(n)** Ni^{2+} and CO_3^{2-}

4.32 Name each substance having the following chemical formula:
  (a) $CdBr_2$ **(aa)** $Mg(HCO_3)_2$
  (b) $Ba(NO_3)_2$ **(bb)** BaO_2

(c) $Al(OH)_3$

(d) FeS

(e) $HgSO_4$

(f) $SrSO_3$

(g) $Ni(ClO_4)_2$

(h) KOH

(i) MgI_2

(j) $Ca(MnO_4)_2$

(k) $Ba(CN)_2$

(l) H_2O_2

(m) H_2SO_4

(n) O_2

(o) $Mg(ClO_4)_2$

(p) $Pb(C_2H_3O_2)_2$

(q) $AgCl$

(r) NH_4I

(s) $CuCl$

(t) $Zn(ClO)_2$

(u) MnO

(v) NiF_2

(w) HCl (aqueous)

(x) Na_2S

(y) Br_2

(z) NBr_3

(cc) N_2H_4

(dd) $NaClO_4$

(ee) N_2

(ff) Cu_2SO_4

(gg) H_2

(hh) CoF_2

(ii) HCN

(jj) $CdSO_4$

(kk) ZnS

(ll) H_3PO_4

(mm) Na_2O

(nn) Na_2O_2

(oo) HF (nonaqueous)

(pp) Fe_2O_3

(qq) $Sr_3(PO_4)_2$

(rr) NH_4NO_2

(ss) PCl_3

(tt) N_2O_5

(uu) $Ca(HSO_3)_2$

(vv) CO

(ww) Ag_2O

(xx) C_2H_2

(yy) $Pb(ClO_3)_2$

(zz) $Co_3(PO_4)_2$

4.33 Give the chemical formula for each of the following substances:

(a) nitric acid

(b) silver nitrate

(c) cadmium phosphate

(d) nickel phosphate

(e) magnesium peroxide

(f) stannic cyanide

(g) ammonia

(h) strontium chloride

(i) mercuric nitrate

(j) hydrogen sulfide

(k) nitrogen

(l) magnesium bisulfite

(m) ammonium bromide

(n) lead(II) perchlorate

(o) chlorine

(p) hydrazine

(q) magnesium sulfate

(r) barium permanganate

(s) cobalt(II) carbonate

(t) aluminum sulfate

(u) strontium iodide

(v) magnesium ammonium phosphate

(w) phosphoric acid

(x) hydrogen chloride

(aa) ammonium bicarbonate

(bb) cobalt(II) iodide

(cc) carbon tetrachloride

(dd) iron(II) perchlorate

(ee) potassium chloride

(ff) ammonium chlorate

(gg) sodium bisulfite

(hh) carbon disulfide

(ii) cupric hydroxide

(jj) boron trifluoride

(kk) mercurous bromide

(ll) zinc hypochlorite

(mm) sodium iodide

(nn) sulfurous acid

(oo) lead(II) fluoride

(pp) acetylene

(qq) sulfur dioxide

(rr) carbon dioxide

(ss) phosphorus pentafluoride

(tt) calcium acetate

(uu) acetic acid

(vv) potassium hydrogen sulfate

(ww) potassium chlorate

(xx) sodium peroxide

(y) carbon monoxide

(z) chromous nitrate

(yy) manganese(II) chloride

(zz) strontium nitrite

Molecular Weights, Formula Weights, and the Mole

4.34 α-Chloroacetophenone is a very potent lacrimator that law enforcement officers discharge as tear gas during riots and other forms of civil unrest. Its chemical formula is C_8H_7OCl. What is the molecular weight of α-chloroacetophenone?

4.35 The substance known in the military as *mustard gas* has been used during combat as a vesicant, a substance that blisters skin and damages the eyes, mucous membranes, and respiratory tract. The most well-known vesicant is the organic compound 2,2′-dichloroethyl sulfide.

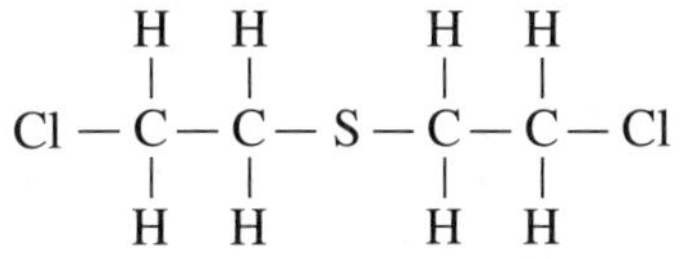

$$Cl - \underset{\underset{H}{|}}{\overset{\overset{H}{|}}{C}} - \underset{\underset{H}{|}}{\overset{\overset{H}{|}}{C}} - S - \underset{\underset{H}{|}}{\overset{\overset{H}{|}}{C}} - \underset{\underset{H}{|}}{\overset{\overset{H}{|}}{C}} - Cl$$

What is the molecular weight of the compound?

4.36 Caproic acid and capric acid are constituents of body perspiration. They are organic compounds having the following chemical formulas:

$$CH_3(CH_2)_4COOH \qquad CH_3(CH_2)_8COOH$$
caproic acid   capric acid

Minute amounts of these and other compounds are detected by bloodhounds when they track the location of fugitives from justice. What are their molecular weights?

4.37 Argon is a monatomic noble gas that can be isolated by liquefying the components of air.

(a) How many moles of argon are contained in 119.98 g of argon?

(b) How many argon molecules are contained in 119.98 g of argon?

4.38 An average penny weighs 3.0 g. Assuming it is composed of 100% copper by mass, determine the number of copper atoms in an average penny.

4.39 When propane burns in the air, 526.3 kcal/mol of heat is released to the environment. How many kcal/g of heat is evolved?

CHAPTER 5 Principles of Chemical Reactions

Once we have properly named substances and written their formulas, we can begin to examine how they interact with one another. Although the chemical reactions of individual hazardous materials will be examined throughout most of the remainder of this textbook, we will note in this chapter some features that are common to all chemical reactions. This includes writing and balancing chemical equations and learning how certain factors affect reaction rates.

In the study of hazardous materials, combustion is a chemical reaction of major concern, especially to fire service personnel. Consequently, we will note what occurs when a substance burns, as well as how other substances effectively function to extinguish fires.

Performance Goals for Section 5.1
- ◆ Describe the nature of a chemical change.
- ◆ Demonstrate a chemical change by means of an equation.

◆ 5.1 THE CHEMICAL REACTION

A substance that has undergone a chemical change is no longer the original substance. In other words, it becomes one or more new substances. To describe a chemical change is to indicate that a substance "reacted" in a particular manner. For instance, we say that "dynamite exploded," "hydrogen burned," "acid corroded metal," or similar expressions. Each of these statements relates to chemical changes that we generally call *chemical reactions*.

In chemistry, it is commonplace to summarize the result of any given reaction in the form of a *chemical equation*. An equation is a shorthand method for expressing a reaction in terms of written chemical formulas. For instance, if we wish to describe the combustion of elemental carbon, the equation may be written as follows:

$$C + O_2 \longrightarrow CO_2$$

TABLE 5.1 ◆ Symbols Used in Chemical Equations

Symbol	Meaning	Examples
(g)[a]	Gaseous reactant or product	$H_2(g), CO_2(g)$
(l)	Liquid reactant or product	$H_2O(l), Br_2(l)$
(s)[b]	Solid reactant or product	$Fe(s), S_8(s)$
(aq)[c]	Reactant or product dissolved in water	$NaCl(aq), KNO_3(aq)$
$(conc)$[d]	Reactant undiluted with water	$HCl(conc)$

[a]An arrow pointing upward ($\uparrow$) is also used when the gas is a product of a reaction.
[b]An arrow pointing downward ($\downarrow$) is also used when a solid precipitates from solution.
[c]Meaning *aqueous.*
[d]Meaning *concentrated.*

The chemical formulas of the substances that enter the chemical reaction are written on the left of the arrow; they are called *reactants*. The formulas of the substances formed as a consequence of the reaction are written on the right of the arrow; they are called *products*. The arrow itself is read as "yields," "produces," "forms," or "gives." Consequently, one way to read this equation is "carbon plus oxygen yields carbon dioxide."

Chemical equations are the basic language of chemistry. They should contain as much information as possible concerning the specific chemical change under consideration. We not only list the formulas of the substances reacting and forming, but sometimes also their physical states under the temperature and pressure conditions of the reaction. Physical states are indicated in parentheses next to the chemical formulas of the reactants and products with the italicized letters *s*, *l*, and *g*, which symbolize solid, liquid, and gas, respectively. Examples of the symbols used in equations are provided in Table 5.1. Using the relevant symbols, the combustion of elemental carbon is denoted by the following equation:

$$C(s) + O_2(g) \longrightarrow CO_2(g)$$

Before a chemical equation can be written, we need to know first the chemical formulas of the reactants and products. To write the more complete form of an equation, we must also know the physical state of the reactants and products. The physical states of reactants and products are determined from chemical reference books or through laboratory experimentation.

Performance Goal for Section 5.2

♦ Describe the process of balancing an equation.

◆ 5.2 BALANCING SIMPLE EQUATIONS

Not only must an equation summarize what occurs qualitatively during a chemical change, but it must also be accountable for other more fundamental observations. In particular, each equation must be written so that the chemical change at issue adheres to the *law of conservation of mass and energy* (Section 2.6). This law states that there is no apparent change in mass during "ordinary" chemical reactions. The term "ordinary" as used here means chemical reactions occurring at the molecular or ionic level, as opposed to the reactions of atomic nuclei.

The conservation of mass and energy requires that during a given chemical change, the atoms of any element are neither created nor destroyed. This means that the number of atoms before and after a reaction remains the same. When chemical equations are written, we make certain that an equal number of atoms for each element have been written on both sides of the arrow. Such an equation is then said to be *balanced*. The simple equation written in Section 5.1 is balanced since it has one carbon atom and two oxygen atoms on each side of the arrow.

Not all equations are directly balanced after writing the chemical formulas of its reactants and products; in fact, they are often unbalanced at this point. We must select a proper coefficient to place *in front of* the appropriate formula so that the equation then becomes balanced. The correct formula of a substance must never be changed when balancing an equation. Furthermore, coefficients are never written in the middle of a formula, such as H_23O.

Most simple equations can be balanced by inspection. Although there are no absolute rules for balancing equations by inspection, the following points are useful when first learning this process:

- Write the correct formula for each reactant and product and separate the reactants from the products with an arrow.
- If known, write the physical state of the reactants and products in parentheses after each formula.
- Choose the formula of the substance containing the greatest number of atoms of an arbitrary element. Insert a number in front of one or both formulas so that the number of atoms for this particular element is balanced.
- If polyatomic ions appear in an equation, balance them as single units only when they retain their identity on both sides of the arrow.
- Balance any remaining atoms or ions.

Let's consider an example. Suppose you wish to write the balanced chemical equation for the reaction that occurs when methane burns in air to form carbon dioxide and water vapor. This is an ordinary combustion reaction involving the chemical combination of methane and atmospheric oxygen. First, we write the chemical formulas of the reactants and products. The formula of methane is CH_4 (from Table 4.8), oxygen is O_2, carbon dioxide is CO_2, and water is H_2O. Under the reaction conditions, each is a gas or vapor. Hence, we initially write the following:

$$CH_4(g) + O_2(g) \longrightarrow CO_2(g) + H_2O(g)$$

Next, we note that this is an unbalanced equation because there are more hydrogen atoms in CH_4 than in H_2O. We balance the hydrogen atoms by inserting a 2 in front of the formula for water, as follows:

$$CH_4(g) + O_2(g) \longrightarrow CO_2(g) + 2H_2O(g)$$

Since there are no ions in this equation, we next balance the number of the oxygen atoms. This is accomplished by inserting a 2 in front of O_2, as follows:

$$CH_4(g) + 2O_2(g) \longrightarrow CO_2(g) + 2H_2O(g)$$

When such exercises are performed, it is usually best to make one final check: one carbon atom, four hydrogen atoms, and four oxygen atoms are on each side of the arrow.

Performance Goals for Section 5.3
- ◆ Identify the different types of simple chemical reactions.
- ◆ Write equations for each type.

◆ **5.3 TYPES OF CHEMICAL REACTIONS**

By now, one point should be apparent: an equation denoting the chemical reaction of a hazardous material cannot be written when the products of the reaction are unknown. Although identifying the reaction products is not always a simple feat, they can frequently be determined by knowing the reaction type. The reactions in which we are interested can be classified as one of four types reviewed below with illustrative examples.

COMBINATION OR SYNTHESIS REACTIONS

In a *combination* or *synthesis reaction,* two or more simpler substances combine to form a more complex substance. Some examples of such reactions are illustrated by the following equations:

$$2H_2(g) + O_2(g) \longrightarrow 2H_2O(l)$$
hydrogen   oxygen   water

$$2Na(s) + Cl_2(g) \longrightarrow 2NaCl(s)$$
sodium   chlorine   sodium chloride

$$C(s) + O_2(g) \longrightarrow CO_2(g)$$
carbon   oxygen   carbon dioxide

DECOMPOSITION REACTIONS

In a *decomposition reaction*, a relatively complex substance is broken down into several simpler substances. Some examples of decomposition reactions are illustrated by the following equations:

$$2H_2O(l) \longrightarrow 2H_2(g) + O_2(g)$$
water   hydrogen   oxygen

$$Na_2CO_3(s) \longrightarrow Na_2O(g) + CO_2(g)$$
sodium carbonate   sodium oxide   carbon dioxide

$$(NH_4)_2Cr_2O_7(s) \longrightarrow Cr_2O_3(s) + N_2(g) + H_2O(g)$$
ammonium dichromate   chromium(III) oxide   nitrogen   water

REPLACEMENT OR DISPLACEMENT REACTIONS

In a *replacement* or *displacement reaction*, an element and a compound react so that the free element replaces an element in the compound. Some examples of replacement reactions are illustrated by the following equations:

$$Mg(s) + 2HCl(aq) \longrightarrow MgCl_2(aq) + H_2(g)$$
magnesium   hydrochloric acid   magnesium chloride   hydrogen

$$Cu(s) + 2AgNO_3(aq) \longrightarrow 2Ag(s) + Cu(NO_3)_2(aq)$$
copper   silver nitrate   silver   copper(II) nitrate

$$2KI(aq) + Cl_2(g) \longrightarrow KCl(aq) + I_2(s)$$
potassium iodide   chlorine   potassium chloride   iodine

DOUBLE REPLACEMENT OR DOUBLE DISPLACEMENT REACTIONS

In a *double replacement reaction*, there is an exchange of the positively charged ions in two compounds. Some examples of this type of chemical reaction are illustrated by the following equations:

$$2NaCN(s) + H_2SO_4(aq) \longrightarrow Na_2SO_4(aq) + 2HCN(g)$$
sodium cyanide sulfuric acid sodium sulfate hydrogen cyanide

$$PbS(s) + 2HCl(aq) \longrightarrow PbCl_2(aq) + H_2S(aq)$$
lead(II) sulfide hydrochloric acid lead(II) chloride hydrogen sulfide

$$Na_2SO_4(aq) + BaCl_2(aq) \longrightarrow BaSO_4(s) + 2NaCl(aq)$$
sodium sulfate barium chloride barium sulfate sodium chloride

Solved Exercise 5.1

Potassium bicarbonate is a useful dry-chemical fire extinguisher. When heated, it produces potassium carbonate, water, and carbon dioxide. Write the balanced chemical equation for this reaction.

Solution: First, it is necessary to write the chemical formulas for the reactant and products associated with this chemical reaction. Following the methods learned in Chapter 3, the chemical formulas of the four relevant compounds are written as follows:

potassium bicarbonate, $KHCO_3(s)$ potassium carbonate, $K_2CO_3(s)$
water, $H_2O(g)$ carbon dioxide, $CO_2(g)$

The physical states of these substances are denoted parenthetically. They are known from experience.

These formulas can now be used to write an unbalanced equation representing the chemical reaction.

$$KHCO_3(s) \longrightarrow K_2CO_3(s) + H_2O(g) + CO_2(g)$$

Finally, we must balance the equation. Since oxygen atoms are more abundant in this equation than any other type of atom, we begin by balancing the number of oxygen atoms. Initially, there are three atoms of oxygen on the left side of the arrow, but six on the right side $(3 + 1 + 2)$. Balance oxygen by inserting a "2" in front of the formula for potassium bicarbonate. The "2" also balances the number of potassium atoms on each side of the arrow. The equation now looks as follows:

$$2KHCO_3(s) \longrightarrow K_2CO_3(s) + H_2O(g) + CO_2(g)$$

Performing a final check on the number of atoms, we see that there are two atoms of potassium, two atoms of hydrogen, two atoms of carbon, and six atoms of oxygen on each side of the arrow. Hence, this equation is now balanced.

Solved Exercise 5.2

Hydrogen cyanide, a highly toxic gas, is produced when hydrochloric acid is mixed with barium cyanide. Identify this type of chemical reaction, and write a balanced equation illustrating the reaction.

Solution: The names of three substances are mentioned in this exercise: hydrogen cyanide, hydrochloric acid, and barium cyanide. Their chemical formulas are HCN, HCl, and $Ba(CN)_2$, respectively. Using them, the following partial equation is written:

$$Ba(CN)_2(s) + HCl(aq) \longrightarrow \underline{\hspace{1cm}} + HCN(g)$$

It is apparent that the nature of the reaction involves an exchange of ions. Consequently, this reaction is an example of a double replacement or double displacement reaction.

Knowing the reaction type, we can now identify the remaining compound as barium chloride. The chemical formula of barium chloride is $BaCl_2$. The unbalanced equation now looks as follows:

$$Ba(CN)_2(s) + HCl(aq) \longrightarrow BaCl_2(aq) + HCN(g)$$

In this form, there is an unequal number of chloride and cyanide ions on each side of the arrow. To overcome this obstacle, we write a "2" in front of the formulas for hydrochloric acid and hydrogen cyanide, as follows:

$$Ba(CN)_2(s) + 2HCl(aq) \longrightarrow BaCl_2(aq) + 2HCN(g)$$

Equal numbers of atoms are now present on each side of the arrow: one barium ion, two cyanide ions, two hydrogen ions, and two chloride ions. This equation is balanced.

Performance Goals for Section 5.4
- ◆ Describe the phenomena of oxidation and reduction.
- ◆ For a given chemical reaction, identify the oxidizing agent, the reducing agent, the substance oxidized, and the substance reduced.

◆ 5.4 OXIDATION–REDUCTION REACTIONS

Chemists also classify a chemical process in terms of whether it represents an oxidation–reduction reaction, frequently called a *redox* reaction. Combination, decomposition, and simple replacement reactions involve oxidation–reduction processes, whereas double replacement reactions do not. Although we shall study redox reactions in more depth in Chapter 11, a basic understanding is now required since we will encounter them frequently.

OXIDATION

Oxidation is any of the following processes:

Elements and compounds *oxidize* when they gain oxygen atoms. When a compound is oxidized, each type of atom within the compound combines with oxygen. For example, carbon, hydrogen, and methane oxidize by combining with oxygen.

$$C(s) + O_2(g) \longrightarrow CO_2(g)$$

carbon   oxygen   carbon dioxide

$$2H_2(g) + O_2(g) \longrightarrow 2H_2O(l)$$

hydrogen   oxygen   water

$$CH_4(g) + 2O_2(g) \longrightarrow CO_2(g) + 2H_2O(g)$$

methane   oxygen   carbon dioxide   water

Compounds also *oxidize* when they lose hydrogen atoms. When methanol decomposes, for instance, formaldehyde and hydrogen form.

$$CH_3OH(g) \longrightarrow HCHO(g) + H_2(g)$$

methanol   formaldehyde   hydrogen

Since methanol loses hydrogen atoms, it is said to be *oxidized*.

A compound *oxidizes* when it becomes less affiliated with its electrons. For ionic substances, this is accomplished by the *loss* of one or more electrons.

$$Na(s) \longrightarrow Na^+(aq) + e^-$$

$$Mg(s) \longrightarrow Mg^{2+}(aq) + 2e^-$$

$$Cu(s) \longrightarrow Cu^{2+}(aq) + 2e^-$$

$$Fe^{2+}(aq) \longrightarrow Fe^{3+}(aq) + e^-$$

$$2Cl^-(aq) \longrightarrow Cl_2(g) + 2e^-$$

In the first three examples, neutral atoms of sodium, magnesium, and copper, respectively, lose either one or two electrons as indicated and become positively charged ions; in the fourth example, the iron(II) ion loses an electron and becomes the iron(III) ion; and in the fifth example, each of two chloride ions loses an electron to form a neutral molecule of chlorine.

REDUCTION

Oxidation is always associated with the accompanying process called *reduction*. Any one of the following processes constitutes reduction:

Compounds *reduce* when they lose oxygen atoms. For example, when sodium perchlorate is heated, it loses oxygen atoms.

$$NaClO_4(s) \longrightarrow NaCl(s) + 2O_2(g)$$

sodium perchlorate   sodium chloride   oxygen

Sodium perchlorate is *reduced*.

Compounds *reduce* when they gain hydrogen atoms. For example, the organic compound ethene combines with hydrogen to become ethane.

$$C_2H_4(g) + H_2(g) \longrightarrow C_2H_6(g)$$

ethene   hydrogen   ethane

Since it gains hydrogen atoms, ethene is *reduced*.

Substances reduce when they become more affiliated with electrons. For ionic systems, reduction is associated with the *gain* of electrons.

$$Cl_2(g) + 2e^- \longrightarrow 2Cl^-(aq)$$

$$S_8(s) + 16e^- \longrightarrow 8S^{2-}(aq)$$

$$Fe^{3+}(aq) + e^- \longrightarrow Fe^{2+}(aq)$$

$$Fe^{2+}(aq) + 2e^- \longrightarrow Fe(s)$$

In the first two examples, neutral elements gain electrons and form negative ions; in the third example, the iron(III) ion gains an electron and becomes the iron(II) ion; and

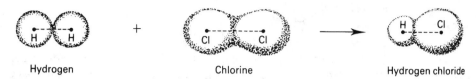

Hydrogen   Chlorine   Hydrogen chloride

FIGURE 5.1 ◆ When the oxidation–reduction phenomenon occurs between covalently bonded substances, electrons are not completely transferred from one reacting species to the next. In the hydrogen and chlorine molecules, shown to the
left of the arrow, the electron pairs are mutually shared between the like atoms;
but in the hydrogen chloride molecule, shown to the right of the arrow, the electronic distribution is asymmetric about the center. This partial loss and gain of
electron density is typical of the oxidation–reduction phenomenon between all
covalent substances.

in the final example, the iron(II) ion gains two electrons and becomes an atom of elemental iron. The molecules and ions on the left of these arrows are said to be *reduced.*

Oxidation and reduction also occur in covalent systems, but here, a total transference of electrons does not occur. For instance, consider the chemical reaction represented by the combination of hydrogen and chlorine.

$$\underset{\text{hydrogen}}{H_2(g)} + \underset{\text{chlorine}}{Cl_2(g)} \longrightarrow \underset{\text{hydrogen chloride}}{2HCl(g)}$$

In the hydrogen and chlorine molecules, the electron pairs in the covalent bonds are
shared equally by their respective atoms. In the hydrogen chloride molecule, however,
the chlorine atom shares the pair of bonding electrons to a greater degree than the
hydrogen atom. This unequal sharing of the electron pair is illustrated in Figure 5.1. It
causes an unsymmetrical electron distribution in the molecule of hydrogen chloride.
This unsymmetrical distribution of electrons is typical of oxidation in covalent systems. Hydrogen has been oxidized and chlorine has been reduced.

In either ionic or covalent systems, the substances oxidized are called *reducing
agents,* and the substances reduced are called *oxidizing agents,* or *oxidizers.* These names
result from the effect the agent has on other substances. In the combination of hydrogen and chlorine, chlorine is the oxidizing agent and hydrogen is the reducing agent.

Consider another example. Decades ago, cameras used flashbulbs to send out a
brilliant blaze to lighten a darkened scene. The brilliance was associated with a chemical
reaction in which metallic magnesium burned to form magnesium oxide.

$$\underset{\text{magnesium}}{Mg(s)} + \underset{\text{oxygen}}{O_2(g)} \longrightarrow \underset{\text{magnesium oxide}}{2MgO(s)}$$

During this reaction, a magnesium atom loses two electrons to become a magnesium ion.
It also combines with oxygen. For both reasons, magnesium is *oxidized.* Each atom of
the oxygen molecule gains two electrons, and each becomes an oxide ion. The oxygen is
reduced. Magnesium is the *reducing agent* and oxygen is the *oxidizing agent,* or *oxidizer.*

Performance Goals for Section 5.5
   ◆ Describe the rate of reaction.
   ◆ Identify the seven factors that affect the rate at which a chemical reaction occurs.
   ◆ Describe the impact each factor has upon the rate of reaction.

◆ 5.5 FACTORS AFFECTING THE RATE OF REACTION

Each chemical reaction occurs at a definite speed called its *rate of reaction*. Sometimes the rate of reaction is referenced to a correlating chemical phenomenon. This results in the use of terms such as the combustion rate, corrosion rate, or explosive rate. Chemists establish these rates of reaction by experimentally noting the change in concentration of a reactant or product over time.

The speed at which a given substance undergoes a chemical change is often associated with its hazardous nature. This is clearly illustrated by the detonation of nitroglycerin. Several grams of nitroglycerin will completely decompose within a millionth of a second. Fortunately, not all chemical reactions occur as rapidly as the explosion of nitroglycerin. Otherwise, we would have even greater problems when responding to emergencies involving hazardous materials. The rate of reaction depends on at least seven factors, each of which will be discussed independently. When appropriate, the influence of each factor shall be noted as it bears upon the rate of combustion.

NATURE OF THE MATERIAL

Some substances do not burn at all. Examples are water, carbon dioxide, nitrogen, and the noble gases. Other substances, like wood, paper, hydrogen, magnesium, and sulfur, do not begin burning until they are first exposed to a spark, flame, or other source of ignition. Still other substances burn spontaneously, even without exposure to an ignition source. An example is elemental phosphorus, which, when exposed to the air, bursts into flame. These rates of combustion vary from zero to some finite value. It is their individual chemical nature that causes some substances not to burn at all, others to burn only when kindled, and still others to burn spontaneously.

SUBDIVISION OF THE REACTANTS

Wooden logs do not burn spontaneously. Initially, they must first be kindled, perhaps by the heat generated from the burning of smaller pieces of wood. By contrast, when the dust from the same type of wood is dispersed or suspended in the air and exposed to an ignition source, it is likely to ignite spontaneously and burn with explosive violence throughout its entire mass. The phenomenon is referred to as a *dust explosion*.

Sawdust, coal dust, grain dust, flour dust, and cotton lint are examples of combustible materials that can be dispersed in the air. Dust explosions involving these materials are well-acknowledged phenomena. The static electricity generated as one particle circulates among the others can serve as a source of ignition.

Why does the dust of a combustible material burn explosively, whereas bulk quantities of the same materials must be kindled before they burn? The answer to this question is associated with particle size. The smaller the particle, the greater the total exposed surface area of a given mass. This means that as particle size decreases, more molecules become available to react.

More molecules are also available to react in configurations that provide an increased surface area. Figure 5.2 shows that a flammable liquid burns fastest in the vessel that allows it to assume the greatest surface area. In general, anytime the surface area of a given substance is increased, the substance reacts at an increased rate. This occurs because its molecules are not internally bound to each other and are more free to react with the molecules of neighboring substances.

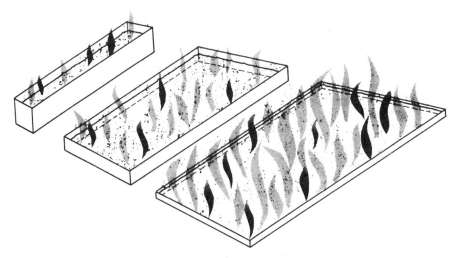

FIGURE 5.2 ◆ The same amount of a flammable liquid has been added to three containers of progressively increasing size. When ignited at the same instant, the liquid burns at the fastest rate in the container that provides the largest surface area.

STATE OF AGGREGATION

The rate at which a substance reacts is also affected by its physical state of matter. This is particularly true for the rate of combustion. Generally, reactions involving gases occur much faster than reactions involving liquids or solids. These differences in reaction rates are affected by the nature of the gaseous, liquid, and solid states of matter. Molecules of gaseous substances are relatively far apart and exert small attractive forces upon each other. This allows diffusion to occur very rapidly. But, in the liquid and solid states of matter, the particles are in contact and held tightly together. This hinders their likelihood of encountering other particles and reduces their reaction rates.

CONCENTRATION OF REACTANTS

Before a chemical reaction occurs, the particles that make up the structure of the reactants must contact each other. On the other hand, particle contact does not signify that a reaction will necessarily occur. The probability of particles contacting each other increases as the number of particles in a given volume increases. In other words, if different concentrations of the same reactants are put into two vessels, the rate of the reaction is generally faster in the vessel containing the greater concentration of reactants.

Imagine that we have four containers of equal volume holding different numbers of two hypothetical molecules, A and B (Figure 5.3). What is the relative number of collisions between unlike molecules? (We ignore the collisions between like molecules, such as A contacting A or B contacting B, since these collisions do not cause a chemical reaction.)

Let's consider each container separately. The first container holds one molecule of A and one molecule of B, while the second container holds two molecules of A and one molecule of B. It follows that the likelihood of an A molecule colliding with a B molecule is twice as great in the second container as compared with the first one. In the third container, which holds two molecules of A and two molecules of B, the probability of collision between unlike molecules is increased to four times the possibility

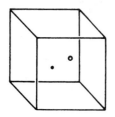 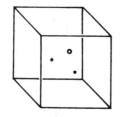 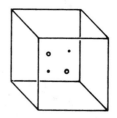 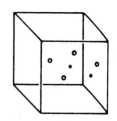

FIGURE 5.3 ◆ The probability of a chemical reaction increases as the number of reactant molecules confined to a container increases. The molecules of the reactants A and B are designated as dots and open circles, respectively.

in the first container and two times that of the second one. Finally, in the fourth container, which holds two molecules of A and four molecules of B, the probability of unlike molecules is increased to eight times that observed in the first container. This shows that an increase in the concentration of the reactants causes the rate of a given reaction to increase.

We also noted in Section 3.1 that the concentration of the reactants has an important bearing on the rate at which combustion occurs. Unless the concentration of flammable gases or the vapors of flammable liquids is within the flammable range of a given material, the material does not burn.

The concentration of atmospheric oxygen also affects the rate of combustion. In clean air at sea level, the concentration of oxygen is about 21% by volume. The majority of the remaining 78% by volume is nitrogen gas. Atmospheric nitrogen serves as a diluent during the combustion of materials in the air and retards their combustion rates. In an atmosphere of pure oxygen, combustion always occurs with an increased intensity. Some substances that are ordinarily stable in the air could burn spontaneously in an atmosphere of pure oxygen.

ACTIVATION ENERGY

While there are some substances that react spontaneously upon contact, most must be supplied with a minimum amount of energy before they chemically react. Consider pieces of combustible solids lying exposed to atmospheric oxygen. They do not normally burst into flame. On the other hand, they readily ignite when exposed to an energy source. This minimum amount of energy that must be supplied to initiate a chemical reaction is called the *activation energy*.

Before the combustion of a solid substance can occur, heat must ordinarily be supplied. The heat raises the temperature of the substance to its *kindling point*, the temperature at which the burning is sufficiently rapid to proceed without the need for additional heat from an external source. In exothermic reactions, the activation energy does not need to be continuously supplied as the reaction proceeds, since it is replaced by the energy released in the process. Once combustion has been initiated, it continues until either the material or the oxygen is exhausted.

An amount of activation energy must also be supplied to initiate endothermic reactions. In this case, the chemical phenomenon ceases if the source of energy is removed. We must supply not only the activation energy, but enough additional energy to replace the energy absorbed by the reactants. Consider *electrolysis*, the process of decomposing a substance through use of an electric current. Electrolysis is an endothermic phenomenon. When we disconnect the current, electrolysis stops.

TEMPERATURE

Heating a mixture of reactants causes its particles to move more rapidly, and this motion increases the probability that the particles will collide. As the speed of the particles increases, the temperature rises accordingly. Thus, a reaction rate increases with a rise in temperature, because more molecules become sufficiently activated at higher temperatures.

According to a classic rule in chemistry, the rate of a chemical reaction doubles for every rise of 18°F (10°C). Thus, twice as many molecules are activated when the temperature of the system has been increased by 18°F (10°C); four times as many are activated when the temperature is increased 36°F (20°C); and eight times as many for a 54°F (30°C) increase in temperature.

CATALYSIS

A *catalyst* is a substance that affects the rate of a reaction and appears to remain unchanged. A familiar example of a catalyzed reaction is the rusting of iron in the presence of atmospheric water vapor. Iron exposed to moisture corrodes much faster than iron stored in a dry atmosphere. An additional example of a catalyzed reaction is the combustion of hydrogen in the presence of platinum. A mixture of hydrogen and oxygen can be kept in a vessel for years without reacting to any noticeable degree; however, if a small amount of platinum is introduced into this vessel, the mixture reacts explosively. At the end of the reaction, hydrogen and oxygen form water while the platinum appears unchanged.

During a given reaction, the catalyst itself could undergo a chemical change, but then react again and return to its original condition. In the end, its overall chemical identity is not lost. A catalyst does not always undergo a chemical change, but could be altered physically by the particles of one or more reactants adhering to its surface.

Solved Exercise 5.3

In rural regions, propane is frequently more cost-effective to use for heating than other fuels. Propane is also popular domestically for cooking food on portable grills. When propane burns completely, carbon dioxide and water vapor are produced as combustion products. Write a balanced equation representing the complete combustion of propane.

Solution: First, we require the chemical formulas of the reactants and products. The chemical formula of propane was provided in Table 4-8 as C_3H_8. The chemical formulas of elemental oxygen, carbon dioxide, and water vapor are O_2, CO_2, and H_2O, respectively. The unbalanced equation representing the combustion of propane is then written as follows:

$$C_3H_8(g) + O_2(g) \longrightarrow CO_2(g) + H_2O(g)$$

As written, there are three atoms of carbon, eight atoms of hydrogen, and two atoms of oxygen on the left side of the arrow, but only one atom of carbon, two atoms of hydrogen, and three atoms of oxygen on the right side of the arrow. To balance this equation, we write a 3 in front of the formula for carbon dioxide. This provides three atoms of carbon on each side of the arrow, but the number of hydrogen and oxygen atoms still remains unbalanced. Hence, we insert a 4 in front of the formula for water. Now, there are eight hydrogen atoms on each side of the arrow, but the number of oxygen atoms remains unbalanced: 2 on the left side and 10 on the right side of the arrow. To overcome this problem, we insert a 5 in front of the formula for oxygen, and the equation is balanced in the following final form:

$$C_3H_8(g) + 5O_2(g) \longrightarrow 3CO_2(g) + 4H_2O(g)$$

Performance Goals for Section 5.6
- ◆ Describe the ordinary combustion of a substance in air.
- ◆ Distinguish between complete and incomplete combustion.
- ◆ Describe the nature of chemical energy and the heat of combustion.
- ◆ Identify the common commercial fuels.

◆ 5.6 THE COMBUSTION PROCESS

Combustion is a chemical reaction that releases energy to the surroundings as heat and light. The light is ordinarily visible to the naked eye and observable as flames. The flames appear during a combustion reaction when minute particles of the fuel or combustion products are heated to incandescence. When combustion is referred to in everyday practice, we say that something is "burning" or "on fire."

Combustion is not always accompanied by the presence of visible flames. Hydrogen and methanol are examples of substances that burn with nearly imperceptible flames. The energy released to the surroundings is still heat and light, but the light is outside the visible range. To locate flames when such substances are leaking from storage and transport vessels during emergency-response actions, experts recommend the use of a thermal-imaging camera (Figure 5.4).

Two or more substances chemically unite during combustion. Since one of them is typically oxygen, we often hear that a substance *oxidizes* when it burns. There are also combustion reactions that involve the union of one substance with a substance that is not oxygen. Elemental phosphorus, for example, burns in an oxygen environment, but it also burns in a chlorine environment. Although oxygen is not involved in the latter process, we still indicate that the phosphorus *oxidizes* when it combines with chlorine.

Combustion manifests itself as fire when an oxidation occurs relatively fast. For this reason, combustion is often described as a rapid oxidation. Combustion is also a self-sustaining reaction; that is, unless the process is intentionally extinguished, combustion continues until the concentration of the substance falls below a minimum value.

The substance that burns is called the *fuel*. When a fuel burns, its atoms are never destroyed. Instead, they unite with the atoms of other substances to form one or more new substances. Since a substance typically unites with atmospheric oxygen when it burns, the products of the combustion are often metallic or nonmetallic oxides. Oxygen itself does not burn; it is said to *support* combustion.

Many flammable and combustible substances contain carbon in their molecular framework. When they burn in air, the constituent carbon atoms become carbon mon*oxide* or carbon di*oxide*. When carbon dioxide forms, the combustion is said to be *complete*, but when carbon monoxide forms, we regard the process as *incomplete*,

FIGURE 5.4 ◆ Emergency responders use a thermal-imaging camera like this Stealth-IR camera to locate objects that are not visible to the naked eye. The camera may also be used to locate points on storage and transport vessels from which vapor escapes and burns with nearly imperceptible flames. *Courtesy of Sierra Pacific Innovations Corporation, Las Vegas, Nevada.*

or *partial oxidation*. We represent the incomplete and complete combustion of carbon as follows:

$$2C(s) + O_2(g) \longrightarrow 2CO(g)$$
carbon   oxygen   carbon monoxide

$$C(s) + O_2(g) \longrightarrow CO_2(g)$$
carbon   oxygen   carbon dioxide

Incomplete combustion can be a relatively complicated phenomenon. It often occurs when items smolder or burn slowly without a visible flame. During the incomplete combustion of carbon-containing fuels, some carbon atoms do not even unite with oxygen. When this occurs, elemental carbon is produced, which disperses in the air as the tiny particulates of smoke.

In Section 2.6 we noted that chemical energy is present in each substance. This energy is composed of two types:

- Some energy is present with a substance by virtue of the motion of its atoms and molecules.
- Energy is also stored in a substance as the result of its unique structure. This energy is associated with the strength of its chemical bonds. During any chemical reaction, some bonds break, new ones form, and any excess energy is either absorbed or released into the environment. This is the source of the energy released during combustion.

TABLE 5.2 ◆ Heat Values of Some Common Fuels

Fuel	Btu/lb	kJ/kg
Hydrogen	61,600	143,000
Methane (natural gas)	24,100	56,000
Acetylene	21,600	50,200
Propane	21,500	50,000
Diesel fuel	20,700	48,000
Ethanol	12,900	30,000
Methanol	9900	23,000
Wood	8500	19,700

When combustion occurs, heat is evolved to the surroundings. It is called the *heat of combustion*. Heats of combustion are generally noted in an energy unit per gram, pound, or mole of substance burned, for example, Btu/lb, kcal/g, kJ/kg, or kJ/mol. Materials that burn and evolve an amount of heat greater than approximately 5000 Btu/lb (11,600 kJ/kg) are useful commercial and industrial fuels. Some common fuels and their heats of combustion are listed in Table 5.2. These limited data indicate that hydrogen outranks other fuels from the viewpoint of the amount of energy liberated per unit mass. We shall note the use of hydrogen as a fuel in Section 7.2.

Chemists sometimes denote the complete combustion of methane by an equation like the following:

$$CH_4(g) + 2O_2(g) \longrightarrow CO_2(g) + 2H_2O(g) + 24{,}100 \text{ Btu/lb (56,000 kJ/kg)}$$

methane   oxygen   carbon dioxide   water

Methane is the fuel and oxygen is the oxidizing agent. When it burns, the constituent carbon and hydrogen atoms in each molecule of methane become molecules of carbon dioxide and water vapor, both of which are routinely discharged into the atmosphere. Chemical bonds are broken and new bonds are formed. The amount of heat released during the combustion of methane is 24,100 Btu/lb, or 56,000 kJ/kg.

▪▪

Performance Goal for Section 5.7
   ◆ Identify the conditions necessary for the occurrence of spontaneous combustion.

▪▪

◆ 5.7 SPONTANEOUS COMBUSTION

Some substances undergo oxidation so slowly that their fires are initially imperceptible. When the oxidation occurs within a confined space where the circulation of the air is poor, these substances often absorb their own heat of reaction. The continued absorption ultimately heats these substances to their autoignition temperatures, at which they self-ignite. Since a source of ignition other than the heat of reaction is absent, the process is called *spontaneous combustion*.

FIGURE 5.5 ◆ Oily rags and improperly stored linseed oil, varnishes, lacquers, and other oil-based paint products are among the materials likely to undergo spontaneous combustion. Each year, millions of dollars in property are needlessly lost because of fires that originated from such sources.

There are flammable and combustible animal and vegetable oils that oxidize slowly. An example is linseed oil, which was formerly a component of commercial paints. As demonstrated in Figure 5.5, animal and vegetable oils should be regarded as potential sources of spontaneous combustion. The improper storage or disposal of rags dampened with linseed oil is a well-acknowledged circumstance in which spontaneous combustion is probable. When such rags are negligently tossed into a pile in a broom closet, for instance, the heat cannot readily dissipate to the surroundings. Instead, the rags absorb the heat; before long, the oil bursts into flame. Since the availability of oxygen is limited, a thick, black smoke usually accompanies these slowly burning fires.

Spontaneous combustion can also occur within stacks of undried agricultural products, such as damp hay and grass. Microorganisms proliferate in these moist materials, and their physiological activities generate heat. This microbiological activity is known as *thermogenesis*. It is supplemented by chemical oxidation until the temperature rises to approximately 160°F (71°C), at which the microorganisms can no longer survive. Biological activity ceases, but chemical oxidation continues. When heat evolves faster than it can dissipate, the internal temperature increases and the product self-ignites. Since the burning product is damp, evaporating water typically evolves from the fire as billows of white smoke.

Performance Goals for Section 5.8
- Describe the greenhouse effect.
- Identify the principal greenhouse gas.
- Describe the association between the greenhouse effect and global warming.
- Describe the intent of the Kyoto protocol.

◆ 5.8 THE GREENHOUSE EFFECT

Our planet is continuously warmed by the radiant energy that enters our atmosphere from the sun. The Earth's surface absorbs this energy, some of which is then emitted back into outer space. Over the past 500,000 years, the rates at which the radiation was absorbed and emitted have been balanced on the average. This balance between the absorption and emission rates produced an average surface temperature fluctuating between 66°F (19°C) and 80°F (27°C).

Today, however, the average temperature is at the high point of this range. Scientists fear the temperature of the Earth's surface may rise even higher, because the absorbed energy is now hindered from effectively radiating back into outer space. This interference is caused in part by an elevated concentration of carbon dioxide in the atmosphere compared to the concentration in the past. Carbon dioxide is released into the atmosphere each time a substance burns. While some carbon dioxide is absorbed into the oceans and other bodies of water, most remains unaltered in the atmosphere. This means that carbon dioxide has been regularly accumulating in Earth's atmosphere in ever-increasing amounts, especially since industrialization began during the 19th century.

Carbon dioxide possesses the comparatively unusual capability of absorbing the energy that would otherwise pass into space. After it has absorbed the energy, carbon dioxide then re-emits it to Earth instead of into outer space. This process has upset the conventional balance between the rates of radiation absorption and emission, which in turn has caused the planet to warm. Using computer modeling, researchers predict that the atmospheric accumulation of carbon dioxide might cause Earth to warm roughly 3 to 9 degrees Fahrenheit (2 to 5 degrees Celsius) over the next century.

The phenomenon associated with increased global warming has been compared to the glass walls of a greenhouse. Accordingly, it is called the *greenhouse effect*. A greenhouse traps radiant energy and reduces the rate at which it dissipates to the outside environment, thereby keeping the enclosure warmer than the surroundings. Each gas that contributes to global warming—such as carbon dioxide—is called a *greenhouse gas*.

Many scientists believe we have been experiencing the global impact of the greenhouse effect for some time. They point to polar glaciers that are melting and sea levels that are rising. Although we are unable to accurately predict the localized impacts of the greenhouse effect, it seems plausible that until methods have been implemented to reduce the atmospheric concentration of the greenhouse gases, severe droughts and wild fires will occur more frequently than in the past—especially in some areas of the world. As we advance into the 21st century, the need to control the spread of these untamed fires is likely to be greater than at any known time in the past.

The greenhouse gases are primarily produced by the industrialized nations of the world. The United States alone produces approximately one-fourth of the world's greenhouse gases. In an attempt to curb the impact of the greenhouse effect, the United Nations spearheaded a policy aimed at producing a worldwide reduction in greenhouse gas emissions among its industrialized members. In the international agreement called the *Kyoto protocol,* the United States was asked to reduce greenhouse gas emissions 7% below the 1990 level by the year 2012.

The concentration of greenhouse gases in the United States is directly linked to numerous ongoing industrial processes, the combination of which affects the American economy. For this reason, the United States did not ratify the protocol. Rather than agreeing to set specific goals for greenhouse gas reductions, the United States elected to develop its own program by correlating greenhouse-gas-emission reduction with economic output and intensifying a search for technologies that reduce or control the buildup of carbon dioxide in the atmosphere.

Global warming is a contested scientific and political topic. It is clear that restraining its consequences will continue to be an ongoing and challenging international effort for some time.

Performance Goals for Section 5.9
- ◆ Describe the concept of reactivity as it is used by EPA in RCRA regulations.
- ◆ Determine whether a chemical waste exhibits the RCRA characteristic of reactivity.

◆ 5.9 THE RCRA CHARACTERISTIC OF REACTIVITY

The RCRA regulations (Section 1.5) denote a material as a hazardous waste if it exhibits one or more characteristics, one of which is reactivity. A material exhibits the characteristic of *reactivity* when it conforms with one or more of the following conditions:

- ◆ It readily undergoes violent change without detonating.
- ◆ It reacts violently with water.
- ◆ It forms potentially explosive mixtures with water.
- ◆ It is a cyanide- or sulfide-bearing substance that, when exposed to acid or alkaline solutions, generates toxic gases, vapors, or fumes in a quantity sufficient to present a danger to human health or the environment.
- ◆ It is capable of detonation or an explosive reaction if it is subjected to a strong initiating source or if it is heated under confinement.
- ◆ It is readily capable of detonation, explosive decomposition, or reaction at standard temperature and pressure.
- ◆ It is a forbidden explosive as this term is used in the DOT regulations (Section 14.4).

A waste that exhibits the characteristic of reactivity is a hazardous waste and is subject to EPA's treatment, storage, and disposal regulations. When a waste exhibits the characteristic of reactivity, it is assigned the hazardous waste number "D003."

Performance Goals for Section 5.10
- ◆ Describe the combustion process by using the fire triangle and the fire tetrahedron.
- ◆ Describe the nature of a free radical.
- ◆ Demonstrate the importance of free radicals in the combustion process.

◆ 5.10 THE FIRE TRIANGLE AND FIRE TETRAHEDRON

Decades ago, we visually represented the combustion process by using the *fire triangle* in Figure 5.6. Each leg of the triangle was assigned a factor required for combustion. The intent of this model was to display the process by noting the interplay

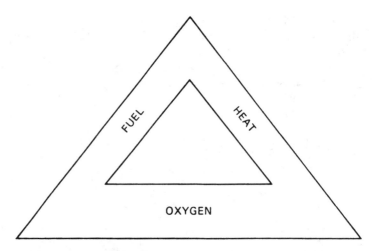

FIGURE 5.6 ◆ The fire triangle. The self-sustenance of an ordinary fire was once considered possible only when ample fuel, oxygen, and heat simultaneously existed, the three components of this triangle. Today, however, this concept has been broadened into the fire tetra-hedron shown in Fig. 5.7, which includes a fourth component, free radicals.

between fuel, oxygen, and heat. Although scientists initially believed that combustion could be described by the use of only these factors, it soon became apparent that the model could not adequately account for everything scientists knew about combustion.

Scientists had also discovered that a fourth factor—the presence of free radicals—was important during combustion. This meant that a four-sided geometrical figure was needed to visually depict the four components. For this purpose, a *fire tetrahedron* is used. Each face of the fire tetrahedron in Figure 5.7 is assigned an element of the combustion process. The message conveyed by this figure is that combustion occurs when adequate fuel, oxygen, heat, and free radicals are simultaneously present.

FIGURE 5.7 ◆ The fire tetrahedron, the four components of which are fuel, oxygen, heat, and free radicals. An ordinary fire is self-sustained only when there is simultaneously ample fuel, oxygen, heat, and free radical propagation. Furthermore, fires are extinguished by removing any one of these components from the other three.

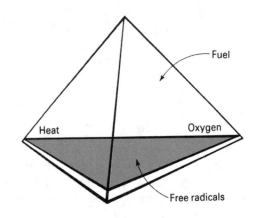

A *free radical* is a molecule that has an unpaired electron. It forms when a covalent bond is broken. Consider the methane molecule. When one of its four chemical bonds is broken or split, we represent the resulting structure as follows:

$$
\begin{array}{c}
\text{H} \\
| \\
\text{H}-\text{C} \cdot \\
| \\
\text{H}
\end{array}
$$

This species is the *methyl free radical.* The dot represents the unpaired electron.

Once produced, a free radical rapidly seeks out an atom, a molecule, or another free radical. Since it possesses an unpaired electron, a free radical is an extraordinarily reactive, short-lived species. Nonetheless, during its transient existence, a free radical undergoes many kinds of chemical reactions, which occur in the following four ways:

- Donating an electron to a stable molecule or atom
- Removing an electron from a stable molecule or atom
- Removing a group of atoms from a molecule
- Adding to a molecule

To illustrate the importance of free radicals during combustion, we need to examine the burning of methane in more detail. Chemists now view the combustion of methane as a series of individual reactions, each of which involves free radical reactions. As the combustion proceeds, the products of one reaction activate additional molecules, which then initiate other reactions. This stepwise description of the manner by which reactants are converted into products is called the *mechanism* of the reaction.

In all, the combustion of methane can be summarized by means of a mechanism consisting of the following steps:

$$
\underset{\text{methane}}{CH_4(g)} \longrightarrow \underset{\text{methyl radical}}{CH_3\cdot(g)} \longrightarrow \underset{\text{formaldehyde}}{HCHO(g)} \longrightarrow
$$

$$
\underset{\text{formyl radical}}{HCO\cdot(g)} \longrightarrow \underset{\text{carbon monoxide}}{CO(g)} \longrightarrow \underset{\text{carbon dioxide}}{CO_2(g)}
$$

It is an example of a *chain reaction*, that is, a series of steps, each of which generates a reactive substance that brings about another step. Let's examine it more closely.

INITIATION STEP

The first mechanistic step involves the production of a methyl free radical and a hydrogen atom. This occurs when a methane molecule first absorbs activation energy from an ignition source.

$$
\underset{\text{methane}}{CH_4(g)} \longrightarrow \underset{\text{methyl radical}}{CH_3\cdot(g)} + \underset{\text{hydrogen atom}}{H\cdot(g)}
$$

It is called the *initiation step* of the mechanism.

CHAIN-PROPAGATING STEPS

The hydrogen atom then reacts with molecular oxygen, which results in the formation of a hydroxyl radical ($\cdot$OH) and an oxygen atom ($\cdot\ddot{O}\cdot$).

$$
\underset{\text{hydrogen atom}}{H\cdot(g)} + \underset{\text{oxygen}}{O_2(g)} \longrightarrow \underset{\text{hydroxyl radical}}{\cdot\ddot{O}H(g)} + \underset{\text{oxygen atom}}{\cdot\ddot{O}\cdot(g)}
$$

It is an example of a *chain-propagating step*, since one reactive species has reacted and generated new ones.

During the chain-propagating steps that occur during the combustion of methane, new reactive intermediates, atoms, and molecules are produced by reactions involving methyl radicals, hydroxyl radicals, hydrogen atoms, and oxygen atoms. Examples of these chain-propagating steps are represented by a group of equations like the following:

$$CH_4(g) \; + \; \cdot\ddot{O}H(g) \longrightarrow H_2O(g) + CH_3\cdot(g)$$
methane   hydroxyl radical   water   methyl radical

$$CH_4(g) + H\cdot(g) \longrightarrow CH_3\cdot(g) + H_2(g)$$
methane   hydrogen atom   methyl radical   hydrogen

$$CH_3\cdot(g) + \cdot\ddot{O}\cdot(g) \longrightarrow \underset{\underset{H}{|}}{\overset{\overset{O}{\parallel}}{H-C}}(g) + H\cdot(g)$$
methyl radical   oxygen atom   formaldehyde   hydrogen atom

$$\underset{\underset{H}{|}}{\overset{\overset{O}{\parallel}}{H-C}}(g) + CH_3\cdot(g) \longrightarrow \underset{\underset{H}{|}}{\overset{\overset{O}{\parallel}}{\cdot C}}(g) + CH_4(g)$$
formaldehyde   methyl radical   formyl radical   methane

$$\underset{\underset{H}{|}}{\overset{\overset{O}{\parallel}}{H-C}}(g) + \cdot\ddot{O}H(g) \longrightarrow \underset{\underset{H}{|}}{\overset{\overset{O}{\parallel}}{\cdot C}}(g) + H_2O(g)$$
formaldehyde   hydroxyl radical   formyl radical   water

$$\underset{\underset{H}{|}}{\overset{\overset{O}{\parallel}}{H-C}}(g) + H\cdot(g) \longrightarrow \underset{\underset{H}{|}}{\overset{\overset{O}{\parallel}}{\cdot C}}(g) + H_2(g)$$
formaldehyde   hydrogen atom   formyl radical   hydrogen

$$\underset{\underset{H}{|}}{\overset{\overset{O}{\parallel}}{H-C}}(g) + \cdot\ddot{O}\cdot(g) \longrightarrow \underset{\underset{H}{|}}{\overset{\overset{O}{\parallel}}{\cdot C}}(g) + \cdot OH(g)$$
formaldehyde   oxygen atom   formyl radical   hydroxyl radical

$$\underset{\underset{H}{|}}{\overset{\overset{O}{\parallel}}{\cdot C}}(g) \longrightarrow CO(g) + H\cdot(g)$$
formyl radical   carbon monoxide   hydrogen atom

$$CO(g) \; + \; \cdot\ddot{O}H(g) \longrightarrow CO_2(g) + H\cdot(g)$$
carbon monoxide   hydroxyl radical   carbon dioxide   hydrogen atom

The combination of the reactions illustrated by these equations allows combustion to continue until the supply of the reactants or the reactive intermediates is exhausted.

TERMINATION STEPS

Finally, the free radicals and atoms combine in *termination steps* similar to the reactions denoted by the following equations:

$$CH_3 \cdot (g) \; + \; H \cdot (g) \; \longrightarrow \; CH_4(g)$$

methyl radical   hydrogen atom   methane

$$CH_3 \cdot (g) + CH_3 \cdot (g) \longrightarrow CH_3 - CH_3(g)$$

methyl radicals   ethane

In the termination steps of the mechanism, two reactive intermediates combine. Since new reactive species are not generated, the combustion process slows or ceases.

Solved Exercise 5.4

During the combustion of methane, formaldehyde is produced during the propagation step of the mechanism. Since it is an intermediate, chemists indicate that formaldehyde is produced by the *partial oxidation* of methane. While only a trace of formaldehyde is present during the normal burning of methane, the reaction can be controlled so as to commercially manufacture formaldehyde.

(a) Write the Lewis formula of formaldehyde.

(b) Write the balanced equation illustrating the production of formaldehyde by the partial oxidation of methane.

Solution:

(a) The chemical formula of formaldehyde is HCHO. The Lewis structure is shown below:

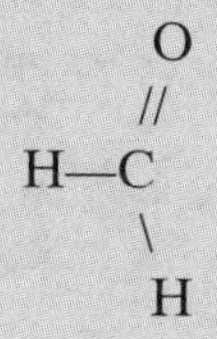

$$
\begin{array}{c}
O \\
\parallel \\
H - C \\
\backslash \\
H
\end{array}
$$

(b) We begin by writing an unbalanced partial equation:

$$CH_4(g) + O_2(g) \longrightarrow HCHO(g) + \underline{\hspace{2cm}}$$

After some consideration, it becomes apparent that the remaining substance is water. When included, the production of formaldehyde by the partial oxidation of methane is depicted as follows:

$$CH_4(g) + O_2(g) \longrightarrow HCHO(g) + H_2O(g)$$

Performance Goals for Section 5.11

- Discuss the manner by which water effectively functions as a fire extinguisher.
- Identify the conditions during which the use of water as a fire extinguisher is inadvisable.
- Identify the nature of an emergency-response action in which the use of either AFFF or AR-AFFF is recommended.

◆ 5.11 WATER AS A FIRE EXTINGUISHER

Although water is generally effective only for extinguishing class A fires, in certain circumstances, it can also be used to extinguish class B fires. Some examples are denoted below:

- When water is immiscible with a burning liquid and floats upon its surface, the water extinguishes the fire by absorbing heat and preventing the escape of vapor from the liquid into the atmosphere.
- Water is effective as a fire extinguisher when it dilutes the burning fuel. If a flammable liquid is soluble in water, it becomes nonflammable when a sufficient amount of water has been mixed with it. Water extinguishes such fires primarily by reducing the amount of available fuel for self-sustenance.

Water is often selected as a fire extinguisher because it is usually available in relatively large quantities. Yet water is not a good all-purpose fire extinguisher for the following reasons:

- The application of water to fires often causes considerable damage; in fact, the damage resulting from the use of water frequently surpasses that caused directly by the fire.
- The fluidity of water causes it to be highly inefficient as a fire extinguisher. Most water that is applied to a fire drains from the scene into adjacent areas, where its exposure to the heat generated by the fire may be so limited it does not rapidly vaporize.
- Many parts of the world are cold enough for water to freeze when exposed to the environment. Although antifreeze agents can be added to keep the water liquid at temperatures as low as $-60°F$ $(-51°C)$, these water solutions are frequently corrosive to metals and require the use of special equipment for mixing and discharging them. Furthermore, when insufficient antifreeze and water are applied to a fire, the water evaporates, allowing the antifreeze to burn.
- Water is often more dense than the burning liquids that constitute class B fires. When applied to them, the water sinks below their surfaces, where it is incapable of absorbing the evolved heat.
- Water ruins delicate electronic circuitry, and thus it is not recommended on many class C fires. Water containing dissolved mineral salts conducts electricity, which can put firefighters at the risk of being electrocuted.
- Water often reacts with the burning metals that constitute class D fires and may aid in sustaining combustion rather than extinguishing such fires.

The efficiency of water as a fire extinguisher is markedly improved by using a low- or high-velocity fog nozzle to discharge it as a mist, spray, or fog (Figure 5.8). The increased surface area of the fog particles leads to more rapid vaporization of water, and thus, to faster removal of heat from the fire scene.

FOAM SYSTEMS

The efficiency of water as a fire extinguisher is also improved when the water has been sealed into a gel or foam called an *aqueous-film-forming foam,* or *AFFF.* Several AFFF concentrates are commercially available for producing foam upon demand. They are composed of a surfactant that increases the wetting and penetrating ability of water. An AFFF concentrate can be mixed with fresh, sea, or brackish water in specialized equipment. Firefighters typically use an aqueous-film-forming foam to extinguish class B fires involving a bulk volume of a flammable liquid that is water-insoluble, like gasoline, but

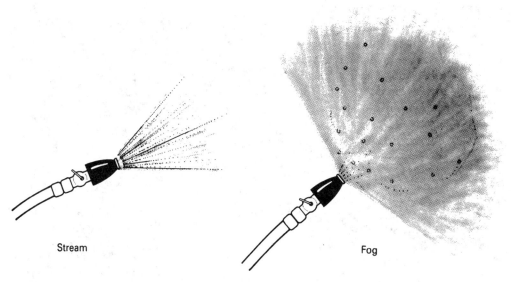

Stream   Fog

FIGURE 5.8 ◆ A comparison between the application of water to a fire as a solid stream and as a fog. In the latter case, the droplets of water present more surface area to the heat source and thus are capable of absorbing heat more effectively than the solid stream.

the use of aqueous film-forming foam is also recommended for extinguishing some stubborn class A fires in upholstery, bedding, paper, hay, and brush.

AFFF foam is an effective fire extinguisher because it functions as a blanketing agent over the burning fuel. Its application produces a barrier between the fuel and the atmosphere that prevents contact between the fuel and the atmospheric oxygen needed for combustion. The foam also cools the fuel and serves as a vapor-sealing film over the surface of the fuel.

Foam concentrate is often stored in the bladder-tank system shown in Figure 5.9. Water pressure squeezes the bladder, providing the concentrate at the same pressure. Although little maintenance is normally needed, firefighters soon learn that considerable skill must be developed to effectively use aqueous-film-forming foam for extinguishing a major fire. An aqueous-film-forming foam is applied improperly by plunging it into burning fuel rather than across its surface. Then, the foam is unable to isolate the burning fuel from atmospheric oxygen, and the fire continues to burn.

There has been some apprehension with certain aqueous-film-forming foams since their use releases hydrofluoric acid (Section 8.9) and other fluorides. Their use also leads to groundwater contamination and failure of the wastewater treatment systems to which they are discharged.

A versatile and popular fire extinguisher is *alcohol-resistant aqueous-film-forming foam,* or *AR-AFFF.* Several AR-AFFF concentrates are available commercially. When they are mixed with water and properly dispensed, the resulting foam is commonly used for extinguishing class B fires involving both water-soluble and water-insoluble flammable liquids.

Another example of a fire-extinguishing foam is *protein foam,* a material prepared from natural protein materials such as fish meal, horn-and-hoof meal, or feather meal. The foam typically contains from 3% to 6% by weight of a protein concentrate as well as a stabilizing agent to provide permanence. Protein foam possesses a lubricating nature; because of this feature, it is often used on airport runways to assist disabled aircraft during landing.

FIGURE 5.9 ◆ Foam concentrate is often stored in a pressure-rated tank containing a nylon-reinforced elastomeric bladder. Water pressure is used to squeeze the bladder and provide the foam concentrate at the same pressure. *Courtesy of Reliable Fire Equipment Company, Alsip, Illinois.*

FIGURE 5.10 ◆ This fire onboard the *Nassia* in the Bosphorus Straits was extinguished in just 12.5 minutes using Pyrocool FEF. The insurer, Lloyd's of London, estimated that a period of 10 days would be required for extinguishing the fire by routine means. As a result of using Pyrocool FEF, firefighters were able to salvage 80% of the ship's cargo — 78,000 tons (70,900 tonnes) of crude oil — and prevent its release into the sea. *Courtesy of Pyrocool Technologies, Inc., Monroe, Virginia.*

Environmental problems associated with the use of firefighting foams can be circumvented by the use of modern-day products such as Pyrocool FEF (fire extinguishing foam). Pyrocool FEF is an environmentally benign fire extinguishing agent. Its use has been associated with two principal advantages. Pyrocool FEF does not pollute the environment, and it is capable of extinguishing major fires very rapidly (Figure 5.10).

Performance Goals for Section 5.12
   ◆ Discuss the manner by which carbon dioxide effectively functions as a fire extinguisher.
   ◆ Identify the nature of the emergency-response action in which the use of carbon dioxide in a total flooding situation is recommended.

◆ 5.12 CARBON DIOXIDE AS A FIRE EXTINGUISHER

At ordinary temperatures and pressures, carbon dioxide is encountered as a gas. Since it is nonflammable, carbon dioxide is often used to extinguish class B fires, class C fires, and even some class A fires. Its use is never recommended on class D fires. The effectiveness of carbon dioxide as a fire extinguisher is linked with its vapor density. Since carbon dioxide possesses a vapor density of 1.52, it is about 1½ times more dense than air. Thus, when it is applied to a fire, carbon dioxide prevents contact between the burning material and atmospheric oxygen.

Figure 5.11 demonstrates the pressure–temperature relationship for carbon dioxide. This is called the *phase diagram* of carbon dioxide, since the sketch provides the temperature and pressure conditions at which the substance exists as a gas, liquid, and solid. This phase diagram illustrates that at normal room conditions, carbon dioxide exists either as a solid or vapor but not as a liquid. This is a relatively unique property. Whereas most solids liquefy before they vaporize, solid carbon dioxide, called *dry ice,* changes directly from the solid state of matter into its vapor at normal atmospheric conditions. The physical transformation of a substance directly from its solid state to its gaseous state without becoming liquid is called *sublimation.*

Solved Exercise 5.5

What is the intended message of the following dry ice pictographs?

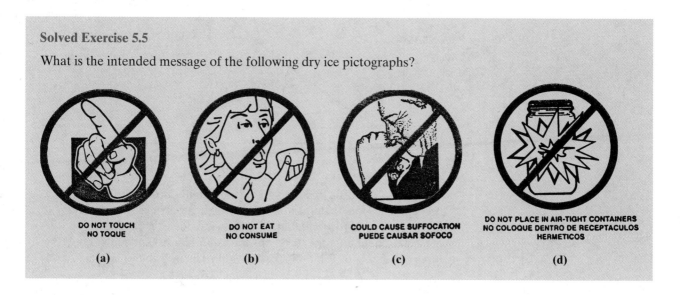

DO NOT TOUCH
NO TOQUE

DO NOT EAT
NO CONSUME

COULD CAUSE SUFFOCATION
PUEDE CAUSAR SOFOCO

DO NOT PLACE IN AIR-TIGHT CONTAINERS
NO COLOQUE DENTRO DE RECEPTACULOS
HERMETICOS

(a)   (b)   (c)   (d)

Solution: Dry ice is extremely cold relative to room temperature and should be handled with the specific precautions noted for cryogens in Section 2.13. Each pictograph successively conveys the following message:

(a) Do not allow dry ice to contact the skin, since it can bond so firmly to the tissue that the flesh is ripped or torn when separation is attempted.

(b) Do not attempt to consume dry ice, since it can bond so firmly to the tongue or cheeks that the flesh is ripped or torn when separation is attempted.

(c) The sublimation of dry ice is simultaneously associated with a displacement of air. Since carbon dioxide does not support life, the inhalation of an atmosphere in which carbon dioxide has displaced the air can cause suffocation and death.

(d) Dry ice sublimates into a substantially larger volume as the gas. When it is confined within a storage vessel, internal pressure develops as sublimation occurs. This can cause the container to rupture.

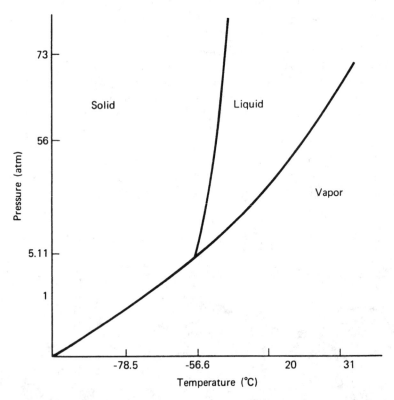

FIGURE 5.11 ◆ The phase diagram (not to a uniform scale) of carbon dioxide. Note that at 1 atm carbon dioxide only exists as either its vapor or solid (dry ice). Liquid carbon dioxide exists only at pressures equal to or greater than 75.1 psi (5.11 atm) and temperatures between −69.9°F (−56.6°) and approximately 88°F (31°C).

FIGURE 5.12 ◆ A portable carbon dioxide fire extinguisher. This extinguisher delivers a quick smothering action to flames, suffocating the fire of atmospheric oxygen. Its operation is simple: remove the locking pin, aim the horn at the base of the fire, squeeze the lever to discharge carbon dioxide, and release the lever to stop the discharge. *Courtesy of Walter Kidde, Mebane, North Carolina.*

When intended for use as a fire extinguishing agent, carbon dioxide is often encountered as a compressed gas within a portable extinguisher (Figure 5.12). Portable hand-held extinguishers containing compressed carbon dioxide in capacities from 2 to 25 lb (0.9 to 10 kg) are available commercially. Carbon dioxide is also discharged as a gas from storage tanks in which it is confined as a liquid under pressure. Discharged as the gas, the carbon dioxide is much colder than the surrounding atmosphere. It functions not only by smothering the burning materials, but by cooling them.

Unlike water, carbon dioxide does not ordinarily damage the fire area. It is considered a "clean" fire extinguisher. Since it is a gas at normal temperature and pressure conditions, carbon dioxide dissipates into the surrounding atmosphere after use without leaving a residue.

Carbon dioxide is generally applied to fires that occur inside buildings where the atmosphere is calm. Even under this circumstance, however, firefighters often need to apply carbon dioxide repeatedly to fires that have been initially extinguished. Heat generated by the hot material causes the carbon dioxide gas to dissipate into the atmosphere. Unless it has cooled sufficiently, the material reignites upon contacting atmospheric oxygen.

Carbon dioxide is often stored in industrial plants for potential use as a fire extinguisher in a total flooding situation. The term "total flooding" means that the

FIGURE 5.13 ◆ Low-pressure carbon dioxide is discharged in a total flooding system into an evacuated manufacturing facility. *Courtesy of Chemetron Fire Systems, Matteson, Illinois.*

impacted area is literally flooded upon demand with carbon dioxide (Figure 5.13). Two systems involving the use of carbon dioxide are commercially available for total flooding:

- In the *high-pressure system,* liquid carbon dioxide is stored in steel cylinders under a pressure of approximately 850 psi$_a$ (5900 kPa).
- In the *low-pressure system,* liquid carbon dioxide is maintained at 0°F (−18°C) by means of refrigeration under a pressure of approximately 300 psi$_a$ (2100 kPa).

The principal danger posed by the use of carbon dioxide as a fire extinguisher is the threat of asphyxiation. When discharged into a confined area, the carbon dioxide plume replaces the ground-level air. Since individuals who inhale carbon dioxide within the treated area could lose consciousness, it is critical to first evacuate everyone from the area *before* the carbon dioxide is discharged.

Some fire extinguishers produce carbon dioxide by chemical action with other substances. The once popular soda-acid fire extinguisher was a system that used a chemical reaction to produce carbon dioxide and water. "Soda" is the common name of sodium bicarbonate. Separate solutions of soda and sulfuric acid were arranged so that when the assembly was inverted, the two mixed together. The resulting chemical reaction produced carbon dioxide.

$$2NaHCO_3(aq) + H_2SO_4(aq) \longrightarrow Na_2SO_4(aq) + 2H_2O(l) + CO_2(g)$$

sodium bicarbonate   sulfuric acid   sodium sulfate   water   carbon dioxide

The carbon dioxide produced in the soda-acid fire extinguisher generated a pressure that forced water out the nozzle. It was the water, not the carbon dioxide, that extinguished the fire. The standard 2½-gallon soda-acid fire extinguisher shown in Figure 5.14 provided 2½ gallons of water. Sodium bicarbonate solutions slowly

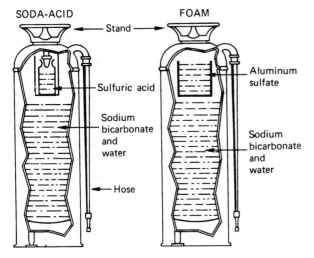

SODA-ACID

FOAM

Stand

Sulfuric acid

Sodium
bicarbonate
and
water

Hose

Aluminum
sulfate

Sodium
bicarbonate
and
water

FIGURE 5.14 ◆ On the left is the soda-acid fire extinguisher, now relatively obsolete, containing sodium bicarbonate and sulfuric acid solutions. When inverted, the acid and bicarbonate solutions mix and chemically react, producing carbon dioxide, which forces the solution mixture out of the nozzle. On the right is a modification of this fire extinguisher, which uses solutions of aluminum sulfate and sodium bicarbonate to produce the carbon dioxide.

deteriorated over time; hence, soda-acid extinguishers required periodic recharging with fresh solutions.

Carbon dioxide can also be contained as bubbles within a *chemical foam*. This is another example of a film-forming foam. Here, sodium bicarbonate and aluminum sulfate are stored as separate water solutions or powders. When they are mixed, these substances react to produce carbon dioxide.

$$6NaHCO_3(aq) + Al_2(SO_4)_3(aq) \longrightarrow 3Na_2SO_4(aq) + 2Al(OH)_3(s) + 6CO_2(g)$$

sodium bicarbonate   aluminum sulfate   sodium sulfate   aluminum hydroxide   carbon dioxide

Sometimes an extract of licorice root is mixed with these solutions to help produce foam. In combination with aluminum hydroxide, a tough coating is formed around each bubble of carbon dioxide. The entire mass emerges as a foam and acts as a wet blanket on a fire, preventing the burning material from contacting atmospheric oxygen.

Some large petroleum refineries and/or their outlying tank farms employ a two-solution wet-foam system for smothering fires that occur in or around their tanks. Solutions of aluminum sulfate and sodium bicarbonate are stored within separate tanks equipped with foam-mixing chambers, usually located at the top ring or roof of these tanks where they are combined to produce the chemical foam. When spread upon the surface of a burning petroleum product, the foam starves the fire by preventing its contact with atmospheric oxygen.

Performance Goals for Section 5.13

+ Discuss the manner by which the halons effectively function as fire extinguishers.
+ Describe why the use of the halons as fire extinguishers is being phased out in the United States.
+ Describe the intent of the Montreal protocol.

The *halons* are nonflammable compounds whose molecules possess one or two carbon atoms and from four to six halogen atoms. They are halogenated derivatives of methane and ethane, that is, one or more hydrogen atoms in the molecules of methane and ethane have been substituted with halogen atoms. Two representative halon formulas are CF_3Br and CF_2ClBr. In the first, the four hydrogen atoms in the methane molecule have been substituted with three fluorine atoms and one bromine atom. In the second, the four hydrogen atoms have been substituted with one chlorine atom, one bromine atom, and two fluorine atoms. These compounds were once very popular fire extinguishers. Their vapors were discharged at room temperature as total flooding and streaming systems.

The word *halon* was first devised by the U.S. Army Corps of Engineers as an abbreviation for <u>hal</u>ogenated hydrocarb<u>on</u>, a class of organic compounds we shall visit again in Section 12.7. The commercial halon products are named using the word Halon and a four-digit number. When the formula of the product is denoted as $C_aF_bCl_cBr_d$, the number is represented as abcd.

The halons used as fire extinguishers are listed in Table 5.3. The most broadly used are Halon 1301 and Halon 1211. Both are more effective at extinguishing class A, class B, and class C fires than either water or carbon dioxide.

The chemical formulas of the halons having the commercial names Halon 1301 and Halon 1211 are CF_3Br and CF_2ClBr, respectively. Chemists name them as the halogenated derivatives of methane. The halogen atoms are named by replacing the *-ine* suffix on the name of the halogen with *-o*. The number of halogen atoms is indicated by the use of *mono-, di-, tri-,* and *tetra-*, as relevant, for 1, 2, 3 and 4, respectively, and when multiple halogens are part of the same substance, they are named alphabetically. Thus, the chemical names of Halon 1301 and Halon 1211 are bromotrifluoromethane and bromochlorodifluoromethane, respectively.

TABLE 5.3 ◆ Some Physical Properties of the Common Halon Agents

Fire Extinguishing Agent	Chemical Formula	Halon No.	Boiling Point	Melting Point	Specific Gravity of Liquid at 68°F (20°C)
Bromochloromethane	CH_2BrCl	1011	151°F(66°C)	−124°F(−87°C)	1.93
Dibromodifluoromethane	CBr_2F_2	1202	76°F(24.5°C)	−223°F(−142°C)	2.28
1,2-Dibromo-1,1,2,2-tetrafluoroethane	$F_2BrC—CBrF_2$	2402	117°F(47°C)	−167°F(−111°C)	2.17
Bromotrifluoromethane	$CBrF_3$	1301	−72°F(−58°C)	−270°F(−168°C)	1.57
Dichlorodifluoromethane	CCl_2F_2	122	−22°F(−30°C)	−252°F(−158°C)	1.31
Bromochlorodifluoromethane	$CBrClF_2$	1211	25°F(−3.9°C)	−257°F(−161°C)	1.83
1,2-Dichloro-1,1,2,2-tetrafluoroethane	$F_2ClC—CClF_2$	242	39°F(3.9°C)	−137°F(−94°C)	1.44

$$
\begin{array}{cc}
\text{F} & \text{Br} \\
| & | \\
\text{F}-\text{C}-\text{F} & \text{F}-\text{C}-\text{F} \\
| & | \\
\text{Br} & \text{Cl} \\
\text{bromotrifluoromethane} & \text{bromochlorodifluoromethane} \\
\text{(Halon 1301)} & \text{(Halon 1211)}
\end{array}
$$

Collectively, the halons possess two properties that cause them to be effective fire extinguishers. First, their vapors are much denser than air. This means that when halons are discharged upon fires, their vapors smother them. Second, the halons decompose at the elevated temperatures experienced during fires, producing free radicals and atoms. These reactive species scavenge the combustion intermediates produced during class B fires. For instance, when Halon 1301 thermally decomposes, trifluoromethyl radicals and bromine atoms are produced.

$$
\text{CF}_3\text{Br}(g) \longrightarrow \;
\begin{array}{c}
\text{F} \\
| \\
\text{F}-\text{C} \cdot (g) \\
| \\
\text{F}
\end{array}
\; + \; :\ddot{\text{B}}\text{r} \cdot (g)
$$

The bromine atoms have a far greater affinity for reacting with combustion intermediates than the combustion intermediates have for reacting with each other.

The halons were once popular fire extinguishers for the reasons summarized below:

- They effectively extinguish class A, class B, and class C fires.
- They dissipate into the atmosphere and leave no residue following their use, thereby avoiding secondary property damage. This was highly desirable when assessing the protection of delicate and expensive electronic equipment, computer circuitry, and aircraft interiors.
- They do not conduct electricity.
- When used with proper exposure controls, they pose only a minimum health risk to firefighters and other personnel.

The halons were suitable for application on fires within vaults, museums, libraries, and hospitals. They were also used in military and civilian planes for extinguishing fires in engines, cargo bays, cockpits, passenger compartments, and washrooms. They were once widely recommended for use in total flooding systems (such as stationary fire suppression systems in large computer facilities) and streaming systems (such as hand-held portable fire extinguishers).

Given these features, why are they not widely used today? The answer to this question is linked with their ability to deplete the ozone that occupies a stratospheric layer lying at the edge of our atmosphere. Ozone is a form of elemental oxygen having three atoms of oxygen per molecule instead of the usual two. As we shall examine in more detail in Section 7.1, ozone exists as a blanket around Earth, where its presence protects life on Earth from overexposure to harmful ultraviolet radiation.

When most substances are released into the atmosphere, they ultimately undergo chemical reactions that convert them into relatively innocuous substances. The halons are unlike them. They are relatively unreactive substances. When released into the environment, their vapors migrate to the stratosphere where they react with the ozone. The combination of such reactions has resulted in depleting the stratospheric ozone.

Scientists fear that the depletion will allow excessive ultraviolet radiation to penetrate to the surface of our planet, where it can contribute to the incidence of skin cancer and cataracts and cause untold harm in other ways.

To ensure the survival of planet Earth, 168 representatives of the world's industrialized nations agreed to phase out the manufacture, import, and use of ozone-depleting substances in an international agreement known as the *Montreal protocol*. This action was accomplished in the United States under the statutory authorities of the Clean Air Act and the Toxic Substances Control Act.

The production and import of ozone-depleting substances such as the halons have been banned in the United States since January 1994, and the use of ozone-depleting substances for all nonessential uses has been banned since 2001. This latter ban has resulted in virtually phasing out the use of halons in hand-held portable fire extinguishers.

Solved Exercise 5.6

The United States Department of Defense (DOD) and the National Institute of Standards and Technology (NIST) have embarked on a program called the "Next Generation Fire Suppression Technology Program," or NGP. Its aim to "develop and demonstrate, by 2005, technology for economically feasible, environmentally acceptable, and user-safe processes, techniques, and fluids that meet the operational requirements currently satisfied by Halon 1301 systems in aircraft." What is the most likely reason DOD and NIST have embarked on this program?

Solution: Halon 1301 has been used successfully to combat fires in military and civilian aircraft dry bays and engine nacelles. Notwithstanding the benefit of this use, the halons are ozone-depleting substances. EPA could ban its use for the referenced purpose at any time. It is likely that DOD and NIST engaged in this program to find a fire suppressant that performs at least as well as Halon 1301 when used to extinguish aircraft fires, but whose use does not simultaneously deplete of stratospheric ozone.

EPA conditionally supports the use of the halon products for certain essential uses, but only until an economically feasible, environmentally friendly alternative has been marketed. A replacement that has received EPA's approval for firefighting purposes is the compound having the following molecular structure:

$$
\begin{array}{ccccccccc}
 & F & & F & & & F & & F \\
 & | & & | & & & | & & | \\
F - & C & - & C & - & C & - & C & - & C & - F \\
 & | & & | & & \| & & | & & | \\
 & F & & F & & O & & CF_3 & & F
\end{array}
$$

It is named 1,1,1,2,2,4,5,5,5-nonafluoro-4-(trifluoromethyl)-3-pentanone and is sold under the trademark Novec 1230. This compound possesses an atmospheric lifetime

FIGURE 5.15 ◆ The fire extinguisher Novec 1230 is marketed as an environmentally friendly replacement for Halon 1301. It may be suitably applied to fires within vaults, museums, libraries, and hospitals to extinguish class A, class B, and class C fires. *Courtesy of 3M Corporation, Minneapolis, Minnesota.*

of only 3 to 5 days, which is far too short to affect stratospheric ozone. It is also nontoxic.

Novec 1230 is a liquid at room temperature. When used to extinguish fires, as in Figure 5.15; it may be discharged as a liquid stream or gas in local and total flooding applications. Since it is a nontoxic substance, it outranks carbon dioxide for extinguishing fires by total flooding.

Performance Goal for Section 5.14
- ◆ Discuss the manner by which dry alkali metal bicarbonates, monoammonium phosphate, and graphite effectively function as fire extinguishers.

◆ 5.14 DRY CHEMICAL FIRE EXTINGUISHERS

The following chemical substances are generally used in the solid state as *dry chemical* fire extinguishers: sodium bicarbonate, potassium bicarbonate, sodium chloride, potassium chloride, and monoammonium phosphate. Some dry chemical fire extinguishers effectively suppress class B and class C fires; others are useful on class D fires;

and some extinguish class A, class B, and class C fires. They are not clean fire extinguishers. After their use, messy residues remain.

Fire extinguishers containing sodium bicarbonate and potassium bicarbonate are sometimes referred to as "regular" dry chemical fire extinguishers. Potassium bicarbonate is usually referred to as the commercial product *Purple K*. Although ineffective on class A fires, both sodium and potassium bicarbonates effectively suppress and extinguish class B flammable liquid fires and are often used during initial tactics when fighting a fire involving a flammable compressed gas.

The effectiveness of the alkali metal bicarbonates as fire extinguishers is related in part to their ability to produce carbon dioxide when heated.

$$2NaHCO_3(s) \longrightarrow Na_2CO_3(s) + CO_2(g) + H_2O(g)$$

sodium bicarbonate   sodium carbonate   carbon dioxide   water

The carbon dioxide smothers the fire. In addition, sodium atoms are produced when these compounds are exposed to high temperatures. The sodium atoms react with other reactive chemical species to produce water.

During fires involving petrochemical fuels, hydroxyl radicals ($\cdot\ddot{O}H$) are among the reactive species produced. During fires involving the burning of petrochemical fuels, sodium atoms react with the hydroxyl radicals to produce water.

$$Na\cdot(g) + \cdot\ddot{O}H(g) \longrightarrow NaOH(g)$$

sodium atoms   hydroxyl radicals   sodium hydroxide

$$NaOH(g) + \cdot H(g) \longrightarrow H_2O(l) + Na\cdot(g)$$

sodium hydroxide   hydrogen atoms   water molecules   sodium atoms

Overall, the reaction of hydrogen atoms and hydroxyl radicals forms water.

$$H\cdot(g) + \cdot\ddot{O}H(g) \longrightarrow H_2O(l)$$

hydrogen atoms   hydroxyl radicals   water molecules

In essence, the sodium atoms serve to catalyze the production of water and remove the free radicals.

Another dry chemical is monoammonium phosphate, known more properly in chemistry as ammonium dihydrogen phosphate. This substance is often employed in the *multipurpose ABC fire extinguisher* illustrated in Figure 5.16. The reason for the effectiveness of this substance as a fire extinguisher is twofold:

- When applied to a burning material, monoammonium phosphate decomposes endothermically, as follows:

$$4NH_4H_2PO_4(s) \longrightarrow P_4O_{10}(s) + 4NH_3(g) + 6H_2O(g)$$

monoammonium phosphate   tetraphosphorus decoxide   ammonia   water

  Since it absorbs heat as it decomposes, monoammonium phosphate cools the combustible material below the minimum temperature at which it can burn.

- The ammonia generated by the thermal decomposition of monoammonium phosphate reacts with certain atoms and free radicals. Ammonia, for example, reacts with hydroxyl radicals produced when petroleum fuels burn. The reaction effectively removes the free radicals.

There are also commercial chemical products, known as *dry powders,* that effectively suppress certain class D fires. An important dry powder fire extinguisher is composed primarily of graphite. The graphite-based dry powders are used for extinguishing the fires of combustible metals, such as magnesium and aluminum (Sec-

FIGURE 5.16 ◆ A portable, multipurpose, dry chemical fire extinguisher. This extinguisher contains monoammonium phosphate as the fire extinguishing agent. It may be used to effectively suppress most class A, class B, and class C fires. *Courtesy of Walter Kidde, Mebane, North Carolina.*

tion 9.3). Although these fire extinguishers are successful for this purpose, attention must be directed to the chemistry of graphite whenever they are used. Applied to a combustible metal fire, graphite produces water-reactive metallic carbides, such as magnesium carbide and aluminum carbide. Upon exposure to water, metallic carbides generate flammable gases including methane and acetylene. Consequently, the water reactivity of metallic carbides must be addressed when removing the residues following the use of graphite-based dry powders on combustible metal fires. We shall revisit this issue in Chapter 9.

■ ■

REVIEW EXERCISES

Writing Chemical Equations

5.1 Why is it necessary to "balance" chemical equations?

5.2 One type of portable oxygen generator is constructed of separate compartments, one containing iron powder and the other containing sodium chlorate.

When the compartment contents are mixed and heated, oxygen is generated as follows:

$$Fe(s) + NaClO_3(s) \longrightarrow FeO(s) + NaCl(s) + O_2(g)$$

Show that this equation is balanced.

5.3 When concentrated perchloric acid is heated to 198°F (92°C), it decomposes into water vapor, chlorine, and oxygen. Write the balanced equation that denotes this decomposition.

5.4 When ammonium ~~......~~ ed, it decomposes to produce ~~......~~ force is generated. Some ex- ~~......~~ e are shown by the following

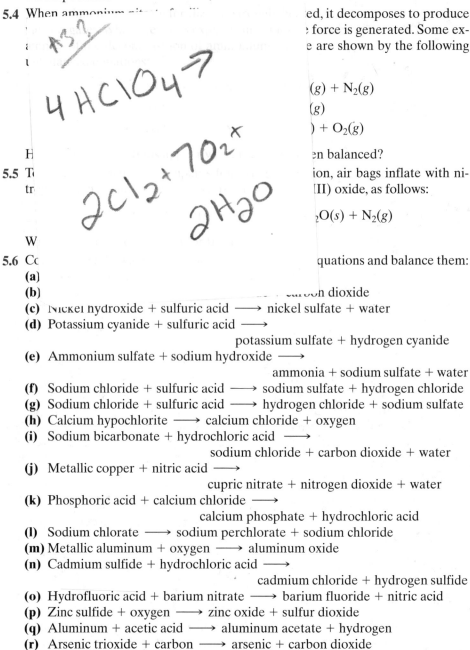

$$132$$
$$4 HClO_4 \longrightarrow$$
$$2 Cl_2 + 7 O_2 \times$$
$$2 H_2O$$

~~......~~ (g) + N_2(g)
~~......~~ (g)
~~......~~) + O_2(g)

H ~~......~~ en balanced?

5.5 T ~~......~~ ion, air bags inflate with ni-
tr ~~......~~ (II) oxide, as follows:

~~......~~ _2O(s) + N_2(g)

W

5.6 Co ~~......~~ quations and balance them:
(a) ~~......~~
(b) ~~......~~ carbon dioxide
(c) Nickel hydroxide + sulfuric acid $\longrightarrow$ nickel sulfate + water
(d) Potassium cyanide + sulfuric acid $\longrightarrow$
potassium sulfate + hydrogen cyanide
(e) Ammonium sulfate + sodium hydroxide $\longrightarrow$
ammonia + sodium sulfate + water
(f) Sodium chloride + sulfuric acid $\longrightarrow$ sodium sulfate + hydrogen chloride
(g) Sodium chloride + sulfuric acid $\longrightarrow$ hydrogen chloride + sodium sulfate
(h) Calcium hypochlorite $\longrightarrow$ calcium chloride + oxygen
(i) Sodium bicarbonate + hydrochloric acid $\longrightarrow$
sodium chloride + carbon dioxide + water
(j) Metallic copper + nitric acid $\longrightarrow$
cupric nitrate + nitrogen dioxide + water
(k) Phosphoric acid + calcium chloride $\longrightarrow$
calcium phosphate + hydrochloric acid
(l) Sodium chlorate $\longrightarrow$ sodium perchlorate + sodium chloride
(m) Metallic aluminum + oxygen $\longrightarrow$ aluminum oxide
(n) Cadmium sulfide + hydrochloric acid $\longrightarrow$
cadmium chloride + hydrogen sulfide
(o) Hydrofluoric acid + barium nitrate $\longrightarrow$ barium fluoride + nitric acid
(p) Zinc sulfide + oxygen $\longrightarrow$ zinc oxide + sulfur dioxide
(q) Aluminum + acetic acid $\longrightarrow$ aluminum acetate + hydrogen
(r) Arsenic trioxide + carbon $\longrightarrow$ arsenic + carbon dioxide
(s) Mercury(II) oxide $\longrightarrow$ metallic mercury + oxygen

(t) Magnesium sulfate + calcium hydroxide $\longrightarrow$
magnesium hydroxide + calcium sulfate

(u) Ammonia + hydrogen sulfide $\longrightarrow$ ammonium sulfide

(v) Carbon disulfide + oxygen $\longrightarrow$ carbon monoxide + sulfur dioxide

(w) Silver nitrate $\longrightarrow$ metallic silver + nitrogen dioxide + oxygen

(x) Nickel(II) chlorate $\longrightarrow$ nickel(II) chloride + oxygen

(y) Sulfur tetrachloride + water $\longrightarrow$ sulfurous acid + hydrogen chloride

(z) Iron(III) oxide + carbon monoxide $\longrightarrow$
metallic iron + carbon dioxide

Balancing Equations

5.7 Balance each of the following chemical equations:

(a) $C_2H_4(g) + O_2(g) \longrightarrow CO_2(g) + H_2O(g)$

(b) $KOH(aq) + H_2SO_4(aq) \longrightarrow K_2SO_4(aq) + H_2O(l)$

(c) $Na(s) + H_2O(l) \longrightarrow NaOH(aq) + H_2(g)$

(d) $NiO(s) + H_2SO_4(aq) \longrightarrow NiSO_4(aq) + H_2O(l)$

(e) $MnO_2(s) + HCl(aq) \longrightarrow MnCl_2(aq) + Cl_2(g) + H_2O(l)$

(f) $As_2O_3(s) + H_2S(g) \longrightarrow As_2S_3(s) + H_2O(l)$

(g) $POCl_3(l) + H_2O(l) \longrightarrow H_3PO_4(aq) + HCl(g)$

(h) $Fe(s) + O_2(g) + H_2O(g) \longrightarrow Fe(OH)_3(s)$

(i) $CS_2(l) + Cl_2(g) \longrightarrow CCl_4(l) + S_2Cl_2(l)$

(j) $Ag_2S(s) + H_2O(l) \longrightarrow Ag(s) + H_2S(g) + O_2(g)$

(k) $KClO_3(s) \longrightarrow KCl(s) + KClO_4(s)$

(l) $S_8(s) + O_2(g) \longrightarrow SO_2(g)$

(m) $Fe_2O_3(s) + HCl(aq) \longrightarrow FeCl_3(aq) + H_2O(l)$

(n) $Al(s) + KClO_4(s) \longrightarrow Al_2O_3(s) + KCl(s)$

(o) $Bi(s) + H_2SO_4(aq) \longrightarrow Bi_2(SO_4)_3(aq) + SO_2(g) + H_2O(l)$

(p) $K_2Cr_2O_7(s) \longrightarrow Cr_2O_3(s) + K_2O(s) + O_2(g)$

(q) $Fe(OH)_2(s) + H_2O(g) + O_2(g) \longrightarrow Fe(OH)_3(s)$

(r) $PCl_3(l) + H_2O(l) \longrightarrow H_3PO_3(aq) + HCl(g)$

(s) $Hg(l) + HNO_3(aq) \longrightarrow Hg_2(NO_3)_2(aq) + NO(g) + H_2O(l)$

(t) $P_4(s) + NaOH(aq) + H_2O(l) \longrightarrow NaH_2PO_2(aq) + PH_3(g)$

(u) $Mg_3N_2(s) + H_2O(l) \longrightarrow Mg(OH)_2(s) + NH_3(g)$

(v) $Al(NO_3)_3(aq) + H_2O(l) + NH_3(g) \longrightarrow NH_4NO_3(aq) + Al(OH)_3(s)$

(w) $Pb(C_2H_3O_2)_2(aq) + BaCl_2(aq) \longrightarrow Ba(C_2H_3O_2)_2(aq) + PbCl_2(s)$

(x) $Zn(OH)_2(s) + H_3PO_4(aq) \longrightarrow Zn_3(PO_4)_2(aq) + H_2O(l)$

(y) $HCN(g) + O_2(g) \longrightarrow CO_2(g) + NO_2(g) + H_2O(g)$

(z) $NaBr(aq) + H_2SO_4(aq) \longrightarrow Br_2(g) + SO_2(g) + Na_2SO_4(aq) + H_2O(l)$

Types of Chemical Reactions

5.8 Identify whether each of the following equations represents a combination, decomposition, single replacement, or double replacement reaction:

(a) $CaO(s) + H_2O(l) \longrightarrow Ca(OH)_2(aq)$

(b) $Al_2(CO_3)_3(s) \longrightarrow Al_2O_3(s) + 3CO_2(g)$

(c) $2Na(s) + 2H_2O(l) \longrightarrow 2NaOH(aq) + H_2(g)$

(d) $Pb(NO_3)_2(aq) + H_2S(g) \longrightarrow PbS(s) + 2HNO_3(aq)$

(e) $SO_3(g) + H_2O(l) \longrightarrow H_2SO_4(aq)$

(f) $AgC_2H_3O_2(aq) + HCl(aq) \longrightarrow AgCl(s) + HC_2H_3O_2(aq)$
(g) $CdSO_3(s) \longrightarrow CdO(s) + SO_2(g)$
(h) $Fe(s) + CuCl_2(aq) \longrightarrow FeCl_2(aq) + Cu(s)$

5.9 Water should only be used for extinguishing a magnesium fire when a deluging volume can be applied within a relatively short period. Otherwise, the water reacts with the hot magnesium to produce hydrogen and magnesium hydroxide. Under fire conditions, the hydrogen ignites, and the heat of reaction causes the magnesium to rekindle.
   (a) What type of chemical reaction produces the hydrogen?
   (b) Write the balanced equation illustrating this reaction.

5.10 Supply the name of the reactant or product that completes each of the following word equations:
   (a) _____ + hydrochloric acid $\longrightarrow$ iron(III) chloride + hydrogen
   (b) Aluminum + bromine $\longrightarrow$ _____
   (c) Potassium oxide + _____ $\longrightarrow$ potassium hydroxide
   (d) _____ + calcium chloride $\longrightarrow$ lead(II) chloride + calcium acetate
   (e) Calcium hypochlorite $\longrightarrow$ calcium chloride + _____
   (f) Cadmium chloride + _____ $\longrightarrow$ cadmium hydroxide + sodium chloride
   (g) Copper(II) sulfite $\longrightarrow$ _____ + sulfur dioxide
   (h) Nickel carbonate $\longrightarrow$ nickel oxide + _____
   (i) Hydrogen + _____ $\longrightarrow$ hydrogen fluoride
   (j) Hydrogen sulfide + _____ $\longrightarrow$ mercury(II) sulfide + nitric acid

Oxidation–Reduction

5.11 In some agricultural regions, butane is more cost-effective as a tractor fuel than diesel fuel. The chemical formula of butane is C_4H_{10}. When butane burns completely, carbon dioxide and water vapor are produced as combustion products.
   (a) Write a balanced equation representing the complete combustion of butane.
   (b) Identify the substance oxidized and the substance reduced.
   (c) Identify the oxidizing agent and the reducing agent.

5.12 Several elements burn with incandescence when exposed to an atmosphere of hot chlorine. Examples of these elements include copper and arsenic.
   (a) Write balanced equations illustrating the formation of cupric chloride and arsenic trichloride from their respective elements.
   (b) In each instance, identify the substance oxidized and the substance reduced.
   (c) In each instance, identify the oxidizing agent and the reducing agent.

Factors Affecting Reaction Rates

5.13 Among the various factors that influence reaction rates, which is involved in each of the following observations:
   (a) Motor fuel ignites more easily during warm weather compared to cold weather.
   (b) A stack of wooden logs lying exposed to the air fails to burn until a single log has been kindled.

(c) Lumps of flour are difficult to ignite directly, but a dispersion of flour dust ignites explosively.

(d) A sheet of magnesium metal is virtually impossible to ignite, but powdered magnesium in a camera flashbulb ignites almost instantaneously.

(e) Metallic calcium reacts slowly with cold water, but it reacts rapidly with hot water.

Combustion Phenomena

5.14 Charcoal briquettes consist almost entirely of elemental carbon. When they burn in the open air, carbon monoxide or carbon dioxide is produced. Write balanced chemical equations illustrating the production of each combustion product.

5.15 Why are metallic oxides incapable of being combustible fuels?

5.16 Sooty plumes generally billow from petroleum tank fire scenes. What does the presence of these sooty plumes generally indicate about the nature of the combustion phenomena?

5.17 Tetraphosphorus trisulfide is the active component of strike-anywhere matches. When it burns in air, each nonmetallic constituent is converted into a nonmetallic oxide.

(a) Identify the combustion product(s) that form when tetraphosphorus trisulfide burns completely in air.

(b) Write balanced equations that denote the combustion phenomenon.

5.18 Hydrazine is a substance sometimes used as a rocket propellant. Its chemical formula is N_2H_4. When hydrazine burns in air, nitrogen and water vapor are produced.

(a) Write a balanced equation denoting the combustion.

(b) Identify the oxidizing agent and reducing agent.

RCRA Characteristic of Reactivity

5.19 Why is each of the following chemical wastes likely to exhibit the RCRA characteristic of reactivity?

(a) An electroplating waste containing dissolved sodium hydroxide, sodium cyanide, and zinc cyanide.

(b) A leather-tanning waste containing sodium hydroxide and sodium sulfide.

Chemistry of Fire Extinguishers

5.20 The DOT regulation at 49 CFR §173.615(d) stipulates that shippers and carriers can transport no more than 440 lb (200 kg) of dry ice in any one cargo pit or bin on any aircraft other than with specific approval. What is the most likely reason DOT enacted this regulation?

5.21 The OSHA regulation set forth at 29 CFR. §1910.162(b)(5) stipulates that employers must provide a predischarge employee alarm system within the workplace to alert employees if a concentration of 4% or greater of carbon dioxide is released from a fixed fire extinguishing system. What is the most likely reason OSHA enacted this regulation?

5.22 Grease fires occurring in kitchens can usually be extinguished with ordinary table salt (sodium chloride). Describe the mechanism by which table salt effectively extinguishes grease fires.

5.23 The first halogenated hydrocarbon discovered to possess firefighting properties was carbon tetrachloride, or tetrachloromethane. During its use, however, the toxic gas phosgene was produced. For this reason, the use of carbon tetrachloride as a fire extinguisher was prohibited.

  (a) Write the chemical formula for carbon tetrachloride.

  (b) What halon name would have been assigned to carbon tetrachloride as a fire extinguisher?

  (c) The vapor density of carbon tetrachloride is 5.3 (air = 1). Suggest two reasons why carbon tetrachloride effectively extinguishes petroleum fires.

Aspects of the DOT Hazardous Materials Regulations

6 CHAPTER

Transportation regulations relating to hazardous material shipments were prepared by the U.S. Department of Transportation (DOT) to protect transportation personnel and equipment from exposure to hazardous materials and to provide fire and emergency-response personnel, the public, and transport workers with communication information in the event of a transportation mishap. Although DOT is limited to regulating the transportation of hazardous materials within the United States and its territories, the DOT regulations have been written to conform with generally accepted international standards. Consequently, their proper implementation can potentially aid fire and emergency-response personnel, the public, and transport workers worldwide.

The material in this chapter is an overview of certain aspects of the DOT regulations that apply to domestic shipments. Since they are subject to change from time to time, the parties affected by them should always consult 49 CFR Parts 171 through 179 for the latest revisions and amendments.

> **Performance Goals for Section 6.1**
> - Identify the information represented by each columnar entry in Table 6.1.
> - Illustrate how shippers and carriers use the Hazardous Materials Table to obtain basic descriptions for specific hazardous materials.
> - Discuss the meaning of a reportable quantity, marine pollutant, hazardous substance, and inhalation hazard.
> - Review the nature of the basic descriptions listed in Table 6.4.

◆ 6.1 THE SHIPPING PAPER

A *hazardous material* is defined by DOT as any designated material "capable of posing an unreasonable risk to health, safety, and property when transported." Each hazardous material designated by DOT is tabulated in the *Hazardous Materials Table,* excerpts of which are provided in Table 6.1. Although there are actually 10 columnar

TABLE 6.1 ◆ Hazardous Materials Table[a]

Symbols (1)	Hazardous Materials Descriptions and Proper Shipping Names (2)	Hazard Class or Division (3)	Identi-fication Numbers (4)	PG (5)	Label Codes (6)	…	Quantity Limitations (9)		Vessel Stowage (10)	
							Passenger Aircraft/ Rail (9A)	Cargo Aircraft Only (9B)	Location (10A)	Other (10B)[b]
	Acetone	3	UN1090	II	3	…	5 L	60 L	B	
	Acetonitrile	3	UN1648	II	3	…	5 L	60 L	B	40
	Acetyl benzoyl peroxide, solid, or with more than 40 percent in solution	Forbidden								
	Acrolein, stabilized	6.1	UN1092	I	6.1, 3	…	Forbidden	Forbidden	D	40
	Alcoholic beverages	3	UN3065	II	3	…	5 L	60 L	A	
I	Ammonia, anhydrous	2.3	UN1005		2.3, 8	…	Forbidden	Forbidden	D	40, 57
D	Ammonia, anhydrous	2.2	UN1005		2.2	…	Forbidden	Forbidden	D	40, 57
D	Ammonia solutions, *relative density less than 0.880 at 15°C in water with more than 50 percent ammonia*	2.2	UN3318		2.2	…	Forbidden	25 kg	D	40, 57
I	Ammonia solutions, *relative density less than 0.880 at 15°C in water with more than 50 percent ammonia*	2.3	UN3318		2.3, 8	…	Forbidden	25 kg	D	40, 57
	Ammonia solutions, *relative density between 0.880 and 0.957 at 15°C in water, with more than 10% but not more than 35% ammonia*	8	UN2672	III	8	…	5 L	60 L	A	40, 85
	Ammonia solutions, *relative density less than 0.880 at 15°C in water, with more than 35% but not more than 50% ammonia*	2.2	UN2073		2.2	…	Forbidden	150 kg	E	40, 57
D	Asbestos	9	NA2212	III	9	…	200 kg	200 kg	A	34, 40
	Barium cyanide, solid	6.1	UN1565	I	6.1	…	5 kg	50 kg	A	26, 40
	Battery fluid, acid	8	UN2796	II	0	…	1 L	30 L	B	
	Calcium	4.3	UN1401	II	4.3	…	15 kg	50 kg	E	

Symbols	Hazardous materials descriptions and proper shipping names	Hazard class or Division	Identification Numbers	PG	Label Codes	Special provisions	Passenger aircraft/rail	Cargo aircraft only	Vessel stowage Location	Other
A	Calcium oxide	8	UN1901	III	8	...	25 kg	100 kg	A	
I	Carbon, activated	4.2	UN1362	III	4.2	...	0.5 kg	0.5 kg	A	12
I	Carbon, *animal or vegetable origin*	4.2	UN1361	II	4.2	...	Forbidden	Forbidden	A	12
		4.2	UN1361	III	4.2	...	Forbidden	Forbidden	A	12
	Caustic soda (etc.) see Sodium hydroxide, etc.									
D	Charcoal *briquettes, shell, screenings, wood, etc.*	4.2	NA1361	III	4.2	...	25 kg	100 kg	A	
	Chlorine	2.3	UN1017		2.3, 8	...	Forbidden	Forbidden	D	40, 51, 55, 62, 68, 69, 90
	Chlorobenzene	3	UN1134	III	3	...	60 L	220 L	A	
	Coal tar distillates, flammable	3	UN1136	II	3	...	5 L	60 L	B	
	Cyclohexane	3	UN1145	II	3	...	5 L	60 L	A	
	Dichloromethane	6.1	UN1593	III	6.1	...	60 L	220 L	E	
	Dinitrogen tetroxide	2.3	UN1067		2.3, 5.1, 8	...	Forbidden	Forbidden	D	40, 89, 90
	Environmentally hazardous substances, liquid, n.o.s.	9	UN3062	III	9	...	No limit	No limit	A	
	Ethyl acetate	3	UN1173	II	3	...	5 L	60 L	B	
	Ethylbenzene	3	UN1175	II	3	...	5 L	60 L	B	
	Ethylene, compressed	2.1	UN1962		2.1	...	Forbidden	150 kg	E	40
	Ethyl mercaptan	3	UN2363	I	3	...	Forbidden	30 L	E	95, 102
	Ethyl methyl ketone *or* methyl ethyl ketone	3	UN1193	II	3	...	5 L	60 L	B	
G	Flammable liquids, n.o.s.	3	UN1993	I	3	...	1 L	30 L	E	
		3	UN1933	II	3	...	5 L	60 L	B	
		3	UN1933	III	3	...	60 L	220 L	A	
G	Flammable liquids, toxic, n.o.s.	3	UN1992	I	3, 6.1	...	Forbidden	30	E	40
		3	UN1922	II	3, 6.1	...	1 L	60 L	B	40
		3	UN1922	III	3, 6.1	...	60 L	220 L	A	
	Fluorine, compressed	2.3	UN1045		2.3, 5.1, 8	...	Forbidden	Forbidden	D	40, 89, 90

TABLE 6.1 ◆ (continued)

Symbols (1)	Hazardous Materials Descriptions and Proper Shipping Names (2)	Hazard Class or Division (3)	Identification Numbers (4)	PG (5)	Label Codes (6)	...	Quantity Limitations (9)		Vessel Stowage (10)	
							Passenger Aircraft/ Rail (9A)	Cargo Aircraft Only (9B)	Location (10A)	Other (10B)[b]
D	Fuel oil (*No. 1, 2, 4, 5, or 6*)	3	NA1993	III	3	...	60 L	220 L	A	
	Gasoline	3	UN1203	II	3	...	5 L	60 L	E	
D	Hazardous waste liquid, n.o.s.	9	NA3082	III	9	...	No limit	No limit	No limit	A
D	Hazardous waste solid, n.o.s.	9	NA3077	III	9	...	No limit	No limit	A	
	Hydrazine, anhydrous or Hydrazine aqueous solutions *with more than 64 per hydrazine by mass*	8	UN2029	I	8, 3, 6.1	...	Forbidden	2.5 L	D	21, 40, 42, 100
	Hydrofluoric acid, *anhydrous see* Hydrogen fluoride, anhydrous									
	Hydrogen, compressed	2.1	UN1049		3	...	Forbidden	150 kg	E	40, 57
	Hydrogen fluoride, anhydrous	8	UN1052	I	8, 6.1	...	Forbidden	Forbidden	D	40
	Hydrogen, refrigerated liquid (*cryogenic liquid*)	2.1	UN1966		2.1	...	Forbidden	Forbidden	D	40
	Hydrogen sulfide	2.3	UN1053		2.3, 2.1	...	Forbidden	Forbidden	D	40
	Lime, unslaked, see Calcium oxide									
	Liquefied petroleum gas, see Petroleum gases, liquefied									
	Lithium aluminum hydride	4.3	UN1410	I	4.3	...	Forbidden	15 kg	E	
	London purple	6.1	UN1621	II	6.1	...	25 kg	100 kg	A	
	Lye, see Sodium hydroxide, solutions									
	Mercury oxide	6.1	UN1641	II	6.1	...	25 kg	100 kg	A	
	Methyl ethyl ketone, see Ethyl methyl ketone									
	Methyl isobutyl ketone	3	UN1245	II	3	...	5 L	60 L	B	
	Methylene chloride, see Dichloromethane									

Symbol	Hazardous materials descriptions and proper shipping names	Hazard class or Division	Identification Numbers	PG	Label Codes	Special Provisions	Passenger aircraft/rail	Cargo aircraft only	Vessel Stowage Location	Other
	Nickel carbonyl	3	UN1259		3, 6.1	...	Forbidden	Forbidden	D	18, 40
+	Nitric acid, red fuming	8	UN2032	I	7, 8.1, 6.1	...	Forbidden	Forbidden	D	40, 66, 74, 89, 90
	Nitrogen, compressed	2.2	UN1066		2.2	...	75 kg	150 kg	A	40
	Nitrogen dioxide, see Dinitrogen Tetroxide									
	Organophosphorus pesticides, solid, toxic	6.1	UN2783	I	6.1	...	5 kg	50 kg	A	40
	Oxygen, compressed	2.2	UN1072		2.2, 5.1[c]	...	75 kg	150 kg	A	
+	Oxygen generator, chemical spent	9	NA3356	III	9	...	Forbidden	Forbidden	A	
	Oxygen, refrigerated liquid (cryogenic liquid)	2.2	UN1073		2.2, 5.1[c]	...	Forbidden	Forbidden	D	
	Petroleum gases, liquefied or Liquefied petroleum gas	2.1	UN1075		2.1	...	Forbidden	150 kg	E	40
	Phosgene	2.3	UN1076	I	2.3, 8	...	Forbidden	Forbidden	D	40
	Phosphoric acid	8	UN1805	III	8	...	5 L	60 L	A	
	Phosphorus, amorphous	4.1	UN1338	III	4.1	...	25 kg	100 kg	A	74
	Phosphorus, white dry or Phosphorus, white, under water or Phosphorus, white, in solution or Phosphorus, yellow dry or Phosphorus, yellow, under water or Phosphorus, yellow, in solution	4.2	UN1381	I	4.2, 6.1	...	Forbidden	Forbidden	E	
	Phosphorus, white, molten	4.2	UN2447	I	4.2, 6.1	...	Forbidden	Forbidden	E	
	Phthalic anhydride with more than 0.05 percent maleic anhydride	8	UN2214	III	8	...	Forbidden	Forbidden	D	
	Sodium	4.3	UN1428	I	4.3	...	Forbidden	15 kg	D	
	Sodium hydroxide, solid	8	UN1823	II	8	...	15 kg	50 kg	A	
	Sodium hydroxide solution	8	UN1824	II	8	...	1 L	30 L	A	
				III	8	...	5 L	60 L	A	
	Sodium peroxide	5.1	UN1504	I	5.1	...	Forbidden	15 kg	B	13, 75, 106
	Stannic chloride, anhydrous	8	UN1827	II	8	...	1 L	30 L	C	
D	Sulfur	9	NA1350	III	9	...	No limit	No limit	A	19, 74
I	Sulfur	4.1	UN1350	III	4.1	...	No limit	No limit	A	19, 74

TABLE 6.1 ◆ (continued)

Symbols (1)	Hazardous Materials Descriptions and Proper Shipping Names (2)	Hazard Class or Division (3)	Identification Numbers (4)	PG (5)	Label Codes (6)	...	Quantity Limitations (9) Passenger Aircraft/ Rail (9A)	Cargo Aircraft Only (9B)	Vessel Stowage (10) Location (10A)	Other (10B)[b]
D	Sulfur, molten	9	NA2248	III	9	...	Forbidden	Forbidden	C	61
I	Sulfur, molten	4.1	NA2248	III	4.1	...	Forbidden	Forbidden	C	61
G	Tear gas substances, liquid, n.o.s.	6.1	UN1693	I	6.1	...	Forbidden	Forbidden	D	40
	Tetrachloroethane	6.1	UN1702	II	6.1	...	5 L	60 L	A	40
	Toluene	3	UN1294	II	3	...	5 L	60 L	B	
	1,1,1- Trichloroethane	6.1	UN2831	III	6.1	...	60 L	220 L	A	40
	Trinitrotoluene, or TNT, *dry or wetted with 30% water by mass*	1.1D	UN0209	II	1.1D	...	Forbidden	Forbidden	10	
	Trinitrotoluene, or TNT, *wetted with not less than 30% water by mass*	4.1	UN1356	I	4.1	...	0.5 kg	0.5 kg	E	28
	Xylenes	3	UN1307	II	3	...	5 L	60 L	B	
		3	UN1307	III	3	...	60 L	220 L	A	

[a]Excerpted from the DOT Hazardous Materials Table, 49 CFR §172.101. Columns 7 and 8 have been redacted.
[b]Examples of the coded provisions are provided in Table 6.7. For the provisions not listed in Table 6.7, the interested reader should consult 49 CFR §176.84.
[c]See also Section 7.1, p. 261.

entries for each designated hazardous material, the entries for columns 7 and 8 have been omitted here, since the tabulated information is not especially instructive to emergency-response personnel.

One or more of the following six symbols are sometimes entered in column 1: "+," "A," "D," "G," "I," and "W." The significance of these symbols is summarized as follows:

- A "+" sign in column 1 specifies the selection of the proper shipping name, hazard class, and packing group for a given entry in columns 2, 3, and 4, respectively. These terms are described shortly. These three entries may not be changed, irrespective of any other information.
- The letter "A" signifies that DOT regulates the transportation of the given material by aircraft only, unless the material is a hazardous waste or hazardous substance.
- The letter "D" specifies the proper shipping name that is appropriate for describing a material in domestic transportation. The name may be inappropriate for describing the material in international transportation.
- The letter "G" specifies n.o.s. (not otherwise specified) and generic proper shipping names for which one or more technical names of the hazardous material must be entered in parentheses, in association with the basic description.
- The letter "I" specifies the proper shipping name that is appropriate for describing a material in international transportation.
- The letter "W" signifies that DOT regulates the transportation of the given material by watercraft only, unless the material is a hazardous waste or hazardous substance.

With few exceptions, DOT requires a hazardous material to be properly described for transportation on a *shipping paper*. A shipping paper can take a number of forms such as a shipping order, bill of lading, manifest, railroad waybill, or similar document. The shipping papers used by Federal Express Corporation and United Parcel Service are shown in Figure 6.1.

For a given consignment, DOT requires shippers and carriers to provide the *basic description* for each hazardous material on the accompanying shipping paper. The basic description of a hazardous material provides each of the following five components in the sequence noted:

- Proper shipping name (including the technical name, when applicable)
- Hazard class or division
- Identification number
- Packing group (when applicable)
- Total quantity by mass, volume, or as otherwise appropriate

DOT requires this sequence to be entered on a shipping paper in English in the specific fashion illustrated in Figure 6.2.

In addition to entering the basic description on a shipping paper, DOT requires shippers and carriers to execute either of the following certification statements:

- "This is to certify that the above-named materials are properly classified, described, packaged, marked, and labeled, and are in proper condition for transportation according to the applicable regulations of the Department of Transportation."
- "I hereby declare that the contents of this consignment are fully and accurately described above by the proper shipping name, and are classified, packaged, marked, and labeled/placarded, and are in all respects in proper condition for transport according to applicable international and national governmental regulations."

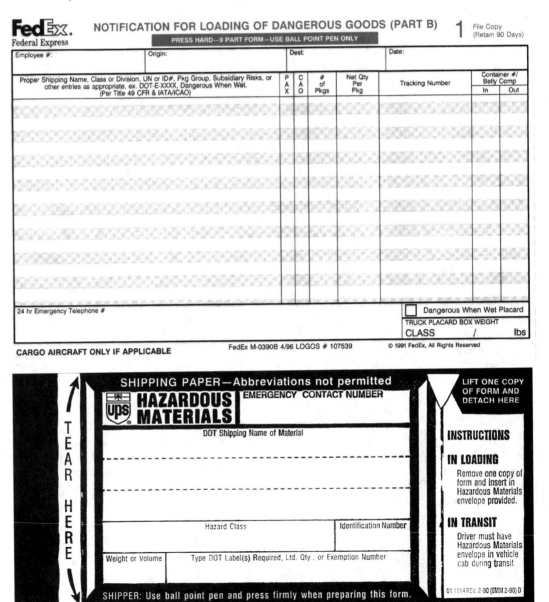

FIGURE 6.1 ◆ Federal Express and United Parcel Service require completion of their respective forms of a shipping paper when a hazardous material package is offered for transportation. Each form requests the information that DOT requires at 49 CFR §172.200.

We shall now briefly note some features of each component of the basic description of a hazardous material.

PROPER SHIPPING NAME

The *proper shipping name* of a hazardous material is the name that appears in plain, non-italic type in column 2 of the Hazardous Materials Table. It is not the alternative

STRAIGHT BILL OF LADING —
SHORT FORM — ORIGINAL — NOT NEGOTIABLE

RECEIVED, subject to the classifications and tariffs in effect on the date of the issue of this Bill of Lading.

DATE	PAGE: OF	FREIGHT PREPAID except when box at right is checked	COLLECT ☐	CARRIER	NO.	ROUTE/DEL. CARRIER

SHIPPER'S NO'S

CUSTOMER'S ORDER NO'S

CONSIGNED TO

[illegible]

Received $ ______ to apply in prepayment of the charges on the property described hereon.

______ Agent or Cashier

Per ______ (The signature here acknowledges only the amount prepaid)

Charges Advanced: $

Subject to Section 7 of conditions, of applicable Bill of Lading, if this shipment is to be delivered to the consignee without recourse on the consignor, the consignor shall sign the following statement:
The carrier shall not make delivery of this shipment without payment of freight and all other lawful charges.

(Signature of Consignor)

The fibre boxes used for this shipment conform to the specifications set forth in the box maker's certificate thereon, and all other requirements of the Consolidated Freight Classification.

HM	DESCRIPTION OF ARTICLES IF HAZARDOUS MATERIAL - PROPER SHIPPING NAME (CHEMICALS AND ACIDS - RVNX - 50¢/LB.)	EXCEPTIONS	EXEMPTIONS	** WEIGHT (SUB TO CORR)	CLASS OR RATE	NO PKGS	TYPE PKG.
x	10 drums, Gasoline, 3, UN1203, PG II			4,500 LBS			
x	40 Cylinders, Nitrogen, compressed, 2.2, UN1066			800 LBS			
x	1 Drum, Flammable solids, n.o.s., 4.1, UN1325, PG II			452 LBS			
	4 Boxes, Advertising Materials, Paper, NOI			60 LBS			
	1 Roll, Paper printing, newsprint			690 LBS			
	12 Sets, Carbon paper			22 LBS			
				LBS			
	Emergency Contact:			LBS			
	1-(__) ___-____			LBS			
				LBS			
				LBS			

Shipper —

Per ______

This is to certify that the above named materials are properly classified, described, packaged, marked and labeled, and are in proper condition for transportation, according to the applicable regulations of the Department of Transportation.

This shipment is within the limitations prescribed for Passenger/Cargo only Aircraft. (Delete nonapplicable).

Per ______

**If the shipment moves between two ports by a carrier by water, the law requires that the Bill of lading shall state whether it is carrier's or shipper's weight.
† NOTE—Where the rate is dependent on value, shippers are required to state specifically, in writing, the agreed or declared value of the property. The agreed or declared value of the property is to be specifically stated by the shipper to be not exceeding RVNX 50¢/lb.

Agent ______ Per ______

AGENT MUST SIGN ORIGINAL BILL OF LADING AND MUST DETATCH AND RETAIN THE SHIPPING ORDER

FIGURE 6.2 ◆ The proper shipping description of a hazardous material must be provided in the manner that DOT prescribes on a shipping paper. When the consignment is involved in a transportation mishap, the components of the proper shipping description allow a team responding to the disaster to readily identify the nature of each hazardous material therein.

name sometimes listed in italics or any other name used to describe the material. The hazardous material is not always described in this table by its chemical name. A technical name of a commercial preparation or mixture may be provided instead.

When a consignment comprises hazardous materials and other materials whose transportation is not regulated by DOT, the proper shipping names of the hazardous materials must be entered *first* on the relevant shipping paper; or a contrasting color or an "x" in the "HM" column can be entered to identify the entries that are hazardous materials. As in Figure 6.2, gasoline, nitrogen, and an unspecified flammable solid are listed on the shipping paper *before* the materials that are not subject to the DOT hazardous materials regulations. The hazardous materials are also specifically identified through use of an "x" in the "HM" column.

Shippers and carriers are required to describe a particular hazardous material by selecting the name from the Hazardous Materials Table that most accurately describes the material to be transported. Suppose a carrier desires to transport ten 55-gallon drums containing industrial heating oil. The carrier must identify this commodity as "Fuel Oil" on a shipping paper accompanying the shipment, since the Hazardous

Materials Table provides only this name as the closest description of the commodity. An arbitrary selection of "Oil," "Heating Oil," or any other similar term does not comply with the intent of the DOT regulations. DOT does allow the use of the name of the hazardous material listed in italics in the table, but only *in addition to* the proper shipping name. In this example, the proper shipping name could also be correctly denoted as "Fuel Oil (Fuel Oil No. 2)," but only when this name constitutes the fuel's closest description.

In the event that the correct technical name of a material is neither listed nor entirely accurate, selection is then made from certain general descriptions; or shippers and carriers can use the letters *n.o.s.* (meaning "not otherwise specified"), *n.o.i.* (meaning "not otherwise indexed"), or *n.o.i.b.n.* (meaning "not otherwise indexed by name").

On occasion, the letters "RQ" are used with the basic description of a hazardous material. "RQ" refers to the amount of a hazardous substance (Section 1.5) that constitutes a reportable quantity. Reportable quantities of hazardous substances are provided in transportation regulations,* several of which are provided in Table 6.2.

There are only five reportable quantities: 5000 lb (2270 kg); 1000 lb (454 kg); 100 lb (45.4 kg); 10 lb (4.54 kg); and 1 lb (0.54 kg). The variation of the reportable quantities is associated with the relative differences in the toxicity of hazardous substances; that is, severely toxic hazardous substances have much lower reportable quantities than less toxic hazardous substances.

When shippers or carriers intend to transport *in one package* a hazardous substance in an amount equal to or greater than its reportable quantity, DOT requires them to enter "RQ" on the shipping paper either before or after its basic description. Suppose, for instance, that a carrier intends to ship 4500 lb of aqueous ammonia containing not less than 12% but not more than 44% ammonia by railroad tankcar. The reportable quantity of aqueous ammonia is listed in Table 6.2 as 1000 lb (454 kg). Since the carrier intends to ship an amount of ammonia exceeding the reportable quantity, the letters "RQ" must be included with the basic description for this commodity on the accompanying waybill as either of the following:

RQ, Ammonium hydroxide, 8, UN2672, III	4500 lb
Ammonium hydroxide, 8, UN2672, III, RQ	4500 lb

The RQ entry is very important. It quickly alerts emergency-response personnel to the presence of a hazardous substance that, when released, adversely affects public health and the environment. First-on-the-scene responders who identify spills or leaks of this hazardous substance can then implement procedures to minimize or reduce the impact of the release.

When shippers and carriers intend to transport a commodity by means of a ship or other form of watercraft, they must determine whether it is a marine pollutant. A *marine pollutant* is a material whose transportation is regulated to reduce or eliminate the impact on the aquatic environment from releases into waterways. Some marine pollutants are not hazardous materials. DOT provides a listing of the marine pollutants,[†] an excerpt of which is reproduced in Table 6.3.

---

*A complete listing of the hazardous substances and their reportable quantities is provided at 40 CFR §302.4 and 49 CFR §172.101, Appendix A, Table 1.

[†]A complete listing of the marine pollutants is tabulated at 49 CFR §172.101, Appendix B.

TABLE 6.2 ◆ **Some Hazardous Substances and Their Reportable Quantities**[a]

Hazardous Substance	Reportable Quantity	
	Pound	Kilogram
Acetone	5000	2270
Acetonitrile	5000	2270
Ammonia solutions (as ammonium hydroxide)	1000	454
Asbestos, white	1	0.454
Chlorine	10	4.54
Chlorobenzene	100	45.4
Dichloromethane	1000	454
Hydrazine	1	0.454
Hydrogen fluoride, anhydrous	100	45.4
Hydrogen sulfide	100	45.4
Mercury oxide		___[b]
Methyl ethyl ketone	5000	2270
Methyl parathion	100	45.4
Nickel carbonyl	10	4.54
Nitric acid	1000	454
Phosgene	10	4.54
Phosphine	100	45.4
Phosphorus	1000	454
Phthalic anhydride	5000	2270
Sodium	10	4.54
Sodium hydroxide	1000	454
1,1,2,2-Tetrachloroethane	100	45.4
Toluene	1000	454
1,1,1-Trichloroethane	1000	454
Xylene(s)	100	45.4

[a]40 CFR §302.4 and 49 CFR §172.101, Appendix A, Table 1.
[b]Although all compounds of mercury are hazardous substances, their reportable quantities are not listed in the references noted in footnote a.

When a material intended for shipment is a marine pollutant, the words "marine pollutant" must be included with the basic description. For example, when shippers and carriers intend to transport 10 lb (4.54 kg) of mercury(II) oxide by watercraft, they convey that the substance is a marine pollutant on the accompanying shipping paper as follows:

10 lb   Mercury oxide, 6.1, UN1641, II (marine pollutant)

TABLE 6.3 ◆ Some Marine Pollutants[a]	
Arsenic chloride	Lead acetate
Barium cyanide	London purple
Cadmium compounds	Mercuric oxide
Carbon tetrachloride	Phosphorus, white or yellow
Chlorine	Sodium cyanide
Coal tar	Zinc bromide
Cupric sulfate	

[a]49 CFR §172.101, Appendix B.

When the name of the commodity does not include the name of the component that causes it to be a marine pollutant, the name of the component must be identified parenthetically. For example, when shippers and carriers transport 6.0 lb (2.7 kg) of the insecticide *London purple* by watercraft, they denote it on the accompanying shipping paper as follows:

> 6 lb   London purple (contains diarsenic trioxide and aniline),
> 6.1, UN1621, II (marine pollutant)

Here, the first parenthetical entry identifies the substances in London purple that cause it to be a marine pollutant.

When providing the basic description of a hazardous material, shippers and carriers must be aware of other DOT regulatory requirements that may be germane to identifying the given consignment. Some examples of these regulations are provided in Table 6.4. They illustrate that valuable information can be gleaned from the entries on a shipping paper.

HAZARD CLASS

Each hazardous material regulated by DOT is associated with a specific *hazard class* that corresponds to a particular defining criterion. Nine hazard classes are symbolized by a number 1 through 9, and an additional hazard class is represented by the letters "ORM," the acronym for "other regulated materials."

Several hazard classes are further divided. Each division is symbolized by a number that follows the hazard class number; thus, "1.2" refers to division "2" of class 1. Table 6.5 lists the DOT class numbers and division numbers that have been assigned to hazardous materials, as well as the name of the label that DOT requires shippers and carriers to affix to the hazardous materials packaging. We shall observe the format of each DOT label in Section 6.4.

For a given hazardous material, the hazard class is determined from the entry in column 3 of the Hazardous Materials Table. When providing the basic description of a hazardous material on a shipping paper, the column 3 designation is entered on a shipping paper immediately adjacent to the proper shipping name. Thus, as represented in Figure 6.2, shippers and carriers desiring to transport gasoline entered a "3" next to "Gasoline."

Sometimes, the word "forbidden" is entered in column 3 of the Hazardous Materials Table. This means that DOT prohibits shippers and carriers from accepting this hazardous material for transportation *by any mode*.

TABLE 6.4 ◆ **Examples of Additional Information Required by DOT in Basic Descriptions**[a,b]

Example	*Relevant Regulation*
Transportation of a Hazardous Material by Any Mode	
RQ, Hydrogen fluoride, anhydrous, 8, UN1052, I *or* Hydrogen fluoride, anhydrous, 8, UN1052, I, RQ	When a shipper or carrier transports a hazardous substance in an amount that equals or exceeds its reportable quantity, the letters RQ are included before or after the basic description.
RQ, Sodium, 4.3, UN1428, I (Dangerous when wet)	When a hazardous material, by chemical interaction with water, is liable to become spontaneously flammable or give off flammable gases in dangerous quantities, the words "Dangerous when wet" are included with the basic description.
Corrosive liquids, n.o.s. (Valeric acid), 8, UN1760, I	If a hazardous material is described as an n.o.s. entry in the Hazardous Materials Table, the basic description includes the name of the substance in parentheses immediately following the proper shipping name.
Flammable liquids, toxic, n.o.s. (contains xylene and methanol), 3, UN1992, I	When a mixture or solution of two or more hazardous materials described by an n.o.s. entry in the Hazardous Materials Table is to be transported, the technical names of at least two components most predominantly contributing to the hazards of the mixture or solution are parenthetically entered immediately next to the basic description.
RQ, Hydrazine aqueous solutions, 8, UN2029 (poison) *or* RQ, Hydrazine aqueous solutions, 8, UN2029 (toxic)	Notwithstanding the hazard class to which a hazardous material is assigned, if a liquid or solid material in a package meets the definition of a Division 6.1, PG I or II, and the fact that it is a poison is not disclosed in the shipping name or class entry, either of the words "poison" or "toxic" is entered in the basic description.
Organophosphorus pesticides, solid, toxic, n.o.s., 6.1, UN2783, I, RQ (Ciodrin)	Notwithstanding the hazard class to which a hazardous material is assigned, if the technical name of the compound or principal constituent that causes a material to meet the definition of a Division 6.1, PG I or II, is not included in the proper shipping name for the material, the technical name is entered in parentheses in association with the basic description.
RQ, Phosgene, 2.3, UN1076 (POISON—Inhalation Hazard, Zone A)	For materials that are poisonous by inhalation, the words "POISON—Inhalation Hazard" and the words "Zone A," "Zone B," "Zone C," or "Zone D" for gases or "Zone A" or "Zone B" (Section 10.1) for liquids, as appropriate, are entered immediately adjacent to the basic description.
Flammable solids, n.o.s., 4.1, UN1325, II (sodium)	If a hazardous material is also a hazardous substance other than a radioactive material, and the name of the hazardous material does not identify the hazardous substance by name, one of the following descriptions is entered in parentheses in association with the basic description of the hazardous material: **(a)** the name of the hazardous substance; *or*
Hazardous waste liquid, n.o.s., 9, UN3082, III (F001)[c]	**(b)** for waste streams, the hazardous waste number; *or*
Hazardous waste liquid, n.o.s., 9, UN3082, III (EPA ignitability)	**(c)** for waste streams that exhibit an EPA characteristic, the letters "EPA" followed by the characteristic or the EPA hazardous waste number.[d]

TABLE 6.4 ◆ (continued)

Example	Relevant Regulation
Transportation of a Hazardous Material by Any Mode	
RQ, London purple (contains diarsenic trioxide and aniline), 6.1, UN1621, II	If a hazardous material is also a hazardous substance other than a radioactive material, and the name of the hazardous material does not identify the hazardous substance by name, the letters RQ are entered either before or after the basic description.
Residue: Last contained ethyl methyl ketone, 3, UN1193, II	The basic description for the residue of a hazardous material contained in packaging other than a railroad tankcar includes the words "Residue: Last contained _____ (name of hazardous material in packaging before it was emptied)."
RQ, Hydrogen fluoride, anhydrous, 8, UN1052, I (DOT-E***)	Each basic description of a hazardous material under a DOT-approved exemption bears the notation "DOT-E" followed by the exemption number assigned and so located that the notation is clearly associated with the description to which the exemption applies. A DOT exemption is issued by DOT. It allows shippers and carriers to transport a hazardous material legally in an unconventional fashion. (*** is replaced with the appropriate exemption number.)
Transportation of a Hazardous Material by Passenger-Carrying Railroad or Aircraft	
Stannic chloride, anhydrous, 8, UN1827, II (Limited Quantity) *or* Stannic chloride, anhydrous, 8, UN1827, II (Ltd Qty)	The basic description for a material transported in a "limited quantity" by passenger-carrying railroad or aircraft as determined through use of the Hazardous Material Table includes the words "Limited Quantity" or "Ltd Qty" following the basic description.
Transportation of a Hazardous Material by Cargo Aircraft	
Ethyl mercaptan, 3, UN2363, I (Cargo aircraft only)	When DOT allows the transportation of a package containing a hazardous material by cargo aircraft and the Hazardous Materials Table prohibits its transportation aboard passenger-carrying aircraft, the words "Cargo aircraft only" are entered after the basic description.
Ethyl mercaptan, 3, UN2363, I (Cargo aircraft only) (Limited Quantity) *or* Ethyl mercaptan, 3, UN2363, I (Cargo aircraft only) (Ltd Qty)	The basic description for a material transported in a "limited quantity" by cargo aircraft as determined through use of the Hazardous Material Table includes the words "Limited Quantity" or "Ltd Qty" following the basic description.
Transportation of a Hazardous Material on Water	
Phosphorus, white, under water, 4.2, UN1381, I (marine pollutant)	When a shipper or carrier transports a marine pollutant on water, the words "marine pollutant" are included with the basic description.
London purple (contains diarsenic trioxide and aniline), 6.1, UN1621, II (marine pollutant)	When a shipper or carrier transports on water a marine pollutant that does not include the name of the component causing it to be a marine pollutant, the name of the component is identified parenthetically.

TABLE 6.4 ◆ (*continued*)

Example	*Relevant Regulation*

Transportation of a Hazardous Material on Water

Environmentally hazardous substances, liquid, n.o.s., 9, UN3082, III (sodium cyanide and copper) (marine pollutant)

When a material designated with an n.o.s. entry in the Hazardous Materials Table is a marine pollutant, the names of at least two of the components most predominantly contributing to the marine pollutant designation appear in parentheses in association with the description.

RQ, Chlorobenzene, 3, UN1134, III (4 steel drums/pallet; gross mass: 515 lb/drum)

When a hazardous material is to be transported on water, the type of packaging, the number of each type of packaging on a pallet or freight container, and the gross mass of each type of package are identified in addition to the basic description.

Transportation of a Hazardous Material by Railroad

RQ, Petroleum gases, liquefied, 2.1, UN1075 (Placarded FLAMMABLE GAS)

The basic description of a hazardous material contained within a railroad tankcar includes the notation "Placarded" followed by the name of the placard that DOT requires to be posted on the railroad car.

RQ, Petroleum gases, liquefied, 2.1, UN1075 (DOT-113A) (Placarded FLAMMABLE GAS) (Do Not Hump or Cut Off Car While in Motion)

The basic description of a flammable gas contained in a DOT specification railroad tankcar contains an appropriate notation, such as "DOT-113A" and the statement "Do Not Hump or Cut Off Car While in Motion."

HOT Phthalic anhydride, 8, UN2214, III, RQ (Maximum Operating Speed 15 mph)

When DOT permits the transportation by railroad of a hazardous material at an elevated temperature pursuant to certain denoted exceptions, the basic description contains an appropriate notation such as "Maximum Operating Speed 15 mph."

HOT Phthalic anhydride, 8, UN2214, III, RQ

If a liquid hazardous material other than molten sulfur and molten aluminum is to be transported at an "elevated temperature," and the fact that it is an elevated-temperature material[e] is not disclosed in the shipping name, the word "HOT" immediately precedes the proper shipping name of the hazardous material.

RESIDUE: Last contained ethyl methyl ketone, 3, UN1193, II (placarded FLAMMABLE RESIDUE)

The basic description for the residue of a hazardous material contained in a railroad tankcar includes the words "Residue: Last contained _____ (name of hazardous material in a railroad tankcar before it was emptied)" together with the appropriate placard notation ("Placarded: _____ RESIDUE").

Transportation of Anhydrous Ammonia by Highway

RQ, Ammonia, anhydrous, 2.2, UN1005 (0.2 percent water) (Inhalation Hazard)

When a carrier transports anhydrous ammonia having a water content equal to or greater than 0.2% by mass in a DOT Specification MC 330 or MC 331 tank truck, the basic description includes "0.2 percent water" to indicate the suitability for shipping anhydrous ammonia in a cargo tank constructed from quenched and tempered steel.

RQ, Ammonia, anhydrous, 2.2, UN1005 (Not for Q and T tanks) (Inhalation Hazard)

When a carrier transports anhydrous ammonia having a water content less than 0.2% by mass in a DOT Specification MC330 or MC 331 tank truck, the basic description includes the phrase "Not for Q and T tanks."

TABLE 6.4 ◆ *(continued)*	
Example	*Relevant Regulation*
Transportation of Anhydrous Ammonia by Highway	
RQ, Ammonia, anhydrous, 2.2, UN1005 (0.2 percent water) (Inhalation Hazard)	When a carrier transports anhydrous ammonia or ammonia solutions, *relative density less than 0.880 at 15 degrees C in water,* the basic description includes the words "Inhalation Hazard" without the word POISON or the designation of a zone.
Transportation of Liquefied Natural Gas by Highway	
RQ, Petroleum gases, liquefied, 2.1, UN1075 (Noncorrosive)	When a carrier transports liquefied petroleum gas in a DOT Specification MC 330 or MC 331 tank truck, the basic description includes either "Noncorrosive" (or "Noncor") or "Not for Q and T tanks," as appropriate.

[a]49 CFR §172.203.

[b]These examples do not include the additional information DOT requires in the basic descriptions of radioactive materials, self-reactive materials, organic peroxides, and explosive materials.

[c]As noted in Section 1.5, the EPA hazardous waste numbers for *listed* hazardous wastes are prefixed with one of the capital letters "F," "K," "P," or "U." Listed hazardous wastes are provided at 40 CFR §§261.31, 261.32, 261.33(e) and 261.33(f).

[d]As set forth in EPA regulations at 40 CFR §§§261.21–261.24, there are four EPA *characteristics* of a hazardous waste: ignitability (Section 3.2); corrosivity (Section 8.13); reactivity (Section 5.8); and toxicity (Section 10.1). Hazardous wastes exhibiting these characteristics have been assigned the EPA hazardous waste numbers "D001," " D002," "D003," and "D004 through D043," respectively.

[e]DOT defines an *elevated-temperature material* as a hazardous material that, when offered for transportation or transported in a bulk packaging exists in the:

- ◆ Liquid phase of matter at a temperature at or above 212°F (100°C);
- ◆ Liquid phase of matter with a equal to or greater than 100°F (37.8°C) that is intentionally heated and offered for transportation or transported at or above its; *or*
- ◆ Solid phase of matter at a temperature equal to or greater than 464°F (240°C).

IDENTIFICATION NUMBER

The identification numbers of the hazardous materials are designated in column 4 of the Hazardous Materials Table. They are three- or four-digit numbers preceded by either the prefix "UN" (United Nations) or "NA" (North America). The identification numbers preceded by "UN" are associated with the proper shipping names for hazardous materials transported domestically and internationally, while those preceded by "NA" are associated with the proper shipping names for hazardous materials transported domestically and within Canada *only*.

DOT requires the identification number for a hazardous material to be entered on a shipping paper immediately following the designation of the hazard class. For example, in the basic description for gasoline in Figure 6.2, shippers and carriers must enter the identification number "UN1203" to the immediate right of the "3" designating the hazard class of gasoline.

An identification number is assigned to a hazardous material so that relevant information concerning the consignment is immediately available to emergency-response teams. When the identification number is known, team members can quickly identify the nature of a hazardous material and respond effectively to the incident in which the material is involved. This is accomplished by referring to a guidebook

TABLE 6.5 ◆ Labels Required on Packaging of Hazardous Materials According to Their Classes and Divisions[a]

Hazard Class or Division	Label Name
1.1	EXPLOSIVE 1.1
1.2	EXPLOSIVE 1.2
1.3	EXPLOSIVE 1.3
1.4	EXPLOSIVE 1.4
1.5	EXPLOSIVE 1.5
1.6	EXPLOSIVE 1.6
2.1	FLAMMABLE GAS
2.2	NON-FLAMMABLE GAS
2.3	INHALATION HAZARD
3	FLAMMABLE LIQUID
Combustible liquid	(None)
4.1	FLAMMABLE SOLID
4.2	SPONTANEOUSLY COMBUSTIBLE
4.3	DANGEROUS WHEN WET
5.1	OXIDIZER
5.2	ORGANIC PEROXIDE
6.1 (inhalation hazard, Zone A or B)	INHALATION HAZARD
6.1 (other than inhalation hazard, Zone A or B)	POISON
6.2	INFECTIOUS SUBSTANCE
7	RADIOACTIVE WHITE-I
7	RADIOACTIVE YELLOW-II
7	RADIOACTIVE YELLOW-III
7 (empty packages)[b]	EMPTY
8	CORROSIVE
9	CLASS 9

[a]49 CFR §172.400.
[b]See Section 15.7.

published by DOT, the *Emergency Response Guidebook*. In the guidebook, each identification number is referenced to a guide number that provides specific instructions relating to appropriate response-actions. We shall examine the use of this guidebook in Section 6.8.

PACKING GROUP

Packaging is defined by DOT as "the receptacle and any other components or materials necessary for the receptacle to perform its containment function and to ensure compliance with the minimum [DOT] packing requirements." Two types of packaging are used to containerize hazardous materials:

- Bulk packaging, such as tankcars, cargo tanks, and portable tanks
- Nonbulk packaging, such as drums, cans, cardboard boxes, fiberboard boxes, cylinders, and bags

Nonbulk packaging may also consist of an inner package that has been overpacked to provide an additional measure of safety during transit.

DOT requires packaging to "be designed, constructed, maintained, filled, its contents so limited, and closed," so that each condition noted below is provided by the manufacturer when the packaging is used to transport a hazardous material:

- The release of hazardous materials to the environment will be minimized.
- The effectiveness of the packaging will not be significantly reduced.
- Through any credible spontaneous increase of heat or pressure, a mixture of gases or vapors in the package could not significantly reduce the effectiveness of the packaging.

DOT's packaging requirements for the type of shipping container are specific to the relative degree of hazard possessed by the material to be transported; that is, materials posing similar hazards are generally packaged in the same manner. Hazardous materials that present the most severe hazards are required to be transported in the most secure and strongest packaging.

DOT often distinguishes the authorized means of packaging hazardous materials for bulk and non-bulk shipments. For most hazard classes, DOT assigns a *packing group* and identifies it in column 5 of the Hazardous Materials Table for each hazardous material. DOT notes only three packing groups: Packing Groups I, II, and III. The packing group affects the selection of packaging when shipping a hazardous material. Packaging in Packing Group I is required when the hazardous material is associated with a great degree of danger; packaging in Packing Group II is required when the hazardous material is associated with a medium degree of danger; and packaging in Packing Group III is required when the hazardous material is associated with a minor degree of hazard. DOT does not assign packing groups to hazard classes 1, 7, and 9 and divisions 2.1 and 2.2, since these hazard classes and divisions require their own specially approved packaging. As Table 6.1 notes, DOT allows some hazardous materials to be shipped in packaging associated with each packing group, generally in conjunction with certain provisions, prohibitions, and exceptions.

Referring again to Table 6.1, we see that DOT has assigned the Packing Group II to gasoline. Shippers and carriers interested in transporting gasoline must consult the appropriate section of the DOT regulations to establish the authorized types of bulk and nonbulk packaging that DOT approves for the transportation of this material.

When a packing group for a specific material is denoted in column 5 of the Hazardous Materials Table, DOT requires it to be listed in the relevant basic description immediately following the identification number. For such purposes, the words "packing group" are generally abbreviated as "PG." Thus, when shippers and carriers intend to transport 5000 gal of gasoline, they provide the basic description of this commodity on a shipping paper by one of the following forms:

Gasoline, 3, UN1203, Packing Group II	5000 gal
Gasoline, 3, UN1203, PG II	5000 gal
Gasoline, 3, UN1203, II	5000 gal

TOTAL QUANTITY

DOT also requires the total quantity or amount of each hazardous material in a consignment to be communicated with its basic description. This is typically provided as a mass or volume. Abbreviations are used to express the common units of measurement. Furthermore, the type of packaging (drums, boxes, etc.) in which the quantity is contained are entered in any appropriate manner. The basic description of each hazardous material represented in Figure 6.2 includes 4500 lb of gasoline in 10 drums, 800 lb of nitrogen in 40 cylinders, and 452 lb of a flammable solid in 1 drum.

EMERGENCY-RESPONSE TELEPHONE NUMBER

DOT requires an emergency-response telephone number to be positioned on the shipping paper as follows:

+ After each basic description of a hazardous material
+ In a readily located position if there are multiple hazardous materials listed on the same paper and a single telephone number applies to all of them, or if there is only a single hazardous material listed on the shipping paper

The word "emergency" must also be readily visible, as in Figure 6.2.

Solved Exercise 6.1

The basic description of a hazardous material is provided on a transportation manifest as follows:

> RQ, Hazardous waste liquid, n.o.s. (contains toluene
> and xylene), 9, NA3082, PGII (EPA ignitability)   10,000 gal

What information does this basic description provide to the first-on-the-scene responders that are called to a transportation mishap involving an overturned, leaking tank truck?

   Solution: There are at least four pieces of information in this basic description that are useful to emergency-response teams:

+ The notation "RQ" tells the first-on-the-scene responders that the commodity is a hazardous substance, the release of which could adversely affect public health and the environment. The emergency-response crew should dam or dike the area to confine the leak and prevent its widespread release into the environment.
+ The reference to "EPA ignitability" indicates that the hazardous waste liquid exhibits the RCRA characteristic of ignitability (Section 3.2). The designation informs emergency-response personnel that the hazardous waste liquid possesses a flashpoint equal to or less than 140°F (60°C). Thus, it poses the risk of fire and explosion.
+ Packing Group II indicates that the relative degree of hazard possessed by the commodity is medium. A medium degree of hazard conveys the message that while the material has hazardous properties, other groups of hazardous materials are both less and more hazardous.
+ In Section 6.8, we shall see that the "NA3082" directs emergency-response personnel to recommended procedures for properly responding to an incident involving the release of this hazardous waste liquid.

Solved Exercise 6.2

A tear gas manufacturer wishes to ship 25 lb (11 kg) of "Riot Away," a liquid tear gas product, to the police department in Cincinnati, OH. The active component of the tear gas is the substance whose chemical name is α-chloroacetophenone. What basic description does DOT require the shipper and carrier to enter on the accompanying shipping paper?

Solution: For "Tear gas substances, liquid, n.o.s.," the entry in column 1 of Table 6.1 is "G," indicating that a technical name must be included in the basic description of this hazardous material. Since the active component of the tear gas is α-chloroacetophenone, the basic description is the following:

> Tear gas substances, liquid, n.o.s. (contains
> α-chloroacetophenone), 6.1, UN1693, PG I   25 lb

Performance Goal for Section 6.2
- Identify the locations at which emergency-responders can expect to locate the shipping paper during a transportation mishap involving hazardous materials.

◆ 6.2 LOCATION OF A SHIPPING PAPER DURING TRANSIT

During a transportation mishap, where should emergency-response personnel expect to locate the shipping paper? DOT requires the shipping paper to be carried on transport vehicles in a specific location. Since it is important to quickly locate this document when responding to incidents involving the release of a hazardous material, these requirements are briefly summarized next.

MOTOR CARRIER

DOT requires the drivers of motor vehicles transporting hazardous materials and each carrier using the vehicles to clearly distinguish the appropriate shipping paper from other papers. When the driver is at the vehicle's controls, DOT requires the shipping paper to be within immediate reach and readily visible to any person entering the driver's compartment. For example, it may be stored in a holder mounted to the inside of the door on the driver's side of the vehicle as in Figure 6.3. When the driver is not at the vehicle's controls, DOT requires the shipping paper to be located either in the holder or on the driver's seat inside the vehicle.

RAIL CARRIER

DOT requires a member of the train crew to be in charge of the shipping paper describing the transportation of a hazardous material consignment by railroad. The train crew is also required to retain in its possession a document indicating the position in the train of each loaded and placarded car containing a hazardous material.

Shipping
papers

FIGURE 6.3 ◆ When the driver of a motor vehicle is at the vehicle's controls, shipping papers describing hazardous materials are required by DOT to be kept within immediate reach. This generally means that the papers are in a holder mounted to the inside of the door on the driver's side of the vehicle.

AIRCRAFT

DOT requires the aircraft carrier to provide a shipping paper describing the transportation of a hazardous material consignment by aircraft to the pilot in command.

WATERCRAFT

DOT requires a carrier transporting hazardous materials onboard watercraft to prepare a dangerous cargo manifest, list, or stowage plan. This document must be contained within a designated holder on or near the vessel's bridge.

Performance Goals for Section 6.3
- ◆ Describe how shippers and carriers determine the maximum quantity limits of a hazardous substance allowed for transportation onboard passenger-carrying aircraft or railcar, cargo aircraft, or watercraft.
- ◆ Describe the nature of the DOT regulations pertaining to the stowage of hazardous materials onboard watercraft.

◆ 6.3 TRANSPORTING HAZARDOUS MATERIALS BY PASSENGER-CARRYING AIRCRAFT OR RAILCAR, CARGO AIRCRAFT, OR WATERCRAFT

When shippers and carriers transport hazardous materials either by passenger-carrying aircraft or railcar or by cargo aircraft, they must acknowledge that certain maximum quantity limits may be applicable. DOT denotes the maximum quantity limits on passenger-carrying aircraft or railcar as the entries in column 9A of the Hazardous Materials Table, and on cargo aircraft as the entries in column 9B. Excerpts from column 9 are provided in Table 6.1. Examination of the table discloses that DOT requires shippers and carriers to transport no more than 5 L of gasoline on passenger-carrying aircraft and railcar, whereas they can transport no more than 60 L on cargo aircraft. Both are subject to applicable packaging requirements.

Solved Exercise 6.3

Does DOT limit or prohibit a cargo aircraft carrier from transporting methylene chloride?

Solution: By reference to the entry in column 2 of Table 6.1, we first see that methylene chloride and dichloromethane are synonyms. By reference to the entry in column 9B, we determine that DOT prohibits the carrier from transporting by cargo aircraft a consignment of dichloromethane whose volume exceeds 220 L.

The manner in which nonbulk quantities of hazardous materials are placed on most transport vehicles is not regulated by DOT. When this shipment occurs by means of watercraft, however, certain stowage requirements apply. These requirements are coded in column 10 of the Hazardous Materials Table, excerpts of which are included in Table 6.1. There are two components of column 10:

* Column 10A, titled "location," specifies the authorized stowage location on cargo-carrying vessels and passenger-carrying vessels by means of five letter codes: "A," "B," "C," "D," and "E." Each letter code specifies the authorized location aboard the watercraft as provided in Table 6.6.

TABLE 6.6 ◆ DOT Authorized Stowage Locations Onboard Watercraft[a]

Stowage Category	Location at Which Hazardous Materials Can Be Stowed
A	"On deck" or "under deck" upon a cargo- or passenger-carrying vessel
B	"On deck" or "under deck" upon a cargo-carrying vessel and upon a passenger-carrying vessel having a number of passengers limited to not more than the *larger* of the following: • 25 passengers; or • One passenger per each 9.8 ft (3 m) of overall vessel length, or "On deck" only upon passenger-carrying vessels on which there are more than 25 passengers
C	"On deck" only on a cargo- or passenger-carrying vessel
D	"On deck" only on a cargo-carrying vessel and on a passenger carrying vessel having a number of passengers limited to not more than the larger of 25 passengers or one passenger per each 9.8 ft (3 m) of overall vessel length. Carriage of the hazardous material is prohibited on passenger-carrying vessels on which this limiting number of passengers is exceeded.
E	"On deck" or "under deck" on a cargo-carrying vessel and on a passenger-carrying vessel having a number of passengers limited to not more than the larger of 25 passengers or one passenger per each 9.8 ft (3 m) of overall vessel length, but carriage of the hazardous material is prohibited on passenger-carrying vessels on which the limiting number of passengers is exceeded.

[a]49 CFR §176.83, Table 176.83(f).

TABLE 6.7 ◆ Examples of DOT Stowage Provisions and Segregation Requirements for Hazardous Materials Onboard Watercraft[a]

Code	Provision
2	Temperature controlled hazardous material
8	Glass carboys not permitted on passenger vessels
12	Keep as cool as reasonably practicable
13	Keep as dry as reasonably practicable
14	For metal drums, stowage permitted under deck on cargo vessels
25	Shade from radiant heat
34	Stow "away from" foodstuffs
40	Stow "clear of living quarters"
42	Stow "away from" nitric acid and perchloric acid not exceeding 50% acid by mass
49	Stow "away from" corrosives
32	Stow "away from" copper, its alloys, and its salts[b]
39	Stow "away from" liquid, halogenated hydrocarbons
61	Stow "separated from" corrosive materials
66	Stow "separated from" flammable solids.
71	Stow "separated from" nitric acid
74	Stow "separated from" oxidizers
85	Under deck stowage must be in mechanically ventilated area
89	Segregation same as for oxidizers
90	Stow "separated from" radioactive materials
91	Stow "separated from" flammable liquids
95	Stow "separated from" foodstuffs
98	Stow "away from" all flammable materials
100	Stow "away from" flammable solids
102	Stow "separated from" all odor absorbing cargoes
106	Stow "separated from" powdered metals

[a]49 CFR §176.84.
[b]The term "away from" applies to all hazardous materials within the same hazard class *and* all hazardous materials for which a secondary hazard label of that class is required. For "on deck" stowage, the term "separated from" refers to a separation by a distance of at least 20 ft (6 m) horizontally. When stowed under deck, the term "separated from" refers to different compartments or holds.

- ◆ Column 10B, titled "other," specifies one or more numerical codes, each designating a specific stowage requirement for certain hazardous materials. Some examples of these special stowage provisions are provided in Table 6.7.

Referring to gasoline in Table 6.1, we note that an "E" has been entered under the heading in column 10A, while a letter entry is not listed under the heading in column 10B. From Table 6.6, we determine the "E" code means that shippers and carriers may transport gasoline, but only "on deck" or "under deck" on a cargo-carrying vessel and on a passenger-carrying vessel having a number of passengers not exceeding either of the following limits:

- Not more than 25 passengers *or*
- One passenger per each 9.8 ft (3 m) of overall vessel length

DOT prohibits the carriage of the hazardous material on passenger-carrying vessels when the limiting number of passengers is exceeded.

Performance Goal for Section 6.4
- Describe the method for determining the codes corresponding to the various labels that DOT requires shippers and carriers to affix to the packaging containing a hazardous material.

◆ 6.4 DOT LABELING REQUIREMENTS

When shippers and carriers intend to transport a hazardous material, DOT requires them to affix one or more warning labels directly to the outside surfaces of the packaging. DOT provides label codes corresponding to the required labels by the entries in column 6 of the Hazardous Materials Table. DOT requires the label corresponding to the relevant code to be affixed on the exterior surface of one of the sides or on the top of the relevant packaging. For instance, when shippers and carriers transport gasoline in the 55-gallon steel drum in Figure 6.4, they affix a FLAMMABLE LIQUID label to the outside surface of the drum.

Solved Exercise 6.4

Identify the label(s), if any, that DOT requires a motor carrier to affix to each of five 55-gallon steel drums containing methyl ethyl ketone.

Solution: We first learn that the substance is listed as ethyl methyl ketone in column 2 of Table 6.1. Then, from the entry in column 6, we determine that DOT requires the carrier to affix a FLAMMABLE LIQUID label on the exterior surface of each drum containing this substance.

Solved Exercise 6.5

Identify the label(s), if any, that DOT requires a motor carrier to affix to five fiberboard boxes, each containing eight 1-lb bottles of white elemental phosphorus under water.

Solution: By reference to the entry in column 6 of Table 6.1, we determine that DOT requires the carrier to affix SPONTANEOUSLY COMBUSTIBLE and POISON labels on the exterior surface of the fiberboard boxes. These labels must appear side by side on the surface of two of the sides or ends (other than the bottoms) of the boxes.

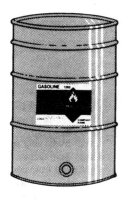

FIGURE 6.4 ◆ When gasoline is shipped in a 42-gallon steel drum, DOT requires the drum to be marked with the proper shipping name and DOT identification number; furthermore, DOT requires the FLAMMABLE LIQUID label to be affixed to the outside surface of the drum.

Each DOT label is diamond-shaped and color-coded to a specific hazard class. To warn observers about the potential hazards of the package contents, the label may illustrate a pictograph. The information conveyed by the label is associated with the principal hazard of the lading being transported. The hazard class or division number is always designated in the lower corner of the warning labels.

Depending upon the unique properties of a hazardous material, the entries in column 6 of the Hazardous Materials Table could direct shippers and carriers to affix multiple labels on packaging. When the hazardous material meets the definition of more than one hazard class, shippers and carriers are directed to display multiple labels on the packaging adjacent to one another on at least two sides or two ends excluding the bottom.

When shippers and carriers transport packages containing hazardous materials authorized solely for transport on cargo aircraft, DOT requires that they display the CARGO AIRCRAFT ONLY label in Figure 6.5 in addition to any other required labels. Shippers and carriers determine whether a hazardous material may be shipped solely by cargo aircraft (instead of passenger-carrying aircraft) by locating the word "forbidden" in column 9A of Table 6.1.

Performance Goal for Section 6.5
   ◆ Describe the information that DOT requires shippers and carriers to mark on packaging containing a hazardous material.

FIGURE 6.5 ◆ The CARGO AIRCRAFT ONLY label required on hazardous material packaging when DOT authorizes transportation only onboard cargo aircraft, but not on passenger-carrying aircraft. The label is black on an orange background.

◆ 6.5 DOT MARKING REQUIREMENTS

With few exceptions, DOT requires shippers and carriers to legibly mark certain required information on the outer surface of packages containing hazardous materials. All markings are written in English. They must be durably, plainly, and legibly displayed, unobstructed by the presence of labels, advertising, and other information on the packaging.

Some commonly encountered DOT marking requirements follow:

NONBULK PACKAGING

- When offered for transportation, nonbulk packages containing a hazardous material must be marked with the name and address of the shipper or receiver of the package.
- Nonbulk packages containing a hazardous material must be marked with the basic description and identification number of the commodity, plus any other applicable specifications or UN marks, instructions, and precautions.
  - When the proper shipping name is described by an "n.o.s." entry, the package must also be marked with the technical name of the material in parentheses immediately following or below the proper shipping name.
  - When the proper shipping name does not reveal that the lading is a toxic substance in either division 2.3 or 6.1, PG I and II, the marking must identify the toxic material in parentheses.
- Each nonbulk package that contains a hazardous substance must be marked with the letters "RQ."
- The words "POISON—INHALATION HAZARD" and the relevant hazard zone (Section 10.14) must be marked on all packages containing a hazardous material that complies with the criteria for a poison inhalation hazard.

> **POISON—INHALATION HAZARD**
> **ZONE B**

- When transporting *liquid* hazardous materials in nonbulk packaging, DOT requires that closure is upward, and that packages other than liquefied compressed gas cylinders must be marked with package orientation markings on two opposite vertical sides of the package with arrows pointing to the correct upright direction as in Figure 6.6.

FIGURE 6.6 ◆ When gasoline is shipped in a 5-gallon metal can inside a cardboard box, DOT requires orientation markings on two opposite vertical sides of the package with arrows pointing in the correct upright direction.

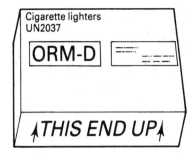

FIGURE 6.7 ◆ The proper marking that DOT requires on packaging containing a hazardous material designated as ORM-D. When solely authorized for transportation by cargo aircraft, the package must be marked "ORM-D-AIR."

- Each nonbulk packaging containing a material classified as ORM-D must be marked on at least one side or an end with the ORM designation immediately following or below the proper shipping name of the material. The appropriate ORM designation is marked inside a rectangle on the surface of a package as in Figure 6.7. The package of an ORM-D material prepared for shipment by air is marked ORM-D-AIR.
- When RCRA-regulated hazardous waste is transported in containers of 110 gallons (416 liters) or less, EPA and DOT require shippers and carriers to mark each container of hazardous waste with the information in Figure 6.8.
- When a marine pollutant is shipped onboard watercraft, DOT requires shippers and carriers to affix the MARINE POLLUTANT symbol in Figure 6.9 on the relevant packages.

Solved Exercise 6.6

A shipper intends to transport six 50-lb cardboard boxes of zinc bromide as cargo by watercraft.

(a) What information does DOT require the shipper to enter on the accompanying dangerous cargo manifest?

(b) What does DOT require the shipper to affix or mark on the six packages?

Solution: (a) Table 6.3 indicates that zinc bromide is a marine pollutant. DOT requires the shipper to communicate this information on the manifest as follows:

Six boxes, Zinc bromide (marine pollutant), 300 lb

(b) Zinc bromide is not listed in Table 6.1 (nor is it listed in the complete Hazardous Materials Table at 49 CFR §172.101). This means that DOT does not require the shipper to affix DOT labels on the six boxes. Nonetheless, DOT requires the shipper to mark each box with the name and address of the shipper or receiver and a description of the package contents as follows:

Zinc bromide (marine pollutant)

DOT also requires the shipper to mark each box with the MARINE POLLUTANT symbol in Figure 6.10.

BULK PACKAGING

- DOT requires shippers and carriers to mark bulk packaging containing an elevated-temperature material other than molten aluminum and molten sulfur with the word

HAZARDOUS WASTE

FEDERAL LAW PROHIBITS IMPROPER DISPOSAL

IF FOUND, CONTACT THE NEAREST POLICE, OR
PUBLIC SAFETY AUTHORITY, OR THE
U.S. ENVIRONMENTAL PROTECTION AGENCY

PROPER D O T
SHIPPING NAME_____ UN OR NA# _____

GENERATOR INFORMATION

NAME_____

ADDRESS_____

CITY_____ STATE_____ ZIP____

EPA   EPA
ID NO _____   WASTE NO _____

ACCUMULATION   MANIFEST
START DATE_____   DOCUMENT NO _____

FIGURE 6.8 ◆ DOT and EPA require this marking on certain containers of "hazardous waste," as this term is used in RCRA (Section 1.5). Before offering hazardous waste for transportation and during its transportation, the generator is required to identify each entry on each container of hazardous waste.

HOT. Molten aluminum and molten sulfur are marked MOLTEN ALUMINUM and MOLTEN SULFUR, respectively.

- DOT requires bulk packaging containing hazardous materials to be marked with the identification number of the lading in one of the following ways:
  - On each side and each end if the capacity of the container equals 1000 gal (3785 L) or more
  - On two opposing sides if the volumetric capacity of the container is less than 1000 gal (3785 L)
  - On each side and each end of the vehicle for cylinders permanently installed on a tube trailer motor vehicle
- DOT requires shippers and carriers to display the appropriate identification number of a hazardous material on the transport vehicle. The identification number is properly displayed either on the relevant placard or the orange panel in Figure 6.10.

MARINE POLLUTANT

FIGURE 6.9 ◆ DOT requires this MARINE POLLUTANT symbol to be marked on packages containing a marine pollutant either when such packages are offered for transportation or during their transportation on watercraft.

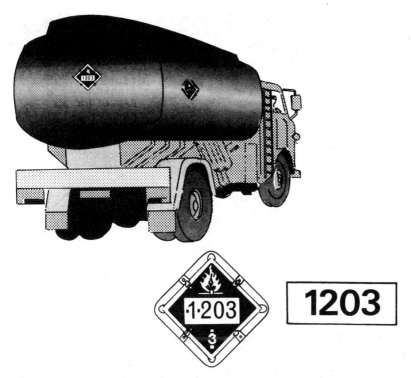

FIGURE 6.10 ◆ DOT requires the identification number of a hazardous material like gasoline to be displayed on certain transport vehicles with or without its UN/NA prefix. The identification number may be displayed on either a placard or an orange panel posted on the vehicle.

Performance Goal for Section 6.6
   ◆ Describe when DOT requires shippers and carriers to post placards on the bulk packaging, freight container, or vehicle used to transport a hazardous material.

◆ 6.6 DOT PLACARDING REQUIREMENTS

Hazardous materials are moved from place to place in multiple ways. The following are the most common modes of transportation:

   ◆ Bulk quantities of hazardous materials are often contained within bulk packaging (e.g., motorized cargo tanks and rail tankcars) that shippers and carriers then move to their destinations.
   ◆ Nonbulk quantities of hazardous materials are loaded into nonbulk packaging (e.g., boxes, cans, and drums) that are loaded into transport vehicles (e.g., railcars and motor vans) or freight containers, which are themselves loaded into transport vehicles. Shippers and carriers then move the transport vehicles to their destinations.

TABLE 6.8 ◆ DOT Hazard Classes That Always Require Placarding[a]

Hazard Class, Division Number, and Additional Information	Placard Name
1.1	EXPLOSIVE 1.1
1.2	EXPLOSIVE 1.2
1.3	EXPLOSIVE 1.3
2.3	INHALATION HAZARD
4.3	DANGEROUS WHEN WET
5.2 (Organic peroxide, Type B, liquid or solid, temperature controlled)	ORGANIC PEROXIDE
6.1 (inhalation hazard, Zone A or B)	INHALATION HAZARD
7 (RADIOACTIVE YELLOW-III label only)	RADIOACTIVE

[a]49 CFR §172.504, Table 1.

TABLE 6.9 ◆ DOT Hazard Classes That Require Placarding Only Under Certain Conditions[a]

Hazard Class, Division Number, and Additional Information	Placard Name
1.4	EXPLOSIVE 1.4
1.5	EXPLOSIVE 1.5
1.6	EXPLOSIVE 1.6
2.1	FLAMMABLE GAS
2.2	NON-FLAMMABLE GAS
3	FLAMMABLE
Combustible liquid	COMBUSTIBLE
4.1	FLAMMABLE SOLID
4.2	SPONTANEOUSLY COMBUSTIBLE
5.1	OXIDIZER
5.2 (other than organic peroxide, Type B, liquid or solid, temperature controlled)	ORGANIC PEROXIDE
6.1 (other than inhalation hazard, Zone A or B)	POISON
6.2	(None)
8	CORROSIVE
9	CLASS 9
ORM-D	(None)

[a]49 CFR §172.504, Table 2.

When warranted, DOT requires shippers and carriers to post one or more placards on each side and each end of the bulk packaging, freight container, or transport vehicle used to transport hazardous materials. The required placards are selected by reference to the appropriate hazard classes in Tables 6.8 and 6.9 as follows:

- Shippers and carriers *always* post the relevant placards when transporting materials whose hazard classes are listed in Table 6.8, regardless of the amounts transported. For example, a shipper or carrier posts RADIOACTIVE placards on a motor van used to transport a wooden box bearing RADIOACTIVE-YELLOW labels, regardless of the amount in the box. The shipper or carrier chooses the RADIOACTIVE placards because hazard class 7 and the RADIOACTIVE YELLOW-III label are listed in Table 6.8.
- Shippers and carriers post the relevant placards when transporting materials whose hazard classes are listed in Table 6.9 when the aggregate gross mass of a hazardous material in any one category equals 1000 lb (454 kg) or more. For example, a shipper or carrier posts FLAMMABLE placards on a railcar used to transport steel drums containing acetonitrile when their aggregate gross mass equals or exceeds 1000 lb (454 kg).

Solved Exercise 6.7

Identify the placard(s), if any, DOT requires a carrier to post on a motor van used to transport six cylinders of compressed fluorine gas, each of which weighs 2.5 lb.

Solution: By reference to the entry in column 3 of Table 6.1, we determine that the hazard class for compressed fluorine is 2.3. By reference to Table 6.8, we determine that when a carrier transports a hazardous material whose hazard class is 2.3, placarding of the transport vehicle is *always* required, irrespective of the amount transported. Consequently, the carrier must display INHALATION HAZARD placards on each side and each end of the transport vehicle.

SHIPMENT OF MULTIPLE PACKAGES OF MATERIALS WHOSE HAZARD CLASSES ARE LISTED IN TABLE 6.9

In lieu of the separate hazard class placards, DOT allows shippers and carriers to post the DANGEROUS placard in Figure 6.11 on freight containers, unit load devices, transport vehicles, and railcars when:

- Multiple nonbulk packages containing materials whose hazard classes are listed in Table 6.9 are transported, and
- The aggregate gross mass of material in any one hazard class that has been loaded at one facility does not exceed 2205 lb (1000 kg).

DANGEROUS placards are never displayed on cargo tanks, portable tanks, or tankcars, and they are never used when placarding materials whose hazard classes are listed in Table 6.8.

Solved Exercise 6.8

A chemical manufacturer intends to transport two chemical substances by motor carrier: four drums of chlorobenzene, each weighing 440 lb, and four drums of solid caustic soda, each weighing 475 lb. Identify the placard(s) DOT requires the carrier to display on the motor van and the basic description of the hazardous materials required on the accompanying shipping paper.

Solution: From the column 3 entries in Table 6.1, we determine that the hazard classes of chlorobenzene and caustic soda are 3 and 8, respectively. Since both hazard classes are listed in Table 6.9, the carrier is required to display FLAMMABLE and CORROSIVE placards on the motor van only when the quantity transported is 1000 lb (454 kg) or more. Since the total quantity of a hazardous material in each hazard class is less than 2205 lb (1000 kg), DOT permits the carrier to display either FLAMMABLE and CORROSIVE placards or DANGEROUS placards on each side and each end of the motor van.

Using Table 6.2, we note that chlorobenzene and sodium hydroxide (caustic soda) are hazardous substances. Nonetheless, the amount of each substance within a single drum does not exceed the reportable quantities for either substance. This means that the letters "RQ" do not need to be entered before or after the basic descriptions of these hazardous materials. Using Table 6.1, the carrier writes the following basic descriptions of the hazardous materials on the accompanying shipping paper as follows:

Description of articles	Mass
Chlorobenzene, 3, UN1134, PG III	1760 lb
Sodium hydroxide, solid, 8, UN1823, PG II	1900 lb

Solved Exercise 6.9

A shipper intends to load 450 lb (205 kg) of compressed nitrogen, 1250 lb (568 kg) of acid battery fluid, and 1750 lb (795 kg) of acetone into an 18-wheel motor van for delivery from the plant site to a customer. Does DOT authorize the shipper to display DANGEROUS placards on each side and each end of the transport vehicle?

Solution: We note from Table 6.1 that the hazard classes of compressed nitrogen, acid battery fluid, and acetone are 2.2, 8, and 3, respectively. Since the aggregate mass of a hazardous material in hazard class 2.2 does not equal 1000 lb (454 kg) or more, DOT does not authorize the shipper to post NON-FLAMMABLE placards on the motor van. Since the aggregate masses of the hazardous materials in hazard classes 3 and 8 equals 1000 lb (454 kg) or more, DOT authorizes the shipper to post FLAMMABLE and CORROSIVE placards on the motor van.

DOT authorizes shippers and carriers to display DANGEROUS placards on a motor vehicle in lieu of the separate hazard class placards when the aggregate gross mass of the materials in any one hazard class in Table 6.9 does not exceed 2205 lb (1000 kg). Since the gross mass of the material in hazard classes 3 and 8 is less than 2205 lb (1000 kg), DOT authorizes the shipper to post DANGEROUS placards on each side and each end of the 18-wheel motor van in lieu of the individual FLAMMABLE and CORROSIVE placards.

FIGURE 6.11 ◆ DOT allows the use of this DANGEROUS placard when two or more categories of hazardous materials in Table 6.9 are transported in the same transport vehicle and when the quantity of a specific category of hazardous material does not exceed 5000 lb (2268 kg). The DANGEROUS placard has red upper and lower triangles with a white rectangular central area and white outer border; the inscription is black.

SHIPMENT OF MULTIPLE PACKAGES OF MATERIALS WHOSE HAZARD CLASSES ARE LISTED IN TABLE 6.8 AND TABLE 6.9

DOT also permits shippers and carriers to transport two or more hazardous materials having the hazard classes listed in Tables 6.8 *and* 6.9 within the same freight container, unit load device, transport vehicle, or railcar. Separate placards for each hazard class are posted when warranted, other than the following:

- When flammable gases are simultaneously transported with 1000 lb (454 kg) or more of oxygen or other nonflammable gas, DOT requires shippers and carriers to post only FLAMMABLE GAS placards.
- When an explosive material in hazard class 1.1 or 1.2 is simultaneously transported with 1000 lb (454 kg) or more of an oxidizer, DOT requires shippers or carriers to post only the relevant EXPLOSIVE 1.1 or EXPLOSIVE 1.2 placards.
- When a transport vehicle or freight container is placarded with POISON GAS placards, DOT does not require shippers or carriers to also post POISON INHALATION HAZARD placards.
- When a transport vehicle or freight container is placarded with POISON INHALATION HAZARD or POISON GAS placards, DOT does not require shippers or carriers to also post POISON placards.

Performance Goals for Section 6.7
- Describe the nature of the nine major DOT hazard classes.
- Describe the nature of the labels and placards that correspond to each DOT hazard class.

◆ 6.7 DOT CLASSIFICATION OF HAZARDOUS MATERIALS

Nine hazard classes have been designated, numbered from 1 through 9, in addition to "other regulated materials." We shall briefly examine the basic features of each class and the DOT labels and placards of each class.

CLASS 1: EXPLOSIVES

An *explosive* is defined by DOT as follows:

> Any chemical compound, mixture or device, the primary or common purpose of which is to function by explosion, that is, with substantially instantaneous release of gas and heat

There are a total of 35 classification codes for explosive materials. Each classification code consists of a division number followed by a compatibility group letter. These letters are used to specify controls for the transportation and storage of explosive materials and to prevent an increase in hazards that might occur if certain types of explosives were stored or transported together. We shall examine the meaning of the compatibility group letter in Section 14.4.

There are six divisions within class 1. Hazardous materials in these divisions are differentiated by the following characteristic features:

Division 1.1	Substances and articles that potentially present a "mass explosive hazard," that is, one that affects almost the entire load instantaneously
Division 1.2	Substances and articles that potentially present a projection hazard but not a mass explosion hazard
Division 1.3	Substances and articles that potentially present a fire hazard and either a minor blast hazard or a minor projection hazard, or both, but not a mass explosion hazard
Division 1.4	Substances and articles that present a minor explosion hazard, but the resulting effect is largely confined to the package and the projection of fragments of appreciable size or range is anticipated; an external fire must not cause an instantaneous explosion of the contents of the package
Division 1.5	Substances so insensitive that there is little probability of initiation or of transition from burning to detonation under normal conditions of transport
Division 1.6	Extremely insensitive articles that do not present a mass explosive hazard and demonstrate a negligible probability of accidental initiation or propagation

DOT requires shippers and carriers to affix the appropriate warning label in Figure 6.12(a) on packaging containing explosive materials. DOT also requires shippers and carriers to display placards like those in Figure 6.12(b) on each side and each end of the bulk packaging, freight container, unit load device, transport vehicle, or railcar used to transport these packages. DOT requires the posting of EXPLO-SIVE 1.1, EXPLOSIVE 1.2, and EXPLOSIVE 1.3 placards whenever any quantity of the corresponding explosive is transported. Posting of the EXPLOSIVE 1.4, EX-PLOSIVE 1.5, and EXPLOSIVE 1.6 placards is required only when 1000 lb (454 kg) or more of the corresponding explosive is transported.

CLASS 2: GASES

Within this hazard class, there are three divisions that relate to the following:

Division 2.1	Flammable gas
Division 2.2	Nonflammable gas
Division 2.3	Poison gas by inhalation

DOT definitions relevant to this hazard class include the following:

A *compressed gas* is any gas or mixture of gases confined within a container and having an absolute pressure exceeding 40 psi (276 kPa) at 70°F (21°C), *or* regardless

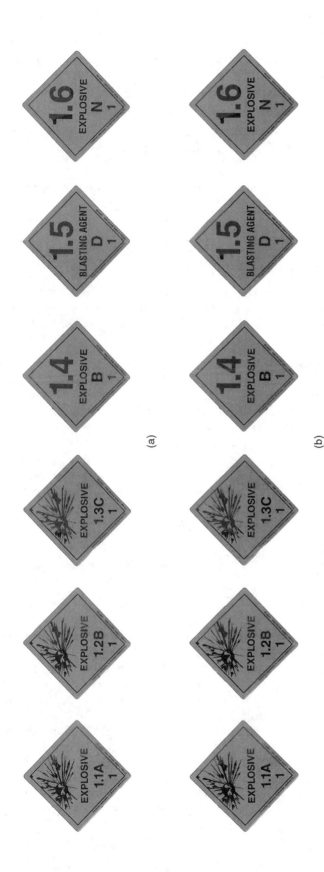

(a)

(b)

FIGURE 6.12 ◆ DOT requires shippers and carriers to affix the relevant label in (a) on containers of division 1.1, 1.2, 1.3, 1.4, 1.5, and 1.6 hazardous materials. The letters that follow these division designations are examples of their compatibility group letters, which we will visit in Section 14.4. Each label is orange with black writing, inscription, and inner border. DOT also requires shippers and carriers to post the relevant placard in (b) on the bulk packaging, freight container, transport vehicle, unit load device, or railcar used to transport class 1 hazardous materials. These placards are orange with a black inner border and inscription.

of the pressure at 70°F (21°C), an absolute pressure exceeding 104 psi (717 kPa) at 130°F (54°C), *or* any liquid flammable material having a vapor pressure exceeding 40 psi$_a$ (276 kPa) at 100°F (37.8°C).

A *flammable gas* is any material that is a gas at 68°F (20°C) or less and 14.7 psi (101.3 kPa) of pressure, *and* is either of the following:

- Ignitable at 14.7 psi (101.3 kPa) when in a mixture of 13% or less by volume with air *or*
- It possesses a flammable range at 14.7 psi (101.3 kPa) with air of at least 12% regardless of the lower limit

A *nonflammable gas* is any material or mixture that does one of the following:

- Exerts in its packaging an absolute pressure of 41 psi (280 kPa), or greater, at 68°F (20°C)
- Does not meet the definition of division 2.1 or 2.3

A *poison gas by inhalation* is a material that is a gas at 68°F (20°C), or less, and a pressure of 14.7 psi (101.3 kPa) *and* is one of the following:

- Known to be so toxic to humans as to pose a health hazard during transportation
- In the absence of adequate data on human toxicity, presumed to be toxic to humans, because when tested on laboratory animals, it has an LC$_{50}$ (Section 10.5) less than 5000 mL/m^3

DOT requires shippers and carriers to affix the relevant warning label in Figure 6.13(a) on packaging containing a class 2 hazardous material. When warranted, DOT requires shippers and carriers to display the relevant placard in Figure 6.13(b) on the bulk packaging, freight container, unit load device, transport vehicle, or railcar used to transport class 2 hazardous materials.

CLASS 3: FLAMMABLE LIQUIDS

DOT distinguishes between a flammable liquid and a combustible liquid. These terms are defined as follows:

With certain exceptions not noted here, a *flammable liquid* is either of the following:

- Any liquid having a flashpoint not more than 141°F (60.5°C)
- Any liquid with a flashpoint at or above 100°F (37.8°C) that is intentionally heated and offered for transportation or transported at or above its flashpoint within bulk packaging

A *combustible liquid* as any liquid that does not meet the definition of any other hazard class and possesses a flashpoint above 141°F (60.5°C) and below 200°F (93°C).

DOT requires shippers and carriers to affix the warning label in Figure 6.14(a) to containers of a flammable liquid. When shippers or carriers transport 1000 lb (454 kg) or more of a class 3 hazardous material, DOT requires the bulk packaging, unit load device, transport vehicle, or railcar to be posted with the FLAMMABLE placard in Figure 6.14(b). When shippers or carriers transport 1000 lb (454 kg) of more of a combustible liquid, the COMBUSTIBLE placard in Figure 6.14(b) is displayed.

When shippers or carriers transport gasoline or fuel oil within a cargo tank or a portable tank moved by motor carrier, DOT permits the words "GASOLINE" or "FUEL OIL," as relevant, to be used in lieu of the word "FLAMMABLE" on the FLAMMABLE placard.

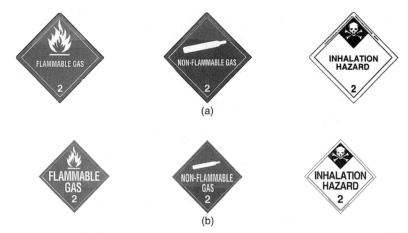

FIGURE 6.13 ◆ DOT requires shippers and carriers to affix the relevant label in (a) on containers of class 2 hazardous materials. The FLAMMABLE GAS label is red with either a black or white symbol and inner border. The NON-FLAMMABLE GAS label is green with either a black or white inscription and inner border. The INHALATION HAZARD label is white with black writing and inner border and a black-and-white skull-and-crossbones symbol. DOT also requires shippers and carriers to post the relevant placard in (b) on the bulk packaging, freight container, transport vehicle, unit load device, or railcar used to transport class 2 hazardous materials. These placards are consistent in design and color scheme with the corresponding labels.

FIGURE 6.14 ◆ DOT requires shippers and carriers to affix the label in (a) on containers of class 3 hazardous materials. The label is red with a white or black border, symbol, and inscription. DOT also requires shippers and carriers to post the FLAMMABLE or COMBUSTIBLE placards in (b) on the bulk packaging, freight container, transport vehicle, unit load device, or railcar used to transport class 3 hazardous materials. Both placards are red with white symbols, inscriptions, and outer borders.

CLASS 4: FLAMMABLE SOLIDS, SPONTANEOUSLY COMBUSTIBLE MATERIALS, AND DANGEROUS-WHEN-WET MATERIALS

There are three divisions within this hazard class:

Division 4.1	Flammable solids
Division 4.2	Spontaneously combustible materials
Division 4.3	Danger-when-wet materials

DOT definitions relevant to this hazard class include the following:

A *flammable solid* is any of the following:

- Any of several wetted explosives
- Any of several self-reactive materials that are thermally unstable and can undergo a strongly exothermic decomposition even without the participation of atmospheric oxygen
- Any readily combustible solids

A *spontaneously combustible material* is either of the following:

- A *pyrophoric material*, that is, a liquid or solid that even in small quantities and without an external ignition source may ignite within five minutes after contacting air. DOT defines a pyrophoric liquid as a liquid that ignites spontaneously in dry or moist air at or below 130°F (54.5°C)
- A *self-heating material*, that is, a material that when in contact with air and without an energy supply is liable to self-heat

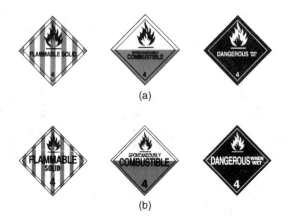

(a)

(b)

FIGURE 6.15 ◆ DOT requires shippers and carriers to affix the relevant label in (a) on containers of a class 4 hazardous material. The FLAMMABLE SOLID label is white with vertical red stripes equally spaced on each side of a vertical red stripe placed in the center of the label: the overprinted symbol, inscription, and inner border are black. The SPONTANEOUSLY COMBUSTIBLE label is red in the lower half and white in the upper half with black or white border, a black symbol, and black printing. The DANGEROUS WHEN WET label is blue with a black or white border line, symbol, and inscription. DOT also requires shippers and carriers to post the relevant placard in (b) on the bulk packaging, freight container, transport vehicle, unit load device, or railcar used to transport class 4 hazardous materials. The FLAMMABLE SOLID, SPONTANEOUSLY COMBUSTIBLE, and DANGEROUS WHEN WET placards are consistent in design and color scheme with their corresponding labels.

A *dangerous-when-wet material* is a material that, by interaction with water, is likely to become spontaneously flammable or to release a flammable or toxic gas or vapor at a rate greater than 28 in.^{3}/lb (1 L/kg) per hour.

DOT requires shippers and carriers to affix the relevant label in Figure 6.15(a) on packaging containing a class 4 hazardous material. When warranted, DOT requires shippers and carriers to display the relevant placard in Figure 6.15(b) on the bulk packaging, freight container, unit load device, transport vehicle, or railcar used to transport class 4 hazardous materials.

CLASS 5: OXIDIZERS AND ORGANIC PEROXIDES

There are two divisions in class 5.

Division 5.1	Oxidizer
Division 5.2	Organic peroxide

DOT defines these terms as follows:

An *oxidizer* is a material that can cause or enhance the combustion of other materials, generally by yielding oxygen.

An *organic peroxide* is an organic compound containing oxygen in the bivalent –O–O– structure.

DOT requires shippers and carriers to affix the relevant DOT label in Figure 6.16(a) on packaging containing a class 5 hazardous material. When warranted, DOT requires shippers and carriers to display the relevant placard in Figure 6.16(b) on the bulk packaging, freight container, unit load device, transport vehicle, or railcar used to transport class 5 hazardous materials.

CLASS 6: POISONOUS MATERIALS AND INFECTIOUS MATERIALS

There are two divisions in class 6.

Division 6.1	Poisonous material
Division 6.2	Infectious material

(a)

(b)

FIGURE 6.16 ◆ DOT requires shippers and carriers to affix the relevant label in (a) on containers of a class 5 hazardous material. DOT also requires shippers and carriers to post the relevant placard in (b) on the bulk packaging, freight container, transport vehicle, unit load device, or railcar used to transport class 5 hazardous materials. These labels and placards are yellow with black printing, symbol, and inner border.

DOT defines these terms as follows:

A *poisonous material* is a material other than a gas that is known to be so toxic to humans such as to afford a hazard to health during transportation, *or that*, in the absence of adequate data on human toxicity is one of the following:

- Presumed to be toxic to humans because of data obtained from tests performed on animals
- An irritating material with properties similar to those of tear gas and causes extreme irritation, especially within confined spaces

An *infectious substance,* or *etiologic agent,* is a viable microorganism or its toxin that causes or can cause disease in humans or animals. An example of an infectious substance is anthrax spores.

DOT requires shippers and carriers to affix the relevant DOT label in Figure 6.17(a) on packages containing a class 6 hazardous material. When warranted, DOT requires shippers and carriers to display the relevant placard in Figure 6.17(b) on the bulk packaging, freight container, unit load device, transport vehicle, or railcar used to transport class 6 hazardous materials.

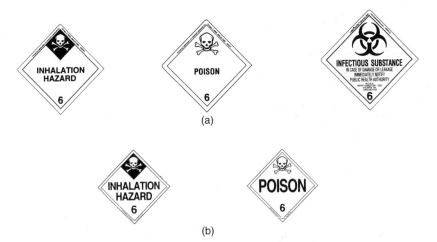

FIGURE 6.17 ◆ DOT requires shippers and carriers to affix the relevant label in (a) on containers of class 6 hazardous materials. DOT also requires shippers and carriers to post the relevant placards in (b) on the bulk packaging, freight container, transport vehicle, unit load device, or railcar used to transport class 6 hazardous materials. The INHALATION HAZARD and POISON (or TOXIC) labels and placards are white with black writing, inscription, and inner border, while the background and symbol of the upper diamond are white. The INFECTIOUS SUBSTANCE label is also white with black writing, inscription, and inner border.

(a)

(b)

FIGURE 6.18 ◆ DOT requires shippers and carriers to affix the relevant label in (a) on containers of class 7 hazardous materials. THE RADIOACTIVE WHITE-I label is white with a black symbol of a three-bladed propeller; the printing of the Roman numeral I is red with all other text written in black. The RADIOACTIVE YELLOW-II and RADIOACTIVE YELLOW-III labels are yellow in the top half and white in the lower half; they bear a black symbol of a three-bladed propeller symbol. The printing of the Roman numerals I and II is red with all other text written in black. DOT also requires shippers and carriers to post the RADIOACTIVE placard in (b) on the bulk packaging, freight container, transport vehicle, unit load device, or railcar used to transport a class 7 hazardous material. The top half of the RADIOACTIVE placard is yellow with a black three-bladed propeller, while the lower half is white with a black inscription.

CLASS 7: RADIOACTIVE MATERIALS

In the DOT regulations, a *radioactive material* is defined as follows:

> any material or combination of materials that spontaneously emits ionizing radiation having a specific activity greater than 70 becquerels per gram (70 Bq/g) [0.002 microcuries per gram (μCi/g)]

Specific activity and the features of nuclear phenomena are noted in Chapter 15.

In general, DOT requires a carrier transporting a package containing a class 7 hazardous material to affix upon the package the relevant DOT label in Figure 6.18(a). These labels are called the *RADIOACTIVE WHITE-I*, *RADIOACTIVE YELLOW-II*, and *RADIOACTIVE YELLOW-III* labels, respectively. The RADIOACTIVE WHITE-I and RADIOACTIVE YELLOW-III labels alert emergency-response personnel to minimum and maximum radiation hazards, respectively. The correct label that DOT requires to be affixed to packages containing a radioactive material is determined in part by the specific activity of the lading, which we shall note further in Chapter 15.

DOT requires shippers and carriers to display the RADIOACTIVE YELLOW placard in Figure 6.18(b) on the bulk packaging, freight container, unit load device, transport vehicle, or railcar used to transport any amount of a radioactive material labeled RADIOACTIVE YELLOW-III.

CLASS 8: CORROSIVE MATERIALS

In the DOT regulations, a *corrosive material* is defined as follows:

- ◆ A liquid or solid that causes visible destruction or irreversible alterations in human skin tissue at the site of contact

FIGURE 6.19 ◆ DOT requires shippers and carriers to affix the label in (a) on containers of class 8 hazardous materials. The label is white in the top-half triangle and black in the bottom-half. The printing is white and the symbols are black and white. DOT also requires shippers and carriers to post the placard in (b) on the bulk packaging, freight container, transport vehicle, unit load device, or railcar used to transport class 8 hazardous materials. The placard is consistent in design and color scheme with the label.

(a)   (b)

◆ A liquid that has a severe corrosion rate on steel or aluminum, as measured in accordance with certain prescribed DOT testing procedures

DOT requires shippers and carriers to affix the warning label in Figure 6.19(a) to packages containing a corrosive material. When warranted, DOT requires shippers and carriers to display the CORROSIVE placard in Figure 6.19(b) on the bulk packaging, freight container, unit load device, transport vehicle, or railcar used to transport class 8 hazardous materials.

CLASS 9: MISCELLANEOUS HAZARDOUS MATERIALS

DOT defines a *miscellaneous hazardous material* as follows:

> any material that presents a hazard during transportation but is not included in any other hazard class

This is referred to as class 9. Included within class 9 is any material that has an anesthetic, noxious, or similar property, as well as elevated-temperature materials, hazardous substances, hazardous wastes, and marine pollutants.

DOT requires shippers and carriers to affix the CLASS 9 label in Figure 6.20(a) on packaging of miscellaneous hazardous materials. When warranted, DOT requires shippers and carriers to display the CLASS 9 placard in Figure 6.20(b) on the bulk packaging, freight container, unit load device, transport vehicle, or railcar used to transport class 9 hazardous materials.

Aside from the hazardous materials previously described in the nine classes, there is one additional group of hazardous materials corresponding to the designation ORM-D. A hazardous material designated as ORM-D is a "material such

FIGURE 6.20 ◆ DOT requires shippers and carriers to affix the label in (a) on containers of class 9 hazardous materials. The label is white with seven black vertical stripes in the upper triangle. DOT also requires shippers and carriers to post the placard in (b) on the bulk packaging, freight container, transport vehicle, unit load device, or railcar used to transport class 9 hazardous materials. The placard is consistent in design and color scheme with the label.

(a)   (b)

as a consumer commodity that . . . presents a limited hazard during transportation due to its form, quantity, and packaging." Hair spray packed in an aerosol container is an example of a consumer commodity that is transported as an ORM-D material.

> **Performance Goals for Section 6.8**
> ◆ When a hazardous material is involved in a transportation mishap, identify the locations at which emergency responders may expect to locate its identification number. Demonstrate how first-on-the-scene responders use the identification number and the *Emergency Response Guidebook* to obtain general response-action information.

◆ 6.8 RESPONDING TO INCIDENTS INVOLVING THE RELEASE OF HAZARDOUS MATERIALS

When shippers and carriers have correctly implemented the DOT regulations, the identification number of a hazardous material can be readily located during an emergency-response action as follows:

◆ Posted upon a placard or an orange panel
◆ Marked upon the packaging containing the hazardous material
◆ Listed on a shipping paper as a component of the basic description of the hazardous material

Having located the identification number, how can first-on-the-scene responders use the number during an incident involving the release of a hazardous material? The answer to this question involves consulting the *Emergency Response Guidebook,** a DOT manual that provides direction to emergency-response crews primarily during the initial phases of a response action.

The guidebook serves as the primary reference source book for first-on-the-scene personnel. It is organized as follows:

◆ All of the hazardous material identification numbers are catalogued on yellow-bordered pages. Each identification number references the hazardous material having that number.
◆ In a second section, having blue-bordered pages, the names of the hazardous materials are listed alphabetically with their corresponding identification numbers.
◆ Each identification number is referenced to a guide number. There are 172 individual guide numbers provided within a section having orange-bordered pages. Each individual guide number is a two-page format that provides information about the potential hazards of the material, recommended actions that relate to public safety, and recommended emergency-response actions.

*The *Emergency Response Guidebook* is available from private commercial companies, the Research and Special Programs Administration, Hazardous Materials Transportation Bureau, U.S. Department of Transportation, Washington, D.C. 20590, and the U.S. Government Printing Office, Superintendent of Documents, Mail Stop: SSOP, Washington, D.C. 20402-9328. From government agencies, request ISBN 0-16-042938-2.

- A fourth section having green-bordered pages provides information concerning initial isolation and protective distances for small and large spills of certain hazardous materials and their classes. It also provides a list of dangerous-when-wet materials.

The guidebook should always be accessible in emergency-response vehicles. Responding personnel should refer to the guidebook, locate the appropriate guide number, and implement the recommended actions.

DOT also requires shippers and carriers to have emergency-response information available to workers during all phases of the transportation of a hazardous material. This requirement is often met by having available a copy of the *Emergency Response Guidebook* during hazardous material loading, unloading, and transfer operations.

Solved Exercise 6.10

Arriving at the scene of a highway transportation mishap, a firefighting team notes a motor van has overturned during an accident. They also observe orange panels bearing the number 1259 and INHALATION HAZARD placards on the visible sides and ends of the motor van. Securing the shipping paper from the driver, the team captain notes that the carrier provided the following basic description of a hazardous material:

8 1-lb bottles overpacked in boxes	Nickel carbonyl, 3, UN1259, PG II (POISON—INHALATION HAZARD, 8 lb Zone B)

Upon cautiously examining the van's interior from a distance, broken boxes marked "Nickel carbonyl, UN1259" and labeled INHALATION HAZARD can be observed. Bottles of the hazardous material were thrown from several boxes and their contents have spilled upon the floor of the van. What actions should be implemented by these first-on-the-scene responders?

Solution: The team learns from the basic description, labeling, placarding, and marking information that nickel carbonyl is a flammable and toxic liquid. The incident involves the release of a substance that poses the risk of fire and inhalation toxicity. To protect public health and the environment, they should immediately cordon off the area, and direct the public away. Although the total quantity at issue is only 8 lb, emergency-response personnel should don total-encapsulating suits with self-contained breathing apparatus. Only then should the intact boxes be removed from the van and set aside.

For more specific guidance in responding to this incident, the captain should contact CHEMTREC. While waiting for CHEMTREC's response, the team members should consult the *Emergency Response Guidebook* for initial response assistance. The identification number 1259 refers responders to Guide No. 57. Here, personnel are provided with direction such as absorbing the spilled liquid and eliminating potential sources of ignition. Guide No. 57 also provides first-aid information to help responders who experience ill effects from inadvertent exposure to nickel carbonyl.

Suppose we need information concerning the appropriate response-action to be taken at a transportation incident involving an initially unidentified hazardous material being transported by motor carrier on a crowded highway. Upon arriving at the scene, personnel note that the carrier posted CORROSIVE placards and orange panels with the number "1805" on the vehicle. Upon securing the shipping paper from the driver, personnel also note that a number of items not regulated by DOT are included in the consignment as well as a hazardous material having the following basic description:

> 485 lb   Phosphoric acid, 8, UN1805, PG III

It is the location of this identification number that is most important to emergency-response personnel.

In the *Emergency Response Guidebook,* the number "1805" directs the reader to guide number 154, which is reproduced in Figure 6.21. This guide number provides general information concerning this substance. Most notably, it denotes that it is a toxic and/or corrosive material, but noncombustible. Any concern that the substance may ignite can be dismissed.

The value of the *Emergency Response Guidebook* is apparent. When properly implemented, each guide provides emergency-response personnel with vital information on how to initially deal with an incident involving the release of a unique hazardous material until more specific information is obtained from the shipper, carrier, manufacturer, or CHEMTREC.

Performance Goal for Section 6.9
- ◆ Describe when carriers are obligated to notify the National Response Center that a hazardous substance has been released into the environment.

◆ 6.9 REPORTING THE RELEASE OF HAZARDOUS SUBSTANCES

As soon as practical, common and contract carriers are required to notify the National Response Center when an amount equal to or greater than the reportable quantity of a hazardous substance is released into the environment. In addition, notice must be given to the National Response Center whenever any of the following occurs:

- ◆ As the result of the release of one or more hazardous materials at least one of the following is true:
  - ◆ A person is killed.
  - ◆ A person receives injuries requiring hospitalization.
  - ◆ The estimated carrier or other property damage exceeds $50,000.
  - ◆ An evacuation of the general public occurs lasting 1 or more hours.
  - ◆ One or more major transportation arteries or facilities are closed or shut down for 1 hour or more.
  - ◆ The operational flight pattern or routine of aircraft is altered.
- ◆ Fire, breakage, spillage, or suspected radioactive contamination occurs involving the shipment of one or more radioactive materials.
- ◆ Fire, breakage, spillage, or suspected radioactive contamination occurs involving the shipment of one or more infectious materials, or etiologic agents.

| GUIDE 154 | SUBSTANCES - TOXIC AND/OR CORROSIVE (NON-COMBUSTIBLE) | ERG2000 |

POTENTIAL HAZARDS

HEALTH

- **TOXIC**; inhalation, ingestion, or skin contact with material may cause severe injury or death.
- Contact with molten substance may cause severe burns to skin and eyes.
- Avoid any skin contact.
- Effects of contact or inhalation may be delayed.
- Fire may produce irritating, corrosive and/or toxic gases.
- Runoff from fire control or dilution water may be corrosive and/or toxic and cause pollution.

FIRE OR EXPLOSION

- Non-combustible, substance itself does not burn but may decompose upon heating to produce corrosive and/or toxic fumes.
- Some are oxidizers and may ignite combustibles (wood, paper, oil, clothing, etc.).
- Contact with metals may evolve flammable hydrogen gas.
- Containers may explode when heated.

PUBLIC SAFETY

- **CALL Emergency Response Telephone Number on Shipping Paper first. If Shipping Paper not available or no answer, refer to appropriate telephone number listed on the inside back cover.**
- Isolate spill or leak area immediately for at least 25 to 50 meters (80 to 160 feet) in all directions.
- Keep unauthorized personnel away.
- Stay upwind.
- Keep out of low areas.
- Ventilate enclosed areas.

PROTECTIVE CLOTHING

- Wear positive pressure self-contained breathing apparatus (SCBA).
- Wear chemical protective clothing which is specifically recommended by the manufacturer. It may provide little or no thermal protection.
- Structural firefighters' protective clothing provides limited protection in fire situations ONLY; it is not effective in spill situations.

EVACUATION

Spill
- See the Table of Initial Isolation and Protective Action Distances for highlighted substances. For non-highlighted substances, increase, in the downwind direction, as necessary, the isolation distance shown under "PUBLIC SAFETY".

Fire
- If tank, rail car or tank truck is involved in a fire, ISOLATE for 800 meters (1/2 mile) in all directions; also, consider initial evacuation for 800 meters (1/2 mile) in all directions.

FIGURE 6.21 ◆ Guide 154 from DOT's *Emergency Response Guidebook* (2000).

ERG2000 SUBSTANCES - TOXIC AND/OR CORROSIVE (NON-COMBUSTIBLE) GUIDE 154

EMERGENCY RESPONSE

FIRE

Small Fires
- Dry chemical, CO_2 or water spray.

Large Fires
- Dry chemical, CO_2, alcohol-resistant foam or water spray.
- Move containers from fire area if you can do it without risk.
- Dike fire control water for later disposal; do not scatter the material.

Fire involving Tanks or Car/Trailer Loads
- Fight fire from maximum distance or use unmanned hose holders or monitor nozzles.
- Do not get water inside containers.
- Cool containers with flooding quantities of water until well after fire is out.
- Withdraw immediately in case of rising sound from venting safety devices or discoloration of tank.
- ALWAYS stay away from tanks engulfed in fire.

SPILL OR LEAK
- ELIMINATE all ignition sources (no smoking, flares, sparks or flames in immediate area).
- Do not touch damaged containers or spilled material unless wearing appropriate protective clothing.
- Stop leak if you can do it without risk.
- Prevent entry into waterways, sewers, basements or confined areas.
- Absorb or cover with dry earth, sand or other non-combustible material and transfer to containers.
- DO NOT GET WATER INSIDE CONTAINERS.

FIRST AID
- Move victim to fresh air.   • Call 911 or emergency medical service.
- Apply artificial respiration if victim is not breathing.
- **Do not use mouth-to-mouth method if victim ingested or inhaled the substance; induce artificial respiration with the aid of a pocket mask equipped with a one-way valve or other proper respiratory medical device.**
- Administer oxygen if breathing is difficult.
- Remove and isolate contaminated clothing and shoes.
- In case of contact with substance, immediately flush skin or eyes with running water for at least 20 minutes.
- For minor skin contact, avoid spreading material on unaffected skin.
- Keep victim warm and quiet.
- Effects of exposure (inhalation, ingestion or skin contact) to substance may be delayed.
- Ensure that medical personnel are aware of the material(s) involved, and take precautions to protect themselves.

FIGURE 6.21 ◆ (*continued*)

- There has been a release of a marine pollutant in a quantity exceeding 119 gal (450 L) for liquids or 882 lb (400 kg) for solids.
- Even when these criteria are not met, the transportation mishap is so severe that, in the judgment of the carrier, it should be reported nonetheless.

As required by regulations at 49 CFR §171.15, the National Response Center can be initially contacted by phone at 800-424-8802 or 202-426-2675. A follow-up report in writing is also required to the Director, Office of Hazardous Materials Regulations, Materials Transportation Bureau, Department of Transportation, Washington, D.C. 20590. DOT also regulates the informational requirements of the follow-up report at 49 CFR §171.16.

■■■

REVIEW EXERCISES

The Shipping Paper

6.1 Identify each hazardous material in Table 6.1 whose transportation is regulated *only* when shipment occurs by each the following means:
   (a) air
   (b) watercraft

6.2 Identify the basic description of oxygen that DOT requires shippers and carriers to enter on a shipping paper, when a chemical company offers it in each of the following states of matter for shipment by motor van:
   (a) 16 cylinders, each containing 10 L of compressed oxygen
   (b) 5000 gallons of cryogenic oxygen

6.3 A mining company intends to separately transport 2500 lb of molten sulfur by motorized tank truck from Louisiana to two locations, California and Peru. Identify the basic description DOT required on the accompanying shipping papers for these domestic and international shipments.

6.4 Dangerous quantities of flammable hydrogen are generated when lithium aluminum hydride reacts with water. When a shipper offers 25 lb (11 kg) of lithium aluminum hydride for shipment by motor carrier, what basic description does DOT require the shipper and carrier to enter on the accompanying shipping paper?

6.5 A liquor company intends to transport twenty 25-gallon kegs of Scotch whiskey by motor van from its distillery to cellars, where it will be stored and aged. When offered for shipment, what basic description does DOT require on the accompanying shipper paper?

6.6 A petroleum company intends to transport 450 m^3 of liquefied petroleum gas in five DOT-113A rail tankcars. What basic description does DOT require the company and rail carrier to provide on the accompanying waybill?

6.7 A chemical incineration and disposal company intends to transport 5000 gal of a spent solvent mixture by motor carrier. If a sample of the waste exhibits the RCRA characteristic of ignitability, what basic description do DOT and EPA require the company and carrier to provide on the accompanying transportation manifest?

6.8 A carrier transports coal tar distillate to a customer's plant site in a cargo tank, from which it is transferred into the customer's storage tank. Once the cargo tank has been emptied, what basic description does DOT require the carrier to enter on a shipping paper before the cargo tank can be returned to the supplier?

Transporting Hazardous Materials by Passenger-Carrying Aircraft or Railcar, Cargo Aircraft, or Watercraft

6.9 A chemistry professor wishes to ship 240 L of acetonitrile, a liquid solvent, to an out-of-state colleague by cargo aircraft. To comply with the DOT hazardous materials regulations:
  (a) May the professor ship the solvent as a single consignment?
  (b) May the professor ship the solvent as four separate consignments?
  (c) What basic description must the professor and airline carrier enter on the accompanying shipping paper?

6.10 A carrier intends to ship six pallets, each holding six 55-gallon drums of methylene chloride, by cargo ship. What is the basic description that DOT requires the carrier to enter on the accompanying manifest? [The density of methylene chloride is 11.07 lb/gal at 68°F (20°C).]

Location of the Shipping Paper During Transit

6.11 The driver of a tank truck hauling hazardous materials stores the relevant shipping paper in the van's glove compartment along with the title to the vehicle and relevant insurance documents. Does this placement of the shipping paper comply with the DOT hazardous materials regulations?

Location of Hazardous Materials Onboard Watercraft

6.12 The length of a cruise ship is 1150 ft (350 m). If 55 couples are onboard, does DOT permit ten 55-gallon steel drums of toluene to be stowed aboard the ship?

DOT Labels, Markings, and Placards

6.13 When offering a 2.5-lb cylinder of compressed fluorine for shipment by motor carrier, the shipper provides the carrier with the following basic description:

> 2.5 lb   Fluorine, compressed, 2.3, UN1045 (POISON—
>   Inhalation Hazard, Zone A)

To comply with the DOT hazardous materials regulations:
  (a) Which labels, if any, must the carrier affix to the fluorine cylinder?
  (b) Which placards, if any, must the carrier post on the motor carrier?

6.14 A chemical manufacturer intends to offer 25 L of ethyl mercaptan to an airline carrier for shipment. To comply with the DOT hazardous materials regulations:
  (a) What basic description must the shipper and carrier enter on the accompanying shipping paper?
  (b) Which labels, if any, must the carrier affix to the packaging?
  (c) What marking information must the carrier provide on the packaging?

6.15 DOT requires that each bulk packaging, including a cargo tank or portable tank, remain placarded, even when the hazardous material has been removed, unless the bulk packaging has been sufficiently cleaned of residue and purged of vapors, or refilled with a material requiring no placards or different placards. What is the most likely reason DOT regulates the emptied bulk packaging?

6.16 A paint manufacturer wishes to ship 5000 gal of a solvent mixture consisting of 15% toluene, 30% methyl ethyl ketone, 25% ethylbenzene, 15% ethyl acetate, and 15% xylene by means of a railroad tankcar. To comply with the DOT hazardous materials regulations:

(a) What basic description must the shipper and rail carrier enter on the accompanying waybill?

(b) Other than accurately describing the solvent mixture on the waybill, what additional information must the train crew document concerning this shipment?

(c) What marking information must the carrier provide on the tankcar?

6.17 An industrial company offers a carrier 10,000 lb (4550 kg) of methylene chloride for shipment by means of a railroad tankcar. To comply with the DOT hazardous materials regulations:

(a) What is the basic description the company and carrier must provide on the accompanying waybill?

(b) What marking information must the carrier provide on the tankcar?

6.18 Which placards, if any, does DOT require a carrier to display on a motor vehicle used solely for the domestic transportation of each hazardous material denoted by the following bill of lading entries:

(a) 2250 lb   RQ, Tetrachloroethane, 6.1, UN1897, PG III
(b) 50 lb   Sodium peroxide, 5.1, UN1504, PG I
(c) 330 lb   Methyl isobutyl ketone, 3, UN1245, PG II
(d) 1500 lb   Calcium oxide, 8, UN1910, PG III
(e) 100 lb   Barium cyanide, 6.1, UN1565, PG I (toxic)

6.19 Determine the total number of placards, if any, that DOT requires to be displayed on a motor van loaded at the same facility with each of the following hazardous materials:

(a) 1000 lb of toluene and 2500 lb of ethylbenzene

(b) 1000 lb of toluene, 2500 lb of ethylbenzene, and 4000 lb of mercuric oxide

6.20 A transportation company is retained to deliver the following mixed load of hazardous materials to a chemical research facility by contract motor carrier:

- Twenty 55-gallon steel drums of 1,1,1-trichloroethane, each weighing 627 lb (285 kg)
- Eight 55-gallon steel drums of acetone, each weighing 383 lb (65 kg)
- Ten 55-gallon steel drums of cyclohexane, each weighing 366 lb (166 kg)
- One cardboard box containing fifteen 1-lb (0.5-kg) bottles of mercuric oxide
- Seven 5-gallon bottles of 1,1,1-trichloroethane, each weighing 60 lb (27 kg) and packaged individually within cardboard boxes

To comply with the DOT hazardous materials regulations:

(a) What is the basic description of each hazardous material the company and carrier must provide on the accompanying shipping paper?

(b) Which labels does DOT require the carrier to affix to the referenced packaging?

(c) Which placards does DOT require the carrier to display on the motor van?

(d) At which location(s) on the motor van does DOT require the carrier to display the placards?

Emergency-Response Actions

6.21 What observations can assist first-on-the-scene responders to establish that a   237
confined flammable gas is present at the scene of a transportation mishap?

6.22 What observations can assist first-on-the-scene responders to establish that a confined flammable or combustible liquid is present at the scene of a transportation mishap?

6.23 A train crew notes that a liquid is leaking from the piping connected to a railroad tankcar containing a hazardous material. This causes the engineer to stop the train and contact the local firefighting team. Upon arriving at the scene, the team captain is provided a waybill by a member of the train crew. The waybill gives the following basic description of the leaking liquid:

> 5000 lb   RQ, Chlorobenzene, 3, UN1134, PG III,
> Placarded FLAMMABLE

Which information in this basic description should the captain immediately convey to these first-on-the-scene responders?

6.24 A firefighting team responds to a fire involving an overturned motor van on which DANGEROUS placards have been posted. The intensity of the fire prevents the team from approaching the vehicle and retrieving the shipping paper. Despite the fact that the DANGEROUS placards are visible, why is it difficult for the team to determine the best action to be taken at this scene?

6.25 Arriving at the scene of a train wreck, a firefighting team notes that the carrier has posted an orange panel bearing the number 3028 upon an overturned, leaking railroad tankcar. The placards and markings on the tankcar are obscured. Using the *Emergency Response Guidebook*:

(a) What is the name of the hazardous material having the identification number 3028?

(b) To which guide number is the identification number 3028 referenced?

(c) Does the relevant guide number indicate that the hazardous material could ignite?

(d) Does the relevant guide number indicate that the hazardous material is poisonous?

7

Chemistry of Some Common Elements

In this chapter, we will examine the characteristic properties of seven elements: oxygen, hydrogen, fluorine, chlorine, phosphorus, sulfur, and carbon. These particular elements have been selected here for significant reasons. The section on fluorine has been included because exposure to this element poses a higher degree of hazard than exposure to any other element. Fluorine is so hazardous that most chemists hesitate before using it. The sections on the other six elements have been included for twofold reasons: These elements possess certain hazardous features of which emergency responders should be aware, and they are used more widely than other elements. This latter distinction increases the likelihood that these elements will be encountered during emergency-response actions involving hazardous materials.

Performance Goals for Section 7.1
- Discuss the physical and chemical properties of elemental oxygen, especially its ability to support the life process and the combustion of matter.
- Describe the manner by which oxygen is separated from the other components of the air for commercial use.
- Identify the industries that use bulk quantities of oxygen.
- Discuss the unique hazards associated with the release of liquid oxygen (LOX).
- Discuss the physical and chemical properties of ozone.
- Describe the manner by which ozone is generated for commercial use.
- Describe the manner by which ozone is generated in the lower atmosphere.
- Describe how the presence of ozone in the lower atmosphere negatively affects the quality of life.
- Describe how ozone depletion in the stratosphere negatively affects life on Earth's surface.

◆ 7.1 OXYGEN AND OZONE

What image is conjured in your mind when someone mentions oxygen? Most individuals first think of the air. It is true that oxygen is a constituent of air. Yet, oxygen makes up only about 21% of the volume of dry air. The major component—78% by

TABLE 7.1 ◆ Physical Properties of Elemental Oxygen	
Specific gravity	1.43
Vapor density (air = 1)	1.1
Freezing point	−360°F (−218°C)
Boiling point	−297°F (−183°C)
Liquid-to-gas expansion ratio	875

volume—is nitrogen. Although oxygen is the most abundant element on Earth, most of it is not found in the free state. Instead, it is combined with other elements in chemical compounds such as water and carbon dioxide. Many oxygen compounds also exist in the chemical structures of the rocks and minerals that make up Earth's crust. A review of Table 4.1 shows that oxygen accounts for approximately 50% by mass of all substances in Earth's crust, the atmosphere, oceans, and other bodies of water.

At the ambient temperatures encountered on Earth, oxygen is a colorless, odorless, and tasteless gas that is only slightly more dense than air. At very low temperatures and under pressure, oxygen exists as a pale blue liquid or solid and has the curious feature of being attracted to a magnet. Other physical properties of oxygen are listed in Table 7.1. Chemists represent the chemical formula of elemental oxygen as O_2.

Oxygen is usually obtained for commercial use by liquefying air. Confined air liquefies when it is compressed at a pressure exceeding 545 psi$_a$ (3760 kPa) while being simultaneously cooled below −318°F (−194°C). To obtain these temperature and pressure conditions, air is passed through a series of compression, expansion, and cooling operations.

By volume, liquid air is essentially a mixture of approximately 43% nitrogen, 54% oxygen, and less than 3% by mass of argon and other gases. When this mixture is allowed to boil at atmospheric pressure, the vapor above the liquid becomes enriched in nitrogen, while simultaneously, the remaining liquid slowly becomes enriched in oxygen. The oxygen-enriched mixture is called *liquid oxygen, or cryogenic oxygen;* in commerce, it is also referred to as *LOX.* When LOX vaporizes, it becomes *gaseous oxygen,* sometimes denoted as *GOX.* Approximately 875 liters of gaseous oxygen form when one liter of liquid oxygen vaporizes at room temperature.

Facilities using large quantities of oxygen often store it as the cryogenic liquid. LOX is typically stored within a tank like that shown in Figure 7.1, from which it can be discharged for use as either the liquid or gas. The discharge of gaseous oxygen is often more desirable from the standpoint of regulating temperature and pressure.

DOT regulates the transportation of compressed oxygen and cryogenic oxygen as a nonflammable gas. For labeling and placarding, it provides two options to shippers and carriers. They may affix NON-FLAMMABLE GAS and OXIDIZER labels to containers of oxygen, or in lieu of these labels, they may affix OXYGEN labels. Shippers and carriers may post either NON-FLAMMABLE GAS placards or OXYGEN placards, when warranted, on the bulk packaging, freight container, transport vehicle, unit load device, or railcar used to transport oxygen. The OXYGEN label and OXYGEN placard are represented in Figure 7.2.

Huge volumes of liquid oxygen are used by the aerospace industry to oxidize rocket fuels and propellants. In today's rockets, such as Apollo 11/Saturn V shown in

FIGURE 7.1 ◆ A demand for a large amount of oxygen occurs at most hospitals, which require the gas for routine medical-care procedures, like oxygen-enriched therapy. In such situations, storing the cryogenic liquid in tanks is a more efficient and economical practice than storing an equal quantity of the compressed gas in cylinders. While stored as the liquid, oxygen is vaporized prior to delivery to hospital lines and other points of use.

Figure 7.3, liquid oxygen and the fuel are pumped into a combustion chamber, where the mixture is ignited. The combustion reaction produces a high-pressure, high-velocity stream of hot gases, which flows through the nozzle and accelerates the rocket to the tremendous velocities needed to reach outer space. Although other oxidizers could be used to oxidize rocket fuels, none has been shown to perform as effectively and economically as liquid oxygen.

Oxygen is also used in a number of less esoteric ways. For example, it is used by cold storage and food processing plants, hospitals, metal fabrication plants, electrical power plants, and the steel industry. Within hospitals, oxygen-enriched air is delivered into oxygen tents and hoods for therapeutic uses. Gaseous oxygen under increased pressure, called *hyperbaric oxygen*, is also used within a chamber to treat various circulatory and respiratory problems, thermal and radiation burns, cerebral palsy, brain injuries, and carbon monoxide poisoning (Section 10.8). Portable cylinders of oxygen are also available for therapeutic uses, especially by individuals afflicted with emphysema.

FIGURE 7.2 ◆ DOT authorizes shippers and carriers to affix the OXYGEN label in (a) to containers of oxygen in lieu of NON-FLAMMABLE GAS and OXIDIZER labels. DOT also authorizes shippers and carriers to post the OXYGEN placard in (b) when warranted, instead of the NON-FLAMMABLE GAS placard on the bulk packaging, freight container, transport vehicle, unit load device, or railcar used to transport oxygen.

(a)   (b)

3rd stage

2nd stage

1st stage

FIGURE 7.3 ◆ The Apollo 11/Saturn V rocket beginning its ascent into space. The three stages of this rocket are depicted on the left. NASA used liquid oxygen to oxidize the fuel (liquid hydrogen) during the second and third stages of this mission. *Courtesy of the National Aeronautics and Space Administration, John F. Kennedy Space Center, Kennedy Space Center, Florida.*

To minimize the risk of fire and explosion, OSHA requires the owners of bulk oxygen storage systems to permanently placard the tanks as follows:

**OXYGEN—NO SMOKING
NO OPEN FLAMES**

OSHA also regulates certain features associated with the installation of bulk oxygen storage systems in the workplace. Examples of these regulations are provided in Solved Exercises 7.1 and 7.2.

Solved Exercise 7.1

OSHA requires the location of bulk oxygen storage systems to be either outdoors aboveground, *or* within a building of noncombustible construction that is adequately vented and used for that purpose exclusively. The selected location must be such that containers and associated equipment are not exposed to electric power lines, flammable or combustible liquid lines, or flammable gas lines. Despite the fact that oxygen is a nonflammable gas, what is the most likely reason OSHA requires bulk oxygen storage systems to be installed in this fashion?

Solution: The most likely reason OSHA enacted this regulation is to minimize the risk of fire and explosion. Although oxygen is a nonflammable gas, it *supports* the combustion of many common materials. Since these materials burn at increased rates in an oxygen-enriched environment, it is prudent to prevent the accumulation of oxygen near flammable or combustible materials. This is accomplished by installing bulk oxygen storage systems in the indicated fashion.

Solved Exercise 7.2

When it is necessary to locate a bulk oxygen storage system on ground lower than adjacent flammable or combustible liquid storage tanks, OSHA requires a suitable means to dike, curb, or grade the nearby structures. What is the most likely reason OSHA enacted this regulation?

Solution: The most likely reason OSHA enacted this regulation is to minimize the risk of fire and explosion. Flammable or combustible liquids that spill or leak from their storage vessels flow toward lower ground. A risk of fire and explosion is posed by the placement of an oxygen storage system on *lower* ground compared to the placement of flammable or combustible liquid storage tanks. To reduce or eliminate the risk, OSHA requires employers to provide suitable diking, curbing, or grading to confine the flammable and combustible liquids when they are inadvertently released from their storage vessels.

Two chemical properties of oxygen are especially important to emergency-response personnel:

- The combustion process is *supported* by atmospheric oxygen; that is, when a substance burns in the air, it unites with atmospheric oxygen. This means that atmospheric oxygen is the oxidizing agent in ordinary combustion reactions. One way to extinguish fires is to smother them, an action that prevents the burning material from contacting the oxygen required for combustion.
- In human and animal organisms, oxygen supports the life process. We assimilate oxygen through the lungs, from which it is carried by the bloodstream to the body's cells. It is then used to oxidize food nutrients. A product of this oxidation is carbon dioxide, which is exhaled from the lungs into the atmosphere. The overall phenomenon releases energy the body uses to maintain life. The process is called *respiration*. Without sufficient oxygen, survival would be impossible.

As the human organism evolved, we learned to adapt to an atmosphere containing oxygen at a concentration of 21% by volume—the concentration we experience in today's atmosphere. When obliged to breathe lower or higher concentrations, the human organism responds to the relevant situation. For instance, oxygen concentrations greater than 21% are commonly administered to patients for therapeutic purposes. In hospitals and clinics, an oxygen concentration between 25% and 50% by volume is often maintained within oxygen tents and hoods. When administered to patients for limited periods, this elevated oxygen concentration typically helps them return to good health more rapidly.

Yet, breathing an elevated concentration of oxygen for long terms does not always provide a positive outcome. Clinical observations demonstrate that adverse health effects result when elevated oxygen concentrations are inhaled for extended periods. For example, individuals who breathe an atmosphere containing 80% oxygen for more than 12 hours suffer coughing, nasal stuffiness, sore throat, chest pain, and other respiratory problems.

It is from the mass of the overlying air that we extract oxygen for survival. We learned in Section 2.8 that the mass of air exerts an average atmospheric pressure on Earth of 14.7 psi, or 101.3 kPa. At elevated heights, the air still contains 21% oxygen, but the atmospheric pressure is reduced. When the atmospheric pressure is less than normal, there is less pressure pushing oxygen into our lungs. This means that we must work harder at elevated heights to take in oxygen. Normal breathing may be unable to provide sufficient oxygen for survival. This occurs when mountaineers climb to great heights and experience *altitude sickness*.

Conversely, we can also experience adverse health effects from breathing oxygen at an *increased* pressure, even for short periods of time. This is the reason clinicians routinely administer hyperbaric oxygen therapy to patients for no more than ½ to 1 hour.

Firefighters are more likely to encounter a situation in which the atmospheric oxygen concentration is *reduced* compared to normal. As Table 7.2 illustrates, we would soon be gasping for breath if the concentration of oxygen in the air fell below 17% by volume. This occurs when fire burns within an enclosed area. Without an alternative air supply, emergency responders could collapse and die if compelled to remain within such an environment for a few short minutes.

Whereas the oxygen in the air is diluted with nitrogen and other gases, compressed oxygen and LOX are concentrated forms of the element. When released from their containers at fire scenes, the oxygen enhances the rate at which combustible materials burn. Put simply, matter burns more rapidly in an atmosphere of pure oxygen than it does in the air. This observation is consistent with our general understanding from Section 5.5, in which we noted that the rate of a reaction increases when the concentration of a reactant is increased.

TABLE 7.2 ◆ Signs and Symptoms Experienced from Inhaling Reduced Levels of Oxygen	
Percent Oxygen in the Air	*Signs and Symptoms*
21	No signs or symptoms (normal atmospheric concentration)
12–15	Muscular coordination for skill movements is lost
10–15	Consciousness continues, but judgment is faulty and muscular effort leads to rapid fatigue
6–8	Collapse occurs rapidly, but quick treatment prevents fatal outcome
<6	Death occurs within 6 to 8 minutes

Liquid oxygen increases the rate of combustion in the following circumstances:

- *The combustion of ordinary matter.* Substances that ordinarily burn slowly may burn spontaneously when in contact with liquid oxygen. Fuels, oils, greases, tar, asphalt, paper, and textiles are examples of matter that does not burn unless it is first kindled. When such substances contact liquid oxygen, however, they are likely to ignite spontaneously.
- *The combustion of metals.* When in contact with liquid oxygen, certain metals are likely to burn. The combination of magnesium shavings and liquid oxygen is so chemically reactive that these elements explode upon contact. The components of aluminum pump parts also react with liquid oxygen, especially when the reaction is initiated by friction. This explains why the use of aluminum is avoided in equipment used for storing and handling cryogenic oxygen.
- *The combustion of porous materials.* Wood, concrete, and asphalt are examples of porous materials. When liquid oxygen contacts them, it readily absorbs into their pores, producing media that can be shock-sensitive. An example is asphalt that has been soaked with liquid oxygen. A mechanical impact may cause it to burn explosively.

CHEMICAL OXYGEN GENERATORS

DOT also regulates the transportation of chemical oxygen generators, portable devices that can be chemically actuated to produce oxygen on demand. They are used in mines and other places where the available oxygen supply may be limited. When the generators are equipped with an attached means of initiation, DOT requires shippers and carriers to obtain approval for their shipment by demonstrating that the generators have been outfitted with two positive means of preventing their unintentional actuation.

OZONE, THE ALLOTROPE OF OXYGEN

Chemists refer to ozone as an allotrope of oxygen. Any *allotrope* is a unique variation of an element possessing a set of physical and chemical properties that are different from the properties of any other form of the element. Four elements that have allotropes are oxygen, phosphorus, sulfur, and carbon.

Ozone is a form of elemental oxygen having three atoms of oxygen per molecule instead of the usual two; thus, we represent its chemical formula as O_3. At room conditions, it is a pale blue gas. With a vapor density of 1.7 relative to air, ozone is more dense than normal oxygen. Whereas it possesses a sweet smell in low concentrations, ozone has an irritating, pungent, "metallic" odor at moderate to high concentrations. The sweet smell of ozone can sometimes be detected around operating electric motors or following a lightning storm.

Two characteristics make ozone one of the most hazardous materials known: a prodigious chemical reactivity and a pronounced toxicity. Ozone is a powerful oxidizing agent—considerably more reactive than oxygen itself. For example, ozone converts lead sulfide into lead sulfate, while oxygen does not react with lead sulfide.

$$\underset{\text{lead sulfide}}{3PbS(s)} + \underset{\text{ozone}}{4O_3(g)} \longrightarrow \underset{\text{lead sulfate}}{3PbSO_4(s)}$$

Since ozone is a toxic substance, every precaution should be taken to prevent breathing it. The ozone damages the scavenger cells of the immune system. These cells, called *macrophages*, customarily destroy foreign bacteria. When they are damaged, the macrophages are unable to effectively accomplish this task, and we fall victim more easily to contracting diseases.

TABLE 7.3 ◆ Adverse Health Effects Associated with Inhalation of Ozone

Ozone Exposure	*Adverse Health Effects*
0.18 ppm for 1 to 3 hours	*In adults:* Reduced exercise performance
0.12 ppm for 1 to 3 hours, or 0.08 ppm for 6 hours	*In heavily exercising adults:* Reduced lung function, cough, shortness of breath, chest pain, and airway inflammation
Repeated exposure to 0.12 ppm	*In laboratory animals:* Changes in lung structure, function, and biochemistry, which could signal the onset of chronic lung disease
0.08 ppm for 3 hours	*In laboratory animals:* Increased susceptibility to bacterial respiratory infections
0.01 to 0.15 ppm for several days	*In children and adolescents playing outdoors:* Reduced lung function, aggravation of asthma, and increased hospital visits

The adverse health effects noted in Table 7.3 may be experienced when an individual inhales an ozone concentration exceeding 0.1 part per million (0.1 ppm) for a period as short as an hour. When the use of ozone is required in the workplace, OSHA requires employers to limit employee exposure to a concentration in air of 0.1 ppm, averaged over an 8-hour workday.

Although ozone is reasonably stable at relatively low temperatures, it decomposes rapidly at room temperature; that is, it spontaneously reverts into "ordinary" oxygen. It is this relative instability that accounts for the fact that ozone is encountered chiefly at very low concentrations.

Ozone is used commercially for several purposes. It is used to bleach oils, fats, textiles, and sugar solutions. It is also used as a microbiocide at water- and sewage-treatment plants. A *microbiocide* is a substance that effectively kills disease-causing microorganisms. Ozone is an especially effective microbiocide since it is capable of destroying the fearsome parasite *Cryptosporidium* that is sometimes found in chlorinated drinking water. *Cryptosporidium* is not killed by chlorination (see Section 7.4). In humans, it causes gastrointestinal diseases that can lead to death. The use of ozone for treating water and sewage has the added advantage that their malodorous sulfides are converted into nonoffensive sulfates (Review Exercise 7.9).

When ozone is used for water treatment, the processes illustrated in Figure 7.4 are implemented. Air is first compressed to separate oxygen and nitrogen. Then, the oxygen is zapped with electricity to produce ozone, which is passed into water under pressure. The ozone reacts with the impurities in the water and kills the constituent microorganisms. The water is then filtered and pumped into the municipal water supply.

Since it is unstable and extremely poisonous, ozone must be synthesized at the point of use in minutely low concentrations. This synthesis is accomplished within an apparatus called an *ozone generator,* or *ozonizer*, which supplies an electrical current to oxygen or air. The ozone generator in Figure 7.5 is familiar to firefighters, since its use is often promoted within buildings that were damaged by fire. While the buildings are enclosed and vacated, the ozone produced by the generator

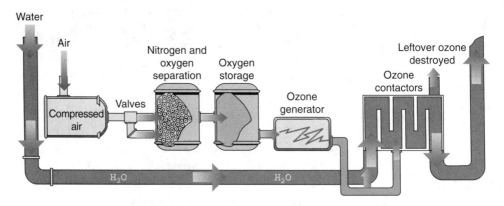

FIGURE 7.4 ◆ A simplified route for producing pure water through the use of ozone. *Adapted from an illustration prepared for the Southern Nevada Water Authority.*

FIGURE 7.5 ◆ Following fires and floods, the use of an ozone generator decontaminates and restores buildings and their furnishings to their original condition. *Courtesy of Medallion Clean Indoor Air Solutions of Las Vegas, Las Vegas, Nevada.*

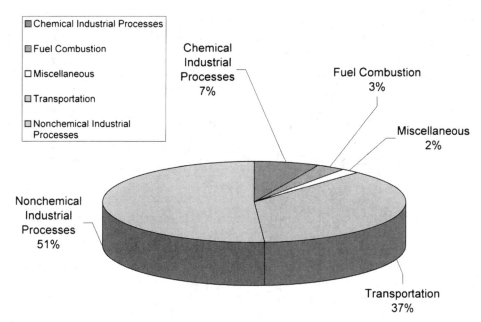

FIGURE 7.6 ◆ Ground-level ozone is produced when the volatile organic compounds in the air react with the nitrogen oxides discharged in vehicular exhaust. *Courtesy of United States Environmental Protection Agency.*

oxidizes the constituents of smoke. This removes odors from furniture, clothing, and other items damaged by fire. Ozone generators are also used to remove the musty odors that persist within walls and carpeting that were damaged by mold or mildew.

In certain metropolitan regions of the United States, ozone is often simultaneously present in the air with nitric oxide, nitrogen dioxide (Section 10.12), and certain organic compounds called *volatile organic compounds,* or *VOCs.* The VOCs are constituents of the air emissions from petroleum refineries, chemical plants, and other industrial facilities. They include the substances in gasoline vapor. Figure 7.6 shows that industrial processes account for the majority of the VOCs in the air. Typically, VOCs are substances that boil at temperatures less than approximately 392°F (200°C).

The oxides of nitrogen are pollutants generated during the operation of automobiles. They are constituents of vehicular exhaust, as well as the plumes emanating from the smoke stacks of fossil-fuel-fired power plants. Within the lower atmosphere, they react photochemically with the VOCs to produce ozone. At ground level where we live and breathe, ozone and the nitrogen oxides are often the *primary* constituents of polluted air, especially during summertime when there is ample sunlight for catalysis.

Solved Exercise 7.3

Scores of industrial facilities, including oil refineries and chemical plants, are located within the ship channel in Houston, TX. Air-sampling studies show that the ozone production rate near the Houston ship channel is extraordinarily high compared to other areas within the city. What is the most logical explanation for this observation?

Solution: Oil refineries and chemical plants discharge prodigious amounts of volatile organic compounds from smokestacks into the air. Ozone is photochemically produced in the air when these compounds react with the nitrogen oxides dispelled to the air in vehicular exhaust. Consequently, the ozone concentration is most elevated in areas where the concentration of VOCs is comparatively high. The Houston ship channel is such an area.

The presence of ozone in the lower atmosphere is linked with the climatic conditions shown in Figure 7.7. When the ozone concentration exceeds approximately 100 ppb, weather forecasters declare an "ozone alert," meaning that the ozone concentration in the air is "approaching unhealthful." At this concentration, affected individuals—especially the elderly—struggle for breath and feel dizzy. Their eyes, noses, throats, and lungs become irritated, and they experience fatigue or a lethargic feeling.

The inhalation of low concentrations of ozone is also likely to exacerbate the illnesses of individuals suffering from heart and lung ailments, emphysema, or chronic bronchitis. The afflicted are likely to cough, experience chest pain and sinus congestion, and suffer severe headaches. Based upon the results of animal studies, researchers have concluded that long-term inhalation of low ozone concentrations contributes to the development of chronic lung disease, cancer, and anemia.

These adverse health effects from exposure to ozone are viewed seriously by governmental agencies charged with evaluating disease and protecting our environment. For example, using the legal authority of the Clean Air Act, EPA established the primary and secondary national ambient air quality standard for this criteria air pollutant (Section 1.5) as 0.12 ppm (235 $\mu g/m^3$) as a 1-hour average and 0.08 ppm (157 $\mu g/m^3$) as an 8-hour average.

Solved Exercise 7.4

On a hot, humid day, the ozone concentration in urban air can exceed 100 ppb. What actions to limit the adverse health effects associated with breathing ozone should be taken by persons who suffer from emphysema?

Solution: Emphysema is a lung ailment associated with the swelling of the alveoli and connecting tissues. Emphysema sufferers cough frequently, experience headaches, and have trouble breathing. When they breathe air contaminated with ozone, these adverse health effects are exacerbated. To limit undue distress when the ozone concentration exceeds 100 ppb, emphysema sufferers are advised to remain within an air-conditioned environment, avoid heavy work and exercise, and breathe oxygen from a portable source.

Although its presence in the lower atmosphere can be a troublesome factor for maintaining good health, ozone otherwise provides a benefit to Earth and its inhabitants.

VERY UNHEALTHY

Ozone concentration 0.201 ppm or more

Very hazy, very hot, and humid
Stationary high atmospheric pressure
Sunny skies
Temperature, 95°F (35°C) or higher

UNHEALTHY

Ozone concentration 0.121 to 200 ppm

Hazy, hot, and humid
Stationary high atmospheric pressure
Sunny skies
Temperature, 90°F (32°C) or higher

APPROACHING UNHEALTHY

Ozone concentration 0.101 to 0.120 ppm

Slow-moving high atmospheric pressure
Sunny skies
Light winds
Temperature, 86°F (30°C) to 92°F (33°C)

MODERATE

Ozone concentration 0.061 to 0.100 ppm

Light to moderate wind
High pressure system
Partly cloudy to sunny skies
Temperature, 75°F (24°C) to 85°F (29°C)

GOOD

Ozone concentration 0.060 ppm or less

Passing cold front
Windy conditions
Partly sunny to cloudy skies or rain
Temperature, 75°F (24°C) to 80°F (27°C)

FIGURE 7.7 ◆ An estimation of the anticipated ozone concentration based upon the prevailing climatic conditions. *Adapted from an illustration prepared by the Delaware Valley Regional Planning Committee for the Ozone Action Partnership.*

Approximately 10 to 19 miles (16 to 30 km) over Earth's surface, ozone occurs naturally in a region of the stratosphere called the *ozone layer* (Figure 7.8). Within the stratosphere, oxygen is bombarded by high-energy particles that originate in the sun. This bombardment causes some oxygen molecules to dissociate into their individual atoms ($\cdot\ddot{O}\cdot$). The oxygen within the ozone layer exists as a mixture of oxygen atoms

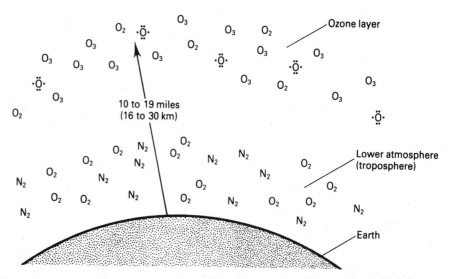

FIGURE 7.8 ◆ At the outer edge of the stratosphere is a region called the ozone layer, where oxygen is constantly converted into ozone. The stratospheric ozone prevents much of the sun's ultraviolet radiation from reaching Earth. Atmospheric scientists have discovered that the amount of ozone is depleted compared to the amount in times past. This depletion is allowing more harmful ultraviolet radiation to penetrate to Earth, increase the instances of skin cancer and cataracts, and weaken our immune systems.

and oxygen molecules. These oxygen species collide to initiate a chemical reaction producing ozone.

$$O_2(g) \longrightarrow 2 \cdot \ddot{O} \cdot (g)$$
oxygen molecule   oxygen atoms

$$O_2(g) \; + \; \cdot \ddot{O} \cdot (g) \longrightarrow O_3(g)$$
oxygen molecule   oxygen atom   ozone molecule

When stratospheric ozone absorbs low-energy ultraviolet radiation, some ozone molecules revert to the ordinary form of oxygen.

$$O_3(g) \longrightarrow O_2(g) \; + \; \cdot \ddot{O} \cdot (g)$$
ozone molecule   oxygen molecule   oxygen atom

$$O_3(g) \; + \; \cdot \ddot{O} \cdot (g) \longrightarrow O_2(g)$$
ozone molecule   oxygen atom   oxygen molecule

Under normal conditions, the continuous formation and photodecomposition of ozone compete with one another. This chemical activity produces a steady-state condition in which the rate at which stratospheric ozone forming from oxygen equals the rate at which ozone undergoes photodecomposition into oxygen.

In the past, the presence of the stratospheric ozone layer protected Earth and its occupants from overexposure to ultraviolet radiation. Today, however, our former use of certain halogen-containing compounds such as the halons is affecting Earth and its occupants. Released into the environment, these compounds diffuse upward to the ozone layer, where chlorine atoms catalyze the decomposition of the ozone and cause its overall depletion. The depleted ozone has been observed as "holes" over both the North and South Poles.

Compared to decades ago, more ultraviolet radiation from the sun now penetrates our atmosphere. The concern among scientists is that we are experiencing serious health and environmental consequences from this increased exposure. There is evidence that the radiation has weakened plant life, which contributes to lesser agricultural output. It may have also caused climatic changes that could ultimately disturb the balance of aquatic and land ecosystems. In humans, overexposure to ultraviolet radiation dramatically increases the incidence of skin cancer and cataracts and weakens our immune systems.

We noted in Section 5.13 that international attempts have been implemented to curb our use of the substances that contribute to the depletion of stratospheric ozone. Most likely, generations will pass before the threat to the ozone layer has entirely lapsed.

Performance Goals for Section 7.2

- ◆ Discuss the physical and chemical properties of elemental hydrogen, particularly noting its flammability, vapor density, rate of diffusion, and heat of combustion.
- ◆ Describe the ways hydrogen is produced for commercial use.
- ◆ Identify the industries that use bulk quantities of hydrogen.
- ◆ Identify the metals that react with acids and/or water to produce hydrogen.
- ◆ Describe the manner by which hydrogen is produced during the charging of storage batteries.
- ◆ Describe why hydrogen is usually selected by the aerospace industry as a rocket fuel.
- ◆ Describe the recommended response actions when hydrogen is released into the environment.

◆ 7.2 HYDROGEN

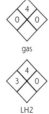

gas

LH2

Although Table 4.1 shows that hydrogen ranks ninth in natural abundance by mass, only traces of this element exist naturally in the free state as the element. Hydrogen is such a reactive element that it is found on Earth only in compounds like water and methane. On the other hand, hydrogen is the most abundant element in the universe, 75.2% by mass.

Elemental hydrogen is an odorless, colorless, tasteless, and nontoxic substance. Its chemical formula is H_2. While hydrogen exists as a gas at ordinary temperatures and pressures, the gas liquefies when compressed under a pressure greater than 294 psi (2030 kPa) and at a temperature less than $-390°F$ ($-234.5°C$). Liquid hydrogen is sometimes denoted as *LH2*.

Elemental hydrogen is commercially available as both the compressed gas and the cryogenic liquid. The chemical industry uses hydrogen as a raw material for the production and manufacture of other chemical substances. The petroleum industry uses hydrogen for the production and manufacture of petroleum fuels. The aerospace industry uses hydrogen as a rocket propellant. At the Kennedy Space Center in Figure 7.9, a staggering 800,000 gal (3000 m^3) of liquid hydrogen is held in a single storage tank awaiting use. Liquid hydrogen is chosen as a rocket fuel since the

FIGURE 7.9 ◆ The largest liquid hydrogen storage tank in the world is located at the John F. Kennedy Space Center in Florida. The tank in the foreground holds 800,000 gal (3000 m^3) of liquid hydrogen. It is dwarfed by the shuttle launch complex in the background. *Courtesy of the National Aeronautics and Space Administration, John F. Kennedy Space Center, Kennedy Space Center, Florida.*

combustion of a very small mass produces a prodigious amount of energy (see below). In the *Columbia* space shuttle shown in Figure 7.10, separate tanks carried 380,000 gal (1438 m^3) of liquid hydrogen and 140,000 gal (530 m^3) of liquid oxygen into space.

In the world of the not-so-distant future, hydrogen could become the fuel used to power automobiles and other surface vehicles. Considerable efforts are underway to identify means of storing compressed hydrogen onboard the cars of tomorrow in a safe, efficient, compact, and economical fashion. Several hydrogen-powered automobiles are already available in the worldwide marketplace. The BMW 745h prototype sedan in Figure 7.11 is powered by a hydrogen internal combustion engine.

The prospective use of hydrogen as a vehicular fuel has the advantage of reducing our reliance on fossil fuel resources from foreign countries and reducing air pollution. Scientists predict that by 2040, the use of hydrogen-powered vehicles could cut annual greenhouse gas emissions by more than 500 million tons (454 million tonnes) and cut oil consumption by 11 million tons per day (10 million tonnes/day). However, the methods now used to produce hydrogen (see p. 276) are associated with the generation of carbon dioxide. This implies that the use of hydrogen-powered vehicles will only be environmentally friendly if the production of the hydrogen fuel does not simultaneously generate carbon dioxide.

Physicists and engineers have also been researching ways to successfully develop a thermonuclear reactor, in which the nuclei of hydrogen atoms fuse together and produce heavier nuclei. These reactions convert mass into energy. Nuclear fusion occurred during the detonation of the hydrogen bomb in Figure 7.12, but the energy release was essentially an uncontrolled phenomenon. To serve as an energy source,

FIGURE 7.10 ◆ The Space Shuttle *Columbia* awaiting its launch into space on January 16, 2003. This was *Columbia*'s 28th and final mission. The shuttle craft was destroyed on February 1, 2002, over east Texas upon reentry into Earth's atmosphere. *Courtesy of the National Aeronautics and Space Administration, John F. Kennedy Space Center, Kennedy Space Center, Florida.*

FIGURE 7.11 ◆ BMW has manufactured a fleet of hydrogen-powered automobiles, including this BMW 745h prototype. The use of hydrogen as a motor fuel is considered to be preferable to the use of gasoline since no harmful combustion products are emitted to the atmosphere. It is for this reason that the words CLEAN ENERGY were written on the side of the prototype car. *Courtesy of 2005 BMW of North America, LLC, Westwood, New Jersey, used with permission. The BMW name and logo are registered trademarks.*

fusion reactions must be controlled on a large scale. If a thermonuclear reactor could be successfully developed,* hydrogen would most likely become *the* energy source of the future, replacing our dependence on modern day fossil-fuel-fired and nuclear power plants.

Hydrogen is produced for commercial use by the following methods:

◆ In the first method, steam is first passed over red-hot coke or coal (Section 7.7). This results in the formation of a mixture of gases called *water gas, synthesis gas,* or *syngas.* This mixture consists mainly of carbon monoxide and hydrogen.

$$C(s) + H_2O(g) \longrightarrow CO(g) + H_2(g)$$
$$\text{carbon} \quad \text{water} \qquad \text{carbon monoxide} \quad \text{hydrogen}$$

The hydrogen is isolated from the carbon monoxide, or the carbon monoxide is converted to carbon dioxide, by passing the water gas with additional steam over a catalyst such as iron(III) oxide.

$$CO(g) + H_2O(g) \longrightarrow CO_2(g) + H_2(g)$$
$$\text{carbon monoxide} \quad \text{water} \qquad \text{carbon dioxide} \quad \text{hydrogen}$$

Here again, the carbon dioxide in the resulting mixture is removed by passing it through an alkaline solution.

---

*A multinational effort, called the International Thermonuclear Experimental Reactor Consortium, aims to have a fusion demonstration power plant in operation by 2014. The United States is a member of the consortium. The underlying technology involves the use of a doughnut-shaped magnetic bottle that will confine hydrogen nuclei as they are heated to 180 million degrees. At this temperature, hydrogen nuclei combine to form helium, neutrons, and sufficient energy to produce electricity. If the initial tests are successful, the consortium estimates that the construction of a commercial-scale demonstration fusion power plant will require an additional 35 years.

FIGURE 7.12 ◆ An experimental thermonuclear device (hydrogen bomb) is exploded on Enewetak on October 31, 1952. *Courtesy of United States Department of Energy, Nevada Operations Office, Las Vegas, Nevada.*

◆ The second industrial method of producing hydrogen involves the chemical action of steam on methane at high temperatures.

$$\underset{\text{methane}}{CH_4(g)} + \underset{\text{water}}{H_2O(g)} \longrightarrow \underset{\text{hydrogen}}{3H_2(g)} + \underset{\text{carbon monoxide}}{CO(g)}$$

The water gas produced by the chemical reaction is treated by either of the methods previously noted for isolating carbon monoxide-free hydrogen.

◆ Hydrogen is also produced by the multistepped, complex reaction between propane and steam, denoted by the following overall equation:

$$\underset{\text{propane}}{C_3H_8(g)} + \underset{\text{water}}{6H_2O(g)} \longrightarrow \underset{\text{carbon dioxide}}{3CO_2(g)} + \underset{\text{hydrogen}}{10H_2(g)}$$

The carbon dioxide in the resulting mixture is removed by passing it through an alkaline solution.

◆ Hydrogen is also produced by the reaction of steam on methanol vapor (Section 12.8) using a copper oxide/zinc oxide catalyst, as follows:

$$\underset{\text{methanol}}{CH_3OH(g)} + \underset{\text{water}}{H_2O(g)} \longrightarrow \underset{\text{carbon dioxide}}{CO_2(g)} + \underset{\text{hydrogen}}{3H_2(g)}$$

Once again, the carbon dioxide is removed by passing the mixture through an alkaline solution.

TABLE 7.4 ◆ Physical Properties of Elemental Hydrogen	
Specific gravity	0.09
Vapor density (air = 1)	0.069
Freezing point	−434°F (−259°C)
Boiling point	−423°F (−253°C)
Lower explosive limit	4%
Upper explosive limit	75%
Liquid-to-gas expansion ratio	865

Several physical properties of elemental hydrogen are listed in Table 7.4. A notable property is its vapor density—only 0.07 when compared to air. Since this value is so low, it gives the gas a natural *lifting power*. When expressed in metric units, the lifting power of 1 L of hydrogen is equal to the difference in mass between 1 L of air and 1 L of hydrogen at a given temperature and pressure. At 32°F (0°C) and 14.7 psi (101.3 kPa), 1 L of air possesses a mass of 1.2930 g, whereas 1 L of hydrogen has a mass of 0.0899 g; hence, the lifting power of hydrogen in air at this temperature and pressure is 1.2930 g/L − 0.0899 g/L, or 1.2031 g/L.

Another important property of hydrogen is the size of its molecules. Hydrogen molecules are extremely small compared to other molecules. They are so tiny that hydrogen molecules easily leak through openings such as pipe joints and valve connections. To prevent loss during storage and transport, the steel cylinders, tubes, and tanks used to hold the compressed gas must be uniquely designed.

Another important property of hydrogen is the relatively high rate at which it diffuses into the air. When released from a tank or container, hydrogen diffuses into the air more rapidly than any other substance; that is, at the same temperature, hydrogen moves faster than other gases or vapors.

Hydrogen is capable of burning when its concentration in air is from 4% to 75% by volume. A concentration within this wide flammable range can be readily achieved when hydrogen is released from its container or tank. Nonetheless, since it dissipates so rapidly, unconfined hydrogen does not remain in any single area for long.

These properties of hydrogen give rise to two potential scenarios. When released indoors or into an enclosure where the gas can accumulate, hydrogen gas poses the risk of fire and explosion. On the other hand, when hydrogen gas is released outdoors or in a manner that enables it to dissipate into the atmosphere, the likelihood that it will form a flammable mixture is relatively small.

Solved Exercise 7.5

The diffusion rate of hydrogen into the air is 0.098 in.^{2}/s (0.634 cm^2/s) at 32°F (0°C). What does the magnitude of this rate indicate about the likelihood as to whether a flammable mixture of hydrogen and air can be produced under most accident scenarios that occur outside buildings?

Solution: The information reveals that hydrogen diffuses very rapidly into the surrounding air. In fact, unconfined hydrogen moves more rapidly into air than any other gas. Hydrogen is also flammable over a wide concentration range (4–75% by volume). Since hydrogen does not remain localized for long, a flammable mixture of hydrogen and air is not ordinarily produced under most accident scenarios that occur outside buildings.

Hydrogen burns in air to form water vapor.

$$2H_2(g) + O_2(g) \longrightarrow 2H_2O(g)$$
hydrogen   oxygen   water

Because of its low vapor density and high diffusion rate, the combustion reaction occurs as the gas moves upward. Hydrogen burns with an almost nonluminous flame, which is difficult to visibly observe during daylight hours.

Since hydrogen rises in the air, hydrogen-sensing devices are often installed near the ceilings of enclosures in which hydrogen could inadvertently be released, such as from a leaking storage tank into a room. These devices activate exhaust fans capable of evacuating the hydrogen into the outside air at very high speeds. When emergency-response crews are called to such incidents, they should enter the room only after the hydrogen has been completely dispelled.

The combustion of hydrogen is unique in the following way. When hydrogen burns in an atmosphere of pure oxygen, the accompanying heat of combustion is 61,600 Btu/lb (143,000 kJ/kg). This is the most energy for the least mass that is evolved during the combustion of any material. We use the magnitude of this energy for launching spacecraft and cutting and welding metals.

The oxyhydrogen torch in Figure 7.13, for example, is used by jewelers to create rings and bracelets made of platinum, a metal that melts at 3191°F (1755°C). In the torch, compressed hydrogen and oxygen are separately stored within steel cylinders and then pressure-fed through tubing into a mixing chamber. This mixture is then discharged into the nozzle, at which point combustion occurs.

Hydrogen is also used commercially in connection with the chemical process called *hydrogenation*. At elevated temperatures and pressures, and generally in the presence of a catalyst, hydrogen combines with certain organic compounds to form commercially

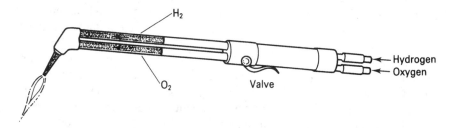

FIGURE 7.13 ◆ The oxyhydrogen torch. Hydrogen and oxygen enter the mixing chamber separately. The mixture discharges from the nozzle at such a rate that the flame burns at the tip of the torch. Using the oxyhydrogen torch, temperatures between 3300°F (1800°C) and 4400°F (2400°C) may be attained.

useful products. When vegetable oils are hydrogenated, for example, they become suitable for use as shortening and raw materials for the manufacture of soaps and lubricants.

In the chemical industry, hydrogen is used to produce metallic and nonmetallic hydrides. Hydrogen combines with sodium, for example, to produce sodium hydride (Section 9.5).

$$2\text{Na}(s) + \text{H}_2(g) \longrightarrow 2\text{NaH}(s)$$
$$\text{sodium} \quad\; \text{hydrogen} \qquad\qquad \text{sodium hydride}$$

Hydrogen also unites with nonmetals to produce compounds such as water, hydrogen chloride, and ammonia.

$$2\text{H}_2(g) + \text{O}_2(g) \longrightarrow 2\text{H}_2\text{O}(g)$$
$$\text{hydrogen} \quad\; \text{oxygen} \qquad\qquad \text{water}$$

$$\text{H}_2(g) + \text{Cl}_2(g) \longrightarrow 2\text{HCl}(g)$$
$$\text{hydrogen} \quad\; \text{chlorine} \qquad\qquad \text{hydrogen chloride}$$

$$3\text{H}_2(g) + \text{N}_2(g) \longrightarrow 2\text{NH}_3(g)$$
$$\text{hydrogen} \quad\; \text{nitrogen} \qquad\qquad \text{ammonia}$$

When it is intended for use at manufacturing and processing plants, hydrogen is generally encountered within cylinders or storage tanks. The capacities of tanks used to store hydrogen range from less than 3000 ft³ (85 m³) to over 15,000 ft³ (425 m³). OSHA regulates the location of these storage tanks in relation to the position of buildings and other storage tanks, so they are readily accessible to delivery equipment and authorized personnel. To alert individuals to the presence of this flammable gas, OSHA also requires the tanks to be permanently placarded, as follows:

> **HYDROGEN—FLAMMABLE GAS**
> **NO SMOKING**
> **NO OPEN FLAMES**

DOT regulates the transportation of compressed hydrogen as a flammable gas, requiring shippers and carriers to affix FLAMMABLE GAS labels to containers and to post FLAMMABLE GAS placards on the bulk packaging, freight container, transport vehicle, unit load device, or railcar used to transport hydrogen.

While hydrogen may be encountered in cylinders and storage tanks, it may also be inadvertently generated as the product of chemical reactions. Certain metals generate hydrogen by displacing it from water and acid solutions. These displacement reactions occur at rates that depend on the nature of the metal. The alkali metals and alkaline earth metals react violently with both water and acids, generating hydrogen at explosive rates. Certain other metals react much more slowly with water and acids. Nonetheless, these reactions pose the risk of fire and explosion, since the hydrogen produced by them often absorbs the heat of reaction and bursts spontaneously into flame. We shall examine this topic more extensively for the alkali metals in Section 9.2.

Many metals possess the capability of releasing hydrogen from acids. Metallic tin and aluminum displace hydrogen from hydrochloric acid and sulfuric acid, respectively, as follows:

$$\text{Sn}(s) + 2\text{HCl}(aq) \longrightarrow \text{SnCl}_2(aq) + \text{H}_2(g)$$
$$\text{tin} \qquad \text{hydrochloric acid} \qquad \text{tin(II) chloride} \quad\; \text{hydrogen}$$

$$2\text{Al}(s) + 3\text{H}_2\text{SO}_4(aq) \longrightarrow \text{Al}_2(\text{SO}_4)_3(aq) + 3\text{H}_2(g)$$
$$\text{aluminum} \quad\; \text{hydrochloric acid} \qquad\; \text{aluminum sulfate} \qquad \text{hydrogen}$$

TABLE 7.5 ◆ Activity Series of the Elements

Elements that release hydrogen from water:	Cesium
	Lithium
	Rubidium
	Potassium
	Barium
	Sodium
	Calcium
	Magnesium
Elements that release hydrogen from acids:	Aluminum
	Manganese
	Zinc
	Chromium
	Iron
	Nickel
	Tin
	Lead
Elements that do not release hydrogen from acids:	Bismuth
	Copper
	Mercury
	Silver
	Platinum
	Gold

The rates at which metals displace hydrogen from water and acidic solutions can be determined experimentally. When the metals are then arranged by decreasing reaction rates, the compilation in Table 7.5 is obtained. This arrangement of the metals is called an *activity series*. The significance of the compilation is summarized as follows:

- The metals listed at the top of the table are so chemically reactive they react with *both* water and acids.
- The metals below magnesium release hydrogen from steam and acid solutions, but not from liquid water.
- The metals below iron in the series do not displace hydrogen from steam, even when the temperature is elevated.
- The metals below lead possess insufficient chemical reactivity to release hydrogen from either water or acids.

Hydrogen is also displaced by certain metals from solutions of sodium hydroxide. The common metals exhibiting this chemistry are aluminum, zinc, and lead, but these reactions occur slowly.

Hydrogen is also produced during the charging of storage batteries. A storage battery produces electric current by chemical action. It operates by means of either of the two reversible chemical reactions commonly referred to as "charging" and "discharging." A battery is said to be "charged" when an outside current has been delivered through it; then, as the battery "discharges," it produces electricity by means of

FIGURE 7.14 ◆ A battery-charging station, such as that illustrated here, should be located in an open, well-ventilated location to prevent the accumulation of hydrogen. Even a mixture containing only 4% by volume of hydrogen in air may explode when ignited by a spark or flame. *Courtesy of Argonne National Laboratory, Argonne, Illinois.*

chemical reactions involving the substances that were formed during the charging process.

To illustrate the manner by which a battery functions, let's consider the common lead storage battery found in most automobiles. When the battery is being "charged," bubbles of oxygen accumulate near its positive plate and hydrogen accumulates near the negative plate. Both gases then dissipate into the atmosphere.

When banks of storage batteries like those shown in Figure 7.14 are simultaneously charged within an enclosure, the concentration of hydrogen is likely to exceed its lower explosive limit. In an enclosed room, for instance, hydrogen rises and concentrates along the length of its ceiling. Since a concentration of only 4% by volume of hydrogen is sufficient to produce a flammable mixture in air, adequate ventilation should be provided within rooms containing banks of storage batteries to prevent the accumulation of hydrogen.

Perhaps the best-known incident involving the burning of hydrogen was the destruction of the *Hindenburg,* the German dirigible shown in Figure 7.15. Although the combustion of diesel fuel powered its engines, hydrogen stored in large gas bags held this airship aloft. Germany had intended to use the *Hindenburg* to inaugurate a new era in fast transatlantic travel, but the airship mysteriously caught fire and exploded while attempting its landing approach on May 6, 1937, at Lakehurst, New Jersey. Given the presence of hydrogen onboard, the immediate sentiment linked the burning of the airship with a hydrogen leak.

Today, airships do not contain hydrogen. Helium has replaced the hydrogen formerly used in dirigibles to provide their lifting power. Helium is nonflammable, but it has only about 93% as much lifting power as hydrogen.

FIGURE 7.15 ◆ The German dirigible *Hindenburg* exploding over its landing site at the Naval Air Station in Lakehurst, New Jersey, on May 6, 1937. This was the first and only time the swastika was permitted to be displayed on a German air vessel entering the United States. Flammable hydrogen gas had held the *Hindenburg* aloft from its departure point in Frankfurt, Germany. *Courtesy of Wide World Photos, New York, New York.*

RESPONDING TO INCIDENTS INVOLVING THE RELEASE OF HYDROGEN

As a practical matter, combating a fire involving hydrogen rarely provides any benefit. It is often best to permit a hydrogen fire to burn itself out without interference. Since the combustion of hydrogen is likely to cause secondary fires, appropriate action should be taken to prevent the spread of these fires.

Attempts should be made to combat an ongoing fire involving hydrogen only when the flow of hydrogen from its containment vessel can be stopped without exposing personnel to risk. At scenes where liquid hydrogen is involved, the use of a water fog is warranted, since it serves to reduce the hydrogen concentration in the nearby atmosphere.

Performance Goal for Section 7.3
- ◆ Discuss the physical and chemical properties of elemental fluorine, especially its pronounced toxicity and unique chemical reactivity.

◆ 7.3 FLUORINE

As noted in Table 4.1, fluorine is found in nature to the extent of 0.027% by mass, but it is never encountered as the free element. Instead, we find it distributed in rocks and minerals. As a hazardous material, the element possesses a certain distinction: it is more chemically reactive than any other element. Fluorine is so reactive that it forms compounds with nearly every other element, the possible exceptions being the lighter noble gases.

At room conditions, elemental fluorine is a pale yellow gas with a pungent, irritating odor. Some important physical properties are provided in Table 7.6. The exceptional chemical reactivity of fluorine is the basis for considerable difficulties both in its preparation and handling. Fluorine can only be prepared by passing an electric current through a molten mixture of potassium fluoride and anhydrous (without water) hydrofluoric acid in a specially designed apparatus.

Although elemental fluorine is available commercially, its pronounced reactivity and poisonous nature generally limit its widespread use. The reactivity of fluorine is associated with its powerful nature as an oxidizing agent. Elemental fluorine does not burn, but like oxygen, it supports the combustion of other substances.

The ability of fluorine to support combustion is apparent from its reaction with hydrogen. Fluorine and hydrogen unite explosively, forming hydrogen fluoride.

$$\underset{\text{hydrogen}}{H_2(g)} + \underset{\text{fluorine}}{F_2(g)} \longrightarrow \underset{\text{hydrogen fluoride}}{2HF(g)}$$

Ordinary materials such as wood, plastics, and metals also burn spontaneously when exposed to fluorine; even "fireproof" asbestos burns in a fluorine-rich atmosphere.

Fluorine also possesses the capability of chemically reacting with the components of water. Although the chemistry is somewhat complicated, it may be represented as follows:

$$\underset{\text{fluorine}}{5F_2(g)} + \underset{\text{water}}{5H_2O(l)} \longrightarrow \underset{\text{hydrogen}}{8HF(aq)} + \underset{\text{oxygen}}{O_2(g)} + \underset{\text{hydrogen peroxide}}{H_2O_2(aq)} + \underset{\text{oxygen difluoride}}{OF_2(g)}$$

The reaction is so violent that its occurrence is sufficient cause to avoid the use of water when combating fires supported by fluorine.

Not only is fluorine a powerful oxidizing agent, but it is also highly toxic when inhaled. Even small traces of fluorine can be irritating, since the element reacts with the moisture in the lungs and respiratory tract to produce other reactive and corrosive substances. Upon contact with the skin, fluorine causes deep, penetrating burns, which can be delayed and progressively worsen with time. To reduce or eliminate these ill effects, OSHA requires employers to regulate employee exposure in the workplace by

TABLE 7.6 ◆ Physical Properties of Elemental Fluorine	
Specific gravity	1.1
Vapor density (air = 1)	1.7
Freezing point	$-360°F$ ($-218°C$)
Boiling point	$-307°F$ ($-188°C$)
Liquid-to-gas expansion ratio	965

limiting the fluorine concentration to no greater than 0.1 part per million (0.1 ppm), averaged over an 8-hour workday.

Although elemental fluorine exists as both a compressed gas and cryogenic liquid, DOT prohibits the shipment of cryogenic fluorine. DOT requires shippers and carriers to transport the compressed gas in DOT-specification cylinders made of a specially designed steel or nickel to which fluorine is relatively unreactive. These cylinders must be seamless and equipped with a valve protection cap, but no safety-relief devices. DOT also prohibits charging of the cylinders with more than 6 lb (3 kg) of fluorine under a pressure not exceeding 400 psi$_g$ (2760 kPa) at 70°F (21°C).

DOT regulates the transportation of fluorine as a poison gas. DOT requires shippers and carriers to affix INHALATION HAZARD, OXIDIZER, and CORROSIVE labels to fluorine packaging and to post INHALATION HAZARD placards on the bulk packaging, freight container, transport vehicle, unit load device, or railcar used to transport fluorine.

Firefighters are cautioned to avoid extinguishing fires supported by elemental fluorine. These fires are generally permitted to burn until the fuel has been totally exhausted. When fluorine cylinders are exposed to flames or excessive heat, water should be used to cool them from unmanned monitors. When they are exposed to prolonged, intense heat, these cylinders rupture since they are unequipped with safety-relief devices.

When fluorine is encountered as a leak from a small cylinder, DOT recommends rapid evacuation of all unauthorized persons to an initial distance in all directions of 100 ft (30 m). DOT further advises their evacuation to a distance downwind from the scene of 0.1 mi (0.2 km) and 0.3 mi (0.5 km) during the day and nighttime hours, respectively.

Performance Goals for Section 7.4

- ◆ Discuss the physical and chemical properties of elemental chlorine, especially its inhalation toxicity.
- ◆ Show that elemental chlorine can act as an oxidizing agent.
- ◆ Identify the industries that use bulk quantities of chlorine.
- ◆ Describe the recommended response actions when chlorine is released into the environment.

◆ 7.4 CHLORINE

Elemental chlorine is not found naturally, but it is found on Earth to the extent of 0.19% by mass in a variety of compounds including sodium chloride, potassium chloride, calcium chloride, and magnesium chloride.

For commercial use, elemental chlorine is prepared by passing an electric current through either molten sodium chloride or an aqueous solution of sodium chloride or magnesium chloride. When aqueous sodium chloride is used, sodium hydroxide and hydrogen are simultaneously produced, as follows:

$$\underset{\text{sodium chloride}}{NaCl(aq)} + \underset{\text{water}}{2H_2O(l)} \longrightarrow \underset{\text{sodium hydroxide}}{2NaOH(aq)} + \underset{\text{hydrogen}}{H_2(g)} + \underset{\text{chlorine}}{Cl_2(g)}$$

The elemental chlorine is encountered commercially as a gas and a liquefied compressed gas.

Throughout the civilized world, large volumes of elemental chlorine are annually required for the following commercial uses:

- A raw material for the production and manufacture of a wide range of chlorine-containing compounds used as solvents, pesticides, dyes, bleaching agents, plastics, refrigerants, and other commercial products
- A microbiocide for treatment of water and sewage
- A bactericide for sterilizing drinking water and pool water against waterborne infectious diseases
- A bleaching agent of paper pulp and certain textiles

At room conditions, elemental chlorine exists as a yellow-green gas with a characteristic penetrating, irritating odor. The element is about 2½ times heavier than air and is highly poisonous when inhaled. Several other physical properties of chlorine are noted in Table 7.7.

Solved Exercise 7.6

Calculate the approximate volume in cubic feet of gaseous chlorine that results when one ton of liquid chlorine is released from its transport vessel. Why does this volume pose the risk of inhalation toxicity?

Solution: Using the specific gravity of the liquid in Table 7.7, the density of liquid chlorine is calculated as follows:

$$1.56 \times 62.3 \, \frac{\text{lb}}{\text{ft}^3} = 97.2 \, \text{lb/ft}^3$$

The volume occupied by 1 ton of liquid chlorine is then calculated.

$$1 \, \text{ton} \times 2000 \, \frac{\text{lb}}{\text{ton}} \times \frac{1 \, \text{ft}^3}{97.2 \, \text{lb}} = 20.6 \, \text{ft}^3$$

From the liquid-to-gas expansion ratio for chlorine listed in Table 7.7, the volume of gaseous chlorine is determined as follows:

$$20.6 \, \text{ft}^3_{\text{(liquid)}} \times 457.6 = 9426 \, \text{ft}^3_{\text{(gas)}}$$

This calculation shows that 1 ton of liquid chlorine occupies a volume of 20.6 ft³, which when released from its transport vessel, expands as the gas to occupy a volume of 9426 ft³.

One ton of chlorine per 9426 ft³ is equivalent to an undiluted concentration of 0.2 lb/ft³. Inhalation of one or two breaths of chlorine at this concentration is fatal.

TABLE 7.7 ◆ Physical Properties of Elemental Chlorine

Specific gravity (gas)	2.48
Specific gravity (liquid)	1.56
Vapor density (air = 1)	2.49
Freezing point	−150°F (−101°C)
Boiling point	−29°F (−34°C)
Liquid-to-gas expansion ratio	457.6

TABLE 7.8 ◆ Signs and Symptoms Resulting from the Inhalation of Various Concentrations of Chlorine

Concentration Inhaled (ppm)	Signs and Symptoms
0.5	No signs or symptoms
3–8	Discomfort; stinging and burning of the eyes
50–250	Severely irritating to eyes and mucous membranes; coughing, choking, nausea, dizziness, headache, and difficulty in breathing
250–500 (30 min)	Delayed pulmonary edema
1000 (few breaths)	Death

The element chlorine should not be confused with solid "chlorine" products used to treat the water in domestic and municipal swimming pools. Although these products are frequently called "chlorine," they are actually oxidizing agents that generate chlorine by chemical action within the swimming pool. The properties of these chlorine-containing products are more appropriately discussed in Chapter 11.

The principal risk associated with exposure to elemental chlorine is inhalation toxicity. The exposure initially causes coughing, dizziness, nausea, headache, and severe inflammation of the eyes, nose, and throat. Extended inhalation is likely to cause congestion of the lungs, which could result in death. The congestion induces the fatal accumulation of fluid within the lungs, a condition known as *pulmonary edema*. The ill effects listed in Table 7.8 are likely to be experienced as well. Individuals who have been exposed to chlorine should be immediately removed from the environment, kept warm with blankets, and provided with immediate medical assistance.

On a somber note, by taking advantage of its highly poisonous nature, chlorine was used as a chemical warfare agent (Section 12.16) during World War I. Germany unleashed the gas against the British and their allies. Discharged with the wind flow, it was carried into the trenches where the chlorine killed unprotected soldiers.

In terms of its chemical reactivity, chlorine is very versatile. Like oxygen and fluorine, it is a nonflammable gas capable of supporting combustion. The oxidizing ability of chlorine is apparent from its reaction with hydrogen. The hydrogen burns to form hydrogen chloride.

$$\underset{\text{hydrogen}}{H_2(g)} + \underset{\text{chlorine}}{Cl_2(g)} \longrightarrow \underset{\text{hydrogen chloride}}{2HCl(g)}$$

Other elements burn when exposed to an atmosphere of chlorine. For example, finely divided, copper, arsenic, antimony, phosphorus, and sulfur burn with incandescence.

$$\underset{\text{copper}}{Cu(s)} + \underset{\text{chlorine}}{Cl_2(g)} \longrightarrow \underset{\text{copper(II) chloride}}{CuCl_2(s)}$$

$$\underset{\text{arsenic}}{2As(s)} + \underset{\text{chlorine}}{3Cl_2(g)} \longrightarrow \underset{\text{arsenic trichloride}}{2AsCl_3(s)}$$

$$\underset{\text{antimony}}{2Sb(s)} + \underset{\text{chlorine}}{3Cl_2(g)} \longrightarrow \underset{\text{antimony trichloride}}{2SbCl_3(s)}$$

$$P_4(s) + 6Cl_2(g) \longrightarrow 4PCl_3(l)$$

phosphorus · · chlorine · · · · · · · · phosphorus trichloride

$$S_8(l) + 4Cl_2(g) \longrightarrow 2S_2Cl_2(l) + 4SCl_4(l)$$

sulfur · · · chlorine · · · · · · disulfur dichloride · · · sulfur tetrachloride

Here, the chlorine reacts as an oxidizing agent. The combination of these reactions causes chlorine to be regarded as a corrosive material.

Chlorine also supports the combustion of certain organic compounds. These combustion reactions occur only in the presence of light, which acts catalytically. For instance, a mixture of elemental chlorine and gasoline vapor is essentially unreactive unless it is exposed to light.

When chlorine atoms are incorporated into a compound, the associated phenomenon is called a *chlorination* reaction. The burning of a substance in a chlorine atmosphere is an example of a chlorination reaction. The term "chlorination" is also used to describe the treatment of drinking water and wastewater with elemental chlorine and chlorine-containing oxidizing agents. When used in this fashion, chlorine acts as a *microbiocide* by killing undesirable microorganisms. The use of chlorine for treating drinking water saves millions of lives each year from such diseases as cholera and dysentery. Notwithstanding this fact, as we noted in Section 7.1, chlorine is incapable of effectively destroying the parasite *Cryptosporidium*.

When dissolved in water, chlorine reacts to form hypochlorous acid, which is even more powerful than chlorine as an oxidizing agent.

$$Cl_2(g) + H_2O(l) \longrightarrow HClO(aq) + HCl(aq)$$

chlorine · · water · · · · · hypochlorous acid · · hydrochloric acid

During the chlorination of drinking water and wastewater, it is the presence of hypochlorous acid that aids in making chlorine an effective microbiocide.

DOT regulates the transportation of chlorine as a liquefied compressed gas in steel cylinders, ton-containers, motorized cargo tanks, and railroad tankcars, several of which are shown in Figure 7.16. The ton-container is less frequently used for transporting chlorine; it is a welded tank having a maximum loaded mass of 3700 lb (1680 kg). As a liquefied compressed gas, elemental chlorine is pressurized at 84 psi (580 kPa) at 70°F (21°C).

Solved Exercise 7.7

What is the most likely reason DOT prohibits shippers and carriers from accepting any quantity of elemental chlorine for domestic transport on passenger-carrying aircraft or railcar?

Solution: The primary hazard associated with exposure to elemental chlorine is inhalation toxicity. The most likely reason DOT prohibits shippers and carriers from transporting chlorine on domestic passenger-carrying aircraft or railcar is to reduce or eliminate the likelihood that passengers and transportation personnel will experience the ill effects caused by chlorine exposure.

DOT regulates the transportation of chlorine as a poison gas. DOT requires shippers and carriers to affix INHALATION HAZARD and CORROSIVE labels to chlorine packaging and to display INHALATION HAZARD placards on the bulk

(a)

(b)

(c)

FIGURE 7.16 ◆ Three transport vessels for shipping chlorine in the United States. (a) Steel cylinders used to transport chlorine as the compressed gas are constructed without seams and have capacities ranging from 1 to 150 lb (0.45 to 68 kg). (b) Insulated tankcars used to transport liquid chlorine are fabricated from certain carbon steels. The manner by which they must be constructed is specified by DOT. These tankcars are nominally of 55- or 90-ton (50- or 82-tonne) capacity and are equipped with safety-relief devices designed to relieve excess internal pressure. (c) Chlorine barges, also used to transport liquid chlorine, are used primarily on inland waters. Each barge has four independent, cylindrical, uninsulated pressure tanks mounted longitudinally. The most common barge capacities are 600 tons (545 tonne); each of the four tanks holds 150 tons (136 tonne). *Courtesy of the Chlorine Institute, Inc., Washington, D.C.*

packaging, freight container, transport vehicle, unit load device, or railcar used to transport chlorine.

The DOT-approved tanks and cylinders used for transporting chlorine are equipped with fusible plugs designed to melt in the temperature range 158–165°F (70–74°C). The plugs are located on the valve just below the valve seat. When exposed to the temperatures routinely associated with fire conditions, the melting of these plugs permits chlorine to be slowly released to the environment and prevents the vessels from rupturing.

DOT also regulates the construction and fabrication of railroad tankcars and tank trucks used to transport chlorine. To maintain the confined chlorine primarily in the liquid state, DOT requires a minimum of 4 in. (10 cm) of insulation about them. As we noted in Section 3.7, thermally insulated tankcars constructed to DOT specifications have been designed to allow external heat to be slowly transferred to the contents under the normal conditions of transport. When exposed to heat, however, these tanks can lose their integrity and rupture.

When exposure to elemental chlorine is required within the workplace, OSHA requires employers to provide their workers with respiratory protective gear. When the exposure involves a relatively small amount of chlorine, goggles and impermeable gloves should be worn. Whenever possible, activities should be conducted within a fume hood. When working on bulk chlorine storage systems, employees should wear total-encapsulating suits and use self-contained breathing apparatus.

To avoid or minimize the ill effects associated with exposure to chlorine, OSHA regulates employee exposure in the workplace by limiting the chlorine concentration in air to no greater than 1 ppm, averaged over an 8-hour workday.

RESPONDING TO INCIDENTS INVOLVING THE RELEASE OF CHLORINE

When an emergency-response team is called to a scene at which chlorine tanks or containers have ruptured or could potentially rupture, it is vital to acknowledge the potential risk of inhalation toxicity. Since inhaling chlorine can be fatal, the use of self-contained breathing apparatus is essential at all times.

The first action of the team responding to a chlorine release is to evacuate all unauthorized persons from the scene to a distance dependent upon the amount of chlorine that has been or could be released into the environment. DOT recommends an isolation distance in all directions of 100 ft (30 m) for a "small" cylinder leak and 900 ft (275 m) for a "large" spill from a bulk container. As additional protection during the occurrence of a "small" spill, DOT also requires the evacuation of unauthorized persons to a distance downwind from the scene of 0.2 mi (0.3 km) and 0.7 mi (1.1 km) during the day and nighttime hours, respectively. For large spills, DOT advises evacuation to a distance of 1.7 mi (2.7 km) and 4.2 mi (6.8 km) during the day and nighttime hours, respectively.

An emergency-response team must also identify those who have been exposed to chlorine. The afflicted individuals should be moved downwind to fresh air where, if necessary, artificial respiration can be supplied. To minimize skin burns, those persons exposed to chlorine should remove their clothing and shower thoroughly. To prevent the impairment of vision, they should irrigate their eyes with flowing water for approximately 30 minutes. Finally, they should be transported to a medical facility for follow-up examinations.

The emergency-response team also needs to examine the condition of the chlorine container or tank. If it has not ruptured, the vessel should be cooled with water as in Figure 7.17. The team needs to pinpoint the specific spots from which chlorine is

FIGURE 7.17 ◆ In this training drill, a chlorine tank was kept cool until the car could be moved away from a nearby brush fire. Note where the firefighters from the Baltimore City Fire Department are directing their fog applicator. It is aimed toward the top housing, which contains the car valves, rather than at the tank itself. All chlorine tank cars are protected by 4 in. (10 cm) of insulation. Hence, the most likely spot at which the tank would rupture is near the valve. *Courtesy of the Chlorine Institute, Inc., Washington, D.C.*

leaking or could potentially leak. Since containers and transport vessels contain *liquid* chlorine, it is often the liquid that drips from valves, fittings, or openings. Liquid chlorine is much more concentrated than its gas, and when unconfined, it readily evaporates. One volume of liquid chlorine evaporates into approximately 460 volumes of gas. For this reason, the area immediately surrounding a chlorine leak becomes a highly toxic environment within seconds.

Locating a chlorine leak is not always simple, especially when the gas has been escaping from its container or transport vessel for some time. One detection method utilizes the chemical reaction between chlorine and ammonia. These two substances react to form ammonium chloride and ammonium hypochlorite, each of which is a white solid.

$$2NH_3(g) + Cl_2(g) + H_2O(l) \longrightarrow NH_4Cl(s) + NH_4ClO(s)$$
$$\text{ammonia} \quad \text{chlorine} \quad \text{water} \quad \text{ammonium chloride} \quad \text{ammonium hypochlorite}$$

The method for detecting a chlorine leak involves tying a rag soaked with household ammonia (Section 10.13) to a broomstick, and then passing the stick along the surface of the chlorine container or tank. Ammonia vaporizes from the liquid and reacts with the chlorine at the point from which it escapes. At this spot, a white cloud drifts into the air. By observing its formation, the source of the leak is easily identified. This procedure is ineffective when the surrounding atmosphere is heavily laden with chlorine, since under this condition, the ammonium compounds are produced virtually throughout the area.

When chlorine is leaking from a tank or container, the exit points must be closed or sealed. When sealing an opening is impossible or impractical, an attempt should be made to prevent the further escape of liquid chlorine into the environment. This can sometimes be accomplished by rolling the vessel so its opening points upward. While chlorine gas continues to escape from the opening, the more concentrated liquid remains confined within the vessel.

When emergency responders arrive at the scene of an incident involving the release of chlorine, specialized equipment should be available for their immediate use. An essential item is a kit that contains a clamping device to seal a leak at the fusible plug and a patching device to seal a leak in the cylinder side-wall. They are components of the so-called Chlorine Institute Emergency Kit "A," which is available from several commercial outlets.*

To improve the speed and effectiveness of a response action at an emergency involving the release of chlorine, the Chlorine Institute formalized a chlorine emergency plan known as CHLOREP. Under this plan, the United States and Canada are divided into regional sectors in which specially trained CHLOREP teams are located. In the event of a chlorine emergency, or when CHEMTREC (Section 1.10) is contacted, the caller is put into immediate contact with the closest CHLOREP team, which then oversees the handling of the chlorine emergency.

Performance Goals for Section 7.5

- Discuss the physical and chemical properties of the red and white allotropes of elemental phosphorus, emphasizing the ability of white phosphorus to spontaneously ignite.
- Describe the manner by which red and white phosphorus is produced.
- Identify the industries that use bulk quantities of white phosphorus.
- Describe the recommended response actions when red and white phosphorus are released into the environment.

◆ 7.5 PHOSPHORUS

Elemental phosphorus has several allotropes, two of which are commercially important: *white phosphorus* and *red phosphorus*. White phosphorus is the unstable form of the element at room conditions. Upon standing, it acquires a yellow coloration due to the partial conversion of the white allotrope to the more stable red allotrope. It is for this reason that white phosphorus is also called *yellow phosphorus*. Red phosphorus is also known as *amorphous phosphorus*.

Both allotropes are important chemical products. They are used to produce special alloys (e.g., phosphor bronze), rodenticides, fireworks, "strike-anywhere" and safety matches (Section 11.17). In the chemical industry, they are used to produce chemical compounds like metallic phosphides (Section 9.6).

The physical and chemical properties of these allotropes are so strikingly different that we shall discuss them separately. Their physical properties are provided in Table 7.9.

---

*The Chlorine Institute is a trade association consisting primarily of company representatives interested in the safe production, distribution, and use of chlorine and other substances associated with the chlor-alkali industry. The offices of the Chlorine Institute are located at 1300 Wilson Boulevard, Rosslyn, Virginia.

TABLE 7.9 ◆ Physical Properties of Elemental Phosphorus

	White Phosphorus	*Red Phosphorus*
Specific gravity	1.82	2.34
Vapor density (air = 1)	4.4	4.8
Melting point	111°F (44°C)	1094°F (590°C)[a]
Boiling point	535°F (280°C)	535°F (280°C)

[a]At 43 atm (4400 kPa)

WHITE PHOSPHORUS

White phosphorus is a waxy, translucent solid at ambient conditions. Since it consists of tetra-atomic molecules, the chemical formula of white phosphorus is P_4.

White phosphorus is industrially prepared by heating calcium phosphate rock with sand and coke in an electric furnace. The principal components of sand and coke are silicon dioxide and carbon, respectively. The production of white phosphorus is denoted by the following equation:

$$2Ca_3(PO_4)_2(s) + 6SiO_2(s) + 10C(s) \longrightarrow 6CaSiO_3(s) + 10CO(g) + P_4(g)$$

calcium phosphate   silicon dioxide   carbon   calcium silicate   carbon monoxide   phosphorus

The phosphorus vapors are vented from the furnace and condensed under water to produce a white solid. Small quantities are generally stored and transported under water in containers, but bulk quantities are often encountered as molten phosphorus under water or blanketed with nitrogen. Both are transported in containers or in railroad tankcars (Figure 7.18).

FIGURE 7.18 ◆ A railroad tank car used for transporting molten elemental phosphorus. *Courtesy of FMC Corporation, Phosphorus Chemicals Division, Pocatello, Idaho.*

When white phosphorus is exposed to air, it fumes and spontaneously ignites. The autoignition temperature of white phosphorus is only 86°F (30°C). This temperature is so low that even body heat initiates ignition. White phosphorus is stored under water or a blanket of nitrogen to reduce or eliminate the risk of spontaneous combustion.

The odor of white phosphorus is often compared to the smell of garlic and rotten fish, but the comparison is misleading. The odor of garlic and rotten fish is more likely associated with the presence of phosphine (Section 9.6), a toxic gas slowly produced by the reaction between phosphorus and cold water.

$$\underset{\text{phosphorus}}{P_4(s)} + \underset{\text{water}}{6H_2O(l)} \longrightarrow \underset{\text{hypophosphorous acid}}{3H_3PO_2(aq)} + \underset{\text{phosphine}}{PH_3(g)}$$

The rate of this reaction increases when the phosphorus is hot, as when it is burning. Since phosphine is also a flammable gas, its production at a fire scene is generally of minor concern.

The combustion of white phosphorus produces two oxides, tetraphosphorus hexoxide and tetraphosphorus decoxide, as follows:

$$\underset{\text{phosphorus}}{P_4(s)} + \underset{\text{oxygen}}{3O_2(g)} \longrightarrow \underset{\text{tetraphosphorus hexoxide}}{P_4O_6(s)}$$

$$\underset{\text{phosphorus}}{P_4(s)} + \underset{\text{oxygen}}{5O_2(g)} \longrightarrow \underset{\text{tetraphosphorus decoxide}}{P_4O_{10}(s)}$$

As implied by these equations, tetraphosphorus hexoxide is the product of incomplete combustion, while tetraphosphorus decoxide is the product of complete combustion. Both are white compounds. Consequently, when bulk quantities of white phosphorus burn, billows of dense white, choking smoke are produced (Figure 7.19).

Although spontaneous combustion is the principal hazard associated with white phosphorus, exposure to the vapor of white phosphorus poses the risk of inhalation toxicity. The vapor is highly poisonous. Even the repeated inhalation of small quantities causes the affliction known as *phossy jaw,* or *phosphorus necrosis,* which in the most egregious instances causes disintegration of the jawbone. In addition, white phosphorus burns the skin and produces wounds that are extremely painful and slow to heal. For this reason, white phosphorus must always be handled with gloves.

DOT regulates the transportation of white phosphorus as a spontaneously combustible material, requiring shippers and carriers to affix SPONTANEOUSLY

FIGURE 7.19 ◆ When elemental phosphorus and certain phosphorus-bearing compounds burn, the smoke accompanying the combustion is white. This coloration is due to the presence of the combustion products tetraphosphorus hexoxide and tetraphosphorus decoxide, both of which are white. *Courtesy of the United States Environmental Protection Agency, Region 5, Chicago, Illinois.*

COMBUSTIBLE and POISON labels to containers and to post SPONTANEOUSLY COMBUSTIBLE placards on the bulk packaging, freight container, transport vehicle, unit load device, or railcar used to transport white phosphorus.

RESPONDING TO INCIDENTS INVOLVING THE RELEASE OF WHITE PHOSPHORUS

When emergency-response crews first arrive at a scene involving the release of phosphorus, the element is generally burning. Although water is an effective fire extinguisher on a phosphorus fire, experts advise the use of dry sand as the acceptable extinguisher, since it blankets the element, prevents reignition, and avoids the potential exposure of personnel to phosphine.

When small quantities are burning, it is best to segregate them from nearby combustible materials. This may not be a simple matter, since phosphorus melts at a relatively low temperature. Burning, molten phosphorus flows readily into low areas. For this reason, appropriate action should be taken to segregate the molten material from combustible materials by constructing dams or dikes.

Firefighters responding to incidents involving elemental phosphorus should wear protective gear and use self-contained breathing apparatus. They should be particularly cautious to prevent inhaling the fumes from a phosphorus fire. Particulates of the phosphorus oxides are contained within the fumes, which when inhaled could seriously irritate the nose, mouth, throat, and lungs.

RED PHOSPHORUS

The red allotrope of phosphorus is a dark red solid consisting of molecules containing numerous P_4 units. Its chemical formula is usually denoted as either P or P_x. Red phosphorus is produced industrially by heating white phosphorus at 482°F (250°C) in an iron container from which air has been excluded.

Red phosphorus is not nearly as chemically reactive as white phosphorus. Furthermore, it is neither poisonous nor *spontaneously* combustible in small quantities. For this reason, small amounts of red phosphorus may be stored in containers without the overlying protection of water.

Red phosphorus burns when it is exposed to an ignition source. The combustion produces a mixture of two oxides, tetraphosphorus hexoxide and tetraphosphorus decoxide. Red phosphorus also combines with atmospheric oxygen in the *absence* of an ignition source. The reaction occurs very slowly and is highly exothermic. For this reason, the element is rarely stored in bulk to prevent its self-ignition from the accumulated heat of reaction.

DOT regulates the transportation of red phosphorus as a flammable solid, requiring shippers and carriers to affix FLAMMABLE SOLID labels to containers and, when warranted, to post FLAMMABLE SOLID placards on the bulk packaging, freight container, transport vehicle, unit load device, or railcar used to transport the red phosphorus.

RESPONDING TO INCIDENTS INVOLVING THE RELEASE OF RED PHOSPHORUS

Although red phosphorus is unlikely to be encountered in bulk, containers holding small quantities of this element have been involved in fires. They can be extinguished by the application of dry sand, foam or dry chemicals. The use of wet sand is inadvisable, since conditions are appropriate for the production of the toxic gas phosphine (Section 9.6).

◆ 7.6 SULFUR

Elemental sulfur occurs naturally, particularly in countries bordering the Gulf of Mexico and in Japan, Mexico, and Italy. Sulfur occurs naturally in minerals and ores too numerous to mention, in which it is combined with metals and other nonmetals. Sulfur accounts for 0.06% by mass of all the elements found in the Earth's crust, atmosphere, and oceans.

Most of the world's supply of elemental sulfur comes from natural deposits of the element, called *brimstone*. To isolate the sulfur, hot water under pressure is pumped into subterranean brimstone-bearing deposits, whereupon the sulfur melts and is brought to the surface by an air lift.

Sulfur vaporizes when brimstone is heated at atmospheric pressure. When the vapor is allowed to condense upon a cold surface, a finely divided powder of solid sulfur is produced. It is called *flowers of sulfur*. The powder is commercially used as a pesticide.

The pure solid state of sulfur occurs in either of two allotropes called *orthorhombic sulfur* and *monoclinic sulfur*. Orthorhombic sulfur is the stable allotrope of solid sulfur at ambient conditions. It is a pale yellow, crystalline solid. Its physical properties are noted in Table 7.10. When orthorhombic sulfur is maintained at a temperature between 204.8°F (96°C) and 235.2°F (112.8°C), it changes into a mass of long transparent needles composed of monoclinic sulfur. Since monoclinic sulfur is not the stable allotrope, it slowly changes back into the orthorhombic form.

Both solid allotropes exist as molecules having eight sulfur atoms bonded together in a puckered-ring arrangement. For this reason, the chemical formula of solid sulfur is S_8.

$$:S-\ddot{S}-S:$$
$$/ \qquad \backslash$$
$$:S: \qquad :S:$$
$$\backslash \qquad /$$
$$:S-\ddot{S}-S:$$

On the other hand, molten sulfur possesses a highly complex molecular arrangement. Chemists denote it as S_x. The chemical formula of sulfur vapor at its boiling point is

TABLE 7.10 ◆ Physical Properties of Elemental Sulfur	
Specific gravity	2.07
Flash point	405°F (207°C)
Melting point	246°F (119°C)
Boiling point	832°F (444°C)

also represented as S_8, but as the vapor is further heated, the cyclic arrangement breaks down and sulfur assumes the formula S_2.

Elemental sulfur burns with a blue flame. To initiate its burning, sulfur must first be heated to at least 405°F (207°C). The combustion produces sulfur dioxide, a poisonous gas having a suffocating, choking odor (Section 10.10).

$$\underset{\text{sulfur}}{S_8(s)} + \underset{\text{oxygen}}{8O_2(g)} \longrightarrow \underset{\text{sulfur dioxide}}{8SO_2(s)}$$

Elemental sulfur is likely to spontaneously ignite under the following conditions:

◆ When flowers of sulfur are dispersed into air, a potentially explosive mixture is produced. The spontaneous ignition of this mixture is triggered by the static electricity generated by the movement of the sulfur in the air. The fear of a dust explosion can be markedly reduced by electrically grounding the vessel in which the sulfur is contained.

◆ Elemental sulfur reacts with many oxidizing agents. When they are activated to react, the resulting heat of reaction is likely to cause the ignition of the residual sulfur. For this reason, all mixtures of elemental sulfur and oxidizing agents pose the risk of fire and explosion.

Sulfur is one of the world's most important raw materials. There is hardly a component of the chemical industry that doesn't use elemental sulfur or one of its compounds during its manufacturing or production processes. Of the total production, approximately 80% is used as a raw material for the manufacture of sulfuric acid (Section 8.5). Sulfur is also used to produce vulcanized rubber products (Section 13.9), fertilizers, dyes and chemicals, drugs and pharmaceuticals, gunpowder, pesticides, and matches. Several of these important uses are illustrated in Figure 7.20.

FIGURE 7.20 ◆ Elemental sulfur is a constituent of many industrial and domestic products, like gunpowder, matches, insecticides, fertilizers, and vulcanized rubber, some of which are noted here. When the sulfur in these products burns, sulfur dioxide is formed.

Sulfur combines with several nonmetals, two of which are commercially important.

◆ Carbon disulfide is produced by passing sulfur vapor over very hot carbon in the absence of air. The presence of air is avoided, because carbon disulfide is a very flammable substance (Section 12.15). It is used as a solvent by the chemical industry.

$$C(s) \; + \; S_2(g) \longrightarrow CS_2(g)$$
$$\text{carbon} \qquad \text{sulfur} \qquad\qquad \text{carbon disulfide}$$

◆ Sulfur hexafluoride is the predominant product formed when sulfur combines with fluorine.

$$S_8(s) \; + \; 24F_2(g) \longrightarrow 8SF_6(g)$$
$$\text{sulfur} \qquad \text{fluorine} \qquad\qquad \text{sulfur hexafluoride}$$

It is used as an insulator in high-voltage electrical equipment.

Elemental sulfur also combines with most metals. For example, when a mixture of mercury and iron is heated, the elements unite, forming mercury(II) sulfide and iron(II) sulfide, respectively.

$$8Hg(l) \; + \; S_8(s) \longrightarrow 8HgS(s)$$
$$\text{mercury} \qquad \text{sulfur} \qquad\qquad \text{mercury(II) sulfide}$$

$$8Fe(s) \; + \; S_8(s) \longrightarrow 8FeS(s)$$
$$\text{iron} \qquad \text{sulfur} \qquad\qquad \text{iron(II) sulfide}$$

DOT regulates the domestic transportation of solid and molten sulfur as Class 9 hazardous materials, requiring shippers and carriers to affix CLASS 9 labels on containers and, when warranted, to display CLASS 9 placards on the bulk packaging, freight container, transport vehicle, unit load device, or railcar used to transport sulfur.

RESPONDING TO INCIDENTS INVOLVING THE RELEASE OF SULFUR

When emergency-response crews first arrive at a scene involving the release of sulfur, the element is generally burning. The use of water is effective as a fire extinguisher when it is applied as a fog. The fog extinguishes the fire by removing heat. Applying the water as a fog also avoids the buildup of steam under layers of the solid sulfur, which could later erupt and splatter the hot material.

Sulfur readily melts under normal fire conditions, thereby flowing into lower adjacent areas where it can initiate secondary fires. For this reason, appropriate action should be taken to segregate the molten material from combustible materials by constructing dams or dikes.

Firefighters responding to incidents involving elemental sulfur should wear protective gear and use self-contained breathing apparatus. They should be particularly cautious to prevent inhaling sulfur dioxide, since it is highly toxic.

Performance Goals for Section 7.7
- Discuss the physical and chemical properties of elemental carbon in its principal allotropic forms, graphite and diamond.
- Illustrate that carbon acts as a reducing agent.
- Identify the industries that use bulk quantities of coal, coke, and charcoal.
- Describe the nature of coal tar distillates.
- Identify the industries that use bulk quantities of coal tar distillates.
- Describe the recommended response actions when coal, coke, and charcoal are released into the environment.

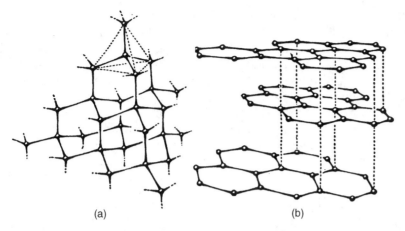

(a)   (b)

FIGURE 7.21 ◆ The crystalline structures of the allotropes of carbon. In the diamond structure shown in (a), each carbon atom is covalently bonded to four carbon atoms in a tetrahedral arrangement. In the graphite structure shown in (b), the covalent bonding is accomplished by means of hexagonal rings that are joined to one another in successive sheets.

◆ **7.7 CARBON**

Elemental carbon occurs naturally as two common allotropes, *graphite* and *diamond*.* It also occurs in *coal, coke, charcoal,* and *carbon black*. In each instance, the elemental carbon is represented by its chemical symbol C.

The natural abundance of carbon is only 0.08% by mass in the Earth's crust, the atmosphere, oceans, and other bodies of water. Yet, carbon ranks much higher in terms of importance since it is a constituent of the compounds needed by all living organisms for survival.

COMMON ALLOTROPES OF CARBON

Graphite and diamond possess substantially different physical properties. Graphite (the "lead" of pencils) is a black, greasy material, but diamond can be cut and polished to a crystalline, transparent luster. Graphite is one of the softest known substances, and diamond is nearly the hardest.

The physical properties of graphite and diamond are linked with the difference in their crystalline structures. The diamond structure in Figure 7.21(a) shows each carbon atom symmetrically surrounded by four other carbon atoms in a tetrahedral arrangement. The graphite structure in Figure 7.21(b) depicts each carbon atom bonded to other carbon atoms in planar hexagonal rings joined to one another in successive sheets.

---

*In 1985, new forms of elemental carbon were discovered by vaporizing graphite with a laser. These forms possess 60, 70, or more carbon atoms per molecule. Collectively, they are known as *fullerenes*. The properties of the fullerenes are not discussed in this book.

The carbon allotrope having the formula C_{60} is called *buckminsterfullerene,* a name derived from the observation that its molecules are shaped like an icosahedron (soccer ball) and resemble the geodesic domes designed by American architect and engineer R. Buckminster Fuller. C_{60} molecules are also called *buckyballs*.

The diamond structure is produced in nature when carbon is subjected to a pressure of approximately 0.8 million psi (5.6 million kPa) within the hot mantle of Earth, over 50 miles (80 km) below the surface. The process has been replicated on Earth's surface by compressing graphite at 1.3 million psi (9 million kPa) in an electric furnace at approximately 3000°F (1650°C) in the presence of a metal catalyst. The resulting synthetic diamonds, called *industrial diamonds,* are used to cut metal and glass, to prepare wire-drawing dies, and to produce drill bits and grinding wheels.

At moderate pressures, graphite is the stable allotropic form of carbon. At the normal temperature and pressure conditions on Earth's surface, diamond converts into graphite at an imperceptibly slow rate. However, when diamond is heated to approximately 1830°F (1000°C) in the absence of air, it converts into graphite at an appreciable rate.

Graphite possesses an extraordinarily high melting point, 6656°F (3680°C). Consequently, it is molded into crucibles that are used to hold molten steel and other high-melting metals. Graphite also possesses an important anomalous property when compared to other nonmetals: It is an extraordinarily good conductor of heat.

Graphite has many industrial applications, of which the following are examples:

- High-temperature operations are often conducted in furnaces and other vessels whose inside walls have been lined with graphite to protect the underlying metal from melting or softening.
- The nose and leading edges of aircraft wings are generally coated with graphite to protect the underlying metal from the heat of friction generated when the aircraft travels through the air.
- Graphite-based dry powder is an effective fire extinguisher for class D fires, because the carbon conducts heat away from the burning metal.

Although graphite and diamond are stable substances at the conditions experienced on Earth's surface, both burn in air. As we first noted in Section 5.6, carbon and the compounds that contain carbon are said to burn either "incompletely" or "completely." The product of the incomplete combustion is carbon monoxide, while the product of complete combustion is carbon dioxide.

$$\text{Incomplete combustion:} \qquad \underset{\text{carbon}}{2C(s)} + \underset{\text{oxygen}}{O_2(g)} \longrightarrow \underset{\text{carbon monoxide}}{2CO(g)}$$

$$\text{Complete combustion:} \qquad \underset{\text{carbon}}{C(s)} + \underset{\text{oxygen}}{O_2(g)} \longrightarrow \underset{\text{carbon monoxide}}{CO_2(g)}$$

In chemical reactions, carbon routinely acts as a reducing agent. This chemical property is put to use during the manufacture of iron and other metals from their naturally occurring ores. In Section 7.5, for example, we observed that the carbon in coke (see next section) is used for manufacturing elemental phosphorus from calcium phosphate rock.

COAL, COKE, AND CHARCOAL

Almost without exception, all forms of carbon found in nature can be traced to the giant plants that grew during the prehistoric period called the carboniferous age. Millions of years ago, these plants grew much more luxuriantly than they do today. As Earth evolved, the remains of these plants were ultimately buried at great depths. The intense temperature and pressure conditions below Earth's surface converted them into coal. The process is called the *carbonization* of vegetable matter.

Solved Exercise 7.8

Why is coal referred to as a *fossil fuel*?

Solution: Plant life that lived millions of years ago became buried deep below the crust of Earth. Here, subjected to high temperature and pressure over time, it decomposed into coal, natural gas, and crude petroleum (Section 12.5). To associate the origin of these materials with past geological ages, they are called *fossil fuels*.

The extent to which carbonization has occurred in nature determines the *rank* of a coal. Five major ranks of coal are recognized, several of which are illustrated in Figure 7.22: lignite, subbituminous, bituminous, semianthracite, and anthracite. Each rank of coal in this listing is progressively older than those before it. Each rank also contains an average higher fixed carbon content that gives it a uniquely different heating value.

Locked within the complex structure of coal are many volatile organic compounds, which burn when their vapors are exposed to an ignition source. Before they can be ignited, a sufficient energy of activation must generally be provided to first vaporize and release them from the inner structure of coal. The evolved heat serves to vaporize more flammable vapor, which subsequently ignites. In this fashion, coal fires are self-sustaining until the flammable compounds in the coal have been entirely exhausted. After these flammable components have burned, a solid residue or ash generally remains. This ash consists of a mixture of mineral oxides.

Solved Exercise 7.9

What physical forms of coal are most likely to ignite without an ignition source?

Solution: Flammable substances are present near the surfaces of freshly mined and recently pulverized coal at concentrations above relevant lower explosive limits. Consequently, the physical forms of coal most likely to ignite spontaneously are freshly mined and recently pulverized coal.

When coal is heated in the absence of air within an assembly like that shown in Figure 7.23, certain volatile gases evolve. The mixture of these gases is called *coal gas*. Methane and ammonia are two constituents of coal gas. Simultaneously, the heating produces a viscous liquid called *coal tar,* which is composed of a mixture of organic compounds including benzene, phenol, cresols, naphthalene, and anthracene, all of which are toxic substances. We shall revisit them in Chapter 12. Coal tar is used as a roofing sealer and a coating for underground pipelines, but the majority of the coal tar produced today is used in the construction of electrodes for the production of elemental aluminum.

coal gas

Coal tar may also be subjected to a separation process called *distillation*, during which the components of the coal tar are first vaporized at specified temperatures and then condensed to liquids. The resulting materials are often called *coal tar distillates*. The latter term is loosely defined, but light and heavy ("pitch") fractions of coal tar

coal tar
light oil

FIGURE 7.22 ◆ Some ranks or classes of coal: peat, lignite, three varieties of bituminous coal, and anthracite. All ranks of coal are thought to have originated from the decay of plants under extreme temperatures and pressures deep below the surface of Earth. Each rank differs from the others by the time necessary to achieve its formation in nature. *Courtesy of the National Coal Association, Washington, D.C.*

coal tar pitch

are commercially recognized. The "light" oil is so named because it floats on water. A commercially important product is isolated from the heavy coal tar distillate by further distillation; it is called *creosote oil*. This product is widely used to preserve railroad cross ties and utility poles against decay and to "cut" asphalt so it is suitable for application as a road and roofing tar.

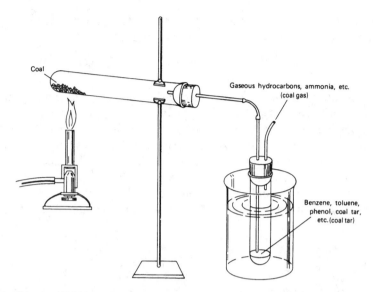

FIGURE 7.23 ◆ Heating coal in the absence of air results in the production of *coal gas* and *coal tar,* both of which contain components that are either flammable or combustible. When coal burns in the presence of air, it is this combination of components that burns.

DOT regulates the transportation of coal gas as a poison gas, requiring shippers and carriers to affix INHALATION HAZARD and FLAMMABLE GAS labels on containers and to post INHALATION HAZARD placards on the bulk packaging, freight container, transport vehicle, unit load device, or railcar used to transport coal gas.

creosote oil

DOT regulates the transportation of coal tar distillate as a flammable liquid, requiring shippers and carriers of the flammable coal tar distillate to affix FLAMMABLE LIQUID labels to containers and, when warranted, to display FLAMMABLE placards on the bulk packaging, freight container, transport vehicle, unit load device, or railcar used to transport coal tar distillates.

Several commercial products are produced when certain carbon-rich materials are heated in the absence of air. For example, when bituminous coal or the residue from treating crude petroleum fractions (Section 12.5) is heated in the absence of air, a solid residue remains that is called *coke.* It resembles the material shown in Figure 7.24. Coke is the form of carbon used to reduce the ores of arsenic, tin, copper, iron, zinc,

FIGURE 7.24 ◆ A sample of coke, the carbonaceous residue that remains when bituminous coal is heated in the absence of air. *Courtesy of the National Coal Association, Washington, D.C.*

phosphorus, and other elements. Most coke produced in the United States is used for reducing the iron oxide in iron ore within blast furnaces.

$$C(s) \ + \ FeO(s) \ \longrightarrow \ Fe(s) \ + \ CO(g)$$

carbon   iron(II) oxide   iron   carbon monoxide

This process is called *smelting,* during which carbon acts as a reducing agent.

When animal bones, nut shells, corn cobs, or peach pits are heated in the absence of air, the product called *charcoal* is produced. Most individuals first experience charcoal briquettes as the fuel for backyard barbeques, but considerable quantities of carbon are also used industrially for adsorbing undesirable substances from products destined for commercial use. *Adsorption* refers to the surface retention of solid, liquid, or gaseous molecules and should not be confused with the term *absorption,* a process involving the physical penetration of one substance into the bulk of another one.

Solved Exercise 7.10

Anesthesiologists sometimes advise their patients to consume lightly burnt toast during the recovery stage following major surgery. What is the chemical basis for this advice?

Solution: When the toast is consumed, the residual anesthetic remaining in the patient's stomach adsorbs to the carbon particles that are present on the surface of the burnt toast. This prevents the further distribution of the anesthetic throughout the body and aids in its elimination.

Charcoal is the most commercially popular of the carbon-containing adsorbing agents. Since charcoal retains the skeletal cellular structure of the material from which it was made, it is highly porous. This porosity gives charcoal a very large surface area per unit weight. Charcoal can also be heated in the absence of air at 1472 to 1652°F (800 to 900°C) to produce a product called *activated charcoal,* or *activated carbon.* This material is often chosen as an adsorbing medium, since its average internal surface area is 284,000 ft^2/oz (929 m^2/g).

Large quantities of activated charcoal are industrially used as adsorbing agents. Activated charcoal is used to purify the atmospheric emissions from gas stacks by reducing or eliminating the concentration of undesirable toxic gases from the gas streams. It is also used in gas-mask canisters and cigarette filter tips for the same purpose.

The transportation of charcoal and activated carbon is regulated domestically and internationally, respectively. Both are regulated as spontaneously combustible materials. DOT requires shippers and carriers to affix SPONTANEOUSLY COMBUSTIBLE labels to their packaging and, when warranted, to display SPONTANEOUSLY COMBUSTIBLE placards on the bulk packaging, freight container, transport vehicle, unit load device, or railcar used to transport them.

CARBON BLACK

When petroleum products are vaporized and burned in a furnace with a limited amount of air, a finely divided form of carbon called *carbon black* is produced. Many grades of carbon black are available commercially, varying primarily in particle size.

These grades are used in a number of consumer products including inks, paints, plastics, and tires, belts, and other abrasion-resistant rubber products (Section 13.8).

RESPONDING TO INCIDENTS INVOLVING THE RELEASE OF COAL, COKE, AND CHARCOAL

Most fires involving coal, coke, and charcoal can be effectively extinguished with water. When bulk quantities of these materials are burning, it is essential to use a deluging volume of water for the following reasons:

- When fires are extinguished only on the surfaces of coal, coke, and charcoal piles, combustion continues within the interior of these piles. Since the liberated heat cannot easily dissipate, the temperature of the entire bulk increases and the coal, coke, or charcoal erupts into flame once again.
- When coal, coke, or charcoal are only lightly moistened with water, the carbon and steam chemically react to produce carbon monoxide and hydrogen.

$$C(s) \ + \ H_2O(g) \ \longrightarrow \ H_2(g) \ + \ \ CO(g)$$
$$\text{carbon} \qquad \text{water} \qquad\qquad \text{hydrogen} \quad \text{carbon monoxide}$$

This is the water gas mixture noted in Section 7.2. The ignition of the water gas serves to rekindle coal, coke, and charcoal fires.

The need to use a deluging volume of water on coal fires is underscored by the occurrence of ongoing fires in several abandoned coal mines in the eastern United States. Although attempts have been made to extinguish them, they are routinely unsuccessful. Firefighters are unable to saturate the coal with water since the risk associated with entering these old mines is too high. The burning coal is exposed to a new source of atmospheric oxygen almost continuously through the countless boreholes that were formerly driven into the mines to provide adequate ventilation to coal miners. Consequently, the coal mine fires are not extinguished and continue to burn, even to today.

■ ■

REVIEW EXERCISES

Oxygen

7.1 Complete and balance each of the following equations, and provide an acceptable name for each product:
   (a) $Mg(s) + O_2(g) \longrightarrow$
   (b) $H_2(g) + O_2(g) \longrightarrow$
   (c) $C(s) + O_2(g) \longrightarrow$

7.2 Why do the suppliers of compressed oxygen warn their customers to refrain from lubricating the valves, regulators, gauges, and fittings associated with oxygen storage containers?

7.3 Why does hot iron melt when it is plunged into a vessel containing liquid oxygen?

7.4 Leadville is a city in Colorado that was once a mining town. Now a tourist location, it is located 10,200 ft (3109 m) above sea level. Leadville's newborns experience one of the highest rates of hospitalization for respiratory ailments in the world. What is the most likely cause of these health problems?

7.5 A steel cylinder containing oxygen is marked with the DOT specification number "DOT-3B500." What is the service pressure in psi of this cylinder?

7.6 Why is the spillage of liquid oxygen upon an asphalt pavement potentially hazardous?

7.7 Which action should be taken by an emergency-response crew to protect lives, property, and the environment at a collision scene involving separate spills of liquid oxygen and automobile fuel?

Ozone

7.8 When ozone generators are used to treat smoke-damaged items following a residential fire, why is it essential for the treatment to be conducted while the home is unoccupied?

7.9 A trace of hydrogen sulfide is a common component of sewage. Its presence is objectionable because it possesses an offensive odor. When sewage is treated with ozone, the hydrogen sulfide is converted into sulfuric acid as the sole product, which at the level produced is benign. Write the balanced equation for this chemical process.

7.10 Why have the countries of the civilized world aggressively curbed the use of halon agents and certain other halogen-containing substances?

Hydrogen

7.11 Complete and balance each of the following equations, and provide an acceptable name for each product:
(a) $Li(l) + H_2(g) \longrightarrow$
(b) $C(s) + H_2(g) \longrightarrow$
(c) $S_8(s) + H_2(g) \longrightarrow$
(d) $FeO(s) + H_2(g) \longrightarrow$

7.12 Why are only traces of elemental hydrogen found in nature?

7.13 When liquid hydrogen spills, why is it likely to spontaneously burst into flame?

7.14 When the use of hydrogen is considered solely from an environmental viewpoint, why is it typically considered an ideal fuel?

7.15 What is the most likely reason DOT prohibits an airline or rail carrier from accepting for domestic transportation any quantity of compressed hydrogen gas by passenger-carrying aircraft or railcar?

7.16 A steel cylinder containing hydrogen is marked "DOT-3XT2000."
(a) What is the DOT specification number of the cylinder?
(b) What is the service pressure of the cylinder?

7.17 The OSHA regulation at 29 CFR §1910.103(b) requires employers to arrange safety-relief devices on all containers of hydrogen in the workplace, other than containers having a capacity of 2 ft³ (0.06 m³) or less, so they discharge upward into the open air in an unobstructed fashion and in such a manner as to prevent the impingement of escaping gas upon its container, adjacent structures, or personnel.
(a) What is the most likely reason OSHA requires employers to arrange the safety-relief devices so they discharge *upward* into the open air?
(b) What is the most likely reason OSHA requires employers to arrange the safety-relief devices so the escaping hydrogen does not impinge upon its container, adjacent structures, or personnel?

7.18 Why do employers post signs like the following in areas where batteries are periodically charged?

> **NO SMOKING**
> **BATTERY-CHARGING AREA**

7.19 Why do experts recommend use of a thermal-imaging camera for detecting a hydrogen leak from a pipe fitting?

Fluorine

7.20 Complete and balance each of the following equations, and provide an acceptable name for each product:
(a) $Cu(s) + F_2(g) \longrightarrow$
(b) $H_2(g) + F_2(g) \longrightarrow$
(c) $P_4(s) + F_2(g) \longrightarrow$

7.21 The DOT regulation at 49 CFR §173.302(d) requires shippers and carriers to transport fluorine as a compressed gas within a specific type of seamless cylinder that is equipped with a valve protection cap but not outfitted with a safety-release device. DOT further requires that each cylinder may not be pressurized to over 400 psi_g (2760 kPa) at 70°F (21°C) and may not contain over 6 lb of gas.
(a) What is the most likely reason DOT requires the use of cylinders that are *not* equipped with safety-relief devices for transporting elemental fluorine?
(b) What is the most likely reason DOT requires the use of seamless cylinders for transporting elemental fluorine?
(c) If a chemical manufacturer compresses 2.5 kg of fluorine into a cylinder at a pressure of 2700 kPa at 21°C, has the manufacturer complied with the mass and pressure requirements of this DOT regulation?

Chlorine

7.22 Complete and balance each of the following equations, and provide an acceptable name for each product:
(a) $Fe(s) + Cl_2(g) \longrightarrow$
(b) $H_2(g) + Cl_2(g) \longrightarrow$
(c) $Al(s) + Cl_2(g) \longrightarrow$

7.23 What is the most likely reason DOT prohibits an airline cargo carrier from accepting any quantity of elemental chlorine for domestic transportation?

7.24 Give examples of the *corrosive* nature of chlorine gas.

Phosphorus

7.25 Complete and balance each of the following equations, and provide an acceptable name for each product:
(a) $P_4(s) + O_2(g) \longrightarrow$
(b) $P_4(s) + Ca(s) \longrightarrow$
(c) $P_4(s) + Cl_2(g) \longrightarrow$

7.26 Why should gloves be worn while handling white phosphorus?

7.27 What is the most likely reason DOT requires a shipper to add an inert gas to railroad tankcars containing residual phosphorus after unloading, or fill them to capacity with water, before accepting the emptied tanks for domestic transportation?

7.28 Using Tables 6.1 and 6.4, identify the basic description that DOT requires a carrier to enter on a manifest accompanying a shipment of 5000 lb (2270 kg) of white phosphorus by railroad tankcar at 250°F (121°C).

7.29 Why are firefighters advised to use dry sand rather than wet sand to extinguish a small red phosphorus fire?

Sulfur

7.30 Complete and balance each of the following equations, and provide an acceptable name for each product:
(a) $C(s) + S_8(s) \longrightarrow$
(b) $Cl_2(g) + S_8(s) \longrightarrow$
(c) $Cu(s) + S_8(s) \longrightarrow$

7.31 When bulk quantities of sulfur are burning, why are nearby secondary fires likely to result?

7.32 When elemental sulfur is shipped onboard watercraft, why does DOT require the shipper to segregate it from oxidizing agents?

Carbon

7.33 Complete and balance each of the following equations, and provide an acceptable name for each product:
(a) $C(s) + F_2(g) \longrightarrow$
(b) $C(s) + CuO(s) \longrightarrow$
(c) $C(s) + H_2O(g) \longrightarrow$
(d) $C(s) + SnO_2(s) \longrightarrow$

7.34 Why are the ends of wooden utility poles intentionally charred prior to their insertion into the ground?

7.35 The DOT regulation at 49 CFR §174.450 stipulates that when fire occurs in a domestic rail shipment of charcoal in transit, water should not be used if it is practicable to locate and remove the burning charcoal. Any charcoal that becomes wet while a fire is being extinguished must be removed from the car and not reshipped, and the remainder of the charcoal must be held under observation in a dry place for at least 5 days before forwarding. What is the most likely reason DOT regulates the domestic transportation of charcoal in this fashion?

7.36 When graphite-based dry powder is used to extinguish a class D fire, which element of the fire tetrahedron illustrated in Figure 5.7 is at issue?

7.37 Why is wood charcoal more likely to spontaneously ignite when it has been ground, crushed, granulated, or pulverized?

7.38 What purpose is served by the granulated carbon used within respirator canisters?

7.39 A cargo carrier intends to ship 1000 lb (454 kg) of activated charcoal from Los Angeles, CA, to Seattle, WA, by motor carrier. What basic description of this commodity is required on the accompanying shipping paper?

7.40 The flashpoint of creosote oil is 165°F (74°C). When 5000 lb (2270 kg) of creosote oil is transported at ambient temperature by means of a motorized tank truck, which placards do shippers and carriers affix to the vehicle?

Chemistry of Some Corrosive Materials

8 **CHAPTER**

To the average nonscientist, corrosion is the undesirable phenomenon that occurs when metals are chemically attacked by components of the air. The most common example of corrosion is the rusting of iron. When iron rusts, the metal combines with atmospheric oxygen and produces a mixture of iron oxides. In a moist environment, these oxides are then converted into their corresponding hydroxides.

Although the associated chemistry is complex, the corrosion of iron may be depicted by the following equation:

$$4Fe(s) + 3O_2(g) + 6H_2O(g) \longrightarrow 4Fe(OH)_3(s)$$
$$\text{iron} \qquad \text{oxygen} \qquad \text{water} \qquad\qquad \text{iron(III) hydroxide}$$

The iron(III) hydroxide is an orange scale on the surface of the iron. It flakes and falls from the surface, thereby exposing the underlying metal to more atmospheric oxygen.

Since the overall integrity of a metallic item is lost, the corrosion of metals is usually a destructive and harmful phenomenon. Notwithstanding this fact, the corrosion of metals by atmospheric oxygen can also be beneficial at times, as when very pure aluminum and titanium metal are corroded in air. In these instances, a film of the relevant oxide is rapidly produced that deposits and strongly adheres to the surface of the metal. The oxide adheres so strongly that the film effectively prevents the further degradation of the underlying metal. Metals like aluminum and titanium, whose surfaces have been protected by an overlying deposit of their corresponding metallic oxide, are said to be *passive* to chemical attack by the constituents of the air.

To chemists, the corrosion phenomenon is not solely limited to the action of atmospheric oxygen on metals. In a broader sense, corrosion occurs whenever any substance "eats into" and destroys the chemical character of metals, minerals, and even body tissues. Within this broader meaning, acids and bases are examples of substances that corrode matter. We shall study their properties in this chapter.

■ **Performance Goals for Section 8.1**
- Describe Arrhenius's theory of acids and bases.
- Distinguish between the nature of strong and weak acids and strong and weak bases.
- Distinguish between mineral acids and organic acids.
- Distinguish between oxidizing and nonoxidizing acids, and name the common acids in each category.
- Distinguish between the nature of diluted and concentrated acids.
- Show that the specific strengths of acid and base solutions are connected to their degree of corrosivity.

◆ 8.1 THE NATURE OF ACIDS AND BASES

Several theories have been proposed to account for the properties of acids and bases, but for simplicity's sake, we shall only use one of them in this book. In 1887, Swedish chemist Svante Arrhenius first advocated a theory that has been modernized here to reflect current scientific knowledge. Arrhenius proposed that an acid is any substance that produces hydrogen ions (H^+) when dissolved in water. Hydrogen ions are nothing more than hydrogen atoms stripped of their electrons. A single hydrogen ion is the same as the nucleus of a hydrogen atom, or a single proton.

Today, chemists know that free hydrogen ions cannot exist alone in aqueous solution due to their high charge density. Instead, they rapidly become "solvated"; that is, they bond loosely to water molecules. These solvated hydrogen ions are very complex in the manner by which they interact; we collectively represent them by the notation $H^+(aq)$.

The ionization of all acids can be represented by Arrhenius' theory. For instance, when hydrogen chloride dissolves in water, hydrogen and chloride ions are produced.

$$\underset{\text{hydrogen chloride}}{HCl(aq)} \longrightarrow \underset{\text{hydrogen ion}}{H^+(aq)} + \underset{\text{chloride ion}}{Cl^-(aq)}$$

The solution of hydrogen chloride in water is called *hydrochloric acid.*

Arrhenius further proposed that a base is a substance that produces hydroxide ions (OH^-) when it is dissolved in water. The sodium and hydroxide ions in sodium hydroxide become solvated and we represent them as $Na^+(aq)$ and $OH^-(aq)$, respectively.

$$\underset{\text{sodium hydroxide}}{NaOH(aq)} \longrightarrow \underset{\text{sodium ion}}{Na^+(aq)} + \underset{\text{hydroxide ion}}{OH^-(aq)}$$

STRONG AND WEAK ACIDS AND BASES

It is the ability of acids and bases to form ions in water that gives rise to their corrosive nature. The relative strength of an acid or base refers to the tendency that an individual substance possesses to form hydrated hydrogen ions and hydroxide ions, respectively, when it is dissolved in water. Those acids and bases that yield a high concentration in water of hydrogen and hydroxide ions are called *strong acids* and *strong bases,* respectively. For instance, hydrochloric acid is an example of a strong acid because the hydrogen chloride in water is almost completely ionized into solvated hydrogen ions and chloride ions. Likewise, sodium hydroxide is an example of a strong base since it almost completely ionizes in water into solvated sodium and hydroxide ions.

On the other hand, some substances retain their unit formulas when they are dissolved in water. Such substances yield relatively low concentrations of hydrogen or hydroxide ions and are called *weak acids* and *weak bases,* respectively. Acetic acid is an example of a weak acid, because it primarily exists as molecules of acetic acid when it is dissolved in water, although some hydrogen and acetate ions are also produced. Ammonium hydroxide is an example of a weak base. When ammonia is dissolved in water, it continues to exist primarily as molecular ammonia and does not appreciably solvate and form ammonium and hydroxide ions.

Each acid in the group listed in Table 8.1 is ranked as a strong or weak acid. Phosphoric acid is regarded as a moderately strong acid, while the acids above and below phosphoric acid are considered strong acids and weak acids, respectively.

MINERAL ACIDS AND ORGANIC ACIDS

An acid can be classified as either a mineral acid or an organic acid. *Mineral acids* consist of molecules having atoms of hydrogen and an identifying nonmetal such as chlorine, sulfur, or phosphorus, and sometimes oxygen. The name is most likely associated with the fact that these acids were initially produced from minerals existing in naturally occurring ores. *Organic acids* are substances whose molecules possess carbon, hydrogen, and oxygen atoms only. All organic acids have molecular structures that contain at least one of the following group of atoms:

$$-\overset{\displaystyle O}{\underset{\displaystyle OH}{\overset{\|}{\underset{\backslash}{C}}}}$$

This group of atoms is characteristic of organic acids and is called the *carboxyl group.* Acetic acid is the only acid noted in this chapter that is an organic acid. The others are mineral acids.

TABLE 8.1 ◆ Relative Strengths of Some Common Acids in Water

Name of Acid	Chemical Formula
Perchloric acid	$HClO_4(aq)$
Sulfuric acid	$H_2SO_4(aq)$
Hydrochloric acid	$HCl(aq)$
Nitric acid	$HNO_3(aq)$
Phosphoric acid	$H_3PO_4(aq)$
Nitrous acid	$HNO_2(aq)$
Hydrofluoric acid	$HF(aq)$
Acetic acid	$CH_3COOH(aq)$
Carbonic acid[a]	$CO_2(aq)$
Hydrocyanic acid	$HCN(aq)$
Boric acid	$H_3BO_3(aq)$

[a]The chemical formula of carbonic acid sometimes appears as $H_2CO_3(aq)$. However, an acid having this chemical composition has never been isolated or identified. The chemical formula of carbonic acid is correctly denoted as $CO_2(aq)$.

OXIDIZING AND NONOXIDIZING ACIDS

An acid chemically reacts as either an *oxidizing acid* or a *nonoxidizing acid*. An oxidizing acid is an acid that can participate in a chemical reaction as an oxidizing agent. In Section 5.4, we noted that an oxidizing agent is a substance that causes oxidation, sometimes by taking electrons from another substance. Nitric acid is an example of an oxidizing acid, since it causes metallic copper, zinc, and other substances to oxidize by taking electrons from them. On the other hand, hydrochloric acid is an example of a nonoxidizing acid. When it participates in chemical reactions, hydrochloric acid does not take electrons from other substances.

Hot sulfuric acid, nitric acid, and perchloric acid are oxidizing acids, while hydrochloric acid, hydrofluoric acid, phosphoric acid, and acetic acid are nonoxidizing acids. The degree to which the oxidizing acids corrode is dependent upon how powerfully they participate as oxidizing agents. We shall note the specific features of acids that act as oxidizing agents when we examine their individual properties.

DILUTED AND CONCENTRATED ACIDS

Chemists often refer to an acid as either *diluted* or *concentrated*. When they refer to a concentrated acid, chemists generally mean the commercially available acid having the most elevated concentration. When referring to a diluted acid, they mean a solution produced by adding water to the concentrated acid.

This is not meant to imply that water is absent from a concentrated acid. A concentrated acid may contain some amount of water. For example, concentrated hydrochloric acid consists of approximately 36% to 38% hydrogen chloride by mass in water. Diluted hydrochloric acid is the solution that results when additional water is added to this concentrated acid. Whereas concentrated hydrochloric acid is a corrosive material, diluted hydrochloric acid may or may not exhibit a corrosive character. When very highly diluted with water, all acids lose their corrosive character.

Performance Goals for Section 8.2

- ◆ Describe the format of the pH scale.
- ◆ Identify the ranges of the pH scale at which the degree of corrosiveness of a solution is greatest.

◆ 8.2 THE pH SCALE

The pH is a number from 0 to 14 that designates the acidity of an aqueous solution. This is expressed in terms of its hydrogen ion concentration. To illustrate, let's begin with a simple example. In pure water, there is a very small hydrogen ion concentration derived by the dissociation of the water molecules, as follows:

$$\underset{\text{water}}{H_2O(l)} \longrightarrow \underset{\text{hydrogen ion}}{H^+(aq)} + \underset{\text{hydroxide ion}}{OH^-(aq)}$$

When the hydrogen ion concentration in pure water is determined experimentally, we find that it is only 0.0000001 mol/L (10^{-7} mol/L). Instead of writing all these zeros, we

TABLE 8.2 ◆ **The pH Values of Some Commonly Encountered Solutions and Mixtures**

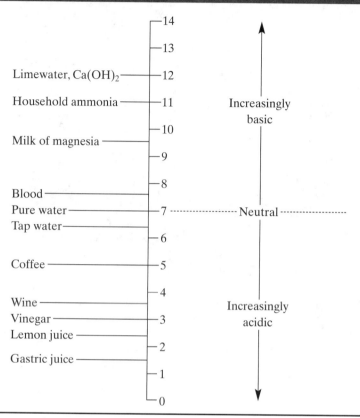

express this concentration by indicating that the pH equals 7; that is, the pH is the degree of the negative exponent. If the hydrogen ion concentration of an aqueous solution is 0.01 mol/L, or 10^{-2} mol/L, the pH equals 2.

Aqueous solutions possessing a pH from 0 to 7 are said to be *acidic*. Aqueous solutions having a pH from 7 to 14 are referred to as *basic, caustic,* or *alkaline*. Since water is neither acidic nor basic, the pH of pure water is 7. Table 8.2 lists the pH of some common solutions and mixtures.

A unit change in a pH value represents a 10-fold difference in the hydrogen ion concentration of a solution, and a difference of two pH units represents a 100-fold difference. For example, the hydrogen ion concentration of an aqueous solution having a pH of 4 is 100 times greater than in a solution having a pH of 6; it is 1000 times greater than in one having a pH equal to 7. In everyday practice, we say that a solution having a pH of 4 is 100 times more acidic than a solution having a pH of 6 and 1000 times more acidic than a solution having a pH of 7. Also, a solution with a pH of 8 is 10 times as alkaline as one with a pH of 7; a solution having a pH of 9 is 10 times as alkaline as one having a pH of 8 and 100 times as alkaline as one having a pH of 7, and so on.

In a chemical laboratory, the pH is frequently determined by means of an electrometric apparatus called a *pH meter*. The pH meter in Figure 8.1 is a voltage-measuring

FIGURE 8.1 ◆ A pH meter used for measuring the pH of aqueous solutions. Electrodes are immersed into the solution (at the right) whose pH is desired. The scale graduated from 0 to 14 is the pH scale; the pH of this solution is 7.8. The other scales shown on the meter are used by chemists for other electrometric measurements. *Courtesy of Orion Research, Inc., Boston, Massachusetts.*

device that has been connected to a pair of electrodes. When the tips of the electrodes are immersed in a solution, the pH of the solution can be readily determined by simply reading the scale.

> **Performance Goals for Section 8.3**
> ◆ Identify the common properties of acids and bases.
> ◆ Define the terms "acidic anhydride" and "basic anhydride."
> ◆ Discuss the manner in which acids and bases are produced by the reactions of acidic anhydrides and basic anhydrides, respectively, with water.
> ◆ Discuss the manner in which salts are produced by the reactions of acidic anhydrides and basic anhydrides.

◆ 8.3 PROPERTIES OF ACIDS AND BASES

All acids are associated with certain common properties. They taste sour; they cause indicator dyes to change to identifiable colors; and they react with bases to form salts and water. On the other hand, bases taste bitter; they feel slippery; they also cause the colors of indicator dyes to change to identifiable colors; and they react with acids to form salts and water.

Acids and bases can be readily differentiated from one another by the colors they impart to pieces of litmus paper. Litmus is a common indicator dye derived from certain *lichens,* any of a group of moss-like plants. A solution of litmus is allowed to impregnate strips of paper that are subsequently dried and then separately moistened with an aqueous solution of an acid and base. Acids turn the litmus paper red, and bases turn the litmus paper blue.

When acids and bases chemically interact, they neutralize each other. An example of the neutralization of an acid and base is represented by the following equation:

$$\underset{\text{hydrochloric acid}}{\text{HCl}(aq)} + \underset{\text{sodium hydroxide}}{\text{NaOH}(aq)} \longrightarrow \underset{\text{sodium chloride}}{\text{NaCl}(aq)} + \underset{\text{water}}{\text{H}_2\text{O}(l)}$$

The compound to the immediate right of the arrow is called a *salt.* Any compound in which the hydrogen in an acid has been replaced by a metallic ion is a salt. NaCl is the chemical formula of sodium chloride, which we commonly know as ordinary table salt. It has dissimilar properties from those of both hydrochloric acid and sodium hydroxide.

THE ANHYDRIDES OF ACIDS AND BASES

When a metallic or nonmetallic element combines with oxygen, the compounds produced are called *metallic oxides* and *nonmetallic oxides,* respectively. Sodium and calcium are two examples of metals that burn in oxygen to form their corresponding metallic oxides; sulfur and phosphorus are two examples of nonmetals that burn in oxygen to form their corresponding nonmetallic oxides.

$$\underset{\text{sodium}}{4\text{Na}(s)} + \underset{\text{oxygen}}{\text{O}_2(g)} \longrightarrow \underset{\text{sodium oxide}}{2\text{Na}_2\text{O}(s)}$$

$$\underset{\text{calcium}}{2\text{Ca}(s)} + \underset{\text{oxygen}}{\text{O}_2(g)} \longrightarrow \underset{\text{calcium oxide}}{2\text{CaO}(s)}$$

$$\underset{\text{sulfur}}{\text{S}_8(s)} + \underset{\text{oxygen}}{8\text{O}_2(g)} \longrightarrow \underset{\text{sulfur dioxide}}{8\text{SO}_2(g)}$$

$$\underset{\text{phosphorus}}{\text{P}_4(s)} + \underset{\text{oxygen}}{3\text{O}_2(g)} \longrightarrow \underset{\text{tetraphosphorus hexoxide}}{\text{P}_4\text{O}_6(s)}$$

A metallic oxide reacts with water to produce a base, whereas a nonmetallic oxide reacts with water to produce an acid. These combination reactions are represented by the following equations:

$$\underset{\text{sodium oxide}}{\text{Na}_2\text{O}(s)} + \underset{\text{water}}{\text{H}_2\text{O}(l)} \longrightarrow \underset{\text{sodium hydroxide}}{2\text{NaOH}(aq)}$$

$$\underset{\text{calcium oxide}}{\text{CaO}(s)} + \underset{\text{water}}{\text{H}_2\text{O}(l)} \longrightarrow \underset{\text{calcium hydroxide}}{\text{Ca(OH)}_2(aq)}$$

$$\underset{\text{sulfur dioxide}}{\text{SO}_2(g)} + \underset{\text{water}}{\text{H}_2\text{O}(l)} \longrightarrow \underset{\text{sulfurous acid}}{\text{H}_2\text{SO}_3(aq)}$$

$$\underset{\text{tetraphosphorus hexoxide}}{\text{P}_4\text{O}_6(s)} + \underset{\text{water}}{6\text{H}_2\text{O}(l)} \longrightarrow \underset{\text{phosphorous acid}}{4\text{H}_3\text{PO}_3(aq)}$$

The word "anhydride" refers to a substance from which the elements of water have been abstracted. It is either a metallic or nonmetallic oxide. A metallic oxide that combines with water to produce a base is called a *basic anhydride.* A nonmetallic oxide that combines with water to produce an acid is called an *acidic anhydride.* Sodium oxide and calcium oxide are examples of basic anhydrides, and sulfur dioxide and tetraphosphorus hexoxide are examples of acidic anhydrides. Some common acids and bases and their respective acidic anhydrides and basic anhydrides are listed in Table 8.3.

TABLE 8.3 ◆ Acidic Anhydrides and Basic Anhydrides

Anhydride	Chemical Formula	Acid/Base	Chemical Formula
Dinitrogen trioxide	$N_2O_3(g)$	Nitrous acid	$HNO_2(aq)$
Dinitrogen pentoxide	$N_2O_5(g)$	Nitric acid	$HNO_3(aq)$
Sulfur dioxide	$SO_2(g)$	Sulfurous acid	$H_2SO_3(aq)$
Sulfur trioxide	$SO_3(g)$	Sulfuric acid	$H_2SO_4(aq)$
Tetraphosphorus hexoxide	$P_4O_6(s)$	Phosphorous acid	$H_3PO_3(aq)$
Tetraphosphorus decoxide	$P_4O_{10}(s)$	Phosphoric acid	$H_3PO_4(aq)$
Calcium oxide	$CaO(s)$	Calcium hydroxide	$Ca(OH)_2(aq)$
Magnesium oxide	$MgO(s)$	Magnesium hydroxide	$Mg(OH)_2(aq)$
Potassium oxide	$K_2O(s)$	Potassium hydroxide	$KOH(aq)$
Sodium oxide	$Na_2O(s)$	Sodium hydroxide	$NaOH(aq)$

Solved Exercise 8.1

Write equations for the following chemical phenomena:

(a) Bubbles of hydrogen are generated as metallic calcium slowly reacts with an aqueous solution of hydrochloric acid.

(b) A surface coating of zinc oxide is removed by an aqueous solution of sulfuric acid.

(c) Bubbles of carbon dioxide are generated when potassium carbonate is mixed into an aqueous solution of phosphoric acid.

Solution: The phenomena in (a), (b), and (c) are representative of the chemical reactions of acids with metals, metallic oxides, and metallic carbonates, respectively.

(a) Acids react with common metals other than copper, silver, gold, and mercury to produce hydrogen and a salt of the metal. Consequently, metallic calcium reacts with hydrochloric acid to produce hydrogen and calcium chloride, as follows:

$$\underset{\text{calcium}}{Ca(s)} + \underset{\text{hydrochloric acid}}{2HCl(aq)} \longrightarrow \underset{\text{calcium chloride}}{CaCl_2(aq)} + \underset{\text{hydrogen}}{2H_2(g)}$$

(b) Acids react with metallic oxides to produce water and a salt of the metal. Consequently, zinc oxide reacts with sulfuric acid as follows:

$$\underset{\text{zinc oxide}}{ZnO(s)} + \underset{\text{sulfuric acid}}{H_2SO_4(aq)} \longrightarrow \underset{\text{zinc sulfate}}{ZnSO_4(aq)} + \underset{\text{water}}{H_2O(l)}$$

(c) Acids react with metallic carbonate to produce carbon dioxide, water, and a salt of the metal. Consequently, potassium carbonate reacts with phosphoric acid as follows:

$$\underset{\text{potassium carbonate}}{3K_2CO_3(s)} + \underset{\text{phosphoric acid}}{2H_3PO_4(aq)} \longrightarrow \underset{\text{carbon dioxide}}{3CO_2(g)} + \underset{\text{water}}{3H_2O(l)} + \underset{\text{potassium phosphate}}{2K_3PO_4(aq)}$$

◆ 8.4 ACIDS AND BASES AS CORROSIVE MATERIALS

In this book, acids and bases are the primary corrosive materials of interest. Acids act as corrosive materials by reacting with metals, metallic oxides, metallic carbonates, and skin tissue. Bases act as corrosive materials by reacting with metals and skin tissue. We shall consider these phenomena separately.

REACTIONS OF ACIDS AND METALS

Diluted acids react with all the commonly encountered metals other than copper, silver, gold, and mercury. These are simple displacement reactions that produce hydrogen and a salt of the metal. Two examples of such reactions are illustrated by the following equations:

$$\underset{\text{magnesium}}{Mg(s)} + \underset{\text{hydrochloric acid}}{2HCl(aq)} \longrightarrow \underset{\text{magnesium chloride}}{MgCl_2(aq)} + \underset{\text{hydrogen}}{H_2(g)}$$

$$\underset{\text{aluminum}}{2Al(s)} + \underset{\text{sulfuric acid}}{3H_2SO_4(aq)} \longrightarrow \underset{\text{aluminum sulfate}}{Al_2(SO_4)_3(aq)} + \underset{\text{hydrogen}}{3H_2(g)}$$

Since acids and metals are chemically incompatible, it is unsafe to store acids in metal containers. They are usually available commercially in plastic containers (Figure 8.2).

REACTIONS OF ACIDS AND METALLIC OXIDES

Acids react with metallic oxides to form a salt of the metal and water. Examples of this type of double-displacement reaction are illustrated by the following equations:

$$\underset{\text{iron(II) oxide}}{FeO(s)} + \underset{\text{hydrochloric acid}}{2HCl(aq)} \longrightarrow \underset{\text{iron(II) chloride}}{FeCl_2(aq)} + \underset{\text{water}}{H_2O(l)}$$

$$\underset{\text{aluminum oxide}}{Al_2O_3(s)} + \underset{\text{nitric acid}}{6HNO_3(aq)} \longrightarrow \underset{\text{aluminum nitrate}}{2Al(NO_3)_3(aq)} + 3H_2O(l)$$

Figure 8.2 ◆ Acids are commercially available in containers like this 55-gal polyethylene drum of hydrochloric acid, 30-gal polyethylene drum of sulfuric acid, 6.5-gal glass carboy of nitric acid (cushioned with polystyrene), and 5-gal polyethylene pail of glacial acetic acid. *Courtesy of Mallinckrodt Baker, Inc., Phillipsburg, New Jersey.*

Acids are sometimes beneficially used to remove metallic oxides and other impurities from the surface of metals. When used in this fashion, the acid is commonly called *pickle liquor* and the associated phenomenon is called *pickling*. Large volumes of pickle liquor are used annually by the steel industry to remove rust from the surface of steel during the manufacture of such products as wire, rod, and nuts and bolts. The common acids used in pickle liquors are sulfuric acid, hydrochloric acid, and phosphoric acid.

Solved Exercise 8.2

Why are sulfur dioxide and sulfur trioxide the respective acidic anhydrides of sulfurous acid and sulfuric acid?

Solution: An acidic anhydride is a compound derived from an acid by elimination of its water. Sulfur dioxide and sulfur trioxide are the acidic anhydrides of sulfurous acid and sulfuric acid, respectively, since they are produced when the water is lost from these acids.

$$H_2SO_3(aq) \longrightarrow SO_2(g) + H_2O(l)$$
sulfurous acid $\qquad$ sulfur dioxide $\quad$ water

$$H_2SO_4(g) \longrightarrow SO_3(g) + H_2O(l)$$
sulfur trioxide $\qquad$ sulfur trioxide $\quad$ water

REACTIONS OF ACIDS AND METALLIC CARBONATES

Acids react with metallic carbonates to produce carbon dioxide, water, and a salt of the metal, as follows.

$$CaCO_3(s) + 2HCl(aq) \longrightarrow CaCl_2(aq) + CO_2(g) + H_2O(l)$$
calcium carbonate $\quad$ hydrochloric acid $\qquad$ calcium chloride $\quad$ carbon dioxide $\quad$ water

$$ZnCO_3(s) + H_2SO_4(aq) \longrightarrow ZnSO_4(aq) + CO_2(g) + H_2O(l)$$
zinc carbonate $\quad$ sulfuric acid $\qquad$ zinc sulfate $\quad$ carbon dioxide $\quad$ water

REACTIONS OF ACIDS AND SKIN TISSUE

The nature of the corrosive effect caused by exposure of skin to an acid is dependent upon the concentration of the acid. Exposure to a diluted acid may only redden the area of exposure, while exposure to a concentrated acid for the same duration could blister skin and produce more extensive tissue damage. Such blistering of the skin tissues caused by exposure to a concentrated acid is illustrated in Figure 8.3. The skin tissue is said to be "burned," since its appearance visually resembles a thermal burn. Very severe burns caused by exposure to concentrated acids result in irreversible tissue damage and permanent disfigurement at the site of contact.

REACTIONS OF BASES AND METALS

Three common metals react with the concentrated solutions of strong bases. They are aluminum, zinc, and lead. During the chemical reactions, hydrogen and a compound

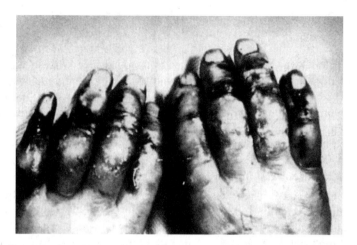

FIGURE 8.3 ◆ An acid burn on the hands. Such damage to the skin from corrosive materials may be irreversible, leaving ugly scars after wounds have healed. To correct the physical appearance of damaged skin, reconstructive cosmetic surgery is often necessary. *Courtesy of Dr. Sophie M. Worobec, M.D., University of Illinois, College of Medicine, and Dr. Paul Lazar, M.D., Chicago, Illinois.*

of the metal are produced. The following equations illustrate the behavior of aluminum, zinc, and lead with a concentrated solution of sodium hydroxide:

$$2Al(s) + 6NaOH(aq) \longrightarrow 2Na_3AlO_3(aq) + 3H_2(g)$$

aluminum   sodium hydroxide   sodium aluminate   hydrogen

$$Zn(s) + 2NaOH(aq) \longrightarrow Na_2ZnO_2(aq) + H_2(g)$$

zinc   sodium hydroxide   sodium zincate   hydrogen

$$Pb(s) + 2NaOH(aq) \longrightarrow Na_2PbO_2(aq) + H_2(g)$$

lead   sodium hydroxide   sodium plumbite   hydrogen

REACTION OF BASES WITH SKIN TISSUE

Aqueous solutions of bases corrode skin tissue in a fashion that depends upon the concentration of the base. When skin is exposed to a diluted solution, it appears reddened at the site of contact. Wounds of this type heal rapidly. On the other hand, when the skin has been exposed to an elevated concentration of the same base, it changes in texture and resembles a thick, sticky liquid. Prolonged exposure can cause the development of deep wounds that are very slow to heal. In this instance, exposure to the base causes irreversible tissue damage and permanent disfigurement at the site of contact.

Performance Goals for Section 8.5

- ◆ Discuss the physical and chemical properties of sulfuric acid.
- ◆ Describe the manner in which sulfuric acid is produced for commercial use.
- ◆ Identify the industries that use bulk quantities of sulfuric acid.
- ◆ Describe the nature of oleum.
- ◆ Identify the principal hazards associated with exposure to concentrated sulfuric acid and oleum.

◆ 8.5 SULFURIC ACID

If we were to list chemical substances by the amounts produced and consumed annually within the United States, we would find sulfuric acid near the top of these lists for the past four decades. In the United States alone, well over 40 million tons of sulfuric acid are produced annually. Sulfuric acid is so important commercially that its production and consumption rates have been used by economists to estimate the degree to which a country has industrialized.

Although the layperson is generally aware of sulfuric acid through its use in lead storage batteries (Figure 8.4), the relative commonplaceness of sulfuric acid is largely connected to its demand within the chemical and petroleum industries. In the chemical industry, sulfuric acid is used for producing and manufacturing other chemical substances, including other acids. In the petroleum industry, it is used as a catalyst for the production of high-octane fuels. Given such widespread usage, sulfuric acid has been called the workhorse of the industrial world. Its popularity suggests that sulfuric acid is likely to be encountered more frequently than other corrosive materials during hazardous materials incidents.

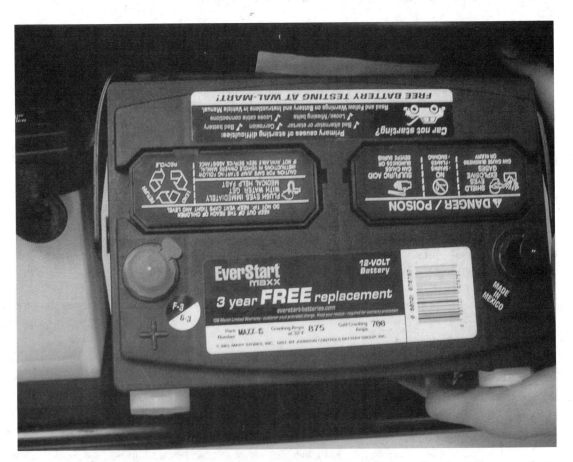

FIGURE 8.4 ◆ An aqueous solution of sulfuric acid is used as the electrolyte in acid storage batteries. The typical sulfuric acid concentration is two moles per liter of solution.

TABLE 8.4 ◆ Physical Properties of Concentrated Sulfuric Acid	
Specific gravity	1.84
Vapor density (air = 1)	2.8
Boiling point	640°F (338°C)
Freezing point	50°F (10°C)
Solubility in water	Infinitely soluble

Sulfuric acid is a mineral acid. It is a colorless, oily liquid having a density approximately twice that of water. Commercially, sulfuric acid is sometimes encountered as a clear-to-brownish-colored liquid, which reflects a lower degree of purity. The chemical formula of sulfuric acid is H_2SO_4. Some of its important physical properties are noted in Table 8.4.

Sulfuric acid is prepared at industrial plants by first burning sulfur to produce sulfur dioxide, and then oxidizing the sulfur dioxide further in the presence of a catalyst to produce sulfur trioxide. This is the acidic anhydride of sulfuric acid. It is united with water to form sulfuric acid.

$$\underset{\text{sulfur}}{S_8(s)} + \underset{\text{oxygen}}{8O_2(g)} \longrightarrow \underset{\text{sulfur dioxide}}{8SO_2(g)}$$

$$\underset{\text{sulfur dioxide}}{2SO_2(g)} + \underset{\text{oxygen}}{O_2(g)} \longrightarrow \underset{\text{sulfur trioxide}}{2SO_3(g)}$$

$$\underset{\text{water}}{H_2O(l)} + \underset{\text{sulfur trioxide}}{SO_3(g)} \longrightarrow \underset{\text{sulfuric acid}}{H_2SO_4(l)}$$

Sulfur trioxide does not unite with pure water readily. When sulfuric acid is industrially manufactured, the sulfur trioxide is absorbed into a solution of sulfuric acid containing 97% sulfuric acid by mass instead of pure water. Sulfur trioxide readily dissolves in this sulfuric acid solution. The final solution boils at 640°F (338°C) at 14.7 psi_a (101.3 kPa) and contains 98.3% sulfuric acid by mass. In commerce, this solution is called concentrated sulfuric acid.

The principal hazards associated with concentrated sulfuric acid are illustrated in Figures 8.5 and 8.6. They are linked with the heat evolved when the acid is diluted, the ability of the acid to extract water from certain materials, and the ability of the acid to react as an oxidizing agent.

LIBERATION OF HEAT

Considerable heat is released when concentrated sulfuric acid is diluted with water. Since this reaction is highly exothermic, extreme caution must be exercised when sulfuric acid is diluted. The recommended practice is to slowly pour the acid into the water while stirring. When concentrated sulfuric acid is diluted in the reverse manner, localized boiling and violent spattering occurs.

EXTRACTION OF WATER

Concentrated sulfuric acid possesses an ability to extract water from materials. Concentrated sulfuric acid so strongly extracts the elements of water from some organic compounds that carbon is often the only visibly remaining residue. It is for this reason that concentrated sulfuric acid completely destroys wood, textiles, and paper. This dehydrating phenomenon also occurs when concentrated sulfuric acid burns body tissues.

FIGURE 8.5 ◆ The label from a 500-mL bottle of concentrated sulfuric acid marketed by Mallinckrodt Baker, Inc. Note the use of the Saf-T-Data System, which provides relative ratings for health, flammability, reactivity, and contact hazards. *Courtesy of Mallinckrodt Baker, Inc., Phillipsburg, New Jersey.*

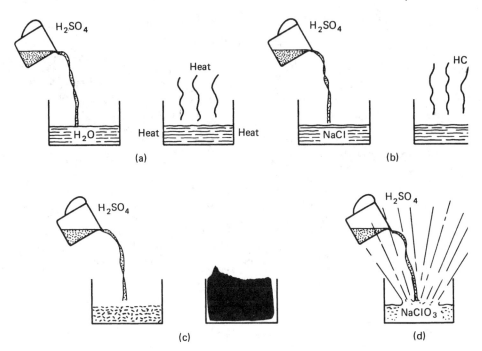

FIGURE 8.6 ◆ Some potentially hazardous features of concentrated sulfuric acid. In (a), the mixture of concentrated sulfuric acid with water causes the evolution of sufficient heat to trigger the self-ignition of some materials. In (b), concentrated sulfuric acid reacts with sodium chloride in a double-replacement reaction. In this instance, toxic vapors of hydrogen chloride are produced. In (c), concentrated sulfuric acid dehydrates sugar; that is, it removes the elements of water from sugar, leaving a residue of carbon. In (d), concentrated sulfuric acid reacts with sodium chlorate, again extracting the elements of water; these two substances are chemically incompatible and explode on contact.

OXIDIZING POTENTIAL

Hot concentrated sulfuric acid reacts as a strong oxidizing agent. Many acids do not generally react with copper, carbon, and lead, but hot, concentrated sulfuric acid oxidizes them.

$$\underset{\text{copper}}{Cu(s)} + \underset{\text{sulfuric acid}}{2H_2SO_4(conc)} \longrightarrow \underset{\text{copper(II) sulfate}}{CuSO_4(aq)} + \underset{\text{sulfur dioxide}}{SO_2(g)} + \underset{\text{water}}{2H_2O(g)}$$

$$\underset{\text{carbon}}{C(s)} + \underset{\text{sulfuric acid}}{2H_2SO_4(conc)} \longrightarrow \underset{\text{carbon dioxide}}{CO_2(g)} + \underset{\text{sulfur dioxide}}{SO_2(g)} + \underset{\text{water}}{2H_2O(g)}$$

$$\underset{\text{lead}}{Pb(s)} + \underset{\text{sulfuric acid}}{3H_2SO_4(conc)} \longrightarrow \underset{\text{lead(II) bisulfate}}{Pb(HSO_4)_2(s)} + \underset{\text{sulfur dioxide}}{SO_2(g)} + \underset{\text{water}}{2H_2O(g)}$$

On the other hand, concentrated sulfuric acid reacts so slowly with these elements at room temperature that the rates are barely perceptible.

Solved Exercise 8.3

Sulfuric acid is one of the most important substances in the chemical industry. It is an example of a strong acid, a mineral acid, and (when hot) an oxidizing acid. What is meant by these terms?

Solution: Each acid is either strong or weak, depending upon the extent to which it ionizes in water. Those that ionize completely are strong acids, while those that do not are weak acids. Sulfuric acid is a strong acid, since it ionizes almost completely when dissolved in water, as follows:

$$H_2SO_4(conc) \longrightarrow H^+(aq) + HSO_4^-(aq)$$
$$HSO_4^-(aq) \longrightarrow H^+(aq) + SO_4^{2-}(aq)$$

Sulfuric acid is a mineral acid, since a nonmetal—sulfur— is a component of its chemical composition.

Hot sulfuric acid is an oxidizing acid, because it participates as a reactant during which it takes electrons from other substances. An example of this oxidizing nature of sulfuric acid is evident in its reaction with carbon.

$$C(s) + 2H_2SO_4(conc) \longrightarrow CO_2(g) + SO_2(g) + 2H_2O(g)$$

OLEUM (FUMING SULFURIC ACID)

Sulfur trioxide is readily soluble in concentrated sulfuric acid. When the acid contains a higher proportion of sulfur trioxide than is contained in ordinary sulfuric acid, the resulting material is called *oleum* or *fuming sulfuric acid*. This acid is available commercially in concentrations having from 10 to 70% sulfur trioxide.

The chemical formula of oleum is often denoted as $xH_2SO_4 \cdot ySO_3$, where x and y are numbers. For example, oleum whose concentration is 65% sulfur trioxide by mass is expressed by the formula $4H_2SO_4 \cdot 9SO_3$.

Like sulfuric acid, oleum is a corrosive material. When oleum burns the skin, ugly scars are produced. In addition, oleum spontaneously releases toxic sulfur trioxide vapors, which pose a high degree of inhalation toxicity. For this reason, oleum should only be stored and used in a well-ventilated location.

Performance Goals for Section 8.6
- Discuss the physical and chemical properties of nitric acid.
- Describe the manner in which nitric acid is produced for commercial use.
- Identify the industries that use bulk quantities of nitric acid.
- Describe the nature of fuming nitric acid.
- Identify the principal hazards associated with exposure to concentrated nitric acid and fuming nitric acid.

◆ 8.6 NITRIC ACID

Nitric acid is second among the acids most commonly used throughout the United States. It is the raw material used for the manufacture of ammonium nitrate fertilizers, explosives, and nitrated organic compounds. It is also required for the production of rayon cloth. The chemical formula of nitric acid is HNO_3. Its important physical properties are provided in Table 8.5.

TABLE 8.5 ◆ **Physical Properties of Concentrated Nitric Acid**	
Specific gravity	1.50
Boiling point	187°F (86°C)
Freezing point	–44°F (–42°C)
Solubility in water	Infinitely soluble

Pure nitric acid is a colorless liquid. It is encountered commercially in concentrated and diluted forms. Concentrated nitric acid is an aqueous solution consisting of 68.2% nitric acid by mass. All concentrations containing less than 68.2% acid are called diluted nitric acid.

When encountered, nitric acid is often yellow to red-brown in color, which indicates that nitrogen dioxide is present. Nitrogen dioxide is a red-brown gas produced by the slow decomposition of nitric acid, a phenomenon catalyzed by sunlight.

$$4HNO_3(l) \longrightarrow 4NO_2(g) + 2H_2O(l) + O_2(g)$$

nitric acid   nitrogen dioxide   water   oxygen

Almost all nitric acid is industrially manufactured from ammonia by means of a series of reactions. Gaseous ammonia is first mixed with about 10 times its volume of air, and then exposed to platinum gauze. The platinum increases the rate of the reaction that converts ammonia into nitric monoxide (NO). Then, additional air is permitted to enter the reaction system so the nitric oxide can be further oxidized to nitrogen dioxide.

$$4NH_3(g) + 5O_2(g) \longrightarrow 4NO(g) + 6H_2O(g)$$

ammonia   oxygen   nitric oxide   water

$$2NO(g) + O_2(g) \longrightarrow 2NO_2(g)$$

nitric oxide   oxygen   nitrogen dioxide

The nitrogen dioxide is then reacted with water to produce nitric acid.

$$3NO_2(g) + H_2O(l) \longrightarrow 2HNO_3(l) + NO(g)$$

nitrogen dioxide   water   nitric acid   nitric oxide

The excess nitrogen monoxide produced in this reaction is recycled through the system.

The hazards of concentrated nitric acid are illustrated in Figure 8.7. Like sulfuric acid, nitric acid is an oxidizing acid. Sulfuric acid is a powerful oxidizing agent only when it is hot, whereas nitric acid is a powerful oxidizing agent even at room temperature.

OXIDATION OF METALS

When nitric acid chemically attacks metals, the metals are oxidized to their corresponding positive ions, while the nitric acid is reduced to one or more of the following: nitrogen, nitric oxide, nitrogen dioxide, dinitrogen monoxide, or the ammonium ion. Nitric acid reacts with some metals to form each of these products under specific conditions. For example, it reacts with zinc to form nitrogen, nitric oxide, nitrogen dioxide, dinitrogen monoxide, or ammonium nitrate in separate reactions.

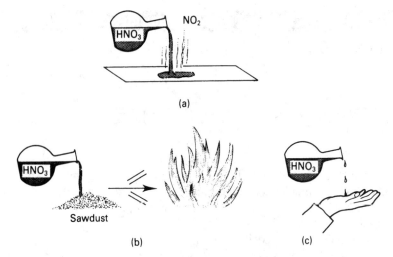

FIGURE 8.7 ◆ Some potentially hazardous features of concentrated nitric acid. In (a), the hot acid acts as a corrosive material; nitric acid corrodes certain metals and nonmetals alike, simultaneously producing an oxide of nitrogen. When nitric acid corrodes these substances, it chemically reacts as a strong oxidizing agent. This is further illustrated in (b), which shows concentrated nitric acid causing the ignition of sawdust. In (c), concentrated nitric acid acts as a corrosive material on skin tissue, leaving ugly, yellow scars.

$$5Zn(s) + 12HNO_3(aq) \longrightarrow 5Zn(NO_3)_2(aq) + 6H_2O(l) + N_2(g)$$
zinc   nitric acid   zinc nitrate   water   nitrogen

$$3Zn(s) + 8HNO_3(aq) \longrightarrow 3Zn(NO_3)_2(aq) + 4H_2O(l) + 2NO(g)$$
zinc   nitric acid   zinc nitrate   water   nitric oxide

$$Zn(s) + 4HNO_3(conc) \longrightarrow Zn(NO_3)_2(aq) + 2H_2O(l) + 2NO_2(g)$$
zinc   nitric acid   zinc nitrate   water   nitrogen dioxide

$$4Zn(s) + 10HNO_3(aq) \longrightarrow 4Zn(NO_3)_2(aq) + 5H_2O(l) + N_2O(g)$$
zinc   nitric acid   zinc nitrate   water   dinitrogen monoxide

$$4Zn(s) + 10HNO_3(aq) \longrightarrow 4Zn(NO_3)_2(aq) + 3H_2O(l) + NH_4NO_3(aq)$$
zinc   nitric acid   zinc nitrate   water   ammonium nitrate

When nitric acid oxidizes a metal, the most common products that form are nitric oxide and nitrogen dioxide. Diluted nitric acid oxidizes metals to produce nitric oxide, and concentrated nitric acid oxidizes metals to produce nitrogen dioxide.

Some finely divided metals react at explosive rates when they contact concentrated nitric acid. On the other hand, aluminum, iron, chromium, and cobalt are not chemically attacked at all by concentrated nitric acid. The corrosion of aluminum and iron by nitric acid does not occur at room temperature if the concentration of nitric acid is greater than 80% and 70% by mass, respectively. Chromium resists chemical attack by nitric acid at any concentration, and stainless steel resists attack by concentrations of nitric acid other than those greater than 78% by mass.

OXIDATION OF NONMETALS

Hot nitric acid also corrodes nonmetals such as carbon and sulfur, as follows:

$$C(s) + 4HNO_3(conc) \longrightarrow CO_2(g) + 4NO_2(g) + 2H_2O(l)$$
carbon   nitric acid   carbon dioxide   nitrogen dioxide   water

$$S_8(s) + 48HNO_3(conc) \longrightarrow 8H_2SO_4(l) + 48NO_2(g) + 16H_2O(l)$$
sulfur   nitric acid   sulfuric acid   nitrogen dioxide   water

OXIDATION OF ORGANIC COMPOUNDS

Many flammable organic compounds are oxidized, sometimes explosively, by nitric acid. This results in their subsequent ignition. For example, the organic compounds turpentine, acetic acid, acetone, ethanol, nitrobenzene, and aniline react so vigorously when mixed with hot concentrated nitric acid that they burst into flame.

NITRIC ACID AND CELLULOSIC MATERIALS

Nitric acid is capable of initiating the spontaneous ignition of wood, excelsior, and other cellulosic materials, especially when these materials have been finely divided. It is for this reason that bottles of nitric acid are not cushioned with sawdust or other cellulosic materials when they are transported.

NITRIC ACID BURNS BODY TISSUES

Nitric acid corrodes body tissues by reacting with the complex proteins that make up their structures. The exposure of nitric acid to the skin results in ugly, yellow burns that heal very slowly. The chemistry associated with this phenomenon involves the production of a yellow-brown colored substance called *xanthoproteic acid*. The discoloration of the skin typically wears away in 2 to 3 weeks.

Solved Exercise 8.4

The hydrogen in nitric acid is not replaced by metallic copper, but nitric acid does react with copper as an oxidizing agent. What are the products of the reaction when metallic copper is oxidized by concentrated nitric acid and diluted nitric acid?

   Solution: Concentrated nitric acid oxidizes metals to produce nitrogen dioxide, while diluted nitric acid oxidizes metals to produce nitric oxide. The metal is oxidized to a positive ion.

$$Cu(s) + 4HNO_3(conc) \longrightarrow Cu(NO_3)_2(aq) + 2NO_2(g) + 2H_2O(l)$$

copper   nitric acid   copper(II) nitrate   nitrogen dioxide   water

$$3Cu(s) + 8HNO_3(dil) \longrightarrow 3Cu(NO_3)_2(aq) + 2NO(g) + 4H_2O(l)$$

copper   nitric acid   copper(II) nitrate   nitric oxide   water

---

FUMING NITRIC ACID

The nitrogen oxides are readily soluble in concentrated nitric acid. When the acid contains a higher proportion of nitrogen oxides than is contained in ordinary nitric acid, the resulting material is called *fuming nitric acid*. In commerce, there are two forms of fuming nitric acid. One form is called *red fuming nitric acid*. It contains more than 85% nitric acid, less than 5% water, and from 6 to 15% nitrogen oxides. The second form is called *white fuming nitric acid*. It contains more than 97.5% nitric acid, less than 2% water, and less than 0.5% nitrogen oxides.

    Contact of the skin with both forms of fuming nitric acid is highly irritating. In addition, the substances spontaneously release toxic nitrogen oxide vapors, which pose a high degree of inhalation toxicity. They are also powerful oxidizing agents. For this reason, fuming nitric acid should be segregated from other chemical substances and should be stored and used in a well-ventilated location.

Performance Goals for Section 8.7

- ◆ Discuss the physical and chemical properties of hydrochloric acid.
- ◆ Describe the manner in which hydrochloric acid is produced for commercial use.
- ◆ Identify the industries that use bulk quantities of hydrochloric acid.
- ◆ Identify the principal hazards associated with exposure to concentrated hydrochloric acid.
- ◆ Identify the principal hazards associated with exposure to concentrated hydrochloric acid and anhydrous hydrogen chloride.

◆ 8.7 HYDROCHLORIC ACID

 Hydrochloric acid is another commercially important acid. It is most familiar to the general public as a constituent of certain household cleaning products and the liquid that maintains the proper acidity of the water in residential swimming pools (Figure 8.8).

Outside the home, hydrochloric acid has dozens of uses. As a component of pickle liquor, it is used in galvanizing, tinning, and enameling processes. Hydrochloric acid is a processing agent used by the food industry during the production of certain food products such as corn syrup. It is used by the petroleum industry to activate petroleum wells, and it is used by the chemical industry for the production of dozens of important compounds.

Pure hydrochloric acid is a colorless, fuming, and pungent-smelling liquid composed of hydrogen chloride dissolved in water. The chemical formulas of hydrochloric acid and hydrogen chloride are $HCl(aq)$ and $HCl(g)$, respectively. Some physical properties of the concentrated acid are noted in Table 8.6.

Hydrochloric acid is prepared by dissolving anhydrous hydrogen chloride (see below) in water. The concentrated acid contains from 36 to 38% hydrogen chloride by mass. An industrial grade of hydrochloric acid, called *muriatic acid,* is a dilute solution that is slightly yellow because of the presence of dissolved compounds of iron.

VAPORIZATION OF HYDROCHLORIC ACID

The principal hazardous feature of hydrochloric acid is associated with the hydrogen chloride vapor released spontaneously from the concentrated acid. Hydrogen

---

TABLE 8.6 ◆ Physical Properties of Concentrated Hydrochloric Acid	
Specific gravity	1.20
Vapor density (air = 1)	1.3
Boiling point	−121°F (−85°C)
Freezing point	−175°F (−115°C)
Solubility in water	85 g/100 g

FIGURE 8.8 ◆ Hydrochloric acid is the active component of several household cleaning products, including hard-water-stain removers and toilet bowl cleaners. It is also the acid used in residential swimming pools.

TABLE 8.7 ◆ Inhalation Effects of Hydrogen Chloride on Humans

Hydrogen Chloride Concentration in Air (ppm)	Symptoms
1–5	Threshold limit for detection by smell
5–10	Mild irritation of mucous membranes, eyes, nose, and throat
35	Distinct irritation of mucous membranes, eyes, nose, and throat
50–100	Barely tolerable effects including severe coughing and panic; possible injury to the bronchial region
1000	Danger of pulmonary edema after 24-hour exposure; potentially fatal

chloride is a poisonous gas. When concentrated hydrochloric acid spills or leaks from its container, the pungent odor of the toxic vapor is immediately detectable. Since the vapor is approximately one-fifth heavier than air, it lingers in low areas before dissipating into the atmosphere. The adverse symptoms associated with the inhalation of various concentrations of hydrogen chloride causes the adverse effects noted in Table 8.7.

HYDROCHLORIC ACID CORRODES METALS

While its concentrated and diluted forms are strong acids, hydrochloric acid is a nonoxidizing acid. When hydrochloric acid chemically corrodes metals and tissue, it does so by means of the general phenomena previously noted in Section 8.4.

HYDROCHLORIC ACID REACTS WITH OXIDIZING AGENTS

Hydrochloric acid is chemically incompatible with oxidizing agents such as metallic chlorates, metallic dichromates, and metallic permanganates. These reactions result in the production of toxic chlorine.

$$\underset{\text{potassium chlorate}}{KClO_3(s)} + \underset{\text{hydrochloric acid}}{6HCl(conc)} \longrightarrow \underset{\text{potassium chloride}}{KCl(aq)} + \underset{\text{water}}{3H_2O(l)} + \underset{\text{chlorine}}{3Cl_2(g)}$$

$$\underset{\text{potassium dichromate}}{K_2Cr_2O_7(s)} + \underset{\text{hydrochloric acid}}{14HCl(conc)} \longrightarrow \underset{\text{potassium chloride}}{2KCl(aq)} + \underset{\text{chromium(III) chloride}}{2CrCl_3(aq)} + \underset{\text{water}}{7H_2O(l)} + \underset{\text{chlorine}}{3Cl_2(g)}$$

$$\underset{\text{potassium permanganate}}{2KMnO_4(s)} + \underset{\text{hydrochloric acid}}{16HCl(conc)} \longrightarrow \underset{\text{manganese(II) chloride}}{2MnCl_2(aq)} + \underset{\text{potassium chloride}}{2KCl(aq)} + \underset{\text{water}}{8H_2O(l)} + \underset{\text{chlorine}}{5Cl_2(g)}$$

To reduce unwanted reactions between hydrochloric acid and chlorine-producing pool chemicals, acid manufacturers often label their acid containers with a warning message like the following:

**DO NOT STORE NEAR
CHLORINE-PRODUCING
POOL CHEMICALS**

ANHYDROUS HYDROGEN CHLORIDE

Anhydrous hydrogen chloride is itself a commercial chemical product that is prepared by the direct combination of elemental hydrogen and chlorine.

$$H_2(g) + Cl_2(g) \longrightarrow 2HCl(g)$$
$$\text{hydrogen} \quad \text{chlorine} \quad \text{hydrogen chloride}$$

Hydrogen chloride is also produced as a by-product during the chlorination of organic compounds.

The inhalation of hydrogen chloride is associated with the ill effects noted in Table 8.7. When inhaled, elevated concentrations of hydrogen chloride could completely deteriorate the tissue cells within the respiratory tract, destroy the lining of the tract, and cause pulmonary edema.

Performance Goals for Section 8.7
   ◆ Discuss the physical and chemical properties of perchloric acid.
   ◆ Describe the manner by which perchloric acid is produced for commercial use.
   ◆ Identify the industries that use bulk quantities of perchloric acid.
   ◆ Identify the principal hazards associated with exposure to concentrated perchloric acid.

◆ 8.8 PERCHLORIC ACID

Perchloric acid is another common acid used primarily by the chemical, electroplating, and incendiary (fireworks) industries. Its chemical formula is $HClO_4$.

Perchloric acid is a colorless, aqueous liquid having an approximate composition of 72% by mass. This is the composition of the concentrated perchloric acid available commercially. Although grades of perchloric acid greater than 72% are known, they are not routinely encountered since they are explosively unstable. Some physical properties of concentrated perchloric acid are noted in Table 8.8.

Concentrated perchloric acid is prepared by distilling a mixture of potassium perchlorate and sulfuric acid at less-than-atmospheric pressure. The chemical reaction associated with its production is represented by the following equation:

$$2KClO_4(s) + H_2SO_4(aq) \longrightarrow 2HClO_4(conc) + K_2SO_4(aq)$$
$$\text{potassium perchlorate} \quad \text{sulfuric acid} \quad \text{perchloric acid} \quad \text{potassium sulfate}$$

DECOMPOSITION OF PERCHLORIC ACID

Although concentrated perchloric acid can be safely heated to 194°F (90°C), it is likely to explosively decompose if heated above this temperature, especially when

TABLE 8.8 ◆ Physical Properties of Concentrated Perchloric Acid	
Specific gravity	1.70
Boiling point	397°F (203°C)
Freezing point	0 °F (−18°C)
Solubility in water	Very soluble

confined. When heated to approximately 198°F (92°C), perchloric acid decomposes, as follows:

$$4HClO_4(conc) \longrightarrow 2Cl_2(g) + 7O_2(g) + 2H_2O(g)$$

$\underset{\text{perchloric acid}}{} \qquad \underset{\text{chlorine}}{} \quad \underset{\text{oxygen}}{} \quad \underset{\text{water}}{}$

OXIDIZING POTENTIAL OF PERCHLORIC ACID

Hot concentrated perchloric acid is a strong oxidizing acid, but perchloric acid is a very weak oxidizing agent when cold and dilute. Hot concentrated perchloric acid violently reacts with organic compounds, including cellulosic materials such as sawdust. A mixture of perchloric acid and sawdust ignites spontaneously.

Performance Goals for Section 8.9
- Discuss the physical and chemical properties of hydrofluoric acid.
- Describe the manner by which hydrofluoric acid is produced for commercial use.
- Identify the industries that use bulk quantities of hydrofluoric acid.
- Describe the physical and chemical properties of anhydrous hydrogen fluoride.
- Identify the principal hazards associated with exposure to concentrated hydrofluoric acid and anhydrous hydrogen fluoride.

◆ 8.9 HYDROFLUORIC ACID

Hydrofluoric acid is a water solution of hydrogen fluoride; hence, the chemical formulas of hydrofluoric acid and hydrogen fluoride are HF(aq) and HF(g), respectively.

Hydrofluoric acid is used for a variety of purposes. In the glass industry, it is used to polish, etch, and frost glass. In the metallurgical and steel industries, it is used to pickle brass, copper, and certain steel alloys. In the computer industry, it is used to produce computer chips. In the chemical industry, it is used as a catalyst and fluorinating agent.

The concentrated hydrofluoric acid of commerce is a solution containing 60% hydrogen fluoride by mass. It is a colorless, fuming liquid whose physical properties are noted in Table 8.9. Hydrofluoric acid containing less than 60% hydrogen fluoride is also commercially available. These forms of the acid are often transported in bulk by means of railroad tankcars and tank trucks like those shown in Figure 8.9.

The preparation of hydrofluoric acid involves the chemical reaction between sulfuric acid and calcium fluoride, a constituent of the naturally occurring ores fluorspar and fluorite. The production reaction is represented by the following equation:

$$CaF_2(s) + H_2SO_4(conc) \longrightarrow CaSO_4(s) + 2HF(aq)$$

$\underset{\text{calcium fluoride}}{} \quad \underset{\text{sulfuric acid}}{} \qquad \underset{\text{calcium sulfate}}{} \quad \underset{\text{hydrofluoric acid}}{}$

The calcium sulfate is filtered and the solution of hydrofluoric acid is boiled, resulting in the production of the concentrated composition containing approximately 60% hydrogen fluoride by mass.

HYDROFLUORIC ACID BURNS BODY TISSUES

Hydrofluoric acid can severely corrode body tissue and produce burns that heal very slowly. Even moderate exposure to the acid causes the development of painful sores.

TABLE 8.9 ◆ **Physical Properties of Concentrated Hydrofluoric Acid**	
Specific gravity	1.0
Vapor density (air = 1)	0.71
Boiling point	68°F (20°C)
Freezing point	−117°F (−83°C)
Solubility in water	Infinitely soluble

The damage to the skin may not be limited to external destruction. The tissues beneath the outermost layer of the skin may be permanently destroyed. Pain from hydrofluoric acid burns may not be experienced immediately, and visible signs may not be apparent until hours following the initial exposure. At this point, the area of

FIGURE 8.9 ◆ Hydrofluoric acid is commercially available as its anhydrous liquid and as aqueous solutions. Railroad tankcars, tank trucks, and other vehicles used to transport the forms of hydrofluoric acid are placarded CORROSIVE in accordance with DOT regulatory requirements. *Courtesy of E. I. du Pont de Nemours & Company, LaPorte, Texas.*

contact may appear blanched and bloodless. When the victim is unattended by a physician, gangrene may develop, and the destructive action may have penetrated into the bones.

HYDROFLUORIC ACID REACTS WITH SILICON COMPOUNDS

The most distinguishing chemical property of concentrated hydrofluoric acid is its ability to slowly react with silicon compounds to produce gaseous silicon tetrafluoride. Silicon compounds are components of ordinary glass. Hydrofluoric acid reacts with them to produce silicon tetrafluoride. For example, hydrofluoric acid reacts with the sodium silicate and calcium silicate in glass as follows:

$$\underset{\text{sodium silicate}}{Na_2SiO_3(s)} + \underset{\text{hydrofluoric acid}}{6HF(conc)} \longrightarrow \underset{\text{sodium fluoride}}{2NaF(aq)} + \underset{\text{silicon tetrachloride}}{SiF_4(g)} + \underset{\text{water}}{3H_2O(l)}$$

$$\underset{\text{calcium silicate}}{CaSiO_3(s)} + \underset{\text{hydrofluoric acid}}{6HF(conc)} \longrightarrow \underset{\text{calcium fluoride}}{CaF_2(s)} + \underset{\text{silicon tetrachloride}}{SiF_4(g)} + \underset{\text{water}}{3H_2O(l)}$$

Since hydrofluoric acid reacts with the components of glass, the acid is routinely stored and transported in polyethylene or other hydrofluoric-acid-resistant plastic bottles and drums.

Solved Exercise 8.5

When an aqueous solution of hydrofluoric acid is stored in a glass bottle, why does the solution appear milky after time?

Solution: When stored in a glass bottle, hydrofluoric acid reacts with the constituent sodium silicate and calcium silicate to produce silicon tetrafluoride (see text for equations). The reaction with calcium silicate forms calcium fluoride, which is insoluble in water. Consequently, an aqueous hydrofluoric acid solution that has been stored in a glass bottle appears milky after time.

ANHYDROUS HYDROGEN FLUORIDE

Anhydrous hydrogen fluoride is itself a commercial chemical product that is prepared from calcium fluoride and sulfuric acid.

$$\underset{\text{calcium fluoride}}{CaF_2(s)} + \underset{\text{sulfuric acid}}{H_2SO_4(conc)} \longrightarrow \underset{\text{calcium sulfate}}{CaSO_4(s)} + \underset{\text{hydrogen fluoride}}{2HF(g)}$$

Although designated here as a gas, anhydrous hydrofluoric acid exists at room temperature as both a gas and compressed liquid.

In the computer industry, hydrogen fluoride and hydrofluoric acid are used to etch silicon wafers during the manufacture of computer chips. In the petroleum industry, it is used as a catalyst in connection with converting petroleum-derived gasoline into high-octane gasoline (Section 12.5). In the chemical industry, it is used to produce chlorofluorocarbons (Section 12.7).

Like its acid counterpart, anhydrous hydrogen fluoride corrodes skin tissue and produces severe burns deep beneath the skin. The corrosion may occur hours after exposure of the acid to the skin, even when no pain is initially experienced. Vapor burns to the eyes result in the formation of lesions and blindness.

■ **Performance Goals for Section 8.10**
  ◆ Discuss the physical and chemical properties of phosphoric acid.
  ◆ Describe the manner in which phosphoric acid is produced for commercial use.
  ◆ Identify the industries that use bulk quantities of phosphoric acid.
  ◆ Identify the principal hazards associated with exposure to concentrated phosphoric acid.
  ◆ Describe the physical and chemical properties of phosphoric anhydride.

◆ 8.10 PHOSPHORIC ACID

There are at least eight acids containing phosphorus, but phosphoric acid is the only one that is commonly encountered. Its chemical formula is H_3PO_4.

Phosphoric acid is used as a raw material for manufacturing a number of commercially important metallic phosphates. For instance, *superphosphate* is produced from phosphoric acid by the fertilizer industry. It is a synthetic fertilizer consisting of a mixture of calcium dihydrogen phosphate and calcium sulfate. Ammonium dihydrogen phosphate, the dry fire extinguisher, is also produced from phosphoric acid. The acid is also used to prepare the surface of steel sheets for painting.

Various grades of phosphoric acid are commercially available. When an aqueous solution of phosphoric acid is allowed to boil at atmospheric pressure, a syrupy solution containing about 85% phosphoric acid by mass is produced. This is the concentrated phosphoric acid of commerce, some physical properties of which are noted in Table 8.10. In addition, a commercially available food-grade phosphoric acid is used as an ingredient of certain soft drinks and other food products.

Phosphoric acid is manufactured from phosphate rock, an ore containing calcium phosphate, by either of the following methods:

  ◆ In the first method, phosphate rock is reacted with sulfuric acid.

$$\underset{\text{calcium phosphate}}{Ca_3(PO_4)_2(s)} + \underset{\text{sulfuric acid}}{3H_2SO_4(aq)} \longrightarrow \underset{\text{calcium sulfate}}{3CaSO_4(s)} + \underset{\text{phosphoric acid}}{2H_3PO_4(aq)}$$

  ◆ In the second method, elemental phosphorus is produced from phosphate rock by the method noted in Section 7.5; then, the phosphorus is burned to form tetraphosphorus decoxide, which is then reacted with water.

$$\underset{\text{phosphorus}}{P_4(s)} + \underset{\text{oxygen}}{O_2(g)} \longrightarrow \underset{\text{tetraphosphorus decoxide}}{P_4O_{10}(s)}$$

$$\underset{\text{tetraphosphorus decoxide}}{P_4O_{10}(s)} + \underset{\text{water}}{6H_2O(l)} \longrightarrow \underset{\text{phosphoric acid}}{4H_3PO_4(aq)}$$

The reaction with water is violent and releases considerable heat, 41 kJ/mol.

TABLE 8.10 ◆ Physical Properties of Concentrated Phosphoric Acid

Specific gravity	1.69
Boiling point	500°F (260°C)
Freezing point	108°F (42°C)
Solubility in water	Very soluble

Phosphoric acid exhibits the hazardous features of all corrosive materials previously noted in Section 8.4.

PHOSPHORIC ANHYDRIDE

Tetraphosphorus decoxide is the acidic anhydride of phosphoric acid and is sometimes known as *phosphoric anhydride*. Its chemical formula is P_4O_{10}. This substance possesses such a prodigious affinity for water that it is used industrially as a drying agent.

Considerable heat is evolved to the surroundings when phosphoric anhydride reacts with water. Given this fact, phosphoric anhydride should be segregated from combustible matter to reduce the likelihood of its ignition.

Solved Exercise 8.6

The DOT regulation at 49 CFR §173.188(a)(1) requires shippers and carriers to package phosphoric anhydride in hermetically sealed (airtight) bottles that are inserted into a wooden box and cushioned with incombustible packing material. What is the most likely reason DOT regulates the shipment of phosphoric anhydride in this fashion?

Solution: Phosphoric anhydride reacts with water—including atmospheric moisture—to produce phosphoric acid, a corrosive material. The most likely reason DOT requires use of hermetically sealed bottles is to prevent entry of water into the containers during their shipment. When phosphoric anhydride reacts with water, the heat evolved to the surroundings is 41 kJ/mol. This is sufficient heat to initiate the combustion of combustible packing material. The most likely reason DOT requires carriers to use incombustible packing material when shipping bottles of phosphoric anhydride is to reduce the likelihood of fire.

Performance Goals for Section 8.11
- ◆ Discuss the physical and chemical properties of acetic acid.
- ◆ Describe the manner in which acetic acid is produced for commercial use.
- ◆ Identify the industries that use bulk quantities of acetic acid.
- ◆ Identify the principal hazards associated with exposure to concentrated acetic acid.

◆ 8.11 ACETIC ACID

Acetic acid is the most commonly encountered organic acid. Its chemical formula is CH_3COOH. Molecules of acetic acid are represented by the following Lewis structure:

$$
\begin{array}{ccc}
 & H & O \\
 & | & \parallel \\
H- & C-C \\
 & | & \backslash \\
 & H & OH
\end{array}
$$

Acetic acid is the substance responsible for the sour taste and sharp odor of vinegar, the common food product in Figure 8.10 containing from 3 to 6% acetic acid by

FIGURE 8.10 ◆ Food-grade vinegar is an aqueous solution that contains approximately 3 to 6% acetic acid. In this range of concentration, acetic acid is not a hazardous material.

volume. It is commercially used by the chemical industry for the synthesis of ethyl acetate, vinyl acetate, cellulose acetate, and other chemical and pharmaceutical products.

At room temperature, concentrated acetic acid is a colorless, pungent liquid. Its important physical properties are noted in Table 8.11. It is manufactured by a number of methods, the most common of which involves the following two-step procedure:

◆ The chemical union of acetylene and water vapor to produce acetaldehyde

$$C_2H_2(g) + H_2O(g) \longrightarrow CH_3 - \overset{\overset{\displaystyle O}{\displaystyle \|}}{\underset{\underset{\displaystyle H(g)}{\displaystyle \backslash}}{C}}$$

acetylene   water   acetaldehyde

TABLE 8.11 ◆ Physical Properties of Concentrated Acetic Acid	
Specific gravity	1.05
Vapor density (air = 1)	2.1
Boiling point	244°F (118°C)
Freezing point	61°F (17°C)
Solubility in water	Infinitely soluble
Lower explosive limit, by volume	4%
Upper explosive limit, by volume	16%
Flashpoint	109°F (43°F)

◆ The gas-phase oxidation of acetaldehyde to acetic acid

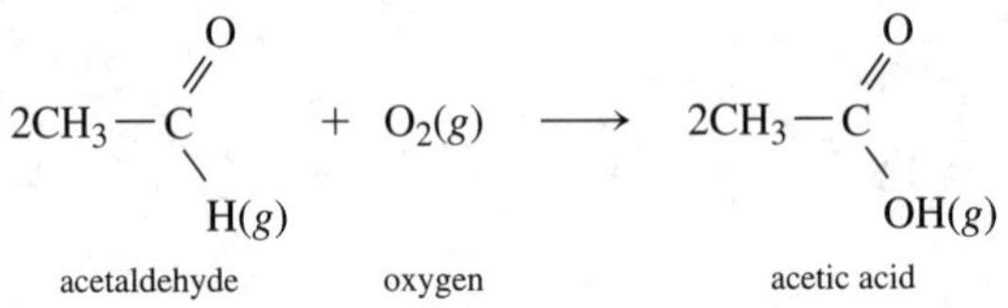

$$
2CH_3\!-\!\overset{\displaystyle O}{\underset{\displaystyle H(g)}{\overset{\|}{C}}} \quad + \quad O_2(g) \quad \longrightarrow \quad 2CH_3\!-\!\overset{\displaystyle O}{\underset{\displaystyle OH(g)}{\overset{\|}{C}}}
$$

acetaldehyde oxygen   acetic acid

When aqueous solutions of acetic acid are cooled to temperatures near 61°F (16°C), a mixture consisting of liquid and solid phases is produced. The liquid phase contains impurities and is recycled or discarded, but the solid phase generally contains greater than 99% acetic acid by mass. This component is the concentrated acetic acid of commerce; it is called *glacial acetic acid*.

Acetic acid is a weak, nonoxidizing acid. Although the concentrated acid is incapable of reacting with oxidizing agents, it is associated with the following hazardous features:

VAPORIZATION OF ACETIC ACID

Concentrated acetic acid releases a vapor that, when inhaled, is choking and suffocating. The vapor can readily damage the bronchial tract. Exposure to other body tissues, particularly the eyes, results in severe burns.

FLAMMABLE NATURE OF ACETIC ACID

Concentrated acetic acid is a class II combustible liquid. Since the aqueous solutions of acetic acid are nonflammable, fires involving the concentrated acetic acid can be extinguished by diluting the acid with water.

Performance Goals for Section 8.12
- Identify the major alkaline metal hydroxides.
- Describe the manner in which they are produced for commercial use.
- Identify the industries that use bulk quantities of the alkaline metal hydroxides.
- Identify the principal hazards associated with exposure to the alkaline metal hydroxides.

◆ **8.12 ALKALINE METAL HYDROXIDES**

There are three commercially important alkaline corrosive materials: sodium hydroxide, potassium hydroxide, and calcium hydroxide. Their chemical formulas are NaOH, KOH, and $Ca(OH)_2$, respectively. We shall refer to them in this section as the "alkaline metal hydroxides." The physical properties of sodium hydroxide, potassium hydroxide, and calcium hydroxide are noted in Table 8.12.

Sodium hydroxide is also known commercially as *lye* and *caustic soda,* while potassium hydroxide is sometimes called *caustic potash,* or *potash lye.* Sodium hydroxide is used in countless applications throughout industry, including the purification of petroleum products, the reclaiming of rubber, and the processing of textiles and paper. The chemical industry uses large amounts of sodium hydroxide as a raw material essential for the manufacturing of soap, rayon, and cellophane.

Sodium hydroxide solutions are used as industrial cleaners at car- and truck-washing facilities and garages. They are also used at wastewater facilities to remove metallic compounds as water-insoluble metallic hydroxides. To the layperson, these solutions are best known as the constituents of commercial products intended to unclog the stoppage in plumbing (Figure 8.11).

Potassium hydroxide is used mainly by the chemical industry for production of compounds used as fertilizers, soft soaps, and pharmaceutical products. It is also the electrolyte utilized in alkaline storage batteries.

Sodium hydroxide and potassium hydroxide are generally manufactured by passing an electric current through solutions of sodium chloride (brine) and potassium chloride, respectively.

$$2NaCl(aq) + 2H_2O(l) \longrightarrow 2NaOH(aq) + H_2(g) + Cl_2(g)$$

sodium chloride   water   sodium hydroxide   hydrogen   chlorine

$$2KCl(aq) + 2H_2O(l) \longrightarrow 2KOH(aq) + H_2(g) + Cl_2(g)$$

potassium chloride   water   potassium hydroxide   hydrogen   chlorine

At room temperature, sodium hydroxide and potassium hydroxide are white solids that are commercially available as flakes, pellets, sticks, granulated forms, and concentrated aqueous solutions.

Calcium hydroxide is known more commonly as *slaked lime,* or *quicklime.* It is primarily used as a raw material for the production of mortar and plaster, but it is widely used commercially for other purposes as well. Calcium hydroxide is also available in a variety of grades including a food grade. Since it is only sparingly soluble in water, calcium hydroxide is not available commercially as a solution.

TABLE 8.12 ◆ Physical Properties of the Alkaline Metal Hydroxides

	Sodium Hydroxide	*Potassium Hydroxide*	*Calcium Hydroxide*
Specific gravity	2.13	2.04	2.50
Boiling point	2534°F (1390°C)	2408°F (1320°C)	decomposes
Melting point	599°F (315°C)	680°F (360°C)	1076°F (580°C)
Solubility	42 g/100 g of H_2O	107 g/100 g of H_2O	0.18 g/100 g of H_2O

FIGURE 8.11 ◆ Lye is a common name for sodium hydroxide. It is the active component of chemical products such as Liquid-Plumr and Drāno that are used to open clogged drains.

Calcium hydroxide is industrially prepared by reacting calcium oxide and water as follows:

$$\underset{\text{calcium oxide}}{CaO(s)} + \underset{\text{water}}{H_2O(l)} \longrightarrow \underset{\text{calcium hydroxide}}{Ca(OH)_2(s)}$$

This reaction is called *slaking*.

The hazardous features of the alkaline metal hydroxides are summarized as follows:

◆ Since sodium hydroxide and potassium hydroxide are strong bases, their concentrated solutions corrode aluminum, zinc, and lead as well as body tissues in the manner previously noted in Section 8.4. The exposure of the eyes to solutions of these caustic substances causes injurious changes in the structure of the cornea, ultimately leading to complete opacification (clouding).

◆ A considerable amount of heat is evolved when the alkaline metal hydroxides dissolve in water. The absorption of this heat can initiate the burning of liquid organic compounds.

Performance Goals for Section 8.13

- ◆ Describe the concept of corrosivity as it is used by EPA in RCRA regulations.
- ◆ Determine whether a chemical waste exhibits the RCRA characteristic of corrosivity.
- ◆ Describe the nature of the labels and placards used by DOT and the accident-prevention tags, warning labels, and worded signs used by OSHA to signal the presence of corrosive materials.

■ ■

◆ 8.13 DOT, EPA, AND OSHA REGULATIONS REGARDING CORROSIVE MATERIALS

The DOT definition of a *corrosive material* was provided earlier in Section 6.7. The components of the DOT definition are illustrated by the pictographs on the COR-ROSIVE label and CORROSIVE placard previously shown in Figure 6.19. These pictographs are internationally used to denote the potentially destructive nature of corrosive materials when they inadvertently contact skin and metals.

The corrosive materials discussed throughout this chapter are provided in Table 8.13. DOT regulates their transportation by requiring carriers and shippers to affix the relevant label to packages containing corrosive materials. When 1000 lb (454 kg) or more of a corrosive material is transported, DOT requires shippers and carriers to post CORROSIVE placards on the bulk packaging, freight container, transport vehicle, unit load device, or railcar used to transport it.

EPA is concerned with the treatment, storage, and disposal of corrosive materials when they are hazardous wastes (Section 1.5). A waste exhibits the RCRA characteristic of *corrosivity* when it possesses either of the following properties:

- ◆ It is aqueous and has a pH less than or equal to 2 or greater than or equal to 12.5.
- ◆ It is a liquid and corrodes steel at a rate greater than 0.250 in. (6.35 mm) per year at a test temperature of 130°F (55°C) using a specified test method.

A waste that exhibits the characteristic of corrosivity is assigned the hazardous waste number "D002."

OSHA is attentive to protecting workers from the ill effects caused by exposure to corrosive materials. To accomplish this aim, it requires employers to post and use accident-prevention tags, warning labels, and worded signs within the workplace to signal the presence of corrosive materials that could damage property or cause accidental injury to workers or the public. Some examples of these warning items are shown in Figure 8.12.

Solved Exercise 8.7

A metal-plating company generates aqueous acid waste. When a pH meter is used to establish the pH of a representative sample of the waste, a pH of 1.8 is recorded.

(a) Based solely upon this information, does the waste exhibit the RCRA characteristic of corrosivity?

(b) What is the proper shipping description that DOT requires a carrier to enter on a shipping manifest when the carrier intends to transport 45 drums of the waste?

Solution: (a) An aqueous waste exhibits the RCRA characteristic of corrosivity when a sample possesses a pH equal to or less than 2.0, or equal to or greater than 12.5. Since the aqueous acid waste possesses a pH equal to 1.8, the waste exhibits the characteristic of corrosivity.

(b) Following the directions in Section 6.1, Table 6.4, DOT requires the carrier to enter either of the following proper shipping descriptions on the relevant manifest: "Hazardous waste liquid, n.o.s., 9, UN3082 (EPA corrosivity)" *or* "Hazardous waste liquid, n.o.s., 9, UN3082 (D002)."

TABLE 8.13 ◆ Some Corrosive Materials Regulated for Transport by DOT

Proper Shipping Name	*Label Code(s)*
Acetic acid, glacial, *with more than 80% acid, by mass*	8, 3
Acetic acid solution, *not less than 50% but not more than 80% acid, by mass*	8
Acetic acid solution, *with more than 10% and less than 50% acid, by mass*	8
Hydrochloric acid	8
Hydrofluoric acid solution *with more than 60% strength*	8
Hydrofluoric acid solution *with not more than 60% strength*	8
Hydrogen chloride, anhydrous	2.3, 8
Hydrogen fluoride, anhydrous	8, 6.1
Nitric acid, *other than red fuming, with more than 70% nitric acid*	8, 5.1
Nitric acid, *other than red fuming, with not more than 70% nitric acid*	8
Nitric acid, red fuming	8, 5.1, 6.1
Perchloric acid, *with more than 50 percent acid but not more than 72 percent acid, by mass*	5.1, 8
Perchloric acid, *with not more than 50% acid, by mass*	8, 5.1
Phosphoric acid	8
Phosphorus pentoxide[a]	8
Potassium hydroxide, solid	8
Potassium hydroxide, solution	8
Sodium hydroxide, solid	8
Sodium hydroxide, solution	8
Sulfuric acid, fuming, *with less than 30 percent free sulfur trioxide*	8
Sulfuric acid, fuming, *with 30 percent or more free sulfur trioxide*	8, 6.1
Sulfuric acid, *with more than 51 percent acid*	8
Sulfuric acid, *with not more than 51 percent acid*	8
Sulfuric acid, spent	8
Sulfur trioxide, stabilized	8, 6.1

[a]Phosphorus pentoxide refers to the substance whose proper chemical name is tetraphosphorus decoxide.

FIGURE 8.12 ◆ When employees are required to use corrosive materials, OSHA requires employers to use accident-prevention tags, warning labels, and worded signs to warn employees of their presence within the workplace.

Performance Goals for Section 8.14
- Describe the nature of the response actions to be taken when corrosive materials have been released into the environment.
- Describe the nature of the action to be taken by paramedics when an individual has swallowed a corrosive material.

◆ 8.14 RESPONDING TO INCIDENTS INVOLVING THE RELEASE OF CORROSIVE MATERIALS

First-on-the-scene responders can generally identify the presence of a corrosive material at a transportation mishap by observing at least one of the following:

- The number 8 as a component of a basic description of a hazardous material listed on a shipping paper
- The word "CORROSIVE" printed upon a black-and-white label affixed to packaging
- The word "CORROSIVE" printed upon a black-and-white placard posted on the relevant transport vehicle

At an emergency scene involving a corrosive material, each member of the response crew should don acid-resistant apparel like the acid-entry suit with clear face shield in Figure 8.13. An acid-resistant suit is designed to protect the wearer from inadvertently splashing a corrosive material upon skin or into the eyes. The use of the acid-resistant suit can minimize or eliminate the primary hazard associated

FIGURE 8.13 ◆ This fully encapsulating suit is worn when responding to incidents involving the spillage of acids and other corrosive materials. The suit is made of a material that is resistant to permeation by corrosive materials and thus prevents bodily contact with them. *Courtesy of Mine Safety Appliances Co., Pittsburgh, Pennsylvania.*

with a corrosive material. When an acid-resistant suit is unavailable at an emergency scene involving the release of a corrosive material, nonpermeable gloves and either protective eyeglasses or a face shield should be used to provide minimal protection.

When called to a scene involving the release of a corrosive material, emergency responders should implement the procedures in Figure 8.14. In addition, either of the following actions can be taken:

- Dilute the corrosive material with a volume of water at least equal to 10 volumes of the material that has been released into the environment. This is usually an adequate response when relatively small quantities of an acid or base have spilled or leaked from its container or storage tank. For instance, suppose that a 1-gal container of liquid sulfuric acid has inadvertently spilled on a laboratory floor and has flowed toward a drain leading to an off-site wastewater treatment plant. In this situation, dilution of the spilled acid with a copious volume of water lessens the acid's corrosive nature by reducing the concentration of hydrogen ions.
- Neutralize the corrosive material. This action is recommended when a crew responds to the environmental release of a relatively large volume of a corrosive material, such as 10,000 gal (38 m³). A large volume of a corrosive material is typically encountered during a transportation mishap involving a tank truck like that shown in Figure 8.15. Diluting such a large volume of a corrosive material with water is generally impractical, since a substantially larger volume of water is needed to effectively reduce its corrosivity.

The spill of a large volume of an acid can be neutralized with solid substances such as either "soda ash" or "lime," the common names for anhydrous sodium carbonate and calcium hydroxide, respectively. As we noted in Section 8.3, an acid reacts with a base to produce a salt of the acid and water; and an acid reacts with a metallic carbonate to produce a salt of the acid, water, and carbon dioxide.

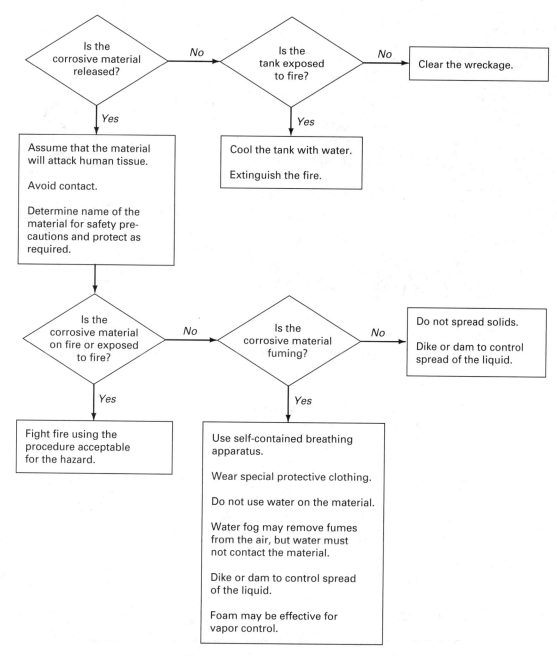

FIGURE 8.14 ◆ The recommended procedures when responding to a disaster involving a corrosive material contained within either non-bulk or bulk packaging. *Adapted with permission of the American Society for Testing and Materials, from a figure in ASTM STP 825, A Guide to the Safe Handling of Hazardous Materials Accidents. Copyright © 1983, American Society for Testing and Materials.*

FIGURE 8.15 ◆ A typical tank truck used to transport a bulk quantity of a corrosive material. This tank truck is transporting ferric chloride solution, the corrosive material whose DOT identification number is 2582. *Courtesy of Bulk Transportation, Walnut, California.*

When lime or soda ash are used to neutralize a spill of hydrochloric acid, the resulting chemical reactions are represented by the following equations:

$$\underset{\text{calcium hydroxide}}{Ca(OH)_2(s)} + \underset{\text{hydrochloric acid}}{2HCl(aq)} \longrightarrow \underset{\text{calcium chloride}}{CaCl_2(aq)} + \underset{\text{water}}{2H_2O(l)}$$

$$\underset{\text{sodium carbonate}}{Na_2CO_3(s)} + \underset{\text{hydrochloric acid}}{2HCl(aq)} \longrightarrow \underset{\text{sodium chloride}}{2NaCl(aq)} + \underset{\text{carbon dioxide}}{CO_2(g)} + \underset{\text{water}}{H_2O(l)}$$

The use of lime and soda ash in this incident results in reducing or eliminating the corrosive nature of hydrochloric acid by chemically converting it into a group of relatively benign substances.

Solved Exercise 8.8

First-on-the-scene responders arriving at a domestic transportation mishap observe that a 5,000-gal (19,000-L) overturned tank truck is leaking its liquid contents. The transportation manifest indicates that the consignment consists solely of concentrated hydrochloric acid. The highway upon which the truck was traveling is located approximately 250 ft (76.2 m) from the edge of a lake. What procedures should these first responders implement to save lives, property, and the environment?

Solution: The principal hazard associated with concentrated hydrochloric acid is exposure to the hydrogen chloride vapor that spontaneously evolves from it. To avoid inhaling the vapor, the use of self-contained breathing apparatus is essential. To avoid contact of the skin with the acid, an acid-impervious suit should be donned.

To save lives, property, and the environment, Figure 8.12 directs the first-on-the-scene responders to implement the following:

- Use self-contained breathing apparatus.
- Wear special protective clothing.
- Use water fog or foam to reduce toxic gas fumes in the air.
- Dike or dam the spilled material to prevent its spread.
- Use lime to neutralize the acid within the diked area.

RESPONDING TO ACID AND ALKALI POISONING INCIDENTS

Paramedics are often called to assist in situations in which individuals have inadvertently been exposed to corrosive materials. In such incidents, a member of the paramedic team should immediately contact the American Association of Poison Control Centers (see footnote, p. 440). In addition, the following actions can be taken:

- Since corrosive materials can irreversibly alter skin tissue at the site of contact, the affected area should be extensively flushed with water.
- When a corrosive material has been inadvertently splashed into an individual's eyes, pain, swelling, cornea erosion, and blindness can rapidly ensue. Consequently, it is vital to immediately flush the eyes with a gentle stream of running water for at least 30 minutes. If the afflicted individual wears contact lenses, the eyes should first be irrigated for several minutes, and then, the lenses should be removed and the eyes again irrigated. The individual should be advised to promptly contact an ophthalmologist for professional eye treatment.
- When a corrosive material has been swallowed, the individual may experience difficulty in swallowing, nausea, intense thirst, shock, difficulty in breathing, and death. Paramedics should *not* induce vomiting unless given direction to do so by the Poison Control Center. The stomach wall is relatively tough and normally capable of withstanding the presence of gastric juices having a pH less than 2. Vomiting should not be induced because the individual's stomach contents could inadvertently be channeled into the bronchial tract, where serious damage could result.
- When a corrosive material has been consumed, paramedics should attempt to neutralize the individual's stomach contents. *Milk of magnesia,* a white suspension of magnesium hydroxide in water, neutralizes acids. Acidic foods such as vinegar and citrus fruit juices neutralize bases. To ensure that neutralization of the corrosive material has been complete, the individual should consume a volume of the neutralizing agent at least equal to the amount swallowed.

Corrosive materials are frequently used in science laboratories and within certain work environments. In all areas where individuals can be exposed to an injurious corrosive material, suitable facilities for quick drenching and flushing of the eyes and body must be available. A suitable eye-wash station and shower are shown in Figure 8.16.

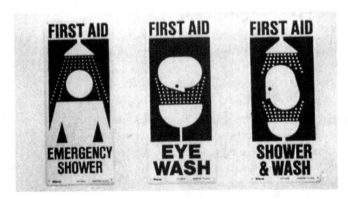

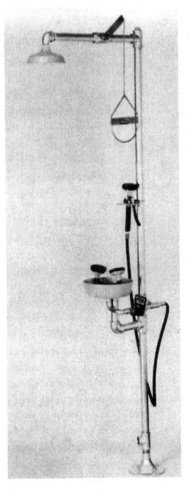

FIGURE 8.16 ◆ When corrosive materials are splashed in the eyes or when they otherwise come in contact with skin tissue, they must be removed immediately to protect the skin against permanent damage by flushing the afflicted area with water. For this reason, eye-wash stations and showers are necessarily located in science laboratories and in other areas where corrosive materials are regularly stored and used. *Courtesy of Lab Safety Supply Company, Janesville, Wisconsin.*

■■

REVIEW EXERCISES

Chemical Nature of Acids and Bases

8.1 By means of an equation, use Arrhenius's theory to account for the acidic character of the solution produced when hydrogen bromide dissolves in water.

8.2 Write the equation that illustrates the salts produced when the acidic anhydride listed under the heading "A" reacts with the basic anhydride listed under the heading "B":

	A	B
(a)	sulfur trioxide	barium oxide
(b)	carbon dioxide	calcium oxide
(c)	tetraphosphorus decoxide	potassium oxide

8.3 Identify two acidic anhydrides that contain nitrogen. Write balanced equations illustrating the production of an acid when each combines with water.

8.4 Complete and balance each of the following equations, and provide an acceptable name for each reactant and product:

(a) $Cr_2O_3(s) + H_2SO_4(aq) \longrightarrow$

(b) $PbCO_3(s) + HNO_3(aq) \longrightarrow$

(c) $Zn(s) + HClO_4(aq) \longrightarrow$

(d) $MgO(s) + HNO_3(aq) \longrightarrow$

(e) $CuCO_3(s) + HCl(aq) \longrightarrow$

(f) $NiO(s) + H_3PO_4(aq) \longrightarrow$

(g) $CaCO_3(s) + CH_3COOH(aq) \longrightarrow$

8.5 Write a balanced equation illustrating the following neutralization reactions:

(a) sulfuric acid with limewater

(b) nitric acid with potassium hydroxide

(c) hydrochloric acid with magnesium hydroxide

pH Scale

8.6 What information is required to determine whether the aqueous solution in Figure 8.1 is acidic, basic, or neutral?

8.7 A solution having a pH of 4 is how many times more acidic than a solution having a pH of 5?

Corroding Nature of Substances

8.8 Write an equation that illustrates the corrosive effect, if any, that occurs when each substance listed under the heading "A" reacts with the substance listed under the heading "B":

	A	B
(a)	atmospheric oxygen	copper
(b)	sulfuric acid	iron
(c)	hydrochloric acid	cadmium oxide
(d)	sulfuric acid	chromium(III) carbonate
(e)	phosphoric acid	tin
(f)	sodium hydroxide	metallic aluminum
(g)	perchloric acid	zinc carbonate

8.9 When sulfuric acid is a component of polluted air, it chemically attacks statues, memorials, and monuments made from limestone. Given that the most common mineral in limestone is calcium carbonate, illustrate how these limestone objects are corroded by the sulfuric acid in polluted air.

8.10 How can the storage of a diluted acid within a steel tank connected to metal fittings and pipes constitute a fire and explosion hazard?

Sulfuric Acid

8.11 Complete and balance each of the following equations, and provide an acceptable name for each reactant and product:

(a) $NaCN(s) + H_2SO_4(aq) \longrightarrow$

(b) $K(s) + H_2SO_4(aq) \longrightarrow$

(c) $Cu(s) + H_2SO_4(conc, hot) \longrightarrow$

(d) $NaCl(s) + H_2SO_4(conc) \longrightarrow$

8.12 What is the most likely reason DOT requires chemical manufacturers to affix POISON and CORROSIVE labels to packages of oleum with a sulfur trioxide content equal to or greater than 30% by mass?

8.13 Why does an aqueous solution of sulfuric acid conduct an electric current, but concentrated sulfuric acid does not?

8.14 What occurs when concentrated sulfuric acid is accidentally spilled on cotton trousers?

Nitric Acid

8.15 Complete and balance each of the following equations, and provide an acceptable name for each reactant and product:

(a) $BaCO_3(s) + HNO_3(aq) \longrightarrow$

(b) $Fe_2O_3(s) + HNO_3(aq) \longrightarrow$

(c) $PbS(s) + HNO_3(aq) \longrightarrow$

(d) $Ag(s) + HNO_3(conc) \longrightarrow$

8.16 The DOT regulation at 49 CFR §173.158(a) stipulates that nitric acid at any concentration exceeding 40% by mass cannot be packaged with any other material. What is the most likely reason DOT enacted this regulation?

8.17 When carriers intend to transport nitric acid at concentrations other than those greater than 78% by mass, the DOT regulation at 49 CFR §173.158(b)(iii) stipulates the use of containers constructed from American Iron and Steel Institute type number "304HT," where "304HT" refers to a heat-treated grade of stainless steel. What is the most likely reason DOT requires carriers to transport nitric acid in stainless steel containers when the acid concentration is less than 78% acid by mass?

8.18 Which placards does DOT require shippers and carriers to post on a vehicle that is solely used to transport drums of red fuming nitric acid?

8.19 Following a transportation mishap involving the release of nitric acid, members of the emergency-response crew note a yellowing of the skin on their feet. What is the most likely cause of the discoloration?

Hydrochloric Acid

8.20 Complete and balance each of the following equations, and provide an acceptable name for each reactant and product:

(a) $BaO(s) + HCl(aq) \longrightarrow$

(b) $CaSO_3(s) + HCl(aq) \longrightarrow$

(c) $SnO_2(s) + HCl(aq) \longrightarrow$

8.21 Write an equation that demonstrates the production of hydrochloric acid by the union of its elements.

8.22 Write an equation that demonstrates the effective removal of rust from an iron tub by the use of pickle liquor containing hydrochloric acid.

Perchloric Acid

8.23 Complete and balance each of the following equations, and provide an acceptable name for each reactant and product:

(a) $ZnO(s) + HClO_4(aq) \longrightarrow$

(b) $Al_2O_3(s) + HClO_4(aq) \longrightarrow$

(c) $CdCO_3(s) + HClO_4(aq) \longrightarrow$

(d) $Zn(s) + HClO_4(aq) \longrightarrow$

8.24 What is the most likely reason DOT prohibits carriers from accepting for transport perchloric acid that contains more than 72% acid by mass?

Hydrofluoric Acid

8.25 Complete and balance each of the following equations, and provide an acceptable name for each reactant and product:

(a) $NiO(s) + HF(aq) \longrightarrow$

(b) $Mg(s) + HF(aq) \longrightarrow$

(c) $Al_2(CO_3)_3(s) + HF(aq) \longrightarrow$

8.26 Why do sealed glass bottles containing hydrofluoric acid often burst?

Phosphoric Acid

8.27 Complete and balance each of the following equations, and provide an acceptable name for each reactant and product:

(a) $Fe_2O_3(s) + H_3PO_4(aq) \longrightarrow$

(b) $Sn(s) + H_3PO_4(aq) \longrightarrow$

(c) $CuCO_3(s) + H_3PO_4(aq) \longrightarrow$

8.28 What is the most likely reason DOT prohibits a railroad carrier from transporting any amount of phosphoric anhydride by passenger-carrying railcar?

Acetic Acid

8.29 Complete and balance each of the following equations, and provide an acceptable name for each reactant and product:

(a) $BaO(s) + CH_3COOH(aq) \longrightarrow$

(b) $Na(s) + CH_3COOH(aq) \longrightarrow$

(c) $MgCO_3(s) + CH_3COOH(aq) \longrightarrow$

8.30 A fire inspector notes that 13-gal (49-L) glass carboys containing glacial acetic acid have been stored within a temperature-controlled warehouse. To prevent the acid from freezing, identify the temperature at which the warehouse should be maintained.

Alkaline Metal Hydroxides

8.31 Lye is often used to remove mixtures of solidified grease and hair from clogged plumbing. Why is lye effective for this purpose?

8.32 When responding to a domestic transportation mishap involving a spill of a caustic soda near a lake, why should emergency responders take precaution to prevent the spilled material from entering the lake?

8.33 When quicklime is exposed to moist air, it absorbs water vapor and swells in size; it then cracks and crumbles to a powder. Quicklime first forms calcium hydroxide and then slowly unites with atmospheric carbon dioxide. The result is a mixture of calcium hydroxide and calcium carbonate that is commercially available as a product called *air-slaked lime*.

(a) Write the equations representing these chemical phenomena.

(b) Why is air-slaked lime an effective neutralizing agent for use when responding to an emergency involving a spill of a large volume of concentrated acetic acid?

Government Regulations Relating to Corrosive Materials

8.34 A carrier intends to transport 5000 lb (2270 kg) of anhydrous hydrogen fluoride by public highway. Using Tables 6.1 and 6.4, identify the basic description that DOT requires the carrier to enter on the accompanying shipping paper.

8.35 A carrier intends to transport 80 drums of a hazardous waste containing sulfuric acid and phosphoric acid, a sample of which exhibits the RCRA characteristic of corrosivity. Using Tables 6.1 and 6.4, determine the basic description that DOT requires the carrier to enter on the accompanying waste manifest.

8.36 Using Table 8.13, identify the nature and number of labels DOT requires a carrier to affix to packaging containing bottles of each of the following:

(a) Hydrochloric acid

(b) 60% aqueous perchloric acid solution

(c) Phosphoric acid

(d) Red fuming nitric acid

8.37 A chemical manufacturer in Chicago, Illinois, intends to ship 5000 lb (2270 kg) of caustic soda solution by railroad tankcar to a customer in Baltimore, Maryland. What basic description does DOT require the manufacturer and carrier to enter on the accompanying shipping paper?

Responding to Incidents Involving the Release of Corrosive Materials

8.38 Physicians recommend administering baking soda to a child who has accidentally swallowed a corrosive material. Baking soda is sodium bicarbonate. Write the equation illustrating the reaction between sodium bicarbonate and hydrochloric acid.

8.39 Acetic acid is the principal ingredient in food-grade vinegar. Through use of an equation, illustrate how milk of magnesia neutralizes an individual's stomach contents following the consumption of an excessive amount of vinegar-based salad dressing.

8.40 When first-on-the-scene responders arrive at a transportation mishap, they discover that concentrated nitric acid is leaking from a ruptured stainless-steel drum. What procedures should be implemented to save lives, property, and the environment?

Chemistry of Some Water-Reactive Substances

 **9** CHAPTER

The use of water as a fire extinguishing agent has always been a common practice; but on occasion, the application of water at an emergency scene involving hazardous materials can be inappropriate and potentially dangerous. Water reacts with certain hazardous materials to produce spontaneously flammable products that generate secondary fires and make it more difficult to extinguish the original one. Furthermore, water reacts with some hazardous materials to produce toxic or corrosive products that could unknowingly endanger emergency responders and the general public. We shall collectively refer to these types of hazardous materials as *water-reactive substances*.

When is it inappropriate for emergency-response crews to use water when responding to an incident involving the release of hazardous materials? To assist in answering this question, this chapter is devoted to the study of substances that react with water at such rapid rates they could pose a risk to the general public as well as to the emergency-response personnel.

Performance Goals for Section 9.1

♦ Discuss the importance to emergency responders of identifying hazardous materials that are water-reactive substances.

♦ Discuss the procedures NFPA and DOT use to warn firefighters of the presence of water-reactive substances at locations where they are stored or transported.

◆ 9.1 WHEN IS A SUBSTANCE WATER-REACTIVE?

In Section 6.7, we noted DOT's definition of a dangerous-when-wet material. To paraphrase, it is any substance that reacts with water to produce a spontaneously flammable and/or toxic product.

DOT regulates the transportation of water-reactive substances as one or more of the following hazardous materials:

- Dangerous-when-wet materials
- Spontaneously combustible materials
- Flammable solids
- Corrosive materials

When they are transported, DOT requires shippers and carriers to comply with labeling, marking, and placarding requirements. In particular, DOT requires shippers and carriers to affix the relevant label to packages containing water-reactive substances. When any amount of a dangerous-when-wet material is transported, DOT requires shippers and carriers to post DANGEROUS WHEN WET placards on the bulk packaging, freight container, transport vehicle, unit load device, or railcar used to transport it. When 1000 lb (454 kg) or more of a spontaneously combustible material, flammable solid, or corrosive material is transported, DOT requires shippers and carriers to post SPONTANEOUSLY COMBUSTIBLE, FLAMMABLE SOLID, or CORROSIVE placards, as relevant.

When water reacts with another substance, the chemical phenomenon is called *hydrolysis*. Chemists represent this process by using the following general equation:

$$A + H_2O(l) \longrightarrow C + D$$

Here, "A" represents a water-reactive substance and "C" and "D" are the substances produced when "A" reacts with water. We should always avoid applying water during emergency-response actions at which "A" is present, when either "C" or "D" is a flammable or toxic substance.

Consider the metals that react chemically with water to produce flammable hydrogen. These metals constitute the fuels of class D fires (Section 1.4). Several of them are so chemically reactive that they spontaneously ignite in air without exposure to an ignition source. These metals form hydrogen when they react with the moisture in the air. The hydrogen absorbs the heat of reaction and self-ignites. It is this combustion of hydrogen that activates the combustion of the metal.

A substance that self-ignites without exposure to an ignition source is said to be *pyrophoric*. This dictionary definition is expanded in OSHA regulations as follows: A substance that ignites spontaneously in air at a temperature of 130°F (54.4°C) or below. Many pyrophoric substances react vigorously with water and in atmospheres of high humidity. To prevent their ignition, pyrophoric substances are generally stored and handled under suitable liquids that prevent their contact with air. When pyrophoric substances are also water reactive, they are processed in enclosed, oxygen-free or inert, dry atmospheres. To avoid their ignition during transit, DOT requires their manufacturers to seal them in airtight containers.

The chemical reaction between a substance and water may also form a product that is toxic or corrosive. Some dangerous-when-wet substances produce a sufficient amount of toxic vapor when in contact with water that they pose a health risk to individuals within 0.3–6.0 miles (0.5–10 km) downwind. A listing of these substances is provided in DOT's *Emergency Response Guidebook* and is reproduced in Table 9.1. Emergency responders should give special attention to these substances, so appropriate action can be taken to prevent illness or death when they are involved in accidents.

TABLE 9.1 ◆ Some Classes of Water-Reactive Substances

Class of Substance	Example	Chemical Formula	Hydrolysis Product
Acetyl halides	Acetyl bromide	CH_3OBr	Hydrogen bromide
	Acetyl chloride	CH_3OCl	Hydrogen chloride
Acids	Fluorosulfonic acid	FSO_3H	Hydrogen fluoride
	Nitrosylsulfuric acid	NO_2SO_3H	Nitrogen dioxide
Chlorosilanes	Methyldichlorosilane	CH_3SiCl_2H	Hydrogen chloride
	Methyltrichlorosilane	CH_3SiCl_3	Hydrogen chloride
	Trichlorosilane	Cl_3SiH	Hydrogen chloride
Metallic amides	Lithium amide	$LiNH_2$	Ammonia
	Magnesium diamide	$Mg(NH_2)_2$	Ammonia
Metallic cyanides	Potassium cyanide	KCN	Hydrogen cyanide
	Sodium cyanide	$NaCN$	Hydrogen cyanide
Metallic halides	Aluminum bromide, anhydrous	$AlBr_3$	Hydrogen bromide
	Aluminum chloride, anhydrous	$AlCl_3$	Hydrogen chloride
	Antimony pentafluoride, anhydrous	SbF_5	Hydrogen fluoride
Metallic hypochlorites	Calcium hypochlorite	$Ca(ClO)_2$	Chlorine, hydrogen chloride
	Lithium hypochlorite	$LiClO$	Chlorine, hydrogen chloride
Metallic oxychlorides	Chromium oxychloride	$Cr(OCl)_3$	Hydrogen chloride
Metallic phosphides	Aluminum phosphide	AlP	Phosphine
	Calcium phosphide	Ca_3P_2	Phosphine
	Magnesium aluminum phosphide	$Mg_3P_2 \cdot AlP$	Phosphine
	Magnesium phosphide	Mg_3P_2	Phosphine
	Potassium phosphide	K_3P	Phosphine
	Sodium phosphide	Na_3P	Phosphine
	Stannic phosphide	Sn_3P_4	Phosphine
	Strontium phosphide	Sr_3P_2	Phosphine
	Zinc phosphide	Zn_3P_2	Phosphine
Nonmetallic halides	Iodine pentafluoride	IF_5	Hydrogen fluoride
	Phosphorus pentachloride	PCl_5	Hydrogen chloride
	Silicon tetrachloride	$SiCl_4$	Hydrogen chloride
	Thionyl chloride	$SOCl_2$	Hydrogen chloride, sulfur dioxide
Sulfides	Ammonium hydrosulfide	NH_4HS	Hydrogen sulfide, ammonia
	Ammonium sulfide	$(NH_4)_2S$	Hydrogen sulfide, ammonia
Others	Chloride dioxide (hydrate)[a]	ClO_2	Chlorine
	Uranium hexafluoride	UF_6	Hydrogen fluoride

[a]See Section 11.8.

Solved Exercise 9.1

Upon observing a posted NFPA diamond on the exterior wall of a burning shed, what information regarding the water reactivity of chemical products stored therein is immediately conveyed to responding firefighters?

Solution: Information regarding the water reactivity of a substance is conveyed on an NFPA diamond in two ways. As we first noted in Section 1.9, the relative degree of the health, fire, and reactivity hazards of a substance is conveyed through the number that has been entered in the three topmost quadrants of a diamond. The relative degree of water reactivity is conveyed by the number that has been entered in the rightmost yellow quadrant, as follows:

- A "3" means the substance reacts violently with water.
- A "2" means the substance may react violently with water or may form potentially explosive mixtures with water.
- A "1" means the substance may react with water with some release of energy, but not violently.
- A "0" means the substance is not reactive with water.

Secondly, entering a "W" with a line drawn through its center in the bottommost quadrant of the diamond cautions firefighters against the application of water.

Responding firefighters use this combination of information for selecting an action appropriate to the incident at hand.

Performance Goals for Section 9.2
- Identify and compare the physical and chemical properties of lithium, sodium, and potassium, especially their water reactivity.
- Identify the industries that use bulk quantities of these alkali metals.
- Discuss the recommended procedures for extinguishing the fires of the alkali metals.

◆ 9.2 ALKALI METALS

Although metallic sodium is the only alkali metal of major commercial importance, we shall consider some properties of three alkali metals in this section: lithium, sodium, and potassium. The properties of these three metals illustrate the uniqueness of their reactions. Some physical properties of these metals are noted in Table 9.2.

Metallic lithium, sodium, and potassium displace hydrogen from water, as follows:

$$2\text{Li}(s) + 2\text{H}_2\text{O}(l) \longrightarrow 2\text{LiOH}(aq) + \text{H}_2(g)$$
$$\text{lithium} \quad\quad \text{water} \quad\quad\quad \text{lithium hydroxide} \quad \text{hydrogen}$$

$$2\text{Na}(s) + 2\text{H}_2\text{O}(l) \longrightarrow 2\text{NaOH}(aq) + \text{H}_2(g)$$
$$\text{sodium} \quad\quad \text{water} \quad\quad\quad \text{sodium hydroxide} \quad \text{hydrogen}$$

$$2\text{K}(s) + 2\text{H}_2\text{O}(l) \longrightarrow 2\text{KOH}(aq) + \text{H}_2(g)$$
$$\text{potassium} \quad\quad \text{water} \quad\quad\quad \text{potassium hydroxide} \quad \text{hydrogen}$$

When lithium reacts with water, the hydrogen produced does not immediately burn; but when sodium and potassium react with water, the hydrogen bursts spontaneously into flame.

DOT regulates the transportation of the alkali metals, their alloys, and dispersions as dangerous-when-wet materials, some examples of which are provided in Table 9.3. DOT requires shippers and carriers to affix DANGEROUS WHEN WET labels on their containers and to post DANGEROUS WHEN WET placards on the bulk packaging, freight container, transport vehicle, unit load device, or railcar used to transport them.

TABLE 9.2 ◆ Physical Properties of the Alkali Metals

	Lithium	*Sodium*	*Potassium*
Specific gravity	0.53	0.97	0.86
Melting point	354°F (179°C)	208°F (98°C)	147°F (64°C)
Boiling point	2437°F (1337°C)	1638°F (892°C)	1425°F (774°C)

LITHIUM METAL

Lithium is a soft, silvery metal and is the least dense solid element at normal conditions. Metallic lithium is so light that its pieces float upon the surface of liquid petroleum products, such as kerosene and gasoline.

Metallic lithium and lithium compounds are valuable raw materials during the manufacturing of porcelain, ceramics, castings, fungicides, bleaching agents, pharmaceuticals, and greases. Metallic lithium is itself a component of a lightweight magnesium alloy.

When lithium metal reacts with water, the hydrogen is slowly displaced from the water. It dissipates into the surrounding environment without ever achieving a concentration equal to or greater than 4% by volume, its lower explosive limit.

During its reaction with water, metallic lithium remains in the solid state of matter. Although the heat of the reaction is initially absorbed by the metal, it is transmitted to the surrounding water. The temperature of the metal remains below the boiling point of water.

When metallic lithium is left exposed to the air at room conditions, it does not spontaneously ignite. Although the metal oxidizes in the air, it does so very slowly. Even molten lithium oxidizes so slowly it can be poured from a container in the open air without losing its bright luster.

In an atmosphere of absolutely dry air, lithium metal does not spontaneously burn. When exposed to an ignition source, however, the metal burns in the air with a characteristic crimson color; forming a mixture of lithium oxide and lithium nitride.

$$4\underset{\text{lithium}}{\text{Li}}(s) + \underset{\text{oxygen}}{\text{O}_2}(g) \longrightarrow \underset{\text{lithium oxide}}{2\text{Li}_2\text{O}}(s)$$

$$6\underset{\text{lithium}}{\text{Li}}(s) + \underset{\text{nitrogen}}{\text{N}_2}(g) \longrightarrow \underset{\text{lithium nitride}}{2\text{Li}_3\text{N}}(s)$$

To reduce its potential for ignition, lithium is generally stored under kerosene.

TABLE 9.3 ◆ Some Alkali Metals Regulated by DOT for Transport

Proper Shipping Name	*Label Code*
Alkali metal alloys, liquid, n.o.s.	4.3
Alkali metal amalgam, liquid	4.3
Alkali metal amalgam, solid	4.3
Alkali metal dispersions	4.3
Lithium	4.3
Potassium	4.3
Potassium metal alloys	4.3
Potassium sodium alloys	4.3
Sodium	4.3

FIGURE 9.1 ◆ Metallic sodium is often encountered commercially in the shape of bricks, like those shown here. Individual sodium bricks should only be handled with dry gloves made of loose-fitting canvas or similar material, previously dusted with dry soda ash to remove any adhering moisture. Such bricks are ordinarily immersed in a neutral oil within steel drums. These containers should be stored in a dry, fireproof building or room used exclusively for sodium storage. *Courtesy of E. I. du Pont de Nemours & Company, Wilmington, Delaware.*

SODIUM METAL

Like metallic lithium, sodium is also a soft, silvery bright metal. Of the alkali metals, it is the most commonly encountered and the only one produced in an amount greater than 1 ton per year. Sodium metal is generally available for commercial use as the solid bricks shown in Figure 9.1.

The majority of the metallic sodium produced in the United States was formerly used as a raw material for the manufacture of the vehicular fuel additives tetraethyllead and tetramethyllead. This use of metallic sodium was sharply curtailed in 1975, when EPA banned the use of leaded gasoline. Today, metallic sodium is primarily used as a raw material for the production of metallic titanium or highly reactive sodium compounds such as sodium peroxide and sodium hydride. In addition, metallic sodium is used as a catalyst during the production of certain types of synthetic rubber.

Metallic sodium is sometimes encountered commercially in alloys such as the sodium/potassium alloy, sodium/lead alloy, and sodium amalgam. An *amalgam* is a special alloy of one or more metals that have been mixed with elemental mercury. When using a sodium alloy instead of sodium, the reactions of metallic sodium proceed less vigorously. Consequently, manufacturers generally use a sodium alloy when the rate of a reaction needs careful control.

Metallic sodium reacts very rapidly with water. This reaction occurs so rapidly that the hydrogen produced is unable to dissipate before it ignites. Instead, it concentrates in the immediate vicinity of the metal where, induced by the heat of reaction, it self-ignites and spontaneously burns. The metallic sodium absorbs the heat of reaction and melts, thereby exposing an underlying surface of the solid metal for further reaction.

Unlike lithium, metallic sodium does not react with atmospheric nitrogen. Metallic sodium burns in an atmosphere of oxygen, producing a mixture of sodium oxide and sodium peroxide.

$$4Na(s) + O_2(g) \longrightarrow 2Na_2O(s)$$
$$\text{sodium} \quad\quad \text{oxygen} \quad\quad\quad \text{sodium oxide}$$

$$2Na(s) + O_2(g) \longrightarrow Na_2O_2(s)$$
$$\text{sodium} \quad\quad \text{oxygen} \quad\quad\quad \text{sodium peroxide}$$

Sodium peroxide is an oxidizing agent.

In air, metallic sodium ignites spontaneously at room temperature with a characteristic yellow flame. In absolutely dry air, however, this oxidation does not occur at an appreciable rate, which suggests that the oxidation of metallic sodium in air is triggered by the reaction of metallic sodium and atmospheric water vapor. To reduce its potential for ignition, sodium is generally stored under kerosene.

POTASSIUM METAL

Potassium is a soft, silvery metal. Although it was formerly used with sodium as a heat-exchanger fluid in nuclear reactors, metallic potassium has such relatively few commercial uses that it is rarely encountered.

Metallic potassium reacts with water even more rapidly than sodium. The vigorous nature of this reaction is most likely due to the presence of minute amounts of potassium superoxide (see the following equation) that are produced when potassium burns. The hydrogen produced by the reaction of potassium and water initially concentrates around the metal, where it self-ignites. Then, the heat of reaction triggers the burning of the potassium metal.

The combustion of potassium metal in air is associated with the production of a characteristic purple flame. The combustion product is primarily potassium oxide.

$$\underset{\text{potassium}}{4K(s)} + \underset{\text{oxygen}}{O_2(g)} \longrightarrow \underset{\text{potassium oxide}}{2K_2O(s)}$$

On the other hand, when potassium burns in an atmosphere of pure oxygen, a mixture of potassium oxide, potassium peroxide, and potassium "superoxide" is produced. A *superoxide,* more properly called a *hyperoxide,* is a compound containing the superoxide ion, O_2^-. The production of potassium superoxide is denoted as follows:

$$\underset{\text{potassium}}{K(s)} + \underset{\text{oxygen}}{O_2(g)} \longrightarrow \underset{\text{potassium superoxide}}{KO_2(s)}$$

Metallic superoxides are extraordinarily reactive oxidizing agents. It is their exceptional chemical reactivity that accounts for the pronounced rate of reaction between metallic potassium and water. Potassium superoxide reacts with atmospheric moisture to produce oxygen and hydrogen peroxide.

$$\underset{\text{potassium superoxide}}{2KO_2(s)} + \underset{\text{water}}{2H_2O(l)} \longrightarrow \underset{\text{potassium hydroxide}}{2KOH(aq)} + \underset{\text{hydrogen peroxide}}{H_2O_2(aq)} + \underset{\text{oxygen}}{O_2(g)}$$

As potassium burns in moist air, this added presence of oxygen within the burning zone enhances the rate at which the metal burns.

The reaction between potassium superoxide and atmospheric moisture occurs with such ease it has been utilized commercially as a means of supplying oxygen in self-contained breathing apparatus.

FIGHTING ALKALI METAL FIRES

It is appropriate to first consider two common fire extinguishers that are ineffective on alkali metal fires:

- *Water.* The alkali metals displace flammable hydrogen from water.
- *Carbon dioxide.* When carbon dioxide is used on an alkali metal fire, carbon particulates are produced.

$$\underset{\text{sodium}}{4Na(s)} + \underset{\text{carbon dioxide}}{CO_2(g)} \longrightarrow \underset{\text{sodium oxide}}{2Na_2O(s)} + \underset{\text{carbon}}{C(s)}$$

FIGURE 9.2 ◆ A commercially available portable fire extinguisher for class D fires (left) and an illustration of its use (right). *Courtesy of Ansul Fire Protection, Marinette, Wisconsin.*

Since the reaction is exothermic, the underlying metal usually erupts into flame as the carbon dioxide dissipates.

The use of a dry-chemical or dry-powder fire extinguisher (Figure 9.2) is normally recommended for extinguishing or controlling the spread of alkali metal fires. Nonetheless, caution needs to be exercised when using graphite-based dry powder to extinguish an alkali metal fire. Graphite effectively extinguishes the fire by a smothering action, thereby limiting the amount of atmospheric oxygen and moisture available to the metal. At the high temperatures accompanying alkali metal fires, the graphite may react with the relevant metal to produce compounds called *metallic carbides*. An example of a metallic carbide is sodium carbide.

$$\underset{\text{sodium}}{2Na(s)} + \underset{\text{carbon}}{2C(s)} \longrightarrow \underset{\text{sodium carbide}}{Na_2C_2(s)}$$

Like the alkali metals, the alkali metal carbides react with water to produce flammable products.

When dry powder has been used to extinguish an alkali metal fire, it is always necessary to carefully dispose of the remaining residue. In addition to the presence of the alkali metal carbide, bits and pieces of unburned alkali metal often remain confined under this bed of residue. The use of water on the entire mass of residue could produce flammable substances, whose ignition could rekindle the alkali metal fire. When dry powder has been used to extinguish an alkali metal fire, small quantities of the residue should be removed from the bulk and reacted with water in a controlled fashion at an isolated location.

FIGURE 9.3 ◆ Lith-X dry powder, a graphite-based fire extinguishing agent. In general, the use of Lith-X is recommended for extinguishing class D fires. *Courtesy of Ansul Fire Protection, Marinette, Wisconsin.*

When combating lithium fires, the following two fire extinguishing agents are uniquely recommended for use, both of which effectively function by smothering the fire:

- Lithium chloride, a dry-chemical that extinguishes a lithium fire without subsequently producing other substances
- *Lith-X,* a graphite-based dry-powder that is suitable for use on lithium fires (Figure 9.3)

The environment of the class D fires can be extremely caustic because of the formation of the corresponding metallic oxides, hydroxides, and carbonates. The particulates of such compounds are constituents of the smoke accompanying alkali metal fires. Their inhalation can cause adverse health effects ranging from minor irritation and congestion of the nose, throat, and bronchi to severe lung injury.

Performance Goals for Section 9.3

- Compare the physical and chemical properties of magnesium, zirconium, titanium, aluminum, and zinc.
- Discuss the nature of the fire hazards associated with these metals when they are produced as dusts, powders, chips, turnings, flakes, punchings, borings, ribbons, and shavings during metal forging and machining operations, and particularly when they are coated with combustible cutting oils.
- Identify the industries that use bulk quantities of these combustible metals.
- Discuss the recommended procedures for extinguishing the fires of the combustible metals.

◆ 9.3 COMBUSTIBLE METALS

Magnesium, zirconium, titanium, aluminum, and zinc possess a common hazardous feature. Although bulk pieces of these metals are typically difficult to ignite, their finely divided forms readily burn. They are called *combustible metals,* and their fires are regarded as class D fires.

The finely divided forms of the combustible metals noted in this section are regarded as pyrophoric and water-reactive substances to varying degrees. They include dusts, powders, chips, turnings, flakes, punchings, borings, ribbons, and shavings. These forms are commonly produced during metal forging and machining operations; often, the heat retained from these processes is sufficient to cause the metals to spontaneously ignite. The finely divided forms of metals generated during machining, grinding, boring, and other fabrication processes are also likely to be coated with cutting oils, which could ignite as the primary fuel.

The finely divided forms of combustible metals react with water to produce hydrogen. The spontaneous ignition of the hydrogen triggers the burning of the underlying metal. The rate of hydrogen production is affected by a number of factors including the particle size, distribution and dispersion, purity, and ignition temperature of the metal, as well as the moisture content of the surrounding atmosphere. Some relevant physical properties of these metals are provided in Table 9.4.

DOT regulates the transportation of combustible metals as flammable solids, spontaneously combustible materials, or dangerous-when-wet materials, some examples of which are provided in Table 9.5. DOT requires shippers and carriers to affix FLAMMABLE SOLID, SPONTANEOUSLY COMBUSTIBLE, or DANGEROUS WHEN WET labels, as relevant, on their containers and to post the appropriate FLAMMABLE SOLID, SPONTANEOUSLY COMBUSTIBLE, or DANGEROUS WHEN WET placards on the bulk packaging, freight container, transport vehicle, unit load device, or railcar used to transport them.

MAGNESIUM METAL

Magnesium is an exceptionally lightweight metal; it is thereby often employed in the construction of aircraft, racing cars, transportable machinery, engine parts, automobile frames and bumpers, and other items for which the mass of the object is pertinent. Because of the popularity of magnesium as a structural metal, it is commercially available in a variety of sizes ranging from powders to large ingots.

TABLE 9.4 ◆ Physical Properties of Several Combustible Metals

	Magnesium	*Zirconium*	*Titanium*
Specific gravity	1.74	6.5	4.5
Melting point	1204°F (651°C)	3362°F (1850°C)	3146°F (1800°C)
Boiling point	2025°F (1107°C)	5252°F (3578°C)	5432°F (3262°C)
	Aluminum	*Zinc*	
Specific gravity	2.7	7.14	
Melting point	1220°F (660°C)	787°F (419°C)	
Boiling point	3733°F (2056°C)	1661°F (905°C)	

TABLE 9.5 ◆ Some Combustible Metals Regulated by DOT for Transport

Proper Shipping Name	Label Code
Aluminum, molten	9
Aluminum powder, coated	4.1
Aluminum powder, uncoated	4.3
Aluminum silicon powder, uncoated	4.3
Magnesium granules, coated *particle size not less than 149 microns*	4.3
Magnesium *or* Magnesium alloys *with more than 50% magnesium in pellets, turnings, or ribbons*	4.1
Titanium powder, dry	4.2
Titanium powder, wetted *with not less than 25% water (a visible excess of water must be present) (a) mechanically produced, particle size less than 53 microns; (b) chemically produced, particle size less than 840 microns*	4.1
Titanium sponge granules *or* Titanium sponge powders	4.1
Zinc ashes	4.3
Zinc powder *or* Zinc dust	4.3, 4.2
Zirconium, dry, *coiled wire, finished metal sheets, strip (thinner than 254 microns but not thinner than 18 microns)*	4.1
Zirconium dry, *finished sheets, strip, or coiled wire*	4.1
Zirconium powder, dry	4.2
Zirconium powder, wetted *with not less than 25% water (a visible excess must be present) (a) mechanically produced, particle size less than 53 microns; (b) chemically produced, particle size less than 840 microns*	4.1
Zirconium scrap	4.2
Zirconium suspended in a liquid	3

Magnesium is a very reactive metal. Whereas it reacts slowly with cold water, magnesium reacts rapidly with warm water, producing hydrogen.

$$\underset{\text{magnesium}}{Mg(s)} + \underset{\text{water}}{2H_2O(l)} \longrightarrow \underset{\text{magnesium oxide}}{Mg(OH)_2(s)} + \underset{\text{hydrogen}}{H_2(g)}$$

When raised to its ignition temperature of 1153°F (623°C), metallic magnesium melts and burns vigorously with brilliant, blinding white flames (Figure 9.4). The liquid metal, burning as it flows, drops to lower levels, where it ignites all combustible materials in its pathway.

As magnesium burns, approximately 75% combines with atmospheric oxygen to form magnesium oxide. The remaining 25% combines with atmospheric nitrogen to form magnesium nitride.

$$\underset{\text{magnesium}}{2Mg(s)} + \underset{\text{oxygen}}{O_2(g)} \longrightarrow \underset{\text{magnesium oxide}}{2MgO(s)}$$

$$\underset{\text{magnesium}}{3Mg(s)} + \underset{\text{nitrogen}}{N_2(g)} \longrightarrow \underset{\text{magnesium nitride}}{Mg_3N_2(s)}$$

FIGURE 9.4 ◆ Bumpers, chassis, wheel frames, and other auto-motive parts are often made of a magnesium alloy. When they become engulfed in a fire, such as the car-dealership structural fire illustrated here, the magnesium itself burns, producing an intensely brilliant, hot flame.

It is the combustion of magnesium powder that was responsible for the brilliant spurt of light produced when camera photoflash bulbs were energized. Magnesium powder is also a component of many fireworks (Section 11.9), in which its reactions contribute to brilliant displays of light.

Solved Exercise 9.2

When magnesium burns in air, a mixture having an approximate composition of 75% magnesium oxide and 25% magnesium nitride is produced. Since the composition of the air is approximately 21% oxygen and 78% nitrogen, why doesn't the combustion of magnesium produce a mixture of 21% magnesium oxide and 78% magnesium nitride?

Solution: A mixture of 21% magnesium oxide and 78% magnesium nitride could only be produced *if* the element possessed the same affinity for atmospheric nitrogen as it does for atmospheric oxygen. Other factors being equal, the rate at which magnesium combines with nitrogen would then equal the rate at which it combines with oxygen.

In reality, magnesium does not possess the same affinity for nitrogen as it does for oxygen. Nitrogen is ordinarily regarded as a semi-inert element. In combustion processes, atmospheric nitrogen combines with only the most reactive elements. On the other hand, oxygen combines not only with the reactive elements, but with many elements possessing lesser reactivities. Given this reluctance of nitrogen to combine with other elements, it is plausible to conclude that the rate at which magnesium combines with nitrogen is much slower than the rate at which it combines with oxygen. This results in the production of a mixture of combustion products having an approximate composition of 75% magnesium oxide and 25% magnesium nitride.

ZIRCONIUM METAL

Today, zirconium is used almost exclusively in the nuclear and steel industries. In the nuclear industry, zirconium is used in reactor cores and to clad uranium fuel rods, whereas in the steel industry, zirconium is used to remove oxygen from molten steel. Zirconium dust was also used commercially in some camera photoflash bulbs.

Zirconium dust possesses an autoignition temperature of only 70°F (21°C). Because of this fact, the presence of zirconium dust can constitute a serious fire and explosion risk. When zirconium dust is transferred from one container to another, the dust particles absorb the heat generated by friction as the particles move against each another. When the temperature of the dust rises to its autoignition point, it spontaneously ignites.

When zirconium dust burns in the air, the resulting fire provides an exceedingly brilliant, white flame. The combustion results in the production of a mixture of zirconium oxide and zirconium nitride, as follows:

$$\underset{\text{zirconium}}{Zr(s)} + \underset{\text{oxygen}}{O_2(g)} \longrightarrow \underset{\text{zirconium oxide}}{ZrO_2(s)}$$

$$\underset{\text{zirconium}}{2Zr(s)} + \underset{\text{nitrogen}}{N_2(g)} \longrightarrow \underset{\text{zirconium nitride}}{2ZrN(s)}$$

The reactions may be initiated by the combustion of hydrogen, produced when the dust reacts with atmospheric moisture.

$$\underset{\text{zirconium}}{Zr(s)} + \underset{\text{water}}{2H_2O(g)} \longrightarrow \underset{\text{zirconium oxide}}{ZrO_2(s)} + \underset{\text{hydrogen}}{2H_2(g)}$$

TITANIUM METAL

Metallic titanium is 45% lighter in mass than steel, but nonetheless, it is just as strong as steel. Since it possesses this combination of lightness and strength, titanium is often alloyed with aluminum and vanadium and used to manufacture military aircraft parts, jet engines, and missiles.

Titanium metal is virtually immune to corrosive attack. This is due to the affinity that titanium has for oxygen. Once exposed to air, the surface of the metal is quickly passivated with a thin layer of titanium(IV) oxide, which serves to protect the underlying metal from further chemical attack. This passivated metal is highly resistant to corrosion by most acids, chlorine, oxidizing agents, and seawater.

The resistance of titanium to chemical attack is put to use in the following ways:

- Since it is resistant to corrosion by seawater, metallic titanium can be used in the hulls of submarines and in underwater machinery.
- Since it is resistant to corrosion by most acids and chlorine, metallic titanium is also used in the construction of vessels in which these raw materials will be stored, transported, or reacted. One notable exception, however, is hydrofluoric acid, which chemically attacks titanium.

The production of metallic titanium is costly, which currently limits its widespread use. Notwithstanding the elevated cost, titanium is found in everyday items such as jewelry, skis, and golf equipment. Osteopathic surgeons also choose titanium for hip and knee replacements.

Whereas none of these bulk forms of titanium is considered hazardous, finely divided titanium is a pyrophoric metal that poses a dangerous fire and explosion risk.

When metallic titanium burns in air, a mixture of titanium(IV) oxide and titanium(III) nitride is produced.

$$\text{Ti}(s) \; + \; \text{O}_2(g) \; \longrightarrow \; \text{TiO}_2(s)$$
titanium   oxygen   titanium(IV) oxide

$$2\text{Ti}(s) \; + \; \text{N}_2(g) \; \longrightarrow \; 2\text{TiN}(s)$$
titanium   nitrogen   titanium(III) nitride

The reactions may be initiated by the combustion of the hydrogen produced when the finely divided metal reacts with atmospheric moisture.

$$\text{Ti}(s) \; + \; 2\text{H}_2\text{O}(l) \; \longrightarrow \; \text{TiO}_2(s) \; + \; 2\text{H}_2(g)$$
titanium   water   titanium(IV) oxide hydrogen

Solved Exercise 9.3

Why is the installation of an automatic sprinkling system recommended in machine shops that use lathes to fabricate titanium parts?

Solution: Various-sized pieces of hot metal are produced when titanium parts are mechanically fabricated. These pieces include titanium dust, powder, chips, turnings, flakes, ribbons, and shavings, all of which are prone to self-ignite. The likelihood of ignition is greatest when the pieces are first generated, since at this time, they retain the heat of friction. Once the metal ignites, it is essential to extinguish the fire as quickly as possible to prevent its spread to adjoining areas. This can be effectively accomplished through use of an automatic sprinkling system that applies water to the fire in an amount sufficient to rapidly cool the burning metal.

ALUMINUM METAL

Table 4.1 lists aluminum as the most abundant metal on Earth's surface. As the element, aluminum is too reactive to be found in an uncombined form in nature. Instead, we find it in such materials as clay and feldspar, in which it is combined with silicon and oxygen.

Aluminum is an example of a *malleable* metal; that is, it can be rolled into a relatively thin sheet or foil. Firefighters encounter aluminum foil bonded to fabric in the aluminized protective coats commonly called "silvers", they are recommended for use when responding to fires that release high levels of radiant heat, such as bulk flammable liquid or flammable gas fires. Aluminum foil is also a popular kitchen item.

Since it is lightweight and durable, aluminum sheeting is used to produce a wide variety of commercial products including the common soda can. In building construction, aluminum sheeting is used as siding, eaves, screens, and window and door frames. During building fires, these aluminum components can melt and collapse, since the temperature achieved often exceeds the melting point of aluminum, 1220°F (660°C).

The principal material employed as the metal skin of most standard aircraft is an aluminum or aluminum alloy sheeting. The metal cannot be used as the outer skin of supersonic aircraft, however, since it becomes too hot and softens from the friction generated by the fast movement through the air.

Aluminum is also a *ductile* substance; that is, it can be drawn into wires. Although aluminum wire is twice as effective as copper wire for conducting electricity, the use of aluminum electrical wiring is largely not favored. The unpopularity is linked with the heavy deposit of aluminum oxide produced on aluminum wiring, which restricts the flow of electrical current and causes the metal to become overly heated. This situation constitutes a fire hazard.

The deposition of aluminum oxide on the surface of aluminum wiring is connected with the extraordinary affinity that aluminum has for oxygen. Aluminum exposed to air is covered with a thin, tenacious coating of aluminum oxide that gives the metal a dull, white luster. While this oxide coating protects the underlying metal from further oxidation, the coating does not protect aluminum from other forms of chemical attack. Seawater, for instance, corrodes aluminum.

This chemical affinity of metallic aluminum for oxygen is evident from the chemical reaction described in Figure 9.5 between powdered aluminum and iron(III) oxide. The mixture of 27% powdered aluminum and 73% iron(III) oxide is commonly called *thermite*. When the mixture is activated, a reaction producing iron and aluminum oxide occurs.

$$2Al(s) \ + \ Fe_2O_3(s) \ \longrightarrow \ 2Fe(l) \ + \ Al_2O_3(s)$$

aluminum   ferric oxide   iron   aluminum oxide

The phenomenon is called the *thermite reaction*. It releases such considerable heat that temperatures of approximately 3990°F (2199°C) result. Since this is above the melting point of iron [2800°F (1538°C)], the iron produced by this chemical reaction is a molten, white-hot liquid. In the days of the old West, the thermite reaction was put to use to weld rails together during the construction of railroads.

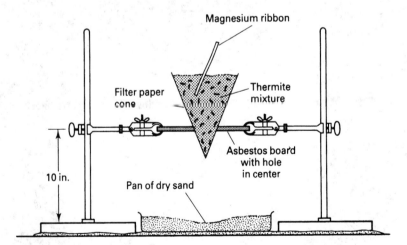

FIGURE 9.5 ◆ The thermite reaction can be illustrated in a chemical laboratory as follows: A mixture of powdered aluminum and iron(III) oxide is inserted into a cone over a pan of dry sand, which protects the tabletop from possible damage. A magnesium ribbon is inserted into the thermite mixture and ignited. The heat of combustion initiates the reaction between aluminum and iron(III) oxide. Molten iron spits from the reaction mixture and drips into the sand.

Whereas elemental aluminum is very stable in the form of foil and sheets, aluminum dust and powder are pyrophoric materials that pose the risk of fire and explosion. The aluminum burns violently in air with an intensely bright, white and orange flame. A mixture of aluminum oxide and aluminum nitride is produced.

$$4Al(s) \ + \ 3O_2(g) \longrightarrow \ 2Al_2O_3(s)$$
$$\text{aluminum} \qquad \text{oxygen} \qquad\qquad \text{aluminum oxide}$$

$$2Al(s) \ + \ N_2(g) \longrightarrow \ 2AlN(s)$$
$$\text{aluminum} \qquad \text{nitrogen} \qquad\qquad \text{aluminum nitride}$$

The reactions may be initiated by the combustion of hydrogen, produced when the dust and powder react with atmospheric moisture.

$$2Al(s) \ + \ 3H_2O(l) \longrightarrow \ Al_2O_3(s) \ + \ 3H_2(g)$$
$$\text{aluminum} \qquad \text{water} \qquad\qquad \text{aluminum oxide} \qquad \text{hydrogen}$$

Upon contact with liquid oxygen, powdered aluminum burns spontaneously as shown in Figure 9.6. Aluminum oxide is the sole product of combustion.

The reactivity of aluminum powder is put to use in many fireworks (Section 11.9), in which the metal reacts to produce a brilliant display of orange light. It is also incorporated into some paints and varnishes, but consideration must be given to their use, since these coatings behave as flammable solids once the paint solvent has evaporated. Aluminum powder is also a component of solid rocket fuels, in which it is mixed with

FIGURE 9.6 ◆ This four-nozzle flamethrower-like device is burning a flowing mixture of liquid oxygen and powdered aluminum. The intensity of the flames illuminates the night sky. *Courtesy of the United States Department of Defense.*

ammonium nitrate and ammonium perchlorate. The mixture of powdered aluminum and ammonium nitrate is an explosive called *ammonal* (Section 11.11).

The climax of the German dirigible *Hindenberg* (Figure 7.8) may have been linked with the combustion of aluminum powder. The exterior surface of the dirigible consisted of a cloth cover impregnated with a doping mixture of aluminum powder and ferric oxide. The presence of aluminum powder provided a surface having high reflectivity. The aluminum particles reflected heat off the vessel and prevented the hydrogen from expanding. The premise is that the aluminum powder first caught fire at an isolated location, perhaps triggered by static electricity. Once initiated, the fire then rapidly spread across the entire covering, ultimately igniting the reserves of hydrogen. The resulting inferno consumed the vessel.

In circumstances where the temperature is substantially elevated compared to the norm, even bulk aluminum acts as a fast-burning fuel. The skin of the Space Shuttle, for example, must be armored with heat shielding to protect the shuttle when it re-enters Earth's atmosphere from outer space. During this reentry it could experience temperatures in excess of 3000°F (1650°C). If this shielding is pierced in any way, the underlying aluminum becomes superheated. Aluminum melts at 1220°F (660°C) and vaporizes at 3733°F (2056°C). At these temperatures, aluminum fires occur.

In 2003, the Space Shuttle *Columbia* exploded on reentry into our atmosphere, killing the seven astronauts aboard. The shuttle was covered with more than 20,000 interlocking ceramic tiles designed to protect the aluminum alloy shell from the heat of reentry. Experts who examined debris from the accident wreckage observed droplets of aluminum and stainless steel. This observation caused them to link the cause of the accident with the loss of tiles on the left wing, especially along its leading edge. Without its protective covering, the underlying aluminum alloy burned, ultimately engulfing the entire shuttle.

ZINC METAL

Zinc is a relatively common metal used frequently for industrial purposes. Most commonly, zinc is used to coat iron and protect it from corrosion. The zinc-coated iron is said to be *galvanized*. Zinc is also used as a component of a number of alloys. For example, zinc and copper combine to form the alloy *brass*. Metallic zinc is also used in dry-cell batteries and a variety of structural materials.

Zinc powder and dust are pyrophoric materials whose presence poses a fire and explosion hazard. They ignite spontaneously in air with a green flame, producing zinc oxide as the sole combustion product.

$$2Zn(s) \ + \ O_2(g) \ \longrightarrow \ 2ZnO\,(s)$$
$$\text{zinc} \qquad \text{oxygen} \qquad\qquad \text{zinc oxide}$$

The reaction may be initiated by the combustion of hydrogen, produced when the dust and powder react with atmospheric moisture.

$$Zn(s) \ + \ H_2O(l) \ \longrightarrow \ ZnO(s) \ + \ H_2(g)$$
$$\text{zinc} \qquad \text{water} \qquad\qquad \text{zinc oxide} \qquad \text{hydrogen}$$

FIGHTING FIRES INVOLVING COMBUSTIBLE METALS

Firefighters are frequently forewarned against using water on combustible metal fires. Notwithstanding this generally sound advice, water can effectively extinguish combustible metal fires when the following two conditions are met:

◆ The water must be discharged in a volume that totally deluges the fire scene and cools the metal.

◆ The water must be discharged rapidly, and soon after the fire first begins.

The fulfillment of these conditions may be impossible in certain circumstances. Consideration must always be given as to whether enough water is available at the fire scene, and whether an appropriate means is available to rapidly apply the water to the fire.

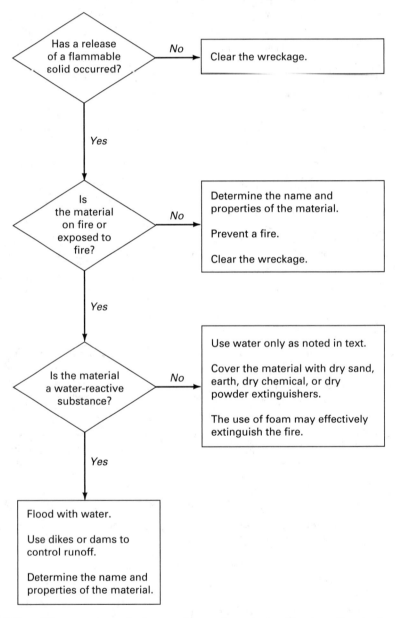

FIGURE 9.7 ◆ The recommended procedures when responding to a disaster involving a flammable solid. *Adapted with permission of the American Society for Testing and Materials, from a figure in ASTM STP 825, A Guide to the Safe Handling of Hazardous Materials Accidents. Copyright © 1983, American Society for Testing and Materials.*

Some combustible metals react with water very slowly at room temperature but very vigorously at elevated temperatures. For example, at ambient temperatures magnesium reacts only slowly with water, but the reaction is vigorous when the magnesium is hot.

$$Mg(s) + 2H_2O(l) \longrightarrow Mg(OH)_2(s) + H_2(g)$$
magnesium   water   magnesium hydroxide   hydrogen

When water is applied too slowly to a magnesium fire, it cannot adequately cool the metal. The evolved hydrogen ignites and causes the metal fire to be periodically rekindled.

The application of carbon dioxide is not recommended for extinguishing combustible metal fires, since hot magnesium reacts with carbon dioxide.

$$2Mg(s) + CO_2(g) \longrightarrow 2MgO(s) + C(s)$$
magnesium   carbon dioxide   magnesium oxide   carbon

Since this reaction is exothermic, the use of carbon dioxide does not cool the burning metal and the fire is not extinguished.

When extinguishing combustible metal fires, experts recommend the implementation of the procedures in Figure 9.7. Experts recommend the use of a deluging volume of water or dry sand, earth, dry chemicals, or dry powders. Special extinguishing agents are commercially available for use on specific combustible metal fires; for example, zirconium silicate is commercially available for extinguishing zirconium fires.

Since combustible metal fires frequently burn with exceptionally brilliant flames, firefighters should be aware that the radiant energy could damage the retinas of their eyes. They should also avoid breathing the smoke evolved during these fires, as it contains tiny particulates of caustic metallic oxides. When inhaled, these particles cause considerable discomfort and localized lung injury.

Performance Goals for Section 9.4
- Discuss the chemistry of the aluminum alkyl compounds, noting especially that they are pyrophoric, water-reactive, spontaneously combustible, and highly toxic substances.
- Identify the industries at which aluminum alkyl compounds are used.

◆ 9.4 ALUMINUM ALKYL COMPOUNDS

The *aluminum alkyls* are a group of compounds whose molecules are composed of an aluminum atom covalently bonded to three carbon atoms of an alkyl group (Section 12.2). Two examples of aluminum alkyl compounds are triethylaluminum and tri(isobutyl)aluminum, in which "ethyl" and "isobutyl" are the alkyl groups, respectively. Some physical properties of these compounds are provided in Table 9.6.

Aluminum alkyls are commercially used by the chemical industry as polymerization catalysts. One such catalyst is a mixture of titanium(IV) chloride and an aluminum alkyl. It is called the *Ziegler–Natta catalyst,* after Karl Ziegler and Giulio Natta, the chemists who discovered it.

Aluminum alkyls are also used as incendiary agents by the military. Trimethylaluminum has been used for production of a luminous trail in the upper atmosphere for tracking the location of rockets. Triethylaluminum has also been used as an incendiary agent in flamethrowers (Figure 9.8) (see Solved Exercise 9.4).

TABLE 9.6 ◆ Physical Properties of Aluminum Alkyl Compounds

	Triethylaluminum $(CH_3)_3Al$	*Tri(isobutyl)aluminum* $[(CH_3)_2CHCH_2]_3Al$
Melting point	−63°F (−53°C)	39°F (4°C)
Boiling point	381°F (194°C)	237°F (114°C)
Specific gravity	0.837	0.788
Flashpoint	−63°F (−53°C)	<32°F (<0°C)

The commercially available aluminum alkyl compounds are pyrophoric, violently water-reactive, and highly toxic liquids. The spontaneous combustion of triethylaluminum, for example, produces aluminum oxide, carbon dioxide, and water vapor.

$$2(C_2H_5)_3Al(g) + 21O_2(g) \longrightarrow Al_2O_3(s) + 12CO_2(g) + 15H_2O(g)$$

triethylaluminum   oxygen   aluminum oxide   carbon dioxide   water

Triethylaluminum also reacts vigorously with water, producing flammable ethane.

$$2Al(C_2H_5)_3(l) + 3H_2O(l) \longrightarrow Al_2O_3(s) + 6C_2H_6(g)$$

triethylaluminum   water   aluminum oxide   ethane

FIGURE 9.8 ◆ A United States Navy Zippo flamethrower is tested from a patrol boat during the Vietnam encounter. The active component in these flamethrowers are pyrophoric substances such as trimethylaluminum. *Courtesy of the United States Department of Defense.*

Once generated, the ethane bursts into flame and burns with the unhydrolyzed triethyl-aluminum. Considerable heat is evolved to the environment as these reactions occur.

Bulk quantities of aluminum alkyl compounds burn so vigorously and persistently that they pose a substantial fire risk. The extraordinary heat of combustion associated with their burning necessitates that firefighters wear special protective gear like the "silvers" in Figure 9.9 when combating these fires.

To warn users that these chemical products possess an inordinately high degree of reactivity, chemical manufacturers label each container of an aluminum alkyl compound as follows:

> **SPONTANEOUSLY FLAMMABLE**
> **IN AIR**
> **HIGHLY FLAMMABLE LIQUID**
> **REACTS VIOLENTLY WITH WATER**
> **MOISTURE SENSITIVE**
> **STORE UNDER INERT GAS**
> **INCOMPATIBLE WITH AIR,**
> **MOISTURE, WATER,**
> **OXIDIZING AGENTS, AND**
> **ALCOHOLS**

To prevent their accidental ignition, the aluminum alkyl compounds must be stored in electrically grounded containers under an atmosphere of nitrogen in a cool, well-ventilated area.

Solved Exercise 9.4

During the Vietnam conflict, the U.S. military used triethylaluminum as the active component of a portable encapsulated flame system for controlling the extent of unwanted vegetation and to rout the enemy from jungles and other densely vegetated areas. What is the most logical reason the Department of Defense selected triethylaluminum as an incendiary agent for this use?

   Solution: When discharged into the open air, triethylaluminum sponta-neously bursts into flame. This property gives rise to its potential use in an item intended to rapidly discharge a stream of fire upon demand, like a portable en-capsulated flame system. The most logical reason the Department of Defense selected triethylaluminum as the active component of a portable flamethrower is associated with the military's desire to use a substance that could sponta-neously burst into flame upon demand.

For extinguishing aluminum alkyl fires, the use of dry chemical powder pressur-ized with nitrogen, vermiculite, or dry sand is recommended.

DOT regulates the transportation of the generic group of aluminum alkyls and their hydrides and halides as spontaneously combustible materials. DOT requires ship-pers and carriers to affix SPONTANEOUSLY COMBUSTIBLE and DANGEROUS

FIGURE 9.9 ◆ Water-reactive aluminum alkyl compounds are highly pyrophoric. Consequently, when responding to fire incidents in which they are involved, special protective clothing must be worn. This specialized aluminum suit reflects heat and provides protection against potential bodily contact with the reactive substance. *Courtesy of Ethyl Corporation, Baton Rouge, Louisiana.*

WHEN WET labels on their containers and to post SPONTANEOUSLY COM-
BUSTIBLE placards, if relevant, on the bulk packaging, freight container, transport
vehicle, unit load device, or railcar used to transport them.

Performance Goals for Section 9.5
- ◆ Discuss the chemistry of the three types of metallic hydrides, especially their water
  reactivity.
- ◆ Identify the industries in which metallic hydrides are used.

◆ 9.5 METALLIC HYDRIDES

Approximately 10 metallic hydrides are encountered commercially. They are used as
powerful reducing agents by the chemical industry. These compounds can be classified
according to their general chemical composition as simple metallic hydrides, metallic
borohydrides, and complex aluminum hydrides.

SIMPLE METALLIC HYDRIDES

These compounds are composed of metallic ions with hydride (H^-) ions. They are
lithium hydride, sodium hydride, calcium hydride, magnesium hydride, and aluminum
hydride, whose chemical formulas are LiH, NaH, CaH_2, MgH_2, and AlH_3, respectively. They are produced from the metal and hydrogen. For example, sodium hydride
is a simple metallic hydride produced by the union of sodium metal and hydrogen.

$$2Na(s) + H_2(g) \longrightarrow 2NaH(s)$$

$\quad\quad\text{sodium}\quad\quad\text{hydrogen}\quad\quad\quad\text{sodium hydride}$

METALLIC BOROHYDRIDES

These compounds are composed of metallic ions with borohydride (BH_4^-) ions. They
are lithium borohydride, sodium borohydride, and aluminum borohydride, whose
chemical formulas are $LiBH_4$, $NaBH_4$, and $Al(BH_4)_3$, respectively. These substances
are produced by relatively complex chemical reactions. Sodium borohydride, for example, is prepared from sodium hydride and trimethyl borate, as follows:

$$4NaH(s) + (CH_3)_3BO_3(l) \longrightarrow NaBH_4(s) + 3NaOCH_3(s)$$

$\quad\text{sodium hydride}\quad\quad\text{trimethyl borate}\quad\quad\quad\text{sodium borohydride}\quad\text{sodium methoxide}$

COMPLEX ALUMINUM HYDRIDES

These compounds are composed of metallic ions with aluminum hydride (AlH_4^-) ions.
They are lithium aluminum hydride and sodium aluminum hydride, whose chemical
formulas are $LiAlH_4$ and $NaAlH_4$, respectively. These substances are produced from
the relevant metallic hydride and anhydrous aluminum chloride (Section 9.8). Lithium
aluminum hydride, for example, is prepared from lithium hydride and anhydrous aluminum chloride.

$$4LiH(s) + AlCl_3(s) \longrightarrow LiAlH_4(s) + 3LiCl(s)$$

$\quad\text{lithium hydride}\quad\text{aluminum chloride}\quad\quad\text{lithium aluminum hydroxide}\quad\text{lithium chloride}$

The reaction is conducted in diethyl ether (Section 12.9), which acts as a solvent.

WATER REACTIVITY OF THE METALLIC HYDRIDES

At room temperature, the metallic hydrides are relatively stable, solid compounds. Nonetheless, they possess several common hazardous features. Of special interest here is the fact that these compounds react with water to produce flammable hydrogen. To avoid contact with atmospheric moisture, all metallic hydrides should be stored in tightly sealed containers.

When the metallic hydrides are encountered commercially, they are often covered with petroleum oil. The presence of the oil lends an element of safety when handling and storing metallic hydrides. The metallic hydrides are also encountered as ethereal solutions; that is, they are dissolved in diethyl ether (Section 12.9), a highly flammable liquid. The combination of diethyl ether and a metallic hydride poses a twofold risk of fire and explosion.

The following equations illustrate the water reactivity of several representative metallic hydrides:

$$\underset{\text{lithium hydride}}{LiH(s)} + \underset{\text{water}}{H_2O(l)} \longrightarrow \underset{\text{lithium hydroxide}}{LiOH(aq)} + \underset{\text{hydrogen}}{H_2(g)}$$

$$\underset{\text{sodium borohydride}}{3NaBH_4(s)} + \underset{\text{water}}{6H_2O(l)} \longrightarrow \underset{\text{sodium borate}}{3NaBO_2(aq)} + \underset{\text{hydrogen}}{12H_2(g)}$$

$$\underset{\text{lithium aluminum hydride}}{LiAlH_4(s)} + \underset{\text{water}}{4H_2O(l)} \longrightarrow \underset{\text{aluminum hydroxide}}{Al(OH)_3(s)} + \underset{\text{lithium hydroxide}}{LiOH(aq)} + \underset{\text{hydrogen}}{4H_2(g)}$$

$$\underset{\text{aluminum borohydride}}{Al(BH_4)_3(s)} + \underset{\text{water}}{12H_2O(l)} \longrightarrow \underset{\text{aluminum hydroxide}}{Al(OH)_3(s)} + \underset{\text{boric acid}}{3H_3BO_3(aq)} + \underset{\text{hydrogen}}{12H_2(g)}$$

As metallic hydrides react with water, the evolved hydrogen absorbs the heat of reaction and spontaneously bursts into flame. To extinguish these fires, chemical manufacturers recommend the use of dry powder (Figure 9.10).

DOT regulates the transportation of metallic hydrides as spontaneously combustible materials or dangerous-when-wet materials, some examples of which are provided in Table 9.7. DOT requires shippers and carriers to affix SPONTANEOUSLY

TABLE 9.7 ◆ Metallic Hydrides Regulated by DOT for Transport

Proper Shipping Name	Label Code
Aluminum borohydride *or* Aluminum borohydride in devices	4.2, 4.3
Aluminum hydride	4.3
Calcium hydride	4.3
Lithium aluminum hydride	4.3
Lithium aluminum hydride, ethereal	4.3, 3
Lithium borohydride	4.3
Lithium hydride	4.3
Lithium hydride, fused solid	4.3
Potassium borohydride	4.3
Sodium aluminum hydride	4.3
Sodium borohydride	4.3
Sodium hydride	4.3

SODIUM ALUMINUM HYDRIDE

FIREFIGHTING MEASURES

No standard methods have been developed for extinguishing large-scale sodium aluminum hydride fires. Small fires may be extinguished using dry powder, provided that a large excess of dry powder is used. Water should be used only on small amounts of sodium aluminum hydride. Water-based foams, chemical foams, and halogenated extinguishers should NOT be used.

The following general procedures are recommended for spills involving portable containers of sodium aluminum hydride:

(a)   Water may be used (if one is thoroughly familiar with the violent reaction and if personnel are properly protected from possible explosion) to decompose sodium aluminum hydride from small spills and to keep adjacent tanks or equipment cool.

(b)   If spillage from a container is burning, move flammables away or move the container to a safe place, if possible. Avoid increasing the spill. Keep water, materials wet with water, and liquid fire extinguishers away. Small fires can be controlled with dry materials or with water, if one is thoroughly familiar with the violent reaction and if personnel are properly protected from possible explosion. If possible use sandbags to contain and isolate the burning material.

(c)   If material is spilled from a container but is not burning, ventilate the area or move the container out in the open; otherwise, proceed as in (b).

The following procedure is recommended for handling sodium aluminum hydride when there are fires from an external source in the vicinity:

If a container is threatened by fire from an external source (not from the chemical substance itself), extinguish the external fire by conventional means, or move the container, whichever can be done quicker and with less danger. If this cannot be done, keep the container cool by spraying water on it. Water reacts violently with sodium aluminum hydride but has been found to be an effective fire control treatment if one is thoroughly familiar with the violent reaction and personnel and properly protected from explosion.

FIGURE 9.10 ◆ This excerpt from the Material Safety Data Sheet for sodium aluminum hydride cautions emergency responders against using water to combat fires involving sodium aluminum hydride. This information applies as well to extinguishing fires involving other metallic hydrides. *Courtesy of Albemarle Corporation, Baton Rouge, Louisiana.*

COMBUSTIBLE and/or DANGEROUS WHEN WET labels on their containers. DOT also requires shippers and carriers to post SPONTANEOUSLY COMBUSTIBLE placards, when relevant, or DANGEROUS WHEN WET placards on the bulk packaging, freight container, transport vehicle, unit load device, or railcar used to transport them.

◆ 9.6 METALLIC PHOSPHIDES

Metallic phosphides are produced by combining a given metal with elemental phosphorus. Calcium phosphide, for example, is formed by heating calcium and phosphorus.

$$6Ca(s) \ + \ P_4(s) \ \longrightarrow \ 2Ca_3P_2(s)$$

$$\text{calcium} \qquad \text{phosphorus} \qquad\qquad \text{calcium phosphide}$$

These compounds were once popular fumigants used on grain and other postharvest crops, but in the United States, their use is now nearly nonexistent.

The metallic phosphides function as fumigants by reacting with atmospheric moisture to produce the toxic gas *phosphine*.

$$Ca_3P_2(s) \ + \ 6H_2O(l) \longrightarrow 3Ca(OH)_2(s) + 2PH_3(g)$$

$$\text{calcium phosphide} \qquad \text{water} \qquad \text{calcium hydroxide} \qquad \text{phosphine}$$

When calcium phosphide is applied within an enclosure used for the storage of crops, it is the phosphine that kills mice and other unwanted pests.

DOT regulates the transportation of metallic phosphides as dangerous-when-wet materials or poisons, some examples of which are provided in Table 9.8. DOT requires shippers and carriers to affix DANGEROUS WHEN WET and/or POISON labels on their containers and to post DANGEROUS WHEN WET or POISON placards on the bulk packaging, freight container, transport vehicle, unit load device, or railcar used to transport them.

Solved Exercise 9.5

What is the most likely reason DOT requires packages of tin(IV) phosphide, or stannic phosphide, to be labeled FLAMMABLE SOLID and DANGEROUS WHEN WET when offered for domestic shipment?

Solution: Metallic phosphides are solid compounds that react with water to produce the flammable and toxic substance phosphine.

$$Sn_3P_4(s) \ + \ 12H_2O(l) \longrightarrow 3Sn(OH)_4(s) + 4PH_3(g)$$

$$\text{stannic phosphide} \qquad \text{water} \qquad \text{stannic hydroxide} \qquad \text{phosphine}$$

The high degree of flammability for phosphine is evident from the fact that its autoignition temperature is only 100°F (37.8°C). The high degree of toxicity of this gas can be estimated from the OSHA regulation that limits employee exposure in the workplace to a concentration of 0.3 ppm, averaged over the 8-hour workday. Atmospheric moisture induces the reaction between stannic phosphide and water, which poses an additional element of concern.

Based on these facts, the most likely reason DOT requires packages of tin(IV) phosphide to be labeled FLAMMABLE SOLID and DANGEROUS WHEN WET for domestic shipment is to quickly forewarn emergency-response and transportation personnel that a flammable and toxic substance could be generated when stannic phosphide is involved in a transportation mishap.

PHOSPHINE

This gas can be readily identified by its exceptionally offensive odor, which some describe as a mixture of garlic and rotten fish. It is fortunate that this putrefying odor is detectable at minutely low concentrations, since phosphine is an extraordinarily poisonous gas. When inhaled, it causes pulmonary edema and massive destruction of the lung tissues. Long-term exposure causes the bones to soften.

Phosphine is also spontaneously flammable. Its autoignition temperature is only 100°F (37.8°C), a value readily achieved in most environments. Phosphine burns in air to produce tetraphosphorus decoxide and water vapor.

$$4PH_3(g) + 8O_2(g) \longrightarrow \underset{\text{tetraphosphorus decoxide}}{P_4O_{10}(s)} + \underset{\text{water}}{6H_2O(g)}$$
$$\underset{\text{phosphine}}{} \underset{\text{oxygen}}{}$$

Phosphine is available commercially although it is not widely used. When it is present in the workplace, OSHA requires employers to limit employee exposure to a concentration of 0.3 ppm, averaged over the 8-hour workday.

DOT regulates the transportation of phosphine as a poison gas. DOT requires shippers and carriers to affix INHALATION HAZARD and FLAMMABLE GAS labels on its containers and to post INHALATION HAZARD placards on the bulk packaging, freight container, transport vehicle, unit load device, or railcar used to transport it.

TABLE 9.8 ◆ Metallic Phosphides Regulated by DOT for Transport

Proper Shipping Name	Label Codes
Aluminum phosphide	4.3, 6.1
Aluminum phosphide pesticides	6.1
Calcium phosphide	4.3, 6.1
Magnesium aluminum phosphide	4.3, 6.1
Magnesium phosphide	4.3, 6.1
Potassium phosphide	4.3, 6.1
Sodium phosphide	4.3, 6.1
Stannic phosphide	4.3, 6.1
Zinc phosphide	4.3, 6.1

◆ 9.7 METALLIC CARBIDES

The compounds consisting of metallic ions with carbon ions that exist as C_2^{2-}, C^{4-}, or C_3^{4-} are called *metallic carbides*. Calcium carbide and aluminum carbide are the only two commercially important metallic carbides. DOT regulates their transportation as dangerous-when-wet materials. DOT requires shippers and carriers to affix DANGEROUS WHEN WET labels on their containers and to post DANGEROUS WHEN WET placards on the bulk packaging, freight container, transport vehicle, unit load device, or railcar used to transport them.

CALCIUM CARBIDE

Calcium carbide is manufactured by heating a mixture of coke and lime in an electric furnace at an elevated temperature.

$$CaO(s) \ + 3C(s) \longrightarrow \ CaC_2(s) \ + \ CO(g) \ [T > 3600°F \ (1982°C)]$$

calcium oxide   carbon   calcium carbide   carbon monoxide

The major use of calcium carbide is as a raw material for the production of welding-grade acetylene.

$$CaC_2(s) \ + 2H_2O(l) \longrightarrow Ca(OH)_2(s) \ + C_2H_2(g)$$

calcium carbide   water   calcium hydroxide   acetylene

Because of its water-reactive nature, calcium carbide that has been exposed to humid air poses a flammable and explosive hazard. For this reason, it must be stored and handled within a dry environment free of ignition sources.

> **Solved Exercise 9.6**
>
> Why do chemical manufacturers recommend that steel drums containing calcium carbide be protected against the buildup of electrostatic charges?
>
> **Solution:** Calcium carbide is a water-reactive material. Atmospheric moisture sealed inside the steel drums could react with the product to produce a concentration of acetylene sufficient to pose the risk of fire and explosion. When the drums are unsealed, the presence of an electrostatic charge could then ignite the acetylene. To prevent such an incident, chemical manufacturers urge users of calcium carbide to ground the drums and prevent the buildup of electrostatic charges.

ALUMINUM CARBIDE

Aluminum carbide is prepared by heating aluminum oxide with coke in an electric furnace. It is used as a catalyst by the chemical industry. Aluminum carbide is also a water-reactive substance, forming flammable methane as a hydrolysis product.

$$\underset{\text{aluminum carbide}}{Al_4C_3(s)} + \underset{\text{water}}{12H_2O(l)} \longrightarrow \underset{\text{aluminum hydroxide}}{4Al(OH)_3(s)} + \underset{\text{methane}}{3CH_4(g)}$$

Aluminum carbide that has been exposed to humid air poses a flammable and explosive hazard.

Performance Goals for Section 9.8
- ◆ Discuss the chemistry of the water-reactive substances that form hydrogen chloride as a product of their hydrolysis.
- ◆ Use Table 8.7 to discuss the adverse health effects caused by the inhalation of various concentrations of hydrogen chloride.
- ◆ Identify the industries at which these water-reactive substances are used.

◆ 9.8 WATER-REACTIVE SUBSTANCES THAT PRODUCE HYDROGEN CHLORIDE

Certain substances react with water to produce hydrogen chloride vapor or hydrochloric acid as products of their hydrolysis. When they are encountered, these substances routinely possess the suffocating, pungent odor of hydrogen chloride, which fumes in the air and limits visibility. As we noted in Section 8.7, hydrogen chloride is a toxic, irritating gas, and hydrochloric acid is a corrosive liquid. In the following discussion, the hydrolysis product is denoted solely as hydrogen chloride vapor.

DOT regulates the transportation of these substances as corrosive materials, poisons, or dangerous-when-wet materials, some examples of which are provided in Table 9.9. DOT requires shippers and carriers to affix CORROSIVE, POISON, and/or

TABLE 9.9 ◆ Some Miscellaneous Water-Reactive Compounds Regulated by DOT for Transport

Proper Shipping Name	Label Codes
Aluminum chloride, anhydrous	8
Methyldichlorosilane	4.3, 8, 3
Methyltrichlorosilane	3, 8
Phosphorus oxychloride	8, 6.1
Phosphorus pentachloride	8
Phosphorus trichloride	6.1, 8
Silicon tetrachloride	8
Stannic chloride, anhydrous	8
Sulfuryl chloride	8, 6.1
Thionyl chloride	8
Titanium tetrachloride	8, 6.1
Trichlorosilane	4.3, 3, 8

DANGEROUS WHEN WET labels, as relevant, on their containers. DOT also requires them to post CORROSIVE, POISON, or DANGEROUS WHEN WET placards on the bulk packaging, freight container, transport vehicle, unit load device, or railcar used to transport these hazardous materials.

ALUMINUM CHLORIDE (ANHYDROUS)

This compound is a white-to-yellow solid whose chemical formula is $AlCl_3$. In the chemical industry substantial quantities are used as catalysts and as a raw material for the production of aluminum alkyl compounds and lithium aluminum hydride. It is also used to produce antiperspirants.

Anhydrous aluminum chloride reacts violently with water to produce hydrogen chloride.

$$2AlCl_3(s) \ + \ 3H_2O(l) \longrightarrow \ Al_2O_3(s) \ + \ 6HCl(g)$$

aluminum chloride   water   aluminum oxide   hydrogen chloride

For this reason, the manufacturers and distributors of this substance caution that it could irritate the skin, eyes, and respiratory tract (Figure 9.11).

PHOSPHORUS OXYCHLORIDE

This substance, also called phosphoryl chloride, is a colorless, fuming liquid whose chemical formula is $POCl_3$. It is used primarily by the chemical industry as a chlorinating agent.

Phosphorus oxychloride reacts violently with water to produce hydrogen chloride.

$$POCl_3(l) \ + \ 3H_2O(l) \longrightarrow H_3PO_4(aq) + \ 3HCl(g)$$

phosphorus oxychloride   water   phosphoric acid   hydrogen chloride

---

ANHYDROUS ALUMINUM CHLORIDE

HAZARDS IDENTIFICATION

CAUTION!

MAY CAUSE IRRITATION TO SKIN, EYES, AND RESPIRATORY TRACT

STABILITY AND REACTIVITY

Stability: Stable under ordinary conditions of use and storage.

Hazardous Decomposition Products: May emit toxic chloride fumes when heated to decomposition.

Hazardous Polymerization: Will not occur.

Incompatibilities: Water and moist air.

Conditions to Avoid: Water

FIGURE 9.11 ◆ These excerpts from the Material Safety Data Sheet for anhydrous aluminum chloride note the hazardous properties of this water-reactive substance. *Adapted from the Material Safety Data Sheet for Aluminum Chloride, Mallinckrodt Baker, Inc., Phillipsburg, New Jersey.*

TABLE 9.10 ◆ Physical Properties of Some Chlorosilanes			
	Methyldichlorosilane	*Methyltrichlorosilane*	*Trichlorosilane*
Specific gravity	1.10	1.27	1.34
Boiling point	106°F (41°C)	149°F (66.4°C)	90°F (32°C)
Flashpoint	−26°F (−15°C)	8°F (−13°C)	7°F (−14°C)

PHOSPHORUS TRICHLORIDE

This substance is a colorless, fuming liquid whose chemical formula is PCl_3. It is used in the chemical industry as a chlorinating agent and catalyst. Phosphorus trichloride is used, for example, as a raw material for producing acetyl chloride (Section 9.9).

Phosphorus trichloride reacts with water to form phosphorous acid and hydrogen chloride.

$$PCl_3(l) + 3H_2O(l) \longrightarrow H_3PO_3(aq) + HCl(g)$$
phosphorus trichloride water phosphorous acid hydrogen chloride

PHOSPHORUS PENTACHLORIDE

This substance is a yellow-to-green solid whose chemical formula is PCl_5. It is primarily used as a chlorinating and dehydrating agent by the chemical industry.

Phosphorus pentachloride is decomposed by water in a multistep process that is summarized as follows:

$$PCl_5(s) + 4H_2O(l) \longrightarrow H_3PO_4(aq) + 5HCl(g)$$
phosphorus pentachloride water phosphoric acid hydrogen chloride

SILICON TETRACHLORIDE AND CHLORINATED SILANES

Silicon tetrachloride is a colorless, fuming liquid whose chemical formula is $SiCl_4$. It is used as a raw material to manufacture liquid or semisolid silicon-containing polymers known as *silicones*, which are widely used as electrical insulation. Silicon tetrachloride is also used in the semiconductor manufacturing industry.

Silicon tetrachloride reacts with water to form silicic acid and hydrogen chloride.

$$SiCl_4(l) + 4H_2O(l) \longrightarrow H_4SiO_4(aq) + 4HCl(g)$$
silicon tetrachloride water silicic acid hydrogen chloride

Other than silicon tetrachloride, several organic compounds containing silicon and chlorine are water-reactive. They are called *chlorosilanes*. Three examples of chlorosilanes are methyldichlorosilane, methyltrichlorosilane, and trichlorosilane. Several representative properties of these substances are noted in Table 9.10.

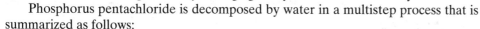

Solved Exercise 9.7

Why should emergency-response personnel use protective gear including self-contained breathing apparatus when responding to a domestic transportation mishap involving a spill of trichlorosilane?

> **Solution:** Trichlorosilane is a flammable and water-reactive substance. Upon reacting with atmospheric moisture, hydrogen chloride is produced. As we noted in Table 8.7, the inhalation of this vapor causes a variety of adverse health effects. Since the prolonged breathing of hydrogen chloride poses an inhalation toxicity hazard, the use of self-contained breathing apparatus is essential.

Exposure to the chlorosilanes irritates the respiratory tract, skin, and eyes. When elevated concentrations are inhaled, pulmonary edema could result. Vapor burns to the eyes cause lesions and blindness.

The chlorosilanes produce hydrogen chloride and hydrochloric acid upon exposure to water. For example, the water reactivity of methyltrichlorosilane is denoted as follows:

$$\underset{\text{methyltrichlorosilane}}{CH_3SiCl_3(l)} + \underset{\text{water}}{3H_2O(l)} \longrightarrow \underset{\text{methylsilanetriol}}{CH_3Si(OH)_3(l)} + \underset{\text{hydrogen chloride}}{3HCl(g)}$$

SULFURYL CHLORIDE

This substance, also called sulfonyl chloride, is a colorless liquid whose chemical formula is SO_2Cl_2. Sulfuryl chloride is mainly used in the chemical industry as a chlorinating and dehydrating agent.

Sulfuryl chloride reacts with water to form sulfuric acid and hydrogen chloride.

$$\underset{\text{sulfuryl chloride}}{SO_2Cl_2(l)} + \underset{\text{water}}{2H_2O(l)} \longrightarrow \underset{\text{sulfuric acid}}{H_2SO_4(aq)} + \underset{\text{hydrogen chloride}}{2HCl(g)}$$

THIONYL CHLORIDE

This is a red-to-yellow liquid whose chemical formula is $SOCl_2$. Thionyl chloride is used mainly within the chemical industry.

Thionyl chloride reacts vigorously with water to form sulfurous acid and hydrogen chloride, as follows:

$$\underset{\text{thionyl chloride}}{SOCl_2(l)} + \underset{\text{water}}{2H_2O(l)} \longrightarrow \underset{\text{sulfurous acid}}{H_2SO_3(aq)} + \underset{\text{hydrogen chloride}}{2HCl(g)}$$

TIN(IV) CHLORIDE (ANHYDROUS)

This substance, also known as stannic tetrachloride, is a colorless, fuming liquid whose chemical formula is $SnCl_4$. It is used to manufacture blueprint and similarly sensitized types of paper.

Anhydrous tin(IV) chloride reacts with water to form tin(IV) oxide and hydrogen chloride, as follows:

$$\underset{\text{tin(IV) chloride}}{SnCl_4(l)} + \underset{\text{water}}{2H_2O(l)} \longrightarrow \underset{\text{tin(IV) oxide}}{SnO_2(s)} + \underset{\text{hydrogen chloride}}{4HCl(g)}$$

TITANIUM(IV) CHLORIDE (ANHYDROUS)

Titanium(IV) chloride, or titanium tetrachloride, is a colorless, volatile liquid whose chemical formula is $TiCl_4$. It is used as a paint pigment and a polymerization catalyst. As we noted in Section 9.4, the mixture of titanium(IV) chloride and an aluminum alkyl compound is an important polymerization catalyst.

Titanium(IV) chloride reacts with water to produce titanium(IV) dioxide and hydrogen chloride.

$$\underset{\text{titanium(IV) chloride}}{TiCl_4(l)} + \underset{\text{water}}{2H_2O(l)} \longrightarrow \underset{\text{titanium(IV) oxide}}{TiO_2(s)} + \underset{\text{hydrogen chloride}}{4HCl(g)}$$

◆ 9.9 WATER-REACTIVE COMPOUNDS THAT PRODUCE ACETIC ACID

When acetic acid is formed as a product of a chemical reaction, its highly irritating and pungent odor is immediately evident. This can pose serious consequences, as the inhalation of acetic acid vapor is suffocating and exposure to the eyes and nose is severely irritating.

We shall briefly note here two organic compounds that react with water to produce acetic acid vapor: acetic anhydride and acetyl chloride. Some physical properties of these compounds are provided in Table 9.11.

ACETIC ANHYDRIDE

This is a colorless, fuming liquid whose chemical formula is $(CH_3CO)_2O$. Acetic anhydride is principally used by the chemical, pharmaceutical, and polymer industries for the manufacture of aspirin, cellulose acetate (Section 13.5), and related products.

When acetic anhydride combines with water, the sole product produced is acetic acid.

$$CH_3-C(=O)-O-C(=O)-CH_3\,(l) + H_2O(l) \longrightarrow 2\,CH_3-C(=O)-OH\,(g)$$

acetic anhydride   water   acetic acid

Since the inhalation of acetic acid is suffocating, breathing becomes difficult when responding to an emergency involving a major spill or leak of acetic anhydride.

When acetic anhydride is used in the workplace, OSHA requires employers to limit employee exposure to a concentration of 5 ppm, averaged over the 8-hour workday.

DOT regulates the transportation of acetic anhydride as a corrosive material. DOT requires shippers and carriers to affix CORROSIVE and FLAMMABLE LIQUID labels on its containers and to post FLAMMABLE placards, if relevant, on

TABLE 9.11 ◆ Physical Properties of Two Water-Reactive Organic Compounds

	Acetic Anhydride	*Acetyl Chloride*
Specific gravity	1.08	1.1
Vapor density (air = 1)	3.5	2.7
Melting point	−99°F (−73°C)	−170°F (−112°C)
Boiling point	284°F (140°C)	124°F (51°C)
Flashpoint	129°F (54°C)	40°F (4°C)

the bulk packaging, freight container, transport vehicle, unit load device, or railcar used to transport it.

ACETYL CHLORIDE

This is a colorless, fuming liquid whose chemical formula is CH_3COCl. Acetyl chloride is used principally within the chemical industry. It is produced by various means, one of which involves the reaction between phosphorus trichloride and acetic acid.

$$PCl_3(l) \ + \ 3CH_3-\overset{\displaystyle O}{\underset{\displaystyle OH}{\overset{\|}{C}}}(l) \ \longrightarrow \ 3CH_3-\overset{\displaystyle O}{\underset{\displaystyle Cl}{\overset{\|}{C}}}(l) \ + \ H_3PO_3(l)$$

phosphorus trichloride   acetic acid   acetyl chloride   phosphorous acid

Upon contact with water, acetyl chloride reacts violently, producing acetic acid *and* hydrochloric acid vapor.

$$CH_3-\overset{\displaystyle O}{\underset{\displaystyle Cl}{\overset{\|}{C}}}(l) \ + \ H_2O(l) \ \longrightarrow \ CH_3-\overset{\displaystyle O}{\underset{\displaystyle OH}{\overset{\|}{C}}}(g) \ + \ HCl(g)$$

acetyl chloride   water   acetic acid   hydrogen chloride

Table 9.11 reveals that acetyl chloride possesses a relatively low flashpoint. Consequently, the presence of acetyl chloride poses both a fire and health risk. DOT regulates the transportation of acetyl chloride as a flammable liquid and corrosive material. DOT requires shippers and carriers to affix FLAMMABLE LIQUID and CORROSIVE labels on its containers and to post FLAMMABLE placards, if relevant, on the bulk packaging, freight container, transport vehicle, unit load device, or railcar used to transport it.

■■■

REVIEW EXERCISES

Chemical Reactions Involving Water

9.1 Identify the labels that DOT requires shippers and carriers to affix on the packaging of water-reactive substances when offering them for domestic transportation.

9.2 Using Table 9.1, write a balanced equation illustrating the water reactivity of fluorosulfonic acid.

9.3 Oleum (Section 8.5) is an acid that strongly fumes in humid air. What substances are present in these fumes?

Alkali Metals

9.4 Using Tables 6.1 and 6.4, provide the basic description DOT requires shippers and carriers to provide on the accompanying shipping paper when they intend to transport by truck twenty 1-lb canisters of elemental sodium.

9.5 When metallic sodium is briefly exposed to moist air, three white compounds form on its surface. Write the balanced equations that illustrate their formation.

9.6 Write balanced equations illustrating the combustion of metallic potassium in oxygen.

9.7 Why does metallic potassium react at an explosive rate with kerosene, while metallic lithium and sodium are stored in bottles of kerosene without incident?

Combustible Metals

9.8 The OSHA regulation at 29 CFR §1910.157(d)(6) stipulates that employers must provide portable fire extinguishers for class D hazards in working areas where combustible metal powders, flakes, shavings, or similarly sized materials are generated at least once every 2 weeks. Why is this regulation likely to be applicable within work areas where magnesium scrap is periodically generated?

9.9 To prevent their ignition, most combustible materials can be stored within an enclosed area containing an atmosphere of nitrogen. Why is this practice inadvisable for storing freshly generated titanium borings?

9.10 Why is the fabrication of zirconium strips frequently accomplished while the metal is submerged under water?

9.11 Why is the installation of an automatic deluge system often recommended in magnesium plants?

9.12 When titanium(III) nitride is used to coat drill and router bits, the tools last several times longer than standard bits. How is titanium(III) nitride produced?

9.13 Why are zirconium dust and powder always kept wet in storage?

9.14 When a fire investigator wishes to establish the temperature achieved during the burning of a high-rise building, why is the condition of the aluminum door frames closely examined?

9.15 The paint formerly used on some storage tanks contained aluminum powder, which reflected heat from the surface of the tanks. Why did experts recommend grounding these tanks and storing only noncombustible material in them?

Aluminum Alkyl Compounds

9.16 Write a balanced equation illustrating the combustion in air of trimethylaluminum, a pyrophoric fuel.

9.17 The DOT regulation at 49 CFR §173.181 requires shippers and carriers to use specific types of packaging for transporting a pyrophoric liquid like triethylaluminum. Two satisfactory types of packaging are metal cans and steel drums that remain sealed by means of friction-free closures. What is the most likely reason DOT requires a friction-free closure on the packaging used for the domestic shipment of triethylaluminum?

Metallic Hydrides

9.18 Write a balanced equation illustrating the water reactivity of lithium hydride.

9.19 What is the most likely reason DOT limits the quantity of sodium aluminum hydride transported on domestic cargo aircraft to 25 lb (11 kg)?

9.20 A manufacturer in Charleston, South Carolina; intends to ship fifty 1-lb bottles containing lithium aluminum hydride by motor carrier to a customer in Baltimore, Maryland. Using Tables 6.1 and 6.4, what basic description does DOT require the manufacturer and carrier to enter on the accompanying shipping paper?

Metallic Phosphides

9.21 Write a balanced equation illustrating the water reactivity of strontium phosphide.

9.22 What is the most likely reason DOT prohibits railroad carriers from accepting for domestic transport any quantity of aluminum phosphide within a passenger-carrying railcar?

9.23 What is the most likely reason DOT limits the quantity of zinc phosphide transported on domestic cargo aircraft to 100 lb (45.4 kg)?

Metallic Carbides

9.24 Why is calcium carbide sometimes called calcium acetylide?

9.25 When relatively small amounts of magnesium burn, the fires are extinguished using commercially available dry powder. Why might this fire rekindle days later, as someone tries to remove the residue with water?

9.26 What is the most likely reason DOT prohibits the domestic transportation of any quantity of calcium carbide within a passenger-carrying railcar?

Substances That React with Water to Produce Hydrogen Chloride

9.27 Using Tables 6.1 and 6.4, provide the basic description DOT requires shippers and carriers to enter on the accompanying shipping paper when transporting 1 L of anhydrous tin(IV) chloride by means of a passenger-carrying railcar.

9.28 During World War II, anhydrous titanium tetrachloride was used in offensive and defensive operations as a means of concealing the movement of troops and installations within a combat zone.
  (a) Describe how the use of anhydrous titanium tetrachloride effectively deceived the enemy and prevented them from locating the military's presence. (Hint: Titanium dioxide is a white solid.)
  (b) When using anhydrous titanium tetrachloride, why were military personnel concerned about the humidity of the air within the combat zone?
  (c) Which compound was responsible for the acrid smell in the air when using anhydrous titanium tetrachloride?

9.29 What is the most logical reason DOT prohibits airline carriers from accepting for transport any quantity of phosphorus oxychloride by passenger-carrying aircraft?

9.30 Write a balanced equation illustrating the combustion of trichlorosilane. The chemical formula of trichlorosilane is $SiHCl_3$.

Substances That React with Water to Produce Acetic Acid

9.31 Why do chemical manufacturers recommend grounding a railroad tankcar containing acetyl chloride before transferring the product into a storage tank?

9.32 When responding to a bulk spill of acetyl chloride, why should emergency responders prevent the liquid from entering nearby streams and lakes?

Chemistry of Some Toxic Substances  10 CHAPTER

Certain substances, when absorbed even in relatively small quantities, can incapacitate or cause death, illness, or injury. They are called *toxic substances,* or *poisons.* We shall use these two terms interchangeably in this chapter.

Toxic substances are routinely encountered by firefighters and other emergency responders. For example, the toxic gas carbon monoxide is encountered whenever a burning building is entered. Other toxic substances such as asbestos and lead are encountered within the dust that is generated as old buildings burn or are otherwise wracked by the forces associated with fires. To firefighters, the properties of these toxic substances comprise an especially important segment of the study of hazardous materials.

In this chapter, we focus primarily on the following four topics relating to the study of toxic substances:

- The general features of all toxic substances, especially the factors that cause them to be poisonous
- The properties of seven toxic gases—carbon monoxide, hydrogen cyanide, sulfur dioxide, hydrogen sulfide, nitrogen dioxide, and ammonia
- The properties of asbestos, lead, anhydrous ammonia, and pesticides, all of which are toxic substances
- The practices recommended for effectively responding to incidents in which toxic gases and toxic substances have been released

Performance Goals for Section 10.1
- Describe the means used by DOT and OSHA to identify a toxic substance when it is encountered during transit and in the workplace, respectively.
- Describe the nature of the DOT regulations as they apply to the transportation of toxic substances.
- Describe the concept of toxicity as the term is used by EPA in the RCRA regulations.

◆ 10.1 TOXIC SUBSTANCES AND GOVERNMENT REGULATIONS

Although the layperson has a general perception of the meaning of a poison, the term is utilized in a specific manner in government regulations. We shall briefly review how toxicity affects transportation, workplace, and environmental regulations.

TOXICITY AND THE DOT REGULATIONS

In Section 6.7, we first noted that DOT regulates the transportation of two types of toxic substances: a "poison gas" and a "poison." A poison gas is denoted as a division 2.3 hazardous material, while a poison is regarded as a division 6.1 hazardous material. DOT distinguishes between them because the poison gas can pose a substantially greater danger during a transportation mishap. Some examples are provided in Table 10.1.

When either a poison gas or poison is transported, DOT requires shippers and carriers to comply with the labeling, marking, and placarding requirements noted in Chapter 6. In addition to the general requirements, DOT obligates shippers and carriers to disclose that a hazardous material is poisonous by entering "Poison—Inhalation Hazard," "poison," or "toxic" in the basic description of the relevant hazardous material.

TOXICITY AND THE OSHA REGULATIONS

When the use of toxic substances is needed in the workplace, OSHA requires employers to ensure that employees are exposed to no more than certain maximum allowable concentrations, averaged over the 8-hour workday.

In addition, when toxic substances are stored within the workplace, OSHA requires employers to tag, label, or otherwise identify their presence by the use of the accident-prevention tags, warning labels, and worded signs in Figure 10.1. Each warning form includes an imprint of the skull-and-crossbones, the internationally recognized symbol denoting a poison.

THE TOXICITY CHARACTERISTIC AND THE RCRA REGULATIONS

As first noted in Section 1.5, EPA uses the authority of RCRA to regulate the treatment, storage, and disposal of a certain type of waste called a *hazardous waste*. A given waste is a hazardous waste if a representative sample exhibits any of the four characteristics

FIGURE 10.1 ◆ Types of accident-prevention tags, warning labels, and worded signs required by OSHA when toxic substances are present within the workplace.

TABLE 10.1 ◆ Some Toxic Substances Regulated by DOT for Transport

Proper Shipping Name	*Label Codes*
Ammonia, anhydrous	6.1, 8 (I); 2.2 (D)
Ammonia solutions, *relative density between 0.880 and 0.957 at 59°F (15°C) in water with more than 10% but not more than 35% ammonia*	8
Ammonia solutions, *relative density less than 0.880 at 59°F (15°C) in water with more than 35% but not more than 50% ammonia*	2.2
Ammonia solutions, *relative density less than 0.880 at 15°C in water with more than 50 percent ammonia*	6.1, 8 (I); 2.2 (D)
Asbestos	9
Carbamate pesticides, liquid, flammable, toxic, *flashpoint less than 73°F (23°C)*	3, 6.1
Carbamate pesticides, liquid, toxic	6.1
Carbamate pesticides, liquid, toxic, flammable, *flashpoint not less than 73°F (23°C)*	6.1, 3
Carbamate pesticides, solid, toxic	6.1
Carbon monoxide, compressed	2.3, 2.1
Carbon monoxide, refrigerated liquid *(cryogenic liquid)*	2.3, 2.1
Copper cyanide	6.1
Dinitrogen tetroxide	2.3, 5.1, 8
Hydrocyanic acid solutions *or* Hydrogen cyanide, aqueous solutions *with not more than 20% hydrogen cyanide*	6.1
Hydrocyanic acid, aqueous solutions *with less than 5% hydrogen cyanide*	6.1
Hydrogen cyanide, solution in alcohol *with not more than 45% hydrogen cyanide*	6.1, 3
Hydrogen cyanide, stabilized *with less than 3% water*	6.1, 3
Hydrogen cyanide, stabilized *with less than 3% water and absorbed in a porous inert material*	6.1
Hydrogen sulfide	2.3, 2.1
Nitric oxide, compressed	2.3, 5.1, and 8
Organochlorine pesticides liquid, flammable, toxic, *flashpoint less than 73°F (23°C)*	3, 6.1
Organochlorine pesticides, liquid, toxic	6.1
Organochlorine pesticides, liquid, toxic, flammable, *flashpoint not less than 73°F (23°C)*	6.1, 3
Organochlorine pesticides, solid, toxic	6.1
Organophosphorus pesticides, liquid, flammable, toxic, *flashpoint less than 73°F (23°C)*	3, 6.1
Organophosphorus pesticides, liquid, toxic	6.1

(continued)

TABLE 10.1 ◆ (*continued*)

Proper Shipping Name	*Label Codes*
Organophosphorus pesticides, liquid, toxic, flammable, *flashpoint not less than 73°F (23°C)*	6.1, 3
Organophosphorus pesticides, solid, toxic	6.1
Potassium cyanide	6.1
Sodium cyanide	6.1
Sulfur dioxide	2.3, 8
Zinc cyanide	6.1

ignitability, corrosivity, reactivity, or toxicity. In this section, we are concerned with the toxicity characteristic.

The determination as to whether a representative sample of a waste exhibits the toxicity characteristic is made by a chemical analyst. In practice, the analyst conducts certain specified test procedures to determine whether the sample contains one or more constituents in Table 10.2 at concentrations equal to or greater than the relevant threshold value. Each of the listed substances is known to adversely affect public health and the environment. When the sample is found to contain at least one constituent at a concentration equal to or above the listed value, the waste is said to exhibit the characteristic of *toxicity*.

DOT requires shippers and carriers to identify a hazardous waste that exhibits the toxicity characteristic in the basic description entered on a waste manifest. This is accomplished by the use of the term "EPA toxicity" or the relevant EPA hazardous

Solved Exercise 10.1

Using required testing procedures, an analyst determines that a representative liquid waste sample contains lead at a concentration of 12 mg/L.

(a) Does the waste exhibit the RCRA characteristic of toxicity for lead?

(b) When transporting 500 gal (1.9 m³) of the waste for domestic treatment or disposal, what basic description do DOT and EPA require on the accompanying transportation manifest?

Solution: (a) The regulatory threshold concentration of lead is noted in Table 10.2 as 5.0 mg/L. Since the lead concentration in the waste sample exceeds the regulatory threshold concentration, the waste exhibits the RCRA characteristic of toxicity for lead.

(b) Based on the information in Table 6.4, DOT and EPA require either of the following as the basic description of this waste on the accompanying transportation manifest:

500 gal   Hazardous waste liquid, n.o.s., 9, UN3082 (EPA toxicity)

500 gal   Hazardous waste liquid, n.o.s., 9, UN3082 (D008)

TABLE 10.2 ◆ Waste Contaminants Examined for the RCRA Toxicity Characteristic[a]

EPA Hazardous Waste Number	Contaminant	Regulatory Level (mg/L)
D004	Arsenic	5.0
D005	Barium	100.0
D018	Benzene	0.5
D006	Cadmium	1.0
D019	Carbon tetrachloride	0.5
D020	Chlordane	0.03
D021	Chlorobenzene	100.0
D022	Chloroform	6.0
D007	Chromium	5.0
D023	*o*-Cresol	200.0[b]
D024	*m*-Cresol	200.0[b]
D025	*p*-Cresol	200.0[b]
D026	Cresol	200.0[b]
D016	2,4-D[c]	10.0
D027	1,4-Dichlorobenzene	7.5
D028	1,2-Dichloroethane	0.5
D029	1,1-Dichloroethylene	0.7
D030	2,4-Dinitrotoluene	0.13
D012	Endrin	0.02
D031	Heptachlor (and its epoxide)	0.008
D032	Hexachlorobenzene	0.13
D033	Hexachlorobutadiene	0.5
D034	Hexachloroethane	3.0
D008	Lead	5.0
D013	Lindane	0.4
D009	Mercury	0.2
D014	Methoxychlor	10.0
D035	Methyl ethyl ketone	200.0
D036	Nitrobenzene	2.0
D037	Pentachlorophenol	100.0
D038	Pyridine	5.0
D010	Selenium	1.0
D011	Silver	5.0
D039	Tetrachloroethylene	0.7
D015	Toxaphene	0.5

(continued)

TABLE 10.2 ◆ (*continued*)

EPA Hazardous Waste Number	Contaminant	Regulatory Level (mg/L)
D040	Trichloroethylene	0.5
D041	2,4,5-Trichlorophenol	400.0
D042	2,4,6-Trichlorophenol	2.0
D017	2,4,5-TP (Silvex)[d]	1.0
D043	Vinyl chloride	0.2

[a]Adapted from 40 CFR §261.24.
[b]If *o-*, *m-*, and *p*-cresol concentrations cannot be differentiated, the total cresol (D026) concentration is used. The regulatory level of total cresol is 200 mg/L.
[c]2,4-D is 2,4-dichlorophenoxyacetic acid (p. 569).
[d]2,4,5-TP is 2,4,5-trichlorophenoxyacetic acid (p. 569).

waste number provided in Table 10.2. For example, when a shipper offers for domestic transportation a waste that exhibits toxicity due to its arsenic concentration, the waste characteristic is properly denoted on the relevant waste manifest by either "EPA toxicity" or "D004." The basic shipping description for this waste is either of the following:

Hazardous waste liquid, n.o.s., 9, UN3082, III (EPA toxicity)

Hazardous waste liquid, n.o.s., 9, UN3082, III (D004)

Performance Goal for Section 10.2
   ◆ Describe the general means by which toxic substances enter the human organism.

◆ 10.2 HOW TOXIC SUBSTANCES ENTER THE BODY

A toxic substance can enter the body by various routes, but we shall be concerned here with only three: ingestion, skin absorption, and inhalation.

INGESTION

This route refers to the swallowing of a substance through the mouth into the stomach and its subsequent movement through the gastrointestinal tract (Figure 10.2). Once ingested, a substance can pass through the intestinal walls and into the circulatory system, where the substance or its metabolic products are further disseminated to the organs and tissues of the body.

SKIN ABSORPTION

The skin, a cross section of which is illustrated in Figure 10.3, constitutes the largest single organ of the human body. The skin helps the body maintain a normal temperature. It also protects the internal organs and prevents direct contact between them and foreign substances. While the skin acts in this way as an organ of defense, it also acts as a permeable membrane. Some foreign substances penetrate through the epidermis,

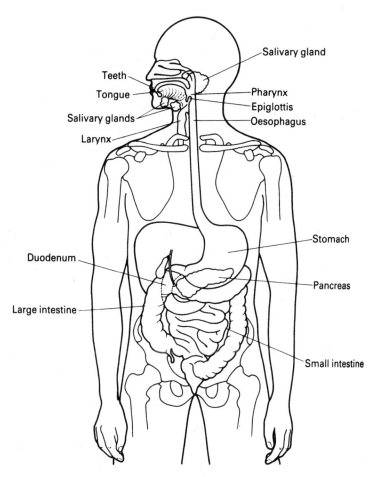

FIGURE 10.2 ◆   The human digestive system, illustrating some of its major components. A substance taken orally passes through the mouth, into the esophagus, and then into the stomach. The substance may undergo chemical alteration to some degree, since partial digestion occurs in the mouth and stomach. It may even pass through the stomach wall directly into the bloodstream. But generally, it is not until the substance or its degradation products have passed into the small intestine that absorption into the bloodstream occurs.

the outermost layer, and enter the underlying dermis, from which it may be further absorbed into the circulatory system and spread throughout the body.

INHALATION

This route is responsible for the movement of gases, vapors, and fumes through the components of the respiratory system illustrated in Figure 10.4. These include the nose, pharynx, larynx, trachea, bronchi, and lungs. When an inhaled substance enters the lungs, it is exposed to blood vessels that cover an average surface area of approximately 90 yd^2 (75 m^2). Given this massive exposure, the substance is absorbed into the bloodstream very rapidly and then circulated to other organs and tissues of the body.

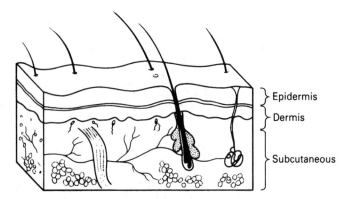

FIGURE 10.3 ◆ The cross section of human skin showing some of its principal components. The outer layer of the skin is the epidermis, a thin surface membrane of dead cells. For a substance to be absorbed through the skin, it must first penetrate through the epidermis. If this occurs, the substance enters the dermis, a collection of cells that acts as a porous diffusion medium. Here the substance may be further absorbed into the bloodstream.

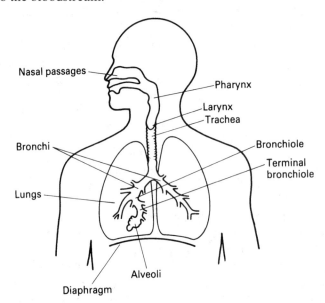

FIGURE 10.4 ◆ The human respiratory system, illustrating its major parts. Inhaled vapors and gases, including air, first pass either through the nose or mouth and then through the pharynx, where they enter the trachea (commonly called the windpipe) at the larynx. Next, inhaled substances enter either of two bronchi, each of which leads to a lung. The individual divisions of each bronchus are called bronchioles.

Performance Goals for Section 10.3

- Describe generally how a toxic substance adversely affects health.
- Identify the nature of the following toxicants: hemotoxicants, hepatotoxicants, nephrotoxicants, neurotoxicants, and reproductive toxicants.
- Discuss the mechanisms that cause asphyxia in humans who have inhaled elevated concentrations of gases.
- Describe how exposure to a substance can result in irritation of the tissues of the respiratory system or skin.

◆ 10.3 SOME COMMON WAYS TOXIC SUBSTANCES ADVERSELY AFFECT HEALTH

Once a toxic substance has been absorbed into the body, it acts in the following general ways:

- It may cause immediate impairment or death.
- It may cause a localized effect at the site of contact.
- It may target the body as a whole. This phenomenon is said to be a *systemic effect*.
- It may target only a specific organ, which experiences localized dysfunction or impairment. When a toxic substance targets a specific organ, the substance is denoted accordingly as one or more of the following:
  - *Hemotoxicant,* a substance that decreases the function of the blood's hemoglobin and deprives the tissues of oxygen
  - *Hepatotoxicant,* a substance that causes liver damage
  - *Nephrotoxicant,* a substance that causes kidney damage
  - *Neurotoxicant,* a substance that affects the central nervous system
  - *Reproductive toxicant,* a substance that adversely affects an individual's reproductive capabilities

ASPHYXIANTS

One way by which a gas or vapor adversely affects health is to induce unconsciousness when inhaled. The substance is called an *asphyxiant.* Common asphyxiants are nitrogen and carbon dioxide.

There are four common means by which an asphyxiant causes unconsciousness in humans. Each means is noted individually below:

- *Denial of sufficient oxygen.* As we first noted in Section 2.5, a gas possessing a vapor density greater than 1 displaces the air within the lower regions of an enclosure. An example of such a gas is carbon dioxide. A person who inhales carbon dioxide loses consciousness since the body is denied sufficient oxygen. When provided fresh air or oxygen, the individual could regain consciousness. If unattended, however, the individual could die from suffocation or experience damage to the central nervous system.

Solved Exercise 10.2

Before removing an underground gasoline storage tank from its location, the American Petroleum Institute recommends purging it with nitrogen or carbon dioxide. Immediately following this purging:

(a) Is it generally safe to implement cutting or welding operations on the tank in connection with its removal?

(b) Is it generally safe for an individual to enter the tank?

Solution: **(a)** When gasoline vapor has been purged from a storage tank with nitrogen or carbon dioxide, its concentration is no longer within the flammable range. For this reason, cutting or welding operations on the tank immediately after conducting the purging can generally be conducted without fear of fire or explosion. Nonetheless, gasoline vapor may enter a supposedly empty tank from connecting pipe lines. For this reason, immediately prior to conducting any work on the tank, it is always prudent to check the vapor concentration therein using a portable combustible gas monitor.

(b) A tank that has been purged with nitrogen or carbon dioxide is deficient in air. An individual who inhales the atmosphere of nitrogen or carbon dioxide experiences asphyxiation. It is rarely safe for an individual to enter the storage tank immediately following the purging operation, since the individual could die from suffocation or experience damage to the central nervous system.

- *Action as a hemotoxicant.* An asphyxiant like carbon monoxide primarily causes unconsciousness by preventing the normal transmission of oxygen in the bloodstream. Although the exposed individual could regain consciousness when provided fresh air or oxygen, we shall see in Section 10.8 that short-term exposures to certain concentrations of carbon monoxide are fatal.
- *Inability to utilize cellular oxygen.* An asphyxiant such as hydrogen cyanide primarily causes unconsciousness by hindering the cells' ability to utilize oxygen. Exposure to fresh air or oxygen may not revive an individual who has inhaled hydrogen cyanide. The victims of hydrogen cyanide poisoning generally survive only when promptly administered an appropriate antidote.

IRRITANTS

The inhalation of a gas or vapor may also impact the body as an *irritant* by injuring the tissues that it contacts. When a low concentration of an irritant is inhaled, the result is often minor inflammation of the tissues that line the respiratory passages. This happens when we inhale a relatively small amount of hydrogen chloride or ammonia. On the other hand, when we inhale a larger concentration of the same substance, we could experience respiratory failure. Whether a substance causes respiratory irritation or more severe harm is often associated with the degree to which it corrodes the nose, pharynx, larynx, trachea, bronchi, and lungs.

Performance Goal for Section 10.4
- Describe the nature of an acute, chronic, short-term, and latent effect caused by exposure of the human organism to a toxic substance.

◆ 10.4 TYPES OF TOXICOLOGICAL EFFECTS

The harmful health effects associated with exposure to substances are often classified as acute, chronic, short-term, or latent. When these substances are constituents of commercial chemical products, these effects are noted by the manufacturer on labels and MSDSs (Section 1.6).

ACUTE HEALTH EFFECT

An *acute health effect* is manifested when an injury, disease, or death that is caused by exposure to a substance develops rapidly and quickly comes to a crisis. An example of an acute effect is the tearing of the eyes that immediately occurs from a short-term exposure to ammonia vapor.

CHRONIC HEALTH EFFECT

A *chronic health effect* is manifested when an injury, disease, or death that is caused by exposure to a substance develops slowly over a few days, weeks, or longer periods. Examples of chronic effects are coronary heart disease and cancer.

SHORT-TERM HEALTH EFFECT

A *short-term health effect* is manifested when an injury or disease of relatively short duration is caused by exposure to a substance *and* from which recovery occurs rapidly. For example, when an individual becomes asphyxiated from exposure to carbon dioxide but recovers when provided with oxygen or air, the asphyxiation is called a short-term health effect.

LATENT HEALTH EFFECT

A *latent health effect* is manifested when an injury or disease caused by exposure to a substance occurs *only* after an extensive time period has passed. For example, the development of liver angiosarcoma by individuals exposed to vinyl chloride vapor is a latent health effect, since the appearance of malignant tumors on the liver occurs decades after they inhale the vapor. A number of cancerous and noncancerous afflictions are also examples of latent effects associated with exposure to certain types of asbestos fibers. These diseases usually emerge 30 or more years after the initial exposure.

Performance Goal for Section 10.5
- ◆ Describe the nature of the following factors that affect the degree of toxicity: the quantity of a substance; the duration of exposure; the rate at which a substance is absorbed into the bloodstream; the age and health of the afflicted individual; and individual sensitivities.

◆ 10.5 FACTORS AFFECTING THE DEGREE OF TOXICITY

The human body is a very complex and delicately balanced system. Each cell assimilates nutrients from foods, resists biological attack, reproduces, and generates the substances that the host organism needs for its survival. Chemical reactions in the cells are responsible for life itself; but when toxic substances are absorbed across cellular membranes, they can upset this delicate chemistry. When the cells malfunction, the host organism experiences impaired health, disease, or death.

Fortunately, the body possesses natural mechanisms for protecting itself against foreign substances. One such mechanism involves the action of specific organs. For instance, the liver is particularly effective at converting many harmful substances into harmless ones or into substances that can be rapidly excreted in urine. As the result of these metabolic changes, foreign substances are modified in chemical structure, temporarily stored in specific organs, and/or directly eliminated.

Notwithstanding these natural protective mechanisms, the body is often incapable of protecting itself from invasion by toxic substances. Whether a toxic substance actually causes death, disease, or injury in the human organism depends upon several factors, the most important of which are the following: the quantity of substance, the duration of exposure, the rate at which a substance is absorbed into the bloodstream, the age and health of the afflicted person, and individual sensitivities. We shall examine each factor independently.

QUANTITY OF SUBSTANCE

While all substances have the potential to be toxic, we are concerned with those substances that elicit an adverse impact upon an organism in relatively small amounts. The amount of the substance administered or absorbed per body weight is called the *dose*. Some substances are so toxic that exposure to the tiniest dose is lethal. For instance, the bacterium *Clostridium botulinum*, the cause of botulism, is a single cell that releases a toxicant so potent that four hundred-thousandths of an ounce can cause death. Fortunately, there are relatively few substances that exhibit this unusually high degree of toxicity.

The impact that a specific substance has on an organism is a function of the dose. Most of us are familiar with the effects that alcoholic beverages exert on the body as a function of dose. When we drink enough alcohol to achieve a blood alcohol concentration of 50 mg/dL, we feel slightly subdued, relaxed, perhaps even elevated in spirit; but when the blood alcohol concentration reaches 150 mg/dL, we are patently intoxicated. [One deciliter (dL) is 100 milliliters, or 1/10 of a liter.]

DURATION OF EXPOSURE

As a general policy, the longer or more frequently an individual is exposed to low doses of a toxic substance, the more that person is likely to experience ill effects when the substance is not eliminated. Consider the mixture of the toxic substances contained within tobacco smoke. Whereas the inhalation of tobacco smoke has been irrefutably shown to cause throat and lung cancer, most of us know people who regularly smoke but have never contracted cancer. If tobacco products cause cancer, why aren't all tobacco smokers afflicted with throat or lung cancer?

Although the answer to this question is not entirely straightforward, it is apparent that many individuals who become afflicted with throat and lung cancer are not infrequent smokers; instead, they have generally smoked tobacco products for many years. The manifestation of cancer occurs only after many repeated exposures to the toxic substances in tobacco smoke. The assimilation of each dose lengthens the total period that the body is exposed to the substance.

Also, individuals do not respond in the same fashion when exposed to a given concentration of a toxic substance over the same length of time. Some individuals are capable of tolerating toxic substances within their environment more than others. Some even appear to adapt to their presence. Individuals respond to toxic substances over a given length of time in the following ways:

- While some individuals experience ill effects from exposure to a toxic substance, others experience the same effects only after they have been exposed to the same concentration of the same substance for a longer period of time.
- While some individuals experience adverse health effects from exposure to a given concentration of a substance, they may not experience the same effects upon subsequent exposure to the same concentration of the same substance.
- Some individuals who do not experience ill effects from a short-term exposure to one concentration of a substance may be unable to tolerate a long-term exposure to a lesser concentration of the same substance. These individuals require this more lengthy period of exposure to attain their "body burden," that is, the concentration they are unable to tolerate without experiencing ill effects.

RATE AT WHICH A SUBSTANCE IS ABSORBED INTO THE BLOODSTREAM

When a toxic substance is inhaled, the adverse effect caused by its presence may not be evident immediately. After inhalation of a lethal concentration of chlorine or carbon monoxide, for example, the fatality results within seconds or minutes; on the other hand, when a lethal concentration of hydrogen chloride or nitrogen dioxide is inhaled, death could occur hours or days later. These observations demonstrate that the manifestation of an effect caused by exposure to a toxic substance depends not only upon the concentration to which an individual is exposed, but on the rate of the mechanism that causes impairment or death.

When a toxic substance is ingested or absorbed through the skin, sufficient time must then elapse for the substance to be absorbed into the bloodstream. When ingested, the substance must usually pass from the stomach into the small intestine, where it is then absorbed into the bloodstream. When exposed to skin, the substance must first pass into the dermis, where it is then absorbed into the bloodstream.

It is fortuitous that time is required for the absorption of a substance into the bloodstream. The time needed for this action to occur provides a window of opportunity during which emergency-response personnel can administer an antidote, induce vomiting, or implement other appropriate first-aid actions.

AGE AND HEALTH OF AFFLICTED INDIVIDUAL

The degree to which toxic substances affect individuals often depends upon their age and general state of health. Young children and the elderly are typically more susceptible than middle-aged individuals to the impact from exposure to toxic substances. Children are especially vulnerable since their average body weight is relatively low and their immune systems are still developing.

Likewise, those persons already weakened from disease are more likely than healthy people to be susceptible to the presence of toxic substances. This sensitivity is often evident in individuals suffering from heart and respiratory ailments. Whereas such people are often incapable of tolerating the presence of air pollutants, such as low concentrations of ozone, healthy people could be oblivious to the pollutants' presence.

A well-recognized group of unhealthy individuals are tobacco smokers. As a general rule, nonsmokers are able to tolerate the presence of toxic substances better than smokers. They are also less likely to fall victim to certain illnesses. For example, nonsmokers do not fall victim as easily as smokers to asbestosis (Section 10.17), a progressive, irreversible lung disease associated with exposure to asbestos fibers. Firefighters who are nonsmokers are less likely than firefighters who smoke to experience the ill effects caused by exposure to carbon monoxide and other toxic gases.

INDIVIDUAL SENSITIVITIES

There are vast differences in the ways in which individuals respond when exposed to toxic substances. Often, these differences are not even associated with how old or healthy the individuals are. For example, some asthmatics are seemingly able to tolerate exposure to ground-level ozone, while others experience breathing problems from exposure to a lesser dose for a shorter period of time. How is this variability explained?

The answer to this question could have something to do with our inherited traits. Our genes are known to influence individual capabilities of contracting certain diseases such as coronary heart disease and some types of cancer. They also could play a role in whether or not we satisfactorily tolerate exposure to a given concentration of a toxic substance. The mechanism by which our genes influence the vulnerability to toxic substances is uncertain and remains an active area of medical research.

Performance Goals for Section 10.6

- Describe the methods by which toxicity is quantified from studies conducted on animals: lethal dose, 50% kill (LD_{50}); lethal concentration, 50% kill (LC_{50}); up-and-down dose; and threshold limit value (TLV).
- Demonstrate the manner by which the lethal concentration and lethal dose of a substance can be approximated for humans from the LC_{50} and LD_{50} values obtained from animal studies.
- Define the short-term exposure limit (STEL), permissible exposure limit (PEL), and immediately-dangerous-to-life-and-health level (IDLH).
- Describe how emergency responders use the information conveyed by these limits when they encounter a toxic substance during the line of duty.

◆ 10.6 MEASURING TOXICITY

Toxicologists have devised procedures for measuring the concentration of a substance that causes an organism to experience injury, disease, or death. These procedures involve exposing groups of laboratory animals to concentrations of a substance, observing the effects caused by the exposure, and extrapolating the results to humans. When exposed to a unique concentration of a substance, animals and humans do not always respond in the same way. Consequently, these parameters may have only limited relevance.

In the workplace, OSHA requires employers to reduce or eliminate the likelihood that employees will experience adverse health effects from exposure to toxic substances. To do so, employers consider three relevant exposure limits: the short-term exposure limit, the permissible exposure limit, and the immediately-dangerous-to-life-and-health level. Firefighters also use these limits at emergency-response scenes to evaluate the relative inhalation hazard posed by a known concentration of a toxic substance.

LETHAL DOSE, 50% KILL (LD_{50})

The LD_{50} is the amount of a substance that, when administered to laboratory animals, kills half of them during a preestablished time. The LD_{50} is expressed in milligrams of administered substance per body mass of the animal in kilograms (mg/kg). When a substance affects a specific organ, the LD_{50} may be measured by considering the mass of that organ.

The lethal dose of a substance to humans is calculated using the LD_{50} measurement obtained from animal studies. For an average person having a mass of w kilograms, the lethal dose is the product of the LD_{50} and the weight:

$$\text{Lethal dose} = LD_{50} \times w$$

LETHAL CONCENTRATION, 50% KILL (LC_{50})

The LC_{50} is the concentration of a substance that, when administered to laboratory animals, kills half of them during a preestablished time. The LC_{50} is typically expressed in parts per million (ppm) by volume. These measurements are used by DOT to characterize materials in the hazard classes 2.3 and 6.1 and to establish the hazard zone for a substance that poses an inhalation hazard (Section 10.14).

The lethal concentration of a substance to humans is calculated using the LC_{50} measurement obtained from animal studies. For an average person having a mass of w kilograms, the lethal concentration is the product of the LC_{50} and the weight:

$$\text{Lethal concentration} = LC_{50} \times w$$

UP-AND-DOWN DOSE

This is the concentration in mg/kg of a substance that kills a laboratory animal within 48 hours. First, a single dose of the substance is administered to the animal. If it remains alive after 48 hours, a single, slightly higher dose is administered to a second animal. If the first animal dies within 48 hours, a slightly lower dose is administered to the second animal. This process is then repeated by administering higher or lower doses to additional animals until the specific concentration that can kill a laboratory animal has been identified.

THRESHOLD LIMIT VALUE (TLV)

The TLV is the upper limit of a concentration to which an average healthy person can be repeatedly exposed on an all-day, everyday basis without suffering adverse health effects. The TLV for gaseous substances in the air is usually expressed in ppm. The TLV for fumes or mists in air is expressed in milligrams per cubic meter (mg/m^3). The value of a TLV is the level of exposure at which the probability of the occurrence of adverse health effects is deemed negligible. They are established by the American Conference of Governmental Industrial Hygienists.*

Solved Exercise 10.3

Experiments conducted at the National Institute of Standards and Technology, Gaithersburg, Maryland, established the following inhalation LC_{50} values for hydrogen cyanide by exposing laboratory animals to this toxic gas for the indicated time periods.

Exposure Time (min)	LC_{50} (ppm)
1	3000
2	1600
5	570
10	290
20	170
30	110
60	90

Animal deaths were determined during both the exposure and postexposure periods. What do these data demonstrate about the inhalation toxicity of hydrogen cyanide?

---

*TLVs are published in ACGIH (American Conference of Governmental Industrial Hygienists), *Documentation of the Threshold Limit Values and Biological Exposure Indices*, Cincinnati, Ohio (2001).

Solution: These data are interpreted as follows: When exposed to hydrogen cyanide, half of the animals died following a 1-minute exposure to a concentration of 3000 ppm; half died following a 2-minute exposure to a concentration of 1600 ppm; half died following a 5-minute exposure to a concentration of 570 ppm; and so forth. Overall, these data demonstrate that death can result not only from a short-term exposure to a high concentration of a toxicant, but also from a long-term exposure to a lesser concentration of the same toxicant.

Solved Exercise 10.4

In 1995, a doomsday cult called the Aum Shinrikyo placed plastic bags of the neurotoxicant *sarin* in commuter train cars of the Tokyo subway system. This misdeed caused the deaths of 12 people. An earlier release of sarin by the same group had killed seven people in Matsumoto, Japan, in 1994. Sarin is an example of a "nerve gas" (Section 12.16), a volatile liquid whose vapor, when inhaled, kills by disrupting the normal function of the central nervous system. Its inhalation LD_{50} is 1.7×10^{-5} mg/kg, or 0.000017 mg/kg. Determine the minimum amount of sarin that, when inhaled, was fatal to a 200-lb subway commuter.

Solution: The lethal dose of sarin for a 200-lb person is calculated as follows:

$$200 \text{ lb} \times \frac{1 \text{ kg}}{2.2 \text{ lb}} \times 0.000017 \text{ mg/kg} = 0.0015 \text{ mg, or } 1.5 \times 10^{-3} \text{ mg}$$

The inhalation of only 0.0015 mg of sarin by a 200-lb person is fatal.

SHORT-TERM EXPOSURE LIMIT (STEL)

The STEL is the concentration of a substance to which workers can be exposed continuously for approximately 15 minutes without suffering irritation, chronic or irreversible tissue damage, or narcosis sufficient to increase the likelihood of accidental injury, impair self-rescue, or materially reduce work efficiency. This assumes that the daily threshold limit value, time-weighted average for the substance has not been exceeded. STELs for specific substances are published when toxicological effects from relatively elevated short-term exposures to either humans or animals have been reported. They are established by the National Institute for Occupational Safety and Health.*

PERMISSIBLE EXPOSURE LIMIT (PEL)

The PEL is the time-weighted average threshold limit value of a toxic substance to which workers can be exposed continuously for 8 hours in conformance with relevant OSHA regulations. Employers calculate the cumulative employee exposure to a listed toxic substance over an 8-hour work shift (see Review Exercise 10.9). They are established by the National Institute for Occupational Safety and Health (NIOSH).

Ceiling values are also denoted for some substances by NIOSH. They represent the maximum concentrations to which workers can be exposed.

*STELs, PELs, and IDLHs are published in *NIOSH Pocket Guide to Chemical Hazards*, U.S. Department of Health and Human Services, Public Health Service, Centers for Disease Control and Prevention, National Institute for Occupational Safety and Health (January 2003), available from the Superintendent of Documents, U.S. Government Printing Office, Washington, DC 20402. PELs are also published at 29 CFR §1910.1000.

TABLE 10.3 ◆ Toxicity Measurements of Some Common Gases

Toxic Substance	*LC_{50} (ppm)*	*Permissible Exposure Limit*	*Short-Term Exposure Limit*	*Immediately-Dangerous-to-Life-and-Health Level [ppm]*
Ammonia	4000	25 ppm (18 mg/m^3)(NIOSH)	50 ppm (35 mg/m^3) (OSHA) 35 ppm (27 mg/m^3) (NIOSH)	300
Carbon monoxide	3760	50 ppm (55 mg/m^3) (OSHA) 35 ppm (40 mg/m^3) Ceiling 200 ppm (229 mg/m^3) (NIOSH)		1200
Chlorine	293	Ceiling 1 ppm (3 mg/m^3) (OSHA) Ceiling 0.5 ppm (1.45 mg/m^3), 15 minutes (NIOSH)		10
Fluorine	185	0.1 ppm (0.2 mg/m^3) (OSHA/NIOSH)		25
Hydrogen chloride	2810	Ceiling 5 ppm (7 mg/m^3) (OSHA/NIOSH)		50
Hydrogen cyanide		10 ppm (11 mg/m^3), skin absorption (OSHA)	4.7 ppm (5 mg/m^3), skin absorption (NIOSH)	50
Hydrogen sulfide	712	Ceiling 20 ppm (28.4 mg/m^3) 50 ppm/10-minute maximum peak (OSHA) 10 ppm (15 mg/m^3), 10-minute maximum peak (NIOSH)		100
Nitric oxide	115	25 ppm (30 mg/m^3) (OSHA/NIOSH)		100
Nitrogen dioxide	115	Ceiling 5 ppm (9 mg/m^3) (OSHA)	1 ppm (1.8 mg/m^3) (NIOSH)	20
Phosgene	5	0.1 ppm (0.2 mg/m^3) (OSHA) 0.1 ppm (0.4 mg/m^3) Ceiling 0.2 ppm (0.8 mg/m^3), 15-minute exposure (NIOSH)		2
Phosphine	20	0.3 ppm (0.4 mg/m^3) (OSHA/NIOSH)	1 ppm (1 mg/m^3) (NIOSH)	50
Sulfur dioxide	2520	5 ppm (13 mg/m^3) (OSHA) 2 ppm (5 mg/m^3) (NIOSH)	5 ppm (13 mg/m^3) (NIOSH)	100

IMMEDIATELY-DANGEROUS-TO-LIFE-AND-HEALTH LEVEL (IDLH)

The IDLH is the atmospheric concentration of any substance that poses an immediate threat to life, causes irreversible or delayed adverse health effects, or interferes with an individual's ability to escape during a 30-minute period from a dangerous atmosphere. They are established by NIOSH.

The relative differences between several of these toxicity measurements are noted for some common gases in Table 10.3.

Solved Exercise 10.5

A chlorine manufacturer intends to ship 5000 lb (2270 kg) of chlorine as a liquefied compressed gas by barge. Using Tables 6.1, 6.2, 6.3, 6.4, 10.3, and 10.14, identify the basic description DOT requires on the accompanying shipping paper.

Solution: Each table listed in this question provides information needed for preparing the basic description of this shipment.

Table 6.2	A reportable quantity is at issue.
Table 6.3	Chlorine is a marine pollutant.
Table 10.3	The LC_{50} of chlorine is 293 ppm.
Table 10.14	The relevant hazard zone is "B."

Using information in Tables 6.1 and 6.4, we then write the basic description as follows:

5000 lb   RQ, Chlorine, 2.3, UN1017 (Poison—Inhalation Hazard, Zone B) (marine pollutant)

Performance Goals for Section 10.7
- Describe the general atmosphere of the typical fire scene, especially the routine presence of carbon monoxide and smoke.
- Describe the physical composition of smoke and note the specific ways it hampers performance when implementing emergency-response activities.
- Describe the manner in which the inhalation of smoke negatively affects the human organism.

◆ 10.7 TOXICITY OF THE FIRE SCENE

The fire scene is usually an extremely dangerous environment. Even if we could eliminate all of its physical hazards, we would still encounter an atmosphere filled with smoke, dust, toxic gases, and other toxic substances. The potential for illness and death at a fire scene often results when individuals are exposed to the smoke and toxic gases produced during the fire.

What is smoke? Put simply, smoke is the gray-to-black plume of matter that emanates from class A and class B fires. It consists of an airborne suspension of water droplets and finely divided particulates of carbon and ash.

During most fires, carbon is initially produced as microscopic particles, which rapidly conglomerate into black, visible-sized particulates collectively referred to as *soot*. These particulates possess diameters less than 10 μm. Consequently, they are small enough to be inhaled into the lungs. The larger particulates in smoke are usually filtered within the nasal passageway, but the smaller ones can be drawn into the bronchi and lungs. The ultrafine soot particles—those having diameters less than 2.5 μm—can penetrate deeply.

Scientists have linked the long-term inhalation of particulate matter with an increase in cardiovascular and pulmonary diseases in susceptible individuals. In particular, the inhalation of these particulates has been linked with hastening the deaths of sick, elderly individuals by contributing to the premature onset of heart attacks, strokes, and emphysema. It is prudent to infer that soot particulates can also contribute to the inception of adverse health effects in those firefighters who regularly inhale soot while combating fires.

Using the authority of the Clean Air Act, EPA regulates the concentration of particulate matter in the ambient air as a criteria air pollutant (Section 1.5). For particulates having a diameter of 10 μm or less, EPA set the primary national air quality 24-hr standard at 150 μg/m^3. For particulates having a diameter of 2.5 μm or less, EPA set the primary national ambient air quality 24-hr standard at 65 μg/m^3 and the annual standard at 15μg/m^3.

Smoke is produced as a consequence of the incomplete combustion of matter (Figure 10.5). At a fire scene, the presence of smoke and combustion products like carbon monoxide often represents a greater hazard to life and a more serious hindrance to firefighting efforts than the fire itself. Individuals who are unable to escape from a fire scene, like those in Figure 10.6, often die from exposure to the smoke and combustion products, even before the fire reaches them.

SMOKE OBSCURES VISION

Zero visibility is common inside burning buildings from which smoke cannot readily escape. When our eyes are exposed to the irritant gases in smoke, they sting and involuntarily tear and close. Certain irritants in smoke are *lacrimators* (Section 12.17). Their vapors act upon the sensitive nerve endings of the mucous membrane of the eye, which causes an involuntary unleashing of a flood of tears. Examples of the lacrimators in smoke are formaldehyde (Section 12.10) and acrolein (see Review Exercise 10.7).

INHALING SMOKE CAUSES CHOKING AND GAGGING

As carbon particulates coat the bronchial passageways and lung surfaces, they interfere with normal respiration and contribute to the onset of adverse health effects. The trachea and bronchi are lined with tiny hair-like projections called *cilia*, which act in a wave-like manner to force deposited particles upwards towards the esophagus, where they are generally swallowed. When smoke is inhaled, it greatly impairs the ability of the cilia to effectively remove carbon particulates from the respiratory passages. This then causes the onset of choking, gagging, and respiratory injuries.

INHALING SMOKE CAUSES LATENT HEALTH EFFECTS

Epidemiologists have linked the inhalation of fine particulates with an increase in deaths caused by lung cancer. Evidence indicates that the risk of lung cancer increases by 8% for those individuals who frequently inhale particulate matter. The onset of latent health effects like cancer may be caused by the presence of certain compounds that are present in smoke. These compounds include the polynuclear aromatic hydrocarbons, which we shall visit in Section 12.4.

CARBON PARTICULATES IN SMOKE ADSORBS TOXIC GASES

The carbon particulates in smoke act as adsorbants to which toxic gases adhere. This means that each time smoke is inhaled, the carbon particulates aid in drawing carbon monoxide and other undesirable gases into the bronchi and lungs. As these particulates deposit upon the surfaces of the bronchi and lung, the expulsion of the gases from the body is simultaneously hindered.

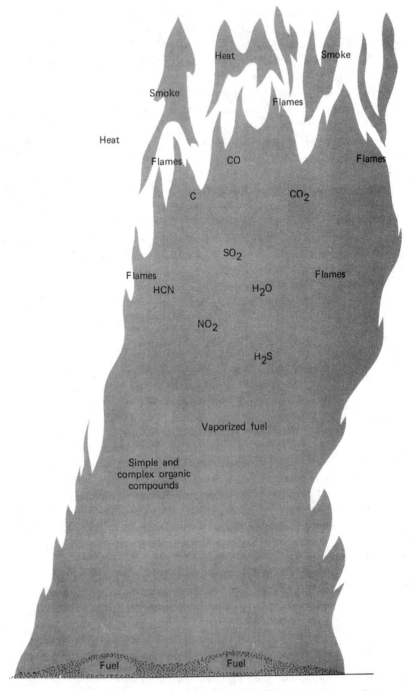

FIGURE 10.5 ◆ Smoke is typically produced during the incomplete combustion of carbon-bearing fuels, along with carbon monoxide and water vapor. Depending on the nature of the fuel, however, the atmosphere surrounding a fire may also contain vapors of the fuel, its decomposition products, and one or more fire gases. When inhaled, these substances are frequently more hazardous to firefighters and fire victims than the accompanying flames or heat.

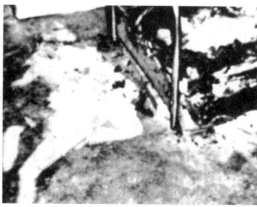

FIGURE 10.6 ◆ A victim of a mattress fire. Note the outline of the body on the floor in the photograph at the right, as well as the lack of observable soot under the body. This suggests that death occurred prior to the onset of a flaming combustion and was probably caused by inhalation of a noxious or poisonous atmosphere, such as one containing carbon monoxide. *Courtesy of I. N. Einhorn, Flammability Research Center, University of Utah, Salt Lake City, Utah.*

Class A and class B fires often involve the combustion of wood, plastics, heating fuels, and other carbon-rich materials. When such materials burn, the constituent carbon unites with oxygen to form carbon monoxide or carbon dioxide, both of which are colorless, odorless gases. The presence of carbon monoxide at a fire scene is always a matter of special concern since carbon monoxide is highly poisonous.

To understand how these gases are produced at fire scenes, it is necessary to re-visit two processes first introduced in Section 7.7: incomplete and complete combustion (Figure 10.7).

♦ *Incomplete combustion* is the burning phenomenon that occurs when the supply of air or oxygen, or access to either air or oxygen, is limited; that is, incomplete combustion occurs during fuel-rich fires. When carbon-bearing fuels burn, incomplete combustion processes

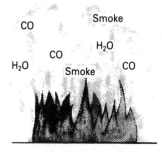
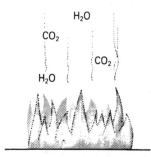

FIGURE 10.7 ◆ The incomplete combustion of carbon-bearing fuels, shown on the left, results in the production of carbon monoxide, carbon particulates (smoke), and water vapor. This situation is typical of a fuel-rich or oxygen-starved fire. On the right, the complete combustion of carbon-bearing fuels results in production of carbon dioxide and water vapor. This combustion is typical of a fuel that had been previously mixed with air or oxygen.

produce carbon monoxide and soot. The following equations illustrate the production of these substances during the incomplete combustion of methane:

$$2CH_4(g) + 3O_2(g) \longrightarrow 2CO(g) + 4H_2O(g)$$

methane oxygen carbon monoxide water

$$CH_4(g) + O_2(g) \longrightarrow C(s) + 2H_2O(g)$$

methane oxygen carbon water

The carbon monoxide produced during incomplete combustion enters the atmosphere, where it slowly oxidizes to carbon dioxide.

• By contrast, *complete combustion* is the burning process that occurs when plenty of air or oxygen is available. When carbon-bearing fuels burn, complete combustion is associated with the production of carbon dioxide. The complete combustion of methane is illustrated by the following equation:

$$CH_4(g) + 2O_2(g) \longrightarrow CO_2(g) + 2H_2O(g)$$

methane oxygen carbon dioxide water

Complete combustion occurs when the flow of fuel and air can be regulated as in a normal heating system.

While carbon monoxide and carbon dioxide are present at virtually all fire scenes, several other gaseous combustion products could also be present. They include hydrogen cyanide, ammonia, sulfur dioxide, nitrogen dioxide, hydrogen chloride, and acrolein (Section 13.4). Whether one or more of them are actually present at a given fire scene depends upon the chemical nature of the material that burns, smolders, or undergoes thermal decomposition.

Research studies demonstrate that exposure to a mixture of toxic gases adversely impacts laboratory animals differently than exposure to a single gas. Our understanding with respect to the manner by which the combination toxicologically interacts when inhaled is an active area of ongoing research.

Performance Goals for Section 10.8

- Discuss the basis for associating the presence of carbon monoxide with virtually all class A and class B fires.
- Identify the adverse effects experienced when humans are exposed to different concentrations of carbon monoxide.
- Describe how carbon monoxide interacts with the blood's hemoglobin to impede the proper transfer of inhaled oxygen.
- Describe how carbon monoxide is produced for commercial use.
- Identify the industries that use bulk volumes of carbon monoxide.

◆ 10.8 CARBON MONOXIDE

Carbon monoxide is an odorless, colorless, tasteless, and nonirritating gas that is present at virtually every fire scene. Given this combination of properties, the presence of carbon monoxide is unlikely to be immediately apparent. Nonetheless, we must always acknowledge that carbon monoxide is likely to be a component of the atmosphere at *all* fire scenes.

The majority of the illnesses and deaths associated with fires is linked with the victims' exposure to carbon monoxide. In the United States, the inhalation of carbon monoxide is also the leading cause of accidental poisoning in the home. Faulty furnaces,

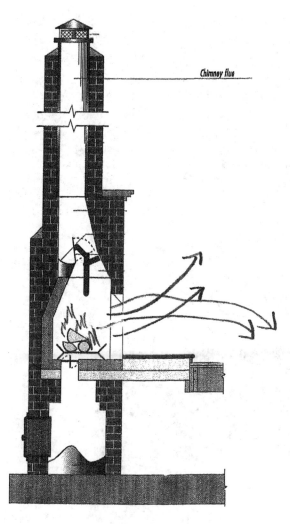

Chimney flue

FIGURE 10.8 ◆ When a chimney flue is blocked, the combustion products produced in an operating fireplace may enter the home. Under this circumstance, the fireplace serves as a potential source of carbon monoxide poisoning. To prevent a poisonous atmosphere from emerging, homeowners should always be assured that chimney flues are unblocked so the combustion products properly vent to the outside.

stoves, and space heaters and blocked chimneys and flues (Figure 10.8) are among the domestic sources of carbon monoxide.

Solved Exercise 10.6

The carbon monoxide concentration in the lower atmosphere normally ranges from 0.2 to 0.5 ppm. Since human activities (e.g., the burning of fuels) regularly produce carbon monoxide, why doesn't the carbon monoxide concentration become so elevated that the lower atmosphere turns poisonous?

   Solution: Carbon monoxide does not permanently remain within the atmosphere. It slowly converts into carbon dioxide by chemical union with atmospheric oxygen.

$$2CO(g) + O_2(g) \longrightarrow 2CO_2(g)$$

While some carbon dioxide is absorbed into the oceans and other bodies of water, most remains unaltered within the atmosphere.

Solved Exercise 10.7

The carbon monoxide concentration in a car's exhaust is approximately 8% by volume when the car is idling, but it reduces to approximately 4% when the car is driven. Why are these concentrations so dissimilar?

Solution: When the supply of air is restricted, incomplete combustion occurs. This situation exists within the car's combustion chambers and is associated with production of carbon monoxide. A comparatively lesser volume of air is drawn into the car's combustion chambers during idling than when the car is driven. This results in the production of a correspondingly lesser amount of carbon monoxide during idling compared to when the engine is operating.

Solved Exercise 10.8

Why are a growing number of cities and states requiring the installation in homes of devices that detect the presence of carbon monoxide?

Solution: A carbon monoxide detector like the simple device shown below can alert individuals to the presence of this odorless, tasteless, colorless, and poisonous gas.

Carbon monoxide detector. *Courtesy of KIDDE Safety, Mebane, North Carolina.*

City and state legislators are hopeful that the widespread use of carbon monoxide detectors will result in saving the lives of family members, just as the requirement for installing smoke detectors in homes has done. The presence of carbon monoxide detectors does not replace the necessity for the installation of smoke detectors; in fact, carbon monoxide detectors are not designed to detect smoke or fire. *Both* should be installed in homes.

To understand why carbon monoxide is a poison, it is necessary to examine the chemistry that occurs during respiration. When we inhale air into our lungs, a supply of atmospheric oxygen is absorbed into the bloodstream. The oxygen is then carried throughout the body by a complex component of the blood called *hemoglobin*. There are about 300 million hemoglobin molecules inside each red blood cell. The molecular structure of hemoglobin in Figure 10.9 indicates its complexity; hence, the hemoglobin molecule is represented here by the symbol *Hb*.

When hemoglobin assimilates oxygen, the compound called *oxyhemoglobin* is produced. It is represented here as O_2Hb. The respiratory process can then be represented in the following simplified fashion:

$$\underset{\text{hemoglobin}}{Hb(aq)} + \underset{\text{oxygen}}{O_2(g)} \longrightarrow \underset{\text{oxyhemoglobin}}{O_2Hb(aq)}$$

Oxyhemoglobin molecules move within the circulatory system and eventually arrive at the various tissues and organs of the body, where they release their oxygen at the cellular level for biochemical use. The hemoglobin molecules then return through the circulatory system to the lungs, where they secure a new supply of oxygen.

Carbon monoxide interrupts this normal act of respiration in part by interfering with the transport of oxygen. Carbon monoxide reacts with hemoglobin, forming a substance called *carboxyhemoglobin* (COHb), as follows:

$$\underset{\text{hemoglobin}}{Hb(aq)} + \underset{\text{carbon monoxide}}{CO(g)} \longrightarrow \underset{\text{carboxyhemoglobin}}{COHb(aq)}$$

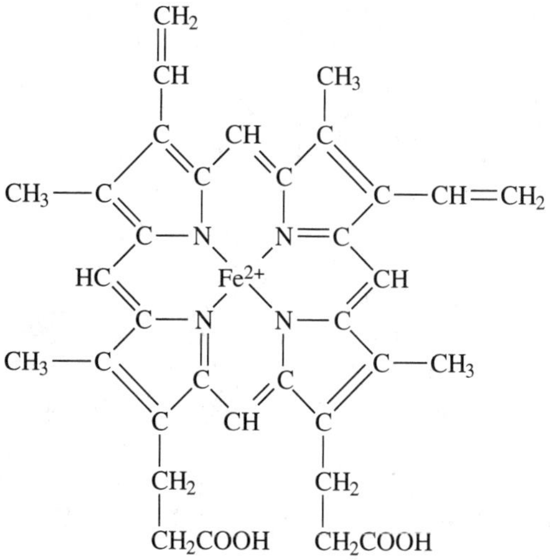

FIGURE 10.9 ◆ The molecular structure of hemoglobin, a complex component of blood whose normal function is to transport oxygen from the lungs to the various tissues and organs of the body.

TABLE 10.4 ◆ Signs and Symptoms Associated with Various Concentrations of Carboxyhemoglobin

Percent Carboxyhemoglobin	Signs and Symptoms
0–10	No signs or symptoms
10–20	Tightness across forehead, possible slight headache, dilation of the cutaneous blood vessels
20–30	Headache and throbbing in the temples
30–40	Severe headache, weakness, dizziness, dimness of vision, nausea, vomiting, and collapse
40–50	Same as above, greater possibility of collapse; cerebral anemia, and increased pulse and respiratory rates
50–60	Cerebral anemia, increased respiratory and pulse rates, coma, and intermittent convulsions
60–70	Coma, intermittent convulsions, depressed heart action and respiratory rate, and possible death
70–80	Weak pulse and slow respiration, leading to death within hours
80–90	Death in less than 1 hour
90–100	Death within a few minutes

The chemical affinity for the formation of carboxyhemoglobin is about 200 times greater than the affinity for forming oxyhemoglobin. When the blood's hemoglobin is bound within the structure of carboxyhemoglobin, it cannot perform its normal bodily function. The carboxyhemoglobin is unable to provide any respiratory benefit.

Individuals who inhale carbon monoxide experience the combination of signs and symptoms listed in Table 10.4 and illustrated in Figure 10.10. An encounter with carbon monoxide may be deadly. Indeed, exposure to an atmosphere containing only 0.1% carbon monoxide converts over half of the hemoglobin in blood to the useless carboxyhemoglobin and causes death within 1 hour. The presence of carboxyhemoglobin in the bloodstream kills the host victim by hindering the transport of oxygen. The host victim suffers from *anoxia,* the lack of oxygen. Anoxia causes the victim's skin to assume a bright pink color. As the tissues are deprived of oxygen, the afflicted person experiences difficulty in breathing, drowsiness, headache, dizziness, and chest pains. If deprived of oxygen for an extended period, the victim may be unable to survive.

The ill effects caused by inhalation of carbon monoxide are not solely associated with the production of carboxyhemoglobin. Research studies show that carbon monoxide inhaled into the lungs does not always bond to the blood's hemoglobin; instead, it assimilates into the cells of the body's tissues where it can interfere with certain enzymatic processes. It is for this reason that long-term exposure to a low concentration of carbon monoxide may cause ill effects that are not otherwise experienced by short-term exposure to the same concentration.

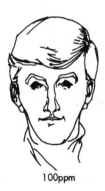

100ppm   1000ppm   1300ppm   >2000ppm

FIGURE 10.10 ◆ Carbon monoxide is a toxic combustion product to which all firefighters are exposed during routine work-related activities. An adult may tolerate carbon monoxide at a concentration in air of 100 ppm without noticeably suffering adverse health effects. However, a 1-hour exposure to an atmosphere containing a concentration of carbon monoxide of 1000 ppm generally causes the individual to experience a mild headache; a reddish coloration of the skin simultaneously develops. A 1-hour exposure to an atmosphere containing 1300 ppm of carbon monoxide usually causes the skin to turn cherry red; the accompanying headache becomes throbbing. A 1-hour exposure to an atmosphere containing 2000 ppm of carbon monoxide is likely to cause either the individual's death or damage to the respiratory and nervous systems of an individual who survives.

An individual who has lost consciousness from inhaling carbon monoxide and other toxic gases should be swiftly moved to an open space where a means of artificial respiration and fresh (100%) oxygen can be administered. Breathing 100% oxygen for 1 hour can reduce the concentration of carbon monoxide in the victim's blood to approximately one-half its initial concentration.

A seriously afflicted individual, especially when comatose, should be rapidly transported to a hospital or clinic capable of administering hyperbaric oxygen therapy. This treatment process involves subjecting the person to a high-oxygen atmosphere within a pressurized chamber of the type shown in Figure 10.11. The inhalation of 100% oxygen at a pressure of 2 atm (202.6 kPa) for ½ to 1 hour results in the faster conversion of carboxyhemoglobin to oxyhemoglobin compared to the use of 100% oxygen at 1 atm (101.3 kPa) for the same period. It is for this reason that hyperbaric oxygen treatment may reverse the impact caused by inhaling carbon monoxide.

Breathing oxygen under increased pressure allows the blood to carry more oxygen deeper into the body's tissues, including the brain. Under pressure, the capillaries diffuse blood farther, allowing oxygen-rich blood to be carried into otherwise inaccessible areas of the tissues. Even with effective treatment, however, a person who has survived exposure to carbon monoxide may still suffer long-term damage to the heart and brain.

Aside from its production at fire scenes, carbon monoxide is also commercially available as a liquefied compressed gas, nonliquefied compressed gas, and cryogenic liquid. Some of its physical properties are listed in Table 10.5. Carbon monoxide is also encountered as a component of water gas, large volumes of which are produced by the chemical industry for the manufacture of hydrogen (Section 7.2) and methanol (Section 12.8). The isolation of carbon monoxide from water gas is a primary means by which carbon monoxide is produced for commercial use.

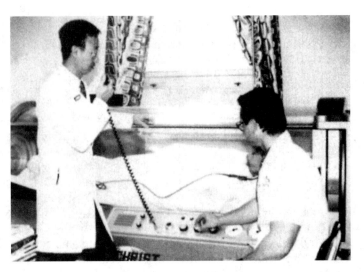

FIGURE 10.11 ◆ A therapist and physician attending to a patient in a hyperbaric oxygen chamber. Hyperbaric oxygen therapy involves administering 100% oxygen at a pressure up to three times the normal atmospheric pressure for 30 to 60 minutes. The utilization of hyperbaric oxygen therapy at 2 atm (202.6 kPa) accelerates the removal of carbon monoxide from the bloodstream. *Courtesy of St. James Hospital, Medical Center, Chicago Heights, Illinois.*

Although the primary risk associated with exposure to carbon monoxide is inhalation toxicity, the gas also poses the risk of fire and explosion. When ignited, carbon monoxide oxidizes to carbon dioxide.

$$\underset{\text{carbon monoxide}}{2CO(g)} + \underset{\text{oxygen}}{O_2(g)} \longrightarrow \underset{\text{carbon dioxide}}{2CO_2(g)}$$

Using the authority of the Clean Air Act, EPA regulates the concentration of carbon monoxide in the ambient air as a criteria air pollutant (Section 1.5). EPA has set the primary national ambient air quality standard for carbon monoxide at 9 ppm (10 mg/m³).

When carbon monoxide is present in the workplace, OSHA requires employers to limit employee exposure to a concentration of 50 ppm (55 mg/m³), averaged over an 8-hour workday.

DOT regulates the transportation of carbon monoxide as a poison gas. DOT requires shippers and carriers to affix INHALATION HAZARD and FLAMMABLE GAS labels on its packaging and to display INHALATION HAZARD placards on

TABLE 10.5 ◆ **Physical Properties of Carbon Monoxide**	
Freezing point	−341°F (−207°C)
Boiling point	−314°F (−192°C)
Vapor density (air = 1)	0.97
Lower explosive limit	12.5%
Upper explosive limit	74%
Liquid-to-gas expansion ratio	700

the bulk packaging, freight container, transport vehicle, unit load device, or railcar used to transport carbon monoxide.

▪▪
Performance Goals for Section 10.9
  ◆ Identify the products that decompose to produce hydrogen cyanide when they smolder or are exposed to intense heat.
  ◆ Identify the adverse effects experienced by the human organism from exposure to hydrogen cyanide.
  ◆ Describe how hydrogen cyanide and hydrocyanic acid are produced for commercial use.
  ◆ Identify the industries that use bulk quantities of hydrogen cyanide, hydrocyanic acid, and metallic cyanides.
▪▪

◆ 10.9 HYDROGEN CYANIDE

Hydrogen cyanide is a colorless gas possessing the odor of bitter almonds. It is a highly toxic gas, possessing a TLV of only 10 ppm. Examples of its other important physical properties are provided in Table 10.6.

Hydrogen cyanide is encountered at some fire scenes. It is generated during the thermal decomposition of products composed of substances whose molecules possess carbon atoms bonded to nitrogen atoms (C≡N). These products include wool, silk, polyacrylonitrile, polyurethane, and nylon. When carpeting and other textiles composed of these materials smolder or are exposed to intense heat, hydrogen cyanide is released into the surrounding environment. Scientists believe that hydrogen cyanide generally does not survive at fire scenes, but is consumed by combustion. The incomplete and complete combustion processes are respectively represented as follows:

$$4HCN(g) + 5O_2(g) \longrightarrow 4CO(g) + 4NO(g) + 2H_2O(g)$$
hydrogen cyanide   oxygen   carbon monoxide   nitric oxide   water

$$4HCN(g) + 9O_2(g) \longrightarrow 4CO_2(g) + 4NO_2(g) + 2H_2O(g)$$
hydrogen cyanide   oxygen   carbon dioxide   nitrogen dioxide   water

Inhalation is the most common means by which hydrogen cyanide can enter the body, although skin absorption is also a viable mechanism. When absorbed into the body, the impaired experience the ill effects listed in Table 10.7. The combination of these effects is known as *cyanosis*. The mechanism of cyanosis is connected with the ability of hydrogen cyanide to inhibit the normal biological activity of *cytochrome oxidase,* an enzyme essential for cellular respiration and energy production. While the supply of oxygen in the cells of the body's tissues may be plentiful, the hydrogen cyanide inhibits the ability to utilize the oxygen effectively.

TABLE 10.6 ◆ Physical Properties of Hydrogen Cyanide

Freezing point	6.8°F (−14°C)
Boiling point	79°F (26°C)
Vapor density (air = 1)	0.93
Lower explosive limit	6%
Upper explosive limit	41%

TABLE 10.7 ◆ Toxicological Properties of Hydrogen Cyanide

HCN Concentration (ppm)	Symptoms
0.2–5.0	Threshold of odor
10	Threshold limit value
18–36	Slight symptoms (headache) after several hours
45–54	Tolerated for $\frac{1}{2}$ to 1 hour without difficulty
100	Death within 1 hour
110–135	Fatal in $\frac{1}{2}$ to 1 hour
181	Fatal after 10 minutes
280	Immediately fatal

Aside from its production during certain fires, hydrogen cyanide is also produced for commercial use by the catalytic reaction between ammonia and air with methane or natural gas.

$$2NH_3(g) + 3O_2(g) + 2CH_4(g) \longrightarrow 2HCN(g) + 6H_2O(g)$$
$$\text{ammonia} \quad \text{oxygen} \quad \text{methane} \quad \text{hydrogen cyanide} \quad \text{water}$$

The hydrogen cyanide molecules produced by the reaction tend to react with other hydrogen cyanide molecules at an explosive rate. To avoid the reaction, the commercial grades of hydrogen cyanide are stabilized with water and 0.05% phosphoric acid.

DOT regulates the transportation of the following three forms of hydrogen cyanide as poisons:

Hydrogen cyanide, stabilized *with less than 3 percent water*

Hydrogen cyanide, stabilized *with less than 3 percent water and absorbed in a porous inert material*

Hydrogen cyanide, solution in alcohol *with not more than 45 percent hydrogen cyanide*

When transporting the first and second forms, DOT requires shippers and carriers to affix INHALATION HAZARD and FLAMMABLE LIQUID labels to their packaging and to post INHALATION HAZARD placards on the bulk packaging, freight container, transport vehicle, unit load device, or railcar used to transport them. When transporting the third form of hydrogen cyanide, DOT requires shippers and carriers to affix INHALATION HAZARD labels to its packaging and to post INHALATION HAZARD placards on the bulk packaging, freight container, transport vehicle, unit load device, or railcar used to transport it.

HYDROCYANIC ACID

Hydrogen cyanide dissolves in water to form a solution known as *hydrocyanic acid, or prussic acid.* The acid solution is so weak it is incapable of turning litmus paper red. It is colorless, but very volatile. It is used to fumigate ships, warehouses, and greenhouses. During warfare and terrorist attacks, it could potentially be used as a blood agent (Section 12.16).

DOT regulates the transportation of the following forms of hydrocyanic acid as poisons:

Hydrocyanic acid solutions *or* Hydrogen cyanide, aqueous solutions
with not more than 20 percent hydrogen cyanide

Hydrocyanic acid, aqueous solutions *with less than 5 percent*
hydrogen cyanide

DOT requires shippers and carriers to affix INHALATION HAZARD labels to their packaging and to post INHALATION HAZARD placards on the bulk packaging, freight container, transport vehicle, unit load device, or railcar used to transport them.

METALLIC CYANIDES

These are substances such as sodium cyanide, potassium cyanide, copper(II) cyanide, and zinc cyanide. Sodium cyanide and potassium cyanide are available commercially as solids and aqueous solutions, while copper(II) cyanide and zinc cyanide are available solely as solids. All are used to prepare solutions from which copper and zinc are respectively electroplated.

In states where the death penalty is allowed, hydrogen cyanide inhalation is sometimes selected as the means of execution. The substance is generated by chemical action within a gas chamber. To generate hydrogen cyanide, pellets of a metallic cyanide are dropped into acid within an enclosed system.

$$NaCN(s) \quad + \quad HCl(aq) \quad \longrightarrow \quad NaCl(aq) \quad + \quad HCN(g)$$

sodium cyanide   hydrochloric acid   sodium chloride   hydrogen cyanide

Given the notoriety associated with cyanides, law enforcement agencies should be provided with an accounting of the cyanide products potentially available within their jurisdictions to terrorists.

When hydrogen cyanide is present in the workplace, OSHA requires employers to limit employee exposure to a dermal concentration of 10 ppm ($11 \, mg/m^3$), both averaged over an 8-hour workday.

DOT regulates the shipment of metallic cyanides as poisons. DOT requires shippers and carriers to affix POISON labels on their packaging and to display POISON placards on the bulk packaging, freight container, transport vehicle, unit load device, or railcar used to transport them.

Performance Goals for Section 10.10

- Identify the nature of the products that produce sulfur dioxide when they burn.
- Identify the potential hazard experienced when humans are exposed to sulfur dioxide.
- Describe how sulfur dioxide is produced for commercial use.
- Identify the industries that use bulk quantities of sulfur dioxide.

◆ 10.10 SULFUR DIOXIDE

Sulfur dioxide is a colorless, toxic gas possessing the sharp, pungent odor associated with burning tires. Some other physical properties of this gas are noted in Table 10.8.

TABLE 10.8 ◆ Physical Properties of Sulfur Dioxide	
Freezing point	$-105°F$ $(-76°C)$
Boiling point	$14°F$ $(-10°C)$
Vapor density (air = 1)	2.3

Sulfur dioxide is produced naturally during the eruption of volcanos. In the United States, the eruption of the Kilauea volcano in Hawaii (Figure 10.12) spews massive amounts of sulfur dioxide into the atmosphere daily.

Other sulfur-containing compounds are also found naturally in coal, natural gas, and crude oil. Complex proteins containing sulfur are also present in wool, hair, and animal hides (Section 13.4). Sulfurous compounds are also found in several natural and synthetic polymers, including vulcanized rubber. When these materials burn, the sulfur in their constituent compounds becomes sulfur dioxide. Firefighters encounter sulfur dioxide when they respond to fires involving the combustion of sulfur-containing substances or products to which sulfur was added.

Since sulfur dioxide is produced when coal and other fossil fuels burn, it is encountered as an air pollutant. Fossil-fuel-fired power plants in the United States emit

FIGURE 10.12 ◆ The eruption of the Kilauea Volcano, the active crater on the island of Hawaii, serves as the major natural source of sulfur dioxide in the United States. *Courtesy of the Hawaiian Volcano Observatory, United States Geological Survey, Hawaii National Park, Hawaii; photograph by J. D. Griggs.*

almost two-thirds of the sulfur dioxide released into the air. Not only is sulfur dioxide toxic, but it slowly oxidizes to sulfur trioxide, which in turn reacts with atmospheric moisture to form sulfuric acid.

$$2SO_2(g) \ + \ O_2(g) \ \longrightarrow \ 2SO_3(g)$$
$$\text{sulfur dioxide} \qquad \text{oxygen} \qquad \quad \text{sulfur trioxide}$$

$$SO_3(g) \ + \ H_2O(g) \ \longrightarrow \ H_2SO_4(aq)$$
$$\text{sulfur trioxide} \qquad \text{water} \qquad \qquad \text{sulfuric acid}$$

The presence of sulfuric acid in the atmosphere has severely damaged building materials made of concrete, marble, mortar, and limestone. A startling example of the impact that sulfuric acid has had on stone is evident from observing Cleopatra's Needle, an obelisk that was moved from Egypt to New York City in the late 19th century. It has deteriorated more in approximately 100 years than in the 2000 years it stood in Egypt.

Aside from its presence at fire scenes and in polluted air, sulfur dioxide is also available commercially as a liquefied compressed gas. It is primarily produced by burning sulfur.

$$S_8(g) \ + \ 8O_2(g) \ \longrightarrow \ 8SO_2(g)$$
$$\text{sulfur} \qquad \text{oxygen} \qquad \quad \text{sulfur dioxide}$$

It is used as a refrigerant and a bleaching agent in the pulp and paper industry, in refining sugar, and in processing dried fruits.

Inhalation toxicity is the primary hazard associated with exposure to sulfur dioxide. Coughing, chest pains, shortness of breath, and constriction of the airways are immediate symptoms of exposure to this gas. As an atmospheric pollutant, very low concentrations of sulfur dioxide are moderately tolerable for short periods of time. However, a sulfur dioxide concentration of 100 ppm in air is considered immediately dangerous to life and health; death could result from only a short exposure.

Various regulatory agencies address the impact of sulfur dioxide on public health and the environment. EPA regulates the concentration of sulfur dioxide in the ambient air as a criteria air pollutant. EPA has set the primary national ambient air quality standard for sulfur dioxide as an annual arithmetic mean equal to 0.030 ppm (80 μg/m^3) and a 24-hour average equal to 0.14 ppm (365 μg/m^3). The secondary standard is a 3-hour average equal to 0.50 ppm (1300 μg/m^3).

When sulfur dioxide is present in the workplace, OSHA requires employers to limit employee exposure to a concentration of 5 ppm (13 mg/m^3), averaged over an 8-hour workday.

DOT regulates the transportation of sulfur dioxide as a poison gas. DOT requires shippers and carriers to affix INHALATION HAZARD and CORROSIVE labels to its packaging and to display INHALATION HAZARD placards on the bulk packaging, freight container, transport vehicle, unit load device, or railcar used to transport it.

Performance Goals for Section 10.11

- ◆ Identify the nature of the products that produce hydrogen sulfide when they smolder or are exposed to intense heat.
- ◆ Discuss the reason hydrogen sulfide is encountered only at trace concentrations at fire scenes.
- ◆ Identify the potential hazard experienced when humans are exposed to hydrogen sulfide.
- ◆ Describe how hydrogen sulfide is produced for commercial use.
- ◆ Identify the industries that use bulk quantities of hydrogen sulfide.

◆ 10.11 HYDROGEN SULFIDE

Hydrogen sulfide is a colorless, toxic gas possessing the disagreeable odor associated with rotten eggs. The odor is offensive even at concentrations as low as 3 to 5 ppm. We sometimes hear individuals describing the odor of a material by saying that it smells like sulfur. Since elemental sulfur is an odorless solid (Section 7.6), what they actually mean to say is that the material smells like hydrogen sulfide.

Hydrogen sulfide occurs naturally. It is produced by the decay of organisms in swamps, sewers, and other anaerobic (non-oxygen) environments from which the gas seeps into the atmosphere. Hydrogen sulfide is referred to by sanitary engineers who work in sewers as *sewer gas*.

Since it possesses a very disagreeable odor, most people immediately become aware of the presence of hydrogen sulfide. Nonetheless, the long-term exposure to hydrogen sulfide can temporarily deaden an individual's sense of smell. The phenomenon is called *olfactory fatigue*. This causes the individual to become oblivious to the presence of the gas. When this occurs, they could inhale dangerous amounts unknowingly.

The presence of hydrogen sulfide at fire scenes, although possible, is generally improbable. Small concentrations could evolve during the thermal decomposition of certain materials manufactured from animal products, such as leather items and wool carpeting. The long-term survival of hydrogen sulfide generated during fires is improbable since the gas is generally consumed by combustion (see the following equation).

Although hydrogen sulfide is available as a liquefied compressed gas, it is not a commercially popular product. For its limited commercial use, it is primarily derived from petroleum refineries and natural gas wells. In commerce, hydrogen sulfide is used to process mineral ores, produce metallic sulfides such as nickel sulfide and molybdenum sulfide, and manufacture phosphors used in television tubes. Hydrogen sulfide is

FIGURE 10.13 ◆ Two specially equipped vehicles used to ship bulk quantities of hydrogen sulfide as a liquefied, compressed gas: a 5000-gal ($1.9\text{-}m^3$) tank truck and a 22,000-gal ($83\text{-}m^3$) rail tankcar. These transport vehicles hold approximately 14 ton (13 tonne) and 64 ton (58 tonne) of hydrogen sulfide, respectively. DOT requires the shipper to display the identification number 1053 and an INHALATION HAZARD placard on both sides and both ends of each vehicle. *Courtesy of Montana Sulphur & Chemical Company, Billings, Montana.*

FIGURE 10.13 ◆ *continued*

available in cylinders and bulk transport containers (Figure 10.13). Examples of some of its important physical properties are provided in Table 10.9.

Inhalation toxicity is the primary hazard associated with exposure to hydrogen sulfide. Exposure to this gas initially gives rise to dizziness and the onset of a headache, but unconsciousness and respiratory paralysis can follow immediately. Death from asphyxiation could occur readily unless breathing is restored before heart action ceases. Most fatalities associated with hydrogen sulfide inhalation occur when safety practices are ignored at petroleum refineries, chemical manufacturing facilities, and other locations at which hydrogen sulfide is stored or generated.

Hydrogen sulfide also poses the risk of fire and explosion. When ignited, the gas burns in air as follows:

$$\underset{\text{hydrogen sulfide}}{2H_2S(g)} + \underset{\text{oxygen}}{3O_2(g)} \longrightarrow \underset{\text{water}}{2H_2O(g)} + \underset{\text{sulfur dioxide}}{2SO_2(g)}$$

It is because of its ease of ignition that the hydrogen sulfide generated during fires is generally consumed by combustion.

When hydrogen sulfide is present in the workplace, OSHA requires employers to limit employee exposure to a concentration of 50 ppm within a 10-minute maximum period.

DOT regulates the transportation of hydrogen sulfide as a poison gas. DOT requires shippers and carriers to affix INHALATION HAZARD and FLAMMABLE GAS labels to its packaging and to display INHALATION HAZARD placards on the

TABLE 10.9 ◆ **Physical Properties of Hydrogen Sulfide**	
Freezing point	−117°F (−83°C)
Boiling point	−76°F (−60°C)
Vapor density (air = 1)	1.2
Lower explosive limit	4.3%
Upper explosive limit	46%

bulk packaging, freight container, transport vehicle, unit load device, or railcar used to transport it.

◆ 10.12 NITROGEN OXIDES

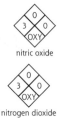

nitric oxide

nitrogen dioxide

Eight oxides of nitrogen are known, but we are concerned here only with the two that are produced when atmospheric nitrogen combines with oxygen. They are nitric oxide and nitrogen dioxide, whose chemical formulas are NO and NO_2, respectively. Since they are often produced in combination, nitric oxide and nitrogen dioxide are represented jointly as NO_x.

The chemical combination of atmospheric nitrogen and oxygen occurs only slowly at ambient temperatures, but the reaction occurs rapidly at elevated temperatures.

$$\underset{\text{nitrogen}}{N_2(g)} + \underset{\text{oxygen}}{O_2(g)} \longrightarrow \underset{\text{nitric oxide}}{2NO(g)}$$

$$\underset{\text{nitric oxide}}{2NO(g)} + \underset{\text{oxygen}}{O_2(g)} \longrightarrow \underset{\text{nitrogen dioxide}}{2NO_2(g)}$$

At fossil-fuel-fired power plants and in the combustion chambers of motor vehicles, NO_x is produced in such significant concentrations (Figure 10.14) that it is classified as one of the major air pollutants in the United States.

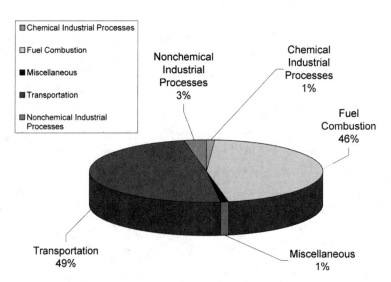

FIGURE 10.14 ◆ Nearly all the NO_x in the ambient air is derived from transportation and combustion activities. *Courtesy of United States Environmental Protection Agency.*

TABLE 10.10 ◆ Inhalation Effects of Nitrogen Dioxide in Humans

Nitrogen Dioxide Concentration in Air (ppm)	Symptoms
5	Threshold limit for detection by smell
10–20	Mild irritation to the eyes, nose, and upper respiratory tract
25–40	No adverse effects in workers exposed over a period of several years
50	Distinct irritation to the eyes, nose, and upper respiratory tract
80	Tightness in the chest after 3- to 5-minute exposure
90	Pulmonary edema occurs following 30-minute exposure
100–200	Threat to life after 30- to 60-minute exposure
250	Danger of death following short-term exposure

The presence of NO_x in the lower atmosphere is directly linked with the production of ground-level ozone (Section 7.1) and smog. Concerned about the adverse impact on public health posed by NO_x, especially within urban areas, EPA required automobile manufacturers to build cars that reduced the amount of NO_x emitted in vehicular exhaust. In response, the manufacturers installed catalytic converters in new automobiles. One chemical reaction that occurs on the surface of the converter is the reduction of the nitric oxide into environmentally friendly nitrogen and oxygen. The use of converters thus reduces the amount of NO_x into the atmosphere as we operate modern-day motor vehicles.

EPA regulates the concentration of nitrogen dioxide in the ambient air as a criteria air pollutant. EPA has set the primary and secondary national ambient air quality standards for nitrogen dioxide at 0.053 ppm (100 mg/m^3) as an annual arithmetic mean.

Nitric oxide is a colorless, odorless gas, but nitrogen dioxide is a dark-red-brown gas having a pungent, acrid odor. Both gases are poisonous. When nitrogen-containing materials burn, each gas is produced as the product of incomplete and complete combustion, respectively. Examples of these nitrogenous materials include the polyurethane products (Section 13.7) used in certain types of building insulation, bedding, and furniture cushioning. While the presence of nitrogen dioxide at a fire scene can be signaled by its red-brown color, it is generally impossible to visually identify this color in a smoke plume that is predominantly black.

Firefighters need to be wary of the presence of the nitrogen dioxide at fire scenes, because the gas poses the risk of inhalation toxicity. When an individual inhales nitrogen dioxide, the ill effects in Table 10.10. are experienced. The gas absorbs into the bloodstream, where it combines with hemoglobin to form *methemoglobin,* as follows:

$$\text{Hb}(aq) \ + \ \text{NO}_2(g) \ \longrightarrow \ \text{(NO}_2\text{)Hb}(aq)$$
$$\text{hemoglobin} \quad \text{nitrogen dioxide} \quad \text{methemoglobin}$$

Like carboxyhemoglobin, methemoglobin cannot effectively transport oxygen to the body's tissues, causing the host to suffer from the ailment called *methemoglobinemia.* Latent effects can be experienced by individuals exposed to nitrogen dioxide, resulting in death days after the initial exposure.

TABLE 10.11 ◆ Physical Properties of Nitrogen Dioxide

Freezing point	12°F (−11°C)
Boiling point	68°F (20°C)
Vapor density (air = 1)	1.59

The toxicity of nitrogen dioxide is also associated with its ability to react with water. A mixture of nitrous acid and nitric acid is produced, as follows:

$$2NO_2(g) + H_2O(g) \longrightarrow HNO_2(aq) + HNO_3(aq)$$
$$\text{nitrogen dioxide} \quad \text{water} \quad \quad \text{nitrous acid} \quad \text{nitric acid}$$

When it is inhaled, nitrogen dioxide reacts with atmospheric moisture to produce a mixture of these acids within the respiratory system. Even in low concentrations, nitric acid can damage the lining of the lungs and connective tissues and produce pulmonary edema.

Aside from its potential presence at fire scenes, nitrogen dioxide is also a commercial chemical product. Some of its important physical properties are provided in Table 10.11. Nitrogen dioxide is produced for commercial use by the oxidation of ammonia, as follows:

$$4NH_3(g) + 5O_2(g) \longrightarrow 4NO(g) + 6H_2O(g)$$
$$\text{ammonia} \quad \text{oxygen} \quad \quad \text{nitric oxide} \quad \text{water}$$

$$2NO(g) + O_2(g) \longrightarrow 2NO_2(g)$$
$$\text{nitric oxide} \quad \text{oxygen} \quad \quad \text{nitrogen dioxide}$$

When confined within a cylinder under pressure at room temperature, the nitrogen dioxide combines with itself, mole-for-mole, to produce dinitrogen tetroxide.

$$2NO_2(g) \longrightarrow N_2O_4(g)$$
$$\text{nitrogen dioxide} \quad \quad \text{dinitrogen tetroxide}$$

The mixture is used in the aerospace industry as a powerful oxidizing agent.

Nitric oxide is also a chemical product. It is available commercially as a liquefied compressed gas. Although it is poisonous, nitric oxide is used in hospitals and clinics to treat emphysema and similar pulmonary diseases. When a very low dose is mixed with oxygen and administered to a respiratory patient, the nitric oxide reduces inflammation and dilates blood vessels. When this mixture of gases is administered to premature babies, the nitric oxide reduces the risk of death and brain damage.

When nitric oxide and nitrogen dioxide are used in the workplace, OSHA requires employers to limit employee exposure to maximum concentrations of 25 ppm (30 mg/m^3) and 5 ppm (9 mg/m^3), respectively.

DOT regulates the transportation of nitric oxide and nitrogen dioxide (dinitrogen tetroxide), each as a poison gas. DOT requires shippers and carriers to affix INHALATION HAZARD, OXIDIZER, and CORROSIVE labels to their packaging and to display INHALATION HAZARD placards on the bulk packaging, freight container, transport vehicle, unit load device, or railcar used to transport them.

Performance Goals for Section 10.13

- Identify the common commercial products that contain ammonia.
- Identify the adverse effects experienced when humans are exposed to ammonia.
- Describe how ammonia is produced for commercial use.
- Identify the industries that use bulk volumes of ammonia.
- Identify the reason ammonia is encountered only at trace concentrations at fire scenes.

Ammonia is a colorless, pungent gas, easily detected and recognized by its familiar odor. It has been known since the days of alchemy, when it was produced by heating either coal or animal hoofs and horns.

For commercial use, ammonia gas is produced from atmospheric nitrogen and hydrogen at elevated temperature and pressure conditions at ammonia manufacturing plants (Figure 10.15).

$$N_2(g) + 3H_2(g) \longrightarrow 2NH_3(g)$$
$$\text{nitrogen} \quad \text{hydrogen} \qquad \text{ammonia}$$

The ammonia produced in the United States is directly used as a fertilizer or a coolant in large refrigeration systems. Over 80% is used for agricultural purposes; less than 2% is used in refrigeration. Ammonia was once the principal refrigerant used for very large refrigeration systems. It remains the most economical coolant choice for meat-packing plants, dairies, and similar units.

FIGURE 10.15 ◆ The major supply of ammonia is produced by the direct synthesis of hydrogen and atmospheric nitrogen under high temperature and pressure. The Phillips Petroleum plant in Beatrice, Nebraska, can produce 575 tons of ammonia daily. The large commercial demand for ammonia is linked with its utilization as a fertilizer for crops, either directly or indirectly, in such products as ammonium nitrate and ammonium phosphate. *Courtesy of the American Petroleum Institute, Washington, D.C.*

TABLE 10.12 ◆ Physical Properties of Anhydrous Ammonia	
Specific gravity (liquid)	0.77
Specific gravity (gas)	0.68
Vapor density (air = 1)	0.6
Freezing point	$-108°F$ $(-78°C)$
Boiling point	$-28°F$ $(-33°C)$
Lower explosive limit	16%
Upper explosive limit	25%

When used for such purposes, it is encountered as the liquefied compressed gas called *anhydrous ammonia.* Some of its properties are listed in Table 10.12. This product is transported by public highway, railway, and watercraft, and transferred by pipeline.

When farmers use anhydrous ammonia, they dispense it directly to soil or irrigation waters. The ammonia is typically discharged into soil from a tractor saddle tank

FIGURE 10.16 ◆ For domestic use, DOT regulates the transportation of anhydrous ammonia as a nonflammable gas (Division 2.2). DOT requires the nurse tank used for the application of anhydrous ammonia to agricultural soils to be marked ANHYDROUS AMMONIA and bear the identification number 1005 on NON-FLAMMABLE GAS placards or orange panels posted on all four sides of the tank. DOT also requires the warning IN-HALATION HAZARD to be marked on two opposing sides of the tank.

or nurse tank mounted behind the tillage tool through a distribution pod (Figure 10.16). When the use of anhydrous ammonia is required on a farm or other workplace, OSHA requires employers to limit employee exposure to a concentration of 50 ppm (35 mg/m^3), averaged over an 8-hour workday.

DOT regulates the domestic transportation of anhydrous ammonia as a nonflammable gas. DOT requires shippers and carriers to affix NON-FLAMMABLE GAS labels to its packaging and to display NON-FLAMMABLE GAS placards on the bulk packaging, freight container, transport vehicle, unit load device, or railcar used to transport it.

When shippers and carriers transport anhydrous ammonia by means of a DOT MC 330 or MC 331 tank truck, DOT requires them to provide information regarding the constituent water in the basic description (Table 6.4). For example, when the anhydrous ammonia contains water at a concentration of 0.2% or more by mass, the basic description is the following:

> 1 t/t   RQ, Anhydrous ammonia, 2.2, UN1005
> (0.2 percent water) (Inhalation hazard)

DOT also requires shippers and carriers to evaluate the nature of the tank truck used for transporting anhydrous ammonia, and to provide this information in the basic description. For example, shippers and carriers may transport anhydrous ammonia in a tank truck only when it has been constructed from a specially "quenched and tempered steel." When transporting anhydrous ammonia containing water at a concentration of 0.2% or more by mass, the basic description is the following:

> 1 t/t   RQ, Anhydrous ammonia, 2.2, UN1005 (0.2 percent
> water) (Not for Q & T tanks) (Inhalation Hazard)

In both instances, DOT also requires "Inhalation Hazard" as a component of the basic description (without the word "POISON" and without a zone designation).

When destined for international locations, DOT regulates the transportation of anhydrous ammonia as a poison gas. DOT requires shippers and carriers to affix INHALATION HAZARD and CORROSIVE labels to its packaging and to display INHALATION HAZARD placards on the bulk packaging, freight container, transport vehicle, unit load device, or railcar used to transport it.

AMMONIA SOLUTIONS

Most individuals have encountered ammonia as the gas that escapes from the commercial cleaning product in Figure 10.17 known as *household ammonia*. This is an aqueous solution of ammonia primarily used for cleaning glass and porcelain surfaces. Chemists refer to any aqueous solution of ammonia as *ammonium hydroxide,* or *aqueous ammonia;* its chemical formula is $NH_3(aq)$, or NH_4OH. The concentration of ammonia dissolved in household ammonia ranges from 2 to 5% by volume. The gas is perceptible at concentrations as low as approximately 5 ppm.

DOT regulates the transportation of the ammonium hydroxide solutions having the following descriptions:

> **Ammonia solutions,** *relative density less than 0.880 at 15°C in*
> *water with more than 50 percent ammonia*
>
> **Ammonia solutions,** *relative density between 0.880 and 0.957 at 15°C*
> *in water, with more than 10% but not more than 35% ammonia*
>
> **Ammonia solutions,** *relative density less than 0.880 at 15°C in water*
> *with more than 35 percent but not more than 50 percent ammonia*

FIGURE 10.17 ◆ Household ammonia is an aqueous solution that contains approximately 2 to 5% ammonia by volume. In this range of concentration, the solution is not a hazardous material. Commercial products containing household ammonia are used for cleaning walls, woodwork, tile and linoleum floors, porcelain, ceramics, and kitchen appliances.

DOT regulates the first solution domestically and internationally. When it is shipped domestically, DOT requires shippers and carriers to affix NON-FLAMMABLE GAS labels to the packaging and to display NON-FLAMMABLE GAS placards, when relevant, on the bulk packaging, freight container, transport vehicle, unit load device, or railcar used to transport it. When shippers and carriers domestically transport ammonia solutions, *relative density less than 0.880 at 15°C in water with more than 50 percent ammonia* by tank truck, DOT requires them to insert the words "Inhalation Hazard" (without the word "POISON" and without a zone designation) as a component of the basic description as follows:

    1 t/t   RQ, Ammonia solutions, 2.2, UN3318 (Inhalation Hazard)

When this solution is shipped internationally, DOT requires shippers and carriers to affix INHALATION HAZARD and CORROSIVE labels to the packaging and to display INHALATION HAZARD placards on the bulk packaging, freight container, transport vehicle, unit load device, or railcar used to transport it.

When the second ammonia solution is shipped, DOT requires shippers and carriers to affix CORROSIVE labels to its packaging and to display CORROSIVE placards, when relevant, on the bulk packaging, freight container, transport vehicle, unit load device, or railcar used to transport it.

When the third ammonia solution is shipped, DOT requires shippers and carriers to affix NON-FLAMMABLE GAS labels to its packaging and to display NON-FLAMMABLE GAS placards, when relevant, on the bulk packaging, freight container, transport vehicle, unit load device, or railcar used to transport it.

DOT does not regulate the transportation of household ammonia.

RESPONDING TO INCIDENTS INVOLVING THE RELEASE OF ANHYDROUS LIQUID AMMONIA

Individuals experience the ill effects listed in Table 10.13 when they inhale ammonia. The eyes, skin, and mucous membranes are particularly susceptible to severe irritation. The inhalation of ammonia acts as a respiratory stimulant. Elevated ammonia concentrations in the blood can cause death by suffocation. Anhydrous liquid ammonia also is corrosive and burns the skin and eyes.

Given the nature of the ill effects likely to be experienced from exposure to anhydrous liquid ammonia, the use of self-contained breathing apparatus is essential when responding to incidents involving the release of ammonia. When released directly from a liquid storage tank into the atmosphere, ammonia is generally visible as a white fog consisting of droplets of condensed atmospheric moisture. This fog can aid workers in identifying the location from which the ammonia is leaking from its storage tank or container. Nonetheless, the presence of the fog severely limits visibility within the immediate environment and hampers the performance of emergency-response personnel when encountering incidents involving the release of ammonia.

TABLE 10.13 ◆ Inhalation Effects of Ammonia on Humans

Ammonia Concentration in Air (ppm)	Symptoms
5–10	Detectable limit by odor
50	No chronic effects
150–200	General discomfort; eye tearing; irritation and discomfort of exposed skin; irritation of mucous membranes
400–700	Pronounced irritation and discomfort to the eyes, ears, nose, and throat
2000	Barely tolerable for more than a few moments; serious blistering of the skin; danger of pulmonary edema, asphyxia, and death within minutes following exposure

As shown in Table 10.13, ammonia is lighter than air. When it is released outdoors, ammonia quickly disperses into the atmosphere, especially under windy conditions. When anhydrous liquid ammonia is released into moist air, however, small liquid droplets of ammonia may be produced along with the gas. Under such conditions, these droplets behave as a dense gas, traveling along the surface of the ground instead of rising into the air and dispersing.

Water absorbs a very large volume of ammonia; at room conditions, 1 volume of water absorbs 1176 volumes of ammonia gas. This physical property is used to advantage when responding to incidents involving the release of ammonia. The recommended practice is to establish a water curtain downwind from the point at which the ammonia is being released. The water dissolves the ammonia and reduces the concentration that can travel further.

Solved Exercise 10.9

To protect public health, safety, and the environment, what immediate action should be taken by first-on-the-scene responders when encountering a strong odor of ammonia within an enclosed room at an ice cream plant?

Solution: Ammonia cannot ignite in air unless a concentration within its flammable range is exposed to an ignition source. The flammable range of ammonia is 16 to 25% by volume. A concentration within this range is readily achieved when ammonia escapes from large refrigeration units into an enclosed room. When the members of an emergency-response team first detect a strong odor of ammonia, their primary concern should be aimed at eliminating the likelihood of its ignition. This can be accomplished by opening doors and windows to vent the ammonia from the enclosure while simultaneously taking precautions to avoid generating a spark.

Since an exposure to ammonia burns the eyes and nasal passageways and the inhalation of elevated concentrations causes immediate asphyxiation, the implementation of any emergency-response action involving ammonia should be undertaken only while wearing a full-body suit and using self-contained breathing apparatus. No activity should ever be conducted within an enclosure in which the ammonia concentration is within its flammable range, as survival is unlikely if the ammonia subsequently ignites.

Although ammonia is a flammable gas, its flammable range extends from 16 to 25% by volume. Ammonia generally dissipates when released outdoors, and a flammable mixture is not achieved in air. In these circumstances, ammonia fires are unlikely. Nonetheless, ammonia concentrations in air *are* achieved within the flammable range during some accident scenarios, such as during the indoor release of ammonia from refrigeration systems into a confined area. In these circumstances, the occurrence of an ammonia fire is highly probable.

When ammonia ignites, it burns rapidly, producing nitrogen dioxide and water vapor, as follows:

$$4NH_3(g) + 7O_2(g) \longrightarrow 4NO_2(g) + 6H_2O(g)$$

ammonia   oxygen   nitrogen dioxide   water

When anhydrous ammonia is present in the workplace, OSHA requires employers to limit employee exposure to a short-term limit of 50 ppm (35 mg/m^3). OSHA also regulates the manner by which anhydrous ammonia is stored and handled in all workplaces other than ammonia manufacturing facilities and refrigeration plants using anhydrous ammonia solely as a refrigerant.

AMMONIA AT FIRE SCENES

Materials manufactured from animal products thermally decompose when they are exposed to intense heat. These materials include leather items and wool carpeting. When they are present at fire scenes, the odor of ammonia is often perceptible. Nonetheless, the ammonia concentration produced by the thermal decomposition of animal products is generally too low to inflict harm upon firefighters.

The odor of ammonia may also be encountered at a fire scene where fertilizers containing ammonium compounds are stored, usually within garages or sheds. When these fertilizers decompose upon exposure to heat, ammonia is produced. For example, when ammonium sulfate fertilizer is heated to high temperatures, it thermally decomposes to ammonia, sulfur dioxide, nitrogen, and water.

$$3(NH_4)_2SO_4(s) \longrightarrow 4NH_3(g) + N_2(g) + 3SO_2(g) + 6H_2O(g)$$
$$\text{ammonium sulfate} \qquad \text{ammonia} \quad \text{nitrogen} \quad \text{sulfur dioxide} \quad \text{water}$$

Performance Goals for Section 10.14

- Describe the nature of the response actions to be implemented when toxic substances have been released into the environment.
- Describe the manner in which emergency responders use the *Emergency Response Guidebook* to establish the initial isolation zone and the protective-action zone associated with large and small spills of a toxic substance.

◆ 10.14 RESPONDING TO INCIDENTS INVOLVING THE RELEASE OF TOXIC SUBSTANCES

At the scene of a transportation mishap, an emergency-response crew may learn that a toxic substance has been released to the environment by observing any of the following:

- The hazard class numbers 2.3 or 6.1 have been included in the basic description of a toxic substance on a shipping paper.
- POISON or POISON—INHALATION HAZARD, Zone A, Zone B, Zone C, or Zone D (see below) is marked on the packaging containing the toxic substance. The significance of the hazard zones shall be noted shortly.
- White POISON GAS (or TOXIC GAS), INHALATION HAZARD, or POISON labels are affixed to the packaging of a hazardous material.
- White POISON GAS (or TOXIC GAS), INHALATION HAZARD, or POISON placards are displayed on the bulk packaging, freight containers, transport vehicles, and rail cars used to transport the toxic material.

TABLE 10.14 ◆ Assignment of the Hazard Zone For Toxic Substances That Pose an Inhalation Hazard[a]

Hazard Zone	Inhalation Toxicity
A	LC_{50} less than or equal to 200 ppm
B	LC_{50} greater than 200 ppm and less than or equal to 1000 ppm
C	LC_{50} greater than 1000 ppm and less than or equal to 3000 ppm
D	LC_{50} greater than 3000 ppm or less than or equal to 5000 ppm

[a]49 CFR §173.116

DOT requires shippers and carriers to establish the hazard zone of a hazardous material that poses an inhalation hazard. The *hazard zone* is designated by one of the four letters "A," "B," "C," and "D" and is established on the basis of the inhalation toxicity criteria provided in Table 10.14. For the four possible descriptions, the greatest health risk is posed by the presence of a hazardous material whose basic description includes the phrase "Inhalation Hazard (Zone A)."

DOT requires shippers and carriers to communicate the hazard zone of a hazardous material that poses an inhalation hazard by marking the relevant packaging with the words "POISON—INHALATION HAZARD" and "Zone A," "Zone B," "Zone C," or "Zone D" for gases or "Zone A" or "Zone B" for liquids. DOT also requires shippers and carriers to enter these precautionary warnings immediately adjacent to the basic description of a toxic substance on the relevant shipping paper. Table 6.4 illustrates this DOT requirement for a shipment of the toxic gas phosgene.

When a toxic substance has spilled during a transportation mishap, DOT provides emergency-response personnel with initial isolation and protective-action distances in tabular form in its *Emergency Response Guidebook*. Excerpts of the relevant table relating to the toxic substances noted in this chapter are provided in Table 10.15. To use this table effectively, first-on-the-scene responders must first determine whether a "small" or "large" spill of the toxic substance has occurred. This determination can ordinarily be made by speaking with the individual most immediately responsible for transportation of the hazardous material. A "small" spill is one that involves a single package such as a 55-gal (208-L) drum, a small cylinder, or a small leak from a large package. A "large" spill involves a release from bulk packaging or multiple releases from several small packages.

Once a determination regarding the size of the release has been confirmed, the guidebook is used to establish each of the following:

- The *initial isolation distance*. This is the distance from an emergency scene to which should be quickly moved in a *crosswind* direction to establish the *initial isolation zone* shown in Figure 10.18(a). The initial isolation zone is the area surrounding the emergency scene in which persons may be exposed to dangerous, life-threatening concentrations of a toxic gas or vapor that poses an inhalation hazard. Emergency-response personnel first direct persons outside the initial isolation zone.
- The *protective-action distance*. This is the distance downwind from the scene that establishes the *protective-action zone* shown in Figure 10.18(b). The protective-action zone is the area in which persons may become incapacitated and unable to take protective action and/or incur serious, irreversible health effects. After they direct persons outside

Solved Exercise 10.10

Using Table 10.15, identify the initial isolation and protective-action distances recommended by DOT for protection of public health when responding to an 11:00 P.M. domestic transportation mishap involving a leaking 75-lb (34-kg) cylinder of hydrogen sulfide.

   Solution: It is first necessary to identify whether this incident involves a "small" or "large" spill of hydrogen sulfide. It is prudent to acknowledge that 75 lb (34 kg) constitutes a "large" spill of a toxic substance. Since it is dark at 11:00 P.M. in most parts of the world, Table 10.14 is examined under the heading "night" for the initial isolation and protective-action distances for potential exposure to hydrogen sulfide. These distances are listed in Table 10.15 as 400 ft (125 m) and 0.9 mi (1.4 km), respectively.

the initial isolation zone, emergency-response personnel direct others to outside the protective-action zone, beginning with those individuals closest to the right-hand semi-circular component of the initial isolation zone.

When emergency-response personnel are required to encounter a toxic substance, they must first don appropriate protective clothing to prevent the possibility of skin contact. This protective clothing includes impermeable coveralls or similar fully en-capsulating body suits, gloves, head coverings, and self-contained breathing apparatus. Since toxic substances can enter the body not only by inhalation but also by skin ab-sorption, the use of the full encapsulating body suit minimizes or eliminates the risk of exposure to the substance by both routes. When clothing has contacted a toxic sub-stance, it should be thoroughly washed with water prior to removal.

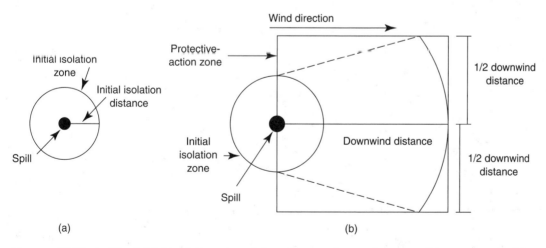

FIGURE 10.18 ◆ The initial isolation zone (a) and protective-action zone (b) associated with the spill of a toxic substance that is an inhalation hazard. Both zones constitute regions in which individuals risk exposure to toxic gases or vapors associated with the spill of a hazardous material in hazard classes 2.3 and 6.1. From a practical viewpoint, the initial isolation zone is a circle whose radius equals the initial isolation distance, and the protective-action zone is a square each of whose sides equal the protective-action (downwind) distance.

TABLE 10.15 ◆ Initial Isolation and Protective-Action Distances[a]

ID No.	NAME OF MATERIAL	SMALL SPILLS (From a Small Package or Small Leak from a Large Package)						LARGE SPILLS (From a Large Package or from Many Small Packages)					
		First ISOLATE in All Directions		Then PROTECT Persons Downwind During-				First ISOLATE in All Directions		Then PROTECT Persons Downwind During-			
				DAY		NIGHT				DAY		NIGHT	
		Meters	(Feet)	Kilometers	(Miles)	Kilometers	(Miles)	Meters	(Feet)	Kilometers	(Miles)	Kilometers	(Miles)
1005	Ammonia, anhydrous	30 m	(100 ft)	0.2 km	(0.1 mi)	0.2 km	(0.1 mi)	60 m	(200 ft)	0.5 km	(0.3 mi)	1.1 km	(0.7 mi)
1005	Ammonia, anhydrous, liquefied												
1005	Ammonia, solution, with more than 50% Ammonia												
1005	Anhydrous ammonia												
1005	Anhydrous ammonia, liquefied												
1016	Carbon monoxide	30 m	(100 ft)	0.2 km	(0.1 mi)	0.2 km	(0.1 mi)	125 m	(400 ft)	0.6 km	(0.4 mi)	1.8 km	(1.1 mi)
1016	Carbon monoxide, compressed												
1051	Hydrocyanic acid, aqueous solutions, with more than 20% Hydrogen cyanide	60 m	(200 ft)	0.2 km	(0.1 mi)	0.5 km	(0.3 mi)	400 m	(1300 ft)	1.3 km	(0.8 mi)	3.4 km	(2.1 mi)
1051	Hydrocyanic acid, liquefied												
1051	Hydrogen cyanide, anhydrous, stabilized												
1051	Hydrogen cyanide, stabilized												
1053	Hydrogen sulfide	30 m	(100 ft)	0.2 km	(0.1 mi)	0.3 km	(0.2 mi)	215 m	(700 ft)	1.4 km	(0.9 mi)	4.3 km	(2.7 mi)
1053	Hydrogen sulfide, liquefied												
1053	Hydrogen sulphide												
1053	Hydrogen sulphide, liquefied												
1062	Methyl bromide	30 m	(100 ft)	0.2 km	(0.1 mi)	0.3 km	(0.2 mi)	95 m	(300 ft)	0.5 km	(0.3 mi)	1.4 km	(0.9 mi)
1067	Dinitrogen tetroxide	30 m	(100 ft)	0.2 km	(0.1 mi)	0.5 km	(0.3 mi)	305 m	(1000 ft)	1.3 km	(0.8 mi)	3.9 km	(2.4 mi)
1067	Dinitrogen tetroxide, liquefied												
1067	Nitrogen dioxide												
1067	Nitrogen dioxide, liquefied												
1067	Nitrogen peroxide, liquid												
1067	Nitrogen tetroxide, liquid												
1079	Sulfur dioxide	30 m	(100 ft)	0.3 km	(0.2 mi)	1.1 km	(0.7 mi)	185 m	(600 ft)	3.1 km	(1.9 mi)	7.2 km	(4.5 mi)

ID No.	Name of Material	SMALL SPILLS First ISOLATE in all Directions	Then PROTECT persons Downwind DAY	Then PROTECT persons Downwind NIGHT	LARGE SPILLS First ISOLATE in all Directions	Then PROTECT persons Downwind DAY	Then PROTECT persons Downwind NIGHT
1079	Sulfur dioxide, liquefied						
1079	Sulfur dioxide						
1079	Sulfur dioxide, liquefied						
1613	Hydrocyanic acid, aqueous solution, with not more than 20% Hydrogen cyanide (when "inhalation Hazard" is on a package or shipping paper)	30 m (100 ft)	0.2 km (0.1 mi)	0.2 km (0.1 mi)	125 m (400 ft)	0.5 km (0.3 mi)	1.3 km (0.8 mi)
1613	Hydrogen cyanide, aqueous solution, with not more than 20% Hydrogen cyanide (when "inhalation Hazard" is on a package or shipping paper)						
1614	Hydrogen cyanide, anhydrous, stabilized (absorbed)	60 m (200 ft)	0.2 km (0.1 mi)	0.5 km (0.3 mi)	400 m (1300 ft)	1.3 km (0.8 mi)	3.4 km (2.1 mi)
1614	Hydrogen cyanide, stabilized (absorbed)						
1660	Nitric oxide	30 m (100 ft)	0.3 km (0.2 mi)	1.3 km (0.8 mi)	155 m (500 ft)	1.3 km (0.8 mi)	3.5 km (2.2 mi)
1660	Nitric oxide, compressed						
1680	Potassium cyanide (when spilled in water)	30 m (100 ft)	0.2 km (0.1 mi)	0.3 km (0.2 mi)	95 m (300 ft)	0.8 km (0.5 mi)	2.6 km (1.6 mi)
1689	Sodium cyanide (when spilled in water)	30 m (100 ft)	0.2 km (0.1 mi)	0.3 km (0.2 mi)	95 m (300 ft)	1.0 km (0.6 mi)	2.6 km (1.6 mi)
1953	Compressed gas, flammable, poisonous, n.o.s. (Inhalation Hazard Zone A)	185 m (600 ft)	1.8 km (1.1 mi)	5.6 km (3.5 mi)	915 m (3000 ft)	10.8 km (6.7 mi)	11.0 + km (7.0 + mi)
1953	Compressed gas, flammable, poisonous, n.o.s. (Inhalation Hazard Zone B)	30 m (100 ft)	0.3 km (0.2 mi)	1.1 km (0.7 mi)	305 m (1000 ft)	3.1 km (1.9 mi)	7.7 km (4.8 mi)
1953	Compressed gas, flammable, poisonous, n.o.s. (Inhalation Hazard Zone C)	30 m (100 ft)	0.2 km (0.1 mi)	1.0 km (0.6 mi)	215 m (700 ft)	2.1 km (1.3 mi)	5.6 km (3.5 mi)

(continued)

437

TABLE 10.15 ◆ (continued)

ID No.	NAME OF MATERIAL	SMALL SPILLS (From a Small Package or Small Leak from a Large Package)					LARGE SPILLS (From a Large Package or from Many Small Packages)					
		First ISOLATE in All Directions		Then PROTECT Persons Downwind During-				First ISOLATE in All Directions		Then PROTECT Persons Downwind During-		
				DAY		NIGHT				DAY		NIGHT	
		Meters	(Feet)	Kilometers	(Miles)	Kilometers	(Miles)	Meters	(Feet)	Kilometers	(Miles)	Kilometers	(Miles)
1953	Compressed gas, flammable, poisonous, n.o.s. (Inhalation Hazard Zone D)	30 m	(100 ft)	0.2 km	(0.1 mi)	0.6 km	(0.4 mi)	185 m	(600 ft)	1.6 km	(1.0 mi)	4.3 km	(2.7 mi)
1953	Compressed gas, flammable, toxic, n.o.s. (Inhalation Hazard Zone A)	185 m	(600 ft)	1.8 km	(1.1 mi)	5.6 km	(3.5 mi)	915 m	(3000 ft)	10.8 km	(6.7 mi)	11.0 + km	(7.0 + mi)
1953	Compressed gas, flammable, toxic, n.o.s. (Inhalation Hazard Zone B)	30 m	(100 ft)	0.3 km	(0.2 mi)	1.1 km	(0.7 mi)	305 m	(1000 ft)	3.1 km	(1.9 mi)	7.7 km	(4.8 mi)
1953	Compressed gas, flammable, toxic, n.o.s. (Inhalation Hazard Zone C)	30 m	(100 ft)	0.2 km	(0.1 mi)	1.0 km	(0.6 mi)	215 m	(700 ft)	2.1 km	(1.3 mi)	5.6 km	(3.5 mi)
1953	Compressed gas, flammable, toxic, n.o.s. (Inhalation Hazard Zone D)	30 m	(100 ft)	0.2 km	(0.1 mi)	0.6 km	(0.4 mi)	185 m	(600 ft)	1.6 km	(1.0 mi)	4.3 km	(2.7 mi)

[a] Adapted from *Emergency Response Guidebook* (Washington, D.C.: U.S. Department of Transportation), 2000.

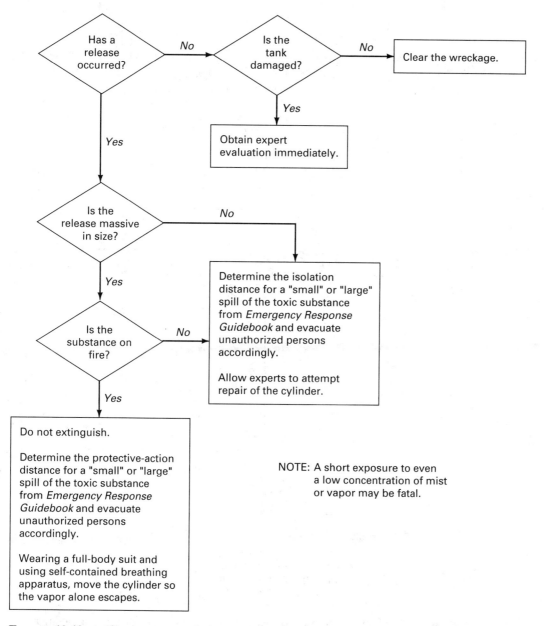

FIGURE 10.19 ◆ The recommended procedures when responding to a disaster involving a poison gas contained in cylinders. *Adapted with permission of the American Society for Testing and Materials, from a figure in ASTM STP 825, A Guide to the Safe Handling of Hazardous Materials Accidents. Copyright © 1983, American Society for Testing and Materials.*

When an emergency-response team is called to a scene involving the release of a toxic substance, the procedures in Figures 10.19, 10.20, and 10.21 should be implemented. Whenever there is a poisoning emergency, information concerning the unique properties of the poison may be essential for the successful protection of lives, property, and the environment. To reduce or eliminate the risk of injury from poison

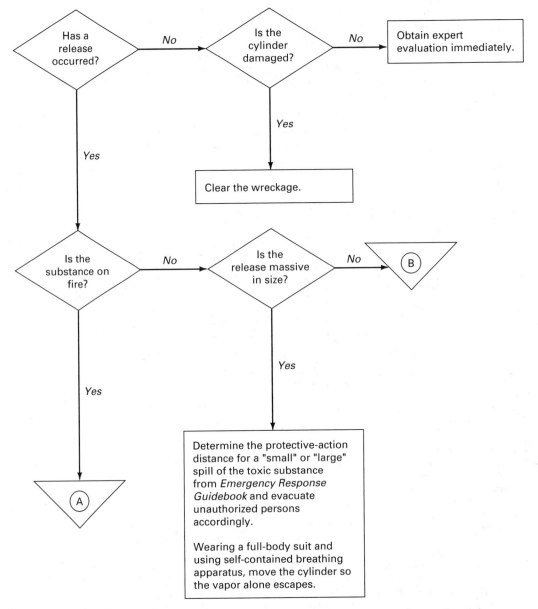

FIGURE 10.20 ◆ The recommended procedures when responding to a disaster involving a poison gas in a bulk container such as a large cylinder. *Adapted with permission of the American Society for Testing and Materials, from a figure in ASTM STP 825, A Guide to the Safe Handling of Hazardous Materials Accidents. Copyright © 1983, American Society for Testing and Materials.*

exposure, the American Association of Poison Control Centers can be utilized as an informational source.*

When fighting fires involving a poison, firefighters should also observe the following basic principles:

---

*The American Association of Poison Control Centers can be accessed by telephone at (800) 222-1222. It is essentially a library staffed by nonmedical personnel. Be prepared to give the name of the poisonous product involved in the emergency and any information provided on the label.

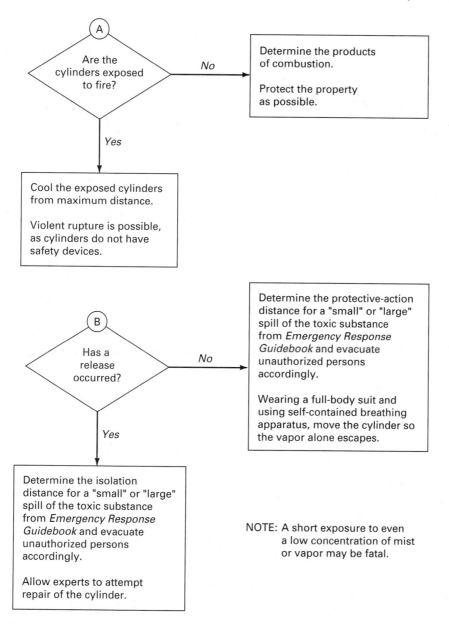

FIGURE 10.20 ◆ *(continued)*

- Use impermeable coveralls or similar full-body suit, gloves, head covering, and self-contained breathing apparatus.
- Avoid direct exposure to the smoke or fumes that evolve from these fires.
- Attack the fire from an upwind location.
- Use fog instead of direct streams of water, thereby limiting the quantity of poisonous dust at the fire scene.
- Keep runoff water to a minimum, and channel runoff water into a temporary reservoir to prevent its entrance into local sewers or waterways.
- Notify the operators of the relevant storm and sanitary sewer system and treatment plant of the ongoing fire.

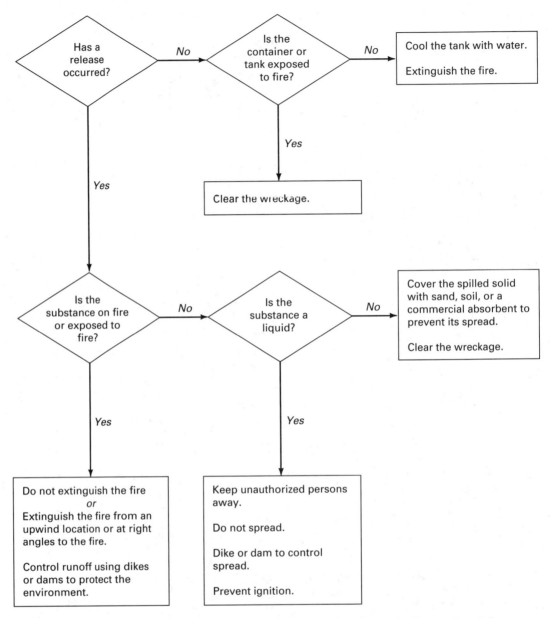

FIGURE 10.21 ◆ The recommended procedures when responding to a disaster involving a poison such as most pesticides. *Adapted with permission of the American Society for Testing and Materials, from a figure in ASTM STP 825, A Guide to the Safe Handling of Hazardous Materials Accidents. Copyright © 1983, American Society for Testing and Materials.*

Performance Goals for Section 10.15
- Describe the nature of a carcinogen.
- Describe the manner by which epidemiologists classify carcinogens.
- Describe the means OSHA uses to signal the presence of carcinogens in the workplace.

Millions of individuals are annually diagnosed with cancer, which in many instances is believed to be associated with exposure to chemical substances. The substances that can cause cancer are called *carcinogens*. The determination as to whether a given substance causes cancer is an active area of epidemiological research.

Epidemiologists are scientists who are concerned with the causes, nature, and transmission of diseases. Since it is considered unethical to conduct research upon humans, they largely perform their research upon animal subjects or cell cultures. As a result of their studies, the process of *carcinogenesis*—the contracting of cancer—has been identified as a multistepped cellular phenomenon. During the *initiation stage,* exposure to a carcinogen permanently alters the genetic material in a cell. Sometimes, repeated exposures to the same substance are needed to induce the cellular alteration, while on other occasions, a single exposure is ample. During the subsequent *promotion stage,* the damaged cell divides into new but similarly damaged cells, each of which subsequently again divides into new damaged cells. When this division occurs over and over, the aggregates of damaged cells create masses of tissue called *polyps,* which can undergo further genetic mutations that transform into malignant tumors.

Epidemiologists recognize the three groups of carcinogens noted below:

- *Known human carcinogens.* These are substances for which there is sufficient evidence from epidemiological studies to conclude unquestionably that they cause cancer in humans.
- *Probable human carcinogens.* These are substances for which there is sufficient evidence to conclude that they cause cancer in experimental animals, but no direct evidence that they cause cancer in humans. Nonetheless, prudence dictates that exposure to a probable human carcinogen can be reasonably anticipated to cause cancer in humans.
- *Possible human carcinogens.* These are substances for which there is reason to suspect that they may cause cancer in animals or humans, but no studies have yet affirmed the suspicion.

The classification of a substance as a known, probable, or possible human carcinogen is a necessary one, since the existing data do not always enable epidemiologists to clearly ascertain whether the substance causes cancer in the human organism. Research studies in carcinogenesis are compounded by the observation that certain substances appear incapable of directly causing cancer themselves, but can affect the rate of cancerous tumor formation when the exposure occurs along with other substances. The biochemical mechanisms by which one substance induces another to initiate or promote cancer are not entirely understood.

Exposure to cancer-causing substances occurs in a variety of ways. Carcinogens are contained within certain commercial products that we use routinely. They include asbestos, crude petroleum, and petroleum products. When the presence of a carcinogen within a commercial product is known, the consumer must be advised. Conveying this information is the responsibility of the seller, who posts the information where the product is sold. For example, the sign in Figure 10.22 advises parties who purchase gasoline that the product contains low levels of benzene and other carcinogens.

When carcinogens are present in the workplace, OSHA requires employers to provide employees with respirators and protective clothing. OSHA also requires

FIGURE 10.22 ◆ The laws in some states require the posting of a sign to warn customers of the presence of cancer-causing substances within commercial products sold within the state. This sign warns drivers that exposure to some substances within gasoline can cause cancer, birth defects, or other reproductive harm.

> **WARNING**
>
> Chemicals known to the state to cause cancer, birth defects, or other reproductive harm are found in gasoline, crude petroleum, and many other petroleum products and their vapors, or result from their use. Read and follow label directions and use care when handling or using all petroleum products.

employers to provide a segregated area in which the work with carcinogens can be safely conducted, and to post this area as follows:

> **CANCER-SUSPECT AGENT EXPOSED IN THIS AREA**
>
> **IMPERVIOUS SUIT INCLUDING GLOVES, BOOTS, AND AIR-SUPPLIED HOOD REQUIRED AT ALL TIMES**
>
> **AUTHORIZED PERSONNEL ONLY**

Emergency-response teams should heed this warning when they are required to enter buildings in which carcinogens are stored or used.

We shall identify the carcinogens of immediate concern to emergency-response personnel as we encounter them in this and future chapters.*

Performance Goals for Section 10.16
- Identify the metals and metalloids required in the diet for good health.
- Identify the negative impact experienced by the human organism from exposure to lead.
- Identify ways by which emergency-response personnel are likely to encounter lead.
- Describe the means OSHA uses to signal the presence of lead in the workplace.

*Various scientific groups conduct research relating to the ability or tendency of substances to produce cancer in normal tissues. Unless otherwise indicated, the designations in this book are those denoted by the U.S. Department of Health and Human Services and the World Health Organization's International Agency for Research on Cancer (IARC).

For our purposes, metallic compounds may be classified as follows:

- *Essential elements.* For the proper functioning and survival of the human organism, trace amounts of 14 metals and metalloids are essential in our diets. For this reason, nutritionists refer to them as the essential elements. They are calcium, potassium, sodium, magnesium, iron, zinc, copper, tin, vanadium, chromium, manganese, molybdenum, cobalt, and nickel. Their proper balance and availability are essential for proper biological function. This is not meant to imply that the essential elements are beneficial in all doses. The ingestion of excessive concentrations of the essential elements can cause illness or death.
- *Toxic metals.* Certain other metals and metalloids are not known to offer any dietary benefit to the human organism. They include arsenic, barium, beryllium, lead, mercury, and thallium. The consumption of even minute amounts of their compounds can cause illness or death.
- Certain other metals and metalloids are either toxic *or* beneficial, depending on conditions. They include the compounds of selenium and tungsten. They can either exert a toxic effect or perform an essential biological function in the human body.

LEAD AND ITS COMPOUNDS

Of the toxic metals, it is lead whose compounds are most likely to be encountered by emergency responders. Firefighters are exposed to lead whenever they inhale lead-laden dust or smoke.

A primary source of lead compounds is the paint pigment formerly used in lead-based paint. Although the use of lead-based paint has been banned in the United States for most purposes since 1978, millions of apartments and other dwellings with lead-based paint still exist. When these structures collapse during building fires, lead-laden dust is dispersed into the atmosphere. Firefighters are then put at risk of inhaling lead particulates.

The use of respiratory protection equipment significantly eliminates or reduces firefighter exposure to contaminated dust. On the other hand, children who live in dwellings formerly coated with lead-based paint are not as fortunate. Lead-based paint is considered the greatest remaining source of childhood lead exposure. Hungry youngsters, especially those forced to live in substandard housing, sometimes eat old, flaking paint and plaster.

When ingested or inhaled, lead primarily affects the human blood-forming, nervous, and kidney systems, but it also harms the reproductive, endocrine, hepatic, cardiovascular, immunologic, and gastrointestinal processes. Exposure to elevated concentrations of lead can cause severe health problems, including brain disease, colic, palsy, and anemia. Damage to the central nervous system in general, and to the brain in particular, is one of the most severe consequences of chronic lead poisoning.

Children are especially susceptible to the adverse effects of exposure to lead. In particular, lead impairs their ability to learn and execute mental processes effectively. Research studies have equated a 7.4-point *drop* in the intelligence quotient (IQ) of children with an increase from 1 to 10 micrograms of lead per deciliter of blood (1 to 10 µg/dL).

EPA regulates the concentration of lead in the ambient air as a criteria air pollutant. EPA has set the primary and secondary national ambient air quality standards for lead at 1.5 $\mu g/m^3$ as a quarterly average.

Exposure to certain lead compounds may cause latent health effects. In particular, lead acetate and lead phosphate are regarded as probable human carcinogens.

When the use of lead compounds is needed within the workplace, OSHA requires employers to limit employee exposure to a concentration of 0.050 mg/m³, averaged over an 8-hour workday. OSHA also requires the posting of the following warning sign in work areas where airborne lead is generated.

> **WARNING**
>
> **LEAD WORK AREA**
> **POISON**
>
> **NO SMOKING OR EATING**

DOT regulates the transportation of nonoxidizing lead compounds as poisons. These compounds include lead acetate, lead arsenate, lead arsenite, and lead cyanide. DOT requires shippers and carriers to affix POISON labels on their packaging and, when warranted, to post POISON placards on the bulk packaging, freight container, transport vehicle, unit containment device, or railcar used to transport them. DOT regulates the transportation of oxidizing lead compounds as oxidizers. We will note the hazardous features of these compounds in Chapter 11.

Performance Goals for Section 10.17

- Identify the chemical nature of asbestos.
- Identify the type of asbestos products likely to contain friable asbestos.
- Describe the latent effects associated with exposure to certain asbestos fibers.
- Describe the means OSHA uses to signal the presence of asbestos in the workplace.

◆ 10.17 ASBESTOS

 Asbestos is the generic name of a complex mineral composed of several naturally occurring magnesium silicates. Several types of asbestos are recognized, all of which are nonflammable and excellent insulators. The specific type of asbestos in Figure 10.23 exists as strong, flexible fibers that can be spun into threads and woven into fireproof fabric.

When mixed with magnesium oxide, asbestos was formerly used for a variety of fireproofing, insulating, soundproofing, and decorative purposes. It was a component of products like asbestos pipe coverings, flooring products, paper products, antifriction materials (brake lining and clutch facings), roofing materials, and coating and patching compounds.

Although asbestos products were once prized for their qualities, the availability of such products has been sharply curtailed today. Their virtual absence in today's marketplace is linked with the recognition that asbestos exposure causes a number of serious health problems including lung cancer. Unless the asbestos has been completely sealed into a product, as in asbestos floor tiles, it can break apart into tiny fibers, much

FIGURE 10.23 ◆ Cummingtonite (top left) and cummingtonite-grunerite asbestos (bottom left); serpentenite chrysotile veins (right). In nature, asbestos-bearing minerals are usually found as rocks. However, under certain geological conditions, they crystallize as asbestos fibers. Cummingtonite-grunerite asbestos, produced only in the Transvaal province of South Africa, is the mineral that may crystallize as the form of asbestos fibers known as *amosite,* from the acronym for the company, Asbestos Mines of South Africa. *Courtesy of the U.S. Department of the Interior, Bureau of Mines, Washington, D.C.*

smaller and more buoyant than ordinary dust. This occurs when the asbestos-containing material is damaged or disintegrates with age. The crumbly, powdery asbestos, resembles the dust from the asbestos-coated heating pipes in Figure 10.24. It is said to be *friable*. It is the friable form of asbestos that poses severe health hazards to those individuals who inhale or swallow its fibers.

Asbestos fibers range from 0.1 to 10 μm in length (Figure 10.25). They float almost indefinitely in the air and may be readily inhaled or swallowed. When they are inhaled, asbestos fibers may be removed by the cilia lining the respiratory tract. Then, they move into the throat where they are swallowed and pose the threat of causing cancer of the esophagus, stomach, intestines, and rectum.

In addition to lung cancer, asbestos fibers potentially cause the following illnesses once they have been inhaled:

- *Asbestosis,* a noncancerous scarring of lung tissue that occurs when sharp asbestos fibers deposit deeply within the respiratory tract.
- *Mesothelioma,* a unique cancer of the membranes lining the chest and abdomen. Mesothelioma is characteristic of asbestos overexposure, since it rarely occurs in individuals who have not been exposed to asbestos. It is always fatal. Since epidemiologists have established that asbestos exposure causes lung cancer and mesothelioma in humans, asbestos is denoted as a known human carcinogen.

FIGURE 10.24 ◆ Asbestos used as insulation around this piping poses a hazardous condition as it crumbles and flakes. The asbestos separates into a dust of extremely tiny fibers too small to be detected visibly. When these fibers are inhaled or swallowed, they may cause asbestosis, lung cancer, and mesothelioma. In the workplace, when airborne concentrations of asbestos fibers exceed exposure limits established by OSHA, this sign is posted so that employees may take necessary protective steps before entering the area. *Courtesy of the U.S. Environmental Protection Agency, Washington, D.C.*

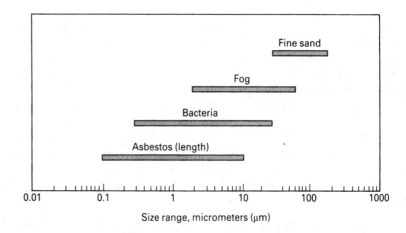

FIGURE 10.25 ◆ The typical size of asbestos fibers, from 0.1 to 10 µm, relative to some other materials. This size is not generally visible to the human eye. *Courtesy of the U.S. Environmental Protection Agency, "Asbestos-Containing Materials in School Buildings: A Guidance Document," Office of Toxic Substances, Washington, D.C., EPA 450/2-78-014, March 1978.*

Asbestos is characteristic of certain substances that are capable of causing a more serious adverse health problem when exposed to the body in combination with a second substance. This phenomenon is called *synergism*. When asbestos fibers are inhaled by cigarette smokers, they act synergistically with the toxic substances in tobacco smoke. The result is that a smoker exposed to asbestos fibers has up to 50 times the chance of developing lung cancer compared to a nonexposed nonsmoker.

Emergency responders are among the individuals who are still likely to be exposed to asbestos fibers. As they fight fires in old buildings, emergency responders are likely to be exposed to airborne asbestos, especially when dust is generated as ceilings and roofs collapse. To reduce or eliminate this exposure, firefighters should combat these fires from upwind locations and use respiratory protection gear.

Various governmental agencies have banned the use of products containing asbestos and have otherwise regulated its exposure. OSHA requires employers to limit employee exposure to an airborne concentration of asbestos fibers of 0.1 fiber per cubic centimeter of air as an 8-hour, time-weighted average. When asbestos is present in the workplace, OSHA requires the area to be posted as follows:

CAUTION

CONTAINS ASBESTOS FIBERS

AVOID CREATING DUST

**BREATHING ASBESTOS DUST MAY
CAUSE SERIOUS BODILY HARM**

FDA has enacted regulations to prevent the contamination of foods, drugs, and cosmetics with asbestos. Furthermore, the Consumer Products Safety Commission has promulgated regulations to eliminate or reduce asbestos in consumer products; for example, the use of asbestos in consumer clothing, hair dryers, and other products is now banned.

Using the authority of several environmental statutes, EPA banned the use of asbestos in pipe and boiler coverings, as well as all uses of asbestos materials that are applied by spraying. To safeguard the health of school children and others who work in schools, EPA launched a program aimed at identifying school buildings that contain asbestos materials, inspecting these buildings to determine if any fibers are being released into the air, and requiring the removal and repair of damaged asbestos material.

EPA also regulates the manner by which the asbestos is stripped from walls, ceilings, and other locations (Figure 10.26). Proper handling practices include sealing a work site before the asbestos is removed, preventing the asbestos fibers from becoming airborne by keeping them wet, and placing the asbestos debris in special bags that are subsequently sealed.

Finally, DOT regulates the domestic transportation of asbestos as a Class 9 hazardous material. DOT requires shippers and carriers to affix CLASS 9 labels to its packaging and to display CLASS 9 placards on the bulk packaging, freight container, transport vehicle, unit load device, or railcar used to transport it.

FIGURE 10.26 ◆ When activities involving the removal of asbestos from walls and ceilings are conducted, OSHA requires employers to limit employee exposure to asbestos fibers by preventing the unnecessary generation of asbestos dust. OSHA also requires employers to further reduce the risk to workers by requiring that they use respirators and wet the dust with water. *Courtesy of Todd Buchanan, photographer.*

Performance Goals for Section 10.18
- Describe the manner by which pesticides are classified.
- Describe the nature of the advisory information pesticide manufacturers provide on their product labels.

◆ 10.18 PESTICIDES

The term *pesticide* is used to denote a commercial product that has been specifically designed to destroy or control the spread of insects, fungi, rodents, plants, and other pests. Although targeted toward nonhuman exposure, pesticides are sometimes encountered under conditions that adversely impact the human organism. To advise the consumer of the hazards posed by exposure to specific pesticides, EPA uses the authority of FIFRA to compel manufacturers to provide advisory information on their pesticide product labels (Section 1.4).

Pesticides can be classified as follows:

- Carbamate pesticides, derivatives of carbamic acid (H_2N—COOH)
- Organochlorine pesticides, chlorinated derivatives of certain complex hydrocarbons
- Organophosphorus pesticides, derivatives of phosphoric acid

DOT regulates the domestic transportation of most pesticides as poisons. DOT requires shippers and carriers to affix POISON labels to its packaging and, when relevant, to display POISON placards on the bulk packaging, freight container, transport vehicle, unit load device, or railcar used to transport them.

Performance Goals for Section 10.19

- ◆ Describe the manner by which terrorists may use biological warfare agents to injure, incapacitate, or cause mass casualties.
- ◆ Identify some common biological warfare agents and the adverse health effects they cause.

◆ 10.19 BIOLOGICAL WARFARE AGENTS

Certain microorganisms and toxins are so highly poisonous to humans that, during wartime, their distribution could serve to injure, incapacitate, or cause mass casualties. They are called *biological warfare agents,* the most common types of which are listed in Table 10.16. Their use during wartime could induce impairment, disease, and death among the enemy on a large scale. Although this practice has been condemned by civilized countries as inhumane, mutual mistrust and fear have prompted the production and stockpiling of biological agents. Today, the threat that terrorists could use them is a widespread fear.

For maximum effectiveness during warfare, biological warfare agents can be sealed into canisters and charged into missile heads or other explosive devices. They can also be disseminated by the following methods:

- ◆ Aerosolized dispersal from aircraft (e.g., crop dusters) over a selected area
- ◆ Contaminating food or a drinking water supply
- ◆ Inoculating fabric, paper, and similar items with infectious material

When biological agents are disseminated, the resulting exposure differs from exposure to chemical agents (Section 12.16) in that hours or days must pass following the initial exposure before observable symptoms develop.

Biological warfare agents are either etiological agents (Section 6.7) or toxins. Etiological agents are living bacteria or viruses, exposure to which causes such diseases as anthrax, botulism, bubonic plague, cholera, and smallpox. On the other hand, *toxins* are poisonous substances produced biochemically by certain plants and animals. Examples of organisms that produce toxins include certain mushrooms, spiders, frogs, toads, fish, and snakes. Upon entry into the body, they typically act upon either the central nervous system or the circulatory system; that is, they act as neurotoxins or cytotoxins, respectively.

The presence of a bacterium or its antibody can only be detected by conducting certain clinical tests on samples of blood, skin lesions, or respiratory sections withdrawn from exposed individuals. Once identified, an appropriate antibiotic can then be administered to protect against death.

Similarly, the presence of a toxin can be detected by conducting certain clinical tests upon infected individuals. Once the toxin has been identified, an appropriate antidote can be administered. In both cases, identification of the bacterium and toxin is

TABLE 10.16 ◆ Potential Biological Warfare Agents[a]

Type of Biological Warfare Agent	Adverse Health Effect Caused by Infection
Bacillus anthracis, a spore-forming bacterium	Anthrax, usually fatal in inhaled form
Clostridium perfringens	Gas gangrene or respiratory distress and failure
Brucella abortus (multiple biotypes)	Brucellosis (an infectious disease often resulting in abortion), bone and joint diseases
Clostridium botulinum (seven biotypes)	Botulism (characterized by vomiting, constipation, thirst, headache, fever, dizziness, blurred vision, and dilation of the pupils)
Clostridium tetani	Tetanus (lockjaw)
Bacillus megaterium, a spore-forming bacterium	Unknown
Bacillus subtilis	Unknown
Bacillus cereus	Food-poisoning-like symptoms
Brucella melitensis	Fatigue, loss of appetite
Franciscella tularensis	Tularemia (a disease often contracted by rodents and transmitted to humans); can lead to fatal pneumonia
Corynebacterium diphtheria	Diphtheria (a respiratory disease, often fatal)
Bacillus licheniformis	Produces a protease enzyme that interferes with protein metabolism
Yersinia pestis	Bubonic plague
Salmonella typhi	Nausea, vomiting, cramps, and diarrhea; in severe cases, typhoid fever (a disease associated with intestinal inflammation and ulceration)
Vibrio cholerae	Cholera, an acute intestinal disease characterized by a profuse watery diarrhea
Salmonella enteritidis	Diarrhea, fever, abdominal cramps, and vomiting
Poxvirus variola	Smallpox (a highly infectious disease that is often fatal)
Plasmodium malaria	Malaria
Streptococcus pneumoniae	Pneumonia
Pneumocystis cornini	Pneumonia
Staphylococcus aureus (multiple biotypes)	Staphylococcus food poisoning (sudden severe nausea, vomiting, stomach cramps, severe diarrhea, headache, fever, and chills)

TABLE 10.16 ◆ *(continued)*	
Type of Biological Warfare Agent	*Adverse Health Effect Caused by Infection*
Coxiella burnetii	Q fever (fever, chills, headache, fatigue, and sometimes chest pain)
Listeria monocytogenes	Fever; muscle ache, nausea, and diarrhea; can lead to meningitis, encephalitis, and intrauterine and cervical infections in women

[a]Iraq identified the first 12 of these agents in December 2002 as types of microorganisms used in its biological weapons program. [Iraq's declaration to the United Nations (December 7, 2002)]

vital to saving lives. If the infected person is untreated, the exposure could well be fatal. The unique features of toxins can be determined by DNA testing.

Emergency-response personnel need to identify broad areas potentially contaminated with biological warfare agents and cordon them from the general public. These areas are then decontaminated by knowledgeable professionals, typically through use of a microbiocide such as chlorine dioxide (Section 11.8). For individual protection, the workers who conduct these decontamination processes need to wear fully encapsulating body suits and use self-contained breathing apparatus.

ANTHRAX

Special concern about potential exposure to anthrax was first raised in the United States in 2001, following terrorist acts in which anthrax-laden mail was discovered in postal areas and in the Hart Senate Office Building. The incidents killed two people and injured 17. Concern was heightened when Iraq informed the United Nations that the anthrax bacillus was among the pathogens used in its biological weapons program.

The bacterium *Bacillus anthracis* inflicts harm on the body by skin absorption, ingestion, and inhalation of the spores of the bacterium. This gives rise to three forms of anthrax infection: skin anthrax, intestinal anthrax, and inhalation anthrax. The most potentially lethal form of the disease is inhalation anthrax, although all three forms can cause death. The initial symptoms of exposure are similar to those of a common cold. Anthrax victims experience fever, fatigue, coughing, and chest discomfort within 6 days of exposure. As time progresses, they develop severe respiratory distress and heavy sweating, followed by the onset of shock. Anthrax victims may be treated by antimicrobial therapy. If untreated, they could die within 24–36 hours after the first onset of symptoms.

RICIN

This is a toxin produced by the common castor plant, *Ricinus communis,* the beans of which also serve as a source of castor oil, the well-known laxative. The chemical analysis of ricin reveals that it is a polymer (Chapter 13). In warfare jargon, it is known as *Agent W.* Its LD_{50} is estimated to be on the order of 3 mg/kg. Ricin can be inhaled as a mist or powder, swallowed as a pellet or dissolved within a liquid, or injected into muscle.

The following incidents involving ricin have been documented:

◆ In 1978, Georgi Markov, a Bulgarian exiled journalist living in London, was jabbed in the thigh with an umbrella by an individual suspected of being a KGB agent. Two days later, he died from cardiac arrest. A follow-up autopsy revealed the presence of a small metal ball with four holes, containing ricin, in his thigh.

- In Paris, a similar platinum–iridium ball was removed from the back of a friend of the journalist. The wax used to seal the holes in the ball had not yet melted. Chemical analysis of the residue in the ball revealed the presence of ricin. The friend recalled being jabbed with an umbrella on the Paris Metro.
- In 1980, a Soviet citizen visiting Tyson Corners, Virginia, was also jabbed. He died en route to the local hospital. A similar platinum–iridium ball was removed from the deceased.
- In 2003, London antiterrorism police and security service agents arrested seven terrorists who were plotting to use ricin to kill and terrify the local population. The details of their methods were never made public.

When ricin is assimilated into the body, its victims experience fever, cough, shortness of breath, chest tightness, and low blood pressure within 8 hours. Death can come within 36 to 72 hours after exposure. There is no specific treatment for ricin exposure.

Ricin may have been used by Iraq in the 1991 Iraq–Iran war. In 2002, plans to produce ricin were identified in Afghanistan.

■ ■

REVIEW EXERCISES

Basic Elements of Toxicology

10.1 Exposure to an adhesive caulk may cause eye irritation and an allergic skin reaction. Prolonged use may cause skin, respiratory, kidney, cardiovascular, and liver damage. Classify each adverse condition as an acute or chronic effect resulting from the exposure to caulk.

10.2 Smallpox is contracted by individuals who have been exposed to certain poxviruses. Following a 7-day incubation period, smallpox victims experience fever, headache, vomiting, and abdominal and back pain. Simultaneously, skin lesions develop on the face, hands, and forearms. The disease is highly contagious; unless properly treated, death could result. Classify each adverse condition as an acute or chronic effect.

10.3 When emergency-response personnel are obliged to work in an atmosphere known to contain an elevated concentration of a toxic gas, why is it essential they don special clothing that prevents contact of the gas with their skin?

10.4 Which factor affecting the onset of ill effects from exposure to a toxic substance is demonstrated by each of the following observations?

(a) Individuals with respiratory ailments and heart ailments find breathing difficult when an ozone alert (Section 7.1) is announced, whereas middle-aged persons may not experience any ill effect.

(b) Massive numbers of mutated cells form in the bronchial passageways of individuals repeatedly exposed to diesel engine exhaust, whereas mutated cells are not observed in individuals who inhale airborne diesel exhaust along transportation routes.

(c) Cigarette smokers are more likely to contract lung cancer the longer they continue to smoke.

(d) Minutes or hours may pass before death occurs from the ingestion of a lethal dose of hydrocyanic acid, whereas death occurs immediately upon inhaling a lethal dose of hydrogen cyanide.

(e) Trace amounts of chromium are essential in our diet to metabolize glucose, but ingestion of hexavalent chromium may cause cancer.

10.5 Experiments conducted at the National Institute of Standards and Technology, Gaithersburg, Maryland, established the following inhalation LC_{50} values for carbon monoxide by exposing laboratory animals to this toxic gas for the indicated time periods:

Exposure Time (min)	LC_{50} (ppm)
1	107,000
2	42,500
5	14,000
10	9800
20	7400
30	6600
60	4900

What do these data illustrate about the relationship between toxicity and the length of time the animals were exposed to carbon monoxide before they died?

Measuring Toxicity

10.6 Two painters work through a normal 40-hour workweek within a spray booth, where they routinely inhale average concentrations of acetone and ethyl acetate of 250 and 450 ppm, respectively. If the TLVs for acetone and ethyl acetate are 200 and 400 ppm, respectively, are the workers likely to experience adverse health effects?

10.7 One constituent of hardwood smoke is the substance called *acrolein*, exposure to which can cause the onset of lachrimation. Acrolein is also a toxic gas whose inhalation LD_{50} is 0.4 mg/kg. Using this information, determine whether the inhalation of 0.05 mg of acrolein by a 180-pound unprotected firefighter is likely to be fatal.

10.8 Governmental hygienists report the IDLH value for sulfur dioxide as 100 ppm.
(a) Is it safe for an emergency-response team without respiratory protection to enter a room in which the concentration of sulfur dioxide is 750 ppm?
(b) For what approximsate period of time can an individual using respiratory protection remain in an environment containing sulfur dioxide at a concentration of 100 ppm without experiencing any escape-impairing symptoms or irreversible health effects?

10.9 The OSHA regulation at 29 CFR §1910.1000(d) requires employers to determine the cumulative exposure of workers to certain gases for an 8-hour work shift through use of the following formula:

$$E = (C_a T_a + C_b T_b + \ldots . C_n T_n)/8$$

E is the equivalent exposure for the work shift, C is the concentration during any period of time in which the concentration remains constant, and T is the duration of time in hours of the exposure. The regulation further stipulates that no employee can be further exposed to the gas when the value of E exceeds the 8-hour, time-weighted average limit set for the given substance. If the 8-hour, time-weighted average limit set by OSHA for anhydrous ammonia is 50 ppm, has the employer complied with this regulation if a worker is exposed to ammonia concentrations throughout the workday as follows: 125 ppm for 4 hr; 75 ppm for 2 hr; and 25 ppm for 2 hr?

10.10 What is the oral lethal dose in ounces of the pesticide *Atrazine* for a 190-lb pesticide applicator. The oral LD_{50} for *Atrazine* is 2000 mg/kg.

10.11 A chemical manufacturer intends to ship 100 lb (45.4 kg) of hydrogen sulfide as a compressed gas by motor van. Using Tables 6.1, 6.2, 6.4, 10.3, and 10.14, identify the basic description DOT requires on the accompanying shipping paper.

Toxicity of the Fire Scene

10.12 Why are active on-duty firefighters likely to experience the adverse health effects from inhaling 300 ppm of carbon monoxide faster than nonactive off-duty firefighters who inhale the same concentration of the same substance?

10.13 The following LC_{50} values were established for carbon monoxide and nitrogen dioxide by exposing laboratory animals for thirty minutes to each gas alone and to a mixture of each gas with 5% carbon dioxide by volume:

	LC_{50} for Gas Alone (ppm)	LC_{50} for Gas With 5% Carbon Dioxide (ppm)
Carbon monoxide	6600	3900
Nitrogen dioxide	200	90

What toxicological information do these data reveal about these gases?

Toxic Gases

10.14 Why is it an unsafe practice to burn charcoal indoors in a hibachi or grill?

10.15 Exposure to elevated concentrations of either carbon monoxide or hydrogen cyanide affects the body as a hemotoxicant, whereas exposure to hydrogen sulfide affects the body as a neurotoxicant. What do these characterizations tell us about the mechanism by which these gases poison?

10.16 During an autopsy, a medical examiner draws blood samples from the body of a fire victim. When subjected to chemical analysis, the samples are found to contain 69% carboxyhemoglobin by volume. Based solely upon this information, is the medical examiner able to properly conclude that carbon monoxide poisoning was the cause of the victim's death?

10.17 Using the following partial sketch of a single-floored home, identify the locations at which carbon monoxide detectors should be installed to provide maximum protection for its occupants.

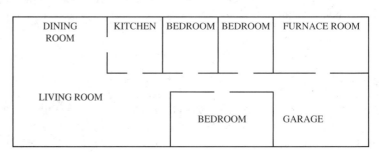

10.18 Why should carbon monoxide detectors be installed at least 15 ft (4.6 m) from fuel-burning appliances?

10.19 Why do charcoal manufacturers often post an advisory warning like the following on bags of charcoal intended for use by consumers?

NEVER BURN CHARCOAL INSIDE HOMES, VEHICLES, OR TENTS

10.20 Use Tables 6.1, 6.2, 6.4, 10.3, and 10.14 to determine the basic description DOT requires a rail carrier to enter on shipping papers when transporting nitrogen dioxide as a liquefied compressed gas.

10.21 Platinum catalytic converters catalyze the oxidation of unwanted carbon monoxide and nitric oxide contained in automobile exhaust. They also reduce the nitrogen dioxide into environmentally benign nitrogen. Write the equations for these reactions.

10.22 Upon entering a laboratory where chemists are working with hydrogen sulfide, an individual immediately recognizes the rotten-egg odor of the gas. Why is this odor no longer apparent to the same person hours later?

10.23 Although ammonia possesses a vapor density of 0.6 (air = 1), why does a release of ammonia sometimes move along the surface of the ground rather than rising into the air and dispersing?

10.24 A chemical manufacturer in Chicago, Illinois, intends to ship 1000 gal (3.8 m^3) of anhydrous ammonia containing less than 0.2% water by mass in a DOT Specification MC 331 tank truck to a farmer in Memphis, Tennessee. Using Tables 6.1 and 6.4, what basic description of this commodity does DOT require on the accompanying shipping paper?

Carcinogens

10.25 Why do decades often pass after an initial exposure to a carcinogen before the onset of cancer becomes evident?

Emergency-Response Actions Involving Toxic Substances

10.26 Why should emergency-response crews who have been exposed to poison gases during the course of duty wash down with water before removing their clothing?

10.27 Upon responding to a warehouse fire, the members of an emergency-response crew locate bags and other containers of a chemical product named *Ciodrin*. The labels indicate that *Ciodrin* is an organophosphate pesticide. With this minimum information, what actions should the crew take to best protect life, property, and the environment?

10.28 Using Table 10.15, identify the initial isolation and protective-action distances recommended by DOT when responding to each of the following domestic transportation mishaps:
(a) A railroad tankcar leaking anhydrous ammonia at 9:30 A.M.
(b) A truck that has overturned at 7:45 P.M. and from which a drum containing aqueous (25%) hydrocyanic acid has burst

10.29 Upon responding to a 2:00 A.M. domestic transportation mishap, a firefighting team determines that a liquefied compressed gas is leaking from a 5-lb (2.3-kg) steel cylinder. The chemical commodity is described on the relevant shipping paper as follows:

> 5 lb   RQ, Compressed gas, flammable, toxic, n.o.s. (POISON—
> Inhalation Hazard, Zone B)

Use Table 10.15 to identify the radius of the initial isolation zone and the magnitude of each side of the protective-action zone associated with the release of this substance.

Metallic Compounds

10.30 Upon chemical analysis using specified test procedures, a representative sample of paint scrapings is found to contain chromium and lead at concentrations of 8.7 and 15.9 mg/L, respectively.
  (a) Do these scrapings exhibit the RCRA characteristic of toxicity?
  (b) When a carrier transports 1 ton of construction debris containing the scrapings for disposal by dump truck, what basic description do DOT and EPA require the carrier to enter on the waste manifest?

10.31 In the 1960s, firefighters were taught to be concerned about the potential for lead poisoning when combating gasoline fires. Why is this concern less probable today?

10.32 In 1992, Congress passed the *Residential Lead-Based Paint Hazard Reduction Act,* which requires housing owners to notify their tenants of the presence or suspicion of the presence of lead-based paint. What is the most likely reason Congress took this action?

Asbestos

10.33 When an individual contracts mesothelioma, why is it reasonably conclusive that the disease was caused by exposure to asbestos?

10.34 The DOT regulation at 49 CFR §173.216 requires that asbestos be transported in "rigid, leak-tight packaging" such as dust- and sift-proof bags. What is the most logical reason DOT requires the use of this special packaging?

Pesticides

10.35 While combating a fire involving the pesticide *Aldicarb,* a 150-lb firefighter inadvertently swallows 0.5 g of pesticide dust, despite the use of protective gear. The oral LD_{50} (rat) for *Aldicarb* is 2.5 mg/kg. The firefighter did not survive.
  (a) What is the most likely reason this fatality occurred?
  (b) Describe the nature of the first-aid action that, if rapidly implemented, could have saved the firefighter's life?

10.36 Once firefighting personnel determine that an ongoing fire involves burning pesticides, why should the operators of the nearby wastewater treatment plant be notified of the fire?

10.37 Why is the recommended practice to allow a small pesticide fire to burn, rather than to extinguish the fire?

Biological Warfare Agents

10.38 Why is it unlikely that terrorists would load a biological warfare agent into a missile warhead with a charge of chemical explosive?

10.39 When civilians suspect they may have been exposed to a biological warfare agent, why is it essential they seek medical attention within hours of the suspected exposure?

Chemistry of Some Oxidizers

11 CHAPTER

Certain chemical reactions provide benefits to civilized societies throughout the world. They include the combustion of fuels, the chlorination of water, the explosion of dynamite, the bleaching of fabrics, and the flaring of fireworks. These phenomena result from *oxidation–reduction reactions,* or *redox reactions* (Section 5.4) for short. These reactions, as well as the hazardous materials that cause them, constitute the subject matter of this chapter.

When oxidation–reduction reactions are conducted in a controlled fashion, the released energy can be harnessed to our advantage. For example, the energy generated in devices such as batteries, dry cells, and fuel cells provides a portable electrical energy supply. However, when redox reactions occur in an uncontrolled fashion, the generated energy is released into the immediate environment, where it could initiate fire and explosion and cause the loss of life and property. This potential for destructiveness necessitates the examination of redox reactions as a component of the study of hazardous materials.

Performance Goals for Section 11.1

- Describe the chemical nature of an oxidizer, especially noting that the presence of oxygen in the chemical composition is unnecessary.
- Describe the nature of the DOT regulations as they apply to the transportation of oxidizers.

◆ 11.1 WHAT IS AN OXIDIZER?

We noted in Chapter 7 that hydrogen burns in oxygen to produce water, in fluorine to form hydrogen fluoride, and in chlorine to produce hydrogen chloride. These combustion reactions are represented by the following equations:

$$2H_2(g) + O_2(g) \longrightarrow 2H_2O(g)$$
$$\text{hydrogen} \quad \text{oxygen} \qquad \text{water}$$

$$H_2(g) + F_2(g) \longrightarrow 2HF(g)$$
$$\text{hydrogen} \quad \text{fluorine} \qquad \text{hydrogen fluoride}$$

$$H_2(g) + Cl_2(g) \longrightarrow 2HCl(g)$$
$$\text{hydrogen} \quad \text{chlorine} \qquad \text{hydrogen chloride}$$

The substances supporting these independent combustion processes are examples of *oxidizers, oxidants,* or *oxidizing agents,* while hydrogen is an example of a *reducing agent.* Oxidizing agents always react with reducing agents in concert.

In Section 6.7, we learned that DOT defines an *oxidizer* as a material that can cause or enhance the combustion of other materials. While elemental oxygen is a common example of such a substance, oxygen is not an essential component of an oxidizer. Furthermore, the occurrence of an oxidation–reduction reaction does not necessarily require the participation of elemental oxygen or compounds containing oxygen. The fact that hydrogen burns in atmospheres of fluorine and chlorine demonstrates that oxygen is not the sole substance that causes or enhances the combustion of other materials. Like oxygen, fluorine and chlorine are oxidizers.

OXIDIZERS AND THE DOT REGULATIONS

DOT regulates the transportation of oxidizers as division 5.1 and 5.2 hazardous materials. When these hazardous materials are transported, DOT requires shippers and carriers to comply with its labeling, marking, and placarding requirements. When a division 5.1 oxidizer is transported, DOT generally requires shippers and carriers to affix OXIDIZER labels to its packaging. When an amount exceeding 1000 lb (454 kg) of a division 5.1 oxidizer is transported, DOT requires shippers and carriers to post OXIDIZER placards on the bulk packaging, freight container, transport vehicle, or railcar used to transport it.

The division 5.2 oxidizers comprise a group of reactive substances called organic peroxides, which we shall study in Section 12.14.

Performance Goals for Section 11.2
- ◆ Describe the concept of the oxidation number.
- ◆ For a given substance, determine the oxidation numbers of its atoms and ions.

◆ 11.2 OXIDATION NUMBERS

Chemists use the concept of an *oxidation number* as a means of describing the combining capability of one ion for another ion or of one atom for another atom. In practice, an oxidation number is assigned to the atoms that make up a substance by the following formal rules:

- ◆ The oxidation number of each atom in an element is zero; thus, the oxidation number of each atom in the elements H_2, O_2, Na, and Mg is zero.
- ◆ The algebraic sum of the oxidation numbers of the atoms in any substance is zero.
- ◆ The hydrogen in hydrogen-containing compounds (other than hydrides) possesses an oxidation number of $+1$. The hydrogen in a hydride (H^-) possesses an oxidation number of -1.
- ◆ The oxygen in oxygen-containing compounds other than peroxides possesses an oxidation number of -2. Each oxygen atom in a peroxide (O_2^{2-}) possesses an oxidation number of -1.

TABLE 11.1 ◆ Examples of Oxidation Numbers

Element	Oxidation Number	Chemical Formula of Representative Compound
F	−1	HF
O	−2	SO_2
	−1	H_2O_2
N	−3	NH_3
	−2	N_2H_4
	−1	NH_2OH
	+1	N_2O
	+2	NO
	+3	HNO_2
	+4	NO_2
	+5	HNO_3
Cl	−1	HCl
	+1	HClO
	+3	$HClO_2$
	+5	$HClO_3$
	+7	$HClO_4$
S	−2	H_2S
	+4	H_2SO_3
	+6	H_2SO_4
P	−3	AlP
	+3	H_3PO_3
	+5	H_3PO_4

◆ The oxidation number of a monatomic ion (i.e., having one atom) is the same as its net ionic charge; thus, the oxidation number of sodium is +1; of magnesium, +2; of the chloride ion, −1; and of the sulfide ion, −2.

◆ The algebraic sum of the oxidation numbers of the atoms in a polyatomic ion is equal to its ionic charge.

The chemical formula of a substance must be known to use these rules and establish oxidation numbers. As an example, let's use them to determine the oxidation number of each atom in sodium chlorate, an oxidizing agent used in fireworks (Section 11.9) and flares (Section 11.10).

The chemical formula of sodium chlorate is $NaClO_3$. This formula indicates that the compound is composed of two ions, the sodium ion (Na^+) and the chlorate ion (ClO_3^-). Using our rules, we establish the oxidation number of each element in sodium chlorate as follows:

Constituent	*Oxidation Number*	*Relevant Rule*
Sodium ion, Na^+	$+1$	The oxidation number of a monatomic ion is the same as its ionic charge.
Oxygen atom	-2	Other than in peroxides, the oxidation number of oxygen is -2.
Chlorine atom	$+5$	The algebraic sum of the oxidation numbers in any substance is zero, that is, $0 = +1 + 5 + [3 \times (-2)]$.
Chlorate ion, ClO_3^-	-1	The algebraic sum of the oxidation numbers of the atoms in a polyatomic ion is equal to its ionic charge, that is, $-1 - +5 + [3 \times (-2)] = -6 + 5$.

Additional examples of determining oxidation numbers are noted in Table 11.1.

Performance Goal for Section 11.3
  • For a given equation, use the change in oxidation number for an atom or ion to identify the oxidizing agent, reducing agent, the substance oxidized, and the substance reduced.

◆ 11.3 OXIDATION–REDUCTION REACTIONS

Since a redox reaction occurs between an oxidizing agent and a reducing agent, the equation illustrating a redox reaction is always written in the following generalized form:

$$\text{Oxidizing agent} + \text{Reducing agent} \longrightarrow \text{Products}$$

A specific example of this is the reaction between iron(III) chloride and tin(II) chloride. We write the equation for this reaction as follows:

$$\underset{\text{iron(III) chloride}}{2FeCl_3(aq)} + \underset{\text{tin(II) chloride}}{SnCl_2(aq)} \longrightarrow \underset{\text{iron(II) chloride}}{2FeCl_2(aq)} + \underset{\text{tin(IV) chloride}}{SnCl_4(aq)}$$

Since the chloride ions do not participate in this redox reaction, they are often called *spectator ions*. We can eliminate them and represent the reaction as an ionic equation:

$$2Fe^{3+}(aq) + Sn^{2+}(aq) \longrightarrow 2Fe^{2+}(aq) + Sn^{4+}(aq)$$

In an ionic equation, only the symbols of the ions contributing to the reaction are written. In this representation, electrons have simply been transferred between the iron and tin ions. The information conveyed by this equation is summarized as follows:

  • The tin(II) ions become tin(IV) ions. The oxidation number of tin increases from $+2$ to $+4$; thus, the tin(II) ions are *oxidized*. Tin(II) chloride is called the *reducing agent,* since it reduces iron(III) chloride.

◆ The iron(III) ions become iron(II) ions. The oxidation number of iron decreases from +3 to +2; thus, the iron(III) ions are *reduced*. Iron(III) chloride is called the *oxidizing agent*, because it oxidizes tin(II) chloride.

Since a simultaneous increase and decrease in oxidation numbers always accompanies oxidation and reduction, respectively, we can summarize the nature of this process in ionic systems as follows:

◆ *Oxidation* is the phenomenon associated with an *increase* in oxidation number and *loss* of electrons from an ion, atom, or group of atoms.
◆ *Reduction* is the phenomenon associated with a *decrease* in oxidation number and *gain* of electrons from an ion, atom, or group of atoms.
◆ The substance that *accepts* electrons during a redox reaction is called the *oxidizing agent*.
◆ The substance that *loses* electrons during a redox reaction is called the *reducing agent*.

Oxidation and reduction phenomena can also be represented separately by equations like the following:

$$Sn^{2+}(aq) \longrightarrow Sn^{4+}(aq) + 2e^-$$
$$Fe^{3+}(aq) + e^- \longrightarrow Fe^{2+}(aq)$$

The phenomenon illustrated by each equation is called a *half-reaction*. One half-reaction represents oxidation, while the other represents reduction. The processes represented by half-reactions always occur simultaneously.

Several other examples of oxidation–reduction phenomena are provided in Table 11.2.

TABLE 11.2 ◆ Some Common Oxidizing Agents

Oxidizing Agent	Element that Changes Oxidation Number	Oxidation Number		Equation Illustrating Half-Reaction
		In Reactant	In Product	
Sodium peroxide	Oxygen	−1	−2	$Na_2O_2(aq) + 2H_2O(l) + 2e^-$ $\longrightarrow 2Na^+(aq) + 4OH^-(aq)$
Metallic hypochlorites	Chlorine	+1	−1	$ClO^-(aq) + 2H^+(aq) + 2e^-$ $\longrightarrow Cl^-(aq) + H_2O(l)$
Metallic chlorates	Chlorine	+5	−1	$ClO_3^-(aq) + 6H^+(aq) + 6e^-$ $\longrightarrow Cl^-(aq) + 3H_2O(l)$
Nitric acid (concentrated)	Nitrogen	+5	+4	$NO_3^-(aq) + 2H^+(aq) + e^-$ $\longrightarrow NO_2(g) + H_2O(l)$
Nitric acid (dilute)	Nitrogen	+5	+2	$NO_3^-(aq) + 4H^+(aq) + 3e^-$ $\longrightarrow NO(g) + 2H_2O(l)$
Metallic peroxydisulfates	Sulfur	+7	+6	$S_2O_8^{2-}(aq) + 2e^-$ $\longrightarrow 2SO_4^{2-}(aq)$

Solved Exercise 11.1

Identify the element oxidized, the element reduced, the oxidizing agent, and the reducing agent in the redox reaction denoted by the following equation:

$$6FeSO_4(aq) + Na_2Cr_2O_7(aq) + 7H_2SO_4(aq) \longrightarrow$$

iron(II) sulfate   sodium dichromate   sulfuric acid

$$3Fe_2(SO_4)_3(aq) + Na_2SO_4(aq) + Cr_2(SO_4)_3(aq) + 7H_2O(l)$$

iron(III) sulfate   sodium sulfate   chromium(III) sulfate   water

Solution: In this equation, the sodium and sulfate ions are identified as spectator ions, because their oxidation numbers are the same on each side of the arrow. Spectator ions can be eliminated, which allows the following ionic equation to be written:

$$6Fe^{2+}(aq) + Cr_2O_7^{2-}(aq) + 14H^+(aq) \longrightarrow 6Fe^{3+}(aq) + 2Cr^{3+}(aq) + 7H_2O(l)$$

To determine the oxidation number of chromium in the dichromate ion, remember that the algebraic sum of the oxidation numbers of the atoms in a polyatomic ion is equal to its ionic charge. We thus determine that the oxidation number of chromium is +6.

$$-2 = [2 \times (+6)] + [7 \times (-2)]$$

The oxidation number of the monatomic ions is the same as their ionic charge.

In the ionic equation, we readily observe that the iron(II) ions become iron(III) ions. The oxidation number of iron increases from +2 to +3 as each iron(II) ion loses an electron. Since an *increase* in oxidation number is associated with oxidation, the iron(II) ions are *oxidized*. Iron(II) sulfate is the *reducing agent*.

The ionic equation also notes that dichromate ions become chromium(III) ions. The oxidation number of chromium decreases from +6 in $Cr_2O_7^{2-}$ to +3 in Cr^{3+} as each chromium ion gains three electrons. Since a *decrease* in oxidation number is associated with reduction, the dichromate ions are *reduced*. Sodium dichromate is the *oxidizing agent*.

Performance Goals for Section 11.4
- Describe the varying degree of hazard associated with different oxidizers.
- Describe the nature of each type of oxidizer listed in Table 11.4.
- Describe the ability of oxidizers containing oxygen atoms or ions to produce oxygen when they thermally decompose.

◆ 11.4 COMMON FEATURES OF OXIDIZERS

Oxidizers are generally perceived as relatively powerful chemical substances since they react rapidly, even at explosive rates. The substances that react with oxidizers include fuels, lubricants, greases, oils, cotton, animal and vegetable fats, paper, coal, coke, straw, sawdust, and wood shavings.

FIGURE 11.1 ◆ The criminal act of detonating a weapon of mass destruction at the Oklahoma City federal building caused the deaths of 169 workers and other individuals. The weapon was a 4000-lb (1800-kg) mixture of ammonium nitrate fertilizer and fuel oil. The detonation occurred at the rate of approximately 13,000 ft/s (4000 m/s). *Courtesy of Lance Moler, photographer, Oklahoma County Newspapers, Inc., Midwest City, Oklahoma.*

Since they are powerfully reactive substances, oxidizers have been chosen by terrorists to cause destruction. In 1995, the oxidizer ammonium nitrate was used with fuel oil to destroy the Murrah Federal Building in Oklahoma City (Figure 11.1). The incident killed 168 people and injured 518.

Different oxidizers possess different strengths. This variation in strength can be approximated by reviewing the listing in Table 11.3. In this series, oxidizing agents are arranged according to their decreasing oxidizing power. Any substance whose name appears on this list is a stronger oxidizing agent than the substances whose names are listed below it. It is instructive to note that many substances are stronger oxidizing agents than oxygen itself.

NFPA distinguishes the degree of hazard posed by specific oxidizers as follows:

Class 1 oxidizers   These are materials whose primary hazard is increasing the burning rate of combustible materials with which they come in contact.

TABLE 11.3 ◆ Relative Strength of Oxidizing Agents[a]

Fluorine
Ozone
Hydrogen peroxide
Hypochlorous acid
Metallic chlorates[b]
Lead(II) dioxide
Metallic permanganates[b]
Metallic dichromates[b]
Nitric acid (concentrated)
Chlorine
Sulfuric acid (concentrated)
Oxygen
Metallic iodates
Bromine
Iron(III) (Fe^{3+}) compounds
Iodine
Sulfur
Tin(IV) (Sn^{4+}) compounds

[a]Listed in descending order of oxidizing power.
[b]In an acidic environment.

Class 2 oxidizers	These are materials that moderately increase the burning rate or cause spontaneous ignition of the combustible material with which they come in contact.
Class 3 oxidizers	These are materials that severely increase the burning rate of combustible materials with which they come in contact; or they may undergo violent self-sustained decomposition when catalyzed or exposed to heat.
Class 4 oxidizers	These are materials that undergo an explosive reaction when catalyzed or exposed to heat, shock, or friction.

The NFPA classification of some specific oxidizers is noted in Table 11.4.

While oxidizers that possess oxygen atoms in their ionic structures do not burn in the traditional sense, they may support combustion by providing a supply of oxygen at the fire scene. For example, when potassium chlorate is exposed to intense heat, it decomposes to produce oxygen.

$$2KClO_3(s) \longrightarrow 2KCl(s) + 3O_2(g)$$
$$\text{potassium chlorate} \qquad \text{potassium chloride} \quad \text{oxygen}$$

The oxygen produced by the reaction supports the combustion of nearby materials at least as well as atmospheric oxygen. When the oxidizer is confined while undergoing decomposition, more oxygen may be produced than is needed to sustain combustion. In such instances, the oxygen generates excessive internal pressure within the confined area. This situation could cause an explosion.

TABLE 11.4   ◆   Some Typical Oxidizers by NFPA Classification

Class 1

Ammonium nitrate	Potassium nitrate
Ammonium persulfate	Potassium persulfate
Barium chlorate	Silver nitrate
Barium nitrate	Sodium carbonate peroxide
Calcium chlorate	Sodium dichloro-*sym*-triazine-2,4,6-trione
Calcium nitrate	dihydrate
Calcium peroxide	Sodium dichromate
Cupric nitrate	Sodium nitrate
Hydrogen peroxide solutions, over 8% but	Sodium nitrite
not exceeding 27.5% by mass	Sodium perborate
Lead nitrate	Sodium perborate tetrahydrate
Lithium hypochlorite	Sodium perchlorate monohydrate
Lithium peroxide	Sodium persulfate
Magnesium nitrate	Strontium chlorate
Magnesium perchlorate	Strontium nitrate
Magnesium peroxide	Strontium peroxide
Nickel nitrate	Thorium nitrate
Nitric acid, 70% or less	Uranium nitrate
Perchloric acid, less than 60% by mass	Zinc chlorate
Potassium dichromate	Zine peroxide

Class 2

Calcium hypochlorite, 50% or less by mass	Sodium chlorite, 40% or less
Chromium trioxide (chromic acid)	Sodium peroxide
Halane (1,3-dichloro-5,5-dimethyl hydantoin)	Sodium permanganate
Hydrogen peroxide, 27.5% to 52% by mass	Trichloro-*sym*-triazine-2,4,6-trione
Nitric acid, more than 70%	(trichloroisocyanuric acid)
Potassium permanganate	

Class 3

Ammonium dichromate	Potassium bromate
Calcium hypochlorite, over 50% by mass	Potassium chlorate
Hydrogen peroxide, 52% to not more than	Potassium dichloro-*sym*-triazine-2,4, 6-trione
91% by mass	(potassium dichloroisocyanurate)
Mono-(trichloro)tetra-(monopotassium	Sodium chlorate
dichloro)-penta-*sym*-triazine-2,4,6-trione	Sodium chlorite, over 40% by mass
Perchloric acid solutions, 60% to 72.5%	Sodium dichloro-*sym*-triazine-2,4,6-trione
by mass	(sodium dichloroisocyanurate)

Class 4

Ammonium perchlorate	Perchloric acid solutions, more than 72.5%
Ammonium permanganate	by mass
Guanidine nitrate	Potassium superoxide
Hydrogen peroxide solutions, more than	
91% by mass	

Solved Exercise 11.2

Magnesium perchlorate and sodium chlorate are mixed in separate containers with equal volumes of turpentine. Which oxidizer poses the greater risk of increasing the rate at which the turpentine spontaneously ignites?

Solution: Using Table 11.4, magnesium perchlorate and sodium chlorate are NFPA class 1 and class 3 oxidizers, respectively. Based on this information, sodium chlorate poses the greater risk of increasing the rate at which turpentine spontaneously ignites.

Performance Goals for Section 11.5
- Describe the hazardous properties of hydrogen peroxide solutions, noting that they possess a degree of hazard that is associated with their concentration.
- Identify the industries that use bulk volumes of hydrogen peroxide.

◆ 11.5 HYDROGEN PEROXIDE

The chemical formula and molecular structure of hydrogen peroxide are H_2O_2 and H—O—O—H, respectively. The substance is encountered commercially in aqueous solutions having the following approximate compositions:

1% to 3% H_2O_2	This solution is a topical antiseptic for use on minor cuts and wounds (Figure 11.2).
6% H_2O_2	This solution is used for bleaching hair. In the process, hydrogen peroxide oxidizes the dark-colored pigment called melanin to colorless products.
30% H_2O_2	This solution is used in the chemical industry for the synthesis of organic peroxides (Section 12.14). The 30% solution is also used by manufacturing and process industries for bleaching cotton, flour, wool, straw, leather, gelatin, and paper. These solutions also appear to be replacing the use of chlorine for treating drinking water.
90% H_2O_2	In the aerospace industry, this solution has been used as a rocket fuel. For example, it helped to launch Apollo 17 (Figure 11.3) and other payloads into space. The solution was also used by the Russian military in torpedo fuel systems.

Although these solutions resemble water in physical appearance, they possess slightly pungent, irritating odors. Hydrogen peroxide without water has also been produced, but it is unavailable commercially.

Bulk quantities of the hydrogen peroxide solutions are transported in tank trucks (Figure 11.4) and railcars.

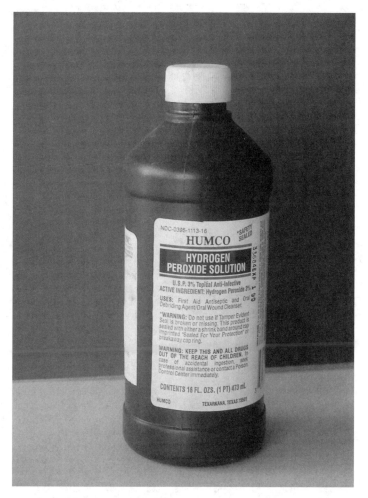

FIGURE 11.2 ◆ A solution containing 3 to 4% hydrogen peroxide by volume is widely used as an antiseptic.

A hydrogen peroxide solution is a member of each NFPA class of oxidizer in Table 11.4. As the hydrogen peroxide concentration increases in the solution, so does its degree of reactivity.

Approximately 95% of the world's hydrogen peroxide is manufactured by the sequential catalytic hydrogenation and oxidation of a suitable organic compound. For simplicity, we shall represent the organic compound here as $X(=O)_2$. The production reactions are accomplished in a solution of an appropriate alcohol (*al*) (Section 12.8), as follows:

$$X(=O)_2(al) + H_2(g) \longrightarrow X(OH)_2(al)$$
$$X(OH)_2(aq) + O_2(g) \longrightarrow X(=O)_2(al) + H_2O_2(al)$$

The hydrogen peroxide is then extracted into water to produce a 25% solution and the organic compound is recycled. The hydrogen peroxide solution may then be concentrated by distillation or diluted to produce the desired concentration.

FIGURE 11.3 ◆ Concentrated hydrogen peroxide was the oxidant used to launch the upper stage of Apollo 17. *Courtesy of the National Aeronautics and Space Administration, John F. Kennedy Space Center, Kennedy Space Center, Florida.*

Hydrogen peroxide is inherently unstable. It slowly decomposes as follows:

$$\underset{\text{hydrogen peroxide}}{2H_2O_2(aq)} \longrightarrow \underset{\text{water}}{2H_2O(l)} + \underset{\text{oxygen}}{O_2(g)}$$

The decomposition reaction is catalyzed by sunlight as well as certain metals, most notably iron, copper, chromium, and silver. An aqueous solution of 8% hydrogen peroxide is completely decomposed following a 10-month exposure to light, while a similar solution kept in darkness for the same length of time remains virtually unaltered. In solutions whose concentrations are less than 30%, the decomposition of hydrogen peroxide occurs so slowly when stored in dark glass bottles that it is virtually imperceptible.

FIGURE 11.4 ◆ In the background is FMC Corporation's hydrogen peroxide manufacturing facility. In the foreground, workers are preparing a cargo tank for transporting hydrogen peroxide to a customer. The container label and Material Safety Data Sheet for this chemical product were previously provided in Figures 1.6 and 1.7, respectively. *Courtesy of FMC Corporation, Hydrogen Peroxide Division, Philadelphia, Pennsylvania.*

Concentrated solutions of hydrogen peroxide ($>6\%$ H_2O_2) become intensely heated when they decompose. These hot solutions then vaporize. To prevent the decomposition reaction from posing a hazard before the intended use of the oxidant, all commercial forms of hydrogen peroxide are stabilized with a small amount of a catalyst that retards the decomposition.

Concentrated hydrogen peroxide solutions possess the following properties:

- Concentrated hydrogen peroxide solutions cause combustible materials to burn spontaneously at explosive rates.
- Hydrogen peroxide solutions having a strength in excess of 20% are highly corrosive. Exposure to the skin causes severe irritation. Exposure to the eyes could cause blindness.

These hazardous properties are routinely identified on container labels as follows:

DANGER

STRONG OXIDIZER
CONTACT WITH OTHER MATERIAL
MAY CAUSE FIRE

CORROSIVE
CAUSES BURNS TO SKIN, EYES,
AND RESPIRATORY TRACT
HARMFUL WHEN SWALLOWED
OR INHALED

Hydrogen peroxide has been linked with a modern-day naval disaster. In 2000, a Russian nuclear submarine, the *Kursk*, exploded and sunk in the Barents Sea. Aboard the *Kursk* were torpedoes, whose fuel systems consisted in part of highly concentrated hydrogen peroxide. Experts have proposed that the cause of the disaster was linked with a leak of hydrogen peroxide from its containment system. The oxidizer interacted with the torpedo's stainless steel casing, which catalyzed the decomposition of the hydrogen peroxide. The subsequent buildup of oxygen resulted in overpressurization of the torpedo and its subsequent explosion. This first explosion then initiated the detonation of other torpedoes in the storage compartment. The hull of the submarine burst, and the *Kursk* floundered and sank. There were no survivors.

When emergency-response teams encounter hydrogen peroxide solutions at accident scenes, an attempt should be made not only to extinguish ongoing fires, but to dam or dike the water runoff generated while combating them. When a fire has not occurred, the solutions should be segregated from combustible materials.

When used in the workplace, OSHA requires employers to regulate employee exposure to an inhalation concentration of 1 ppm, averaged over an 8-hour workday.

DOT regulates the hydrogen peroxide solutions in Table 11.5 as oxidizers. DOT requires shippers and carriers to affix OXIDIZER labels, or OXIDIZER *and* CORROSIVE labels, on their packaging and, when warranted, to post OXIDIZER placards on the bulk packaging, freight container, transport vehicle, unit load device, or railcar used to transport them.

TABLE 11.5 ◆ **Hydrogen Peroxide Solutions Regulated by DOT for Transport**

Proper Shipping Name	Label Codes
Hydrogen peroxide, aqueous solutions *with more than 40% but not more than 60% hydrogen peroxide (stabilized as necessary)*	5.1, 8
Hydrogen peroxide, aqueous solutions *with not less than 8% but less than 20% hydrogen peroxide (stabilized as necessary)*	5.1
Hydrogen peroxide, aqueous solutions *with not less than 20% but not more than 40% hydrogen peroxide (stabilized as necessary)*	5.1, 8
Hydrogen peroxide, stabilized *or* Hydrogen peroxide aqueous solutions, stabilized *with more than 60% hydrogen peroxide*	5.1, 8

Solved Exercise 11.3

When offered for transportation, what proper shipping name does DOT require for an aqueous solution containing 30–32% hydrogen peroxide?

Solution: DOT requires the following as the proper shipping name of an aqueous solution containing 30–32% hydrogen peroxide solution: "Hydrogen peroxide, aqueous solution (with 30–32% hydrogen peroxide)." DOT requires shippers and carrier to affix OXIDIZER and CORROSIVE labels to its packaging and, when warranted, to post OXIDIZER placards to the bulk packaging, freight container, transport vehicle, unit containment device, or railcar used to transport it. Placarding is warranted when shippers or carriers transport 1000 lb (454 kg) or more of the solution.

Performance Goals for Section 11.6
- ◆ Describe the meaning of the term "available chlorine."
- ◆ Describe the hazardous properties of sodium hypochlorite and calcium hypochlorite.
- ◆ Identify the industries that use bulk quantities of sodium hypochlorite and calcium hypochlorite.

◆ 11.6 HYPOCHLORITE OXIDIZERS

Sodium hypochlorite and calcium hypochlorite are most commonly encountered as the active components of household and commercial bleaching and sanitation products. Their chemical formulas are $NaClO$ and $Ca(ClO)_2$, respectively. The purpose of a bleaching agent is to remove undesirable stains from fabrics and other textiles. Hypochlorite bleaching agents can only be safely used on cotton and linen fabrics, not on wool, silk, and nylon.

The hypochlorite bleaching agents are solids at room temperature, but they are also available commercially as aqueous solutions. Non-chemists generally contend that their bleaching action is accomplished by chlorine, but it is the hypochlorite ion that possesses the oxidizing power and accomplishes the bleaching action.

The effectiveness of hypochlorite bleaching agents is commercially described by their *available chlorine*. This term is used to compare the effectiveness of bleaching agents to elemental chlorine. A bleaching agent having 99.2% available chlorine has the same bleaching power as a solution containing 99.2% chlorine by mass. Substances having 99.2% available chlorine are useful for sanitizing water, while substances having an available chlorine of less than approximately 35% are useful as laundry bleaches.

The available chlorine in a bleaching agent is determined by utilizing the reaction between its active ingredient and an acid; for instance, the available chlorine in a bleaching agent containing calcium hypochlorite is determined by reacting the substance with hydrochloric acid, as follows:

$$\underset{\text{calcium hypochlorite}}{Ca(ClO)_2(s)} + \underset{\text{hydrochloric acid}}{4HCl(aq)} \longrightarrow \underset{\text{calcium chloride}}{CaCl_2(aq)} + \underset{\text{water}}{2H_2O(l)} + \underset{\text{chlorine}}{2Cl_2(g)}$$

A chemical analyst then determines the percentage of chlorine produced from the given amount of calcium hypochlorite.

When heated, all metallic hypochlorites decompose to produce oxygen. For instance, calcium hypochlorite decomposes as follows:

$$\underset{\text{calcium hypochlorite}}{Ca(ClO)_2(s)} \longrightarrow \underset{\text{calcium chloride}}{CaCl_2(s)} + \underset{\text{oxygen}}{O_2(g)}$$

When the decomposition reaction occurs at a fire scene, oxygen is supplied to support combustion reactions.

DOT regulates the transportation of the hypochlorites in Table 11.6 as oxidizers. When shipping barium hypochlorite, DOT requires shippers and carriers to affix OXIDIZER and POISON labels on its packaging and, when warranted, to post OXIDIZER placards on the bulk packaging, freight container, transport vehicle, unit load device, or railcar used to transport it. When shipping the remaining hypochlorites, DOT requires shippers and carriers to affix OXIDIZER or CORROSIVE labels on their packaging and, when warranted, to post OXIDIZER placards on the bulk packaging, freight container, transport vehicle, unit load device, or railcar used to transport them.

TABLE 11.6 ◆ Metallic Hypochlorites Regulated by DOT for Transport

Proper Shipping Name	*Label Codes*
Barium hypochlorite *with more than 22% available chlorine*	5.1, 6.1
Calcium hypochlorite, dry *or* Calcium hypochlorite mixtures dry with more than 39% available chlorine (8.8% available oxygen)	5.1
Calcium hypochlorite, hydrated *or* Calcium hypochlorite, hydrated mixtures, *with not less than 5.5% but not more than 10% water*	5.1
Calcium hypochlorite mixtures dry, *with more than 10% but not more than 39% available chlorine*	5.1
Hypochlorite solutions *with more than 5% but less than 16% available chlorine*	8
Hypochlorite solutions *with 16% or more available chlorine*	8
Hypochlorites, inorganic, n.o.s.	5.1
Lithium hypochlorite, dry *or* Lithium hypochlorite mixtures, dry	5.1

FIGURE 11.5 ◆ Clorox is an aqueous solution containing 6% sodium hypochlorite. It is widely used as a laundry bleach. *CLOROX is a registered trademark of The Clorox Company © The Clorox Company. Reprinted with permission.*

SODIUM HYPOCHLORITE

We are most familiar with this substance as the active ingredient of the household laundry product known as *Clorox* (Figure 11.5). This is an aqueous solution containing 6% sodium hypochlorite. Solutions that contain from 10 to 12.5% sodium hypochlorite are also available commercially, including the product known as *Hasachlor* (Figure 11.6). These solutions are useful disinfectants and sanitizers for treating water supplies, sewage effluents, swimming pools, spas, hot tubs, and dairies.

Sodium hypochlorite is manufactured for commercial use by passing gaseous chlorine into an aqueous solution of sodium hydroxide. A mixture of sodium hypochlorite and sodium chloride is produced, as follows:

$$\underset{\text{sodium hydroxide}}{2NaOH(aq)} + \underset{\text{chlorine}}{Cl_2(g)} \longrightarrow \underset{\text{sodium hypochlorite}}{NaClO(aq)} + \underset{\text{sodium chloride}}{NaCl(aq)} + \underset{\text{water}}{H_2O(l)}$$

The sodium chloride does not interfere with the sodium hypochlorite when the latter is intended for use as a bleaching or sanitation agent.

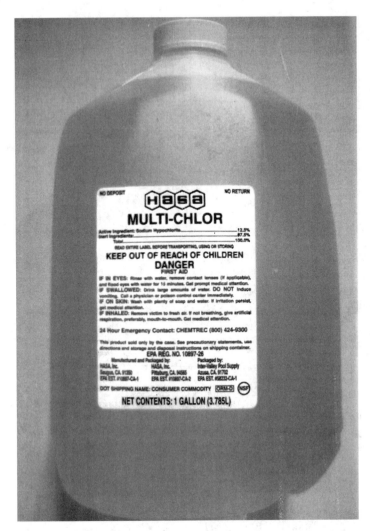

FIGURE 11.6 ◆ Hasachlor is an aqueous solution containing 10 to 12.5% sodium hypochlorite. It is commonly used as a disinfectant and sanitizer. *Courtesy of R. Scott Williams, Grandview, Washington.*

Sodium hypochlorite may also be generated for use upon demand. In the generator in Figure 11.7, it is prepared by electrolyzing a brine solution.

$$\underset{\text{sodium chloride}}{\text{NaCl}(aq)} + \underset{\text{water}}{\text{H}_2\text{O}(l)} \longrightarrow \underset{\text{sodium hypochlorite}}{\text{NaClO}(aq)} + \underset{\text{hydrogen}}{\text{H}_2(g)}$$

CALCIUM HYPOCHLORITE

Calcium hypochlorite is commonly used for large-scale bleaching operations, sanitizing municipal drinking water, disinfecting domestic and municipal swimming pools, and sewage treatment. It is encountered in products known commercially by the tradename *high-test-hypochlorite* ("HTH").

FIGURE 11.7 ◆ Using this generator (shown in its front and back positions), an aqueous solution of sodium hypochlorite is generated for use upon demand. *Courtesy of USFilter/Wallace and Tiernan Products, Vineland, New Jersey.*

Calcium hypochlorite is manufactured by reacting chlorine with a lime slurry. A mixture of calcium hypochlorite and calcium chloride is produced.

$$2Ca(OH)_2(aq) + 2Cl_2(g) \longrightarrow Ca(ClO)_2(aq) + CaCl_2(aq) + 2H_2O(l)$$

calcium hydroxide   chlorine   calcium hypochlorite   calcium chloride   water

Sodium chloride is then added to the solution. This causes the calcium hypochlorite to precipitate as the solid.

Performance Goals for Section 11.7

- Describe the hazardous properties of dichloroisocyanuric acid, trichloroisocyanuric acid, and 1,3-dichloro-5,5-dimethylhyantoin.
- Identify the industries that use bulk quantities of these chlorinating agents.

◆ **11.7 OTHER CHLORINATION AGENTS**

Although the use of calcium hypochlorite continues to be popular in industrial sanitation products, the compound has been largely displaced in household products with certain chlorinated organic compounds. The molecular structures of these compounds

FIGURE 11.8 ◆ Two commercially available products containing trichloro-*sym*-triazine-2,4,6-trione. They are used as chlorination agents, primarily for sanitizing the water in swimming pools.

are relatively complex. We shall study their nature more fully in Chapter 12. The following are representative:

dichloroisocyanuric acid, or
dichloro-*sym*-triazine-2,4,6-trione

trichloroisocyanuric acid, or
trichloro-*sym*-triazine-2,4,6-trione

1,3-dichloro-5,5-dimethylhyantoin,
or halane

The di- and trichloroisocyanuric acids are also available commercially as their sodium and potassium salts.

These chlorination compounds are commercially available as white tablets or powders having 90 to 99% available chlorine. They are components of dry bleaches, detergents, dishwashing compounds, and scouring powders primarily intended for use in the home. The chlorination compounds in Figure 11.8 are used to chlorinate swimming pools.

DOT regulates the transportation of these chlorination compounds as oxidizers. DOT requires shippers and carriers to affix OXIDIZER labels on their packaging and, when warranted, to post OXIDIZER placards on the bulk packaging, freight container, transport vehicle, unit load device, or railcar used to transport them.

Performance Goals for Section 11.8
- ◆ Describe the hazardous properties of chlorine dioxide.
- ◆ Describe the use of chlorine dioxide for decontaminating areas at which biological agents are located.

◆ 11.8 CHLORINE DIOXIDE

There are three oxides of chlorine, but only chlorine dioxide has commercial importance. Its chemical formula is ClO_2. Chlorine dioxide is a red-yellow gas. It is highly unstable, decomposing into its elements.

$$2ClO_2(g) \longrightarrow Cl_2(g) + 2O_2(g)$$

chlorine dioxide   chlorine   oxygen

Although the gas can be satisfactorily stabilized for use by mixing it with carbon dioxide, chlorine dioxide is routinely produced at the point of its consumption. In the generator in Figure 11.9, it is produced by adding sodium chlorite to a mixture of sodium hypochlorite and hydrochloric acid.

$$NaClO(aq) + HCl(aq) \longrightarrow HClO(aq) + NaCl$$

sodium hypochlorite   hydrochloric acid   hypochlorous acid   sodium chloride

$$2NaClO_2(s) + HClO(aq) + HCl(aq) \longrightarrow 2ClO_2(g) + 2NaCl(aq) + H_2O(l)$$

sodium chlorite   hypochlorous acid   hydrochloric acid   chlorine dioxide   sodium chloride   water

When chlorine dioxide is dissolved in water, an aqueous solution is produced called *chlorine dioxide hydrate*. In commerce, this solution is shipped frozen to its point of use. When kept cold or frozen, the aqueous solution survives for a long time without decomposition. Even gentle heating, however, causes the chlorine dioxide to decompose.

As with chlorine, the principal risk associated with exposure to chlorine dioxide is inhalation toxicity. Exposure to chlorine dioxide causes severe respiratory and eye damage.

Chlorine dioxide and elemental chlorine are generally used for the same purposes. By comparison, however, chlorine dioxide is approximately 2½ times more powerful than chlorine as an oxidizing agent. For this reason, chlorine dioxide is sometimes selected over chlorine for bleaching paper pulp and producing white wheat flour, even though chlorine dioxide is more costly.

Chlorine dioxide is sometimes also selected over chlorine for use as a microbiocide. It is used to disinfect water in the dairy, beverage, and other food industries. At water treatment plants, the use of chlorine dioxide is generally supplementary to the use of chlorine.

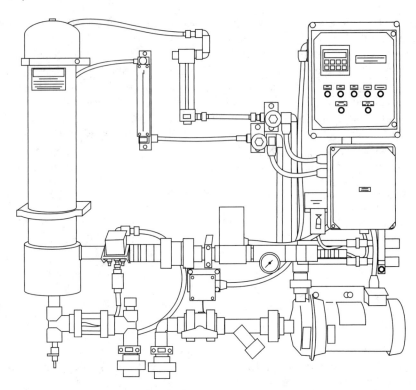

FIGURE 11.9 ◆ In this vacuum-based system, chlorine dioxide gas is generated by means of chemical reactions involving sodium chlorite, sodium hypochlorite, and hydrochloric acid. The gas is used commercially as a microbiocide. *Courtesy of Tramfloc, Inc., Tempe, Arizona.*

In 2002, anthrax-laden mail was discovered in postal areas and in the Hart Senate Office Building. To destroy the anthrax bacterium, EPA contractors chose chlorine dioxide as the disinfectant. The decontamination of mail packages and Senate offices was accomplished by discharging the gas into sealed-off areas, where it remained for several hours before neutralization. The hard surfaces within the Senate building were disinfected using liquid chlorine dioxide.

DOT regulates the transportation of frozen chlorine dioxide hydrate as an oxidizer. DOT requires shippers and carriers to maintain chlorine dioxide in the solid state using dry ice, to affix OXIDIZER and POISON labels on its packaging, and, when warranted, to post OXIDIZER placards on the bulk packaging, freight container, transport vehicle, unit load device, or railcar used to transport it. DOT does not allow the transportation of chlorine dioxide gas by any mode.

Performance Goals for Section 11.9

- ◆ Describe the general chemical composition of all fireworks.
- ◆ Identify the common oxidizers used in fireworks.
- ◆ Describe the nature of the chemical reactions that occur when fireworks are activated.

Items designed primarily for the purpose of producing audible and visible pyrotechnic effects by combustion are called *fireworks*. Familiar examples include firecrackers, Roman candles, pinwheels, flares, serpents, sparklers, and toy torpedoes. The pyrotechnic effects associated with their activation results from a carefully manipulated set of redox reactions, each succinctly timed to provide lavish aerial scenes.

Although state and local laws prevent the widespread sale of fireworks to unauthorized individuals, incidents involving their misuse cause numerous incidents leading to secondary fires, blindness, deafness, the mutilation of fingers, and death annually. To reduce this likelihood, only a licensed retail outlet that posts a permit is allowed to sell fireworks to the public. To further reduce the likelihood of these incidents, state and local laws require that most types of fireworks be energized only by experts in accordance with the manufacturer's warnings and instructions. Even under these conditions, however, the activation of fireworks constitutes a potential risk of fire and explosion (Figure 11.10).

One component of all fireworks is a pyrotechnic mixture of an oxidizing agent and a reducing agent. The most commonly encountered oxidizers are sodium chlorite, sodium chlorate, and sodium perchlorate, all of which are also components of vehicular air bags, solid-rocket fuels, and certain fertilizers. The reducing agents in fireworks include charcoal, sulfur, pulverized magnesium, and aluminum flakes. To provide the colors, one or more metallic chlorides are also added to fireworks formulations.

sodium chlorite

When heated, sodium chlorite, sodium chlorate, and sodium perchlorate decompose to produce oxygen.

sodium chlorate

$$NaClO_2(s) \longrightarrow NaCl(s) + O_2(g)$$
$$\text{sodium chlorite} \qquad \text{sodium chloride} \quad \text{oxygen}$$

$$2NaClO_3(s) \longrightarrow 2NaCl(s) + 3O_2(g)$$
$$\text{sodium chlorate} \qquad \text{sodium chloride} \quad \text{oxygen}$$

$$NaClO_4(s) \longrightarrow NaCl(s) + 2O_2(g)$$
$$\text{sodium perchlorate} \qquad \text{sodium chloride} \quad \text{oxygen}$$

sodium perchlorate

While the oxygen produced by these reactions aids in sustaining the display, more complex redox reactions are responsible for the brilliant lighting and sound effects associated with fireworks. When the mixture of an oxidizer, charcoal, sulfur, and finely divided magnesium or aluminum is activated, the resulting redox reactions occur explosively. The presence of magnesium powder enhances the brilliance of a fireworks display, while the addition of coarse aluminum flakes produces luminous tails. The sets of equations in Table 11.7 illustrate the specific redox reactions associated with the activation of these substances.

The unique colors observed during firework displays are furnished by certain compounds as they vaporize. These compounds are almost exclusively metallic chlorides, which fluoresce strongly in the visible wavelengths. Some examples and the colors they provide are noted below:

barium chloride	yellow-green	cupric chloride	blue-green
lithium chloride	crimson red	calcium chloride	orange red
potassium chloride	lavender	sodium chloride	golden yellow
strontium chloride	carmine red		

FIGURE 11.10 ◆ During a July 4, 1997, fireworks celebration in Alton, Illinois, a misfired 8-in (20-cm) mortar shell rose approximately 10 ft (3 m) over the Mississippi River before dropping upon the barge from which five people were launching the display. The shell subsequently ignited numerous other shells, which rapidly engulfed the entire barge in violent bursts of flame and smoke. This photograph captures not only a shell bursting properly in the air, but also the first of two deadly blasts from the barge (lower right) that resulted in the deaths of two workers. *Courtesy of John Badman, photographer, The Telegraph, Alton, Illinois.*

DOT regulates the transportation of metallic chlorites, chlorates, and perchlorates in Table 11.8 as oxidizers. When shipping barium chlorate, barium perchlorate and lead perchlorate, DOT requires shippers and carrier to affix OXIDIZER and POISON labels to their packaging and, when warranted, to post OXIDIZER placards on the bulk packaging, freight container, transport vehicle, unit load device, or railcar used to transport them. When shipping the other oxidizers in Table 11.8, DOT requires shippers and carriers to affix OXIDIZER labels on their packaging and, when warranted, to post OXIDIZER placards on the bulk packaging, freight container, transport vehicle, unit load device, or railcar used to transport them.

TABLE 11.7 ◆ Chemical Phenomena Occurring During the Activation of Fireworks

Oxidizing Agent	Equations Representing the Relevant Phenomena
Sodium chlorite	$C(s) + NaClO_2(s) \longrightarrow NaCl(s) + CO_2(g)$
	$S_8(s) + 8NaClO_2(s) \longrightarrow 8NaCl(s) + 8SO_2(g)$
	$2Mg(s) + NaClO_2(s) \longrightarrow 2MgO(s) + NaCl(s)$
	$4Al(s) + 3NaClO_2(s) \longrightarrow 2Al_2O_3(s) + 3NaCl(s)$
Sodium chlorate	$3C(s) + 2NaClO_3(s) \longrightarrow 2NaCl(s) + 3CO_2(g)$
	$3S_8(s) + 16NaClO_3(s) \longrightarrow 16NaCl(s) + 24SO_2(g)$
	$3Mg(s) + NaClO_3(s) \longrightarrow 3MgO(s) + NaCl(s)$
	$2Al(s) + NaClO_3(s) \longrightarrow Al_2O_3(s) + NaCl(s)$
Sodium perchlorate	$2C(s) + NaClO_4(s) \longrightarrow NaCl(s) + 2CO_2(g)$
	$S_8(s) + 4NaClO_4(s) \longrightarrow 4NaCl(s) + 8SO_2(g)$
	$4Mg(s) + NaClO_4(s) \longrightarrow 4MgO(s) + NaCl(s)$
	$8Al(s) + 3NaClO_4(s) \longrightarrow 4Al_2O_3(s) + 3NaCl(s)$

TABLE 11.8 ◆ Metallic Chlorites, Chlorates, and Perchlorates Regulated by DOT for Transport

Proper Shipping Name	Label Codes
Barium chlorate	5.1, 6.1
Barium perchlorate	5.1, 6.1
Calcium chlorate	5.1
Calcium chlorate, aqueous solution	5.1
Calcium chlorite	5.1
Calcium perchlorate	5.1
Lead perchlorate, solid	5.1, 6.1
Lead perchlorate, solution	5.1, 6.1
Magnesium chlorate	5.1
Magnesium perchlorate	5.1
Perchlorates, inorganic, aqueous solution, n.o.s.	5.1
Perchlorates, inorganic, n.o.s.	5.1
Potassium chlorate	5.1
Potassium chlorate, aqueous solution	5.1
Potassium perchlorate, solid	5.1
Potassium perchlorate, solution	5.1
Sodium chlorate	5.1
Sodium chlorate, aqueous solution	5.1
Sodium chlorite	5.1
Strontium chlorate	5.1
Strontium perchlorate	5.1
Zinc chlorate	5.1

◆ 11.10 OXIDIZERS IN FLARES AND SIGNALING SMOKES

The military and civilian police regularly use flares and signaling smokes to cordon off emergency sites and coordinate activities during local assault operations. The military also uses signaling smokes by prearrangement to identify friendly units and targets and control the laying and lifting of artillery.

The active chemical component of flares and signaling smokes is a pyrotechnic mixture of an oxidizing agent and reducing agent that has been combined with sodium bicarbonate and an organic dye. Most commonly, the oxidizing and reducing agents are potassium chlorate and elemental sulfur, respectively. The mixture of components is compressed into cartridges, hand grenades, and canisters.

When the mixture is activated by ignition, two chemical reactions occur:

   ◆ The oxidation–reduction reaction between potassium chlorate and sulfur:

$$16KClO_3(s) \; + \; 3S_8(s) \longrightarrow \; 16KCl(s) \; + \; 24SO_2(g)$$

   potassium chlorate   sulfur   potassium chloride   sulfur dioxide

   ◆ The thermal decomposition of sodium bicarbonate:

$$2NaHCO_3(s) \longrightarrow \; Na_2CO_3(s) \; + \; CO_2(g) \; + H_2O(g)$$

   sodium bicarbonate   sodium carbonate   carbon dioxide   water

Carbon dioxide and sulfur dioxide evolve into the air and disperse the dye in the immediate area.

The pyrotechnic mixtures in flares and smoking signals are potentially explosive. DOT regulates their transportation as explosive materials, requiring shippers and carriers to affix the relevant EXPLOSIVE 1.1G, EXPLOSIVE 1.2G, EXPLOSIVE 1.3G, or EXPLOSIVE 1.4S labels to their containers and, when warranted, to post corresponding EXPLOSIVE placards on the bulk packaging, freight container, transport vehicle, or railcar used for their transport.

◆ 11.11 OXIDIZING AMMONIUM COMPOUNDS

All compounds containing the ammonium ion (NH_4^+) are thermally unstable. When these compounds are heated, they decompose in either of the following ways:

TABLE 11.9 ◆ Some Potentially Hazardous Ammonium Compounds

Oxidizer	*Equation Illustrating Thermal Decomposition*
Ammonium bromate	$2NH_4BrO_3(s) \longrightarrow 2NH_4Br(s) + 3O_2(g)$
Ammonium chlorate	$2NH_4ClO_3(s) \longrightarrow 2NH_4Cl(s) + 3O_2(g)$
Ammonium dichromate	$(NH_4)_2Cr_2O_7(s) \longrightarrow Cr_2O_3(s) + 4H_2O(g) + N_2(g)$
Ammonium nitrate[a]	$NH_4NO_3(s) \longrightarrow N_2O(g) + 2H_2O(g)$
Ammonium nitrite	$NH_4NO_2(s) \longrightarrow N_2(g) + 2H_2O(g)$
Ammonium perchlorate	$2NH_4ClO_4(s) \longrightarrow N_2(g) + Cl_2(g) + 4H_2O(g) + 2O_2(g)$
Ammonium permanganate	$2NH_4MnO_4(s) \longrightarrow 2MnO(s) + N_2(g) + 4H_2O(g) + O_2(g)$
Ammonium peroxydisulfate	$3(NH_4)_2S_2O_8(s) \longrightarrow 4NH_3(g) + N_2(g) + 6SO_2(g) + 6H_2O(g) + 3O_2(g)$

[a]See text also.

- ◆ Ammonium compounds that are not oxidizing agents decompose to form ammonia. For instance, ammonium chloride is not an oxidizing agent. When heated, ammonium chloride thermally decomposes to form ammonia and hydrogen chloride.

$$NH_4Cl(s) \longrightarrow NH_3(g) + HCl(g)$$
$$\text{ammonium chloride} \qquad \text{ammonia} \qquad \text{hydrogen chloride}$$

- ◆ Ammonium compounds that are oxidizing agents generally decompose to form nitrogen, oxygen, and either a metallic or nonmetallic oxide. These thermal decomposition reactions are illustrated by the equations in Table 11.9.

Ammonium nitrate is by far the most important commercial chemical product containing the ammonium ion. It is used directly as a fertilizer and a component of certain explosives like *ammonal,* a mixture of ammonium nitrate and powdered aluminum. It is also used to produce dinitrogen monoxide, or nitrous oxide, commonly called *laughing gas*.

$$NH_4NO_3(s) \longrightarrow N_2O(g) + 2H_2O(g)$$
$$\text{ammonium nitrate} \qquad \text{dinitrogen monoxide} \qquad \text{water}$$

It is used as an anesthetic by dentists.

Ammonium nitrate is commonly encountered at farms and distribution centers like those in Figure 11.11, as well as at locations where explosives are used. The ammonium nitrate available today for use as a fertilizer is referred to as *fertilizer-grade* ammonium nitrate. It is a formulation of ammonium nitrate with ammonium sulfate or calcium carbonate, each of which dilutes the oxidizer and reduces the potential risk of its decomposition. Other grades of ammonium nitrate are dynamite-grade, nitrous oxide–grade, and technical-grade. All are available as crystals, flakes, grains, and prills (small round aggregates).

When heated, ammonium nitrate decomposes in a variety of ways. The following equations illustrate some of its methods of decomposition:

$$NH_4NO_3(s) \longrightarrow N_2O(g) + 2H_2O(g)$$
$$\text{ammonium nitrate} \qquad \text{dinitrogen monoxide} \qquad \text{water}$$

$$4NH_4NO_3(s) \longrightarrow 3N_2(g) + 2NO_2(g) + 8H_2O(g)$$
$$\text{ammonium nitrate} \qquad \text{nitrogen} \quad \text{nitrogen dioxide} \quad \text{water}$$

$$2NH_4NO_3(s) \longrightarrow 2N_2(g) + O_2(g) + 4H_2O(g)$$
$$\text{ammonium nitrate} \qquad \text{nitrogen} \quad \text{oxygen} \quad \text{water}$$

FIGURE 11.11 ◆ A farm distribution center where ammonium nitrate can be stored in bulk as well as in bags. Ammonium nitrate is an important source of nitrogen for crops and thus serves as an important fertilizer. *Courtesy of the Fertilizer Institute, Washington, D.C.*

When the atmospheric oxygen has been depleted at a fire scene, the combustion processes are sustained by the dinitrogen monoxide, nitrogen dioxide, and oxygen produced as decomposition products.

Certain precautionary measures must be taken to safely transport, store, and use ammonium nitrate. When storing 1000 lb (455 kg) or more of ammonium nitrate, OSHA requires employers to comply with the following regulations:

- Containers of ammonium nitrate shall not be stored under conditions that could cause the temperature of the ammonium nitrate to exceed 130°F (54°C).
- Bags of ammonium nitrate shall not be stored within 30 in. (76 cm) of the storage building walls and partitions.
- Ordinarily, the height of ammonium nitrate piles shall not exceed 20 ft (6 m), the width 20 ft (6 m), and the length 50 ft (15 m). When the storage is in a building of noncombustible construction or protected with automatic sprinklers, the length of piles shall not be limited. In no case shall the ammonium nitrate be stacked closer than 36 in. (91 cm) below the roof or supporting and spreader beams.
- Aisles shall be provided to separate piles by a clear space of not less than 3 ft (0.9 m) in width. At least one service or main aisle in the storage area shall be not less than 4 ft (1.2 m) in width.

When storing 1000 lb (455 kg) or more of ammonium nitrate in bulk, the following OSHA regulations apply:

- Warehouses shall have adequate ventilation or be capable of adequate ventilation in case of fire.

- Unless constructed of noncombustible material or unless adequate facilities for fighting a roof fire are available, bulk storage structures shall not exceed a height of 40 ft (12 m).
- Bins shall be clean and free of materials that can contaminate ammonium nitrate. (When ammonium nitrate becomes mixed with a combustible substance at a concentration as small as 0.2% by mass, its explosive nature increases dramatically.)
- Due to the corrosive and reactive properties of ammonium nitrate, and to avoid contamination, galvanized iron, copper, lead, and zinc shall not be used in the construction of a bin unless they are suitably protected. Aluminum and wooden bins protected against impregnation by ammonium nitrate are permissible. The partitions dividing the ammonium nitrate storage from other products that could contaminate the ammonium nitrate shall be of tight construction (Figure 11.12).
- The ammonium nitrate storage bins or piles shall be clearly identified by signs reading **"AMMONIUM NITRATE"** with letters at least 2 in. (5 cm) high.
- Piles or bins shall be so sized and arranged that all material in the pile is moved out periodically in order to minimize possible caking of the stored ammonium nitrate.
- The height or depth of piles of ammonium nitrate shall be limited by the pressure-setting tendency of the product; however, in no case shall the ammonium nitrate be piled higher at any point than 36 in. (91 cm) below the roof or the supporting and spreader beams overhead.
- Ammonium nitrate shall not be accepted for storage when the temperature of the product exceeds 130°F (54°C).
- Dynamite, other explosives, and blasting agents shall not be used to loosen ammonium nitrate that has caked.

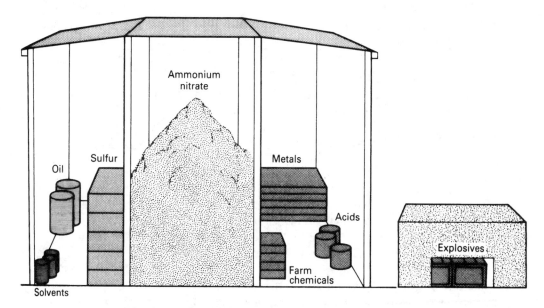

FIGURE 11.12 ◆ Recommended storage for bulk quantities of ammonium nitrate. When ammonium nitrate is stored in bulk, the storage bin should be clean and free of contaminants. The pile within the bin should be sized so that all material in the pile is moved periodically. The bin should be separated by approved fire walls from flammable and combustible materials, corrosive materials, other oxidizers, and substances with which the ammonium nitrate may chemically react. Explosives or blasting agents should never be stored in the same building as ammonium nitrate. *Courtesy of the Fertilizer Institute, Washington, D.C.*

Why has OSHA stringently regulated the manner in which ammonium nitrate is stored in the workplace, but has not regulated the manner in which other oxidizers are stored? The answer to this question is associated with the bulk quantities of this substance that farmers and other users routinely store. No other oxidizer is stored in such a massive quantity at one time at a given location. In today's world, the answer also is connected with the notoriety associated with the incidents in which ammonium nitrate/fuel mixtures have been detonated. Given this notoriety, law enforcement agencies should be provided with an accounting of the ammonium nitrate potentially available to terrorists within their jurisdictions.

Solved Exercise 11.4

The OSHA regulation set forth at 29 CFR §1910.109 stipulates that ammonium nitrate cannot ordinarily be stored in the same building with either sulfur or finely divided metals. What is the most likely reason OSHA regulates the segregation of ammonium nitrate and these substances?

Solution: When exposed to an ignition source, a mixture of an oxidizer and elemental sulfur is often chemically incompatible. The same is true of oxidizers that have been mixed with finely divided metals. With ammonium nitrate, the resulting combustion reactions are violent and occur at explosive rates. Selecting magnesium as an example of a combustible metal, these redox reactions are denoted as follows:

$$S_8(s) + 32NH_4NO_3(s) \longrightarrow 8SO_2(g) + 8NO_2(g) + 64H_2O(g) + 28N_2(g)$$

sulfur　　　ammonium nitrate　　　　sulfur dioxide　nitrogen dioxide　　water　　　　nitrogen

$$12Mg(s) + 16NH_4NO_3(s) \longrightarrow 12MgO(s) + 2NO_2(g) + 32H_2O(g) + 15N_2(g)$$

magnesium　ammonium nitrate　　　magnesium oxide　nitrogen dioxide　　water　　　　nitrogen

OSHA most likely requires the segregation of ammonium nitrate from sulfur and finely divided metals in order to reduce or eliminate the risk of fire and explosion.

Long before it became associated with terrorist activities, ammonium nitrate was the focus of one of the worst maritime incidents associated with a chemical substance. In 1947, the S.S. *Grandcamp*, a French cargo ship docked in Texas City, Texas, caught fire with nearly 2200 tons (2 million kg) of fertilizer-grade ammonium nitrate and 1500 tons (1.4 million kg) of fuel oil aboard (Figure 11.13). Heat caused the ammonium nitrate to melt and then decompose. While fires involving oxidizing agents can generally be effectively extinguished through application of water, this attempt was never made aboard the *Grandcamp*. Instead, the hatches to the hold were sealed with the intent of eliminating the fire's air supply. The fire continued to burn in the absence of atmospheric oxygen because it was supported by an oxidizing agent. The ship and its cargo subsequently exploded. This incident serves to dramatically illustrate the potentially hazardous nature of oxidizers in general and ammonium nitrate in particular.

DOT regulates the transportation of several oxidizers that are ammonium compounds, examples of which are provided in Table 11.10. DOT regulates the transportation

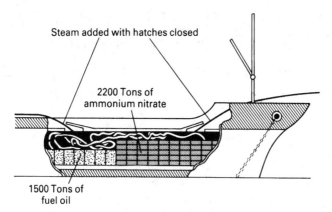

Steam added with hatches closed

2200 Tons of
ammonium nitrate

1500 Tons of
fuel oil

FIGURE 11.13 ◆ On April 16, 1947, 2200 tons of ammonium nitrate was stored on board the S.S. *Grandcamp* along with 1500 tons of fuel oil. While docked at Texas City, Texas, a fire was detected in the hold. An order was given to close the hatches and to apply steam throughout the hold. Under these conditions, the fire could not extinguish. Instead, the internal pressure built up at an uncontrollable rate, causing the cargo vessel to explode. Six hundred people were killed and another 3500 were injured. The property damage was comparable to that experienced during a major wartime bombing incident. The total property loss was estimated as $33 million based on 1947 costs.

of the following as oxidizers: ammonium dichromate, certain ammonium nitrate fertilizers, ammonium perchlorate (see also footnote "a" to Table 11.10), and ammonium persulfate. DOT requires shippers and carriers to affix OXIDIZER labels to their packaging and, when warranted, to post OXIDIZER placards to the bulk packaging, freight container, transport vehicle, unit containment device, or railcar used to transport them.

DOT regulates the transportation of certain ammonium nitrate fertilizers as Class 9 hazardous materials. DOT requires shippers and carriers to affix CLASS 9 labels to their packaging and, when warranted, to post CLASS 9 placards to the bulk packaging, freight container, transport vehicle, unit containment device, or railcar used to transport them.

Finally, DOT regulates the transportation of ammonium nitrate–fuel oil mixtures and ammonium nitrate containing combustible carbon as explosive materials. DOT requires shippers and carriers to affix EXPLOSIVE 1.5D and EXPLOSIVE 1.1D labels to their packaging, respectively, and to post the corresponding EXPLOSIVE placards to the bulk packaging, freight container, transport vehicle, unit containment device, or railcar used to transport them.

DOT prohibits shippers and carriers from accepting any amount of the following ammonium compounds for transportation: ammonium azide, ammonium bromate, ammonium chlorate, ammonium fulminate, ammonium nitrite, and ammonium permanganate.

TABLE 11.10 ◆ Oxidizing Ammonium Compounds Regulated by DOT for Transport

Proper Shipping Name	Label Codes
Ammonium dichromate	5.1
Ammonium nitrate fertilizers	5.1
Ammonium nitrate fertilizers: *uniform non-segregating mixtures of ammonium nitrate that is inorganic and chemically inert towards ammonium nitrate, with not less than 90% ammonium nitrate and not more than 0.2% combustible matter (including organic matter calculated as carbon), or with more than 70% but less than 90% ammonium nitrate and not more than 0.4% total combustible matter*	5.1
Ammonium nitrate fertilizers: *uniform non-segregating mixtures of nitrogen/phosphate or nitrogen/potash types or complete fertilizers of nitrogen/phosphate/potash type, with not more than 70% ammonium nitrate and not more than 0.4% total added combustible material or with not more than 45% ammonium nitrate with unrestricted combustible material*	9
Ammonium nitrate–fuel oil mixture *containing only prilled ammonium nitrate and fuel oil*	1.5D
Ammonium nitrate, *with more than 0.2% combustible substances, including any organic substance calculated as carbon, to the exclusion of any other added substance*	1.1D
Ammonium nitrate, *with not more than 0.2% of combustible substances, including any organic substance calculated as carbon, to the exclusion of any other added substance*	5.1
Ammonium perchlorate[a]	1.1D
Ammonium perchlorate	5.1
Ammonium persulfate	5.1

[a]The hazard class for ammonium perchlorate is either 1.1D or 5.1. The determination of the hazard class depends upon the material's particle size and packaging; hence, experimental testing must be conducted upon a sample of this substance to determine the relevant hazard class.

Performance Goals for Section 11.12
- ◆ Identify the oxidizing chromium compounds.
- ◆ Describe their hazardous properties.
- ◆ Identify the industries that use bulk quantities of these oxidizing chromium compounds.

◆ 11.12 OXIDIZING CHROMIUM COMPOUNDS

Chromium assumes the +6 oxidation state in metallic chromates, metallic dichromates, chromium trioxide, and chromyl chloride. When chromium assumes the +6 oxidation state, it is called hexavalent chromium.

Metallic chromates are compounds composed of a metallic ion and the chromate ion (CrO_4^{2-}); potassium chromate, or K_2CrO_4, is a specific example. Metallic dichromates are compounds composed of a metallic ion and the dichromate ion ($Cr_2O_7^{2-}$); a specific example is potassium dichromate, or $K_2Cr_2O_7$. These compounds are manufactured for commercial use from chromite ore, which we represent here as $FeCr_2O_4$. The ore is heated at kiln temperatures with potassium carbonate to first produce potassium chromate, as follows:

$$4FeCr_2O_4(s) + 8K_2CO_3(s) + 7O_2(g) \longrightarrow 8K_2CrO_4(s) + 2Fe_2O_3(s) + 8CO_2(g)$$

chromite · · · · potassium carbonate · · · oxygen · · · · · · · potassium chromate · · iron(III) oxide · · carbon dioxide

Potassium dichromate is then produced by the reaction of sulfuric acid on potassium chromate.

$$2K_2CrO_4(aq) + H_2SO_4(aq) \longrightarrow K_2Cr_2O_7(aq) + H_2O(l) + K_2SO_4(aq)$$

potassium chromate · · hydrochloric acid · · · · potassium dichromate · · · water · · · · potassium chloride

Metallic chromates and metallic dichromates are solid yellow and orange compounds, respectively. Potassium chromate can be produced from potassium dichromate, and vice versa, as follows:

$$K_2Cr_2O_7(aq) + 2KOH(aq) \longrightarrow 2K_2CrO_4(aq) + H_2O(l)$$

potassium dichromate · · potassium hydroxide · · · potassium chromate · · · water

$$2K_2CrO_4(aq) + 2HCl(aq) \longrightarrow K_2Cr_2O_7(aq) + H_2O(l) + 2KCl(aq)$$

potassium chromate · · hydrochloric acid · · · · potassium dichromate · · · water · · · · potassium chloride

Hexavalent chromium compounds have been used for decades as pigments in industrial paints. Examples are lead chromate, strontium chromate, and zinc chromate. Hexavalent chromium compounds were also used to remove the mill scale that forms within cooling towers (Figure 11.14). Cooling towers are located at petroleum refineries, power plants, and chemical plants. The waste generated in connection with this use was sometimes discharged into unlined surface impoundments, from which the hexavalent chromium seeped into the subsurface and contaminated the underlying groundwater aquifer.

EPA has used the authority of the Toxic Substances Control Act to ban the use of compounds containing hexavalent chromium in the cooling towers used for air conditioners.

Scientific research has revealed that the absorption of a hexavalent chromium compound into the body can cause a number of health ailments. In particular, the inhalation of hexavalent chromium compounds has been linked with lung cancer in humans; thus, epidemiologists rank these compounds as known human carcinogens (Section 10.15) by the inhalation route of exposure.

Research studies concerned with establishing a link between various diseases and the consumption of hexavalent chromium compounds are in progress. Erin Brockovich, the heroine in the movie of the same name, alleged in the film that long-term consumption of water contaminated with excessive levels of hexavalent chromium caused stomach lesions, stomach and intestinal cancers, and other health ailments. Although the evidence may appear to lean in her direction, there are currently no

FIGURE 11.14 ◆ Water is cooled in these tower structures constructed from reinforced concrete, steel, and other materials. As the cooling towers are used, a mineral scale forms on their walls. To remove the scale, compounds containing hexavalent chromium were formerly used as cleaning agents. Their improper disposal subsequently caused groundwater contamination.

scientific studies that definitively support a link between these diseases and the consumption of hexavalent-chromium-contaminated water.

POTASSIUM DICHROMATE

Potassium dichromate is frequently used commercially as a strong oxidizer; for example, it is a component of the bath solutions used for chromium plating. When heated, potassium dichromate decomposes to generate oxygen as follows:

$$2K_2Cr_2O_7(s) \longrightarrow 2Cr_2O_3(s) + 2K_2O(s) + 3O_2(g)$$

potassium dichromate   chromic oxide   potassium oxide   oxygen

Since it is a strong oxidizer, potassium dichromate is chemically incompatible with many other substances. For example, potassium dichromate and hydrochloric acid react to produce chlorine.

$$K_2Cr_2O_7(aq) + 14HCl(aq) \longrightarrow 2CrCl_3(aq) + 2KCl(aq) + 7H_2O(l) + 3Cl_2(g)$$

potassium dichromate   hydrochloric acid   chromic chloride   potassium chloride   water   chlorine

The emergency incidents involving this hazardous material are generally associated with its tendency to oxidize the substances with which it contacts.

CHROMIUM TRIOXIDE

The acids associated with chromates and dichromates exist only in aqueous solutions, but when the water is evaporated from them, a compound called chromium(VI) oxide remains. This compound is also known as chromium trioxide, chromium anhydride, and chromic acid; its chemical formula is CrO_3.

Chromium trioxide is a red solid prepared industrially by adding concentrated sulfuric acid to a saturated solution of potassium dichromate.

$$K_2Cr_2O_7(aq) \ + 2H_2SO_4(conc) \longrightarrow 2KHSO_4(aq) + H_2O(l) + \ 2CrO_3(s)$$
potassium dichromate   sulfuric acid   potassium bisulfate   water   chromium trioxide

Chromium trioxide is used for metal-treatment processes such as chrome plating, copper stripping, aluminum anodizing, and corrosion prevention.

CHROMIUM OXYCHLORIDE

This is a dark-red liquid that forms when a mixture of concentrated hydrochloric acid and sulfuric acid is added to a saturated solution of potassium dichromate. It is also called chromyl chloride; its chemical formula is CrO_2Cl_2.

Not only is chromium oxychloride an oxidizer, but it also reacts very violently with water, forming chromium(III) chloride, chromic acid, and chlorine.

$$6CrO_2Cl_2(l) \ + 4H_2O(l) \longrightarrow \ 2CrCl_3(aq) \ + 4H_2CrO_4(aq) + 3Cl_2(g)$$
chromium oxychloride   water   chromium(III) chloride   chromic acid   chlorine

In this instance, the chromic acid is the "true" chromic acid (not chromium trioxide), which exists only in solution.

AMMONIUM DICHROMATE

This is an orange solid used in pyrotechnics, dyeing fabrics, preparing leather, and photography. Its chemical formula is $(NH_4)_2Cr_2O_7$. When heated, ammonium dichromate decomposes as follows:

$$(NH_4)_2Cr_2O_7(s) \longrightarrow \ Cr_2O_3(s) \ + 4H_2O(g) + N_2(g)$$
ammonium dichromate   chromic oxide   water   nitrogen

TRANSPORTING THE OXIDIZING CHROMIUM COMPOUNDS

DOT regulates the transportation of the chromium compounds in Table 11.11 as oxidizers or corrosive materials. DOT requires shippers and carrier to affix OXIDIZER or CORROSIVE labels, or both, on the packaging used to transport them and, when warranted, to post OXIDIZER or CORROSIVE placards on the bulk packaging, freight container, transport vehicle, unit containment device, or railcar used to transport them.

TABLE 11.11 ◆ Oxidizing Chromium Compounds Regulated by DOT for Transport

Proper Shipping Name	Label Codes
Chromic acid, solid	5.1, 8
Chromic acid solution	8
Chromium oxychloride	8
Chromium trioxide, anhydrous	5.1, 8

◆ 11.13 SODIUM PERMANGANATE AND POTASSIUM PERMANGANATE

Metallic permanganates are compounds in which manganese possesses the +7 oxidation state. They are generally encountered as dark-violet, iridescent crystals.

   The most commercially popular members of this class of compounds are the alkali metal permanganates, sodium permanganate and potassium permanganate, whose chemical formulas are $NaMnO_4$ and $KMnO_4$, respectively. These substances are manufactured from manganese ore containing about 70% manganese dioxide. As the ore is heated with either sodium hydroxide or potassium hydroxide, the manganese undergoes oxidation to produce the corresponding alkali manganate. For example, potassium manganate is formed from the ore and potassium hydroxide.

$$2MnO_2(s) \quad + \quad 4KOH(l) \quad + \quad O_2(g) \longrightarrow \quad 2K_2MnO_4(s) \quad + 2H_2O(g)$$
$$\text{manganese dioxide} \quad \text{potassium hydroxide} \quad \text{oxygen} \quad \text{potassium manganate} \quad \text{water}$$

A solution of the potassium manganate is then oxidized electrolytically to produce potassium permanganate, as follows:

$$2K_2MnO_4(aq) + 2H_2O(l) \longrightarrow \quad 2KMnO_4(aq) \quad + \quad 2KOH(aq) \quad + \quad H_2(g)$$
$$\text{potassium manganate} \quad \text{water} \quad \text{potassium permanganate} \quad \text{potassium hydroxide} \quad \text{hydrogen}$$

The potassium permanganate is then separated from the potassium hydroxide by crystallization.

   Whereas the dilute solutions of sodium permanganate and potassium permanganate are used medicinally to cure dermatitis having a bacterial or fungal origin, the concentrated solutions are used to remove objectionable matter from chemical and biological process wastes by oxidation. The effluent gases from many industrial sources are also often eliminated or reduced in concentration through redox reactions involving the use of sodium permanganate or potassium permanganate.

   DOT regulates the transportation of the metallic permanganates in Table 11.12 as oxidizers. For barium permanganate, DOT requires shippers and carriers to affix OXIDIZER and POISON labels to its packaging and, when warranted, to post OXIDIZER placards on the bulk packaging, freight container, transport vehicle, unit containment device, or railcar used to transport it. For the remaining metallic

TABLE 11.12 ◆ Metallic Permanganates Regulated by DOT for Transport

Proper Shipping Name	Label Codes
Barium permanganate	5.1, 6.1
Permanganates, inorganic, aqueous solution, n.o.s.	5.1
Permanganates, inorganic, n.o.s.	5.1
Potassium permanganate	5.1
Sodium permanganate	5.1

permanganates in Table 11.12, DOT requires shippers and carriers to affix OXIDIZER labels to their packaging and, when warranted, to post OXIDIZER placards on the bulk packaging, freight container, transport vehicle, unit containment device, or rail-car used to transport them.

Performance Goal for Section 11.14

♦ Describe the hazardous properties of the metallic nitrites and metallic nitrates.

◆ 11.14 METALLIC NITRITES AND METALLIC NITRATES

Metallic nitrites and metallic nitrates are compounds composed of metallic ions and the nitrite ion (NO_2^-) and nitrate ion (NO_3^-), respectively. In metallic nitrites, the nitrogen atom assumes an oxidation number of $+3$, whereas in nitrates, the oxidation number is $+5$. Sodium nitrate and potassium nitrate are common components of blasting agents and other explosive articles (Section 14.2). Until the 1980s, sodium nitrite was added to raw meat, such as bacon, hot dogs, and luncheon meats, to preserve color. This use was banned by the U.S. Food and Drug Administration when scientists demonstrated that nitrites produce *N-nitrosamines* in the stomach. Nitrites react with stomach acid to form nitrous acid, which in turn reacts with the protein in meat to form *N*-nitrosamines. The *N*-nitrosamines are organic compounds containing the group of atoms $=N-N=O$. Since their exposure to test animals causes cancer, epidemiologists classify *N*-nitrosamines as probable human carcinogens.

METALLIC NITRITES

Metallic nitrites can act as either oxidizing or reducing agents. They are oxidized to metallic nitrates and reduced to nitric oxide (NO), as follows:

$$NaNO_2(aq) + Na_2O_2(aq) + H_2O(l) \longrightarrow NaNO_3(aq) + 2NaOH(aq)$$

sodium nitrite   sodium peroxide   water   sodium nitrate   sodium hydroxide

$$2NaNO_2(aq) + Na_2SO_3(aq) + 2HCl(aq) \longrightarrow$$

sodium nitrite   sodium sulfite   hydrochloric acid

$$Na_2SO_4(aq) + 2NaCl(aq) + 2NO(g) + H_2O(l)$$

sodium nitrate   sodium chloride   nitric oxide   water

Although metallic nitrites are generally stable to heat, they decompose at elevated temperatures. For example, sodium nitrite decomposes as follows:

$$2NaNO_2(s) \longrightarrow Na_2O(s) + NO_2(g) + NO(g)$$

sodium nitrite   sodium oxide   nitrogen dioxide   nitric oxide

METALLIC NITRATES

Metallic nitrates react only as oxidizing agents. In the presence of acids, they are converted to nitric oxide, nitrogen dioxide, or ammonia. This chemical behavior is exemplified by the chemical reactions of zinc with concentrated nitric acid noted in Section 8.6.

When heated, metallic nitrates decompose in the following ways:

♦ The alkali metal nitrates decompose to form the respective alkali metal nitrite and oxygen. For example, sodium nitrate decomposes as follows:

$$2NaNO_3(s) \longrightarrow 2NaNO_2(s) + O_2(g)$$

sodium nitrate   sodium nitrite   oxygen

◆ Nitrates of the noble metals decompose to produce the metal, nitrogen dioxide, and oxygen. For example, silver nitrate thermally decomposes as follows:

$$2AgNO_3(s) \longrightarrow 2Ag(s) + 2NO_2(s) + O_2(g)$$

silver nitrate   silver   nitrogen dioxide   oxygen

◆ Metallic nitrates other than the alkali metal nitrates and the noble metal nitrates decompose to form the respective metallic oxide, nitrogen dioxide, and oxygen. For example, lead(II) nitrate decomposes as follows:

$$2Pb(NO_3)_2(s) \longrightarrow 2PbO(s) + 4NO_2(g) + O_2(g)$$

lead nitrate   lead oxide   nitrogen dioxide   oxygen

Solved Exercise 11.5

Write equations that illustrate the thermal decomposition of the oxidizers copper(II) nitrate and chromium(III) nitrate.

Solution: Metal nitrates other than the alkali metal nitrates and noble metal nitrates decompose when heated to form the respective metallic oxide, nitrogen dioxide, and oxygen. The thermal decompositions of copper(II) nitrate and chromium(III) nitrate are thus denoted as follows:

$$2Cu(NO_3)_2(s) \longrightarrow 2CuO(s) + 4NO_2(g) + O_2(g)$$

copper(II) nitrate   copper(II) oxide   nitrogen dioxide   oxygen

$$4Cr(NO_3)_3(s) \longrightarrow 2Cr_2O_3(s) + 12NO_2(g) + 3O_2(g)$$

chromium(III) nitrate   chromium(III) oxide   nitrogen dioxide   oxygen

---

DOT regulates the transportation of the metallic nitrates and nitrites in Table 11.13 as oxidizers. When shipping barium nitrate, beryllium nitrate, and lead nitrate, DOT requires shippers and carriers to affix OXIDIZER and POISON labels to their packaging and to post OXIDIZER placards on the bulk packaging, freight container, transport vehicle, unit containment device, and railcar used to transport them. For the remaining metallic nitrates and nitrites in Table 11.13, DOT requires shippers and carriers to affix OXIDIZER labels to their packaging and to post OXIDIZER placards on the bulk packaging, freight container, transport vehicle, unit containment device, or railcar used to transport them.

Performance Goal for Section 11.15

◆ Describe the hazardous properties of the metallic peroxides.

◆ 11.15 METALLIC PEROXIDES

sodium peroxide

Compounds composed of metallic ions and peroxide ions [O_2^{2-} or $(-O-O-)^{2-}$] are called *metallic peroxides*. The commercially important metallic peroxides are those composed of alkali metal ions and alkaline earth ions. Examples are sodium peroxide and barium peroxide, respectively, whose chemical formulas are Na_2O_2 and BaO_2.

TABLE 11.13 ◆ Metallic Nitrates and Nitrites Regulated by DOT for Transport

Proper Shipping Name	Label Codes
Aluminum nitrate	5.1
Barium nitrate	5.1, 6.1
Beryllium nitrate	5.1, 6.1
Calcium nitrate	5.1
Lead nitrate	5.1, 6.1
Lithium nitrate	5.1
Nickel nitrate	5.1
Nickel nitrite	5.1
Nitrates, inorganic, aqueous solution, n.o.s.	5.1
Nitrates, inorganic, n.o.s.	5.1
Nitrites, inorganic, aqueous solution, n.o.s.	5.1
Nitrites, inorganic, n.o.s.	5.1
Potassium nitrate	5.1
Potassium nitrate and sodium nitrite mixtures	5.1
Potassium nitrite	5.1
Silver nitrate	5.1
Sodium nitrate	5.1
Sodium nitrate and potassium nitrate mixtures	5.1
Sodium nitrite	5.1
Strontium nitrate	5.1
Zinc nitrate	5.1

When heated, the metallic peroxides decompose to the corresponding metallic oxide and oxygen.

barium peroxide

$$2Na_2O_2(s) \longrightarrow 2Na_2O(s) + O_2(g)$$

sodium peroxide   sodium oxide   oxygen

$$2BaO_2(s) \longrightarrow 2BaO(s) + O_2(g)$$

barium peroxide   barium oxide   oxygen

They also oxidize finely divided combustible metals. For example, sodium peroxide oxidizes powdered aluminum.

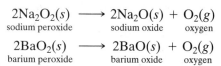

$$2Al(s) + 3Na_2O_2(s) + 3H_2O(l) \longrightarrow Al_2O_3(s) + 6NaOH(s)$$

aluminum   sodium peroxide   water   aluminum oxide   sodium hydroxide

The metallic peroxides are also water-reactive compounds.

$$2Na_2O_2(s) + 2H_2O(l) \longrightarrow 4NaOH(aq) + O_2(g)$$

sodium peroxide   water   sodium hydroxide   oxygen

To prevent metallic peroxides from absorbing atmospheric moisture, they should be stored in tightly closed, moisture-proof containers.

TABLE 11.14 ◆ Metallic Peroxides Regulated by DOT for Transport

Proper Shipping Name	Label Codes
Barium peroxide	5.1, 6.1
Magnesium peroxide	5.1
Peroxides, inorganic, n.o.s.	5.1
Potassium peroxide	5.1
Sodium peroxide	5.1
Strontium peroxide	5.1
Zinc peroxide	5.1

DOT regulates the transportation of the metallic peroxides in Table 11.14 as oxidizers. When shipping barium peroxide, DOT requires shippers and carriers to affix OXIDIZER and POISON labels to their packaging and to post OXIDIZER placards on the bulk packaging, freight container, transport vehicle, unit containment device, and railcar used to transport them. For the remaining metallic peroxides in Table 11.14, DOT requires shippers and carriers to affix OXIDIZER labels to their packaging and to post OXIDIZER placards on the bulk packaging, freight container, transport vehicle, unit containment device, or railcar used to transport them.

Performance Goal for Section 11.16
- Describe the hazardous properties of sodium persulfate and potassium persulfate.

◆ 11.16 SODIUM PERSULFATE AND POTASSIUM PERSULFATE

sodium
peroxide

Metallic persulfates are compounds composed of a metallic ion and the *persulfate* ion, also called the *peroxydisulfate* ion ($S_2O_8^{2-}$). Two examples of metallic persulfates are sodium persulfate and potassium persulfate, both of which are white solids; neither is widely used in commerce.

When sodium persulfate is heated, oxygen is generated, as follows:

$$2Na_2S_2O_8(s) \longrightarrow 2NaO_2(s) + 4SO_2(g) + 3O_2(g)$$

sodium persulfate   sodium oxide   sulfur dioxide   oxygen

DOT regulates the transportation of the metallic persulfates in Table 11.15 as oxidizers. DOT requires shippers and carriers to affix OXIDIZER labels to their

TABLE 11.15 ◆ Metallic Persulfates Regulated by DOT for Transport

Proper Shipping Name	Label Codes
Persulfates, inorganic, aqueous solution, n.o.s.	5.1
Persulfates, inorganic, n.o.s.	5.1
Potassium persulfate	5.1
Sodium persulfate	5.1

packaging and to post OXIDIZER placards on the bulk packaging, freight container, transport vehicle, unit containment device, or railcar used to transport them.

◆ 11.17 CHEMISTRY OF MATCHES

Matches served as one of the earliest commercial products in which redox reactions were used to provide an instant source of fire. Even today, a match is the most common item used to intentionally initiate a fire. It is appropriate to review the chemistry associated with the burning of matches, especially insofar as it involves the use of oxidizers.

The two types of matches in Figure 11.15 are commercially available. They are called *strike-anywhere matches* and *safety matches*. DOT regulates their transportation as flammable solids when the matches are of a type that cannot spontaneously ignite or undergo marked decomposition when subjected for 8 consecutive hours to a temperature of 200°F (93°C). DOT requires shippers and carriers to affix FLAMMABLE SOLID labels to their packaging and, when warranted, to post FLAMMABLE SOLID placards on the bulk packaging, freight container, transport vehicle, unit containment device, or railcar used to transport them.

STRIKE-ANYWHERE MATCHES

The head of a strike-anywhere match consists of a mixture of tetraphosphorus trisulfide, sulfur, lead(IV) oxide, powdered glass, and glue. This mixture is mounted on a small stick of wood and covered with paraffin wax. Tetraphosphorus trisulfide ignites as the match head is struck against a hard surface. The evolved heat of combustion initiates the combustion of the sulfur and wood.

$$\underset{\text{tetraphosphorus trisulfide}}{P_4S_3(s)} + \underset{\text{oxygen}}{8O_2(g)} \longrightarrow \underset{\text{tetraphosphorus decoxide}}{P_4O_{10}(s)} + \underset{\text{sulfur dioxide}}{3SO_2(g)}$$

SAFETY MATCHES

Safety matches ignite by means of friction against a prepared surface. The active components of a safety match consists of a mixture of antimony sulfide, sulfur, and potassium chlorate, held to a piece of cardboard by means of glue. The prepared surface upon which this mixture is struck consists of red phosphorus and powdered glass. This surface is usually on the box, book, or card. When the match head is rubbed upon this surface, the heat initiates a redox reaction as follows:

$$\underset{\text{potassium chlorate}}{16KClO_3(s)} + \underset{\text{sulfur}}{3S_8(s)} \longrightarrow \underset{\text{potassium chloride}}{16KCl(s)} + \underset{\text{sulfur dioxide}}{24SO_2(g)}$$

Then, the energy evolved from the oxidation of the sulfur initiates the ignition of the antimony trisulfide.

$$\underset{\text{antimony sulfide}}{2Sb_2S_3(s)} + \underset{\text{oxygen}}{9O_2(g)} \longrightarrow \underset{\text{antimony oxide}}{2Sb_2O_3(g)} + \underset{\text{sulfur dioxide}}{6SO_2(g)}$$

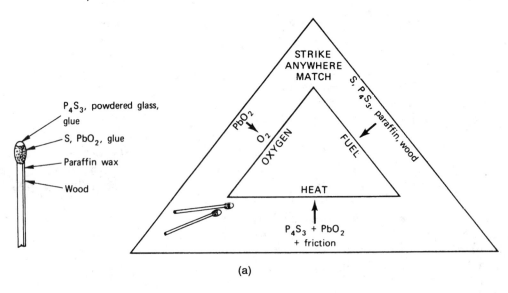

(a)

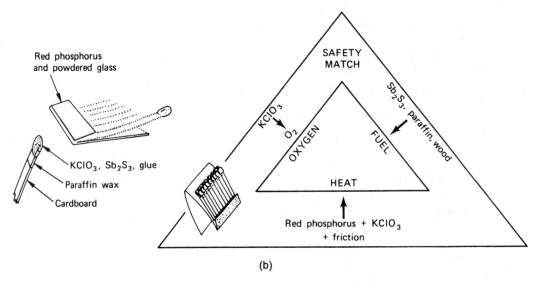

(b)

FIGURE 11.15 ◆ (a) A strike-anywhere match ignites when the surface of the match head is rubbed against a hard surface. (b) A safety match ignites when the surface of a match head is struck against a surface containing red phosphorus and powdered glass. In each instance, a unique redox reaction occurs that results in the production of a flame. The fire triangle relevant to each reaction notes the identity of the fuel, the source of oxygen, and the source of heat.

◆ 11.18 RESPONDING TO INCIDENTS INVOLVING THE RELEASE OF OXIDIZERS

First-on-the-scene responders can generally identify the presence of a simple oxidizer at a transportation mishap by observing at least one of the following:

- ◆ The number 5.1 as a component of a basic description of a hazardous material listed on a shipping paper
- ◆ The word "OXIDIZER" printed upon a yellow label affixed to packaging
- ◆ The word "OXIDIZER" printed upon a yellow placard posted on the relevant transport vehicle

Figure 11.16 indicates that the recommended method of extinguishing fires supported by most liquid or solid oxidizers is to deluge them with water. Since oxidizers are generally soluble in water, the application of water dilutes them. Then, their potential for chemical reactivity is sharply reduced or eliminated.

Caution should be exercised when emergency responders use water on oxidizers. Many solid oxidizers melt before they thermally decompose. This hot molten material flows to adjoining areas, where it may mix with combustible material and cause it to spontaneously ignite. Although water can be effectively used on fires supported by oxidizers, it should never be applied to bulk quantities of molten oxidizers. Steam may rapidly generate and cause the hot, molten material to splatter. When fires are supported by bulk quantities of these molten oxidizers, experts recommend the use of sand to extinguish them.

Solved Exercise 11.6

Why is it essential to dam or dike the water runoff generated during a response action involving an oxidizer?

Solution: When water is used during a response action involving an oxidizer, its primary function is to dissolve the oxidizer so appreciably that it is unable to oxidize nearby matter. As it dissolves in the water, the oxidizer subsequently moves with the flow of the water. When left unattended, the water evaporates and leaves the oxidizer in a dry state, whereupon it is again susceptible to contact combustible materials.

The water runoff generated during a response action involving an oxidizer should be dammed or diked, so the oxidizer can first be neutralized prior to its ultimate discharge. Sodium sulfite is a typical neutralizing agent.

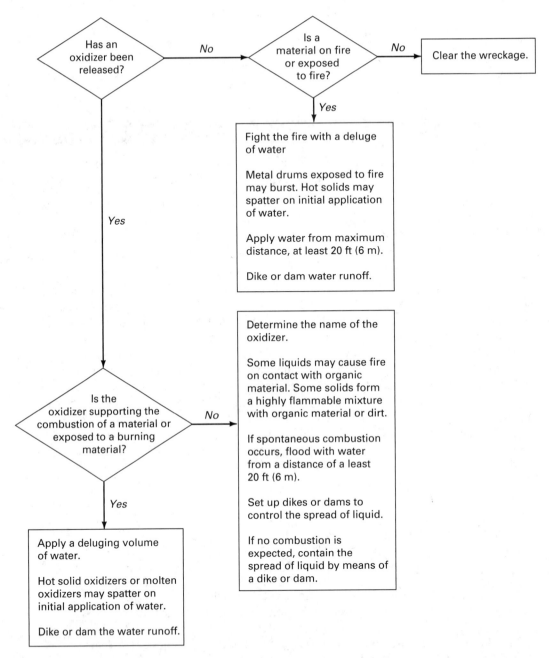

FIGURE 11.16 ◆ The recommended procedures when responding to a disaster involving a combustible material fire supported by an oxidizer other than an organic peroxide. *Adapted with permission of the American Society for Testing and Materials, from a figure in ASTM STP 825, A Guide to the Safe Handling of Hazardous Materials Accidents. Copyright © 1983, American Society for Testing and Materials.*

■■

REVIEW EXERCISES

Basic Concepts Regarding Oxidation–Reduction

11.1 To some, the word *oxidizer* suggests a substance that possesses a certain allegiance with elemental oxygen. Demonstrate that this perception is incorrect by identifying two oxidizers other than oxygen that support the combustion of hydrogen.

11.2 What is the oxidation number of each underlined atom or ion in the following chemical formulas?

(a) $Ba\underline{O}$ and $Ba\underline{O}_2$ (e) $\underline{F}_2$ and $H\underline{F}$

(b) $\underline{N}O$ and $\underline{N}O_2$ (f) $\underline{O}_2$ and $H_2\underline{O}$

(c) $H_2\underline{S}O_3$ and $H_2\underline{S}O_4$ (g) $\underline{P}_4$, $H_3\underline{P}O_3$, $H_3\underline{P}O_4$, and $\underline{P}H_3$

(d) $\underline{Pb}O$ and $\underline{Pb}O_2$ (h) $\underline{Cl}_2$, $K\underline{Cl}$, $K\underline{Cl}O$, $K\underline{Cl}O_2$, $K\underline{Cl}O_3$, and $K\underline{Cl}O_4$

11.3 Identify the oxidizing agent, the reducing agent, the substance oxidized, and the substance reduced in the redox reactions denoted by the following equations:

(a) $4Zn(s) + HNO_3(aq) + 9HCl(aq) \longrightarrow$
$$4ZnCl_2(aq) + NH_4Cl(aq) + 3H_2O(l)$$

(b) $H_2O_2(aq) + 2FeCl_2(aq) + 2HCl(aq) \longrightarrow 2FeCl_3(aq) + 2H_2O(l)$

(c) $6KOH(aq) + 3Cl_2(g) \longrightarrow KClO_3(aq) + 5KCl(aq) + 3H_2O(l)$

(d) $8Al(s) + 3NaClO_4(s) \longrightarrow 4Al_2O_3(s) + 3NaCl(s)$

(e) $Fe(s) + 2HCl(aq) \longrightarrow FeCl_2(aq) + H_2(g)$

11.4 Using Table 11.4, provide the technical basis for concluding that each of the following incidents is potentially hazardous:

(a) Bottles containing potassium dichloro-*s*-triazine-2,4,6-trione are cushioned with sawdust inside a box and transported by railcar.

(b) Potassium permanganate is combined with kerosene.

(c) Bales of unprocessed cotton are shipped with a carboy containing concentrated perchloric acid.

(d) Hot magnesium chips are mixed with sodium nitrate.

(e) Sodium perchlorate is sprinkled on hot charcoal briquettes.

(f) A pinch of potassium superoxide is added to a quart-container of motor oil.

(g) A farmer stores a sulfur-containing pesticide with ammonium nitrate in the same shed.

(h) Hydrogen peroxide solution is washed into a drain.

(i) To aid in their ignition, sodium nitrate is added to piles of cardboard.

(j) Nitric acid is dripped upon wool.

11.5 Which substance in each of the following pairs possesses the greater affinity to spontaneously decompose when exposed to heat?

(a) Potassium chlorate or zinc chlorate

(b) An aqueous solution containing 70% perchloric acid or an aqueous solution containing 70% nitric acid

(c) Sodium peroxide or lithium peroxide

11.6 Using Table 11.4, identify the oxidizer in each of the following pairs that possesses the greater affinity for increasing the combustion rate of a burning material:

(a) Potassium dichromate or ammonium dichromate

(b) Magnesium nitrate or calcium hypochlorite

(c) Lithium peroxide or potassium chlorate

11.7 When magnesium burns in air, a mixture consisting of 75% magnesium oxide and 25% magnesium nitride is produced. Identify the oxidizing agents, the reducing agent, the substance oxidized, and the substances reduced during these combustion reactions.

Hydrogen Peroxide

11.8 What is the most likely reason DOT prohibits the transportation on passenger-carrying aircraft and railcars of any amount of a hydrogen peroxide solution that contains more than 40% hydrogen peroxide by mass?

11.9 Why is the use of water generally the best means of extinguishing fires in which hydrogen peroxide is involved?

11.10 Why are hydrogen peroxide solutions generally stored in bottles made of darkened glass or plastic?

11.11 When hydrogen peroxide solutions soak into cardboard and other combustible materials, why are these materials likely to spontaneously erupt into flame?

11.12 Why is it an unsafe practice to discharge waste hydrogen peroxide solutions into a sewer system?

Hypochlorite Oxidizers

11.13 Why do the manufacturers of hypochlorite bleaching agents identify the available chlorine on their product labels?

11.14 Why is the following label typically affixed to containers of muriatic acid intended for use in home swimming pools?

> **DO NOT STORE NEAR
> CHLORINE-PRODUCING
> POOL CHEMICALS**

Chlorination Agents Other Than Metallic Hypochlorites

11.15 Why must carriers take precaution to segregate trichloroisocyanuric acid from combustible matter when loading it onboard railcars for shipment?

Oxidizers in Fireworks, Flares, and Signaling Smokes

11.16 Strontium perchlorate is the component of railroad flares responsible for the carmine-red light emitted when they are ignited. Write the equation denoting this thermal decomposition.

11.17 The components of a simple firecracker are sulfur, powdered elemental aluminum, and potassium perchlorate. These components are mixed and loosely packed within a cardboard tube. When the mixture is ignited, the redox reaction denoted by the following equation occurs:

$$\underset{\text{sulfur}}{S_8(s)} + \underset{\text{aluminum}}{8Al(s)} + \underset{\text{potassium perchlorate}}{7KClO_4(s)} \longrightarrow \underset{\text{aluminum oxide}}{4Al_2O_3(s)} + \underset{\text{potassium chloride}}{7KCl(s)} + \underset{\text{sulfur dioxide}}{8SO_2(g)}$$

Identify the oxidizing agent, the reducing agents, the substances oxidized, and the substance reduced during this reaction.

Oxidizing Ammonium Compounds

11.18 The OSHA regulation at 29 CFR §1910.109(c)(4)(iii)(*a*) permits employers to store ammonium nitrate in a building having a basement *only* when the basement is open on at least one side. What is the most likely reason OSHA requires one side of the basement to be open?

11.19 The OSHA regulation at 29 CFR §1910.109(c)(4)(iii)(*d*) requires employers to protect the flooring and handling areas in buildings against impregnation by ammonium nitrate, and requires that the floors be constructed without open drains, traps, tunnels, pits, or pockets into which any molten ammonium nitrate could flow and be confined in the event of fire. What is the most likely reason OSHA mandates these structural features?

11.20 Why should the use of ammonium nitrate be avoided in areas where fuel oil is stored?

11.21 Since the 9/11 highjacking incidents, why have law enforcement agencies expressed special interest in identifying the areas within their jurisdictions where bulk quantities of ammonium nitrate are stored?

11.22 When a mixture of ammonium nitrate and fuel oil is activated, an explosion occurs during which nitrogen, carbon dioxide, and water vapor are produced. Using $(CH)_n$ as the chemical formula for the fuel oil, write a balanced equation depicting the reaction between ammonium nitrate and fuel oil.

11.23 Write the equation denoting the oxidation–reduction reaction between the components of ammonal. (Hint: See Solved Exercise 11.4.)

Other Oxidizers

11.24 A chemical manufacturer in Denver, Colorado intends to ship sixteen 0.5-lb bottles of sodium peroxide to a customer in Dallas, Texas by cargo aircraft. Using Table 6.1, what basic description does DOT require the manufacturer and carrier to enter on the accompanying shipping paper?

11.25 The components of ordinary gunpowder are elemental sulfur, charcoal, and potassium nitrate. When gunpowder explodes, these components appear to explode as follows:

$$\underset{\text{sulfur}}{3S_8(s)} + \underset{\text{carbon}}{16C(s)} + \underset{\text{potassium nitrate}}{32KNO_3(s)} \longrightarrow$$

$$\underset{\text{potassium oxide}}{16K_2O(s)} + \underset{\text{carbon dioxide}}{16CO_2(g)} + \underset{\text{nitrogen}}{16N_2(g)} + \underset{\text{sulfur dioxide}}{24SO_2(g)}$$

Identify the oxidizing agent, the reducing agents, the substances oxidized, and the substance reduced during the explosion of gunpowder.

11.26 Write the equation illustrating the thermal decomposition of each of the following metallic nitrates:
   (a) lithium nitrate
   (b) zinc nitrate
   (c) nickel nitrate

11.27 Why is occupational exposure to carcinogens likely to be elevated compared to the norm for workers in the chromate pigment, chromate electroplating, and chromium-alloy manufacturing industries?

Fighting Fires Involving Oxidizers

11.28 When properly posted on the exterior wall of a building, how does the observation of an NFPA diamond convey to on-duty firefighters that an oxidizer is stored nearby?

11.29 Why is the use of carbon dioxide generally ineffective for extinguishing fires supported by oxidizers?

11.30 Why do firefighters recommend the use of a deluging volume of water when extinguishing an oxidizer-supported fire?

Chemistry of Some Hazardous Organic Compounds  **12** CHAPTER

Organic compounds are constituents of many commercial products: heating and motor fuels, solvents, plastics, resins, fibers, paints, varnishes, refrigerants, propellants, aerosols, textiles, and explosives. Their relative commonplace underscores the need to study organic compounds in some detail.

The principal concern about organic compounds to firefighters is that they are usually flammable substances. DOT often regulates their transportation as flammable gases or flammable liquids. It is hardly surprising to learn that when they are encountered at emergency scenes, organic compounds are already burning.

Organic compounds can also cause a variety of detrimental health effects. In the human organism, repeated exposure to certain organic compounds can damage the liver, kidneys, and heart, depress the central nervous system, or cause cancer.

In this chapter, we learn about the properties of simple organic compounds that are commonly encountered commercially, whereas in Chapters 13 and 14 we learn about the properties of more complex organic compounds: polymers and explosives.

Performance Goals for Section 12.1
- Discuss the manner by which a carbon atom covalently bonds to nonmetallic atoms including other carbon atoms.
- Describe the nature of carbon–carbon single bonds, carbon–carbon double bonds, and carbon–carbon triple bonds.

◆ 12.1 WHAT ARE ORGANIC COMPOUNDS?

The molecules of all organic compounds have one common feature: the presence of one or more carbon atoms. In most instances, the carbon atoms share electrons with other nonmetallic atoms. As shown by their molecular structures in Figure 12.1, methane, carbon tetrachloride, carbon monoxide, and carbon disulfide are examples of organic compounds having molecules in which the carbon atoms are bonded to other nonmetallic atoms.

$$\underset{\text{Methane}}{\overset{\displaystyle H}{\underset{\displaystyle H}{H-\overset{\displaystyle |}{\underset{\displaystyle |}{C}}-H}}} \qquad \underset{\text{Carbon tetrachloride}}{\overset{\displaystyle Cl}{\underset{\displaystyle Cl}{Cl-\overset{\displaystyle |}{\underset{\displaystyle |}{C}}-Cl}}} \qquad \underset{\text{Carbon monoxide}}{C\equiv O} \qquad \underset{\text{Carbon disulfide}}{S=C=S}$$

FIGURE 12.1 ◆ Carbon atoms may share their electrons with the electrons of other nonmetallic atoms, like hydrogen, chlorine, oxygen, and sulfur. The compounds that result from such electron sharing are methane, carbon tetrachloride, carbon monoxide, and carbon disulfide, respectively.

Carbon atoms can also mutually share electrons with other carbon atoms. When the molecular structures of such compounds are examined, we find that two carbon atoms can share electrons to form any of the following: carbon–carbon single bonds (C—C), carbon–carbon double bonds (C=C), and carbon–carbon triple bonds (C≡C). Each bond is a shared pair of electrons. Figure 12.2 illustrates the bonding in molecules of ethane, ethene, and acetylene, each of which has two carbon atoms. Molecules of ethane possess carbon–carbon single bonds, molecules of ethylene possess carbon–carbon double bonds, and molecules of acetylene possess carbon–carbon triple bonds.

Covalent bonds between carbon atoms in the molecules of more complex organic compounds can be linked into chains, including branched chains, or into rings. Consider just the carbon skeleton of the following two molecules noted as (a) and (b):

$$-C-C-C-C-C-C- \qquad \underset{\qquad\qquad\quad |\qquad\qquad\quad |}{-C-C-C-C-C-C-}$$
$$ \underset{\qquad\qquad\quad C \qquad\qquad\quad C}{}$$

(a)   **(b)**

In **(a)**, the carbon atoms are bonded to one another in a continuous chain, whereas in **(b)**, the carbon atoms are bonded to one another in a branched pattern. This means that carbon atoms not only bond to other carbon atoms in long chains, they also bond to groups of other carbon atoms as side chains attached to the main chain.

Aside from being bonded to other carbon atoms, each carbon atom in either a straight-chain or branched-chain arrangement is typically bonded to one or more atoms of hydrogen, oxygen, nitrogen, sulfur, or the halogens. For instance, when the carbon atoms in the compound having the carbon skeleton **(a)** are bonded to hydrogen atoms, the molecular structure of the compound is written as follows:

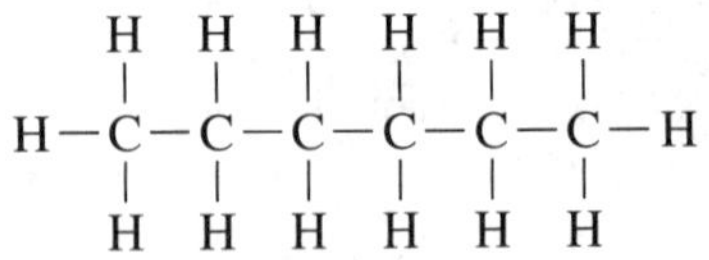

$$H-\overset{\overset{\displaystyle H}{|}}{\underset{\underset{\displaystyle H}{|}}{C}}-\overset{\overset{\displaystyle H}{|}}{\underset{\underset{\displaystyle H}{|}}{C}}-\overset{\overset{\displaystyle H}{|}}{\underset{\underset{\displaystyle H}{|}}{C}}-\overset{\overset{\displaystyle H}{|}}{\underset{\underset{\displaystyle H}{|}}{C}}-\overset{\overset{\displaystyle H}{|}}{\underset{\underset{\displaystyle H}{|}}{C}}-\overset{\overset{\displaystyle H}{|}}{\underset{\underset{\displaystyle H}{|}}{C}}-H$$

When the carbon atoms in **(b)** are bonded to hydrogen atoms, the molecular structure of the compound is written as follows:

$$H-\overset{\overset{\displaystyle H}{|}}{\underset{\underset{\displaystyle H}{|}}{C}}-\overset{\overset{\displaystyle H}{|}}{\underset{\underset{\displaystyle H-C-H}{\underset{\displaystyle H}{|}}}{C}}-\overset{\overset{\displaystyle H}{|}}{\underset{\underset{\displaystyle H}{|}}{C}}-\overset{\overset{\displaystyle H}{|}}{\underset{\underset{\displaystyle H}{|}}{C}}-\overset{\overset{\displaystyle H}{|}}{\underset{\underset{\displaystyle H-C-H}{\underset{\displaystyle H}{|}}}{C}}-\overset{\overset{\displaystyle H}{|}}{\underset{\underset{\displaystyle H}{|}}{C}}-H$$

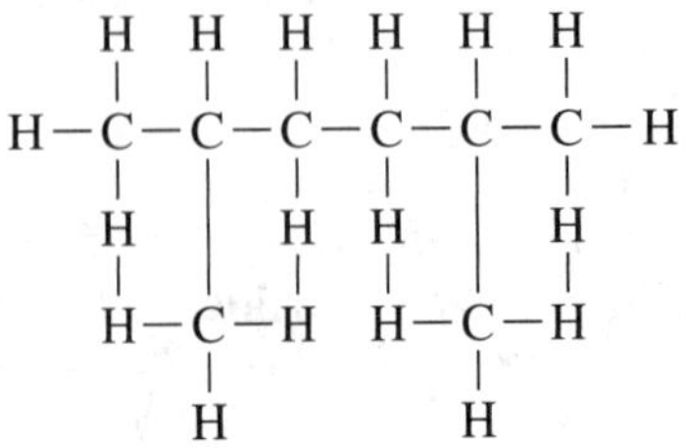

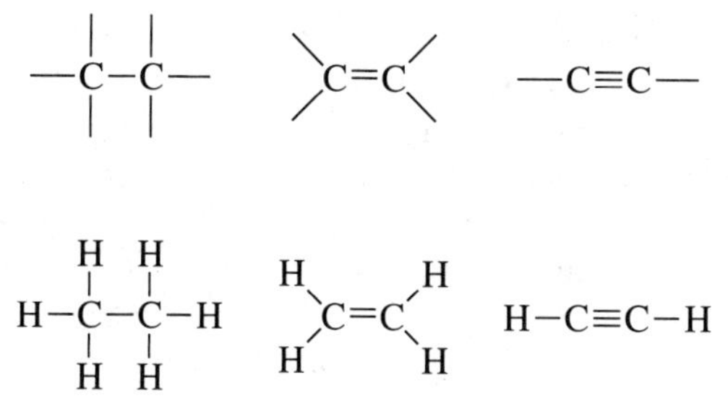

$$-\overset{\displaystyle |}{\underset{\displaystyle |}{C}}-\overset{\displaystyle |}{\underset{\displaystyle |}{C}}- \qquad \diagup\hspace{-0.3em}\underset{}{C}\hspace{-0.3em}=\hspace{-0.3em}\underset{}{C}\hspace{-0.3em}\diagdown \qquad -C\equiv C-$$

$$H-\overset{\displaystyle H}{\underset{\displaystyle H}{\underset{\displaystyle |}{\overset{\displaystyle |}{C}}}}-\overset{\displaystyle H}{\underset{\displaystyle H}{\underset{\displaystyle |}{\overset{\displaystyle |}{C}}}}-H \qquad \overset{H}{\diagdown}C\hspace{-0.3em}=\hspace{-0.3em}C\overset{H}{\diagup} \qquad H-C\equiv C-H$$

FIGURE 12.2 ◆ Two carbon atoms may share their own electrons in either of the three ways noted. This results in the formation of carbon–carbon single bonds, carbon–carbon double bonds, and carbon–carbon triple bonds. When these carbon atoms further bond to hydrogen atoms, the resulting compounds are ethane, ethene, and acetylene, respectively.

This illustrates that many organic compounds consist of continuous or branched chains of carbon atoms. The actual number of such compounds appears to be unlimited. More than one million organic compounds are already known. This abundance of compounds can be intimidating. How can anyone remember the properties of so many substances? Fortunately, it is possible to systematically organize organic compounds into a number of classes, where the properties of the members of each class are largely identified or dictated by certain atoms or groups of atoms. These are called *functional groups* and are identified later in Section 12.6. The study of organic chemistry then becomes the study of the classes of compounds that possess unique functional groups.

We shall begin the study of organic chemistry by examining the simplest organic compounds, the *hydrocarbons;* these are compounds whose molecules are composed of only carbon and hydrogen atoms. All hydrocarbons are broadly divided into two groups: aliphatic and aromatic hydrocarbons. We shall note some of their features in the following two sections.

Performance Goals for Section 12.2
  ◆ Describe the nature of the chemical bonds that exist in the molecules of the alkanes and cycloalkanes.
  ◆ Illustrate that some alkanes have structural isomers.
  ◆ Identify the rules for naming simple alkanes.
  ◆ Describe the nature of the chemical bonds that exist in the molecules of the alkenes, dienes, trienes, and cycloalkenes.
  ◆ Illustrate that some alkenes have structural isomers and geometrical isomers.
  ◆ Identify the rules for naming simple alkenes.
  ◆ Describe the nature of the chemical bonds that exist in molecules of the alkynes.
  ◆ Illustrate that some alkynes have structural isomers.
  ◆ Identify the rules for naming simple alkynes.

◆ 12.2 ALIPHATIC HYDROCARBONS

The word "aliphatic" means "fat-like," but the *aliphatic hydrocarbons* are best characterized by the common features of their molecular structures. We shall examine the following classes of aliphatic hydrocarbons independently.

ALKANES AND CYCLOALKANES

The *alkanes* are hydrocarbons having only carbon–carbon single bonds. The general chemical formula of an alkane is C_nH_{2n+2}, where n is a nonzero integer. When the number of carbon atoms is only one, the number of hydrogen atoms is 4; the corresponding compound is named methane, and its chemical formula is CH_4. When the number of carbon is two, the number of hydrogen atoms is 6; the corresponding compound is named ethane, and its chemical formula is C_2H_6. The alkanes are called *saturated hydrocarbons* since each bonding electron of the carbon atoms is shared with a valence electron from other atoms. Several examples of alkanes having carbon atoms from one to eight are noted in Table 12.1. The names of these simple alkanes and their formulas should be memorized.

The Lewis structures of alkanes having from one to eight carbon atoms are also provided in Table 12.1; for simplicity, only the structures having straight-chain arrangements have been noted. Beginning with butane, however, the molecular structure can be correctly written in several ways. There are two Lewis structures for the formula C_4H_{10}, as follows:

n-butane   *iso*butane

In the first structure, the carbon atoms are bonded to one another in a continuous chain. In the second structure, only three of the carbon atoms are bonded to one another; the fourth carbon atom is bonded to the carbon atom in the middle of the chain. Since there are two correct ways to write the Lewis structures for the formula C_4H_{10}, there must be two distinctly different compounds having this formula. To distinguish them by name, the compound having the carbon atoms bonded in a continuous chain is called *n*-butane, while the compound having the branched structure is named isobutane.

Two or more compounds that have the same molecular formula, but different structural arrangements of their atoms, are called *structural isomers*. One way by which isomers are identified is to use an *n*- (for "normal") in front of the name of the compound that contains a continuous chain of carbon atoms, and *iso*- (meaning "the same") for the compound that has a methyl group (CH_3—) bonded to a carbon atom next to the terminal (end) carbon atom. This system of nomenclature is generally referred to as the *common system*, and the compound names are called *common names*.

Although a molecular structure can be written for any alkane, it is generally convenient to condense it by simply writing the symbols of the atoms next to the symbol of the carbon atom to which they are bonded. For an alkane, we write the symbols of the hydrogen atoms next to the symbols of the carbon atoms to which they are bonded. Such formulas are called *condensed formulas*. They convey the same information as the more complete Lewis structures, but the dashes are either entirely omitted or used only in a limited fashion. The condensed formulas for *n*-butane and isobutane are noted below:

$$CH_3CH_2CH_2CH_3 \qquad CH_3-CH-CH_3$$
$$| $$
$$CH_3$$

TABLE 12.1 ◆ Simple Alkanes Having a Continuous Chain of Carbon Atoms

Name	Molecular Formula	Lewis Structure	Condensed Formula
Methane	CH_4	$H-\overset{\displaystyle H}{\underset{\displaystyle H}{C}}-H$	CH_4
Ethane	C_2H_6	$H-\overset{\displaystyle H}{\underset{\displaystyle H}{C}}-\overset{\displaystyle H}{\underset{\displaystyle H}{C}}-H$	CH_3CH_3
Propane	C_3H_8	$H-C-C-C-H$ (with H above and below each C)	$CH_3CH_2CH_3$
Butane	C_4H_{10}	$H-C-C-C-C-H$ (with H above and below each C)	$CH_3CH_2CH_2CH_3$
Pentane	C_5H_{12}	$H-C-C-C-C-C-H$ (with H above and below each C)	$CH_3CH_2CH_2CH_2CH_3$
Hexane	C_6H_{14}	$H-C-C-C-C-C-C-H$ (with H above and below each C)	$CH_3CH_2CH_2CH_2CH_2CH_3$
Heptane	C_7H_{16}	$H-C-C-C-C-C-C-C-H$ (with H above and below each C)	$CH_3CH_2CH_2CH_2CH_2CH_2CH_3$
Octane	C_8H_{18}	$H-C-C-C-C-C-C-C-C-H$ (with H above and below each C)	$CH_3CH_2CH_2CH_2CH_2CH_2CH_2CH_3$

While the simple hydrocarbons are generally referred to by their common names, the complex hydrocarbons are difficult to name using the common system of nomenclature. To overcome this hurdle, a second system of nomenclature was devised that is both simple and systematic. It was adopted by the International Union of Pure and Applied Chemistry (IUPAC) and is called the *IUPAC system of nomenclature*. The names of organic compounds derived from use of the IUPAC system are called *IUPAC names*.

TABLE 12.2 ◆ Some Common Alkyl Substituents

Name[a]	*Chemical Formula*
Methyl	CH_3-
Ethyl	CH_3CH_2- or C_2H_5-
n-Propyl	$CH_3CH_2CH_2-$ or C_3H_7-
Isopropyl	$CH_3-\underset{\underset{CH_3}{\vert}}{\overset{\overset{H}{\vert}}{C}}-$ or $(CH_3)_2CH-$
n-Butyl	$CH_3CH_2CH_2CH_2-$ or C_4H_9-
Secondary butyl	$CH_3CH_2-\underset{\underset{CH_3}{\vert}}{\overset{\overset{H}{\vert}}{C}}-$ or $CH_3CH_2-\underset{\underset{CH_3}{\vert}}{CH}-$
tert-Butyl	$CH_3-\underset{\underset{CH_3}{\vert}}{\overset{\overset{H}{\vert}}{C}}-$ or $(CH_3)_3-C-$
n-Pentyl[b]	$CH_3CH_2CH_2CH_2CH_2-$ or $C_5H_{11}-$

[a]The alkyl groups are named by replacing the *-ane* suffix of the parent hydrocarbon with *-yl*.
[b]Also called *n*-amyl.

The IUPAC system applies the following rules to the naming of alkanes from their molecular structures:

- Find the longest chain of carbon atoms in the structure. This is called the "main chain." It is not necessary that the main chain be written horizontally in order to be continuous.
- Assign numbers consecutively to each carbon atom in the main chain starting from the end that gives the groups attached to the chain the smaller numbers. In alkanes, the groups are called *alkyl substituents*; their general chemical formula is C_nH_{2n+1}. The names of some common alkyl substituents are provided in Table 12.2. The names and formulas of these alkyl substitutents should be memorized.
- Designate the position of each substituent by the number of the carbon atom along the main chain to which it is attached.
- Name the substituents alphabetically (ethyl before methyl, and so on) and place the names of the substituents as prefixes on the name of the main chain.
- If several substituents occur in the same compound, indicate the number of identical groups by the use of the following prefixes before the name of the substituent: *di-* for two identical groups, *tri-* for three identical groups, *tetra-* for four identical groups, and so on.

The use of these rules in naming several alkanes by the IUPAC system of nomenclature is illustrated in Table 12.3.

There are also alkanes in which the first and last carbon atoms in a continuous chain are joined to each other in a cyclic arrangement. These compounds are called *cycloalkanes*. Their general chemical formula is C_nH_{2n}. They are named by placing the prefix *cyclo-* in front of the name of the parent hydrocarbon to yield cyclopropane, cyclobutane, cyclopentane, and so on. Sometimes the structural formulas for

TABLE 12.3 ◆ Examples of Naming Alkanes

Chemical Formula	Name
$CH_3CH_2CH_2CH_2CH_3$	*n*-Pentane

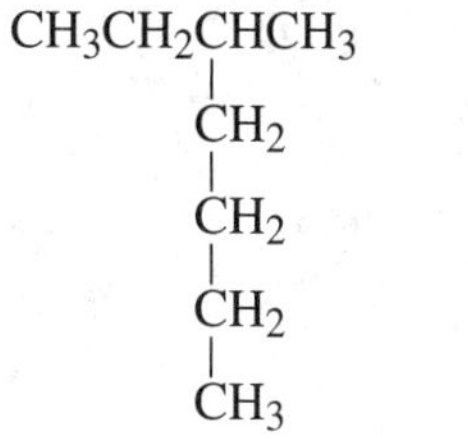 $CH_3CH_2CHCH_3$
    CH_2
    CH_2
    CH_2
    CH_3
	3-Methylheptane

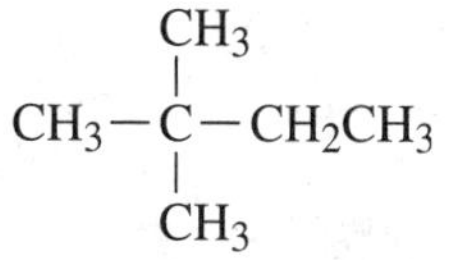 $\quad CH_3$ $CH_3-C-CH_2CH_3$ $\quad CH_3$	2,2-Dimethylbutane

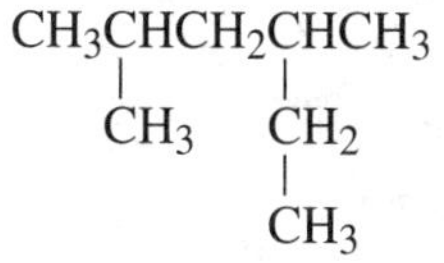 $CH_3CHCH_2CHCH_3$
  $CH_3\quad CH_2$
      CH_3
	2,4-Dimethylhexane

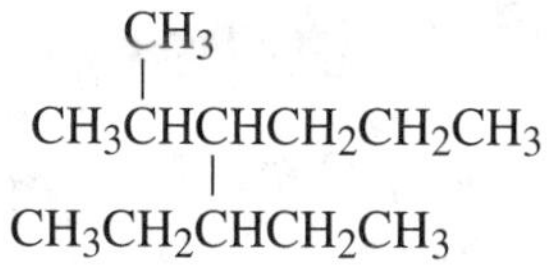 $\quad CH_3$ $CH_3CHCHCH_2CH_2CH_3$ $CH_3CH_2CHCH_2CH_3$	3-Ethyl-4-isopropylheptane

cyclopropane, cyclobutane, and cyclopentane are represented by writing a triangle, square, and pentagon for C_3H_6, C_4H_8, and C_5H_{10}, respectively. The structural formula of cyclohexane is represented by two unique conformations of C_6H_{12}. They are called the *boat* and *chair* conformations and are represented in Table 12.4.

Solved Exercise 12.1

Using the IUPAC system, name the alkane having the following condensed formula:

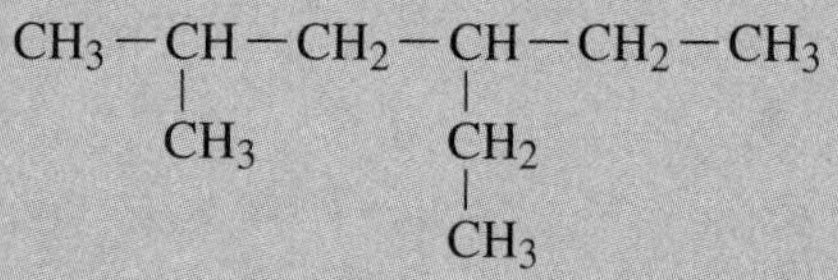

$$CH_3-CH-CH_2-CH-CH_2-CH_3$$
$$\quad\quad CH_3 \quad\quad\quad CH_2$$
$$\quad\quad\quad\quad\quad\quad\quad\quad CH_3$$

Solution: First, identify the longest continuous chain of carbon atoms. By examination, we see that the longest chain contains six carbon atoms, which signifies that the compound is a derivative of hexane. Next, assign a number to each carbon atom of this longest chain.

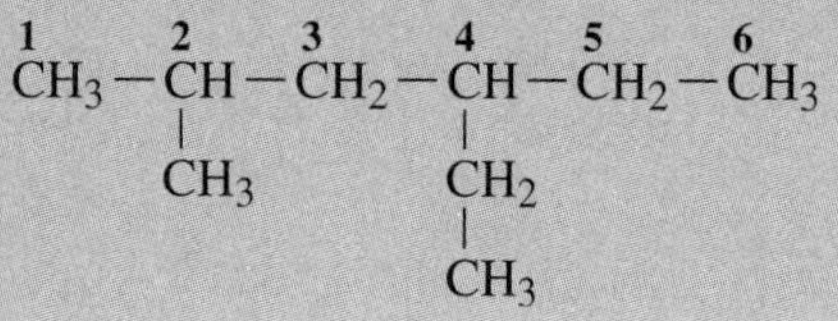

$$\overset{1}{CH_3}-\overset{2}{CH}-\overset{3}{CH_2}-\overset{4}{CH}-\overset{5}{CH_2}-\overset{6}{CH_3}$$
$$\quad\quad CH_3 \quad\quad\quad CH_2$$
$$\quad\quad\quad\quad\quad\quad\quad\quad CH_3$$

We see that the methyl group is bonded to the carbon atom numbered 2 and the ethyl group is bonded to the carbon atom numbered 4. By remembering to name the alkyl substituents in alphabetical order, the compound is correctly named 4-ethyl-2-methylhexane.

It is incorrect to number the carbon atoms from right to left along the horizontal chain of carbon atoms, since this numbering scheme gives larger numbers to the alkyl groups bonded to the main chain (5 for the methyl group and 3 for the ethyl group.) However, it is correct to number the longest chain of carbon atoms in either of the following ways:

$$CH_3 - \overset{2}{C}H - \overset{3}{C}H_2 - \overset{4}{C}H - \overset{5}{C}H_2 - \overset{6}{C}H_3$$
$$\underset{1}{\underset{|}{C}H_3} \qquad \underset{|}{C}H_2$$
$$CH_3$$

$$CH_3 - \overset{2}{C}H - \overset{3}{C}H_2 - \overset{4}{C}H - CH_2 - CH_3$$
$$\underset{1}{\underset{|}{C}H_3} \qquad \underset{5}{\underset{|}{C}H_2}$$
$$\underset{6}{C}H_3$$

The use of either of these numbering schemes also provides 4-ethyl-2-methylhexane as the correct name of this compound.

Solved Exercise 12.2

The detailed analysis of a commercial rubber solvent provides the following approximate chemical composition: **(a)** Fifty percent by volume is a mixture of 2,3-dimethylbutane, 2,3-dimethylpentane, and 3,3-dimethylpentane; **(b)** thirty percent is a mixture of 2,2-dimethylbutane, 3-methylpentane, 2,2-dimethylpentane, methylcyclopentane, methylcyclohexane, 1,2-dimethylcyclopentane, *n*-hexane, and *n*-heptane; and **(c)** twenty percent is a mixture of *n*-butane, *n*-pentane, and 2,4-dimethylhexane. Write the chemical formula for each substance.

Solution: The solvent's constituents have the condensed formulas noted below:

$$\textbf{(a)} \quad CH_3 - \underset{|}{\overset{\overset{\displaystyle CH_3}{|}}{C}}H - \underset{|}{\overset{\overset{\displaystyle CH_3}{|}}{C}}H - CH_3 \qquad CH_3 - \underset{|}{\overset{\overset{\displaystyle CH_3}{|}}{C}}H - \underset{|}{\overset{\overset{\displaystyle CH_3}{|}}{C}}H - CH_2CH_3 \qquad CH_3CH_2 - \underset{|}{\overset{\overset{\displaystyle CH_3}{|}}{C}} - CH_2CH_3$$
$$\underset{|}{CH_3}$$

(2,3-dimethylbutane) (2,3-dimethylpentane) (3,3-dimethylpentane)

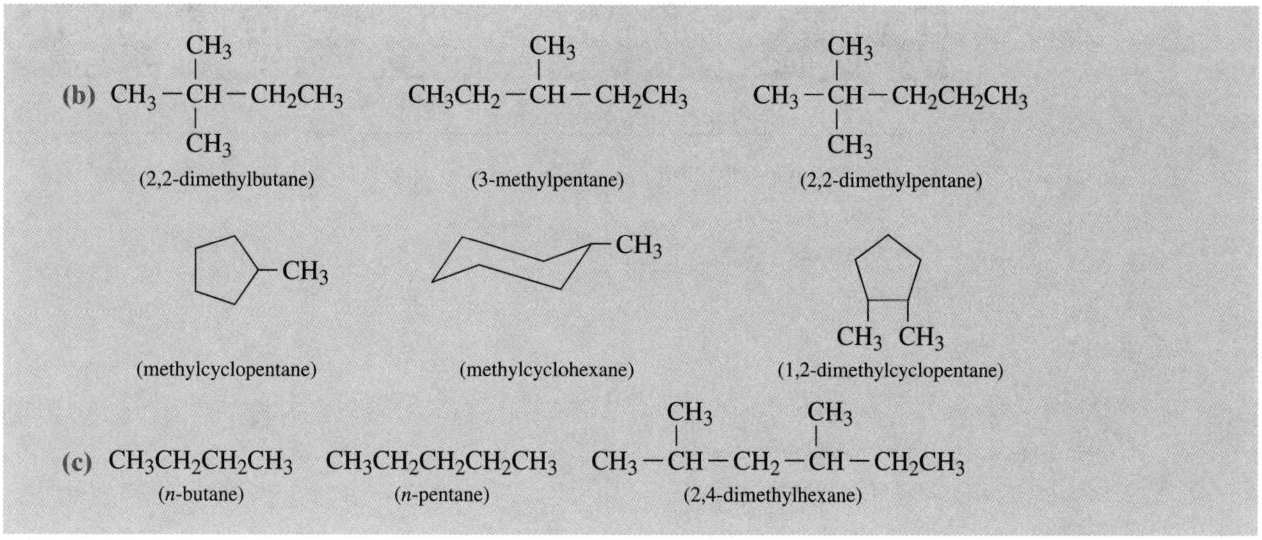

(b)

$$CH_3-\underset{\underset{CH_3}{|}}{\overset{\overset{CH_3}{|}}{CH}}-CH_2CH_3$$

(2,2-dimethylbutane)

$$CH_3CH_2-\underset{\underset{}{}}{\overset{\overset{CH_3}{|}}{CH}}-CH_2CH_3$$

(3-methylpentane)

$$CH_3-\underset{\underset{CH_3}{|}}{\overset{\overset{CH_3}{|}}{CH}}-CH_2CH_2CH_3$$

(2,2-dimethylpentane)

—CH₃

(methylcyclopentane)

—CH₃

(methylcyclohexane)

CH₃ CH₃

(1,2-dimethylcyclopentane)

(c) $CH_3CH_2CH_2CH_3$

(*n*-butane)

$CH_3CH_2CH_2CH_2CH_3$

(*n*-pentane)

$$CH_3-\overset{\overset{CH_3}{|}}{CH}-CH_2-\overset{\overset{CH_3}{|}}{CH}-CH_2CH_3$$

(2,4-dimethylhexane)

DOT regulates the transportation of the alkanes and cycloalkanes in Table 12.5 as flammable gases or flammable liquids, as relevant. DOT requires shippers and carriers to affix FLAMMABLE GAS or FLAMMABLE LIQUID labels on their packaging and, when warranted, to post FLAMMABLE GAS or FLAMMABLE placards on the bulk packaging, freight container, transport vehicle, unit containment device, or railcar used to transport them.

TABLE 12.4 ◆ Some Simple Cycloalkanes

Chemical Formula	*Notation*	*Name*
CH_2 / \ CH_2-CH_2	△	Cyclopropane
CH_2-CH_2		CH_2-CH_2	□	Cyclobutane
CH_2 / \ CH_2 CH_2 \ / CH_2-CH_2	⬠	Cyclopentane
CH_2 CH_2	CH_2-CH_2	CH_2 —— CH_2		Cyclohexane (boat)
CH_2 CH_2 CH_2 CH_2 CH_2 CH_2		Cyclohexane (chair)

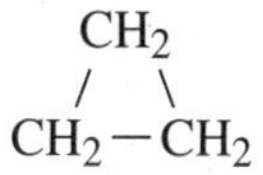

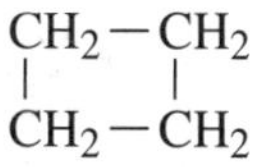

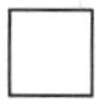

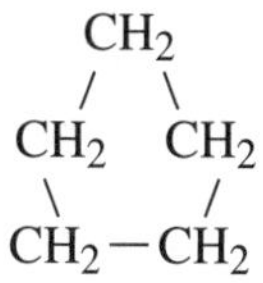

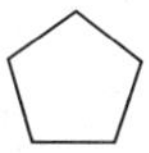

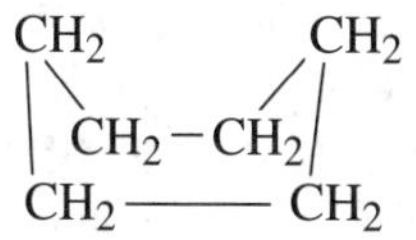

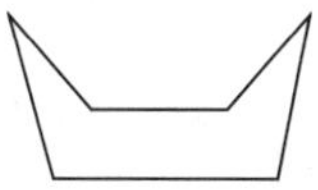

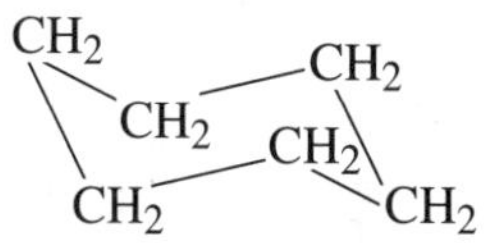

TABLE 12.5 ◆ Some Alkanes and Cycloalkanes Regulated by DOT for Transport

Proper Shipping Name	Label Codes
Butanes *or* Butane mixtures	2.1
Cyclobutane	2.1
Cyclohexane	3
Cyclopentane	3
Cyclopropane, liquefied	2.1
n-Decane	3
2,3-Dimethylbutane	3
Dimethylcyclohexanes	3
2,2-Dimethylpropane	2.1
Ethane, compressed	2.1
Ethane–Propane mixture, refrigerated liquid	2.1
Ethane, refrigerated liquid	2.1
Heptanes	3
Hexanes	3
Isobutane *or* Isobutane mixtures	2.1
Methane, compressed *or* Natural gas, compressed	2.1
Methane, refrigerated liquid *or* Natural gas, refrigerated liquid	2.1
Methylcyclohexane	3
Methylcyclopentane	3
Nonanes	3
Octanes	3
Pentamethylheptane	3
Pentanes	3
Propane *or* Propane mixtures	3
Undecane[a]	3

[a]Undecane is the alkane having the chemical formula $C_{11}H_{24}$.

ALKENES, DIENES, TRIENES, AND CYCLOALKENES

Hydrocarbons containing one or more carbon–carbon double bonds are called *alkenes,* or olefins. Since the molecular structure of each alkene is deficient in hydrogen atoms relative to the molecular structure of its corresponding alkane, the alkenes are said to be *unsaturated hydrocarbons*. The general chemical formula of this group of organic compounds is C_nH_{2n}. The simplest alkene is named ethene, or ethylene; its molecular formula is C_2H_4, and its Lewis structure is the following:

$$\begin{array}{ccc} H & & H \\ \diagdown & & \diagup \\ & C = C & \\ \diagup & & \diagdown \\ H & & H \end{array}$$

The presence of the carbon–carbon double bond in the structure restricts the movement of the atoms about the bond. This means that the six atoms in the ethylene molecule spend their average time in a plane.

Ethene and propene do not possess structural isomers. For example, propene can be correctly represented in either of the following ways:

$$CH_2{=}CHCH_3 \quad CH_3CH{=}CH_2$$

Either formula is the other one turned around end for end. However, we can write two condensed formulas for butene, in which the double bond is positioned differently, as follows:

$$CH_2{=}CHCH_2CH_3 \quad CH_3CH{=}CHCH_3$$

These formulas represent the two structural isomers of butene.

Alkenes are named much like alkanes, except that the *-ene* suffix is used to identify them. Consequently, the first four members of the alkene series are named ethene, propene, butene, and pentene. The common names for these four alkenes are ethylene, propylene, butylene, and amylene, respectively.

In the IUPAC system, we also indicate the position of the double bond in the main chain. This is accomplished by using the following rules:

- Number the main chain of continuous carbon atoms consecutively by beginning at the end that is *nearer* to the double bond. The carbon–carbon double bond must always be included within the main chain of continuous carbon atoms.
- Indicate the position of the double bond by the appropriate numerical prefix.
- Designate the position of each substituent by the number of the carbon atom along the main chain to which it is bonded.

For example, the compound having the formula $CH_2{=}CHCH_2CH_3$ is named 1-butene; the compound having the formula $CH_3CH{=}CHCH_3$ is named 2-butene; and the compound having the following formula is named 2-methyl-1-butene:

$$CH_2{=}C{-}CH_2CH_3$$
$$\overset{\displaystyle |}{CH_3}$$

The relative rigidity of the carbon–carbon double bond leads to a new kind of isomerism, called *geometrical isomerism*. This phenomenon arises when the two substituents or atoms are different on each carbon atom that comprises the carbon–carbon double bond. For example, we can write two Lewis structures for 2-butene, as follows:

$$
\begin{array}{cc}
\begin{array}{ccc}
CH_3 & & H \\
\backslash & & / \\
 & C{=}C & \\
/ & & \backslash \\
H & & CH_3 \\
\end{array}
&
\begin{array}{ccc}
CH_3 & & CH_3 \\
\backslash & & / \\
 & C{=}C & \\
/ & & \backslash \\
H & & H \\
\end{array}
\\[4pt]
\textit{trans}\text{-2-butene} & \textit{cis}\text{-2-butene}
\end{array}
$$

These structures illustrate that a hydrogen atom and a methyl group are bonded to each carbon atom in the double bond. Since the same atoms are bonded to the same atoms in each structure, they do not represent structural isomers. Yet the arrangement of the atoms differs in space. When two compounds have the same structural formula, but differ by the spatial arrangement of their atoms, they are called *geometrical isomers*.

Geometrical isomers are named by using either of the prefixes, *trans-* (meaning on the opposite side of the double bond) or *cis-* (meaning on the same side of the bond). Thus, the two geometrical isomers of 2-butene are named *trans*-2-butene and *cis*-2-butene, respectively, as noted.

The molecules of alkenes can also possess multiple carbon–carbon double bonds. Those having two and three double bonds are called *dienes* and *trienes,* respectively. When these compounds are named, the positions of the double bonds and the use of *-diene* or *-triene* as the relevant suffix are denoted. The following examples illustrate the naming of dienes and trienes:

$$CH_2{=}CH{-}CH{=}CH_2 \qquad\qquad CH_3{-}CH{=}CH{-}CH{=}CH{-}CH{=}CH_2$$

1,3-butadiene   1,3,5-heptatriene

There are also alkenes in which the first and last carbon atoms in a continuous chain are joined to each other in a cyclic arrangement; they are called *cycloalkenes.* They are named by placing the prefix *cyclo-* in front of the name of the parent alkene, such as cyclobutene, cyclopentene, or cyclohexene. Their chemical formulas are usually denoted by using geometrical designs. For example, the chemical formula of cyclohexene is denoted as follows:

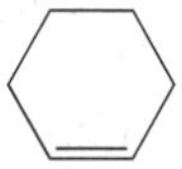

Each alkyl group bonded to the main chain is named with a number identifying its location; hence, the chemical formula of 1-methylcyclopentene is denoted as follows:

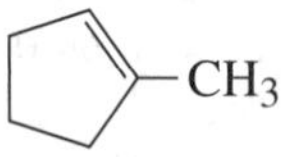

DOT regulates the transportation of the alkenes, dienes, trienes, and cycloalkenes in Table 12.6 as either flammable gases or flammable liquids. DOT requires their shippers and carriers to affix FLAMMABLE GAS or FLAMMABLE LIQUID labels on the packaging and, when warranted, to post FLAMMABLE GAS or FLAMMABLE placards on the bulk packaging, freight container, transport vehicle, unit containment device, or railcar used to transport them.

Solved Exercise 12.3

Using the IUPAC system, name the alkene having the following condensed formula:

$$CH_3CH_2{-}C{=}CH_2$$
$$\vert$$
$$CH_3CH_2$$

Solution: First, identify the longest continuous chain of carbon atoms that contains the carbon–carbon double bond. The longest chain contains four carbon atoms, which signifies that the compound is a derivative of butene. By assigning a number to each carbon atom from the right to the left of this longest chain, the two carbon atoms in the carbon–carbon double bond are numbered 1 and 2, respectively.

$$\overset{4}{C}H_3\overset{3}{C}H_2-\overset{2}{C}=\overset{1}{C}H_2$$
$$|$$
$$CH_3CH_2$$

This means that the compound is a derivative of 1-butene. The ethyl group is bonded to the carbon atom numbered 2. The compound is then correctly named 2-ethyl-1-butene. (We do not assign a number to each carbon atom from the left to the right of the longest chain since then the carbon atoms in the carbon–carbon double bond would be numbered 3 and 4. We always assign numbers to the carbon atoms so that the *smallest* numbers are assigned to the carbon atoms in the carbon–carbon double bond.)

Solved Exercise 12.4

A carrier wishes to transport 5000 lb of compressed ethylene by tank truck.

(a) What information does DOT require the carrier to enter on the accompanying shipping paper?

(b) Identify the DOT placards that DOT requires the carrier to post on the truck.

Solution: (a) DOT requires the basic description to be entered on the accompanying shipping paper. Using Table 6.1, the basic description is entered as follows:

1 T/T (5000 lb) Ethylene, compressed, 2.1, UN1962
Emergency Contact: (_____) _____-_____

(b) DOT requires the carrier to post FLAMMABLE GAS placards on each of the four sides of the truck.

ALKYNES

Hydrocarbons possessing one or more carbon–carbon triple bonds are called *alkynes*. Like the alkenes, they are unsaturated hydrocarbons. The general chemical formula of an alkyne is C_nH_{2n-2}. The simplest member of this series has the formula C_2H_2; it is named ethyne in the IUPAC system, but we know it more generally by its common name, acetylene. The Lewis structure of acetylene is denoted as follows:

$$H-C\equiv C-H$$

In the IUPAC system, alkynes are named by replacing the *-ane* or *-ene* suffix on the associated alkane or alkene, respectively, with the suffix *-yne*.

Although an alkyne can possess structural isomers, it does not possess geometrical isomers. For example, the following Lewis structures designate 1-butyne and 2-butyne:

$$H-C\equiv C-CH_2CH_3 \qquad CH_3-C\equiv C-CH_3$$
1-butyne 2-butyne

TABLE 12.6 ◆ Some Alkenes, Dienes, Trienes, and Cycloalkenes Regulated by DOT for Transport

Proper Shipping Name	Label Codes
n-Amylene	3
Butadienes, inhibited[a]	2.1
Butylene	2.1
1,5,9-Cyclododecatriene	6.1
Cycloheptatrienes	3, 6.1
Cycloheptene	3
Cyclohexene	3
Cyclooctadienes	3
Cyclopentene	3
Dicyclopentadiene	3
Diisobutylenes, isomeric compounds	3
Dipentene	3
Ethylene, compressed	2.1
Hexadienes	3
1-Hexene	3
Isobutylene	2.1
Isoheptenes	3
Isohexenes	3
Isooctenes	3
Isopentenes	3
2-Methyl-1-butene	3
3-Methyl-1-butene	3
Methylpentadienes	3
2,5-Norbornadiene[b]	3
Octadiene	3
Propylene	2.1

[a]DOT requires the addition of a substance to the butadienes to inhibit their autopolymerization. See Section 13-2.
[b]2,5-Norbornadiene is the common name for 2,5-dicycloheptadiene.

When naming alkynes, a numerical prefix is used to indicate the position of the triple bond in the main chain of continuous carbon atoms. Thus, 1-butyne is the alkyne having the molecular formula C_4H_6 and having the carbon–carbon triple bond between a terminal carbon atom and the one immediately adjacent to it; on the other hand, 2-butyne is the C_4H_{10} isomer in which the triple bond is located between two nonterminal carbon atoms.

In the IUPAC system, alkynes are named by using the following rules:

- Number the main chain of continuous carbon atoms consecutively by beginning at the end that is *nearer* to the triple bond. The carbon–carbon triple bond must always be included in the main chain of continuous carbon atoms.
- Indicate the position of the triple bond by the appropriate numerical prefix.
- Designate the position of each substituent by the number of the carbon atom along the main chain to which it is bonded.

Solved Exercise 12.5

Using the IUPAC system, name the alkyne having the following condensed formula:

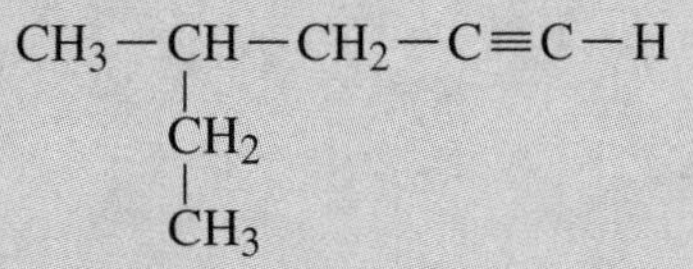

$$CH_3-CH-CH_2-C\equiv C-H$$
$$\mid$$
$$CH_2$$
$$\mid$$
$$CH_3$$

Solution: First, identify the longest continuous chain of carbon atoms that contains the carbon–carbon triple bond. The longest chain contains six carbon atoms, which signifies that the compound is a derivative of hexyne. Next, assign numbers to the carbon atoms from the right to the left downward. The two carbon atoms in the carbon–carbon triple bond are then numbered 1 and 2, respectively.

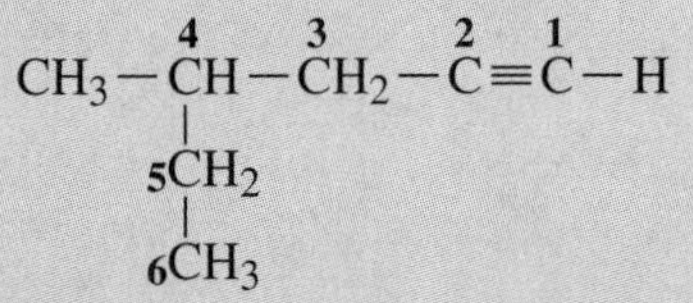

$$CH_3-\overset{4}{C}H-\overset{3}{C}H_2-\overset{2}{C}\equiv\overset{1}{C}-H$$
$$\mid$$
$$_5CH_2$$
$$\mid$$
$$_6CH_3$$

The compound is a derivative of 1-hexyne. The methyl group is bonded to the carbon atom numbered 4. The compound is correctly named 4-methyl-1-hexyne.

Performance Goals for Section 12.3

- Identify the hazardous properties of methane, liquefied petroleum gas, ethylene and propylene, 1,3-butadiene, and acetylene.
- Discuss how these gaseous hydrocarbons are obtained for commercial use.
- Identify the industries that use bulk volumes of methane, liquefied petroleum gas, ethylene and propylene, 1,3-butadiene, and acetylene.

◆ 12.3 GASEOUS HYDROCARBONS

The simple aliphatic hydrocarbons are most commonly encountered as gaseous domestic and industrial fuels. We will discuss them independently.

METHANE

Methane is the simplest hydrocarbon. Its chemical formula is CH_4. Methane is the primary component of natural gas. Although pure methane is odorless, colorless, and tasteless, natural gas is commercially encountered with a slightly offensive smell

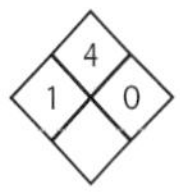

FIGURE 12.3 ◆ At this point near Houston, Texas, products from Phillips Petroleum Company's Sweeny refinery are transported by pipeline to distant locations, like New York. In 1964, a 1531-mi (2465-km) pipeline was completed that linked the Sweeny refinery with the New York City harbor area. *Courtesy of Phillips Petroleum Company, Borger, Texas.*

caused by the presence of organic sulfides, which suppliers purposely add to assist in detecting gas leaks.

Methane occurs naturally as a result of the decay and alteration of animal and plant remains. Consequently, it often accompanies nearby deposits of crude petroleum (Section 12.5). Methane is also found in the atmosphere within coal mines where it is called *firedamp*. When the mud at the bottom of stagnant pools is disturbed, it is methane that bubbles to the surface. It is called *marsh gas*.

Cattle and other bovines also produce methane. There are 1.8 billion bovines worldwide. On the average, a single cow daily exhales 634 qt (600 L) of methane into the air. EPA has estimated that approximately 25% of the methane emitted into the air is linked with the belching of livestock. As amusing as this may sound, the production of methane by livestock has become a serious matter because methane is a greenhouse gas (Section 5.8). The ability of methane to warm the atmosphere is exceeded only by that of carbon dioxide. To reduce the impact caused by methane, agricultural scientists are testing ways of altering the volume of methane that livestock produce.

The natural gas available commercially is generally retrieved from the underground, porous reservoirs where it accumulates. Commercial supplies are also produced from crude petroleum and coal. The production from crude oil involves a process called *cracking*. During cracking, methane is produced by subjecting more complex hydrocarbons to high temperatures under moderate pressure, usually in the presence of a catalyst.

Natural gas is available commercially as a compressed gas. It is directly transferred from petroleum fields or refineries by pipeline to homes and businesses (Figure 12.3). Approximately 60% of American households use natural gas to provide the heat on the kitchen stovetop and fire the home furnace and water heater. Even all-electric homes rely indirectly upon natural gas, since it has become the fuel of choice for many power plants. This situation could soon change, as the American resources of natural gas are seriously declining.

The pipelines that deliver natural gas appear to be buried almost everywhere. Firefighters are often called to scenes at which a pipeline has been inadvertently perforated during a routine landscaping operation. To avoid such events, the local gas company cautions that we notify them of our intention to conduct digging, so it can flag the places where the pipeline is buried.

Natural gas is also available commercially as a cryogenic fluid called *liquefied natural gas* (LNG). When cooled to −258°F (−161°C), the gas reduces to a fraction of its size as it is converted into a liquid. The LNG is transported by tank trucks, and tankers ship it by barge from overseas to domestic ports. The LNG is then delivered to local facilities, where it is vaporized and again sent by pipeline to homes and businesses for use.

Methane burns with a slightly luminous flame, releasing 24,100 Btu/lb (56,000 kJ/kg) to the surroundings. It is a desirable fuel when compared to gasoline not only because it provides this high heat value, but because its combustion produces a lesser number of VOCs (Figure 7.5) compared to the combustion of gasoline. It is for this reason that municipalities sometimes elect to use compressed natural gas as the fuel in city-operated vehicles like the bus in Figure 12.4.

An atmosphere of methane can pose an inhalation hazard as an asphyxiant. Individuals who inhale methane lose consciousness because they have been denied sufficient oxygen. Thus, exposure to an atmosphere of methane can be a direct threat to life by suffocation.

FIGURE 12.4 ◆ Compressed natural gas is now used to fuel some motorized buses in densely populated municipalities where air pollution is potentially problematic. This bus is one of several that operate in southern California. *Courtesy of SunLine Transit Agency, Thousand Palms, California.*

TABLE 12.7 ◆ Physical Properties of Propane and Butane

	Propane	*Butane*
Melting point	−305°F (−187°C)	−216°F (−138°C)
Boiling point	−49°F (−45°C)	31°F (−0.5°C)
Specific gravity	0.58	0.60
Vapor density (air = 1)	1.56	2.04
Lower explosive limit	2.2%	1.9%
Upper explosive limit	9.5%	8.5%

LIQUEFIED PETROLEUM GAS

propane

butane

Liquefied petroleum gas (LPG), also known as *bottled gas,* is produced as a by-product from rectifiers treating natural gas. Its main constituents are propane and butane, but small amounts of ethane, ethene, propene, butene, isobutane, isobutene, and isopentene can also be present.

At most ambient conditions, propane and butane are colorless, odorless, and tasteless gases, but under moderate pressure, they readily liquefy. Both possess heating values similar to that of methane; thus, they are highly desirable as domestic and industrial fuels. Some important physical properties of these two hydrocarbons are noted in Table 12.7.

Propane and butane are shipped under pressure in tankcars from petroleum gas wells and refineries. These liquids are then transferred into a storage tank like that in Figure 12.5; from here, they are often transferred into far smaller storage vessels (that is, they are "bottled"). The bottled gas is then delivered to locations where natural gas cannot be supplied economically by pipeline, such as to rural areas. The cylinders are always stored outside buildings and connected by means of copper tubing to heaters and stoves inside. Although it is stored as a liquid, the fuel is delivered to its outlet as a gas.

When LPG is transported in a DOT Specification M 330 or M 331 tank truck, DOT requires shippers and carriers to determine whether the commodity is corrosive or noncorrosive and to indicate this information in its basic description (Table 6.4) as either of the following:

> 1 t/t   RQ, Petroleum gases, liquefied, 2.1, UN1075
>   (Non-corrosive) *or* Petroleum gases, liquefied, 2.1,
>   UN1075 (Noncor)
>
> 1 t/t   RQ, Petroleum gases, liquefied, 2.1, UN1075
>   (Not for Q and T tanks)

Noncorrosive LPG may be transported in tank trucks that have been constructed from a special "quenched and tempered steel."

ETHYLENE AND PROPYLENE

ethylene

Ethylene and propylene are the common names of the simplest alkenes. As noted earlier, their chemical formulas are C_2H_4, or $CH_2{=}CH_2$, and C_3H_8, or $CH_3{-}CH{=}CH_2$, respectively.

FIGURE 12.5 ◆ These refrigerated spheroidal storage tanks resemble giant silver basketballs. Each tank holds the equivalent of 25,000 barrels of alkylation feedstock chemicals. A thick blanket of glass wool covered by aluminum sheathing insulates the shell of each tank. An adjacent refrigeration plant keeps the alkylate feedstock in the liquid state. *Courtesy of Marathon Oil Co., Robinson, Illinois.*

Ethylene and propylene are produced in the United States in larger amounts than all other substances. Within the chemical industry, ethylene is primarily used for the manufacture of polyethylene (Section 13.6), but it is also used as a raw material for the manufacture of a variety of chemical products including vinyl chloride (Section 13.2), vinylidine chloride (Section 13.2), styrene (Section 13.2), and ethylene glycol (Section 12.8). Ethylene is also used outside the chemical industry; for example, it is used for welding and to hasten the ripening of some produce including apples, bananas, berries, tomatoes, and citrus fruits.

propylene

Propylene is primarily used for the manufacture of polypropylene (Section 13.6), but it is also used to produce isopropyl alcohol, propylene glycol (Section 12.8), and other substances.

The major hazard associated with ethylene and propylene is their flammable nature. Ethylene ignites when a concentration ranging from 3.1% to 36% by volume is exposed to an ignition source. Propylene ignites when a concentration ranging from 2 to 10% is exposed to an ignition source. A fire and explosion hazard is posed when either gas is discharged into an enclosed area.

1,3-BUTADIENE

This substance is a colorless, flammable gas with a mild odor. It is often manufactured at petroleum refineries by the *dehydrogenation* of 1-butene and 2-butene.

$$CH_3CH_2CH{=}CH_2 \longrightarrow CH_2{=}CH{-}CH{=}CH_2 + H_2$$
$$\text{1-butene} \qquad\qquad\qquad \text{1,3-butadiene} \qquad \text{hydrogen}$$

$$CH_3CH{=}CHCH_3 \longrightarrow CH_2{=}CH{-}CH{=}CH_2 + H_2$$
$$\text{2-butene} \qquad\qquad\qquad \text{1,3-butadiene} \qquad \text{hydrogen}$$

Following production, it is transported by pipeline to nearby petrochemical plants. Generally called butadiene (without the numerical prefixes), it is widely used in the rubber industry. For example, approximately 80% of the butadiene produced in the United States is used to manufacture various types of rubber (Section 13.9).

Exposure to butadiene occurs primarily within the workplaces of the rubber and plastics industries. The inhalation of low concentrations causes irritation of the eyes, nose, and throat. When butadiene is present in the workplace, OSHA requires employers to limit employee exposure to a concentration of 1 ppm, averaged over an 8-hour workday.

Research studies link butadiene exposure with the incidence of cancer in humans; hence, butadiene is classified as a known human carcinogen.

ACETYLENE

Acetylene is the simplest alkyne. As noted earlier, its chemical formula is C_2H_2, or $H{-}C{\equiv}C{-}H$. Acetylene is mainly produced in the United States by cracking natural gas, as follows:

$$2CH_4(g) \longrightarrow C_2H_2(g) + 3H_2(g)$$
$$\text{methane} \qquad \text{acetylene} \quad \text{hydrogen}$$

In this instance, the cracking involves dehydrogenation. Acetylene is a colorless gas possessing an ethereal odor when pure. However, industrial grade acetylene can be foul-smelling.

When sparked, a mixture of acetylene and oxygen burns with an intensely hot flame, achieving temperatures as high as 5400°F (2982°C). This process yields 21,600 Btu/lb (50,200 kJ/kg), which is sufficient heat to weld and cut steel and clad metals. The heat derived from use of the oxyacetylene torch in Figure 12.6 is often used for these purposes.

Heat that is sufficient to weld steel can also trigger the ignition of many other commonly encountered flammable and combustible materials. Careless welding practices have been the cause of many major fires. When acetylene is used to generate heat for welding and similar purposes, the likelihood that its heat of combustion will be transmitted elsewhere should always be acknowledged.

Acetylene is also an innately unstable substance. When subjected to thermal or mechanical shock, compressed acetylene decomposes at an explosive rate before it liquefies.

$$C_2H_2(g) \longrightarrow 2C(s) + H_2(g)$$
$$\text{acetylene} \qquad \text{carbon} \quad \text{hydrogen}$$

The decomposition occurs without exposure of the acetylene to air or oxygen. To counteract this likelihood, a special method is utilized to safely compress and store acetylene within steel cylinders. The method takes advantage of the solubility of acetylene in acetone (Section 12.10). One volume of acetone dissolves approximately 25 volumes of acetylene at 1 atm (101.3 kPa) and 300 volumes at 12 atm (1216 kPa). When gas manufacturers dissolve acetylene in acetone, the amount of the gas that can be safely compressed into a cylinder can be increased. First, the acetone is absorbed into a porous

FIGURE 12.6 ◆ Upon combustion, a mixture of acetylene with air or oxygen produces very high temperatures. This feature accounts for the popular use of acetylene for cutting and welding metals. Caution should always be exercised, however, when using acetylene for such purposes. The relatively large amount of heat produced by burning acetylene may be conducted or radiated to nearby flammable and combustible materials, thereby causing their unintended ignition.

medium contained within a cylinder. The porous medium in Figure 12.7 consists of a mixture of monolithic filler and balsa wood. Acetylene is then charged into the cylinder, whereupon it dissolves within the acetone.

When acetylene is released from its container, as during an equipment failure, the principal hazard is its inherent flammability. This is evident from the magnitude of its lower and upper explosive limits. When sparked, acetylene ignites when its concentration is within the range of 3% to 82% by volume.

When acetylene is inhaled, it acts as an asphyxiant. Individuals who inhale adequate acetylene lose consciousness because they have been denied sufficient oxygen. A mixture of acetylene and oxygen can also act as an anesthetic. Mixtures of acetylene of 40% or more by volume with oxygen were once used in anesthesia, although this practice is now obsolete.

Solved Exercise 12.6

The DOT regulation at 40 CFR §173.303 stipulates that when offered for domestic shipment, the pressure in cylinders containing acetylene cannot exceed 250 psi (1720 kPa) at 70°F (21°C). What is the most likely reason DOT limits the pressure under which acetylene can be confined in cylinders during domestic transport?

Solution: Acetylene is an innately unstable substance. Its tendency to undergo self-decomposition is greater when the acetylene has been compressed, such as when it is stored under excessive pressure within cylinders. DOT most likely limits the pressure under which acetylene can be confined within cylinders in order to reduce the likelihood of its self-decomposition.

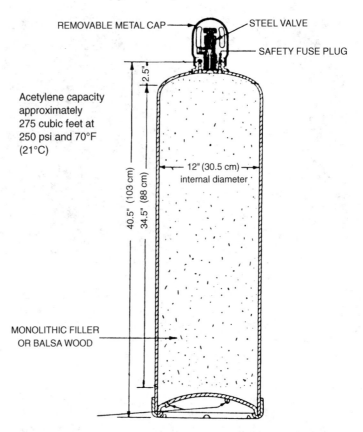

REMOVABLE METAL CAP
STEEL VALVE
SAFETY FUSE PLUG

Acetylene capacity
approximately
275 cubic feet at
250 psi and 70°F
(21°C)

2.5"

40.5" (103 cm)
34.5" (88 cm)

12" (30.5 cm)
internal diameter

MONOLITHIC FILLER
OR BALSA WOOD

FIGURE 12.7 ◆ The cutaway of a typical acetylene cylinder, illustrating its components. The monolithic filler or balsa wood is a porous material that is charged with a suitable solvent, like acetone.

Performance Goals for Section 12.4
- Describe the molecular structure of benzene and other aromatic hydrocarbons.
- Identify the rules for naming aromatic hydrocarbons.
- Discuss the hazardous properties of benzene, toluene, xylene, and the polynuclear aromatic hydrocarbons.

◆ 12.4 AROMATIC HYDROCARBONS

The word "aromatic" suggests that aromatic hydrocarbons are compounds possessing fragrant odors. Some simple aromatic hydrocarbons are actually fragrant, but otherwise, this perception is misleading. *Aromatic hydrocarbons* are regarded as compounds whose molecules are composed of one or more special rings of carbon atoms. These compounds are typified by the organic compound called *benzene,* the simplest aromatic hydrocarbon. The chemical formula of benzene is

C_6H_6. While the molecular structure of benzene is shown by either formula drawn below to the left, it is more commonly represented by a hexagon with a circle inside as shown on the right:

Aromatic and aliphatic hydrocarbons are thus differentiated by the fact that their molecular structures either resemble or do not resemble the molecular structure of benzene. Hereafter, we shall represent the molecular structure of benzene by the hexagonal structural formula with the inscribed circle. Although the symbols of the carbon atoms are not written as part of the hexagon, it is always understood that a carbon atom occupies each corner of the hexagon, where it is bonded to a hydrogen atom.

BENZENE

Benzene is a colorless, water-insoluble, highly volatile liquid. Although it was once a very popular industrial solvent, benzene is no longer used for this general purpose. This reduction in use is associated with the observation that exposure to benzene causes various human blood disorders, including leukemia (see below). Notwithstanding this fact, benzene is still used in substantial amounts as a raw material by the chemical industry for the manufacture of other organic compounds such as the linear alkylbenzene sulfonates, which are popular laundry powders and dish detergents.

Benzene is a flammable compound. When exposed to an ignition source, its vapors burn in the air. The flashpoint of benzene is 12°F (−11°C) and its flammable range is 1.2% to 7.8% by volume.

The principal adverse health effect associated with exposure to benzene is irreversible bone marrow injury. As a result of this injury, the body becomes incapable of properly manufacturing blood cells. Research studies have correlated benzene exposure with the incidence of leukemia in humans; hence, epidemiologists classify benzene as a known human carcinogen (Section 10.14).

When benzene is used in the workplace, OSHA requires employers to limit employee exposure to a vapor concentration of 1 ppm, averaged over an 8-hour workday.

TOLUENE

When any one of the six hydrogen atoms in benzene is substituted with a methyl group, the resulting compound is commonly called *toluene*. The chemical formula of toluene is C_6H_5—CH_3. The molecular structure of toluene is represented as follows:

Toluene is a colorless, water-insoluble, highly volatile liquid, commonly encountered as a solvent in the manufacturing and process industries. Considerable quantities of toluene are also used in solvent-based paint products. Toluene is available commercially in relatively small containers like the 5-gal metal can in Figure 12.8, as well as in bulk containers and tanks.

Toluene is flammable. When exposed to an ignition source, toluene vapors burn readily in the air. The flashpoint of toluene is 40°F (4.4°C) and its flammable range is

FIGURE 12.8 ◆ Toluene is one of the most commonly encountered industrial organic solvents. Chemical laboratories often dispense it from a 5-gal metal container, like that illustrated here. When offered for transportation, the FLAMMABLE LIQUID label is affixed to the container. *Courtesy of J. T. Baker, Inc., Phillipsburg, New Jersey.*

1.1% to 7.1% by volume. Although ignitability is the principal hazard associated with toluene, the occasional inhalation of toluene vapors can irritate the respiratory system and cause dizziness and nausea.

When toluene is present in the workplace, OSHA requires employers to limit employee exposure to a vapor concentration of 200 ppm, averaged over an 8-hour workday.

XYLENE

Two hydrogen atoms in the benzene molecule can also be substituted with methyl groups. Since it is possible to position them in three different ways, there are three structural isomers for the formula $C_6H_4(CH_3)_2$. These three compounds are the structural isomers of xylene. Their molecular structures are written below:

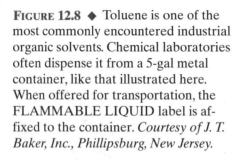

In the common system of nomenclature, the prefixes *ortho-, meta-,* and *para-* are employed to differentiate these structural isomers from one another. *Ortho* means "straight ahead," *meta-* means "beyond," and *para-* means "opposite." The individual compounds and their isomeric mixtures are sometimes referred to as either "xylene" or "xylene(s)" with total disregard of their isomeric distinctions.

The molecular structures of the three compounds in Figure 12.9 illustrate two arbitrary substituents, A and B, positioned on the benzene ring; B is located in the

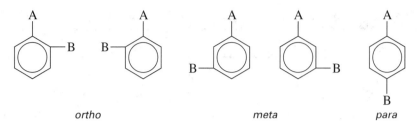

FIGURE 12.9 ◆ The *ortho, meta,* and *para* positions of B, an arbitrary substituent, relative to a fixed position on the benzene ring of another arbitrary sustituent, A. (A and B may be identical or different substituents.) The two configurations noted for the *ortho* and *meta* positions are equivalent ways of denoting the same compound.

ortho-, meta-, and *para-* positions relative to the position of A. When naming these derivatives of benzene, the italicized letters *o-, m-,* and *p-* are used as prefixes.

In the IUPAC system, when two or more alkyl groups are bonded to the benzene ring, the resulting hydrocarbon is named by listing the alkyl groups alphabetically and numbering the first with a 1. Each of the six carbon atoms in the benzene ring is then numbered from 1 to 6. In this manner, *o*-xylene is named 1,2-dimethylbenzene; *m*-xylene is named 1,3-dimethylbenzene; and *p*-xylene is named 1,4-dimethylbenzene.

Like benzene and toluene, the xylene isomers are colorless, water-insoluble, highly volatile liquids. Like toluene, the mixture of xylene isomers is commonly utilized as an industrial solvent. In the chemical industry, xylene competes with naphthalene (see p. 533) as the raw material needed to manufacture phthalic anhydride, a substance widely employed for the production of certain resins and polyesters such as poly(ethylene terephthalate) (Section 13.2).

When exposed to an ignition source, xylene vapors burn readily in air. The flashpoints of *o-, m-,* and *p*-xylene are 90°F (32°C), 84°F (29°C), and 81°F (27°C), respectively. The flammable range of *o*-xylene is 0.9% to 6.7% by volume, while the flammable range for both *m*-xylene and *p*-xylene is 1.1% to 7.0% by volume. While ignitability is the principal hazard associated with the xylene isomers, the inhalation of their vapors may irritate the respiratory system and cause dizziness and nausea.

When xylene is present in the workplace, OSHA requires employers to limit employee exposure to a vapor concentration of 100 ppm, averaged over an 8-hour workday.

OTHER SIMPLE AROMATIC HYDROCARBONS

Individual aromatic hydrocarbons are sometimes named by identifying the benzene ring as a substituent having the formula, C_6H_5-, called the *phenyl* group. Thus, the compound having the following molecular structure is named phenylethylene.

$$\langle\bigcirc\rangle-CH=CH_2$$

The phenyl group and other similar groups of atoms are called *aryl substituents*. The only other aryl substituent we shall encounter in this book is $C_6H_5CH_2-$, called the benzyl group. Additional examples of naming simple aromatic hydrocarbons are provided in Table 12.8.

TABLE 12.8 ◆ Naming Aromatic Hydrocarbons

Chemical Formula	Name
⬡—CH$_2$CH$_3$	Ethylbenzene
⬡—CH$_2$CH$_2$CH$_3$	n-Propylbenzene
CH$_3$—⬡—CH$_2$CH$_2$	p-Ethyltoluene
⬡—C≡C—H	Phenylacetylene
⬡—CH$_2$Cl	Benzyl chloride
CH$_3$CH=CHCH$_2$—⬡	1-Phenyl-2-butene

DOT regulates the transportation of the aromatic hydrocarbons in Table 12.9 as flammable liquids. DOT requires shippers and carriers to affix FLAMMABLE LIQUID labels on their packaging and, when warranted, to post FLAMMABLE placards on the bulk packaging, freight container, transport vehicle, unit containment device, or railcar used to transport them.

TABLE 12.9 ◆ Some Aromatic Hydrocarbons Regulated by DOT for Transport

Proper Shipping Name	Label Codes
Benzene	3
Butylbenzenes	3
Butyltoluenes	6.1
Cymenes[a]	3
Diethylbenzene	3
Ethylbenzene	3
Isopropylbenzene (cumene)	3
n-Propylbenzene	3
Toluene	3
1,3,5-Trimethylbenzene	3
Xylenes	3

[a]Cymene is the common name for isopropyltoluene.

POLYNUCLEAR AROMATIC HYDROCARBONS

These substances are commonly called *PAHs*. The simplest PAHs comprise a group of structurally similar hydrocarbons having two or more mutually fused benzene rings per molecule. Benzene rings are said to be "mutually fused" when they share a pair of carbon atoms and the bond between them. The PAHs having two and three mutually fused benzene rings are naphthalene and anthracene and phenanthrene, respectively.

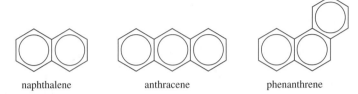

naphthalene   anthracene   phenanthrene

The molecules of more complex PAHs may consist of benzene rings fused to cyclopentenyl (C_5H_7) and other rings.

Naphthalene is the only PAH having commercial importance. It is the simplest PAH from a molecular viewpoint, since each naphthalene molecule is composed of only two mutually fused benzene rings. Naphthalene is a white to colorless solid having the odor of mothballs. As noted earlier, it is primarily used in the chemical industry as a raw material for the manufacture of phthalic anhydride.

When naphthalene is used in the workplace, OSHA requires employers to limit employee exposure to a vapor concentration of 10 ppm (50 mg/m^3), averaged over an 8-hour workday.

DOT regulates the transportation of naphthalene as a flammable solid. DOT requires shippers and carriers to affix FLAMMABLE SOLID labels on its packaging and, when warranted, to post FLAMMABLE SOLID placards on the bulk packaging, freight container, transport vehicle, unit containment device, or railcar used to transport it.

Although naphthalene is the only PAH likely to be encountered commercially, it is possible to unknowingly experience the impact of many other PAHs. Many processes that involve the incomplete combustion of organic materials generate mixtures of PAHs. This implies that many manufacturing and process industries spew PAHs from their smokestacks into the air we breathe. Polynuclear aromatic hydrocarbons are also dissolved constituents of coal tar distillates (Section 7.7), crude petroleum, and certain petroleum fractions including diesel fuel, heating oils, heavy naphtha, and asphalt. They are also components of the emissions associated with the burning of coal, petroleum oil, wood, tobacco, and peat.

The presence of PAH mixtures is especially prominent in diesel-powered vehicular exhaust. The PAHs adsorb to the constituent particulates of soot. Since many heavy-duty trucks and buses are often diesel-powered, the PAHs produced during their use become components of the environments in which they are driven.

During incomplete combustion processes, PAHs are produced simultaneously with soot by a multistep mechanism similar to that depicted in Section 5.10. When simple hydrocarbons are heated to high temperatures, molecular fragments—such as free radicals—are produced. These fragments undergo reactions that result in the production of new molecules. The original molecules and their fragments dehydrogenate, while the new molecules similarly fragment, amalgamate, and continue to dehydrogenate. When dehydrogenation has occurred to its maximum extent, all that remains is soot, or carbon black.

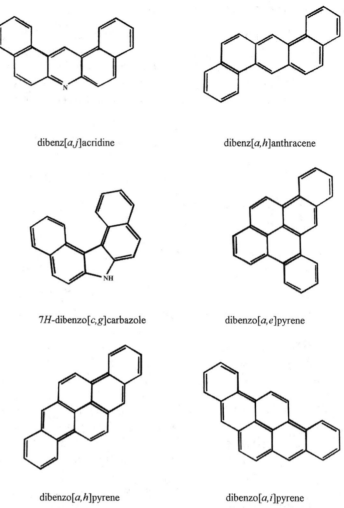

dibenz[*a,j*]acridine   dibenz[*a,h*]anthracene

7*H*-dibenzo[*c,g*]carbazole   dibenzo[*a,e*]pyrene

dibenzo[*a,h*]pyrene   dibenzo[*a,i*]pyrene

FIGURE 12.10 ◆ The molecular structures of the fifteen polynuclear aromatic hydrocarbons classified as probable human carcinogens.

All PAHs are solid compounds at room temperature. The most common way we become exposed to them is by inhalation of the particulate matter in polluted air, wood smoke, tobacco smoke, and vehicular exhaust. Scientists have established that when inhaled, the PAHs collectively pose a chronic respiratory hazard to humans by inducing asthma and activating allergies.

Moreover, the following evidence persuasively links the exposure of 15 PAHs with the onset of cancerous growths:

+ Massive numbers of mutated cells have been observed in the bronchial passageways of individuals repeatedly exposed to diesel engine exhaust within their occupational settings.
+ Epidemiological tests conducted upon animals have demonstrated that malignant tumors are produced systemically and at the physiological sites where the PAHs were directly applied.

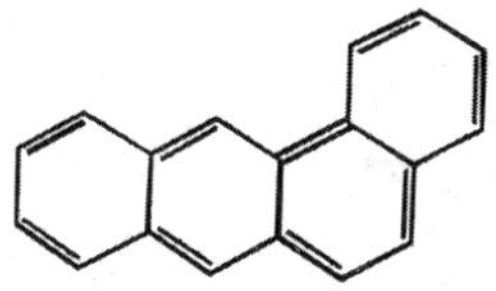

benzo[*a*]anthracene

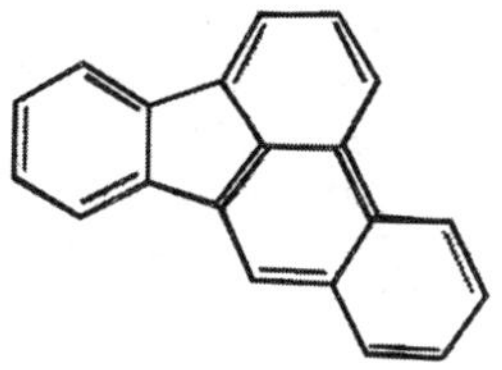

benzo[*b*]fluoranthene

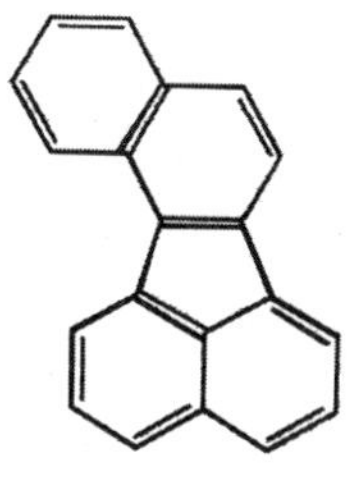

benzo[*j*]fluoranthene

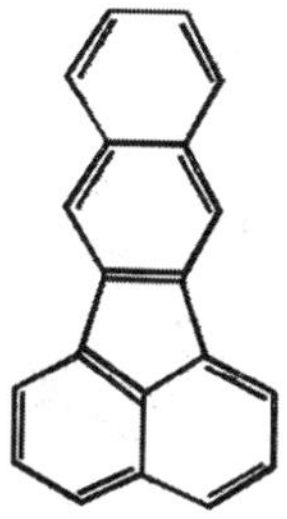

benzo[*k*]fluoranthene

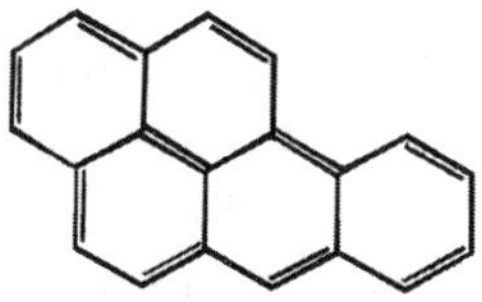

benzo[*a*]pyrene

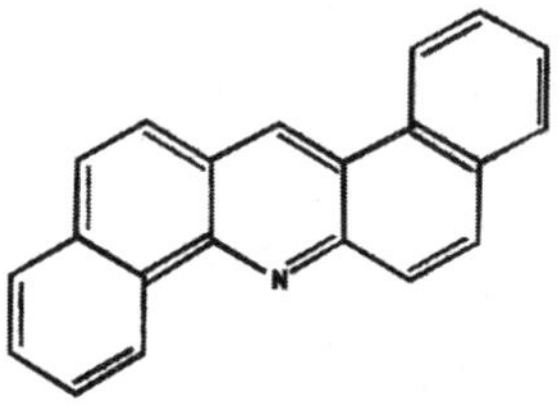

dibenz[*a,h*]acridine

FIGURE 12.10 ◆ *(continued)*

These observations have caused scientists to classify soot as a known human carcinogen and the 15 PAHs in Figure 12.10 as probable human carcinogens. The molecules of 7*H*-dibenzo[*c,g*]carbazole and the isomers of dibenzacridine possess a ring nitrogen atom (that is, a nitrogen atom substitutes for a ring carbon atom). Compounds whose molecules have one or more ring atoms other than carbon are called *heterocyclic compounds*.

Solved Exercise 12.7

Why should firefighters avoid inhaling the smoke that evolves when tar paper and roofing shingles smolder and burn?

Solution: Tar paper and roofing shingles are often manufactured from asphalt, the residue that remains after crude petroleum has been fractionated. The constituents of asphalt include the group of high-molecular-weight compounds

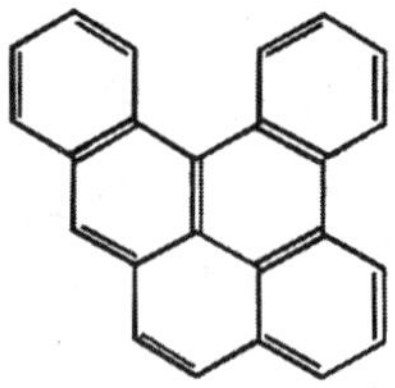

dibenzo[*a,l*]pyrene

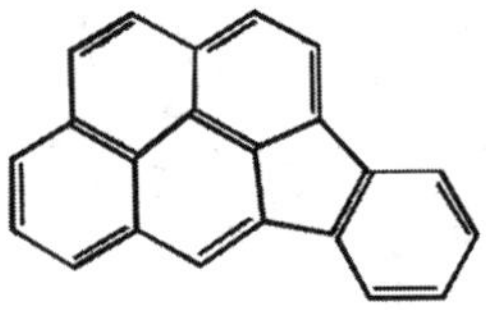

indeno[1,2,3-*cd*]pyrene

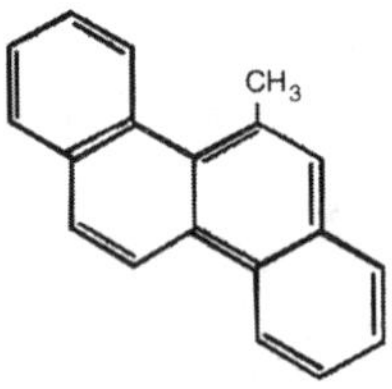

5-methylchrysene

FIGURE 12.10 ◆ (*continued*)

called polynuclear aromatic hydrocarbons (Section 12.4). When asphalt-containing products burn, these compounds either oxidize or become components of the particulate matter evolved in the associated smoke plume. Firefighters should avoid breathing this smoke to reduce the risk of stimulating respiratory ailments or of contracting throat and lung cancer.

(Scientists have verified the presence of dibenzo[*a,h*]pyrene, dibenzo[*a,i*]-pyrene, and indeno[1,2,3-*cd*]pyrene in the matter that volatilizes from molten asphalt. These three are among the 15 PAHs in Figure 12.10 that have been shown to cause cancer in experimental animals.)

Scientists have also concluded that a lifetime exposure to diesel engine exhaust could result in the onset of 450 cases of lung cancer within an exposed population of one million people. This finding has generated concern regarding the potential harmful effects caused by inhaling PAHs and other hazardous components of diesel engine exhaust.

Within occupational settings, career-oriented firefighters are one of the major groups of individuals most vulnerable to *routinely* inhaling PAHs. It is prudent to conclude that they are regularly exposed to airborne PAHs while combating fires. This repeated exposure could constitute a long-term health concern.

■■
Performance Goals for Section 12.5
   ◆ Identify the common products produced by the petroleum industry.
   ◆ Describe the octane number assigned to gasoline products.
   ◆ Describe the chemical nature of petrochemicals.
■■

◆ 12.5 PETROLEUM AND PETROLEUM PRODUCTS

The word "petroleum" is derived from the Greek *petra,* meaning rock, and Latin *oleum,* meaning oil. The name "rock oil" is reminiscent of our earliest contacts with this material: a liquid that oozed from fissures in rocks.

Petroleum is a highly complex mixture consisting of many thousands of organic compounds, approximately 75% of which are hydrocarbons whose molecules possess from three to 60 carbon atoms. The actual number of individual hydrocarbons in petroleum has been estimated to be between 50,000 and 2,000,000. Petroleum occurs naturally in certain areas around the world, deep below Earth's surface. Wells are drilled through the rock to the oil-bearing stratum, through which the petroleum is then pumped to the surface. In this form, it is called *crude petroleum, crude oil,* or *crude.* The crude petroleum is often transferred by pipeline from the oil fields to plant sites known as *petroleum refineries,* where it is stored in large-volume tanks.

Although the mechanism by which crude oil formed in the Earth is open to some debate, most scientists believe that it originated from the partial decomposition of animals and plants that lived millions of years ago. Changes in the position of Earth's crust over the passing millennia buried these materials at great depths. The pressure accompanying these depths caused them to decompose into petroleum.

FIRES INVOLVING CRUDE PETROLEUM

Since the flashpoint of crude petroleum varies from 20 to 90°F (-7 to 32°C), crude petroleum is a highly flammable liquid. Fires involving this material have occurred at oil fields, in transit, and during storage. Capping burning oil wells is a job for specially trained experts, but regular firefighters are called upon to combat fires involving crude oil in transit and storage.

While small crude oil fires can be extinguished using a deluging volume of water, most experts recommend the use of aqueous film-forming foam (AFFF) (Section 5.12) as the most practical means of extinguishing crude oil fires within a bulk storage tank. Here, the use of water is recommended solely for cooling purposes. Water and petroleum are immiscible liquids, and petroleum is less dense than water. When discharged on a petroleum fire, the water sinks to the bottom of the storage tank, where it settles as a separate layer.

The presence of a water layer at the bottom of a bulk storage tank represents a special concern when combating a petroleum fire. While the fire occurs at the tank's surface where flammable vapor is emitted, the accompanying heat of combustion can be slowly transmitted by convection and radiation through the underlying petroleum to the water layer. Absorption of the intense heat causes the temperature of the petroleum and the underlying water to rise until ultimately, the water boils. The water vapor then forces its way upward, pushing the burning petroleum up and over the walls of the storage tank. This phenomenon is appropriately called a *boil-over.*

When a crude petroleum fire occurs within a storage tank, every attempt should be made to extinguish the fire before the boil-over occurs. The burning, frothing petroleum that spills outside the storage tank has been known to flow substantial distances from the tank, triggering numerous secondary fires. The firefighters who combat these fires should be particularly wary, as the flowing, burning petroleum could overtake unsuspecting firefighters in its pathway.

FRACTIONATION OF CRUDE PETROLEUM

One of the major operations occurring at petroleum refineries is the fractional isolation of materials from crude petroleum. The process is called *fractional distillation*, or *fractionation*. It is accomplished by heating the crude within preestablished temperature ranges. The vapors of the constituent compounds thus formed are then condensed and collected within separate receivers.

Figure 12.11 shows that the fractions most commonly isolated during the fractional distillation of crude petroleum consist of the following:

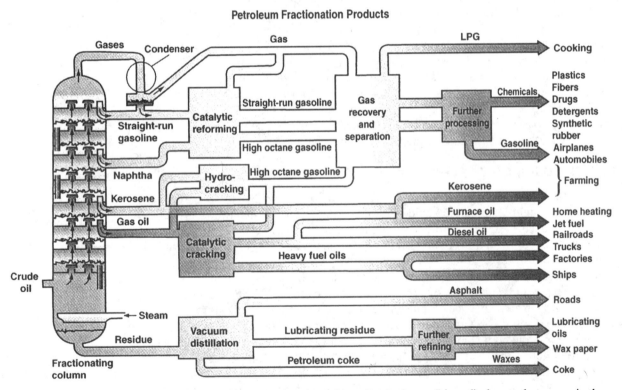

FIGURE 12.11 ◆ Some products derived by the fractionation of petroleum. Liquefied petroleum gas is the only fraction that is marketed in the form that comes directly from the distillation tower. The other fractions require further refinement to remove undesirable components, most notably sulfurous compounds. The refinement processes include catalytic reforming, hydrocracking, catalytic cracking, and vacuum distillation. Additives are also blended with gasoline and other petroleum products to improve their performance. *Chemistry, by Robert C. Smoot, Richard G. Smith, and Jack Price, copyright 1995 by Glencoe/McGraw-Hill; used by permission of the publisher.*

+ *A gaseous mixture of methane (65 to 90%), ethane, propane, and butane.* This fraction separates from crude petroleum at a temperature less than 70°F (21°C). It can be used directly as a feedstock for the production of other substances, or its components can be isolated to produce petroleum products such as bottled gas.

+ *A liquid mixture consisting mainly of straight-chained and cyclic aliphatic hydrocarbons having from five to nine carbon atoms per molecule.* This fraction, called *light naphtha* or *ligroin,* is generally isolated within the boiling point range 158 to 284°F (70 to 140°C). At one time, light naphtha was used directly as the form of gasoline called *straight-run gasoline.* However, light naphtha is now considered environmentally undesirable as a fuel, because it contains a relatively large amount of benzene. Today, light naphtha is used as the feedstock for production of ethylene and propylene.

+ *A liquid mixture consisting mainly of straight-chained and cyclic aliphatic hydrocarbons having from seven to nine carbon atoms per molecule.* This fraction, called *heavy naphtha,* is generally isolated within the boiling point range of 284 to 392°F (140 to 200°C). It is typically used for the production of motor gasoline and jet fuel.

+ *A liquid petroleum product consisting of a mixture of hydrocarbons having from 9 to 16 carbon atoms per molecule.* This fraction, called *kerosene,* is often isolated within the boiling point range 347 to 527°F (175 to 275°C). Kerosene is directly used as a tractor fuel, jet fuel, and heating fuel.

+ *A liquid mixture of compounds having from 15 to 25 carbon atoms per molecule.* This fraction, called *diesel oil* or *diesel fuel,* is usually isolated within the boiling point range 392 to 698°F (200 to 370°C). Sulfurous and nitrogenous compounds are constituents of this untreated fraction of crude oil. To meet today's clean air standards, it is often subjected to a process called *hydrotreating,* during which hydrogen is used to eliminate or reduce the sulfur and nitrogen concentrations in fuels. Diesel oil is used to power some trucks, buses, ships, and certain equipment. It can also be used directly as a heating fuel.

+ *A liquid to semisolid fraction isolated at temperatures above 698°F (370°C).* This is a mixture of compounds that typically contain from 30 to 45 carbon atoms per molecule. The fraction is generally called *lubricating oil.* Although it is used directly as a lubricant, it is also the feedstock for preparing other petroleum-based lubricants, hydraulic fluids, and transmission fluids. Typically, this fraction is hydrogenated and subjected to other chemical processes to remove undesirable compounds containing metals, nitrogen, and sulfur, and then blended to achieve the desired viscosity, stability, and lubricity.

+ *The solid residue of the distillation process.* The fractionation residue is called *asphalt,* a mixture of compounds whose molecules possess 40 or more carbon atoms. Asphalt is transported in portable tanks (Figure 12.12) for use during paving and road construction and as a component of roofing, coatings, and adhesive products. The fractionation residue is also treated to produce *coke.*

A single petroleum fraction may be sold directly as a commercial product, or it may be stored until further processing can be accomplished. The storage occurs in a tank as in Figure 12.13. Bulk volumes of petroleum products are transported to major distribution centers, where they are often stored in multiple tanks within a tank farm as in Figure 12.14, where they await customer demand.

DOT regulates the transportation of crude petroleum and the petroleum products in Table 12.10 as either flammable gases or flammable liquids. DOT requires shippers and carriers to affix FLAMMABLE GAS or FLAMMABLE LIQUID labels on their packaging and, when warranted, to post FLAMMABLE GAS or FLAMMABLE placards on the bulk packaging, freight container, transport vehicle, unit containment device, or railcar used to transport them.

FIGURE 12.12 ◆ A bulk transport tank serves to move large volumes of asphalt cement to locations where it is used to pave and coat roadways. *Courtesy of CEI Enterprises, Inc., Albuquerque, New Mexico.*

FIGURE 12.13 ◆ At this terminal in Morris, Illinois, butane is stored in a spheroidal tank prior to distribution to customers. *Courtesy of Chicago Bridge and Iron Company, Oakbrook, Illinois.*

FIGURE 12.14 ◆ The Los Angeles Texaco refinery at Wilmington, California, where numerous floating-roof tanks are used to store petroleum products prior to their commercial distribution. Over 1 billion gallons of petroleum products may be stored at such locations at any time. *Courtesy of Texaco, Inc., Bellaire, Texas; photographer Thomas Carroll.*

TREATMENT OF PETROLEUM FRACTIONS

An example of a chemical treatment process that is conducted at petroleum refineries is *alkylation*. During alkylation, branched-chain hydrocarbons are produced by the chemical union of an alkane and an alkene. When the alkane is isobutane and the alkene is isobutene, the hydrocarbon commonly called "isooctane" is produced.

The name "isooctane" is actually a misnomer; the correct name of the compound is 2,2,4-trimethylpentane. Alkylation is generally conducted in the presence of hydrofluoric acid, which acts catalytically.

Alkylation processes produce trimethylpentanes, dimethylhexanes, and other branched-chain hydrocarbons. These compounds are desirable additives in gasoline, since they improve the completeness of its combustion. Without their presence, pings and rattling noises are generated within the engine as the fuel burns. This phenomenon is called *knocking*. Fuel additives that serve to reduce the degree of knocking are called *antiknock agents*.

TABLE 12.10 ◆ Some Petroleum Products Regulated by DOT for Transport	
Proper Shipping Name	*Label Codes*
Asphalt, *at or above its flashpoint*	3
Diesel fuel	None
Fuel, aviation, turbine engine	3
Fuel oil	3
Gas oil *or* Diesel oil *or* Heating oil, light	3
Gasoline	3
Hydrocarbon gases, compressed, n.o.s., *or* Hydrocarbon gases mixtures, compressed, n.o.s.	3
Hydrocarbon gases, liquefied, n.o.s. *or* Hydrocarbon gases mixtures, liquefied, n.o.s.	2.1
Hydrocarbons, liquid	3
Kerosene	3
Petroleum crude oil	3
Petroleum distillates, n.o.s. *or* Petroleum products, n.o.s.	3
Petroleum gases, liquefied *or* Liquefied petroleum gas	2.1
Petroleum oil	3
Tars, liquid *including asphalt and oils, bitumen and cut backs*	3

GASOLINE

Different types of gasoline are available commercially for use as motor fuels. They are distinguishable by their inherent octane numbers. The *octane number,* or *octane rating,* is a representation of the antiknock properties of a fuel under laboratory or test conditions. Octane numbers of zero and 100 are arbitrarily assigned to *n*-heptane (a high "knocker") and 2,2,4-trimethylpentane (a low "knocker"), respectively. When a sample of a fuel is tested and found to possess an octane number of 85, its degree of knocking is that of a mixture of 85 parts 2,2,4-trimethylpentane and 15 parts *n*-heptane.

At the pumps in Figure 12.15, gasolines having minimum octane numbers of 87, 89, and 91 are available for sale as motor fuels. Gasolines having octane numbers of approximately 100 or more are sold at airfields as aviation fuels.

Solved Exercise 12.8

In reference to vehicular and aviation fuels, what is meant by the term "high octane"?

Solution: "High octane" refers to a type of gasoline that was most likely manufactured at a petroleum refinery by blending a base stock with certain substances that increase the octane rating of the fuel. These substances include 2,2,4-trimethylpentane, other trimethylpentane isomers, dimethylhexanes, and other branched-chain hydrocarbons, all of which serve as antiknock agents.

2,2,4-Trimethylpentane is commonly called "isooctane" in the petroleum industry. Since the trimethylpentanes and dimethylhexanes are isomers of octane, the
resulting gasoline is elevated in the concentration of these compounds. This is
the origin of the expression "high octane."

Solved Exercise 12.9

Several chemical processes conducted at petroleum refineries involve the production of antiknock agents, which are subsequently blended with base stocks to
increase the octane rating of the resulting fuels. For example, propene that has
been isolated from crude petroleum by fractionation is induced to react with
itself to produce the antiknock agents 2-methyl-1-pentene, 4-methyl-1-pentene,
and 4-methyl-2-pentene. Write the equations that illustrate the production of
these branched alkenes.

Solution: Propene reacts *with itself* to produce 2-methyl-1-pentene, 4-methyl-
1-pentene, and 4-methyl-2-pentene. Consequently, propene is the reactant and 2-
methyl-1-pentene, 4-methyl-1-pentene, and 4-methyl-2-pentene are the reaction
products. The equations illustrating the production reactions are the following:

$$2CH_3CH{=}CH_2 \longrightarrow CH_3CH_2CH_2{-}\overset{\overset{\displaystyle CH_3}{|}}{C}{=}CH_2$$
(propene)   (2-methyl-1-pentene)

$$2CH_3CH{=}CH_2 \longrightarrow CH_2{=}CH{-}CH_2{-}\overset{\overset{\displaystyle CH_3}{|}}{C}H{-}CH_3$$
(propene)   (4-methyl-1-pentene)

$$2CH_3CH{=}CH_2 \longrightarrow CH_3{-}CH{=}CH{-}\overset{\overset{\displaystyle CH_3}{|}}{C}H{-}CH_3$$
(propene)   (4-methyl-2-pentene)

The chemical composition of petroleum fuels varies not only in different parts of
the world, but also at different times of the year. The chemical composition of a given
fuel produced at a refinery could even change daily. Despite this variability, motor
gasoline typically consists of a mixture of hydrocarbons having the following approximate composition by volume: 55% normal and branched alkanes; 34% aromatic
hydrocarbons; 5% alkenes; and 5% cycloalkanes and cycloalkenes.

HEATING FUELS

Some petroleum products are useful as heating fuels. They are manufactured by fractionation, followed by blending with specified additives. For example, diesel fuel and
heavy naphtha are often processed to produce heating oils, either by generating fractions within a narrower distillation range or by adding a specified amount of other petroleum fractions until the final blend possesses desirable ignition temperatures and
heat (Btu) values. These products are utilized not only as home heating oils, but also
as the fuels in boilers for factories, apartment buildings, and schools.

Figure 12.15 ◆ Gasoline fuels are distinguished by their octane numbers, which are established by subjecting a sample to test conditions. In the test, octane numbers of zero and 100 are assigned to *n*-heptane and 2,2,4-trimethylpentane, respectively.

Six grades of heating oils are commercially recognized in the United States, each designated by the word "Fuel Oil" followed by a number from 1 to 6. The grades differ by the temperature at which each fuel ignites, progressively increasing from 444°F (229°C) to 765°F (407°C). These temperatures are the ignition temperatures of Fuel Oil No. 1 and Fuel Oil No. 6, respectively.

PETROCHEMICALS

During refinery operations, a petroleum fraction can be subjected to further distillation to isolate its individual components, or it can be treated to yield substances called *petrochemicals*. Broadly speaking, petrochemicals are substances produced in bulk directly from natural gas, crude petroleum, or a petroleum fraction. The most familiar petrochemicals are ethylene, propylene, 1,3-butadiene, benzene, toluene, xylene(s), and methanol. They serve as indispensable raw materials for the plastics, synthetic rubber, and synthetic fiber industries.

In the United States, by far the most important petrochemicals are ethylene and propylene. They are manufactured by cracking the ethane and propane extracted from natural gas.

$$\underset{\text{ethane}}{C_2H_6(g)} \longrightarrow \underset{\text{ethylene}}{C_2H_4(g)} + \underset{\text{hydrogen}}{H_2(g)}$$

$$\underset{\text{propane}}{2C_3H_8(g)} \longrightarrow \underset{\text{ethylene}}{C_2H_4(g)} + \underset{\text{methane}}{CH_4(g)} + \underset{\text{propylene}}{C_3H_6(g)} + \underset{\text{hydrogen}}{H_2(g)}$$

■ ■

Performance Goal for Section 12.6
- ◆ Identify the functional group characteristic of an alcohol, ether, aldehyde, ketone, organic acid, ester, amine, hydroperoxide, and organic peroxide.

◆ 12.6 FUNCTIONAL GROUPS

One or more hydrogen atoms in the molecular structure of a hydrocarbon can be substituted with another atom or group of atoms. When this substitution has occurred, a new organic compound is produced. The atom or group of atoms that substitutes for the hydrogen atom is called the *functional group* since this is the part of the molecule where chemical reactions often occur.

Table 12.11 lists the functional groups that are components of the molecular structures of many important organic compounds. The symbols R and R′ represent any arbitrary alkyl or aryl group. We shall identify compounds containing these functional groups in Sections 12.7 through 12.14.

TABLE 12.11 ◆ Some Important Functional Groups in Organic Compounds

Name	General Formula	Functional Group
Haloalkane or alkyl halide or aryl halide	R—X	—X (X = halogen)
Alcohol	R—OH	—OH, hydroxyl
Ether	R—O—R′	—O—
Aldehyde	R—C(=O)H	—C(=O)H
Ketone	R—C(=O)—R′	—C(=O)—, carbonyl
Acid	R—C(=O)OH	—C(=O)OH, carboxyl
Ester	R—C(=O)OR′	—C(=O)O—
Amine	R—NH$_2$	—NH$_2$
Hydroperoxide	H—O—O—R	—O—O—, peroxo
Peroxide	R—O—O—R′	—O—O—

◆ 12.7 HALOGENATED HYDROCARBONS

When one or more hydrogen atoms in the molecules of hydrocarbons are substituted with halogen atoms, the resulting compounds are called *halogenated hydrocarbons*. The general chemical formula of the halogenated hydrocarbons is R—X, where R is the formula of an arbitrary alkyl or aryl group and X is the symbol of a halogen. When the halogen is chlorine, the compounds are called *chlorinated hydrocarbons*. For illustration, consider the following Lewis structures of the chlorinated derivatives of methane:

$$
\begin{array}{cccc}
\text{H} & \text{H} & \text{H} & \text{Cl} \\
| & | & | & | \\
\text{H}-\text{C}-\text{Cl} & \text{Cl}-\text{C}-\text{Cl} & \text{Cl}-\text{C}-\text{Cl} & \text{Cl}-\text{C}-\text{Cl} \\
| & | & | & | \\
\text{H} & \text{H} & \text{Cl} & \text{Cl}
\end{array}
$$

As we proceed across the page from left to right, each structure identifies a substance in which one, two, three, and four hydrogen atoms have been respectively substituted with chlorine atoms. In the common system, these compounds are named methyl chloride, methylene chloride, chloroform, and carbon tetrachloride, respectively.

In the IUPAC system, the halogen atoms are named as substituents of the compound having the longest continuous chain of carbon atoms. The halogen atoms are named as substituents by replacing the *-ine* suffix on the name of the halogen with *-o*. The number of halogen atoms is indicated by the use of *mono-*, *di-*, *tri-*, *tetra-*, and so forth for one, two, three, and four atoms, respectively.

Using the IUPAC system, the chlorinated derivatives of methane are named chloromethane, dichloromethane, trichloromethane, and tetrachloromethane, respectively. Similarly, the chlorinated derivatives of ethene are named as follows:

$$
\begin{array}{cccc}
\text{Cl} \quad \text{H} & \text{Cl} \quad \text{Cl} & \text{Cl} \quad \text{H} & \text{Cl} \quad \text{H} \\
\diagdown \quad \diagup & \diagdown \quad \diagup & \diagdown \quad \diagup & \diagdown \quad \diagup \\
\text{C}=\text{C} & \text{C}=\text{C} & \text{C}=\text{C} & \text{C}=\text{C} \\
\diagup \quad \diagdown & \diagup \quad \diagdown & \diagup \quad \diagdown & \diagup \quad \diagdown \\
\text{H} \quad \text{H} & \text{H} \quad \text{H} & \text{H} \quad \text{Cl} & \text{Cl} \quad \text{H}
\end{array}
$$

chloroethene   *cis*-1,2-dichloroethene   *trans*-1,2-dichloroethene   1,1-dichloroethene

$$
\begin{array}{cc}
\text{Cl} \quad \text{Cl} & \text{Cl} \quad \text{Cl} \\
\diagdown \quad \diagup & \diagdown \quad \diagup \\
\text{C}=\text{C} & \text{C}=\text{C} \\
\diagup \quad \diagdown & \diagup \quad \diagdown \\
\text{Cl} \quad \text{H} & \text{Cl} \quad \text{Cl}
\end{array}
$$

trichloroethene   tetrachloroethene

These simple halogenated hydrocarbons are colorless, water-insoluble, highly volatile liquids. Their physical properties once caused them to be highly desirable commercial solvents and aerosol propellants. Trichloroethylene was once a common *degreaser,* that is, a solvent that effectively dissolves oil, grease, wax, and other undesirable matter from metallic, textile, and glass surfaces. Tetrachloroethylene, also called perchloroethylene and "perc," remains a common dry-cleaning solvent.

Certain chlorinated hydrocarbons are capable of depleting stratospheric ozone. Most notable among them are methyl bromide, carbon tetrachloride, 1,1,1-trichloroethane, the halons (Section 5.13), and the chlorofluorocarbons (see p. 549). To minimize their impact upon the ozone layer, EPA used the authority of the Clean Air Act to ban the domestic manufacture of carbon tetrachloride, 1,1,1-trichloroethane, and the chlorofluorocarbons.

When the vapors of the simple chlorinated hydrocarbons are inhaled, they affect the human organism in varying ways. The substance commercially known as *Halothane* is used at hospitals and medical clinics as a general anesthetic; its chemical name is 1-bromo-1-chloro-2,2,2-trifluoroethane.

$$\begin{array}{c} \quad\ \text{Cl}\quad\ \text{F} \\ \quad\ | \quad\quad | \\ \text{Br}-\text{C}-\!\!-\text{C}-\text{F} \\ \quad\ | \quad\quad | \\ \quad\ \text{H}\quad\ \text{F} \end{array}$$

However, when low concentrations of the vapors of other chlorinated ethanes are inhaled, they cause lightheadedness, dizziness, and fatigue. When concentrations above the permissible exposure limits (Section 10.5) are inhaled, the vapors cause a lowering of consciousness that could be life-threatening.

From animal studies, epidemiologists have determined that exposure to the following halogenated hydrocarbons can cause cancer: bromodichloromethane, carbon tetrachloride, chloroform, 1,2-dichloroethane, hexachloroethane, methylene chloride, tetrachloroethylene, tetrafluoroethylene, and trichloroethylene. These substances are probable human carcinogens.

DOT regulates the transportation of the simple halogenated hydrocarbons, some examples of which are provided in Table 12.12. DOT requires shippers and carriers to affix labels that correspond to the hazard codes of the hazardous materials on their packaging and, when warranted, to post placards appropriate to their hazard classes on the bulk packaging, freight container, transport vehicle, unit containment device, or railcar used to transport them.

ORGANOCHLORINE PESTICIDES

Some halogenated hydrocarbons have been used as organochlorine pesticides (Section 10.18). The most recognizable of them is dichlorodiphenyltrichloroethane, more commonly known by its acronym *DDT*.

$$\begin{array}{c} \quad\quad\ \text{Cl} \\ \quad\quad\ | \\ \text{Cl}-\text{C}-\text{CH}-(\langle\bigcirc\rangle-\text{Cl})_2 \\ \quad\quad\ | \\ \quad\quad\ \text{Cl} \end{array}$$

During World War II, DDT was used to eradicate legions of the insects that carried typhoid fever, malaria, and typhus. The insecticide is still used for this purpose in some Third-World countries.

Notwithstanding its insecticidal value, the presence of DDT in the environment had a negative impact on the survival of certain forms of wildlife. This observation

TABLE 12.12 ◆ Some Halogenated Hydrocarbons Regulated by DOT for Transport

Proper Shipping Name	Label Codes
Amyl chlorides	3
Benzyl chloride	6.1, 8
1-Bromo-3-methylbutane	3
Bromobenzene	3
Bromotrifluoroethylene	2.1
n-Butyl bromide	3
Carbon tetrachloride	6.1
1-Chloro-3-bromopropane	6.1
1-Chloro-1,1-difluoroethanes	2.1
1-Chloro-1,2,2,2-tetrafluoroethane	2.2
1-Chloro-2,2,2-trifluoroethane	2.2
Chlorobenzene	3
Chlorobutanes	3
Chlorodifluorobromomethane	2.2
Chlorodifluoromethane	2.2
Chloroform	6.1
2-Chloropropane	3
Chlorotrifluoromethane	2.2
o-Dichlorobenzene	6.1
Dichlorodifluoromethane	2.2
Dichloroethylene	3
Dichloromethane	6.1
Ethylene dichloride[a]	3, 6.1
Hexachlorobenzene	6.1
Hexachlorobutadiene	6.1
Methyl bromide	2.3
Methyl chloride	2.3
Polychlorinated biphenyls	9
Propylene dichloride[b]	3
Tetrachloroethane	6.1
Tetrachloroethylene	6.1
1,1,1-Trichloroethane	6.1
Trichloroethylene	6.1

[a]Ethylene dichloride is the common name of 1,2-dichloroethane.
[b]Propylene dichloride is the common name of 1,2-dichloropropane.

caused EPA to ban the use of DDT in 1972. Over time, the use and manufacture of other organochlorine pesticides have also been discontinued.

CHLOROFLUOROCARBONS

A special group of halogenated hydrocarbons are the *chlorofluorocarbons,* which are often designated as CFCs. These substances were formerly important commercial compounds, but for reasons we shall soon visit, their use has been sharply curbed worldwide. The commercially important chlorofluorocarbons constituted the following classes of substances:

- *Chlorofluoromethanes.* In these compounds, one or more hydrogen atoms in the methane molecule have been substituted with chlorine or fluorine atoms. Their general chemical formula is CF_nCl_{n-x}, where n and x are whole numbers less than 4.
- *Chlorofluoroethanes.* These compounds have had chlorine or fluorine atoms substituted for one or more hydrogen atoms in the ethane molecule. Their general chemical formula is $C_2F_nCl_{n-x}$, where n and x are whole numbers less than 6.

As a class of organic compounds, the chlorofluorocarbons were once highly desirable commercial products. They are relatively inert substances. They either are nonflammable or do not readily ignite.

The chlorofluorocarbons were typically prepared by reacting hydrofluoric acid with a commercially available chlorinated hydrocarbon. For example, a mixture of chlorofluoromethanes was prepared from carbon tetrachloride, as follows:

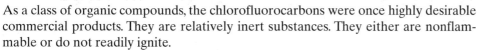

$$CCl_4(g) \quad + \quad HF(l) \quad \longrightarrow$$
carbon tetrachloride   hydrofluoric acid

$$CCl_2F_2(g) \quad + \quad CCl_3F(g) \quad + \quad 3HCl(g)$$
dichlorodifluoromethane   trichlorofluoromethane   hydrogen chloride

The two chlorofluorocarbons were then separated by distillation.

As we first noted in Section 5.13, some chlorofluorocarbons were formerly used as Halon fire extinguishers. These substances were also used as foaming agents by polymer manufacturers in plastic products. As the CFCs vaporized, the plastic expanded and hardened into a foam. CFCs were also used as the fluids in residential and commercial refrigeration and air-conditioning equipment.

Examples of the commercially important chlorofluorocarbons are noted in Table 12.13. Their names were styled as *CFC-wxyz,* where "*w,*" "*x,*" "*y,*" and "*z*" are digits. Here, "*w*" is the number of carbon–carbon double bonds per molecule; "*x*" is the number of carbon atoms per molecule minus one; "*y*" is the number of hydrogen atoms per molecule plus one; and "*z*" is the number of fluorine atoms per molecule. When "*w,*" "*x,*" "*y,*" or "*z*" is zero, the digit is omitted. For example, "*w,*" "*x,*" "*y,*" and "*z*" for the chlorofluorocarbon having the formula $CFCl_3$ are 0, 0, 1, and 1, respectively. It is denoted as CFC-11.

Solved Exercise 12.10

The major component in the fire suppressant known as FM (Fire Master) 200 is 1,1,1,2,3,3,3-heptafluoropropane. Write the structural formula for this substance.

Solution: The name of this fire suppressant indicates that it is a fluorinated derivative of propane, whose chemical formula is C_3H_8. The use of "hepta" means that *seven* of the eight hydrogen atoms have been replaced with fluorine atoms in the molecules of this substance. The notation "1,1,1,2,3,3,3" identifies precisely which hydrogen atoms have been replaced: three that are bonded to the first and third carbon atoms and one that is bonded to the second carbon atom. Consequently, the structural formula of FM 200 is the following:

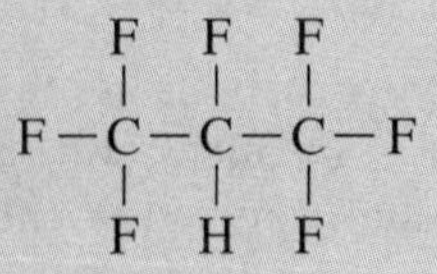

$$
\begin{array}{ccccccc}
 & F & & F & & F & \\
 & | & & | & & | & \\
F- & C & - & C & - & C & -F \\
 & | & & | & & | & \\
 & F & & H & & F & \\
\end{array}
$$

This formula may be condensed to C_3HF_7.

TABLE 12.13 ◆ Some Commercially Important Chlorofluorocarbons

CFC or HCFC Designation	Chemical Name	Molecular Formula
CFC-11	Trichlorofluoromethane	$\begin{array}{c} Cl \\ | \\ Cl-C-Cl \\ | \\ F \end{array}$
CFC-12	Dichlorodifluoromethane	$\begin{array}{c} F \\ | \\ Cl-C-Cl \\ | \\ F \end{array}$
CFC-13	Chlorotrifluoromethane	$\begin{array}{c} F \\ | \\ Cl-C-F \\ | \\ F \end{array}$
CFC-21	Dichlorofluoromethane	$\begin{array}{c} H \\ | \\ Cl-C-Cl \\ | \\ F \end{array}$
HCFC-22	Chlorodifluoromethane	$\begin{array}{c} F \\ | \\ Cl-C-H \\ | \\ F \end{array}$
CFC-113	1,1,2-Trichloro-1,2,2-trifluoroethane	$\begin{array}{c} F \quad F \\ | \quad | \\ Cl-C-C-F \\ | \quad | \\ Cl \quad Cl \end{array}$
CFC-114	1,2-Dichloro-1,1,2,2-tetrafluoroethane	$\begin{array}{c} F \quad F \\ | \quad | \\ Cl-C-C-Cl \\ | \quad | \\ F \quad F \end{array}$
HCFC-123	1,1-Dichloro-2,2,2-trifluoroethane	$\begin{array}{c} Cl \quad F \\ | \quad | \\ Cl-C-C-F \\ | \quad | \\ H \quad F \end{array}$

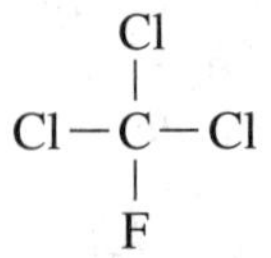

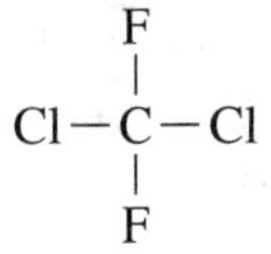

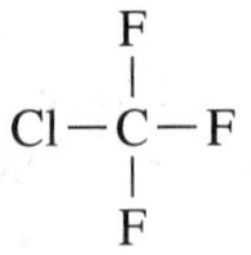

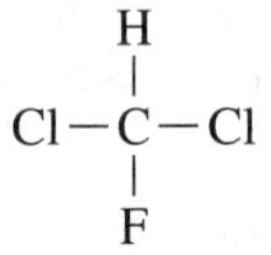

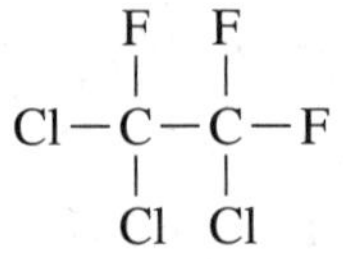

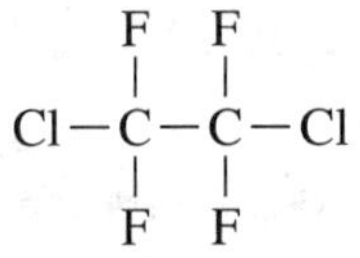

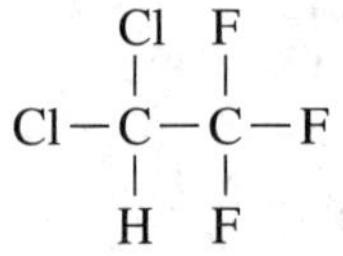

Scientists discovered that the chlorofluorocarbons are carried in the atmosphere upward by air currents to the stratospheric ozone layer (Section 7.1), where they contribute to its depletion. The CFCs absorb certain wavelengths of the ultraviolet radiation from the sun. Upon doing so, carbon-to-chlorine bonds are ruptured, resulting in the production of chlorine atoms (Cl·). When a chlorofluoromethane molecule absorbs ultraviolet radiation, for example, chlorine atoms are produced as follows:

$$CF_nCl_{n-x}(g) \longrightarrow CF_nCl_{n-x-1} \cdot (g) + Cl \cdot (g)$$

When they are produced within the ozone layer, these chlorine atoms are exposed to ozone molecules, with which they unite by a step-wise process that forms molecular oxygen.

$$O_3(g) + Cl \cdot (g) \longrightarrow ClO \cdot (g) + O_2(g)$$
$$ClO \cdot (g) + O \cdot (g) \longrightarrow Cl \cdot (g) + O_2(g)$$

This process contributes to the depletion of stratospheric ozone, and accordingly, more ultraviolet radiation penetrates into the troposphere where we live.

To protect the integrity of the ozone layer, EPA used the authority of the Toxic Substances Control Act to ban the use of chlorofluorocarbons as propellants, foaming agents, and refrigerants. EPA initially allowed manufacturers to use CFC-22 (chloro-difluoromethane) in refrigerators and air conditioners. This tolerance was based upon the belief that CFCs whose molecules possess C—H bonds, denoted *hydrochlorofluorocarbons* (HCFCs), were destroyed at lower altitudes and did not diffuse to the ozone layer. When scientists were unable to substantiate this position, EPA banned the production of HCFCs and phased out their use by 2030.

It remains to be seen whether the imposition of these bans can rectify the damage caused to the stratospheric ozone layer.

Solved Exercise 12.11

While combating a fire at an auto service shop, firefighters encounter a large storage tank labeled as follows:

> **Warning: Contains dichlorofluo-**
> **romethane (CFC-12), a substance**
> **which harms public health and the en-**
> **vironment by destroying ozone in the**
> **upper atmosphere**

What special action should the firefighters take to protect public health, safety, and the environment?

Solution: Dichlorofluoromethane is an example of a "chlorofluorocarbon," a substance which, when released into the environment, is known to contribute to the destruction of the stratospheric ozone layer. The EPA regulation at 40 CFR §82.106 requires the owner of this product to post the referenced label on the storage tank in such a fashion as to be clearly legible and conspicuous to observers. To avoid the potential impact upon destruction of the ozone layer, firefighters should take precautionary action to prevent the rupture of the tank and the subsequent unintended release of the product into the environment. This can be accomplished by cooling the tank with a prodigious volume of water until the fire in the auto service shop has been successfully extinguished.

POLYCHLORINATED BIPHENYLS

Another group of chlorinated hydrocarbons are the *polychlorinated biphenyls,* or PCBs. They are the chlorinated derivatives of the complex hydrocarbon called *biphenyl,* which has the following molecular structure:

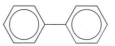

There are 209 structural isomers of the PCBs, each of which has the chemical formula $C_{12}H_xCl_y$, where x ranges from 0 to 9 and y equals $10 - x$. Approximately 130 of these isomers were components of the products formerly manufactured and marketed under trade names such as *Aroclor, Askarel, Inerteen, Pydraul,* and *Pyranol.*

At ordinary ambient conditions, the products containing PCBs are viscous, sticky liquids. Even at elevated temperatures, their individual components are nonflammable, incredibly stable compounds that can be heated to their respective boiling points without undergoing decomposition or bursting into flame. This combination of properties was formerly considered ideal for implementing certain industrial practices that required high temperatures and fire resistance.

From 1930 to 1977, products containing PCBs were marketed for commercial utilization within specialized electrical equipment, including transformers, voltage regulators, circuit breakers, capacitors, motors, and fluorescent light ballasts; they were also used in certain heat transfer and hydraulic systems; and they were incorporated into certain plastic, wax, and rubber products. In all instances, the PCBs were used to provide commercial products with a lower degree of fire resistance.

In the 1960s, however, the PCBs became the focus of several scientific studies conducted on animals. From these studies, scientists concluded that exposure to PCBs could cause cancer, birth defects, liver damage, acne, impotence, and death. They also discovered that once ingested, PCBs dissipate into various receptor tissues of the thymus, lungs, spleen, kidneys, liver, brain, muscle, and testes. The PCBs primarily bioaccumulate in fatty tissues, but their periodic release causes a variety of neurobehavioral disorders and birth abnormalities. Scientists have also discovered that over long periods, the PCBs induced the formation of malignant liver tumors.

Epidemiological studies have shown that the PCBs cause physiological disorders by mimicking the action of the body's own hormones or by interfering with the hormone's normal function. Hormones are substances secreted by the endocrine glands, the organs that control such activities as growth, development, metabolism, and reproduction. When absorbed into the body, the PCBs function like these hormones; hence, they are sometimes referred to as *endocrine disrupters.* Given this combination of adverse properties, epidemiologists rank PCBs as probable human carcinogens.

Because of the potential risk they pose to public health and the environment, EPA banned the manufacture, sale, and distribution of PCBs in 1979, using the legal authority of the Toxic Substances Control Act (Section 1.5). Nonetheless, PCBs are still produced in Russia and used in several other countries. Although their manufacture was banned in the United States, EPA permits the indefinite use of PCBs in systems that were operating prior to 1979, albeit under specific regulatory conditions. Even today, PCBs can be legally used in a "totally enclosed manner," that is, within equipment that has been designed and constructed so as to ensure no exposure of PCBs to humans or the environment. When this equipment is withdrawn from service or repaired, EPA requires replacement of the PCB fluid with a non-PCB-containing product.

An electrical transformer is a totally enclosed system, one type of which can be observed near the tops of electric poles. It is a static device whose function is to transfer electrical energy between circuits of different voltage. Each transformer has a magnetic core around which there are several windings or coils of a metal such as copper. This assembly of core and current-carrying coils is then immersed in a fluid within a tank. The purpose of the fluid, called a *dielectric fluid,* is to increase the resistance of the unit to sustained arcing and serve as a heat-transfer medium.

A transformer containing a commercial PCB fluid is called a *PCB transformer.* A typical PCB transformer contains approximately 200 gal (1500 L) of fluid with a PCB concentration of approximately 50% to 60% by volume, but a larger PCB transformer could hold as much as 1000 gal (7500 L) of PCB fluid. The remaining volume consists of diluents, typically a mixture of trichlorobenzene and tetrachlorobenzene isomers.

The vast majority of transformers in use today were manufactured with mineral oil, a petroleum oil, as their dielectric fluid. They are called *mineral oil transformers.* PCBs were never intentionally added to the mineral oil used in transformers, but the oil often became contaminated when workers conducted routine equipment maintenance on both transformer types using the same hoses and pumping equipment.

Although PCB fluids do not readily ignite, transformer fluids may ignite when high-energy arcing, electrical overloading, or short-circuiting occurs within energized transformers. These fires are characterized by the presence of large amounts of oily, black soot in the accompanying smoke plume. When the energized transformer is located within a building, soot and PCB vapors travel with the smoke plume through ventilation shafts, building ductwork, and open construction areas.

Today's firefighters respond rarely to transformer fires involving PCB dielectric fluids, but since the service lifetime of a transformer is 40 to 85 years, such an action is not entirely outside the realm of possibility. Acknowledging that exposure to the smoke from these class C fires presents firefighters with an unwarranted health risk, EPA requires the transformer owner and operator to post the marking illustrated in Figure 12.16 in such a manner that it can be easily read by the personnel who respond to transformer fires.

EPA has also instituted regulatory controls on facilities that still use PCB transformers. One regulation requires their owners to register PCB transformers with the appropriate local fire department. Through this registration, firefighters are informed of the precise location of the PCB transformers within their areas of jurisdiction. One of the most common locations at which transformers are likely to be found is an electrical substation operated by a local utility (Figure 12.17).

Solved Exercise 12.12

Once fire personnel have been informed of an ongoing fire within a PCB transformer, why does EPA strongly urge the department, on a voluntary basis, to quickly notify the operators of the relevant storm and sanitary sewer system and treatment plants of the ongoing incident?

Solution: Combating a PCB transformer fire can generate a relatively large volume of contaminated water, the discharge of which can subsequently cause widespread pollution of sewer systems and associated treatment plants. The subsequent cleanup of this contamination may be very costly. For these reasons, the operators of sewer systems and treatment plants should be quickly notified of an ongoing PCB transformer fire and of the likelihood that the water used to combat the fire is contaminated with PCBs and their incomplete combustion products. The operators of these plants can then implement appropriate measures to confine the contaminated water and minimize the adverse environmental impact.

PCBs may be released into the environment during transformer and other electrical-equipment fires. In addition, when PCB transformer fluids burn, their constituents undergo incomplete combustion and produce the highly toxic compounds called *polychlorinated dibenzofurans* and *polychlorinated dibenzo*-p-*dioxins* (Section 12.9). These substances may also be released into the environment during electrical-equipment fires.

Firefighters should be especially wary when combating class C fires since exposure to these toxic substances could cause long-term ill effects. Regardless of their magnitude, firefighters should always utilize self-contained respiratory devices and protective clothing to eliminate or reduce the potential for their exposure to PCBs, polychlorinated dibenzofurans, and polychlorinated dibenzo-*p*-dioxins. Public health can also be effectively protected by constructing dikes and dams to confine the spread of the contaminated water generated during these firefighting operations and prevent the spread of the contaminants into storm sewers and elsewhere.

Why are the PCBs regarded as ultrahazardous substances? The answer to this question is linked with the horrific impact they could have not only on public health but also on a wide segment of the environment. When PCBs are consumed by an animal, they adversely affect the health of the animal throughout its lifetime. When the animal is eaten by another animal, the PCBs pass progressively by means of the food

FIGURE 12.16 ◆ Examples of warning signs required by EPA on equipment containing polychlorinated biphenyls pursuant to 40 CFR §761.45.

FIGURE 12.17 ◆ Electrical substations are located in all major municipalities. Here, transformers step-up or step-down the voltage of electric current for distribution on demand to nearby neighborhoods.

chain from animal to animal: worms to fish and birds, fish to birds, and birds and fish to mammals. The PCBs can thereby affect the well-being of many individual species, including humans who consume fish or meat contaminated with PCBs. In the United States alone, EPA has identified 703 bodies of water where health advisories have been posted to warn people against consuming PCB-ridden fish.

Extensive studies have been conducted on fish and birds to identify how the PCBs affect the health of these organisms. They demonstrate that the immune systems of fish who have consumed PCBs become dysfunctional. The number of mature ovaries in these fish are reduced; even when the ovaries are mature, the fish experience a reduced ability to spawn. Both fish and birds who have consumed PCBs experience growth and reproductive impairment. Problems observed in birds include eggshell thinning, lower egg production, decreased egg hatching, deformities in the juvenile birds that do emerge from eggs, and reduction in their growth and survival rates.

In the ocean hierarchy, killer whales are contaminated with PCBs at concentrations higher than those observed in other marine animals since they rank unrivaled at the top of the food chain. Herring may carry PCBs at an average concentration of only 1 ppm, but seals that eat them may contain 20 ppm, and the killer whales that eat these contaminated seals may have PCB levels as high as 250 ppm. The male killer whales carry the PCBs throughout their lives, while the females periodically rid these pollutants from their bodies by passing them to their firstborn calves. This could explain why the males live only half as long as the females, and why many calves die within several months of their birth.

Solved Exercise 12.13

The killer whale, *Orcinus orca,* lives in the waters of the Pacific Northwest. When tissue samples of male killer whales are collected and analyzed, they are found on the average to be contaminated with PCBs at substantially higher concentrations than those observed in the tissues collected from the females or other animals within the same region.

(a) How did the killer whale come to possess this unique distinction?

(b) Why are the PCB concentrations in tissues of males of the species higher than in tissues of females?

Solution: (a) Within the natural environment, contaminant concentrations progressively increase in the tissues of animals at each level of the food chain. (In the case of PCBs, a herring may carry PCBs at a concentration of only 1 ppm, but seals that eat the contaminated herring may contain PCBs at a concentration of 20 ppm, and the killer whales that eat these contaminated seals may have PCB levels of 250 ppm.) Since killer whales rank unrivaled at the top of the food chain in the ocean hierarchy, their tissues become contaminated with a higher concentration of contaminants than other animal species. This statement holds for all contaminants such as PCBs that bioaccumulate in body fat.

(b) Male killer whales can carry the PCBs throughout their lives, whereas the females periodically rid their bodies of some amount of the PCBs by passing these contaminants to their calves in milk. (This may also explain why the male lives only half as long as the female. It may also explain why many calves die within several months of their birth.)

Performance Goals for Section 12.8
- ◆ Identify the functional group that characterizes alcohols.
- ◆ Describe the rules for naming the simple alcohols.
- ◆ Discuss the hazardous properties of methanol, ethanol, phenol, and the cresols.
- ◆ Describe the effects that result from abusing the use of alcoholic beverages.
- ◆ Identify the industries that use methanol, ethanol, phenol, and the cresols.

◆ 12.8 ALCOHOLS

Alcohols are organic compounds that are derived by substituting one or more hydrogen atoms in the hydrocarbon molecule with the hydroxy group (—OH). Thus, the general chemical formula of the simple alcohols is R—OH, where R is the formula of an arbitrary alkyl or aryl group.

When one hydrogen atom is substituted with the hydroxy group in molecules of methane and ethane, the resulting compounds are methyl alcohol and ethyl alcohol, respectively.

$$H-\underset{\underset{H}{|}}{\overset{\overset{H}{|}}{C}}-OH \qquad\qquad H-\underset{\underset{H}{|}}{\overset{\overset{H}{|}}{C}}-\underset{\underset{H}{|}}{\overset{\overset{H}{|}}{C}}-OH$$

methyl alcohol   ethyl alcohol

These are the common names of the two simplest alcohols.

In the IUPAC system, the simple alcohols are named by replacing the *-e* in the name of the corresponding alkane with *-ol*; hence, the compounds having the formulas CH_3OH and CH_3CH_2OH are also named methanol and ethanol, respectively.

When exposed to an ignition source in the air, the vapors of the simple aliphatic alcohols burn. Since oxygen is a component of their molecular structures, the combustion of these alcohols is characterized by the production of nearly imperceptible flames. When methanol and ethanol burn, for example, virtually invisible, pale blue flames are produced without the accompaniment of soot. Their combustion evolves 9900 Btu/lb (23,000 kJ/kg) and 12,900 Btu/lb (30,000 kJ/kg), respectively, to the surroundings.

$$CH_3OH(g) + 2O_2(g) \longrightarrow CO_2(g) + 2H_2O(g)$$

methanol   oxygen   carbon dioxide   water

$$CH_3CH_2OH(g) + 3O_2(g) \longrightarrow 2CO_2(g) + 3H_2O(g)$$

ethanol   oxygen   carbon dioxide   water

Aside from the simple alcohols, compounds that also burn without visible flames are the simple aliphatic aldehydes (Section 12.10), ketones (Section 12.10), and ethers (Section 12.9). Since the flames cannot be detected with the naked eye, firefighters encounter relatively unique difficulties when combating their fires (see Solved Exercise 12.15).

DOT regulates the transportation of the simple alcohols in Table 12.14. DOT requires shippers and carriers to affix labels corresponding to the relevant hazard codes on their packaging and, when warranted, to post placards appropriate to their hazard class on the bulk packaging, freight container, transport vehicle, unit containment device, or railcar used to transport them.

Solved Exercise 12.14

When methanol and ethanol burn in air, why is very little soot produced? How does the absence of soot affect observation of the accompanying flames?

Solution: Methanol and ethanol are examples of organic compounds whose molecules contain oxygen atoms in addition to carbon and hydrogen atoms. This oxygen is utilized together with the available atmospheric oxygen during the combustion of these compounds. Provided with supplemental oxygen, methanol and ethanol burn predominantly in the manner of "complete combustion," producing carbon dioxide instead of carbon monoxide and soot. Since flames are manifested only when minute particulate matter can be heated to incandescence (Section 5.6), the absence of soot causes the flames accompanying the combustion of these simple alcohols to be nearly imperceptible.

Solved Exercise 12.15

Fires involving bulk quantities of methanol are among the most difficult to successfully combat.

(a) What unique characteristic of these fires contributes to this problem?

(b) What special procedures can be implemented to assist firefighters in effectively extinguishing them?

Solution: (a) Bulk methanol fires are often difficult to successfully combat because the evolving flame is virtually invisible to the naked eye. This situation hinders the firefighter's ability to observe the magnitude of these fires and rapidly assess their severity.

(b) The vapor of a flammable liquid generally ignites near the point where the liquid initially leaks from its bulk container or tank. Firefighters must locate and seal the leak before these fires can be permanently extinguished with water or foam. Verging upon a bulk storage or transportation vessel, however, could constitute an unwarranted risk due to the likely occurrence of a BLEVE (Section 3.4). Unable to directly view the flames evolving from methanol fires, firefighters could search from a suitable distance for telltale signs of combustion, such as blistering paint or weakened metal. Specialists recommend the use of a thermal-imaging camera (Section 5.4) to locate the leak more rapidly.

Since the flames associated with methanol fires are nearly invisible, it is difficult to determine not only *where* liquid methanol is leaking, but whether efforts to extinguish the fire were successful. Specialists again recommend using a thermal-imaging camera to establish that the fire has been successfully extinguished.

TABLE 12.14 ◆ Some Alcohols Regulated by DOT for Transport

Proper Shipping Name	*Label Codes*
Alcoholic beverages	3
Alcohols, n.o.s.	3
Alcohols, toxic, n.o.s.	3, 6.1
Alkyl phenols, liquid, n.o.s.	8
Alkyl phenols, solid, n.o.s.	8
Amyl alcohols	3
Butanols	3
Cresols	6.1, 8
Cresylic acid	6.1, 8
Cyclopentanol	3
Denatured alcohol	3 *or* 3, 6.1
Ethanol *or* Ethyl alcohol *or* Ethanol solutions *or* Ethyl alcohol solutions	3
2-Ethylbutanol	3
Isopropanol *or* Isopropyl alcohol	3
Methanol *or* Methyl alcohol	3, 6.1
Methyl cyclohexanols	3
Methyl isobutyl carbinol[a]	3
Phenol, molten	6.1
Phenol, solid	6.1
Phenol solutions	6.1
n-Propanol *or* Propyl alcohol, normal	3
Xylenols	6.1

[a]Methyl isobutyl carbinol is an alternate common name for methyl amyl alcohol.

METHANOL

At room conditions, methanol is a colorless, water-soluble, and highly volatile liquid. It was once called *wood alcohol,* a term that acknowledges methanol as a constituent of the mixture that results when wood is heated in the absence of air.

In the United States, methanol is mainly manufactured by the high-temperature, high-pressure hydrogenation of carbon monoxide in the presence of an appropriate catalyst.

$$\underset{\text{carbon monoxide}}{CO(g)} + \underset{\text{hydrogen}}{2H_2(g)} \longrightarrow \underset{\text{methanol}}{CH_3OH(g)}$$

Industrially, methanol is used as a solvent for shellac and gums; as a raw material for the manufacture of formaldehyde (Section 12.10) and methyl *tert*-butyl ether (Section 12.9); and directly as an antifreeze and aircraft fuel-injection fluid.

The principal hazard associated with methanol is its flammability. A secondary hazard is its poisonous nature. When ingested, as little as 0.06 pint (30 mL) can cause death. Lesser amounts have been known to cause irreversible blindness.

When methanol is present within the workplace, OSHA requires employers to limit employee exposure to a vapor concentration of 200 ppm, averaged over an 8-hour workday.

ETHANOL

Ethanol is also a colorless, water-soluble, highly volatile liquid. It is most commonly known as the active constituent of beer, wine, and the so-called "hard liquors." The ethanol in wine, champagne, and various brandies is produced by fermenting the sugars naturally present in ripe fruits. For example, the ethanol in wine is produced by fermenting the sugar in grapes. When the wine is charged with carbon dioxide, champagne is produced. When the ethanol is distilled from fermented peaches, apricots, apples, and other fruits, brandies are produced.

Ethanol intended for consumption is also produced by fermenting the sugars generated by the enzymatic conversion of the starches in corn, potatoes, barley, rye, and wheat. The ethanol in beer is produced by fermenting malted barley. The ethanol in whiskey is produced by fermenting corn, barley, or wheat; the ethanol in vodka is produced by fermenting potato mash; and the ethanol in gin is produced by fermenting rye. Since it is produced by fermenting cereal grains, the ethanol present in beers, whiskies, gin, and vodka is sometimes called *grain alcohol.*

The concentration of ethanol in alcoholic beverages is frequently expressed by use of the term "proof." In the 17th century, people believed that beverages contained spirits, which caused the physiological effects associated with their consumption. A procedure devised in England to test for the presence of the spirits consisted of preparing a mixture of gunpowder and a sample of the liquor being tested. If the gunpowder ignited after the alcohol had burned away, the event was taken as "proof" or confirmation that spirits were actually present. If the gunpowder did not ignite, the event was taken to mean that spirits were absent. It probably also meant that the liquor had been diluted with too much water.

The percentage by volume is about half the proof of an alcoholic solution. Absolute alcohol is 200 proof; 95% alcohol by volume is approximately 190 proof. A "proof-gallon" of alcohol contains 4.5 lb (2 kg) of ethanol.

When alcoholic beverages are consumed, the ethanol is absorbed into the bloodstream, principally from the small intestine. The rate of absorption into the bloodstream

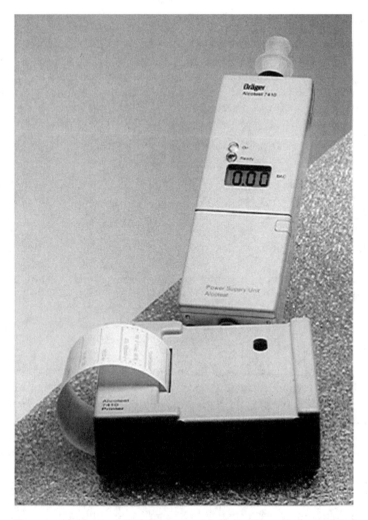

FIGURE 12.18 ◆ A DOT-approved digital Breathalyzer used by law enforcement personnel to determine the concentration of alcohol in the blood. *Courtesy of Dräger Safety, Inc., Breathalyzer Division, Durango, Colorado.*

is faster than the rates of metabolism and subsequent elimination from the body. An estimate of the amount of alcohol in the blood can be approximated through use of a Breathalyzer like the model shown in Figure 12.18, which measures the concentration of alcohol vapors emitted from the lungs.

The metabolism of alcohol occurs primarily in the liver. The metabolic process involves the stepwise oxidation to acetaldehyde and acetic acid.

$$\underset{\text{ethanol}}{CH_3CH_2OH(aq)} \longrightarrow \underset{\text{acetaldehyde}}{CH_3CHO(aq)} \longrightarrow$$

$$\underset{\text{acetic acid}}{CH_3COOH(aq)} \longrightarrow \underset{\text{carbon dioxide}}{CO_2(g)} + \underset{\text{water}}{H_2O(l)}$$

The average rate of oxidation is equivalent to approximately two-thirds to three-fourths of an ounce of whiskey, or about a glass of beer, per hour. When more alcohol

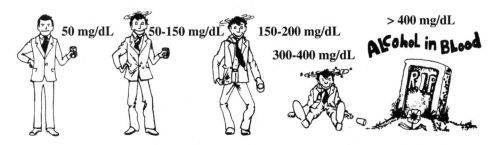

FIGURE 12.19 ◆ The effects experienced by a person having different blood alcohol concentrations: sedation or tranquility, 50 mg/dL; lack of coordination, 50–150 mg/dL; intoxication (delirium), 150–200 mg/dL; unconsciousness, 300–400 mg/dL; and death >400 mg/dL. *Adapted from an illustration by Joe Orlando from Chemistry, by Harvey A. Yablonsky, copyright 1973 by Thomas T. Crowell; used by permission of the publisher.*

is ingested than the liver oxidizes, the blood alcohol concentration, abbreviated BAC, increases until the consumer becomes intoxicated.

In the body, ethanol acts simultaneously as a brain stimulant and a central nervous system depressant. The stimulation is responsible for the pleasurable, sociable, and deinhibiting effects commonly associated with drinking alcohol, while the depression accounts for the anxiety, tension, and dulling of normal cognitive and motor processes.

When alcoholic beverages are misused, the consumed ethanol is responsible for physical, behavioral, and speech changes. The combination of the changes illustrated in Figure 12.19 can be correlated with the blood alcohol concentration. Individuals having blood alcohol concentrations between 50 and 150 mg/dL typically experience a complete lack of coordination. Since they are unable to safely drive a vehicle, 80 mg/dL has been selected as the legal limit above which a person is prohibited from driving on roadways in many states. [One deciliter (dL) is 100 milliliters, or 1/10 of a liter.]

Alcohol should only be consumed in moderation—between one and three drinks per day. "Binge drinking," the indiscriminate consumption of multiple drinks, one after the other, is known to have caused death within a few short hours of consumption.

Since alcohol consumption dulls the senses and causes reflexes to be sluggish, emergency responders who have consumed alcohol in amounts beyond moderation cannot expect to perform at the top of their form. They should take precautionary measures to consume alcohol only in modest amounts at all times.

When individuals consume alcohol in relatively large amounts over long periods of time, they often become afflicted with long-term health problems, the most serious of which is *alcoholism.* This disease results from addiction to alcohol. Such chronic alcohol misuse forces the liver to use alcohol as an energy source. Following aggravated misuse, the liver is unable to function properly. These heavy alcohol drinkers develop *alcoholic hepatitis,* an inflammation of the liver, or *cirrhosis*, a disease in which normal liver tissue is replaced by scar tissue. Since cirrhosis causes serious impairment of liver function, it is often fatal.

Studies have also linked alcohol abuse with mouth, larynx, esophagus, liver, and breast cancers. They show that women who consume two or more drinks a day are 41% more likely to experience breast cancer compared to women who are nondrinkers. Studies have also shown that alcohol enhances the cancer-causing effects of other substances, particularly those found in tobacco smoke. People who smoke and drink are

35 times more likely to contract mouth, tracheal, or esophageal cancers than people who neither smoke nor drink.

In combination with medications containing barbiturates, tranquilizers, or narcotics, ethanol produces cumulative effects, which could slow the cardiac and respiratory systems to the point of causing coma and death. For this reason, individuals on medications should avoid drinking alcoholic beverages. Women who are pregnant should be especially wary, since during pregnancy, alcohol consumption can cause birth defects in their offspring. One third of the population of the mentally handicapped children in the Western world is believed to have resulted from alcohol consumption in combination with medications during pregnancy. Even moderate or social drinking can be harmful to the developing fetus. The U.S. Surgeon General and Britain's Royal College of Psychiatrists uncompromisingly advise total abstinence from alcohol during pregnancy. So alcohol consumers are apprised of this potential danger, the Bureau of Alcohol, Tobacco, and Firearms requires alcohol manufacturers to label their beverages with the message shown in Review Exercise 1.4.

Notwithstanding these adverse effects, research studies reveal that *moderate* drinking contributes to the prevention of heart attacks, regardless of the nature of the alcoholic beverage consumed. Preliminary studies show that individuals who drink at least three times a week experience approximately one-third fewer heart attacks than nondrinkers. Cardiologists are quick to point out, however, that the results of these studies require corroboration.

For industrial use, ethanol is produced by the acid-catalyzed reaction of ethylene and water.

$$\underset{\text{ethylene}}{CH_2{=}CH_2(g)} + \underset{\text{water}}{H_2O(g)} \longrightarrow \underset{\text{ethanol}}{CH_3CH_2OH(g)}$$

Two forms of industrial-grade alcohol are available commercially:

- *Absolute alcohol*, which is ethanol without water.
- *Denatured alcohol*, which is an aqueous solution containing approximately 95% ethanol by volume. Denatured alcohol is intended for use as a solvent. To impede its consumption, a small amount of methanol, pyridine, or aviation gasoline is added to denatured alcohol. The presence of these substances makes the alcohol not only unpalatable but poisonous. The additives are called *denaturants*.

Whereas the sale of absolute alcohol is closely regulated by governments throughout the world, denatured alcohol is sold without the imposition of an excise tax. Manufacturing and process industries use large volumes of denatured ethanol, most typically as a solvent in toiletries, cosmetics, pharmaceuticals and surface coatings and a raw material in the manufacture of other substances. When ethanol is present in the workplace, OSHA requires employers to limit employee exposure to a vapor concentration of 1000 ppm, averaged over an 8-hour workday.

In 1990, following the mandate of the Clean Air Act, EPA required petroleum refineries in certain areas of the United States to blend gasoline with a substance that improved the completeness of the ensuing combustion. This substance is a type of antiknock agent called an *oxygenate*, and the fuel is called *reformulated gasoline*. Ethanol is one of two oxygenates selected as an oxygenate. [The other is methyl *tert*-butyl ether (Section 12.9).]

The combustion of reformulated gasoline produces smaller quantities of the pollutants that are routinely components of vehicular exhaust (such as carbon monoxide and benzene). EPA requires the use of reformulated gasoline in certain urban areas where compliance with the mandates of the Clean Air Act is otherwise difficult. It is anticipated that the use of ethanol as an oxygenate in reformulated gasoline will increase dramatically in the next decade.

GLYCOLS

Glycols are alcohols having two hydroxy groups per molecule. The two most commonly encountered glycols are ethylene glycol and propylene glycol. Both are used as antifreeze agents in motor vehicles. Propylene glycol is also used as a raw material for the manufacture of certain polyester resins.

$$\underset{\text{ethylene glycol}}{HO-CH_2CH_2-OH} \qquad \underset{\text{propylene glycol}}{HO-CH_2CH_2CH_2-OH}$$

PHENOLS

The *phenols* are derivatives of benzene in which one or more hydrogen atoms have been substituted with the hydroxy group of atoms (—OH). They are also called *phenolic compounds*.

The simplest phenolic compound is called *phenol,* or *carbolic acid.*

At room conditions, it is a colorless to white crystalline solid that often darkens to red when exposed to light. Phenol readily absorbs atmospheric moisture; thus, it is often encountered as a liquid. Phenol possesses the sharp distinctive odor characteristic of disinfectants.

Phenol is an important industrial substance since it is the raw material from which a number of derivatives and phenolic resins are manufactured. Phenolic derivatives are used in surgical antiseptics and mouthwashes. Phenol itself was the *first* surgical antiseptic. A solution of one part phenol in 850 parts of water by mass prevents the multiplication of certain bacteria; for this reason, phenol derivatives are common constituents of mouthwashes, gargles, and sprays.

Although phenolic compounds burn, their flashpoints exceed 172°F (78°C). Consequently, they do not pose a risk of fire and explosion. On the other hand, when exposed to skin, mucous membranes, and the eyes, phenol is highly irritating. When ingested, it impairs liver and kidney function and disturbs the central nervous system.

When phenol is present in the workplace, OSHA requires employers to limit employee exposure by skin contact to a concentration of 5 ppm, averaged over an 8-hour workday.

DOT regulates the transportation of molten and solid phenol as poisons. DOT requires shippers and carriers to affix INHALATION HAZARD labels on the packaging of molten phenol and to post INHALATION HAZARD placards on the bulk packaging, freight container, transport vehicle, unit containment device, or railcar used to transport it. DOT requires shippers and carriers to affix POISON labels on the packaging of solid phenol and to post POISON placards on the bulk packaging, freight container, transport vehicle, unit containment device, or railcar used to transport it.

A commercially important group of phenols are the *cresols.* They are the hydroxy derivatives of toluene. There are three isomeric cresols, whose formulas and names follow:

o-cresol

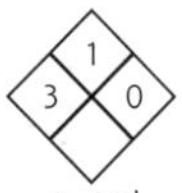

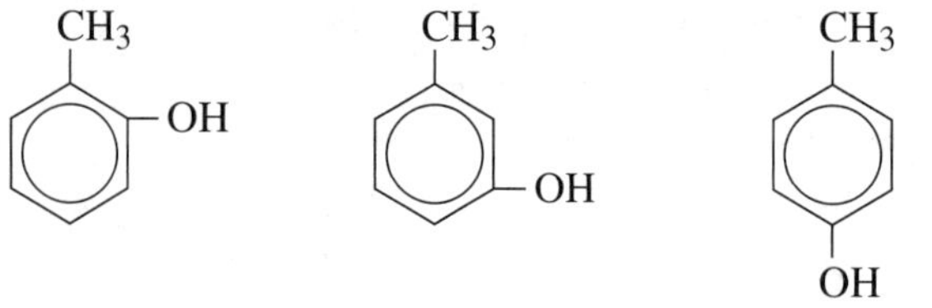

ortho-cresol   *meta*-cresol   *para*-cresol

Their mixture is called *cresylic acid*.

The individual cresol isomers are industrially important. For instance, *o*-cresol is used as a raw material for the manufacture of the fire retardants tricresyl phosphate and cresyl diphenyl phosphate.

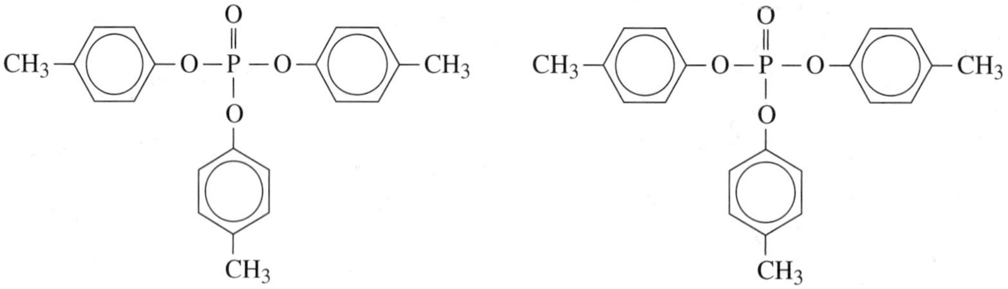

tri-*p*-cresyl phosphate   *p*-cresyldiphenyl phosphate

When used in the workplace, OSHA requires employers to limit employee exposure to *o*-, *m*-, and *p*-cresol to a maximum concentration of 5 ppm ($22 \ mg/m^3$) by skin contact.

DOT regulates the transportation of cresols and cresylic acid as poisons. DOT requires shippers and carriers to affix POISON labels on their packaging and, when warranted, to post POISON placards on the bulk packaging, freight container, transport vehicle, unit containment device, or railcar used to transport them.

Performance Goals for Section 12.9
- Identify the functional group that characterizes ethers.
- Discuss the hazardous properties of the simple ethers.
- Describe the rules for naming them.
- Discuss the likelihood that some commercial ethers may contain organic peroxides, which contribute to their risk of fire and explosion.
- Discuss the hazardous properties of the polychlorinated dibenzofurans and polychlorinated dibenzo-*p*-dioxins, especially noting that they are regarded as major environmental pollutants.

◆ 12.9 ETHERS

The *ethers* are organic compounds whose molecules possess one or more oxygen atoms bridged between two alkyl or aryl groups. Simple ethers have the general chemical formula R—O—R′, where R and R′ are the formulas of arbitrary alkyl or aryl groups.

The common names of the ethers are determined by alphabetically noting the names of the groups bonded to the oxygen atom, followed by the word "ether." For example, the substance having the compound having the following chemical formula has the common name ethyl methyl ether.

$$CH_3—O—CH_2CH_3$$

In the IUPAC system, ethers are named by replacing the *-yl* suffix of the relevant alkyl group with *-oxy* and naming the ether as a substituted alkane. Thus, ethyl methyl ether is also named methoxyethane.

The simple ethers are highly volatile, flammable liquids; hence, they constitute a dangerous fire and explosion hazard. They are hazardous materials not only because they are flammable, but since they produce organic hydroperoxides and organic peroxides (Section 12.14) by reacting with atmospheric oxygen. For example, successive reactions occur between diethyl ether and oxygen to produce diethyl peroxide.

$$CH_3CH_2—O—CH_2CH_3(l) \longrightarrow CH_3CH_2—O—CH_2CH_2—O—OH(s)$$

diethyl ether   2-ethoxyethyl hydroperoxide

$$\longrightarrow CH_3CH_2—O—O—CH_2CH_3(s)$$

diethyl peroxide

These compounds are highly reactive, potentially explosive substances.

The six ethers having the following molecular formulas are especially susceptible to organic peroxide formation:

$$CH_3CH_2—O—CH_2CH_3$$
diethyl ether

$$CH_3—CH—O—CH—CH_3$$
$$\qquad\ |\qquad\quad\ |$$
$$\qquad CH_3\qquad\ CH_3$$
diisopropyl ether

$$\begin{array}{c} CH_2-CH_2 \\ |\qquad\ | \\ CH_2\ \ CH_2 \\ \diagdown\ \diagup \\ O \end{array}$$
tetrahydrofuran

$$\begin{array}{c} O \\ \diagup\ \diagdown \\ CH_2\ \ CH_2 \\ |\qquad\ | \\ CH_2\ \ CH_2 \\ \diagdown\ \diagup \\ O \end{array}$$
dioxane

$$CH_3—O—CH_2CH_2—O—CH_3$$
ethylene glycol dimethyl ether (glyme)

$$CH_3—O—CH_2CH_2—O—CH_2CH_2—O—CH_3$$
diethylene glycol dimethyl ether (diglyme)

To forewarn of the potential reactivity of these ethers, their manufacturers typically affix warning labels like the following on ether containers:

> **DISCARD 30 DAYS AFTER OPENING**
> **OR**
> **AFTER ONE YEAR IF UNOPENED**

The reaction between ethers and atmospheric oxygen is catalyzed by light. Consequently, ethers are often stored in metal cans to prevent the penetration of light. Since the formation of organic peroxides occurs slowly, ethers purchased years ago are likely to be more susceptible to decomposition than those recently acquired. Ethers are also very volatile, a property that causes the organic peroxides to concentrate within the final residual volume of the liquid. For this reason, the likelihood of an explosive reaction may be greatest in containers having only small liquid residues.

DOT regulates the transportation of the simple ethers in Table 12.15 as flammable liquids. DOT requires shippers and carriers to affix FLAMMABLE LIQUID labels on their packaging and, when warranted, to post FLAMMABLE placards on the bulk packaging, freight container, transport vehicle, unit containment device, or railcar used to transport them.

TABLE 12.15 ◆ Some Ethers Regulated by DOT for Transport

Proper Shipping Name	*Label Codes*
Anisole[a]	3
Ethers, n.o.s.	3
Butyl methyl ether	3
Diethoxymethane	3
Diethyl ether *or* Ethyl ether	3
Diisopropyl ether	3
1,1-Dimethoxycthane	3
1,2-Dimethoxyethane	3
Dimethyl ether	3
Dioxane	3
Dipropyl ether	3
Divinyl ether, inhibited[b]	3
Ethyl butyl ether	3
Ethyl methyl ether	2.1
Methyl *tert*-butyl ether	3
Tetrahydrofuran[c]	3

[a]Anisole is the common name for methoxybenzene.
[b]Divinyl ether is the compound having the formula $CH_2{=}CH{-}O{-}CH{=}CH_2$.

[c]Tetrahydrofuran is the cyclic ether having the formula O.

DIETHYL ETHER

This is a colorless, water-soluble, highly volatile liquid. Diethyl ether was once a well-known anesthetic used in medical clinics and hospitals, where it was commonly known simply as *ether*. Today, we are more likely to detect its odor when using automotive starting fluids. The flammable hazard of diethyl ether has contributed to its limited usage.

The physical properties of diethyl ether are provided in Table 12.16. The profoundly low flashpoint, wide flammable range, and relatively high vapor density attest to the flammable nature of diethyl ether. As demonstrated in Figure 12.20, its vapors flow in a downward direction. If they ignite, the resulting fire flashes up to the initial

TABLE 12.16 ◆ Physical Properties of Diethyl Ether

Melting point	−189°F (−123°C)
Boiling point	94°F (34°C)
Specific gravity	0.71
Vapor density (air = 1)	2.55
Lower explosive limit	1.85%
Upper explosive limit	48%

FIGURE 12.20 ◆ Vapors of diethyl ether are heavier than air (vapor density = 2.55), as is illustrated by this experiment. A cloth saturated with the ether is placed at the top of a trough that has been arranged at a 45° angle. A lighted candle is positioned near its lower end, as shown at the left. Ether vapors slowly move down the trough. The candle flame serves as an ignition source. Fire rushes upward to the source of the vapors, as noted at the right.

source. The ether burns with the production of a virtually invisible, pale blue flame and no accompanying soot.

$$CH_3CH_2-O-CH_2CH_3(g) + 6O_2(g) \longrightarrow 4CO_2(g) + 5H_2O(g)$$

diethyl ether   oxygen   carbon dioxide   water

It is the presence of oxygen in the molecular structure that accounts for the generation of the virtually imperceptible flame.

METHYL *tert*-BUTYL ETHER

This is the only other ether of commercial importance. It is commonly denoted by its acronym *MTBE*.

$$CH_3-O-\overset{\displaystyle CH_3}{\underset{\displaystyle CH_3}{\overset{|}{\underset{|}{C}}}}-CH_3$$

Methyl *tert*-butyl ether was formerly one of the major organic compounds manufactured in the United States. It was first introduced in the U.S. fuel market in 1995 to boost the oxygen content of gasoline. It possesses an antiknock rating of 116. On the positive side, the use of MTBE as an oxygenate (Section 12.8) causes the resulting reformulated gasoline to burn more clearly and produce less carbon monoxide than the petroleum fuels containing non-oxygenate antiknock agents. On the negative side, the burning of this reformulated gasoline contributes to an increase in the atmospheric concentration of formaldehyde (Section 12.10), the inhalation of which has been linked with nasal cancer in animals.

In the late 1990s, the presence of MTBE was identified in a drinking water source in California, where its origin was linked to leaking gasoline storage tanks. When it is dissolved in water, even at very low concentrations, MTBE makes the water smell and taste foul. To ensure that our drinking water supplies were simultaneously palatable and unlikely to have harmful constituents, EPA has scheduled the elimination of MTBE as a gasoline additive.

ALKYL ETHERS OF ETHYLENE GLYCOL

The molecules of mono- and dialkyl ethers of ethylene glycol possess the functional groups of ethers and alcohols, —O— and —OH, respectively. Several of the related compounds are commercially known by the trademark *Cellosolve*, some examples of which are noted in Table 12.17. They are used as solvents. For example, ethylene glycol monoethyl ether is a common solvent used in the printing industry. Several are also components of cleaning fluids.

Scientists have linked the inhalation of the vapors of alkyl ethers of ethylene glycol with an increase in miscarriages. For this reason, pregnant women should avoid the use of products containing these organic compounds.

POLYCHLORINATED DIBENZOFURANS AND POLYCHLORINATED DIBENZO-*p*-DIOXINS

These substances are respectively the chlorinated derivatives of the heterocyclic compounds (Section 12.4) dibenzofuran and dibenzo-*p*-dioxin.

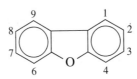

dibenzofuran   dibenzo-*p*-dioxin

TABLE 12.17 ◆ Some Common Ethers of Ethylene Glycol

Trademark	Name	Chemical Formula
Butyl Cellosolve	Ethylene glycol monobutyl ether (2-butoxyethanol)	C_4H_9—O—CH_2CH_2OH
Butyl Cellosolve acetate	Ethylene glycol monobutyl ether acetate	CH_3COO—CH_2CH_2—O—C_2H_9
Cellosolve acetate	Ethylene glycol monoethyl ether acetate	CH_3COO—CH_2CH_2—O—C_2H_5
Cellosolve solvent	Ethylene glycol monoethyl ether (2-ethoxyethanol)	C_5H_5—O—CH_2CH_2OH
Dibutyl Cellosolve	Ethylene glycol dibutyl ether	C_4H_9—O—CH_2CH_2—O—C_4H_9
Methyl Cellosolve	Ethylene glycol monomethyl ether (2-methoxyethanol)	CH_3—O—CH_2CH_2OH
Methyl Cellosolve acetate	Ethylene glycol monomethyl ether acetate	CH_3COO—CH_2CH_2—O—CH_3
Phenyl Cellosolve	Ethylene glycol monophenyl ether (2-phenoxyethanol)	(phenyl)—O—CH_2CH_2OH

Using the indicated numbering system, we shall be primarily interested in the compounds having chlorine atoms bonded to the benzene rings at the 2, 3, 7, and 8 positions. The polychlorinated dibenzofurans and polychlorinated dibenzo-*p*-dioxins that possess at least four chlorine atoms bonded to the benzene rings at the 2, 3, 7, and 8 positions are associated with high degrees of toxicity.

For convenience, we shall designate the polychlorinated dibenzofurans and polychlorinated dibenzo-*p*-dioxins as PCDFs and PCDDs, respectively. There are 135 PCDFs and 75 PCDDs. The PCDF and PCDD having the following molecular structures possess the highest degree of toxicity:

2,3,7,8-tetrachlorodibenzofuran   2,3,7,8-tetrachlorodibenzo-*p*-dioxin

2,3,7,8-Tetrachlorodibenzo-*p*-dioxin is commonly denoted as *dioxin,* or *TCDD.*

The PCDFs and PCDDs have never been intentionally produced or manufactured, but they are generated as unwanted by-products during certain uncontrolled incineration, paper pulp bleaching, and chemical manufacturing operations. The latter operations include the manufacture of trichlorophenol, tetrachlorophenol, pentachlorophenol, and 2,4,5-trichlorophenoxyacetic acid.

Interest in dioxin was first stimulated by public health officials when the substance was identified as a trace contaminant in certain herbicides formerly used as weed and brush killers. These herbicides had been primarily used by the U.S. military during the Vietnam encounter to defoliate jungle terrain and remove cover for the enemy. Approximately 18 million gal (68,100 m^3) of one such substance, *Agent Orange,* was sprayed on southern Vietnam. This herbicide was a 50:50 mixture of 2,4-dichlorophenoxyacetic acid and dioxin-contaminated 2,4,5-trichlorophenoxyacetic acid.

2,4-dichlorophenoxyacetic acid   2,4,5-trichlorophenoxyacetic acid

Military personnel and local civilians who became exposed to this herbicide subsequently contracted horrific diseases. In addition, their children were at times born with serious birth defects. Researchers have now linked Agent Orange exposure in Vietnam veterans with the inception of 10 different diseases, as well as spina bifida (a congenital birth defect) in their offspring. These diseases include cancers of the bronchus, larynx, lung, prostate, and trachea.

Epidemiologists have conducted studies to characterize the high level of toxicity associated with dioxin. These studies show that dioxin, like the PCBs, concentrates in the fatty tissues of humans and animals; also like the PCBs, the dioxin functions adversely as an endocrine disrupter by upsetting the natural order of the neurological, immunological, and reproductive systems. The studies also note that humans who have been exposed to dioxin often suffer from skin diseases, muscle dysfunctions, and nervous system disorders, and that their offspring could be born with birth defects. Since dioxin causes immune-system disorders, the natural defenses toward contracting other diseases are weakened when the host organism has been exposed to dioxin.

Even today, there are ways in which exposure to PCDFs and PCDDs is possible. In Section 12.7, we noted that these substances are generated during certain electrical-equipment fires. In addition, they are generated as products of the incomplete combustion of chlorinated plastics such as poly(vinyl chloride) (PVC) (Section 13.6). It is under such circumstances that emergency responders are at risk from exposure to these highly toxic substances.

The U.S. Department of Health and Human Services has ranked dioxin as a known human carcinogen and the remaining PCDDs and PCDFs with chlorine atoms in the 2, 3, 7, and 8 positions as probable human carcinogens. To avoid adverse health effects from dioxin exposure, EPA has designated 6.4 femtograms per kilogram of body weight as the acceptable daily dosage for humans. [One femtogram is one quadrillionth (10^{-15}) of a gram.]

POLYBROMINATED DIBENZOFURANS AND POLYBROMINATED DIBENZO-*p*-DIOXINS

The polybrominated derivatives of dibenzofuran and dibenzo-*p*-dioxin are also toxic substances. Like the PCDFs and PCDDs, the compounds associated with elevated toxicities are those whose molecules possess at least four bromine atoms bonded to the benzene rings in the 2, 3, 7, and 8 positions. The polybrominated dibenzofurans and polybrominated dibenzo-*p*-dioxins with bromine atoms in the 2, 3, 7, and 8 positions are regarded as probable human carcinogens.

Emergency responders are at risk from exposure to polybrominated dibenzofurans and polybrominated dibenzo-*p*-dioxins. Brominated compounds are used as fire retardants within certain commercial plastic formulations. When these plastics undergo incomplete combustion, they are assumed to release polybrominated dibenzofurans and polybrominated dibenzo-*p*-dioxins to the surroundings. This supposition could place firefighters and others at risk from exposure.

Despite the potential for production of these toxic compounds, brominated fire retardants remain in use in plastic formulations. Their use will most likely continue until manufacturers identify alternative compounds that provide a commensurate degree of fire safety.

Performance Goals for Section 12.10
- ◆ Identify the functional groups that characterize aldehydes and ketones.
- ◆ Describe the rules for naming simple aldehydes and ketones.
- ◆ Describe the hazardous properties of formaldehyde and acetone.
- ◆ Identify the industries that use aldehydes and ketones commercially.

◆ 12.10 ALDEHYDES AND KETONES

Aldehydes and *ketones* are organic compounds whose molecules possess the carbonyl group of atoms ($\overset{\backslash}{\underset{/}{C}}{=}O$). In aldehydes, the carbonyl group is located at the end of a chain of carbon atoms, whereas in ketones, it is located at a nonterminal position.

Thus, aldehydes have the general chemical formula R—CHO, whereas ketones have the formula R—CO—R′. R and R′ are arbitrary alkyl or aryl groups.

In the common system, an aldehyde is named by combining the word "aldehyde" with the prefix of the name of the acid to which it is oxidized. The aldehydes having one, two, three, and four carbon atoms per molecule are named formaldehyde, acetaldehyde, propionaldehyde, and *n*-butyraldehyde, respectively, since they become formic acid, acetic acid, propionic acid, and *n*-butyric acid when oxidized.

$$
\begin{array}{cccc}
\overset{\displaystyle O}{\underset{\displaystyle H}{\overset{\|}{H-C}}} &
\overset{\displaystyle O}{\underset{\displaystyle H}{\overset{\|}{CH_3-C}}} &
\overset{\displaystyle O}{\underset{\displaystyle H}{\overset{\|}{CH_3CH_2-C}}} &
\overset{\displaystyle O}{\underset{\displaystyle H}{\overset{\|}{CH_3CH_2CH_2-C}}} \\
\text{formaldehyde} & \text{acetaldehyde} & \text{propionaldehyde} & \textit{n}\text{-butyraldehyde}
\end{array}
$$

In the IUPAC system, an aldehyde is named by replacing the -*e* in the name of the corresponding hydrocarbon with -*al*. The four aldehydes previously noted are then named methanal, ethanal, propanal, and butanal, respectively, by replacing the -*e* in methane, ethane, propane, and butane with -*al*.

In the common system, a ketone is named by identifying the alkyl or aryl groups of which the substance is composed. Thus, the compound having the following formula is named methyl ethyl ketone.

$$
\underset{\displaystyle O}{\overset{}{CH_3-\overset{\|}{C}-CH_2CH_3}}
$$

In the IUPAC system, a ketone is named by replacing the -*e* in the name of the corresponding hydrocarbon with -*one;* then, the chain of continuous carbon atoms is numbered so as to assign the *smallest possible number* to the carbonyl group. Hence, the IUPAC name for methyl ethyl ketone is 2-butanone.

DOT regulates the transportation of benzaldehyde and two formaldehyde solutions as a Class 9 material, flammable liquid, and corrosive material, respectively. DOT requires shippers and carriers to affix CLASS 9, FLAMMABLE LIQUID, and/or CORROSIVE labels, as relevant, on their packaging and, when warranted, to post CLASS 9, FLAMMABLE, or CORROSIVE placards on the bulk packaging, freight container, transport vehicle, unit containment device, or railcar used to transport them. DOT regulates the remaining aldehydes and ketones in Table 12.18 as flammable liquids. DOT requires shippers and carriers to affix FLAMMABLE LIQUID labels on their packaging and, when warranted, to post FLAMMABLE placards on the bulk packaging, freight container, transport vehicle, unit containment device, or railcar used to transport them.

Solved Exercise 12.16

Identify the functional group in the compound having the following condensed formula, and name the compound.

$$
\underset{\displaystyle O \quad CH_3}{\overset{}{CH_3CH_2-\overset{\|}{C}-\overset{|}{C}HCH_2CH_3}}
$$

Solution: By examination of Table 12.11, we determine that the functional group represented as C=O is characteristic of the group of compounds called ketones. By assigning a number to each carbon atom in the longest continuous chain of carbon atoms containing the carbonyl group, we see that this particular ketone is a "hexanone." The name is derived by replacing the -*e* with -*one* in hexane, the alkane having six carbon atoms per molecule.

$$\overset{1}{C}H_3\overset{2}{C}H_2-\overset{3}{C}-\overset{4}{C}H\overset{5}{C}H_2\overset{6}{C}H_3$$
$$\underset{O}{\overset{\|}{}}\quad\underset{CH_3}{\overset{|}{}}$$

The carbon atom numbered 3 is a component of the carbonyl group, and the carbon atom numbered 4 is bonded to a methyl group. Consequently, in the IUPAC system, the correct name of this compound is 4-methyl-3-hexanone.

It is incorrect to number the longest chain of continuous carbon atoms from right to left, as follows, since a higher number (4) would then be assigned to the carbonyl group.

$$\overset{6}{C}H_3\overset{5}{C}H_2-\overset{4}{C}-\overset{3}{C}H\overset{2}{C}H_2\overset{1}{C}H_3$$
$$\underset{O}{\overset{\|}{}}\quad\underset{CH_3}{\overset{|}{}}$$

In the common system, the compound is named ethyl *sec*-butyl ketone.

FORMALDEHYDE

formalde-
hyde, 37%,
methanol

The simplest aldehyde is formaldehyde, or methanal. Its chemical formula is HCHO.

$$H-\overset{\overset{\textstyle O}{\|}}{C}_{\textstyle \diagdown H}$$

Formaldehyde is a colorless, flammable gas at room conditions, possessing a pungent, highly irritating odor.

In the presence of water vapor, formaldehyde produces a solid substance called paraformaldehyde. This substance possesses an indefinite composition. Its chemical formula is $HO(CH_2O)_nH$, where *n* ranges from approximately 8 to 100.

formalde-
hyde aque-
ous solution

$$_nH-\overset{\overset{\textstyle O}{\|}}{C}_{\textstyle \diagdown H(g)} + H_2O(g) \longrightarrow HO-[-(CH_2)_n-O-]-H(s)$$

formaldehyde   water   paraformaldehyde

Because of its pronounced chemical reactivity, formaldehyde gas is unavailable commercially. On the other hand, formaldehyde solutions are widely available. Formaldehyde is very soluble in water and alcohol, producing a corrosive solution whose properties vary with the formaldehyde concentration. The solution called *form-alin* is an approximately 40% aqueous solution containing 5 to 12% methanol. It is

TABLE 12.18 ◆ Some Aldehydes and Ketones Regulated by DOT for Transport

Proper Shipping Name	Label Codes
Acetaldehyde	3
Acetone	3
Aldehydes, n.o.s.	3
Aldehydes, toxic, n.o.s.	3, 6.1
Amyl methyl ketone	3
Benzaldehyde	9
Butyraldehyde	3
Cyclohexanone	3
Diethyl ketone	3
Diisobutylketone	3
2-Ethylbutryaldehyde	3
Formaldehyde solutions	3, 8
Formaldehyde solutions, *with not less than 25% formaldehyde*	8
n-Heptaldehyde	3
Isobutryaldehyde	3
Ketones, liquid, n.o.s.	3
Methyl ethyl ketone	3
Methyl isobutyl ketone	3
Propionaldehyde	3
Valeraldehyde	3

widely used as a disinfecting, sterilizing, and embalming agent. Formalin containing 37% water and 15% methanol flashes at 122°F (50°C), while an aqueous solution of formaldehyde without methanol flashes at 185°F (85°C).

For industrial use, formaldehyde may be generated as the gas from the following sources:

◆ *Paraformaldehyde.* When it is heated, paraformaldehyde decomposes into formaldehyde and water.

$$HO-[-(CH_2)_n-O-]-H(s) \longrightarrow nH-\overset{\overset{O}{\|}}{C}-H(g) + H_2O(g)$$

paraformaldehyde   formaldehyde   water

DOT regulates the transportation of paraformaldehyde as a flammable solid. DOT requires shippers and carriers to affix FLAMMABLE SOLID labels to its packaging and, when warranted, to post FLAMMABLE SOLID placards on the bulk packaging, freight container, transport vehicle, unit containment device, or railcar used to transport it.

♦ *sym-Trioxane.* This substance is a combustible solid whose molecular structure resembles three united formaldehyde molecules. When mixed with a strong acid, *sym*-trioxane decomposes into formaldehyde.

sym-trioxane   formaldehyde

DOT does not regulate the transportation of *sym*-trioxane as a hazardous material.

Aside from its intentional production, formaldehyde is released into the air during certain combustion processes. Aldehydes are produced as incomplete combustion products when organic substances burn within confined areas. Materials that produce formaldehyde upon burning include tobacco, wood, diesel fuel, kerosene, and natural gas. Researchers have identified aldehydes as the major class of organic compounds comprising diesel engine exhaust, with formaldehyde being the most abundant among them. Individuals who are required to work with diesel-powered equipment are at the highest risk to formaldehyde (and PAH) exposure.

Exposure to formaldehyde can also occur within newly constructed residential and commercial buildings, especially when the inside air is not regularly vented outdoors. The presence of formaldehyde is especially prominent within newly constructed mobile homes. There are at least four potential sources from which formaldehyde can leach within a mobile home. They are the following:

♦ The plastic used to manufacture kitchen and bathroom cabinets is often produced from formaldehyde.
♦ The glue or adhesive in particleboard, hardwood plywood, and fiberboard is often produced from formaldehyde.
♦ The foam insulation used in mobile homes is often manufactured from formaldehyde.
♦ The synthetic carpets installed in mobile homes are often manufactured from formaldehyde.

Figure 12.21 shows that low concentrations of formaldehyde often leach from these sources into the surrounding environment. The rate of release decreases as the age of these products increases and depends upon the prevailing temperature and humidity. Formaldehyde is released into the air at the fastest rate when the temperature and humidity are elevated.

The concern about formaldehyde exposure is linked with the ill effects it causes. When present in the air at a concentration of only 4 to 5 ppm, formaldehyde causes the eyes to tear. When higher concentrations are inhaled, formaldehyde causes respiratory problems ranging from mild irritation of the throat and bronchial passageways to the more severe ailment bronchopneumonia. Exposure to formaldehyde also produces localized sores within the nose, throat, and lungs and causes allergic dermatitis in susceptible individuals.

In research studies conducted on animals, the inhalation of formaldehyde gas has been shown to cause nasal cancer. Based upon these results, epidemiologists rank formaldehyde as a probable human carcinogen by inhalation.

When formaldehyde is present in the workplace, OSHA requires employers to limit employee exposure to a concentration of 0.75 ppm in air, averaged over an 8-hour workday.

FIGURE 12.21 ◆ In a mobile home, formaldehyde very slowly leaches from (1) kitchen cabinets; (2) glue and adhesives used in particleboard, hardwood plywood, and fiberboard; (3) foam insulation; (4) synthetic carpeting. When the home is unvented or highly insulated, formaldehyde becomes sufficiently elevated in concentration to cause respiratory discomfort in sensitive individuals.

ACETONE

The simplest ketone is acetone, or propanone. Its chemical formula is $(CH_3)_2CO$.

$$CH_3 - \underset{\underset{O}{\|}}{C} - CH_3$$

Acetone is a colorless, water-soluble, and highly volatile liquid with a sweet odor. Since its flashpoint is 15°F (−9°C) with a flammable range from 2.6% to 12.8% by volume, the presence of acetone poses a dangerous fire and explosion risk.

Acetone is largely utilized commercially as a solvent in varnishes, lacquers, paints, and fingernail polish remover and as a raw material for the manufacture of methyl methacrylate (Section 13.6), methyl isobutyl ketone (see below), and other substances.

OTHER KETONES

Other simple ketones possess physical properties similar to those of acetone. Two commercially important ketones are methyl ethyl ketone and methyl isobutyl ketone.

$$CH_3 - \underset{\underset{O}{\|}}{C} - CH_2CH_3$$

methyl ethyl ketone
2-butanone

$$CH_3 - \underset{\underset{O}{\|}}{C} - CH_2 - \underset{\underset{CH_3}{|}}{CH} - CH_3$$

methyl isobutyl ketone
4-methyl-2-pentanone

In commerce, they are known more widely by their acronyms, *MEK* and *MIBK*, respectively. Both are flammable, water-soluble, highly volatile liquids at room conditions. They are commonly encountered as solvents or constituents of solvent mixtures, especially those used in solvent-based paints. Their presence constitutes a dangerous fire risk.

Performance Goals for Section 12.11
 + Identify the functional group that characterizes organic acids.
 + Describe the hazardous properties of the simple organic acids.
 + Describe the rules for naming them.
 + Identify the industries that use organic acids commercially.

◆ 12.11 ORGANIC ACIDS

Organic acids are organic compounds containing the carboxyl group, —COOH; hence, they are sometimes called *carboxylic acids*. The general chemical formula of the organic acids is R—COOH, where R is an arbitrary alkyl or aryl group. These simple organic acids have the common names they received historically. In the IUPAC system, they are named by replacing the final *-e* in the name of the corresponding alkane with *-oic acid*.

The organic acids having from one to five carbon atoms per molecule are usually encountered by their common names. The common name of the organic acid having one carbon atom per molecule is formic acid. Its chemical formula is HCOOH.

$$H-\overset{\displaystyle \overset{O}{\|}}{\underset{\displaystyle OH}{C}}$$

The IUPAC name is methanoic acid. The common name of the organic acid having two carbons per molecule is acetic acid. Its chemical formula is CH_3COOH.

$$CH_3-\overset{\displaystyle \overset{O}{\|}}{\underset{\displaystyle OH}{C}}$$

Its IUPAC name is ethanoic acid. Some properties of acetic acid were previously noted in Section 8.11.

The common and IUPAC names of the organic acids having three, four, and five carbon atoms per molecule are listed below:

$$CH_3CH_2-\overset{\displaystyle \overset{O}{\|}}{\underset{\displaystyle OH}{C}} \qquad CH_3CH_2CH_2-\overset{\displaystyle \overset{O}{\|}}{\underset{\displaystyle OH}{C}} \qquad CH_3CH_2CH_2CH_2-\overset{\displaystyle \overset{O}{\|}}{\underset{\displaystyle OH}{C}}$$

propionic acid   *n*-butyric acid   valeric acid (common)
propanoic acid   butanoic acid   pentanoic acid (IUPAC)

Like acetic acid, these simple organic acids are colorless, water-soluble liquids with characteristic odors. Formic acid, acetic acid, and propionic acid possess pungent but not disagreeable odors, but the odors of butyric acid and valeric acids are highly disagreeable.

TABLE 12.19 ◆ Some Organic Acids Regulated by DOT for Transport

Proper Shipping Name	Label Codes
Acetic acid, glacial *or* Acetic acid solution *with more than 80% by mass*	8, 3
Butyric acid	8
Caproic acid	8
Corrosive liquid, acidic, organic, n.o.s.	8
Formic acid	8
Isobutyric acid	3, 8
Propionic acid	8

The simplest aromatic acid is called benzoic acid. Its chemical formula is C_6H_5COOH.

Benzoic acid is a white solid.

Acetic acid and isobutyric acid are the only organic acids that pose a risk of fire and explosion. DOT regulates their transportation as a corrosive liquid and flammable liquid, respectively. DOT requires shippers and carriers to affix CORROSIVE and FLAMMABLE LIQUID labels on their packaging and, when warranted, to post either CORROSIVE or FLAMMABLE placards, as relevant, on the bulk packaging, freight container, transport vehicle, unit containment device, or railcar used to transport them.

DOT regulates the transportation of the organic acids other than isobutyric acid in Table 12.19 as corrosive materials. DOT requires shippers and carriers to affix CORROSIVE labels on their packaging and, when warranted, to post CORROSIVE placards on the bulk packaging, freight container, transport vehicle, unit containment device, or railcar used to transport them.

Performance Goals for Section 12.12
- ◆ Identify the functional group that characterizes esters.
- ◆ Describe the rules for naming simple esters.
- ◆ Describe the hazardous properties of ethyl acetate and di(2-ethylhexyl)phthalate.
- ◆ Identify the industries that use esters commercially.

◆ **12.12 ESTERS**

Organic compounds having the general chemical formula R—CO—OR′ are called *esters*. An ester is the compound produced when an organic acid reacts with an alcohol. This type of chemical reaction is called *esterification*. Esters are named by changing the *-ic* suffix of the organic acid to *-ate*, preceded by the name of the alkyl or aryl group of the alcohol.

TABLE 12.20 ◆ Some Esters Regulated by DOT for Transport	
Proper Shipping Name	*Label Codes*
Amyl acetates	3
Amyl butyrates	3
Butyl acetates	3
n-Butyl formate	3
Cyclohexyl acetate	3
Esters, n.o.s.	3
Ethyl acetate	3
Ethyl butyrate	3
Ethyl formate	3
Isobutyl isobutyrate	3
Isopropyl acetate	3
Isopropyl propionate	3
Methyl acetate	3
Methyl formate	3
Methyl isovalerate	3
n-Propyl acetate	3

The simple esters are colorless, highly volatile liquids that are only partially soluble in water. Many possess pleasant, fruity odors. Esters occur naturally in apples (as ethyl-2-methyl butyrate), bananas (as isoamyl acetate), pineapples (as ethyl butyrate), and other fruits. The vapors of esters are responsible for the unique odors of fruits. Synthetic esters are used commonly as food additives.

DOT regulates the transportation of the esters in Table 12.20 as flammable liquids. DOT requires shippers and carriers to affix FLAMMABLE LIQUID labels on their packaging and, when warranted, to post FLAMMABLE placards on the bulk packaging, freight container, transport vehicle, unit containment device, or railcar used to transport them.

ETHYL ACETATE

This is a commonly encountered constituent of nitrocellulose-based lacquers and other protective coatings. It is produced by reacting ethanol and acetic acid, as follows:

$$CH_3CH_2OH(aq) \ + \ CH_3\overset{O}{\underset{OH}{\overset{\|}{C}}}(aq) \ \longrightarrow \ CH_3-\overset{O}{\underset{O-C_2H_5}{\overset{\|}{C}}}(aq) \ + \ H_2O(l)$$

ethanol   acetic acid   ethyl acetate   water

Ethyl acetate is a colorless liquid with a pleasing, fragrant odor. Since the flashpoint of ethyl acetate is only $24°F(-4°C)$, the presence of this substance constitutes a dangerous fire and explosion risk.

DI(2-ETHYLHEXYL) PHTHALATE

This ester is produced by reacting *o*-phthalic acid with 2-ethylhexanol, as follows:

COOH

⬡—COOH(l)   +   2CH$_3$(CH$_2$)$_3$CHCH$_2$OH(l) $\longrightarrow$

$\overset{\text{CH}_2\text{CH}_3}{\overset{|}{}}$

o-phthalic acid   2-ethylhexanol

COOCH$_2$CH(C$_2$H$_5$)C$_4$H$_9$

⬡—COOCH$_2$CH(C$_2$H$_5$)C$_4$H$_9$ +   H$_2$O

di-(2-ethylhexyl) phthalate   water

Di(2-ethylhexyl) phthalate is commonly denoted as its acronym *DEHP*.

DEHP is a component of various rubber and other polymeric formulations. Its function is to increase the flexibility and toughness of the related products. When used in this fashion, the DEHP is called a *plasticizer*. Since it does not chemically bond with polymers, DEHP is capable of leaching from them.

Health concerns have been raised over this leaching from medical products such as intravenous bags, tubing, and syringes. When blood is stored in plastic bags, DEHP leaches from the plastic and dissolves in the blood. This has been the cause of considerable apprehension, since DEHP has been shown to cause adverse changes in the reproductive systems of animals and to interfere with normal kidney and liver function.

Whether the alterations noted in animals could similarly occur in humans who are exposed to DEHP is a controversial topic. The U.S. Food and Drug Administration has concluded that patient exposures to DEHP are generally well below the levels that could be expected to cause adverse effects, although it concedes that children, especially infants, may represent a special population at increased risk. In recognition of this guarded view, the FDA has proposed voluntary restrictions on the use of DEHP-plasticized devices by pregnant women and newborn infants.

Solved Exercise 12.17

Why is it a prudent practice for pregnant women and nursing mothers to request the use of devices that do not contain DEHP during medical treatment?

Solution: Scientists have determined that exposure to DEHP is linked with health problems in animals, but they do not presently concur as to whether DEHP causes health problems in humans. Until this medical debate has been settled, the prudent practice for pregnant women and nursing mothers, when they are medically treated, is to request the use of devices that were manufactured without DEHP. The primary aim of this practice is to protect the health of the unborn and infants, who represent the groups at the highest risk from potential exposure to DEHP.

◆ 12.13 AMINES

The *amines* are organic derivatives of ammonia, compounds that contain one or more alkyl or aryl groups bonded to nitrogen. Their formulas amines are $R—NH_2$, R_2NH, and R_3N, where R is an arbitrary alkyl or aryl group.

$$R-N\begin{smallmatrix}H\\\\H\end{smallmatrix} \qquad R-N\begin{smallmatrix}H\\\\R\end{smallmatrix} \qquad R-N\begin{smallmatrix}R\\\\R\end{smallmatrix}$$

Simple amines are named by identifying the names of the alkyl or aryl group bonded to the nitrogen atom, and then adding the suffix *-amine*. For example, the names of the compounds having molecules in which one, two, and three ethyl groups are bonded to the nitrogen atom follow:

$$CH_3CH_2-N\begin{smallmatrix}H\\\\H\end{smallmatrix} \qquad CH_3CH_2-N\begin{smallmatrix}CH_2CH_3\\\\H\end{smallmatrix} \qquad CH_3CH_2-N\begin{smallmatrix}CH_2CH_3\\\\CH_2CH_3\end{smallmatrix}$$

ethylamine   diethylamine   triethylamine

When two or three identical groups are bonded to the nitrogen atom, they are sometimes named with the prefixes *bis* and *tris,* respectively, instead of *di* and *tri*.

Complex amines are also named as *N*-substituted derivatives. For example, the compounds having the following molecular structures are named as indicated:

$$CH_3-\underset{\underset{\displaystyle H}{|}}{N}-CH_2CH_2CH_3 \qquad CH_3-\underset{\underset{\displaystyle CH_3}{|}}{N}-CH_2CH_2CH_3$$

N-methyl-*n*-propylamine   *N,N*-dimethyl-*n*-propylamine

In the IUPAC system, the amines are named by replacing the *-e* in the name of the alkane with *-amine*. The simplest amines are also named as aminoalkanes, alkyl-aminoalkanes, and dialkylaminoalkanes. The compounds whose formulas are $CH_3CH_2CH_2NH_2$, $(CH_3)CH—NH_2$, and $(CH_3)_2CH—NHCH_3$ are named as follows:

$$CH_3CH_2CH_2-N\begin{smallmatrix}H\\\\H\end{smallmatrix} \qquad CH_3-\underset{\underset{\displaystyle NH_2}{|}}{C}-CH_3 \qquad \underset{\underset{\displaystyle CH_3}{|}}{CH_2}-\underset{\underset{\displaystyle CH_3}{|}}{CH}-N\begin{smallmatrix}H\\\\H\end{smallmatrix}$$

1-aminopropane   2-aminopropane   2-methylaminopropane

The common name of the simplest aromatic amine is aniline; its chemical formula is $C_6H_5—NH_2$. The substances having the chemical formulas $C_6H_5—NHCH_3$ and $C_6H_5—N(CH_3)_2$ are named *N*-methylaniline and *N,N*-dimethylaniline, respectively.

aniline *N*-methylaniline *N,N*-dimethylaniline

The amines are often used as coagulants, flocculating agents, corrosion inhibitors, bactericides, and fungicides. Among the commercially important amines, methylamine, dimethylamine, ethylamine, and trimethylamine are flammable gases at ambient conditions. The more complex amines are flammable and corrosive liquids. Their corrosivity is linked with the similarity in chemical behavior that they possess with ammonia. The liquid amines also possess the characteristic odor of rotten fish.

DOT regulates the transportation of the simple amines, some examples of which are provided in Table 12-21. DOT requires shippers and carriers to affix labels corresponding to the hazard codes of the hazardous materials on their packaging and,

TABLE 12.21 ◆ Some Amines Regulated by DOT for Transport

Proper Shipping Name	*Label Codes*
Amines, flammable, corrosive, n.o.s. *or* Polyamines, flammable, corrosive, n.o.s.	3, 8
Amines, liquid, corrosive, flammable, n.o.s. *or* Polyamines, liquid, corrosive, flammable, n.o.s.	8, 3
Amines, liquid, corrosive, n.o.s. *or* Polyamines, liquid, corrosive, n.o.s.	8
Amines, solid, corrosive, n.o.s., *or* Polyamines, solid, corrosive, n.o.s.	8
Amyl amines	8, 3
Aniline	6.1
n-Butylamine	3, 8
N-Butylaniline	6.1
Cyclohexylamine	8, 3
Di-*n*-amylamine	3, 6.1
Di-*n*-butylamine	8, 3
Diethylamine	3, 8
N,N-Diethylaniline	6.1
Diisobutylamine	3, 8
Dimethylamine, anhydrous	2.1
Dipropylamine	3, 8
Ethylamine	2.1
Methylamine	2.1
N-Methylbutylamine	3, 8
Tributylamine	8
Triethylamine	3, 8
Trimethylamine, anhydrous	2.1
Tripropylamine	3, 8

when warranted, to post placards appropriate to their hazard classes on the bulk packaging, freight container, transport vehicle, unit containment device, or railcar used to transport them.

Solved Exercise 12.18

Identify the combustion products that are produced when methylamine burns in the air.

Solution: The chemical formula of methylamine is CH_3NH_2, which indicates that each methylamine molecule is composed of carbon, hydrogen, and nitrogen atoms. When methylamine burns in the air, the carbon, hydrogen, and nitrogen atoms unite with atmospheric oxygen. Upon incomplete combustion, carbon monoxide, nitrogen monoxide, and water vapor are produced. Upon complete combustion, carbon dioxide, nitrogen dioxide, and water vapor are produced. These respective processes are denoted as follows:

$$4CH_3NH_2(g) + 9O_2(g) \longrightarrow 4CO(g) + 4NO(g) + 10H_2O(g)$$
methylamine   oxygen   carbon monoxide   nitric oxide   water

$$4CH_3NH_2(g) + 13O_2(g) \longrightarrow 4CO_2(g) + 4NO_2(g) + 10H_2O(g)$$
methylamine   oxygen   carbon dioxide   nitrogen dioxide   water

---

Performance Goals for Section 12.14

- Identify the functional groups that characterize hydroperoxides and organic peroxides.
- Describe the rules for naming these compounds.
- Identify the primary industry that uses hydroperoxides and organic peroxides commercially.
- Discuss the nature of the DOT regulations that uniquely apply to the transportation of peroxo-organic compounds.
- Describe the nature of the response actions to be taken when peroxo-organic compounds have been released into the environment.

◆ 12.14 PEROXO-ORGANIC COMPOUNDS

There are two classes of organic compounds whose molecules contain the peroxy group of atoms (—O—O—). They are called *peroxo-organic compounds*. In commerce, the most commonly encountered peroxo-organic compounds are the *organic hydroperoxides* and *organic peroxides*. The compounds in both classes are derivatives of hydrogen peroxide (Section 11.5), in which one or both of the hydrogen atoms in the H_2O_2 molecule have been substituted with alkyl or aryl groups. Their chemical formulas are H—O—O—R and R—O—O—R′, respectively.

An organic hydroperoxide is identified by the name of the alkyl or aryl group bonded to the H—O—O— group followed by the word "hydroperoxide." Examples are *tert*-butyl hydroperoxide and α,α-dimethylbenzyl hydroperoxide, whose chemical formulas are $(CH_3)_3$—C—O—OH and C_6H_5—C$(CH_3)_2$—O—OH, respectively.

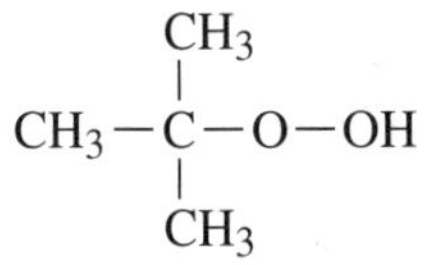

$$CH_3-\overset{\overset{\displaystyle CH_3}{|}}{\underset{\underset{\displaystyle CH_3}{|}}{C}}-O-OH$$

tert-butyl hydroperoxide

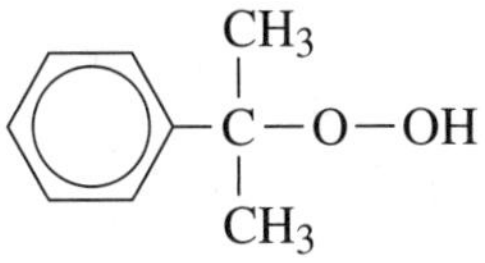

$$\langle\bigcirc\rangle-\overset{\overset{\displaystyle CH_3}{|}}{\underset{\underset{\displaystyle CH_3}{|}}{C}}-O-OH$$

α,α-dimethylbenzyl hydroperoxide
(cumene hydroperoxide)

An organic peroxide is identified by the name of the alkyl or aryl group bonded to the —O—O— group followed by the word "peroxide." When the alkyl or aryl group is the same, the name of the group is preceded with *di*. Examples are diacetyl peroxide and di-*tert*-butyl peroxide, whose chemical formulas are $(CH_3CO_2)_2$ and $[(CH_3)_3-C-]_2O_2$, respectively.

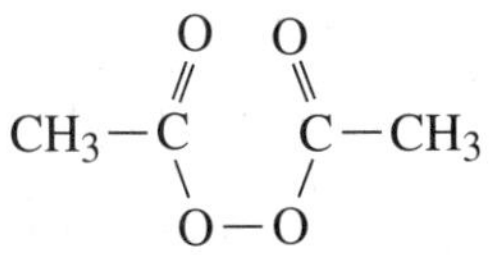

$$CH_3-\overset{\overset{\displaystyle O}{\parallel}}{C}\underset{\underset{\displaystyle O-O}{\diagdown\diagup}}{}\overset{\overset{\displaystyle O}{\parallel}}{C}-CH_3$$

diacetyl peroxide

$$CH_3-\overset{\overset{\displaystyle CH_3}{|}}{\underset{\underset{\displaystyle CH_3}{|}}{C}}-O-O-\overset{\overset{\displaystyle CH_3}{|}}{\underset{\underset{\displaystyle CH_3}{|}}{C}}-CH_3$$

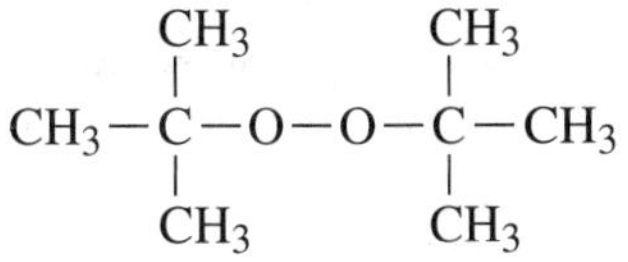

di-*tert*-butyl peroxide

Peroxo-organic compounds are commonly utilized as raw materials to induce polymerization (Section 13.2), a process essential to the production and manufacture of plastics. While they are either liquids or solids at room conditions, peroxo-organic compounds are often encountered dissolved in water or an appropriate organic solvent.

The peroxo-organic compounds are flammable substances, but they are also oxidizers containing active oxygen within their molecular structures. When they burn, peroxo-organic compounds support their own combustion. This combination of properties poses a particularly pronounced risk of fire and explosion. When ignited, the peroxo-organic compounds often burn furiously and more intensely than other combustible substances. This phenomenon is called *deflagration*. An example of deflagration is illustrated by the following equation:

$$[(CH_3)_3-C-]_2O_2(s) + 9O_2(g) \longrightarrow 3CO_2(g) + 5CO(g) + 9H_2O(g)$$

di–*tert*–butyl peroxide   oxygen   carbon dioxide   carbon monoxide   water

Many peroxo-organic compounds are unstable. When ignited, they are just as likely to undergo rapid, autoaccelerated decomposition as they are to burn. Some are so sensitive to friction, heat, or shock that they cannot be safely handled unless maintained at low temperatures and diluted within an inert solid material or solvent.

A typical peroxo-organic compound is dibenzoyl peroxide. The excerpts in Figure 12.22 from the Material Safety Data Sheet for this substance clearly set forth its hazardous features.

TRANSPORTING PEROXO-ORGANIC COMPOUNDS

DOT regulates the transportation of all peroxo-organic compounds as organic peroxides.* When these compounds are transported, DOT requires their shippers and carriers to affix ORGANIC PEROXIDE labels to their packaging and, when warranted, to post ORGANIC PEROXIDE placards on the bulk packaging, freight container, transport vehicle, unit containment device, or railcar used to transport them.

---

* The DOT regulations that apply uniquely to the transportation of organic peroxides are published at 49 CFR §173.225.

DIBENZOYL PEROXIDE

<u>FIREFIGHTING MEASURES</u>

Fire:

Flash point 104°F (40°C)
Autoignition temperature: 176°F (80°C)
Extremely flammable!
Substance is a strong oxidizer and a strong supporter of combustion. Its heat of reaction with combustibles and reducing agents can cause ignition.

Explosion:

Explosive!   Extremely explosion-sensitive to shock, heat, and friction.   May explode spontaneously when dry. Sensitive to mechanical impact. Sensitive to static discharge.

<u>HANDLING AND STORAGE</u>

Protect containers against physical damage. Isolate in a well-detached, fire-resistant, cool and well-ventilated building with no other materials stored therein. Shield containers from direct sunlight and maintain their temperature at less than 100°F (38°C). Employ grounding, bonding, venting, and explosion-relief provisions in accord with accepted engineering practices in any process capable of generating an explosion due to static discharge, shock, impact, heat, friction, or blows. Provide explosion-venting in a safe direction and prohibit any electrical installation or heating facilities. Dibenzoyl peroxide should be stored in and used from original containers. Do not return product that has been taken out of original container. Never mix unless at least 33% water is present. KEEP PURE! Impurities are hazardous in peroxides. Do not add accelerators. Containers of this material may be hazardous when empty since they retain product residues (dust, solids); observe all warnings and precautions listed for the product. DO NOT attempt to clean empty containers since residue is difficult to remove. Do not pressurize, cut, weld, braze, solder, grill, grind or expose such containers to het, sparks, flame, static electricity or other sources of ignition: They may explode and cause injury or death.

FIGURE 12.22 ◆ Two excerpts from the Material Safety Data Sheet for Dibenzoyl Peroxide: Firefighting Measures and Handling and Storage. *Adapted from the Material Safety Data Sheet for Benzoyl Peroxide, Mallinckrodt Baker, Inc., Phillipsburg, New Jersey.*

In connection with transporting organic peroxide, DOT identifies two temperatures:

- The *control temperature,* which is the temperature above which DOT prohibits the transportation an organic peroxide.
- The *emergency temperature,* which is the temperature at which DOT requires the implementation of emergency measures due to the imminent danger of explosion.

When shippers and carriers prepare the relevant shipping paper, DOT requires the entry of certain information with the basic description of an organic peroxide. The nature of this information is summarized below:

- When an organic peroxide is described by a generic type, DOT identifies the shipper or carrier to identify the generic type listed in Table 12.22, each designated by a capital letter, in the basic description. Two examples are the following:

TABLE 12.22 ◆ **Generic Types of Organic Peroxides**

Type A	This is an organic peroxide that can detonate or rapidly deflagrate as packaged for transport. The transportation of Type A organic peroxides is forbidden by DOT.
Type B	This is an organic peroxide that neither detonates nor deflagrates rapidly when correctly packaged for transportation, but it could undergo a thermal explosion.
Type C	This is an organic peroxide that neither detonates nor deflagrates rapidly when correctly packaged for transportation and cannot undergo a thermal explosion.
Type D	This is an organic peroxide that does one of the following: ◆ Detonates only partially, but does not deflagrate rapidly and is not affected by heat when confined ◆ Does not detonate, deflagrates slowly, and manifests no violent effect if heated when confined ◆ Does not detonate or deflagrate, and manifests a medium effect when heated under confinement
Type E	This is an organic peroxide that neither detonates nor deflagrates and manifests either a low or no effect when heated under confinement.
Type F	This is an organic peroxide that will not detonate in a cavitated state, does not deflagrate, manifests only a low or no effect if heated when confined, and possesses either a low or no explosive power.
Type G	This is an organic peroxide that will not detonate in a cavitated state, will not deflagrate at all, shows no effect when heated under confinement, and manifests no explosive power.

Organic peroxide type C, liquid, 5.2, UN3103, II

Organic peroxide type C, solid (dibenzoyl peroxide), 5.2, UN3104, II

◆ When the concentration of an organic peroxide affects the degree of hazard, DOT requires shippers and carriers to include the concentration or the concentration range in the basic description. For example, 50 lb (110 kg) of dibenzoyl peroxide may be transported with either of the following basic descriptions:

50 lb	Organic peroxide type B, solid, 5.2, UN3012, PG II (60% dibenzoyl peroxide)
50 lb	Organic peroxide type B, solid, 5.2, UN3102, PG II (50–72% dibenzoyl peroxide)

◆ When certain organic peroxides are offered for transportation, DOT requires shippers and carriers to include the control temperature and emergency temperature in their basic descriptions. An example of a basic description of an organic peroxide that warrants the inclusion of these temperatures is the following:

Diacetyl peroxide, 20%, 5.2, UN3115, II [Control temperature 68°F (20°C), Emergency temperature, 77°F (25°C)]

First-on-the-scene responders to a transportation mishap should be wary of the information provided by these basic descriptions. For example, in the case of diacetyl peroxide, they become aware that diacetyl peroxide explosively decomposes when its temperature exceeds 77°F (25°C).

RESPONDING TO INCIDENTS INVOLVING THE RELEASE OF PEROXO-ORGANIC COMPOUNDS

Emergency responders identify the presence of a peroxo-organic compound at a transportation mishap or similar emergency scene by observing any of the following:

- The number 5.2 as a component of a proper shipping description of a hazardous material on a shipping paper
- The words "ORGANIC PEROXIDE" on labels affixed to packaging of a hazardous material
- The words "ORGANIC PEROXIDE" on placards posted on the vehicle used to transport the hazardous material

Since many organic compounds are flammable gases or flammable liquids, their fires are fought using the general techniques noted in Sections 3.5 and 3.8. However, this statement does not apply ordinarily to fires involving the peroxo-organic compounds. Since they are potentially explosive substances, the generally recommended practice for fighting fires involving peroxo-organic compounds is to implement either of the following procedures:

- Use unmanned monitors to cool the area where these reactive materials are stored and assure that fire does not reach them.
- When a fire has engulfed the immediate area, evacuate and do not combat the fire.

These recommended procedures are illustrated in Figure 12.23.

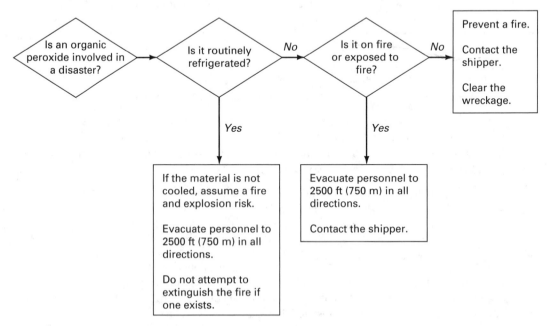

FIGURE 12.23 ◆ The recommended procedures when responding to a disaster involving an organic peroxide. *Adapted with permission of the American Society for Testing and Materials, from a figure in ASTM STP 825, A Guide to the Safe Handling of Hazardous Materials Accidents. Copyright © 1983, American Society for Testing and Materials.*

◆ 12.15 CARBON DISULFIDE

Carbon disulfide is a colorless, water-insoluble, and highly volatile liquid whose chemical formula is CS_2. Whereas pure carbon disulfide possesses a pleasant odor, the industrial grades of carbon disulfide are generally yellow and exhibit disagreeable odors. Some other physical properties of carbon disulfide are listed in Table 12.23.

Carbon disulfide is prepared by reacting methane with vaporized sulfur at 1200°F (650°C).

$$\underset{\text{methane}}{2CH_4(g)} + \underset{\text{sulfur}}{S_8(g)} \longrightarrow \underset{\text{carbon disulfide}}{2CS_2(g)} + \underset{\text{hydrogen sulfide}}{4H_2S(g)}$$

The compound is used commercially as a solvent and as a raw material for the manufacture of viscose rayon, cellophane, and other textiles.

The presence of carbon disulfide at a emergency scene constitutes cause for grave concern to firefighters. Since its flashpoint is −22°F (−30°C), carbon disulfide is a highly flammable liquid. Its flammable range extends from 1 to 44% by volume, and its vapor is 2.6 times heavier than air. The autoignition temperature of carbon disulfide is extremely low, 212°F (100°C). The ignition of carbon disulfide vapor may be initiated by its exposure to a hot steam pipe or an electric light bulb located a considerable distance away.

When carbon disulfide burns, sulfur dioxide and carbon monoxide form as products of incomplete combustion.

$$\underset{\text{carbon disulfide}}{2CS_2(g)} + \underset{\text{oxygen}}{5O_2(g)} \longrightarrow \underset{\text{sulfur dioxide}}{4SO_2(g)} + \underset{\text{carbon monoxide}}{2CO(g)}$$

To protect against inhaling these toxic substances, the use of self-contained breathing apparatus is essential when firefighters respond to major fires involving carbon disulfide.

The inhalation of carbon disulfide vapor is also harmful. The repeated inhalation of the vapor damages the liver and kidneys and permanently affects the central nervous system. The prolonged contact of carbon disulfide with the skin is also harmful, as carbon disulfide absorbs through the skin. In the most egregious case, the absorption is fatal.

When carbon disulfide is present in the workplace, OSHA requires workers to limit employee exposure to a vapor concentration of 20 ppm, averaged over an 8-hour workday.

TABLE 12.23 ◆ Physical Properties of Carbon Disulfide

Melting point	−169°F (−112°C)
Boiling point	115°F (46°C)
Specific gravity	1.26
Vapor density (air = 1)	2.6
Lower explosive limit	1%
Upper explosive limit	44%

DOT regulates the transportation of carbon disulfide as a flammable liquid. DOT requires shippers and carriers to affix FLAMMABLE LIQUID labels on their packaging and, when warranted, to post FLAMMABLE placards on the bulk packaging, freight container, transport vehicle, unit containment device, or railcar used to transport them.

Performance Goals for Section 12.16
- ◆ Describe the nature of chemical warfare agents.
- ◆ Identify the general types of chemical warfare agents.

◆ 12.16 CHEMICAL WARFARE AGENTS

Some substances are so highly toxic to humans that, during wartime, their use could serve to injure, incapacitate, or cause mass casualties among the enemy. They are called *chemical warfare agents*. Their use during wartime has been disdained by civilized countries as an inhumane military act, but in the past, mutual mistrust and fear prompted their production and stockpiling (Figure 12.24). Signatories of the 1997 Chemical Weapons Convention Treaty are now prohibited from producing or stockpiling chemical warfare agents.

An example of a chemical warfare agent is chlorine (Section 7.4). Since it is poisonous and possesses a vapor density of 2.49, the discharge of chlorine during wartime activities could cause mass casualties by maintaining a ground-level vapor concentration capable of perpetrating death among the enemy.

Most substances potentially useful as chemical warfare agents are liquids that can be sealed into canisters or charged into missile warheads and other explosive devices. They vaporize when released, readily providing a lethal vapor concentration.

NERVE GASES

Although referred to as gases, these substances are volatile liquids at room conditions. The most well-known are complex organophosphorus compounds. When their vapors are inhaled, they cause the nerves to stop transmitting impulses to nearby muscles. This paralyzes the muscles, which in turn causes convulsions and organ failure. The nerve gases thus act as neurotoxicants.

Although the U.S. military has produced and stockpiled nerve gases, their actual use has never been authorized. On the other hand, the nerve gas sarin *was* used by Iraq in its war against Iran, as well as against its own citizens in 1988.

Several representative molecular formulas of nerve gases are noted below:

$$
\begin{array}{cc}
\underset{\substack{\text{sarin, GB, or isopropyl} \\ \text{methylphosphonofluoride}}}{
\begin{array}{c}
CH_3 \quad F \\
\backslash \; / \\
P \\
/\!/ \; \backslash \\
O \quad CH-CH_3 \\
\qquad | \\
\qquad CH_3
\end{array}}
&
\underset{\text{GE, or isopropyl ethylphosphonofluoridate}}{
\begin{array}{c}
CH_3CH_2 \quad F \\
\backslash \; / \\
P \\
/\!/ \; \backslash \\
O \quad CH-CH_3 \\
\qquad | \\
\qquad CH_3
\end{array}}
\end{array}
$$

Figure 12.24 ◆ These munitions containing nerve gas are stored in an igloo at the Anniston Army Depot in Anniston, Albama. They are scheduled for destruction into substances that are comparatively less hazardous by means of high-temperature incineration. *Courtesy of Anniston Army Depot and Anniston Chemical Activity, Anniston Alabama.*

$$CH_3 \quad O-CH_2CH_3$$

VX agent, or *O*-ethyl-*S*-(2-diisopropylamino-
ethyl)methyl phosphonothiolate

tabun, or ethyl *N,N*-dimethylphosphoramidocyanidate

VESICANTS

These are substances that, upon exposure to the body, blister skin and damage the eyes, mucous membranes, and respiratory tract.

The best-known vesicant is *mustard gas,* a yellowish-brown, oily liquid having a faint odor of garlic or mustard. Its chemical name is 2,2′-dichloroethyl sulfide.

$$ClCH_2CH_2-S-CH_2CH_2Cl$$

It can be poured on the ground, sprayed into the air, or loaded into artillery shells.

The grisly success of mustard gas as a vesicant is due in part to its low vapor density of 5.4 (air = 1). The vapor hovers for a relatively long period at ground level, where it harms the enemy by causing painful, long-lasting blisters.

The use of 2,2′-dichloroethyl sulfide against enemy troops during World War I is well documented. Discharged as a liquid, its vapor moved along the surface of the ground and into the trenches where soldiers were seeking a level of protection against live artillery. Here, the deadly vapor inflicted the maximum harm upon unsuspecting troops.

Other vesicants include compounds known in the military as the *nitrogen mustards.* These compounds are similar in molecular structure to 2,2′-dichloroethyl sulfide in that an amino nitrogen atom replaces the sulfur atom. There are three well-known nitrogen mustards: methyl bis(2-chloroethyl)amine, ethyl bis(2-chloroethyl)amine, and tris(2-chloroethyl)amine.

methyl bis(2-chloroethyl)amine

ethyl bis(2-chloroethyl)amine

tris(2-chloroethyl)amine

These compounds are liquids possessing especially low vapor densities, 5.9, 5.4, and 7.1 (air = 1), respectively.

BLOOD AGENTS

These are highly volatile liquids that cause seizures, respiratory failure, and cardiac arrest by interfering with the blood's ability to absorb atmospheric oxygen. An example of a blood agent is hydrocyanic acid (Section 10.8), whose presence in the bloodstream inhibits the effective utilization of oxygen at the cellular level.

PULMONARY AGENTS

These are also highly volatile liquids. When their vapors are inhaled, they damage the respiratory tract and cause pulmonary edema (Section 7.4). An example of a pulmonary agent is chloropicrin, whose chemical formula is CCl_3NO_2.

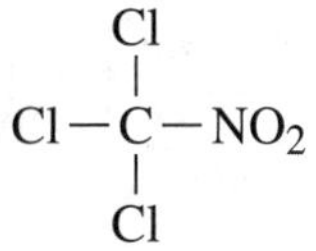

$$Cl-\underset{\underset{Cl}{|}}{\overset{\overset{Cl}{|}}{C}}-NO_2$$

Aside from its potential military use, chloropicrin has also been mixed with the vapors of odorless pesticides to aid in their detection.

Performance Goal for Section 12.17
   ◆ Describe the nature of a lacrimator.

◆ **12.17 LACRIMATORS**

We noted in Section 10.6 that lacrimators are substances that cause the eyes to involuntarily tear and close. Although lacrimators have been used against the enemy during warfare, they are also dispersed by law enforcement personnel to control unruly crowds and discourage their unlawful acts. In law enforcement and military actions, they are commonly known as *tear gases*. When used defensively, exposure to their vapor results in temporarily impaired vision, so the actions of activists and troops can be controlled. It is this *temporary* impact that causes the discriminate use of lacrimators by authorized individuals to be considered humane.

Some representative molecular formulas of lacrimators are shown below:

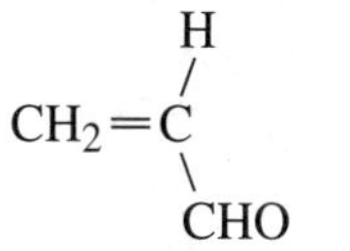

acrolein, or 2-propenal 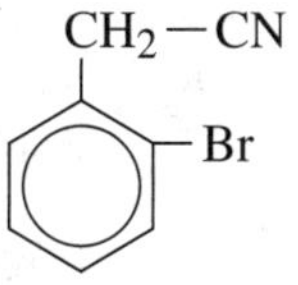   *o*-bromobenzyl cyanide 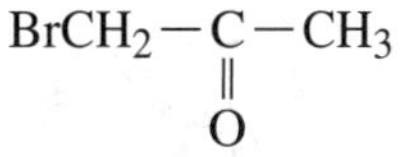   bromoacetone

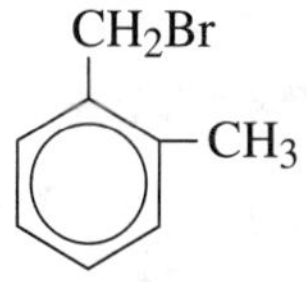

α-bromoxylene   α-chloroacetophenone 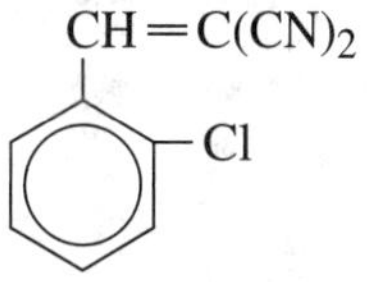   *o*-chlorobenzylidene malononitrile

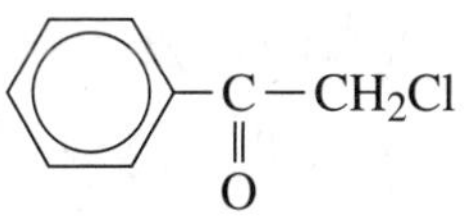

During World War I, *o*-bromobenzyl cyanide was extensively used as a lacrimator, most likely because uncontrollable tearing resulted from an exposure to a concentration in air of only 0.0000003 g/L (0.3×10^{-7} g/L).

Performance Goal for Section 12.18
   ◆ Describe the nature of an incendiary agent such as napalm.

◆ **12.18 INCENDIARY AGENTS**

Certain substances are regularly used during warfare to intentionally initiate fires. They are called *incendiary agents*. The use of triethylaluminum as an incendiary agent in flamethrowers was previously noted in Section 9.4.

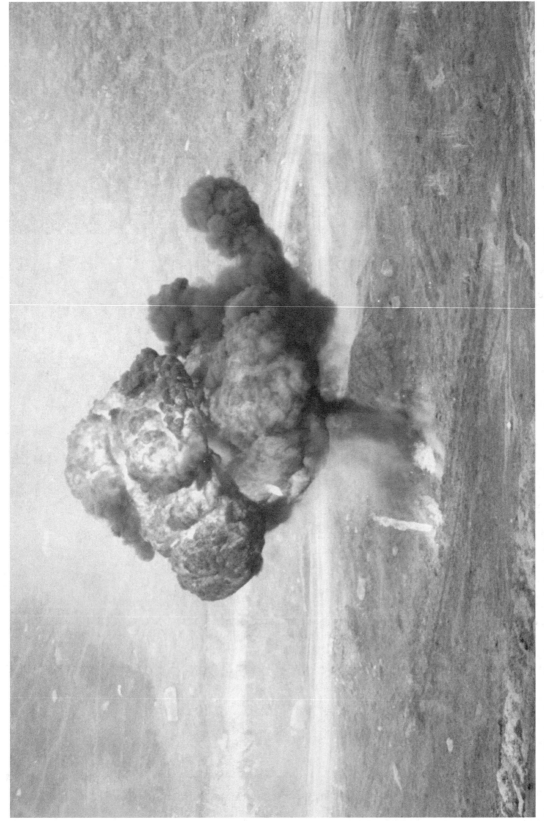

FIGURE 12.25 ◆ A napalm bomb explodes on Nightmare Range after being dropped from an A-10 Thunderbolt II aircraft during a live-fire exercise. *Courtesy of the United States Department of Defense.*

In warfare, several materials have been used to thicken petroleum products for use as incendiary agents. The most commonly known incendiary agent is *napalm*, a mixture of aluminum compounds made from several substances including those found naturally in coconut oil. The following are representative:

$$CH_3(CH_2)_8COO-CH_2 \qquad CH_3(CH_2)_{10}COO-CH_2 \qquad CH_3(CH_2)_{12}COO-CH_2$$
$$CH_3(CH_2)_8COO-CH \qquad CH_3(CH_2)_{10}COO-CH \qquad CH_3(CH_2)_{12}COO-CH$$
$$CH_3(CH_2)_8COO-CH_2 \qquad CH_3(CH_2)_{10}COO-CH_2 \qquad CH_3(CH_2)_{12}COO-CH_2$$

glyceryl tricaprate   glyceryl trilaurate   glyceryl trimyristate

$$CH_3(CH_2)_{14}COO-CH_2 \qquad\qquad CH_3(CH_2)_7CH=CH(CH_2)_7COO-CH_2$$
$$CH_3(CH_2)_{14}COO-CH \qquad\qquad CH_3(CH_2)_7CH=CH(CH_2)_7COO-CH$$
$$CH_3(CH_2)_{14}COO-CH_2 \qquad\qquad CH_3(CH_2)_7CH=CH(CH_2)_7COO-CH_2$$

glyceryl tripalmitate   glyceryl trioleate

The conversion of these substances into napalm is accomplished by their reaction with aluminum hydroxide and excess sodium hydroxide. The reaction of glyceryl tripalmitate is typical.

$$CH_3(CH_2)_{14}COO-CH_2 \qquad\qquad\qquad\qquad\qquad\qquad CH_2OH$$
$$3CH_3(CH_2)_{14}COO-CH(l) \;+\; Al(OH)_3(l) \longrightarrow Al[(CH_3(CH_2)_{14}COO]_3(s) + 3CHOH(l)$$
$$CH_3(CH_2)_{14}COO-CH_2 \qquad\qquad\qquad\qquad\qquad\qquad CH_2OH$$

glyceryl tripalmitate   aluminum hydroxide   aluminum palmitate   glycerol

During wartime, napalm is used to thicken gasoline until a mixture containing approximately 4% by volume is produced. The resulting material is jelly-like in consistency. It is used as the incendiary agent in fire bombs (Figure 12.25) and portable and mechanical flamethrowers.

When conducting military activities, the use of fuel thickened with napalm is associated with results provided by no other material. Napalm increases the range of flamethrowers, imparts slower burning properties, renders a "clinging" feature, and causes the burning flames to move around corners and rebound off walls and other surfaces.

During the Vietnam encounter, flamethrowers were used as formidable weapons. They controlled the extent of unwanted vegetation and routed the enemy from their hiding places in jungles and other densely vegetated terrains.

■ ■

REVIEW EXERCISES

Aliphatic Hydrocarbons

12.1 Determine whether each of the following chemical formulas represents an alkane, alkene, or alkyne: C_4H_8; $CH_3(CH_2)_4CH_3$; $CH_3(CH_2)_4-CH=CH-CH_2CH_3$; and $CH_3-C\equiv C-CH_3$.

12.2 Write the condensed chemical formula of each aliphatic hydrocarbon represented by the following carbon skeletons:

(a) C—C—C—C—C—C—C—C

(b)
$$C-C-C-C-C-C-C$$
$$\quad\quad\quad\quad\quad\quad\quad\;|$$
$$\quad\quad\quad\quad\quad\quad\quad C$$
$$\quad\quad\quad\quad\quad\quad\quad\;|$$
$$\quad\quad\quad\quad\quad\quad\quad C$$

(c)
$$C-C-C-C-C-C$$
$$\quad\;|\quad\quad\;|$$
$$\quad C\quad\quad C$$

(d)
$$C=C-C-C-C$$
$$\quad\quad\;|$$
$$\quad\quad\;C$$
$$\quad\quad\;|$$
$$\quad\quad C-C$$

12.3 Using the IUPAC system, name the alkanes having the following condensed formulas:

(a)
$$CH_3CH_2-CH-CH_2CH_3$$
$$\quad\quad\quad\quad\;|$$
$$\quad\quad\quad\quad CH_3$$

(b)
$$CH_3-CH-CH_2-CH-CH_3$$
$$\quad\quad\quad\;|\quad\quad\quad\quad\;|$$
$$\quad\quad\quad CH_2\quad\quad\; CH_3$$
$$\quad\quad\quad\;|$$
$$\quad\quad\quad CH_3$$

(c)
$$\quad\quad\quad\quad CH_2CH_2CH_3$$
$$\quad\quad\quad\quad\quad\;|$$
$$CH_3CH_2-CH-CH_2CH_3$$

(d)
$$\quad\quad\quad\quad\quad\quad\quad\quad CH_3$$
$$\quad\quad\quad\quad\quad\quad\quad\quad\;|$$
$$CH_3CH_2CH_2CH_2-C-CH_2CH_2CH_3$$
$$\quad\quad\quad\quad\quad\quad CH_3-C-CH_3$$
$$\quad\quad\quad\quad\quad\quad\quad\quad\;|$$
$$\quad\quad\quad\quad\quad\quad\quad\quad CH_3$$

(e)
$$\quad\quad\quad\quad\quad CH_3$$
$$\quad\quad\quad\quad\quad\;|$$
$$CH_3CH_2-C-H$$
$$\quad\quad\quad\quad\;|$$
$$\quad\quad\quad\quad CH_2CH_2-CH-CH_3$$
$$\quad\quad\quad\quad\quad\quad\quad\quad\;|$$
$$\quad\quad\quad\quad\quad\quad\quad\quad CH_3$$

(f)
$$\quad\quad\quad\quad CH_2CH_3$$
$$\quad\quad\quad\quad\;|$$
$$CH_3-CH-CH-CH_2CH_2CH_3$$
$$CH_3CH_2-C-H$$
$$\quad\quad\quad\quad\;|$$
$$\quad\quad\quad\quad CH_3$$

12.4 Using the IUPAC system, name the alkenes having the following condensed formulas:

(a)
$$CH_3\quad H$$
$$\quad\;\backslash\quad /$$
$$\quad\; C=C$$
$$\quad\;/\quad\;\backslash$$
$$\; H\quad\;\; H$$

(b)
$$CH_3CH_2\quad H$$
$$\quad\quad\;\backslash\quad /$$
$$\quad\quad\; C=C$$
$$\quad\quad\;/\quad\;\backslash$$
$$CH_3CH_2\quad H$$

(c)
$$CH_3CH_2CH_2\quad CH_2CH_2CH_2CH_3$$
$$\quad\quad\quad\;\backslash\quad\quad /$$
$$\quad\quad\quad\; C=C$$
$$\quad\quad\quad\;/\quad\;\backslash$$
$$\quad\quad\quad CH_3\quad H$$

(d)
$$CH_3\quad CH_2CH_3$$
$$\quad\;\backslash\quad /$$
$$\quad\; C=C$$
$$\quad\;/\quad\;\backslash$$
$$\; H\quad\;\; H$$

12.5 Using the IUPAC system, name the alkynes having the following condensed formulas:

(a) $H-C\equiv C-CH_3$

(b) $CH_3CH_2-C\equiv C-CH_2CH_3$

(c)
$$\quad\quad\quad CH_3$$
$$\quad\quad\quad\;|$$
$$CH_3-CH-CH_2-C\equiv C-CH_2CH_3$$

(d)
$$CH_3-CH-CH_2-C\equiv C-H$$
$$\quad\quad\;|$$
$$\quad\quad CH_2$$
$$\quad\quad\;|$$
$$\quad\quad CH_2$$
$$\quad\quad\;|$$
$$\quad\quad CH_2$$
$$\quad\quad\;|$$
$$\quad\quad CH_3$$

12.6 Write the condensed formulas for each of the following substances:

(a) propyne

(b) 2-pentene

(c) *n*-hexane

(d) acetylene

(e) 3-methylpentane

(f) cyclohexane

(g) isobutane

(h) 1,2-butadiene

(i) 2,3-dimethylbutane

(j) 2,3,3-trimethylpentane

(k) 3-hexyne

(l) 2,2-dimethylhexane

(m) *trans*-2-butene

(n) 2-ethyl-1-butene

(o) 2-methylcyclopentene

(p) 3-ethyl-3-methylheptane

(q) 3-ethylheptane

(r) cyclobutane

(s) 3-methyloctane

(t) methylcyclopentane

(u) isopentane

(v) 2,3-dimethyl-2-butene

(w) 4-isopropyl-2-methylheptane

(x) *trans*-4-methyl-2-pentene

(y) cyclopentadiene

(z) 1-heptene

(aa) 2,2-dimethylpentane

(ab) 2,2,4,4-tetramethylpentane

(ac) 2-methyl-1,3-butadiene

(ad) diphenylmethane

Gaseous Hydrocarbons

12.7 Use Table 2.5 to determine whether exhaust fans linked to sensing devices should correctly be positioned near the floors or the ceilings of buildings in which a storage tank containing each of the following flammable gases is located:

(a) acetylene

(b) natural gas

(c) propane

12.8 Why are natural deposits of methane and crude petroleum often located in near proximity?

12.9 Which label(s) does DOT require shippers and carriers to affix to containers of each of the following substances when offered for shipment:

(a) butane

(b) cyclobutane

(c) *n*-decane

(d) cryogenic ethane

(e) isononane

(f) isopentane

12.10 Why have most major cities enacted an ordinance requiring individuals to notify the local natural gas company *before* they conduct landscaping or other operations involving the digging of soil?

12.11 Why have many municipal transportation departments elected to power their city vehicles with natural gas rather than petroleum fuels?

12.12 The DOT regulation at 49 CFR §173.315(b)(1) stipulates that shippers or carriers must odorize liquefied petroleum gas by the addition of a small amount of a substance such as ethyl mercaptan, thiophane, or amyl mercaptan. What is the most likely reason DOT requires liquefied petroleum gas to be odorized?

Petroleum Products

12.13 Arrange the following petroleum products according to their increasing risk of fire and explosion at room temperature: asphalt, aviation gasoline, lubricating oil, and kerosene.

12.14 Why are trimethylpentanes, dimethylhexanes, and other branched-chain hydrocarbons commonly produced at petroleum refineries?

12.15 Why are trimethylpentanes and dimethylhexanes sometimes referred to as "isooctanes?"

12.16 Upon standard testing, *n*-butane is found to possess an "octane number" of 99. What does this mean?

Aromatic Hydrocarbons

12.17 Name the derivatives of benzene having the following chemical formulas:

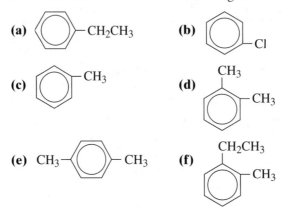

(a) $-CH_2CH_3$ **(b)** $-Cl$

(c) $-CH_3$ **(d)** CH_3 $-CH_3$

(e) CH_3- $-CH_3$ **(f)** CH_2CH_3 $-CH_3$

12.18 Write the molecular formula for each of the following compounds:
- **(a)** benzene
- **(b)** toluene
- **(c)** isopropylbenzene
- **(d)** *o*-ethyltoluene
- **(e)** diphenylacetylene
- **(f)** *m*-xylene
- **(g)** *n*-propylbenzene
- **(h)** *p*-diethylbenzene
- **(i)** 1,3,5-trimethylbenzene
- **(j)** benzyl bromide
- **(k)** 1,2,4,5-tetramethylbenzene

12.19 A chemical manufacturer in Houston, TX, intends to ship 5000 gal (19 m³) of ethylbenzene by rail to a customer in Corpus Christi, TX. Using Tables 6.1 and 6.4, what basic description does DOT require the manufacturer and carrier to enter on the accompanying shipping paper?

12.20 The OSHA regulation at 29 CFR §1910.1028 requires employers to limit employee exposure to a benzene vapor concentration of 1 ppm, averaged over an 8-hour workday. What is the most likely reason OSHA established this especially low worker-exposure limit?

12.21 Which PAHs listed in Table 12.10 generate nitrogen dioxide when subjected to complete combustion within an incinerator?

12.22 What is the most likely reason the U.S. Department of Health and Human Services denotes diesel exhaust particles as "reasonably anticipated to be human carcinogens"?

12.23 Why is occupational exposure to carcinogens for workers within the coal gasification industry likely to be elevated compared to the norm for workers in most other industries?

12.24 Traces of benzo[*a*]pyrene have been identified on the surface of charcoal-broiled meats. Identify the origin of this polynuclear aromatic hydrocarbon.

Functional Groups

12.25 Identify the functional groups in the following molecular structures:

(a) $CH_3CH_2-O-CH_2CH_3$ **(b)** $CH_3CH_2CH_2-\overset{\text{O}}{\underset{\|}{C}}-CH_3$

(c) 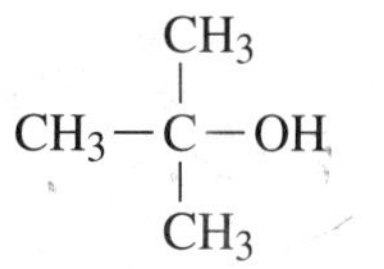 $CH_3-\overset{\overset{\displaystyle CH_3}{|}}{\underset{\underset{\displaystyle CH_3}{|}}{C}}-OH$

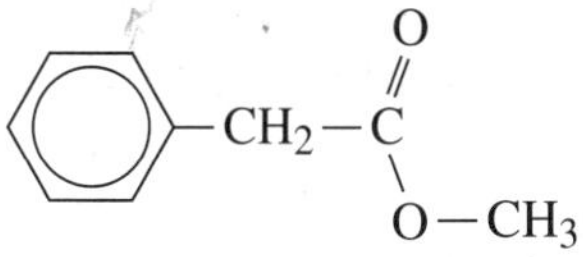

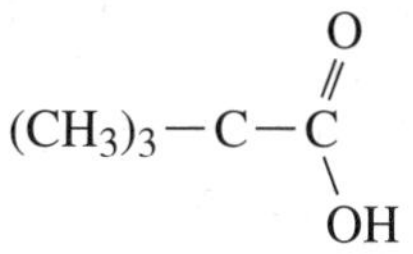

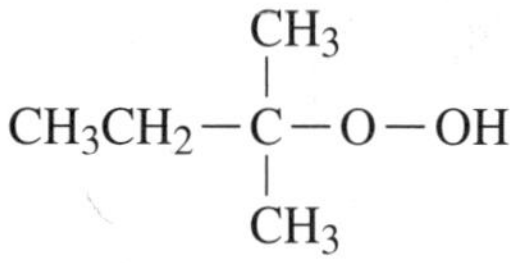

(d) 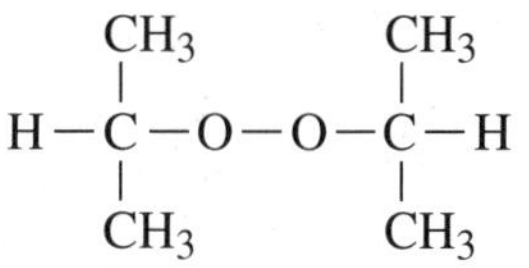 $H-\overset{\overset{\displaystyle CH_3}{|}}{\underset{\underset{\displaystyle CH_3}{|}}{C}}-O-O-\overset{\overset{\displaystyle CH_3}{|}}{\underset{\underset{\displaystyle CH_3}{|}}{C}}-H$

(e) ⟨benzene ring⟩$-CH_2-\overset{\overset{\displaystyle O}{\|}}{C}\diagdown_{O-CH_3}$

(f) $(CH_3)_3-C-\overset{\overset{\displaystyle O}{\|}}{C}\diagdown_{OH}$

(g) $CH_3CH_2-\overset{\overset{\displaystyle CH_3}{|}}{\underset{\underset{\displaystyle CH_3}{|}}{C}}-O-OH$

(h) ⟨benzene ring⟩$-\underset{\underset{\displaystyle O}{\|}}{C}-CH_3$

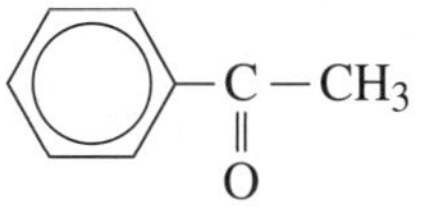

(i) 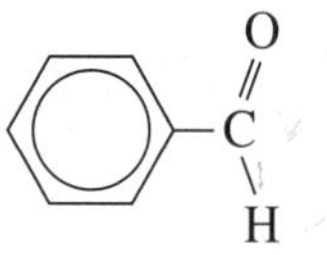 ⟨benzene ring⟩$-\overset{\overset{\displaystyle O}{\|}}{\underset{\underset{\displaystyle H}{\diagdown}}{C}}$

(j) 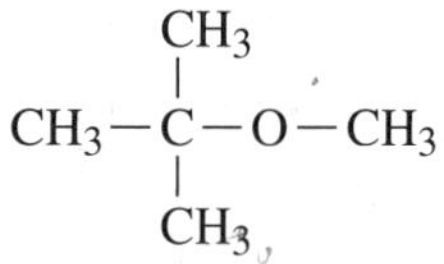 $CH_3-\overset{\overset{\displaystyle CH_3}{|}}{\underset{\underset{\displaystyle CH_3}{|}}{C}}-O-CH_3$

(k) 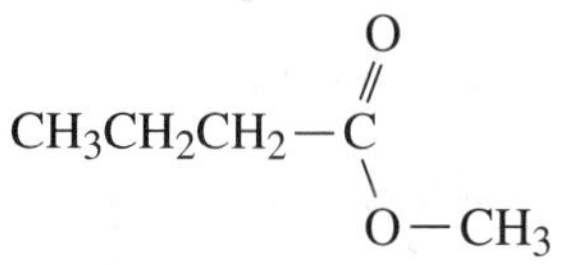 $CH_3CH_2CH_2-\overset{\overset{\displaystyle O}{\|}}{C}\diagdown_{O-CH_3}$

(l) 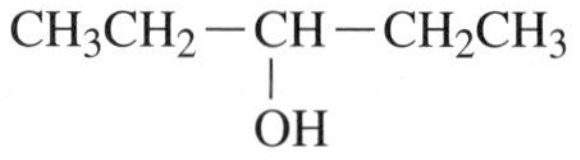 $CH_3CH_2-\overset{}{\underset{\underset{\displaystyle OH}{|}}{CH}}-CH_2CH_3$

Halogenated Hydrocarbons

12.26 Using the IUPAC system, name each compound having the following condensed formula:

(a) 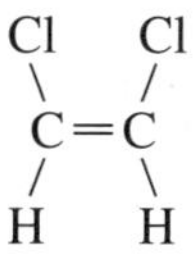 $\overset{\overset{\displaystyle Cl}{\diagdown}\quad\overset{\displaystyle Cl}{\diagup}}{\underset{\underset{\displaystyle H}{\diagup}\quad\underset{\displaystyle H}{\diagdown}}{C=C}}$

(b) 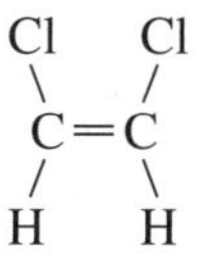 $\overset{\overset{\displaystyle Cl}{\diagdown}\quad\overset{\displaystyle Cl}{\diagup}}{\underset{\underset{\displaystyle H}{\diagup}\quad\underset{\displaystyle H}{\diagdown}}{C=C}}$

(c) CH_3CH_2Br

(d) 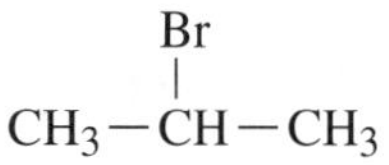 $CH_3-\overset{\overset{\displaystyle Br}{|}}{CH}-CH_3$

(e) 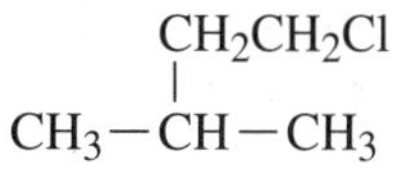 $CH_3-\overset{\overset{\displaystyle CH_2CH_2Cl}{|}}{CH}-CH_3$

(f) 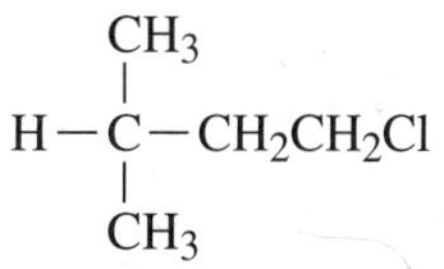 $H-\overset{\overset{\displaystyle CH_3}{|}}{\underset{\underset{\displaystyle CH_3}{|}}{C}}-CH_2CH_2Cl$

12.27 Write the condensed formulas for each of the following substances:

(a) chloroethane
(b) *cis*-2,3-dichloro-2-butene
(c) 2-iodopropane
(d) 2,2-dichlorobutane
(e) dichlorodifluoromethane
(f) trichlorofluoromethane
(g) hexafluoroethane
(h) 1,2-dichloro-1,1,2,2-tetrafluoroethane
(i) 1,3,5-trichlorobenzene
(j) decachlorobiphenyl
(k) 1,2,4,5-tetrachlorobenzene
(l) benzyl bromide

12.28 Why has EPA banned the manufacture of carbon tetrachloride and 1,1,1-trichloroethane in the United States?

12.29 Show that CFC-114 is the commercial name for the chlorofluorocarbon having the chemical name dichlorotetrafluoroethane.

12.30 Formulations of polybrominated biphenyls were once used as fire retardants. The commercial products known as FM BP-6 and FF-1 were composed primarily of hexabromobiphenyls. Why are these fire retardants no longer available commercially in the domestic marketplace?

12.31 The EPA regulation at 40 CFR §761.40(j) stipulates that the owners of electrical transformers containing PCB dielectric fluid must accomplish the following: Mark the vault door, machinery room door, fence, hallway, or other means of access to them with a marking shown in Figure 12.16; coordinate with the primary fire department; and document that the fire department knows, accepts, and recognizes the meaning of this marking. What is the most likely reason EPA has taken this action?

Alcohols

12.32 Using the IUPAC system, name the alcohols having the following condensed formulas:

(a) CH_3CH_2OH

(b) $CH_3-\overset{\overset{\displaystyle CH_3}{|}}{\underset{\underset{\displaystyle OH}{|}}{C}}-H$

(c)

(d) $CH_3CH_2CH_2CH_2OH$

(e) $CH_3-\overset{|}{\underset{\underset{\displaystyle CH_3}{|}}{CH}}-CH_2CH_2CH_2OH$

(f) $CH_2CH_2CH_2CH_2CH_2OH$

(g) $CH_3CH_2-\overset{|}{\underset{\underset{\displaystyle OH}{|}}{CH}}-CH_2CH_3$

12.33 Write the chemical formulas for the following alcohols:
(a) methanol   **(b)** *sec*-butyl alcohol
(c) *m*-cresol   **(d)** 2-methyl-2-propanol
(e) 2-butanol   **(f)** 2-ethyl-2-butanol
(g) 2-methylcyclohexanol   **(h)** *o*-ethylphenol
(i) *n*-amyl alcohol

12.34 What is the percent ethanol by volume in 80-proof gin?

12.35 What is the alcoholic proof of a brandy containing 45% ethanol by volume?

12.36 Why is the consumption of an entire fifth of whisky within a 15-minute period ("binge drinking") likely to be fatal?

12.37 A Breathalyzer test reveals that a truck driver possesses a blood alcohol concentration of 160 mg/dL. What is the most likely reason the driver is subsequently arrested?

12.38 Why is the flame that evolves during the burning of methanol almost invisible to the naked eye?

12.39 Wooden railroad ties were formerly soaked in a solution of pentachlorophenol to kill the microorganisms responsible for decay. Why is this procedure no longer recommended?

Ethers

12.40 Provide an acceptable name for the ethers having the following condensed formulas:

(a) $CH_3CH_2-O-CH_2CH_3$

(b) 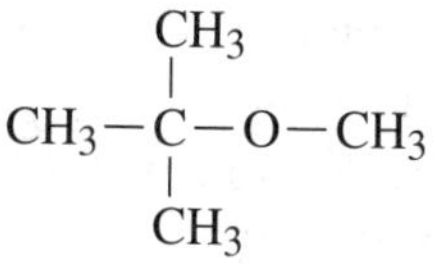 $CH_3-\overset{\overset{\displaystyle CH_3}{|}}{\underset{\underset{\displaystyle CH_3}{|}}{C}}-O-CH_3$

(c) $CH_3CH_2-O-CH_3$

(d) 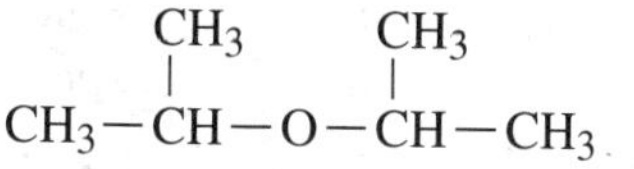 $CH_3-\overset{\overset{\displaystyle CH_3}{|}}{CH}-O-\overset{\overset{\displaystyle CH_3}{|}}{CH}-CH_3$

12.41 Write the chemical formulas for the following ethers:
   (a) dimethyl ether
   (b) ethyl methyl ether
   (c) 2-methoxybutane
   (d) 1,2-dimethoxyethane

12.42 Why should ethers be purchased in small quantities, kept in tightly sealed containers, and used promptly?

12.43 Why do chemical manufacturers recommend marking ether containers with the date on which they are first opened?

12.44 To protect public health and the environment, what special action should firefighters take when responding to a smoldering fire at a metal salvage yard, where electrical transformers are known to have been dismantled to retrieve their copper content?

Aldehydes and Ketones

12.45 Provide an acceptable name for the aldehydes and ketones having the following condensed formulas:

(a) 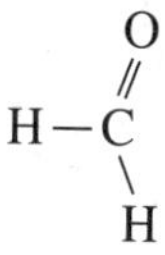 $H-\overset{\overset{\displaystyle O}{\|}}{\underset{\underset{\displaystyle H}{\backslash}}{C}}$

(b) 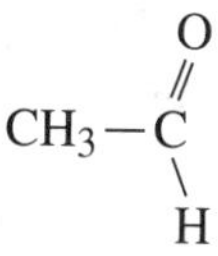 $CH_3-\overset{\overset{\displaystyle O}{\|}}{\underset{\underset{\displaystyle H}{\backslash}}{C}}$

(c) $CH_3CH_2CH_2-\overset{\overset{\displaystyle O}{\|}}{\underset{\underset{\displaystyle H}{\backslash}}{C}}$

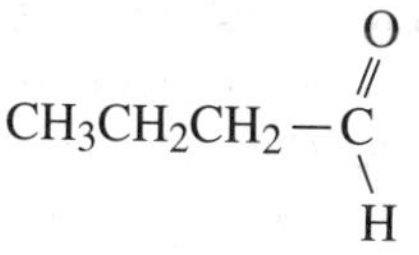

(d) 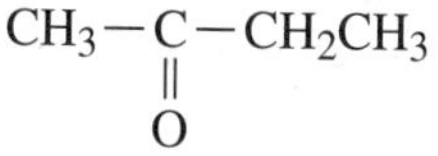 $CH_3-\overset{}{\underset{\underset{\displaystyle O}{\|}}{C}}-CH_2CH_3$

(e) 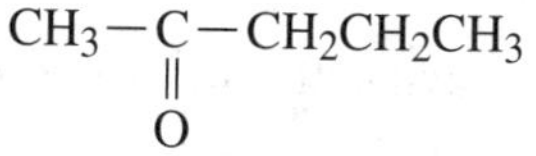 $CH_3-\overset{}{\underset{\underset{\displaystyle O}{\|}}{C}}-CH_2CH_2CH_3$

(f) 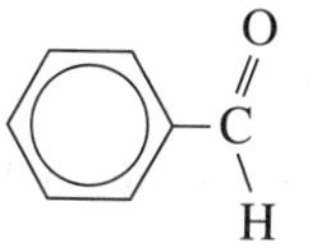  with benzene ring attached to $\overset{\overset{\displaystyle O}{\|}}{\underset{\underset{\displaystyle H}{\backslash}}{C}}$

(g) $CH_3-\overset{\overset{\displaystyle CH_3}{|}}{CH}-CH_2-\overset{}{\underset{\underset{\displaystyle O}{\|}}{C}}-CH_2$

(h) $CH_3CH_2-\overset{}{\underset{\underset{\displaystyle CH_2}{|}}{CH}}-\overset{\overset{\displaystyle H}{/}}{\underset{}{C}}\overset{}{\underset{\displaystyle O}{\backslash\!\!\backslash}}$ with $\overset{}{\underset{\displaystyle CH_3}{|}}$

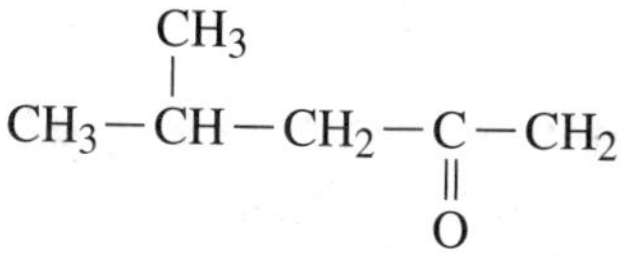

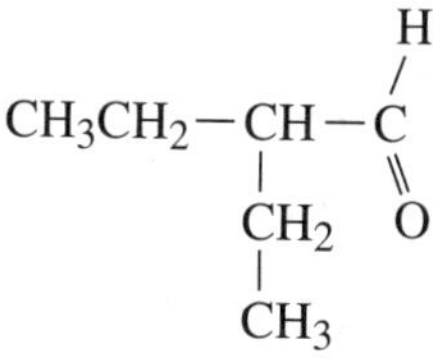

(i)

12.46 Write the chemical formulas for the following aldehydes and ketones:
(a) butanal   (b) propanone
(c) benzaldehyde   (d) methyl propyl ketone
(e) 4-methyl-3-hexanone   (f) 2-pentenal
(g) phenylpropanone   (h) 3-methylcyclopentanone
(i) isobutyraldehyde

12.47 The OSHA regulation at 29 CFR §1910.1048 stipulates that employees cannot be exposed to formaldehyde at a concentration greater than 0.75 ppm when averaged over an 8-hour workday. What is the most likely reason OSHA established this especially low worker exposure limit to formaldehyde?

12.48 What practices can be implemented to reduce formaldehyde emissions within a new home or office building?

12.49 Why do solvent-based paints pose a risk of fire and explosion?

Organic Acids

12.50 Provide an acceptable name for the organic acids having the following condensed formulas:

(a) CH_3-C with $=O$ and $-OH$ (acetic acid structure)

(b) CH_3-CH-C with CH_3 branch, $=O$ and $-OH$

(c) $CH_3CH_2CH_2CH_2-C$ with $=O$ and $-OH$

(d) CH_3-(benzene ring)$-C$ with $=O$ and $-OH$

12.51 Write the chemical formulas for the following organic acids:
(a) formic acid   (b) propanoic acid
(c) 2-propenoic acid   (d) benzoic acid
(e) 3-methylbutanoic acid   (f) *o*-hydroxybenzoic acid

12.52 Why is the risk of fire and explosion for the organic acids having five or fewer carbon atoms per molecule generally regarded as low?

Esters

12.53 Provide an acceptable name for the esters having the following condensed formulas:

(a) $H-C$ with $=O$ and $O-CH_2CH_3$

(b) CH_3-C with $=O$ and $O-CH_2-$(benzene ring)

(c) $CH_3CH_2CH_2-C$ with $=O$ and $O-CH_2CH_3$

(d) (benzene ring)$-C$ with $=O$ and $O-CH_3$

12.54 Write the chemical formulas for the following esters:
    (a) ethyl acetate   **(b)** *n*-butyl acetate
    (c) *n*-propyl formate   **(d)** isopropyl formate
    (e) methyl butyrate   **(f)** isopropyl benzoate
    (g) dimethyl phthalate

12.55 Upon entering an enclosed warehouse where ethyl butyrate is known to be stored, firefighters note an intensely strong odor of pineapples. Why does this odor forewarn them that the risk of fire exists?

Amines

12.56 Provide an acceptable name for the amines having the following condensed formulas:
    (a) $CH_3CH_2CH_2NH_2$

    (b) $CH_3-CH-CH_3$ with NH_3 below the CH

    (c) CH_3CH_2-N with H above and CH_3 below

    (d) CH_3CH_2-N with H above and CH_2CH_3 below

    (e) (benzene ring)$-CH_2NH_2$

    (f) CH_3-CH-N with CH_3 and H above and $CH(CH_3)_2$ below

12.57 Write the chemical formulas for the following amines:
    (a) *n*-butylamine   **(b)** *tert*-butylamine
    (c) 2-aminobutane   **(d)** *N*-*n*-butylaniline
    (e) phenylamine   **(f)** 1-phenylethylamine
    (g) 2-ethylaminobutane

12.58 Write the chemical formula for each amine whose domestic transportation is regulated by DOT as flammable gas.

12.59 What immediate action should be taken when a liquid amine is spilled upon skin tissue?

Peroxo-Organic Compounds

12.60 Provide an acceptable name for the peroxo-organic compounds having the following condensed formulas:

    (a) $CH_3-C-O-OH$ with CH_3 above and CH_3 below the central C

    (b) (benzene ring)$-C(=O)-$ ... $-C(=O)-$(benzene ring) joined by $O-O$

    (c) $CH_3-C-O-C-CH_2-O-OH$ with CH_3, CH_3 above and CH_3, CH_3 below the two central C atoms

12.61 Write the chemical formulas for the following peroxo-organic compounds:
    (a) di-*n*-hexyl peroxide   **(b)** bis-*p*-chlorobenzoyl peroxide
    (c) diisopropylbenzene hydroperoxide   **(d)** acetyl benzoyl peroxide

12.62 During a fire, why are the containers of peroxo-organic compounds likely to explode?

12.63 What information is conveyed on an NFPA diamond to rapidly inform fire-fighters that the peroxo-organic compounds stored in an isolated building are particularly hazardous?

12.64 Why should first-on-the-scene responders to a transportation mishap exercise special caution upon reading the following entry on the related manifest?

> **Organic peroxide type C, solid, temperature controlled, 5.2, UN3114, II**

Carbon Disulfide

12.65 Why is the use of self-contained breathing apparatus essential when firefighters respond to fires involving carbon disulfide?

12.66 Why do carbon disulfide manufacturers recommend that users prevent carbon disulfide vapor from reaching hot water pipes, light bulbs, hot plates, heating mantles, and other hot objects?

Chemical Warfare Agents

12.67 Why is the onset of semiblindness one of the first symptoms of nerve gas poisoning?

Responding to Incidents Involving the Release of Organic Compounds

12.68 Assume that fires erupt in separate containers of the following liquids. When using water as a fire extinguisher, identify which fire is more likely to be extinguished first.
   (a) methanol or toluene   **(b)** acetone or *o*-xylene
   (c) diethyl ether or carbon disulfide

12.69 A railroad waybill bears the following basic description of a chemical consignment:

> 1 t/c   Methyl bromide, 2.3, UN1062 (Poison—Inhalation Hazard, Zone A) (Placarded INHALATION HAZARD)

Use Table 10.15 to identify the isolation and protective-action distances recommended by DOT to be enforced by emergency responders when a domestic transportation mishap occurs at 11:30 P.M.

Chemistry of Some Polymeric Materials

13 **CHAPTER**

Over the past 50 years, our way of life has been dramatically changed by the polymer industry. It is unlikely that a civilized society could long survive without the wares this industry provides. In today's world, we regularly use products manufactured from both natural and synthetic polymers. Our clothing has been made from polymeric fibers, including cotton, polyesters, nylon, and polyacrylics. Our homes are constructed from wood, insulated with polystyrene, carpeted with polypropylene, coated with polyacrylic paints, and decorated with polyacrylonitrile and other polymeric fabrics.

An appreciable portion of our automobiles have also been manufactured from polymers. Often, the bumpers are made of an acrylonitrile–butadiene–styrene copolymer, the roofs are made of poly(vinyl chloride), the upholstery is cushioned with polyurethane, and the rubber tires are manufactured from styrene–butadiene copolymer.

Since they are stable at ambient conditions and do not routinely pose a health risk, polymers are not ordinarily considered hazardous materials. On the other hand, most polymeric products burn and generate toxic gases upon combustion. Not only is it relevant that most polymeric products burn, but their combustion is involved in virtually all common fires. For these reasons, the burning of polymers is a topic of great concern to firefighters.

In this chapter, to understand why they pose special hazards during fires, we will examine the features and structural characteristics of several commonly encountered polymers.

Performance Goals for Section 13.1

- ◆ Discuss the commonplace of polymeric materials in commercial products and relate the importance of this fact to emergency responders.
- ◆ Distinguish between thermoplastic and thermosetting plastics.
- ◆ Describe the method used to name simple polymers.

◆ 13.1 WHAT ARE POLYMERS?

Polymers are substances that are best characterized by the relatively sizable nature of their molecules. Since their molecules are larger than routinely encountered, chemists call them *macromolecules*. Each macromolecule of a polymer comprises a number of repeating smaller units; hence, a polymer can be envisioned as a compound typically composed of hundreds or thousands of repeating units.

Polymers are sometimes described by their response to heat. Some polymers soften when exposed to heat but then return to their original condition when cooled; these polymeric materials are said to be *thermoplastics*. The following are examples of thermoplastic polymers: polyethylene; polypropylene; polystyrene; poly(ethylene terephthalate); and poly(vinyl chloride).

Other polymers solidify or set irreversibly when heated. These are called *thermosetting polymers*. Polyurethane is an example of a thermosetting polymer.

Solved Exercise 13.1

Both the white and yolk of an uncooked egg consist of a mixture of various proteins, all of which are polymers. Why is a cooked egg an example of a natural thermosetting polymer?

Solution: The uncooked egg white and yolk consist of protein strands in a semiliquid state. The protein strands are tightly curled and entwined together. As they are heated, the strands relax and unfold and then interlink to one another until they are firmly set. This process is called *coagulation*. [Typically, the protein strands in the uncooked egg begin to coagulate in the temperature range of 145–150°F (63–66°C).] This is an irreversible process. No physical activity can return the linked strands to their initial semiliquid condition. For this reason, a cooked egg is an example of a thermosetting polymer.

Polymers are often classified according to the ways they are used. One way that some polymers are used is to produce plastic products. Plastics are polymers that can be shaped. The processes used to shape them are molding, casting, extrusion, calendering, laminating, foaming, and blowing. The polymers in plastics are usually combined with other materials including fillers and reinforcing agents. Examples of some common plastics are poly(vinyl chloride), polystyrene, and poly(methyl methacrylate).

Some natural and synthetic polymers are used as threads or yarns collectively called *fibers*. These polymers are characterized by a high tenacity and a high ratio of length to diameter (several hundred to one). They differ widely in form, flexibility, durability, and porosity. Natural fibers are a class of polymers derived from vegetable and animal sources. They include cotton and wool, respectively. Two examples of synthetic fibers are nylon and poly(acrylonitrile). They are used to make numerous goods including rope, woven cloth, matted fabrics, brushes, shingles, and other building and insulating materials, as well as stuffing for pillows and upholstery. Fibers are usually woven or knitted to produce *textiles*. These are items such as garments, carpets, carpet padding, towels, curtains, blankets, mattresses, and upholstery fabrics.

Another way a polymer is used is as an *elastomer*. This type of polymer is characterized by the ability of its molecules to elongate when strained and to reversibly assume the original shape when the tension has been released. Products made from elastomers include rubber bands, belts, footwear, vehicular tires, and the inner tubes of tires. Examples of synthetic elastomers are neoprene and polybutadiene.

Various means have been used to name polymers. Although several common names are used, chemists often place the prefix *poly-* before the name of the substances used to produce specific polymers. Polyethylene, poly(vinyl chloride), and polystyrene are so named, since they are respectively produced from ethylene, vinyl chloride, and styrene. When a polymer is manufactured from a number of substances, all their names may be included within the name of the polymer. The acrylonitrile–butadiene–styrene copolymer commonly used in plastic sewer pipes is so named because it is made from acrylonitrile, butadiene, and styrene. In this instance, the name has thankfully been abbreviated for simplicity to the *ABS* copolymer.

Performance Goals for Section 13.2
  ◆ Distinguish between the chemical nature of a polymer and copolymer.
  ◆ Distinguish between the nature of addition and condensation polymerization.

◆ 13.2 POLYMERIZATION

Polymerization is a unique type of chemical reaction involving the union of certain substances called *monomers*. The equations for such reactions are generally denoted as follows, where "A" and "B" are arbitrary monomers:

$$A + B \longrightarrow Polymer$$

Only a select number of compounds can be induced to polymerize. The polymerization reaction is characterized by the nature of the macromolecules it produces. These macromolecules can be visualized as follows:

  ◆ One type of polymeric macromolecule resembles the mid-section of a freight train having identical railcars as in Figure 13.1. Like a railcar, the repeating unit in this portion of the macromolecule has couplings, here symbolizing chemical bonds, at its front and rear ends.
  ◆ A second type of polymeric macromolecule resembles the mid-section of a freight train that has alternating dissimilar railcars as in Figure 13.2. The substance having such macromolecules is called a *copolymer*. Although the portion of the macromolecule shown in Figure 13.2 possesses regularly repeating units, a copolymer may also be composed of units that alternate irregularly.

When chemists examine the three-dimensional structures of polymers, they find that these chains of repeating units are invariably cross-linked in various ways, as in

—CH$_2$—CH$_2$—CH$_2$—CH$_2$—CH$_2$—CH$_2$—

FIGURE 13.1 ◆ The mid-section of a freight train, each of whose railcars is identical. This assemblage resembles a component of a polymer's macromolecular structure, each of whose units is also identical. This partial structure is that of polyethylene.

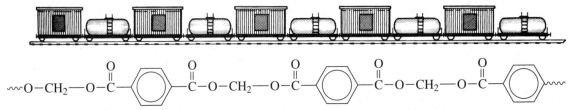

FIGURE 13.2 ◆ The mid-section of a freight train composed of regularly alternating dissimilar rail-cars. This assemblage resembles a component of a copolymer's macromolecular structure, whose alternating units are also dissimilar. The alternation of dissimilar units may occur either regularly or irregularly. This partial structure is that of polyethylene terephthalate.

Figure 13.3. During cross-linking, one chain latches onto another chain. Plastics manufacturers often attempt to increase the degree of cross-linking between the chains in thermosetting plastics. The resulting polymers are more dense, and thus stronger, than those whose macromolecules possess unlinked chains of atoms.

The chains that compose macromolecules can also be folded, coiled, stacked, looped, or intertwined into definite three-dimensional shapes. Although these complex configurations provide polymers with unique properties, we shall require only the information conveyed in their one-dimensional patterns.

The production of polymers is a major activity within the chemical industry. These processes are primarily accomplished by chemical reactions called *addition* and *condensation*. The polymers resulting from these processes are called *addition polymers* and *condensation polymers,* respectively. We shall review both chemical reactions independently.

ADDITION POLYMERIZATION

Polystyrene is an example of an addition polymer; it is produced by the polymerization of the monomer styrene, or phenylethene, as follows:

styrene

Polystyrene

FIGURE 13.3 ◆ A cross-linked polymer. The blackened circles designate an arbitrary monomer, not its atoms. Cross-linking gives the polymer extra strength and durability.

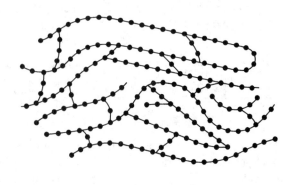

The chemical formula on the left of the arrow represents styrene, the monomer. The formula on the right of the arrow represents a section of the polystyrene macromolecule. The portion of this formula represented in parenthesis is repeated over and over again (*n* times, where *n* is a very large integer). The symbol ~~~ denotes that the unit is repeated.

In the chemical industry, polymerization is initiated in a controlled fashion. First, the reaction must be carefully initiated. One way to initiate polymerization is by using substances capable of forming free radicals (Section 5.10) when they are exposed to heat or light. Examples of such chemical initiators are organic peroxides (Section 12.14) and the Ziegler–Natta catalysts (Section 9.4). Once free radicals are produced by the thermal decomposition of an organic peroxide, they combine with neutral monomer molecules to form more complex free radicals that react with other monomer molecules until the supply of the monomer has been exhausted.

For illustrative purposes, consider the polymerization of styrene induced by free radicals resulting from the dissociation of dibenzoyl peroxide. This polymerization can be envisioned as consisting of a number of independent steps, some of which are represented as follows:

dibenzoyl peroxide   benzoylperoxy radical

styrene   benzoylperoxy radical   polymer fragment

polymer fragment   styrene

polymer fragment

In the first equation, the oxygen–oxygen bond in dibenzoyl peroxide is ruptured, resulting in the production of benzoylperoxy free radicals. In the second equation, a benzoylperoxy free radical reacts with a styrene molecule forming a more complex free radical. In the third equation, this free radical reacts with another molecule of styrene to form a still more complex free radical. Additional steps beyond those illustrated add

FIGURE 13.4 ◆ The interior walls of most refrigerators are produced from polystyrene.

successively more units to the chain in a self-propagating fashion until a long chain of the polymer has been produced and the monomer source has been consumed.

The substance produced by the polymerization of styrene is called polystyrene. The repeating unit in this polymer is the following:

$$\sim\sim\sim CH - CH_2 \sim\sim\sim$$

Polystyrene is commercially used to manufacture products such as brushes, combs, disposable coffee cups, thermally insulated equipment, building and electrical insulation, coaxial television cable, and refrigerator interiors (Figure 13.4).

Polystyrene is also recyclable. Its recycling symbol is an arrowed triangle with an enclosed number 6, beneath which appears the letters PS.

PS

Several other examples of addition polymers are provided in Table 13.1.

Solved Exercise 13.2

Poly(methyl methylacrylate) is produced by the addition polymerization of methyl methyacrylate, a substance having the following molecular formula:

$$CH_2 = C \begin{array}{l} CH_3 \\ | \\ \\ \backslash \\ C = O \\ / \\ O \\ \backslash \\ CH_3 \end{array}$$

Dibenzoyl peroxide is used to initiate polymerization. Write the stepwise equations for this polymerization reaction.

Solution: Addition polymerization reactions occur by a mechanism that includes the production of free radicals. First, free radicals are produced when an organic peroxide such as dibenzoyl peroxide is heated, which causes the dibenzoyl peroxide to melt and decompose.

$$C_6H_5 - C \overset{O}{\underset{\backslash}{\overset{\parallel}{}}} \quad \overset{O}{\underset{/}{\overset{\backslash\backslash}{}}} C - C_6H_5(l) \longrightarrow 2C_6H_5 - \overset{O}{\overset{\parallel}{C}} \cdot (g) \ + \ O_2(g)$$
$$O - O$$

dibenzoyl peroxide   benzoylperoxy free radical oxygen

Then, by adding to a molecule of methyl methacrylate, a benzoylperoxy radical initiates polymerization.

$$C_6H_5 - \overset{O}{\overset{\parallel}{C}} \cdot (g) + CH_2 = C \begin{array}{l} CH_3 \\ | \\ \\ \backslash \\ C = O \\ / \\ O \\ \backslash \\ CH_3 \end{array}(g) \longrightarrow C_6H_5 - \overset{O}{\overset{\parallel}{C}} - CH_2 - C \begin{array}{l} CH_3 \\ | \\ \cdot (g) \\ \backslash \\ C = O \\ / \\ O \\ \backslash \\ CH_3 \end{array}$$

This radical reacts with another molecule of methyl methacrylate to produce an even larger free radical.

$$C_6H_5 - \overset{O}{\overset{\parallel}{\underset{|}{C}}} - CH_2 - \overset{CH_3}{\underset{\backslash}{\overset{|}{C}}} \cdot (g) \ + \ CH_2 = C \begin{array}{l} CH_3 \\ | \\ \\ \backslash \\ C = O \\ / \\ O \\ \backslash \\ CH_3 \end{array}(g) \longrightarrow$$
$$CH_3 \quad\quad C = O$$
$$/$$
$$O$$
$$\backslash$$
$$CH_3$$

$$C_6H_5 - \overset{O}{\overset{\parallel}{C}} - CH_2 - \overset{CH_3}{\underset{\backslash}{\overset{|}{C}}} - CH_2 - \overset{CH_3}{\underset{\backslash}{\overset{|}{C}}} \cdot (g)$$
$$C = O \quad C = O$$
$$/ \quad\quad /$$
$$O \quad\quad O$$
$$\backslash \quad\quad \backslash$$
$$CH_3 \quad CH_3$$

This process repeats until hundreds or thousands of the repeating unit are produced within the macromolecule.

TABLE 13.1 ◆ Examples of Addition Polymers

Monomer	Repeating Unit

Ethylene

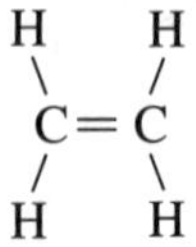

Polyethylene

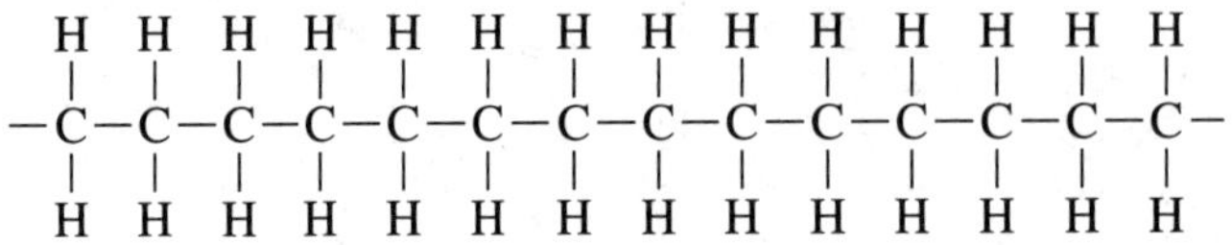

Vinyl chloride

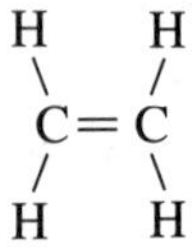

Poly(vinyl chloride)

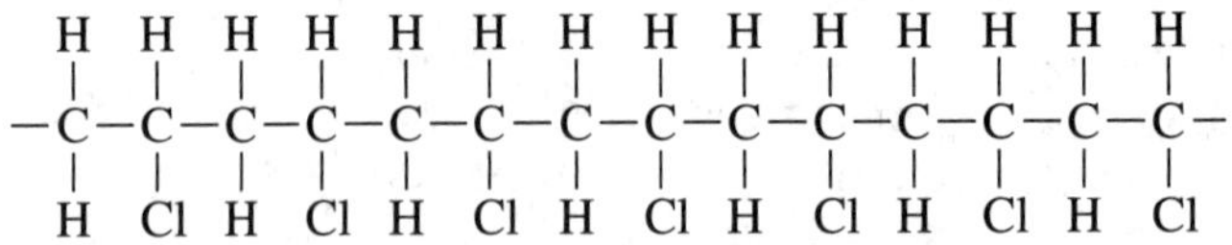

Acrylonitrile

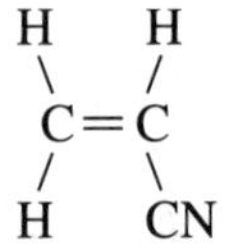

Polyacrylonitrile

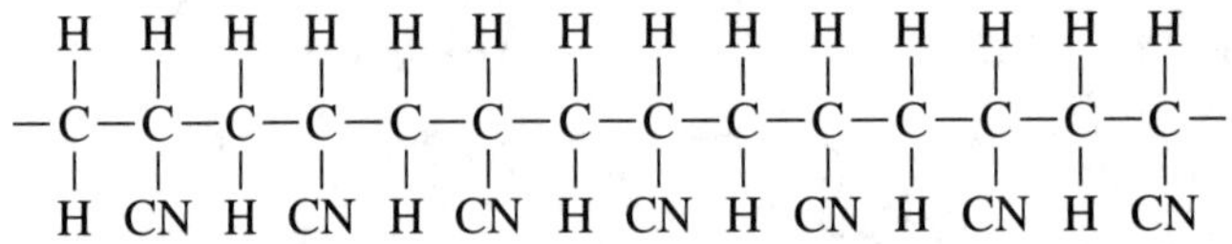

Tetrafluoroethylene

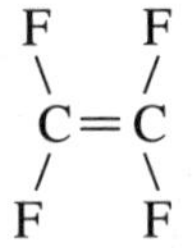

Poly(tetrafluoroethylene)

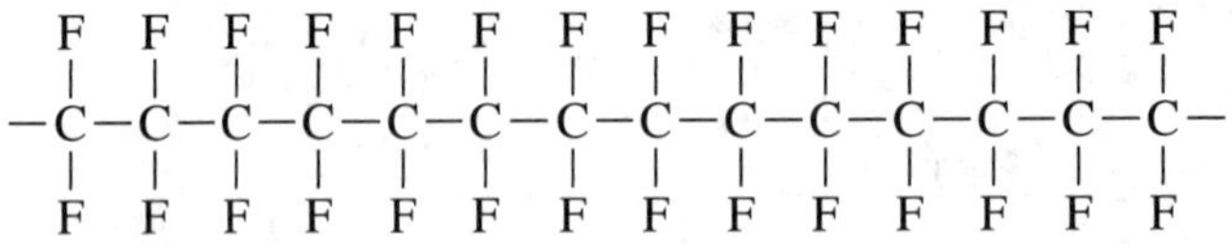

Styrene

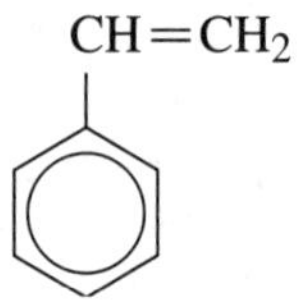

$CH=CH_2$

Polystyrene

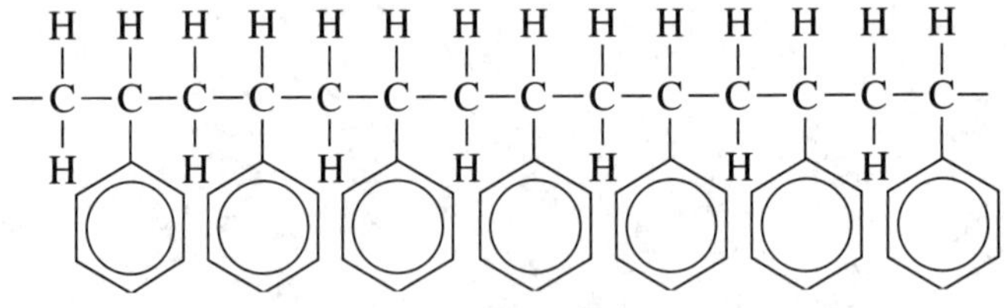

Methyl methacrylate

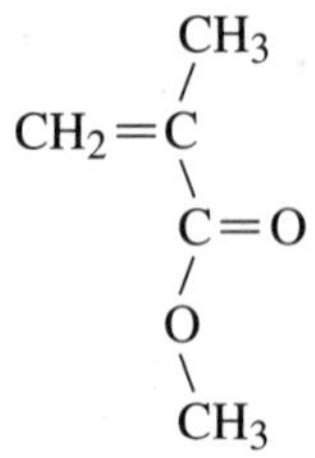

Poly(methyl methacrylate)

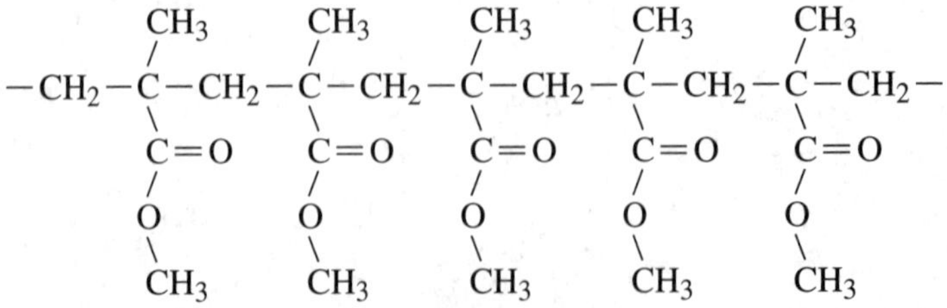

CONDENSATION POLYMERIZATION

Condensation polymers are produced by a chemical reaction called a *condensation polymerization reaction.* A common example of their production involves the reaction between certain alcohols and organic acids. Specifically, a glycol (Section 12.8) is reacted with an acid whose molecules have two carboxyl groups. When heated with a catalyst, they combine with the simultaneous elimination of water. The polymer that results is called a *polyester*.

Consider the chemical reaction between the monomers ethylene glycol and succinic acid. The first step in this reaction is illustrated by the following equation:

$$
\begin{array}{ccccc}
\text{OH} & & \text{COOH} & & \text{COOCH}_2\text{CH}_2\text{OH} \\
| & & | & & | \\
\text{CH}_2 & & \text{CH}_2 & & \text{CH}_2 \\
| & + & | & \longrightarrow & | \qquad + \; \text{H}_2\text{O}(l) \\
\text{CH}_2 & & \text{CH}_2 & & \text{CH}_2 \\
| & & | & & | \\
\text{OH}(l) & & \text{COOH}(s) & & \text{COOH}(s)
\end{array}
$$

ethylene glycol   succinic acid   an intermediate, not yet a polymer

The intermediate compound resulting from this reaction possesses potentially reactive groups at both ends of the carbon–carbon chain. It can react with two molecules of ethylene glycol, eliminating two molecules of water and producing an even more complex intermediate as follows:

$$
\begin{array}{ccccc}
\text{COOCH}_2\text{CH}_2\text{OH} & & \text{OH} & & \text{COOH} \\
| & & | & & | \\
\text{CH}_2 & & \text{CH}_2 & & \text{CH}_2 \\
| & + & | & + & | \\
\text{CH}_2 & & \text{CH}_2 & & \text{CH}_2 \\
| & & | & & | \\
\text{COOH}(s) & & \text{OH}(l) & & \text{COOH}(s)
\end{array}
$$

$$
\longrightarrow
\begin{array}{c}
\text{COOCH}_2\text{CH}_2\text{OH} \\
| \\
\text{CH}_2 \\
| \\
\text{CH}_2 \\
| \\
\text{COOCH}_2\text{CH}_2\text{COOCH}_2\text{CH}_2\text{COOH}(s) + 2\text{H}_2\text{O}(l)
\end{array}
$$

Both ends of this intermediate molecule possess reactive groups. This intermediate can also combine with more ethylene glycol and succinic acid, thus increasing the length of the macromolecular chain. Finally, when the amount of either monomer is exhausted, the polyester that remains can be described by the following formula:

$$
\text{H}\;\text{—}\!\!\!\!\sim\!\!\!\!\left[\!\!\!\sim\!\! \text{O}-\text{CH}_2\text{CH}_2-\text{O}-\overset{\displaystyle\overset{\text{O}}{\|}}{\text{C}}-\text{CH}_2\text{CH}_2-\overset{\displaystyle\overset{\text{O}}{\|}}{\text{C}}-\!\!\sim\!\!\!\!\right]_n\!\!\!\!\sim\!\! \text{OH}
$$

Additional examples of some condensation polymers are provided in Table 13.2.

The tongue-twister poly(ethylene terephthalate) is another polyester more commonly known as *PET, PETE,* or its DuPont trademark *Mylar.* The mechanism for its production is noted in Solved Exercise 13.3. PET is used to manufacture a wide range of industrial and consumer products. Globally, it is predominantly used to manufacture bottles and fibers, but it is also found in audiotapes, videotapes, computer tapes, sailboat sails, telephone and electric cable wires, and dozens of types of food packaging.

Poly(ethylene terephthalate) is also recyclable. Its recycling symbol is an arrowed triangle with an enclosed number 1, beneath which appears the letters PETE.

PETE

Solved Exercise 13.3

Poly(ethylene terephthalate) is a polyester produced by the condensation polymerization of ethylene glycol and terephthalic acid. The chemical formulas of these substances are provided in Table 13.2. Write the stepwise equations that illustrate how this condensation polymerization occurs.

Solution: The chemical reaction between an alcohol and an acid produces an ester. The production of a polyester occurs by a stepwise process, each of which involves the elimination of water, as follows:

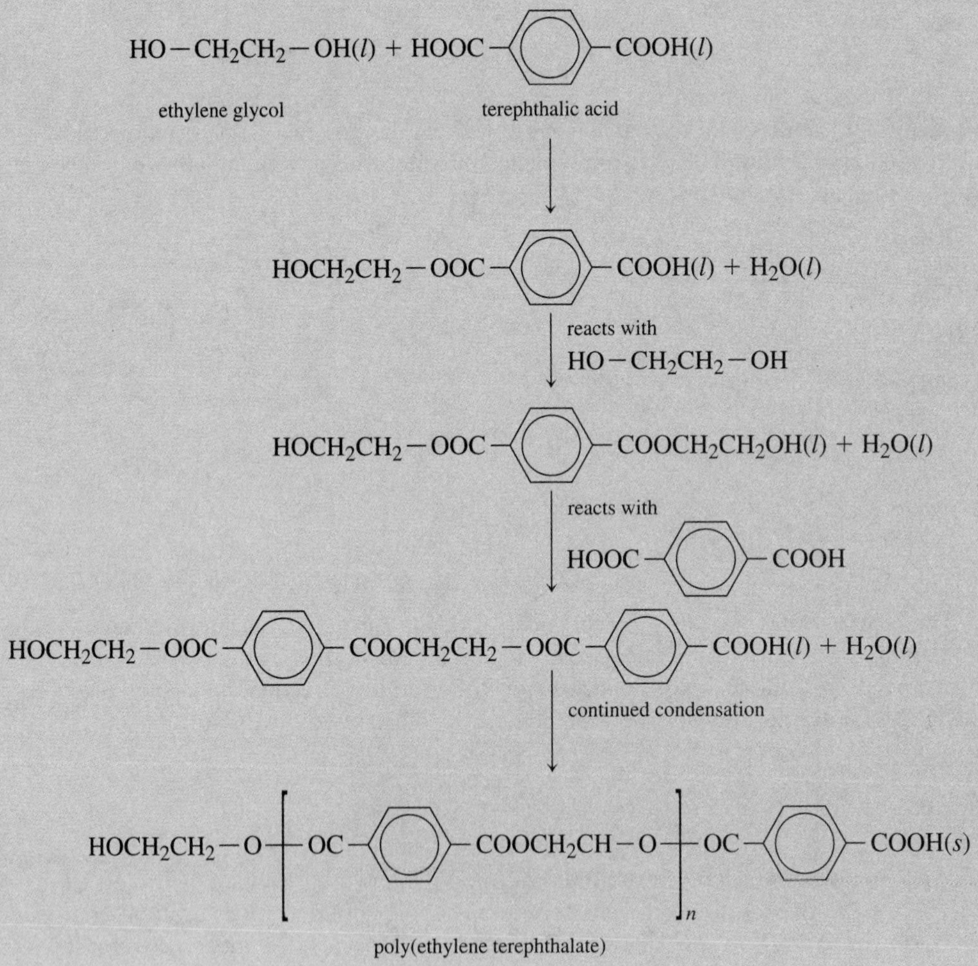

$$HO-CH_2CH_2-OH(l) + HOOC-\langle\!\langle\bigcirc\rangle\!\rangle-COOH(l)$$

ethylene glycol ⟶ terephthalic acid

$$\downarrow$$

$$HOCH_2CH_2-OOC-\langle\!\langle\bigcirc\rangle\!\rangle-COOH(l) + H_2O(l)$$

reacts with
$$HO-CH_2CH_2-OH$$
$$\downarrow$$

$$HOCH_2CH_2-OOC-\langle\!\langle\bigcirc\rangle\!\rangle-COOCH_2CH_2OH(l) + H_2O(l)$$

reacts with
$$HOOC-\langle\bigcirc\rangle-COOH$$
$$\downarrow$$

$$HOCH_2CH_2-OOC-\langle\!\langle\bigcirc\rangle\!\rangle-COOCH_2CH_2-OOC-\langle\!\langle\bigcirc\rangle\!\rangle-COOH(l) + H_2O(l)$$

continued condensation
$$\downarrow$$

$$HOCH_2CH_2-O\left[\!OC-\langle\!\langle\bigcirc\rangle\!\rangle-COOCH_2CH-O\right]_n OC-\langle\!\langle\bigcirc\rangle\!\rangle-COOH(s)$$

poly(ethylene terephthalate)

▪ **Performance Goals for Section 13.3**
   ◆ Describe the nature of autopolymerization.
   ◆ Describe the need for the use of inhibitors to reduce or eliminate the degree of autopolymerization during the storage and transportation and storage of monomers.

TABLE 13.2 ◆ Examples of Condensation Polymers

Reactants	*Repeating Unit*

Ethylene glycol

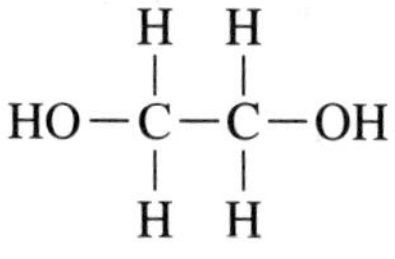

Terephthalic acid

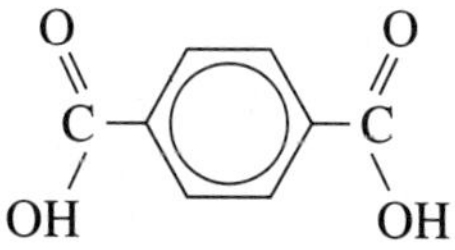

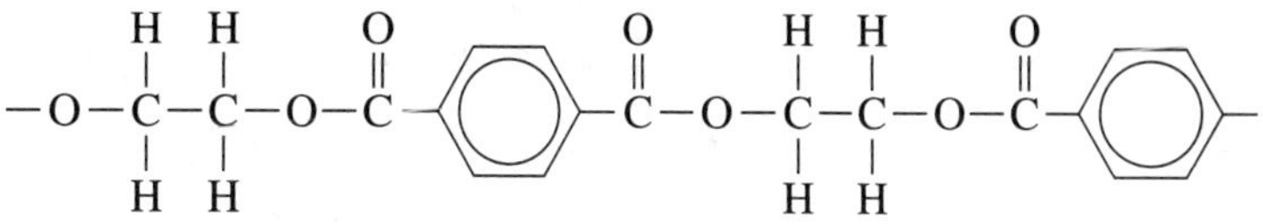

Polyethylene terephthalate

phenol

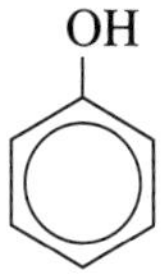

Formaldehyde

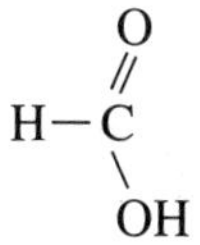

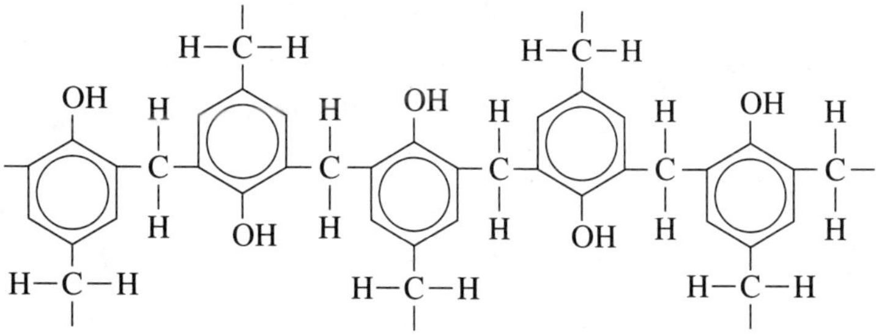

Phenol–formaldehyde

Hexamethylenediamine

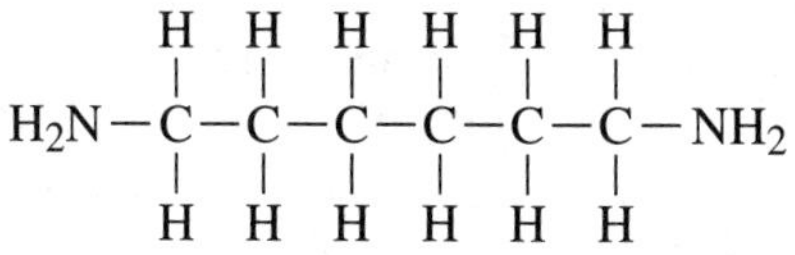

Adipic acid

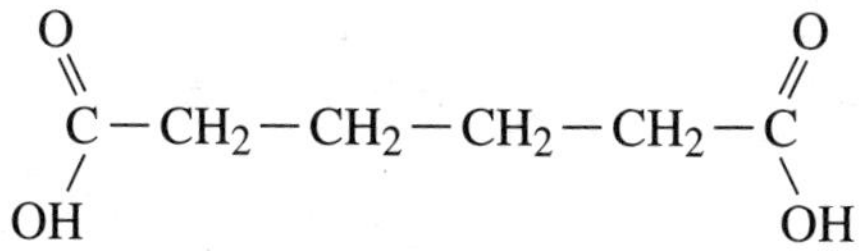

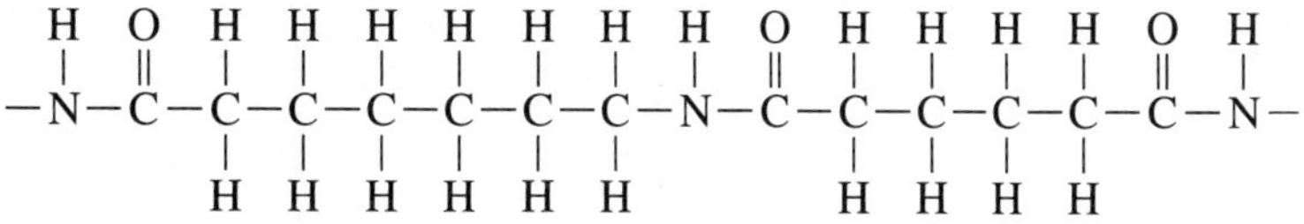

Nylon 66

◆ 13.3 AUTOPOLYMERIZATION

Some monomers are capable of undergoing spontaneous polymerization. This phenomenon is called *autopolymerization*. The autopolymerization of a substance within a tank or container can result in a rise in temperature or pressure, which increases the risk of rupture.

The autopolymerization of a substance can be prevented so the substance can ultimately be utilized for its intended purpose. When the substance is gaseous,

TABLE 13.3 ◆ Some Substances that Autopolymerize Unless Effectively Stabilized

Proper Shipping Name	Label Codes
Acrylonitrile, stabilized	3, 6.1
Chloroprene, stabilized	3, 6.1
Ethyl acrylate, stabilized	3
Ethylene oxide *or* Ethylene oxide with nitrogen *up to a total pressure of 1MPa (100 bar) at 122°F (50°C)*[a]	2.3, 2.1
Isoprene, stabilized	3
Methyl acrylate, stabilized	3
Methyl methacrylate monomer, stabilized	3
Styrene monomer, stabilized	3
Vinyl acetate, stabilized	2.1
Vinyl bromide, stabilized	3
Vinyl butyrate, stabilized	3
Vinyl chloride, stabilized	2.1
Vinylidene chloride, stabilized	3

[a]DOT also regulates the domestic transportation of ethylene oxide mixtures with carbon dioxide, chlorotetrafluoroethane, pentafluoroethane, dichlorodifluoroethane, and tetrafluoroethane.

autopolymerization is hindered by the addition of nitrogen, carbon dioxide, or other inert diluents. Although inert compounds may also be selected to dilute liquids that are prone to autopolymerize, the common practice used by chemical manufacturers is to add an *inhibitor* to their products.

Although there is not a consistent mechanism by which substances autopolymerize, some are initiated by an organic peroxide produced as they react with dissolved atmospheric oxygen. In this instance, the inhibitor selected for retarding autopolymerization is an antioxidant, which immediately reacts with the organic peroxide as it forms.

DOT regulates the transportation of substances that are prone to autopolymerize, some examples of which are provided in Table 13.3. DOT also requires that shippers indicate that their products were inhibited or stabilized against autopolymerization. Examples of the proper shipping name for such polymers are "ethyl acrylate, stabilized" and "vinylidene chloride, stabilized."

To warn firefighters of the presence of a substance susceptible to autopolymerization, NFPA recommends the insertion of a "P" in the bottom quadrant of the NFPA diamond posted in the relevant storage area.

Performance Goals for Section 13.4

- Discuss the general phenomena that occur when polymers are exposed to heat and burn.
- Describe the nature of a flashover.
- Identify the toxic gases produced when polymers are subjected to heat and burn.

Most products from natural and synthetic polymers are combustible when exposed to an ignition source. Their combustion is associated with the following general features:

- Polymeric products often melt and thermally decompose into the monomers from which they were made or a mixture of simpler substances.
- The surfaces of some polymeric products tend to char as they burn.
- Burning polymeric products release considerable heat.
- Burning polymeric products can evolve voluminous amounts of smoke, carbon monoxide, and other hazardous gases, vapors, and fumes.

The melting of polymeric products at the typical fire scene is associated with both beneficial and detrimental issues. The melting often causes the polymer to drip from its source, as from ceiling tile to an underlying floor. Dripping, molten polymer closely resembles dripping, hot candle wax. On the one hand, the dripping serves as a cooling mechanism, removing heat from the immediate site of combustion and hindering the continued combustion of the polymer at that site. On the other hand, when they remain in the fire zone, polymers begin to thermally decompose in their molten state. Then, their decomposition products ignite and the fire is spread.

Prior to its ignition, a polymer frequently undergoes thermal degradation into simpler chemical species; that is, when the polymer is exposed to heat, it decomposes into relatively simpler substances. The decomposition of unique organic polymers occurs by different mechanisms, several of which involve scission of the macromolecular chains. Two types of thermal decomposition are characteristic of polymers:

- When heated, some produce an array of decomposition products. For example, when polypropylene is heated, its decomposition products include methane, ethane, propane, butane, pentane, propene, 2-methylpropene, 1- and 2-pentene, 2-methyl-1-butene, and other hydrocarbons.
- Other polymers produce only the monomers from which they were initially produced. They are said to *depolymerize*. For example, when poly(methyl methacrylate) thermally decomposes, its monomer methyl methacrylate accounts for 91 to 98% by mass of the substances produced. The same is true of poly(ethylene terephthalate). When heated, it reverts into the substances from which is was produced, dimethyl terephthalate and ethylene glycol. Polymers that depolymerize are desirable candidates for recycling.

The vapors of the monomer or mixture of thermal decomposition products initially diffuse to the surface of the polymer, where they mix with atmospheric oxygen and ignite. As the heat increases in intensity, these vapors can also migrate from the immediate burning area and accumulate elsewhere, such as near the ceiling of a room, where they mix with air and burn.

Heat can also be conducted or radiated through a polymeric material, thereby causing the polymer to decompose at a location isolated from the heat source. Consider the section of a wall shown in Figure 13.5. It is constructed of wooden support beams to which polymeric paneling has been applied. Although the heat from a fire impinges on only one side of the wall, it can conduct or radiate *through* the wall, causing the polymer in the paneling to decompose. A mixture of simple combustible substances subsequently produced by thermal decomposition can readily ignite. This generation of flammable vapor at a location isolated from the source of heat gives rise

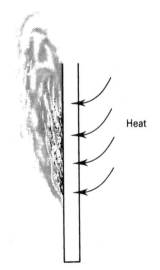

Heat

FIGURE 13.5 ◆ Heat that is conducted or radiated through this section of a wall causes the thermal decomposition of the polymeric paneling on its opposite side. The decomposition of a polymer is associated with the production of simple organic substances, which slowly migrate from their point of origin and mix with surrounding air. When the concentration of this mixture is within its flammable range, an ignition source causes the mixture to ignite.

to *flashover,* the phenomenon responsible in part for the spread of fire from one room into an adjacent room.

Fires pose special problems in large public buildings constructed in part from plastic products, since these polymers usually burn differently than burning wood or other natural materials. The thermal characteristics in air of some common polymers are provided in Table 13.4. By comparison, a hardwood self-ignites in air in the average temperature range of 781°F (416°C) and possesses an average heat of combustion of 8500 Btu/lb (19,700 kJ/kg). Wood and synthetic polymers begin to burn at approximately the same temperature, but most common polymers evolve nearly twice as much heat when they burn as an equivalent amount of wood.

Although the number of deaths caused by fires has declined in recent decades, fatalities still occur. These fatalities often result when individuals are obligated to inhale the gases produced during fires. Since each macromolecule of a polymer is

TABLE 13.4 ◆ Thermal Characteristics of Some Common Polymers

Polymer	Specific Gravity	Decomposition Range	Self-Ignition Temperature	Heat of Combustion
Polyethylene (high-density)	0.965	644–824°F (340–440°C)	662°F (350°C)	20,050 Btu/lb (46,500 kJ/kg)
Polypropylene	0.91	626–770°F (330–410°C)	734–770°F (390–410°C)	19,800 Btu/lb (46,000 kJ/kg)
Polystyrene	1.05	572–752°F (300–400°C)	914°F (490°C)	18,100 Btu/lb (42,000 kJ/kg)
Poly(methyl methacrylate)	1.18	338–572°F (170–300°C)	842°F (450°C)	11,210 Btu/lb (26,000 kJ/kg)
Poly(vinyl chloride)	1.40	392–572°F (200–300°C)	851°(455°C)	8620 Btu/lb (20,000 kJ/kg)

generally composed of hundreds of carbon atoms, burning polymers can produce massive concentrations of carbon monoxide. The larger the fire, the more carbon monoxide produced. Within an enclosure, carbon monoxide and other gases can soar within a matter of seconds to life-threatening concentrations.

Fires involving burning polymers can have a far-reaching impact in areas distant from where the fire originated. This results when gases and thermal degradation products travel by convection through ventilation systems, trash chutes, and similar routes. While this movement spreads the fire, it also causes people to be unsuspectingly exposed to toxic fumes generated elsewhere within a building.

The burning of products made from synthetic polymers often produces a different mixture of gases and fumes than that generated from the burning of nonplastic products. Carbon monoxide is still the most prevalent gas at a fire scene, but other gases associated with burning plastics are also produced. They include hydrogen chloride, ammonia, hydrogen cyanide, sulfur dioxide, and nitrogen dioxide.

At a fire scene, the origin of the nitrogenous and sulfurous gases can be traced to polymeric materials that have thermally decomposed or burned. Thermal decomposition occurs primarily when they are exposed to the heat evolved during slow-burning processes. The polymers that produce hydrogen cyanide, ammonia, nitric oxide, and nitrogen dioxide are nitrogenous organic compounds. They are natural and synthetic polymers that possess one or more of the functional groups listed in Table 13.5. Their thermal decomposition generates hydrogen cyanide and ammonia, and their combustion produces nitric oxide and nitrogen dioxide.

TABLE 13.5 ◆ Some Organic Compounds Containing Nitrogen

Class of Organic Compound	Functional Group	Example
Amine	$-NH_2$	$CH_3CH_2CH_2-NH_2$ *n*-propylamine, or 1-aminopropane
Amide	$-\overset{\displaystyle O}{\overset{\displaystyle \parallel}{C}}\diagdown NH_2$	$CH_3-\overset{\displaystyle O}{\overset{\displaystyle \parallel}{C}}\diagdown NH_2$ acetamide
Amino acid	$-\underset{NH_2}{CH}-\overset{\displaystyle O}{\overset{\displaystyle \parallel}{C}}\diagdown OH$	$CH_3-\underset{NH_2}{CH}-\overset{\displaystyle O}{\overset{\displaystyle \parallel}{C}}\diagdown OH$ 2-aminopropanoic acid
Nitrile	$-C\equiv N$	$CH_2=\overset{\displaystyle H}{\underset{\displaystyle CN}{C}}$ acrylonitrile, or propenenitrile
Nitro compounds	$-NO_2$	$-NO_2$ nitrobenzene
Isocyanate	$-N=C=O$	$CH_3CH_2-N=C=O$ ethyl isocyanate

Polymers that produce hydrogen sulfide and sulfur dioxide are sulfurous organic compounds. They are usually natural polymers having an animal origin. The nonwater matter of vertebrates is composed of proteins, which in turn are made up of amino acids. Although there are 21 amino acids that make up the structures of nearly all proteins, we are concerned here only with two, methionine and cysteine.

$$CH_3-S-CH_2CH_2-\underset{\underset{NH_2}{|}}{CH}-\overset{\overset{O}{\|}}{C}\underset{OH}{\diagdown}$$

methionine

$$HS-CH_2-\underset{\underset{NH_2}{|}}{CH}-\overset{\overset{O}{\|}}{C}\underset{OH}{\diagdown}$$

cysteine

Since sulfur atoms are constituents of these amino acids, they are also constituents of the proteins biologically produced from them. They are contained within animal products made from leather, wool, and animal hair. Their thermal decomposition produces hydrogen sulfide and ammonia, and their combustion produces carbon monoxide, carbon dioxide, sulfur dioxide, nitric oxide, nitrogen dioxide, and water.

Another substance whose presence has been detected in wood smoke is the unsaturated aldehyde known as *acrolein,* or *acrylic aldehyde;* its chemical formula is CH_2=$CHCHO$. This is a pungent-smelling, intensely irritating lacrimator. A 1-minute exposure to air containing as little as 1 ppm of acrolein causes nasal and eye irritation. A concentration of acrolein in air ranging from 2500 to 5000 ppm has been established as the lethal dose to laboratory animals. During World War I, acrolein was offensively used as a lacrimator.

The character of the smoke produced in fires involving polymeric materials varies according to the chemical nature of the polymer. More smoke is typically produced in fires involving polymers that were produced from aromatic monomers. A fire involving burning polystyrene, for instance, produces considerably more soot than one that involves polyethylene.

As Figure 13.6 dramatically illustrates, fire rapidly spreads in untreated fabrics produced from synthetic polymers. To improve the safety of these textiles, they are treated

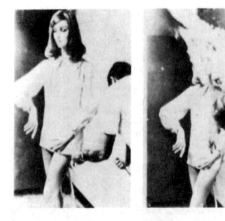

FIGURE 13.6 ◆ A grim demonstration. The manikin shown to the far left is successively photographed 30 seconds, 60 seconds, and 90 seconds after application of a flame to its blouse. This illustrates how quickly fire may spread when certain synthetic fabrics burn. *Courtesy of I. N. Einborn, Flammability Research Center, University of Utah, Salt Lake City, Utah.*

with fire retardants or their fibers are coated with flameproofing agents. Although there is no doubt that these procedures have improved their safety, the treatment is only partially effective. The technology for treating plastics and textiles to make them more resistant to ignition remains an active area of research. Following treatment with fire retardants, polymeric materials still burn, especially when exposed to the intense heat experienced during major fires. Typical fire retardants and flameproofing agents used with modern-day textiles and plastics include antimony trioxide and tricresyl phosphate (Section 12.8).

Aside from the voluntary gestures of plastics and textile manufacturers, state and federal laws have also helped provide materials that are safer for consumers from a fire-related perspective. At the federal level, the *Flammable Fabrics Act* was initially enacted in response to public concern over a number of serious accidents involving brushed rayon high-pile sweaters and children's cowboy chaps that flash-burned when ignited. The statute requires manufacturers to submit apparel fabrics to a specified 45° ignition test and rate-of-burn test. Today, apparel and interior furnishings are subject to such regulations. Regulatory standards applying to carpets and rugs, children's sleepwear, and mattresses have now been adopted.

The Departments of Consumer Affairs in several states have also adopted flammability regulations applicable to a variety of interior furnishings. Even though the hazard can never be totally eliminated, these combined efforts have assisted in untold ways to reduce fire-related injuries when plastics and textiles are involved in fires.

Performance Goal for Section 13.5
   ◆ Describe and compare the macromolecular structures of cellulosic and other vegetable fibers and the common animal fibers.

◆ 13.5 VEGETABLE AND ANIMAL FIBERS

Many common textiles have been produced from naturally occurring vegetable and animal fibers. Cotton and linen are examples of vegetable fibers, whereas wool and silk are examples of animal fibers. These naturally occurring fibers can be directly used to produce textiles, or they can be chemically altered to produce synthetic fibers.

As they occur naturally, vegetable and animal fibers are often mixed with combustible oils; for example, cotton contains cottonseed oil and wool contains lanolin. DOT regulates the transportation of the vegetable and animal fibers in Table 13.6 as hazardous materials. We shall note the properties of some common fibers and pay special attention to their combustible nature.

CELLULOSE AND CELLULOSIC DERIVATIVES

When the natural binding agent *lignin* is removed from wood, the principal substance that remains is a polymer called *cellulose*. This polymer serves as the primary structural component of the cell walls in plants, and thus is generally regarded as nature's most important polymer. Wood is approximately 50% cellulose by mass, whereas cotton and linen are nearly 100% cellulose.

TABLE 13.6 ◆ Some Animal and Vegetable Products Regulated by DOT for Transport

Proper Shipping Name	Label Codes
Cotton	9
Cotton waste, oily	4.2
Cotton, wet	4.2
Fibers or fabrics, animal *or* vegetable *or* synthetic, n.o.s. *with animal or vegetable oil*	4.2
Fibers or fabrics impregnated with weakly nitrated nitrocellulose	4.1
Paper, unsaturated oil treated *incompletely dried (including carbon paper)*	4.2

Cellulose is the raw material of the paper, wood, cotton textiles, and cellulose plastics industries. The natural fiber is encountered in several forms, two of which are cotton and linen.

- *Cotton* comes from any of four species of *Gossypium* that grow in warm climates throughout the world. To produce it as a yarn, raw cotton is first boiled in a dilute solution of sodium hydroxide to remove any wax that naturally adheres to the fibers. Then, it is bleached with chlorine, sodium hypochlorite, or a similar substance. Next, it is passed through a vat of diluted sulfuric acid to neutralize any remaining alkaline materials. Finally, the cotton is washed with water. At this point, it is ready to be spun into a yarn and woven into cloth.
- *Linen* is nearly pure cellulose. Its fibers are derived from the stalk of the flax plant *Linum usitatissimum*. They are among the strongest naturally occurring fibers. Linen fabric absorbs moisture faster than any other fabric. However, linen lacks resiliency, the ability to spring back when stretched. This lack of resiliency causes it to wrinkle easily.

The difference between the physical properties of cotton and linen is associated with the nature of their fibers. When examined under a microscope, individual cotton fibers resemble short, twisted flattened tubes; linen fibers resemble long transparent tubes that possess periodic junctions for cross-linking.

The chemical formula of cellulose is often abbreviated $(C_6H_{10}O_5)_n$ or $[(C_6H_7O_2(OH)_3]_n$, where n is around 5000. The more complete chemical formula is denoted as follows:

The repeating unit pictured in this structure is a substance called *β-glucose*.

Cotton is commercially available in a form called *mercerized cotton*. Cotton fibers swell when they are immersed in a concentrated solution of sodium hydroxide, a process called *mercerizing,* after the discoverer John Mercer. When cotton's flattened fibers are mercerized, they become rounded and more lustrous and acquire additional strength. Mercerizing also allows dyes to penetrate the fabric more easily.

Synthetic fibers have also been derived from the cellulose originating in wood. Cellulose acetate is the fiber produced when wood cellulose is reacted with acetic acid or acetic anhydride. In commerce, the fiber is called *acetate rayon*. Although it possesses a pronounced strength, acetate rayon can be ignited with a hot iron and destroyed by some dry cleaning solvents. Given these adverse features, acetate rayon soon lost its popularity as a fabric of choice, but it is still used today in fireproof motion-picture film, airplane wings, and safety glasses.

Other synthetic products have been derived from cellulose. An example is *cellulose xanthate*, commonly called *rayon*. Highly purified wood pulp is reacted with sodium hydroxide, followed by chemical treatment with carbon disulfide. The resulting solution is extruded through a spinneret, a metal disk having numerous tiny holes, and then into an acid solution, which regenerates the cellulose as tiny transparent fibers.

Another product derived from cellulose is *nitrocellulose*. This substance is manufactured by reacting cellulose with nitric acid. Whereas highly nitrated cellulose is a chemical explosive (Section 14.8), the lesser grades of nitrated cellulose can be dissolved in a solvent and applied to cloth to produce the fabric known as *patent leather*. The products produced from this artificial leather are strong and flexible. They include vehicle seat covers and convertible tops. Nitrocellulose is also used as a film-producing agent in lacquers, printing inks, and enamel nail polishes.

Cotton textiles can be chemically treated to produce fabrics that are wrinkle proof, as well as fabrics that can be washed and drip-dried with few wrinkles. Cotton is often combined with synthetic fibers to produce fabrics that are commercially available as blends with names such as "50% cotton, 50% acrylics."

When triggered by an ignition source, all cellulosic materials burn; however, they do not melt. Since the fires of cellulosic materials are class A fires, they are routinely extinguished with water.

WOOL AND SILK

The most common animal fabrics employed in textiles are wool and silk. Wool is the curly hair of sheep, goats, and llamas. Under a microscope, wool fibers resemble tiny, overlapping scales, much like those of fish. These fibers bend and conform to a variety of physical shapes. They also possess resiliency and tend to hold their shape.

Silk is the soft, shiny fiber produced by silkworms to form their cocoons. Silk fibers are very strong, elastic, and smooth. Under a microscope, silk fibers appear semi-transparent, which accounts for their lustrous sheen. Silk is resilient; its fibers readily spring back to their original position when stretched, bent, or folded. These qualities have made silk one of the most useful fibers for the textile market. Unwinding the long, delicate silk threads of a cocoon is a tedious process; hence, the fabric woven from silk fibers is relatively expensive.

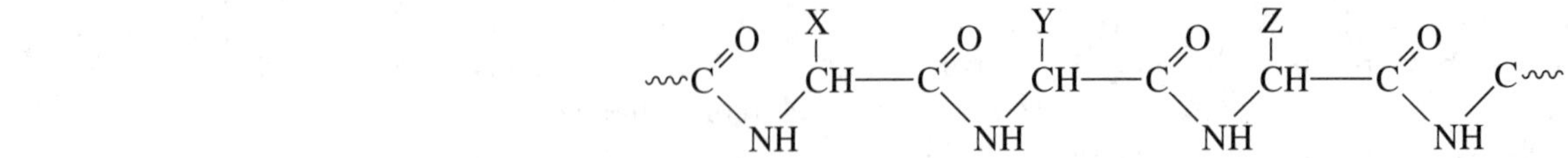

FIGURE 13.7 ◆ A portion of the macromolecular structure of wool. In this protein, X, Y, and Z represent specific substances called *amino acids*, compounds having the chemical formula $R-CHNH_2COOH$. Only 21 amino acids are commonly found in naturally occurring proteins. The interested reader should consult more advanced chemistry texts on this subject.

Like all other forms of animal hair, wool and silk are composed of proteins, which are biological substances whose molecules possess recurring amino groups. The condensed formulas of wool and silk are $C_{42}H_{157}N_5SO_{15}$ and $C_{15}H_{23}N_5O_6$, respectively. A portion of the macromolecular structure of wool is shown in Figure 13.7.

Silk consists of a mixture of two relatively simple proteins called *silk fibroin* and *sericin*. The macromolecular structure of silk fibroin resembles the structure of wool in which X, Y, and Z are invariably $-CH_3$, $HO-C_6H_5-CH_2-$, or $HO-CH_2-$. Sericin possesses a similar macromolecular structure, but the primary recurring group in the protein is the following:

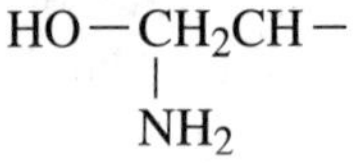

$$HO-CH_2CH-$$
$$|$$
$$NH_2$$

Since wool and silk are proteins, both possess similar properties. Each has an ignition temperature in excess of approximately 1058°F (570°C); hence, wool and silk are difficult to ignite and when ignited, burn very slowly. When tightly woven as in rugs, woolen textiles tend to smolder and char when burning. They absorb great quantities of water, thus permitting their fires to be easily extinguished.

Since the macromolecules of animal fibers contain bonding groups such as $-C\overset{\displaystyle O}{\underset{\displaystyle NH_2}{}}$,

$-NH-$, and $-S-S-$, the presence of ammonia, hydrogen cyanide, and sulfur dioxide at fire scenes involving textiles is possible. Although they may be produced, ammonia and hydrogen cyanide do not appear to survive long at fire scenes, most likely because they burn.

Performance Goals for Section 13.6

- ◆ Identify the commercially important vinyl polymers.
- ◆ Discuss the general properties of polyethylene, polypropylene, poly(vinyl chloride), polyacrylonitrile, and poly(methyl methacrylate) and identify the common products made from each of them.
- ◆ Identify the toxic gases produced when products made from the vinyl polymers are exposed to intense heat and burn.

◆ **13.6 VINYL POLYMERS**

A *vinyl compound* is any substance having the group of atoms $CH_2{=}CH{-}$. Examples of vinyl compounds are listed in Table 13.1 under the heading "Monomer." A polymer made from one or more vinyl compounds is called a *vinyl polymer*.

The following equations illustrate the production of several commercially important vinyl polymers from their respective vinyl compounds:

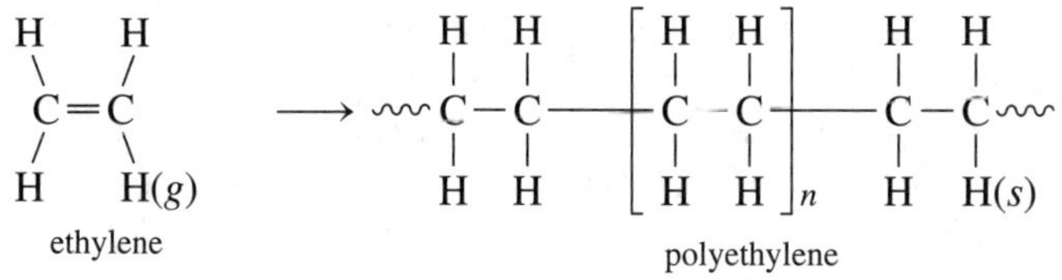

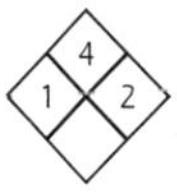

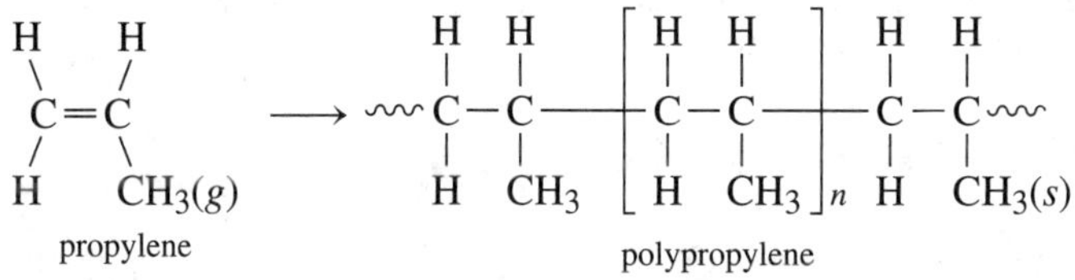

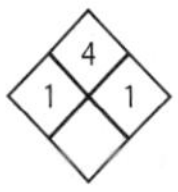

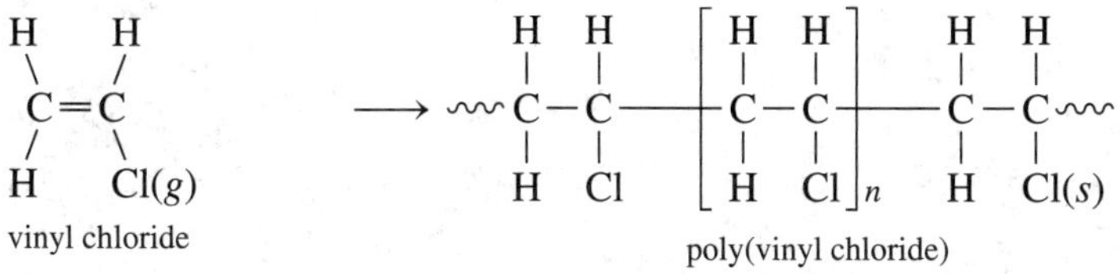

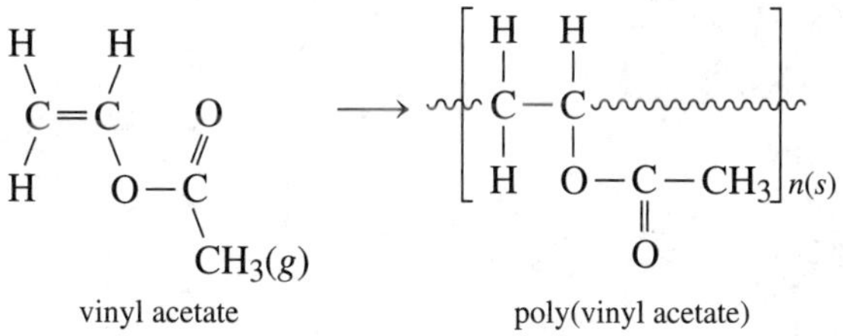

The referenced NFPA diamonds designate the relevant properties of the monomers from which the vinyl polymers are produced.

A vinyl polymer can also be produced from multiple vinyl compounds. For example, the vinyl chloride–vinylidene chloride copolymer is produced from vinyl chloride and vinylidene chloride. Vinyl chloride and vinylidene chloride are synonyms for chloroethene and 1,1-dichloroethene, respectively.

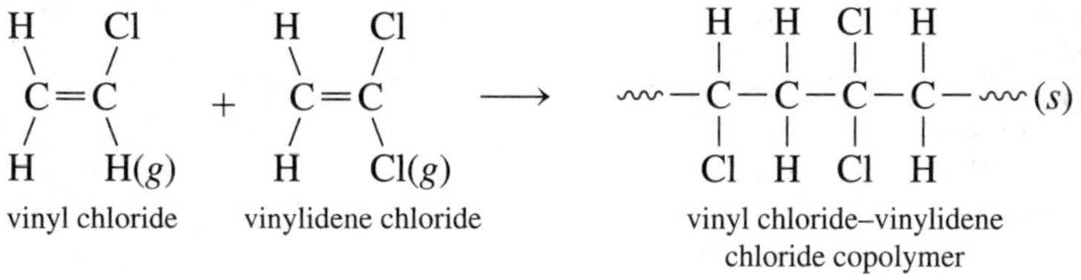

Commercially, the vinyl chloride–vinylidene chloride copolymer is known as *Saran*. Manufacturers form it into sheets, tubes, rods, fibers, and other molded items. Its fibers are used to produce a number of textiles, including carpets, curtains, and upholstery fabrics.

POLYETHYLENE

As we noted in Section 12.3, ethylene is prepared in the petrochemical industry, primarily by cracking ethane and propane. When ethylene is polymerized, the product is polyethylene. It is the most widely manufactured polymer in the United States. Polyethylene macromolecules possess the repeating unit $\sim\sim\sim CH_2 - CH_2 \sim\sim\sim\sim$.

Polyethylene is primarily encountered in two forms: low-density (cross-linked) (LDPE), and high-density (linear) (HDPE). A third form, low-molecular-weight polyethylene, is uniquely used in coatings and polishes. Both LDPE and HDPE are white solids. The low-density form is a thermosetting polymer, whereas the high-density form is a thermoplastic polymer. The density of LDPE may be as low as 0.915 g/mL, and of HDPE as high as 0.965 g/mL.

FIGURE 13.8 ◆ These containers were manufactured from high-density polyethylene. *Courtesy of Mayfair Plastics, Carson, California.*

FIGURE 13.9 ◆ Film manufactured from high-density polyethylene is converted into packaging for the institutional, grocery, and food-service markets.

Low-density polyethylene is largely used for making molded products such as toys, whereas high-density polyethylene is mainly used to manufacture filmed products and hardy containers (Figures 13.8 and 13.9). Included among the latter are plastic milk jugs, detergent containers, sandwich bags, and liners for trash cans, drums, and other containers. These films are used in the building industry as vapor and moisture barriers and in the agricultural industry for mulching, silage covers, greenhouse glazings, pond liners, and animal shelters.

Low-density and high-density polyethylene can be recycled. Their recycling symbols are arrowed triangles with the enclosed numbers 4 and 2, respectively, and beneath which appear the letters LDPE and HDPE, as follows:

Products made from both LDPE and HDPE burn when they are exposed to fire. The data in Table 13.4 show that polyethylene is distinctive among burning polymers insofar as it releases more heat per unit mass than any other commercially popular polymer. During combustion, polyethylene products tend to disintegrate into numerous burning, molten globules or a pool of liquid. Since the macromolecules of polyethylene are composed of only carbon and hydrogen atoms, carbon monoxide, carbon dioxide, and water vapor are produced as combustion products.

POLYPROPYLENE

We noted in Section 12.3 that the cracking of propane produces methane, ethylene, and propylene. In the petrochemical industry, the propylene is separated from this mixture of gases and polymerized to manufacture polypropylene. Its macromolecules possess the repeating unit ~~~CH_2—CH~~~ .
$$|$$
$$CH_3$$

Polypropylene is used to manufacture commercial products ranging from silky fibers to rigid containers. Examples are outdoor tables and chairs, shatterproof glasses, artificial grass and turf, pipes, ropes, nets, twines, carpets, and carpet padding. The plastic items in motor vehicles are largely made from polypropylene. They include the dashboards, bumpers, carpeting, and occasionally, the upholstery (Figure 10.9).

By using the Ziegler-Natta catalyst to initiate polymerization, manufacturers have been able to produce polypropylene so the branching methyl group is arrayed regimentally along the carbon–carbon backbone. This gives rise to the following three types of polypropylene macromolecules:

• *Isotactic,* in which the methyl groups are all pointing in the same direction

~~~CH_2—CH—CH_2—CH—CH_2—CH—CH_2—CH—CH_2—CH—CH_2—CH~~~
      | | | | | |
     CH_3 CH_3 CH_3 CH_3 CH_3 CH_3

• *Syndiotactic,* in which alternate methyl groups point in opposite directions

   CH_3 CH_3 CH_3
   | | |
~~~CH_2—CH—CH_2—CH—CH_2—CH—CH_2—CH—CH_2—CH—CH_2—CH~~~
           | | |
          CH_3 CH_3 CH_3

• *Atactic,* in which the methyl groups are randomly oriented

                   CH_3 CH_3
                   | |
~~~CH_2—CH—CH_2—CH—CH_2—CH—CH_2—CH—CH_2—CH—CH_2—CH~~~
      | | | |
     CH_3 CH_3 CH_3 CH_3

These individual polypropylene types can be cast into shapes, drawn into sheets, or extruded into fibers, thus producing a range of diversified products.

Polypropylene is a recyclable polymer. The recycling symbol for polypropylene is an arrowed triangle with an enclosed number 5, beneath which appear the letters PP.

Polypropylene does not ignite easily. On the other hand, when it is exposed to intense heat, polypropylene thermally decomposes into a mixture of hydrocarbons, which catches fire and burns.

POLY(VINYL CHLORIDE)

Poly(vinyl chloride) is produced when vinyl chloride polymerizes. Vinyl chloride, or chloroethene, is primarily manufactured by the catalytic dehydrochlorination of 1,2-dichloroethane.

$$ClH_2C-CH_2Cl(g) \longrightarrow CH_2{=}\overset{\displaystyle H}{\underset{\displaystyle Cl}{C}}(g) \quad + \quad HCl(g)$$

    1,2-dichloroethane   vinyl chloride   hydrogen chloride

It is a flammable gas and subject to autopolymerization. It is also recognized by epidemiologists as a known human carcinogen.

Poly(vinyl chloride is commonly recognized by its acronym *PVC*. The recurring unit in its macromolecules is $\sim\sim\sim CH_2\underset{\displaystyle |}{\underset{\displaystyle Cl}{CH}}\sim\sim\sim$. The presence of chlorine atoms gives PVC some unique properties. For example, the data in Table 13.4 show that when PVC burns, less heat is evolved compared to other common polymers that burn.

PVC is the polymer used in many construction products, including floor tiles, lighting fixtures, vinyl panels and siding, pipe and pipe fittings (Figure 13.10), conduits, and wall coverings. PVC is now the plastic insulation used in virtually all modern electrical wiring. We also encounter PVC in imitation leather, shower curtains, upholstery material, plastic packaging materials, and medical products. Approximately 70% of the poly(vinyl chloride) produced in the United States is used in building construction materials. Copolymers of PVC and other polymers yield products whose properties vary from elastomeric to rigid. They are fabricated into films, fibers, sheeting, and moldings.

PVC is recyclable. Its recycling symbol is an arrowed triangle with an enclosed number 3, beneath which appears the letter V.

Pure PVC is highly susceptible to degradation, which results in unsightly discoloring and loss of mechanical properties. To inhibit its degradation in construction products, stabilizers are mixed during the manufacturing process with the PVC. These stabilizers are usually compounds of lead and zinc.

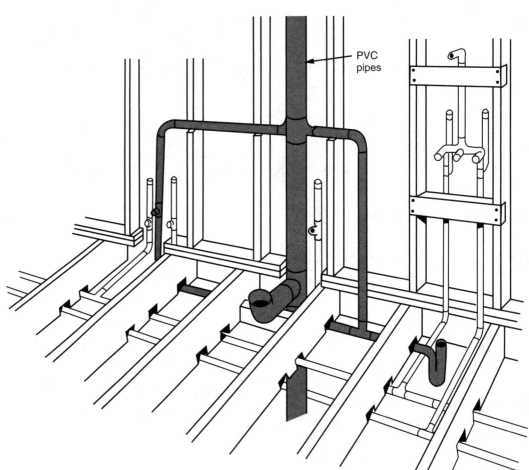

PVC
pipes

FIGURE 13.10 ◆ The use of PVC plumbing pipes in a bathroom, in which they are often used for hot and cold water delivery to fixtures, to carry drainage and waste, and to vent odors. The PVC pipes are usually joined to copper or galvanized steel pipes with transition fittings.

Although PVC does not easily ignite, the same is untrue of the plasticizers added to PVC products. The plasticizer most commonly added to PVC products is dioctyl phthalate (Section 12.12). The ease of ignition of the PVC product increases as more and more plasticizer is incorporated into it.

At elevated temperatures, PVC thermally decomposes and burns. Given the abundance of PVC products that firefighters are likely to encounter, this constitutes a potential health concern for the following reasons:

◆ Hydrogen chloride is produced when PVC burns. We noted in Section 8.7 that the hazard of inhalation toxicity is posed by exposure to hydrogen chloride.

◆ Polychlorinated dibenzofurans and polychlorinated dibenzo-*p*-dioxins are produced during the incomplete combustion of poly(vinyl chloride). We noted in Section 12.9 that exposure to minutely low dioxin concentrations causes a variety of illnesses in individuals and their offspring.

POLYACRYLONITRILE

Acrylonitrile is the monomer from which polyacrylonitrile is made. It is produced from propylene, ammonia, and oxygen, as follows:

$$2C_3H_6(g) \ + \ 2NH_3(g) \ + \ 3O_2(g) \ \longrightarrow \ \underset{\substack{\displaystyle \\ CN}}{\overset{\substack{H \\ /}}{CH_2{=}C(g)}} \ + \ 6H_2O(g)$$

propylene   ammonia   oxygen   acrylonitrile   water

It is a very volatile, toxic, and flammable liquid.

In polyacrylonitrile macromolecules, the repeating unit is ~CH$_2$CH~~~. This
$$\overset{|}{\underset{CN}{}}$$
synthetic polymer was the first to become commercially popular in the form of acrylic fibers. The term *acrylic fibers* refers to those fibers composed of at least 85% by mass of acrylonitrile units. Today, the fibers are primarily used in textiles, since they possess an outstanding resistance to sunlight, are quick drying, and are easy to launder. They are popularly known by the tradenames *Orlon* and *Acrilan*. Polyacrylonitrile is also used to manufacture plastics and synthetic rubber.

As we noted in Section 10.8, hydrogen cyanide is produced when materials made of polyacrylonitrile undergo thermal decomposition. Although life-threatening concentrations of hydrogen cyanide could be generated at fire scenes, the available evidence indicates that the presence of hydrogen cyanide does not significantly contribute to many fire-related fatalities. It is possible that the hydrogen cyanide is produced, but does not itself survive within the fire environment.

Solved Exercise 13.4

Which combustion products are produced when materials made from polyacrylonitrile smolder and burn?

Solution: The macromolecules of polyacrylonitrile possess hundreds or thousands of the following unit:

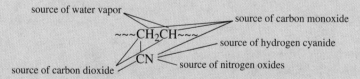

The constituent atoms of these units serve as potential source of combustion products. When the polymer "smolders," it is exposed to the intense heat evolved as the polymer slowly burns. Hydrogen cyanide is produced during the smoldering process. Carbon, hydrogen, and nitrogen atoms are the only constituents that oxidize when polyacrylonitrile burns. When the carbon atoms oxidize, carbon monoxide and carbon dioxide are produced; when the nitrogen atoms oxidize, nitric oxide and nitrogen dioxide are produced; and when the hydrogen atoms are oxidized, water is produced. Consequently, the combustion products generated during the combustion of acrylonitrile are hydrogen cyanide, carbon monoxide, carbon dioxide, nitric oxide, nitrogen dioxide, and water.

POLY(METHYL METHACRYLATE)

Several polyvinyl polymers are produced from the esters of acrylic acid and methacrylic acid. An example is the commercially popular poly(methyl methacrylate), which is frequently designated as *PMMA*. This polymer is produced by polymerizing methyl methacrylate. The repeating unit in PMMA is the following:

$$\sim\sim\sim CH_2 - \underset{\underset{\underset{\underset{CH_3}{|}}{O}}{\overset{\overset{CH_3}{|}}{\underset{\overset{|}{C=O}}{C}}}{C} \sim\sim\sim$$

Methyl methacrylate is manufactured by the following three-step process:

$$CH_3 - \underset{\overset{\|}{O}}{C} - CH_3(l) \;+\; HCN(g) \longrightarrow (CH_3)_2 - \underset{\overset{|}{OH}}{C} - CN(l)$$

acetone   hydrogen cyanide   acetone cyanohydrin

$$(CH_3)_2 - \underset{\overset{\|}{OH}}{C} - CN(l) \;+\; H_2SO_4(l) \longrightarrow CH_2{=}\underset{\underset{\underset{NH_3{\cdot}HSO_4(l)}{|}}{C=O}}{\overset{\overset{CH_3}{|}}{C}}$$

acetone cyanohydrin   sulfuric acid   methylacrylamide bisulfate

$$CH_2{=}\underset{\underset{\underset{NH_3{\cdot}HSO_4(l)}{|}}{C=O}}{\overset{\overset{CH_3}{|}}{C}} \;+\; CH_3OH(l) \longrightarrow CH_2{=}\underset{\underset{\underset{\underset{CH_3(l)}{|}}{O}}{C=O}}{\overset{\overset{CH_3}{|}}{C}} \;+\; NH_4HSO_4(aq)$$

methylacrylamide bisulfate   methanol   methyl methylacrylate   ammonium bisulfate

Methyl methacrylate is a very volatile, flammable liquid.

PMMA is a component of the plastic products known commercially as Plexiglas and Lucite. It is clear as glass, but can be manufactured as a transparent, translucent, or opaque material. Since it is virtually unbreakable under normal conditions of stress and tension, it is useful in windshields, windows, and other products simultaneously requiring weather resistance, strength, and transparency. The largest use of PMMA is associated with the manufacture of advertising signs (Figure 13.11) and displays, but the polymer is also used to make lighting fixtures, building panels, and plumbing and bathroom fixtures. The military uses PMMA in cockpit canopies, windows, gun turrets, and bombardier enclosures. It is also used as a component of latex and enamel paints, a drying oil for varnishes, and a finishing compound on leather.

As we noted earlier, when PMMA is exposed to heat, it depolymerizes and forms its monomer methyl methacrylate. Within a fire environment, the monomer readily burns.

FIGURE 13.11 ◆ Many outdoor advertising signs are manufactured from panels of poly(methyl methacrylate), which offer a combination of elasticity, toughness, and tear-resistance.

- -

Performance Goals for Section 13.7
- Describe the general properties of polyurethane and identify the products produced from it.
- Describe the nature of the combustion of polyurethane.
- Identify the potential hazard associated with the use of untreated polyurethane as insulation material.

- -

◆ 13.7 POLYURETHANE

A *polyurethane* is a substance produced by the condensation of a glycol and an organic diisocyanate. The chemical formula of a simple isocyanate is $R-N=C=O$. Organic diisocyanates are compounds having two isocyanate groups ($-N=C=O$).

A popular commercial polyurethane product is a foam produced by adding a low-boiling liquid to the reaction mixture of toluene 2,4-diisocyanate and ethylene glycol, as follows:

$$O=C=N-\bigcirc\text{-}N=C=O(l) \ + \ HO-CH_2CH_2-OH(l) \longrightarrow$$
$$\hspace{2cm}CH_3$$

toluene 2,4-diisocyanate   ethylene glycol

$$\sim\sim\overset{O}{\underset{\|}{C}}-NH-\bigcirc-NH-\overset{O}{\underset{\|}{C}}\left[O-CH_2CH_2-O-\overset{O}{\underset{\|}{C}}-NH-\bigcirc-NH-\overset{O}{\underset{\|}{C}}\right]O-CH_2CH_2\sim\sim(s)$$
$$\hspace{1cm}CH_3 \hspace{6cm} CH_3 \hspace{1.5cm} n$$

an example of a polyurethane

FIGURE 13.12 ◆ Fire marshals warn that the polyurethane foam in most upholstered furniture poses the most serious fire hazard found in American homes. This loveseat sleeper, chair, ottoman, and pillows have been cushioned with polyurethane foam to simultaneously provide flexibility and resiliency. Furniture manufacturers reduce their ease of ignition by mixing fire retardants into the foam as it solidifies. Notwithstanding the presence of fire retardants, polyurethane burns at the elevated temperatures generated during most fires to produce the deadly gases carbon monoxide and nitrogen dioxide.

The heat released during the polymerization causes this liquid to vaporize. Bubbles of vapor are subsequently captured within the matrix of the polymer. Polyurethane foam results when the frothy mixture solidifies.

Polyurethane foam is available commercially as products that are rigid or flexible plastics. The rigid foam is primarily used as insulation, sound-deadening boards, and wall panels. The insulation is conveniently sprayed upon the surface of concrete, between wall and roof joists, and around window and door frames. The flexible polyurethane foams are used in carpet padding and bedding, furniture (Figure 13.12), and automobile seat cushioning.

Solved Exercise 13.5

Which toxic gases are produced when piles of polyurethane jackets smolder?

Solution: Polyurethane macromolecules possess hundreds or thousands of a unit such as the following:

$$\text{~C−NH−}\bigcirc\!\!\!\!_{CH_3}\!\!\text{−NH−}\overset{O}{\underset{\|}{C}}\!\left[\!\!-O−CH_2CH_2−O−\overset{O}{\underset{\|}{C}}−NH−\bigcirc\!\!\!\!_{CH_3}\!\!\text{−NH−}\overset{O}{\underset{\|}{C}}\!\right]_n\!\!-O−CH_2CH_2\text{~}(s)$$

When the polyurethane jackets smolder, the toxic gases produced are hydrogen cyanide, carbon monoxide, nitric oxide, and nitrogen dioxide.

Polyurethane fires can be effectively extinguished with the application of water, but since they retain considerable heat, it is essential to check for total fire extinguishment within the foam products to prevent their reignition. In the past, it was well acknowledged that caution had to be exercised when fighting fires involving polyurethane products. Their combustion was associated with the production of prodigious quantities of nitrogen dioxide, the presence of which posed an increased risk of inhalation toxicity to building occupants and firefighters. This risk was considered so high that safety engineers required a warning label such as the following to be affixed to polyurethane building products:

> **FIRE HAZARD WHEN USED INSIDE BUILDINGS USE ONLY ON EXTERIOR APPLICATIONS, OR WHEN CONFINED BETWEEN WALLS OF MATERIALS CONSIDERED ACCEPTABLE WHEN USED IN CONSTRUCTION**

The polyurethane products encountered in the contemporary market have been formulated with fire retardants or treated with additives to reduce their ease of combustion. As Figure 13.13 shows, the burning pattern of polyurethane products can be altered by special designing. But does this mean that the concern over the use of

FIGURE 13.13 ◆ Foamed rigid plastic materials made from polyurethane are recognized for their insulating quality and light weight. The polyurethane rigid foam in this "Corner Test" at the Factory Mutual Test Center has been specially formulated and protected, so its contribution to the burning is minimal. *Courtesy of Factory Mutual Engineering & Research, Norwood, Massachusetts.*

polyurethane products is primarily a matter of the past? The answer is "no," because many buildings built with untreated polyurethane still exist today.

Performance Goals for Section 13.8
- Identify the common heat- and fire-resistant polymers and the common products produced from them.

◆ 13.8 HEAT- AND FIRE-RESISTANT POLYMERS

In their zest to identify new and better products, chemists have devoted considerable time to learning to make polymers that are heat- and fire-resistant. Their research efforts have led to the discovery of several polymers possessing these sought-after qualities. Heat- and fire-resistant polymers are very desirable for potential use in certain types of commercial products, especially building construction materials and products for home use.

POLY(TETRAFLUOROETHYLENE)

The commercial name for this polymer is *Teflon,* although it is also known as *PTFE.* The recurring unit in its macromolecules is $\sim\sim\sim CF_2 - CF_2 \sim\sim\sim$.

Teflon is produced by the polymerization of tetrafluoroethylene, which in turn is produced by a two-step process. First, chloroform is reacted with hydrofluoric acid.

$$\underset{\text{chloroform}}{CHCl_3(g)} + \underset{\text{hydrofluoric acid}}{2HF(l)} \longrightarrow \underset{\text{chlorodifluoromethane}}{CHClF_2(g)} + \underset{\text{hydrogen chloride}}{2HCl(g)}$$

Then, the chlorofluorocarbon is passed through a tube heated to 1300°F (700°C), whereupon a mixture of tetrafluoroethylene and hexafluoropropylene is produced.

$$\underset{\text{chlorodifluoromethane}}{2CHClF_2(g)} \longrightarrow \underset{\text{tetrafluoroethylene}}{CF_2{=}CF_2(g)} + \underset{\text{hydrogen chloride}}{2HCl(g)}$$

$$\underset{\text{chlorodifluoromethane}}{2CHClF_2(g)} \longrightarrow \underset{\text{hexafluoropropylene}}{F_3C{-}CF{=}CF_2(g)} + \underset{\text{hydrogen chloride}}{3HCl(g)}$$

Tetrafluoroethylene is isolated from this mixture and used to produce Teflon, or poly(tetrafluoroethylene).

Teflon is well known as the polymer used to produce nonstick cookware. It is also used to produce virtually indestructible tubing, gaskets, and valves. Chemists often use Teflon tubing in laboratories, but cardiologists also use it as artificial veins and arteries.

The heat resistance of Teflon is extraordinary. Teflon products perform well when exposed to temperatures ranging from −400°F (−240°C) to 482°F (250°C). Nonstick cookware coated with Teflon can be heated to temperatures in excess of 932°F (500°C) without burning.

Teflon is extraordinarily unreactive with hot corrosive acids. For this reason, Teflon is coated on the interior walls of tanks intended for acid storage. It is also the polymer used in the waterproof fabric called *Gore-Tex,* which golfers may recognize as the material commonly used in sportswear.

FIGURE 13.14 ◆ Firefighters usually escape from fires with only minor burns when they wear uniforms made of Nomex, a fire-resistant polymer. *Courtesy of Total Fire Group, Dayton, Ohio.*

Nomex

Nomex is the commercial name for a polymer made from *m*-diaminobenzene and isophthaloyl chloride, as follows:

m-phenylene diamine   isophthaloyl chloride   Nomex

From this polymer, a heat-resistant fabric is produced that has been used in the uniforms of firefighters (Figure 13.14) and race car drivers. The fabric provides an element

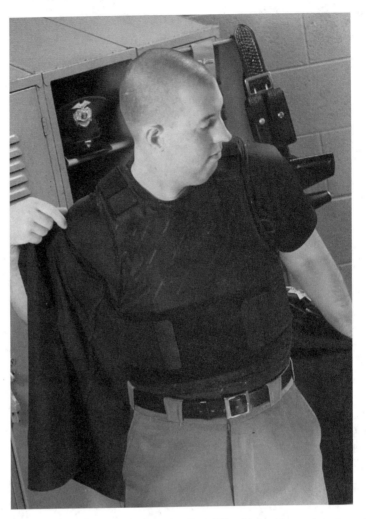

FIGURE 13.15 ◆ A bullet-resistant vest made of Kevlar, a super-strong synthetic fiber. When a bullet is fired at Kevlar body armor, it flattens on impact within the multiple-layer web of the fabric. The Kevlar fabric absorbs the energy of the bullet's impact and disperses it to other fibers in the weave. *Courtesy of Second Chance Body Armor, Inc., Traverse City, Michigan.*

of thermal protection to the wearer in that it carbonizes and thickens when exposed to intense heat. This increases the protective barrier between the heat source and the skin, thus minimizing burn injury.

Kevlar

This is the commercial name for a polymer made from *p*-diaminobenzene and terephthaloyl chloride.

terephthaloyl chloride   *p*-phenylene diamine   Kevlar

Kevlar is flameproof and can be drawn into fibers having a strength five times stronger than steel. Given its astounding resiliency, fabrics made from Kevlar fibers are used as the reinforcement in bullet-resistant vests (Figure 13.15) and helmets. A 16.4-lb (7.5-kg) vest lined with Kevlar and ceramic plates can stop armor-piercing bullets shot from high-powered rifles. A 4-lb (1.8-kg) helmet lined with up to 24 layers of Kevlar is approximately 40% more resistant to shrapnel than the steel helmets formerly used by the military.

No single product compares with Kevlar in terms of the number of lives saved through its use. Nearly 3000 police officers saved their lives because they wore Kevlar vests and helmets during the line of duty; during Operation Iraqi Freedom, the lives of at least two American servicemen were saved because artillery fired at them could not pierce their Kevlar vests.

Aside from their use in military and police body armor, Kevlar is also used in manufacturing rip-resistant jeans, work gloves, skis, hockey sticks, golf balls, ropes, cables, boat hulls, and aircraft structural parts. Some tire manufacturers are using Kevlar to reinforce the tread of radial automobile tires.

Performance Goals for Section 13.9

- ◆ Describe the general nature of the macromolecules in natural rubber.
- ◆ Describe the vulcanization of natural rubber.
- ◆ Identify the substances commonly used in rubber formulations.
- ◆ Identify the general nature of the macromolecules in synthetic rubber produced from 1,3-butadiene and styrene.
- ◆ Identify the toxic gases produced when rubber products burn.
- ◆ Describe the nature of the response actions to be implemented when emergency responders encounter burning rubber products.

◆ 13.9 NATURAL AND SYNTHETIC RUBBER

The term "rubber" refers to any of the natural or synthetic polymers having two main properties: deformation under strain and elastic recovery after vulcanization (defined below). As noted in Section 13.1, these polymers are called *elastomers*.

Natural rubber is isolated from *natural latex*, a white fluid that exudes from the bark of the South American rubber tree *Hevea brasiliensis*. Natural latex consists of approximately 30 to 35% by mass of *cis*-1,4-polyisoprene. Its general chemical formula is $(C_5H_8)_n$. A macromolecular component of natural rubber can be represented as follows:

$$
\begin{array}{ccccccc}
\text{H} & \text{CH}_3 & \text{H} & \text{CH}_3 & \text{H} & \text{CH}_3 \\
\diagdown & \diagup & \diagdown & \diagup & \diagdown & \diagup \\
\text{C}=\text{C} & & \text{C}=\text{C} & & \text{C}=\text{C} \\
\diagup & \diagdown & \diagup & \diagdown & \diagup & \diagdown \\
{\sim}\text{CH}_2 & \text{CH}_2\text{CH}_2 & \text{CH}_2\text{CH}_2 & \text{CH}_2{\sim}
\end{array}
$$

As this structure illustrates, nature selectively produces only the *cis*-isomer of 1,4-polyisoprene in which the methyl groups are oriented in one direction about the carbon-carbon double bonds.

By itself, natural rubber is not entirely desirable for production of commercial products. It is soft and sticky, especially in warm climates. In 1839, however, Charles

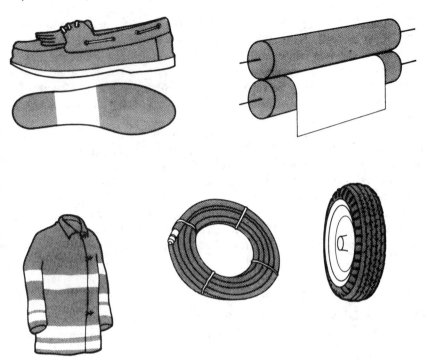

FIGURE 13.16 ◆ Examples of some common products made either directly from rubber or from polymers with rubber additives.

Goodyear accidentally discovered that these undesirable features could be eliminated by heating the natural rubber with elemental sulfur. This discovery led to the development of synthetic rubber and the industrial process called *vulcanization*. Virtually all synthetic rubber formulations today include elemental sulfur or sulfur-bearing compounds. The resulting products are called *vulcanized rubbers*. The elastomeric forms of vulcanized rubbers are tougher and more elastic than natural rubber and are capable of retaining firm shapes over a relatively wide temperature range.

Vulcanization also permits the casting of natural rubber into shapes such as automobile tires. Rubber mixed with sulfur is inserted into a mold, which when closed and heated produces a tire carcass.

A variety of rubber products are manufactured in commerce. Some are shown in Figure 13.16. Their production requires controlling the amount of sulfur used during the vulcanization of natural rubber, as shown by the following:

- Natural rubber mixed with only 3% sulfur by mass is soft and elastic. It is used for manufacturing inner tubes and rubber bands.
- With additional sulfur, vulcanized rubber becomes hard and loses some of its elasticity. Natural rubber having up to 10% sulfur by mass is used to manufacture automobile tires.
- When rubber has been vulcanized to the extent of 68% by mass, it becomes a black solid called *ebonite*, or *hard rubber*. It is used to make black piano keys and the casings for automobile storage batteries.

At the macromolecular level, vulcanization results in cross-linking the *cis*-1,4-polyisoprene chains with disulfide (—S—S—) bonds, as follows:

$$\text{CH}_3 \qquad\qquad\qquad \text{CH}_3$$
$$\text{~~~CH—C}\text{=}\text{CH—CH}_2\text{—CH—C}\text{=}\text{CH—CH}_2\text{~~~}$$
$$| \qquad\qquad\qquad\qquad |$$
$$\text{S} \qquad\qquad\qquad\qquad \text{S}$$
$$| \qquad\qquad\qquad\qquad |$$
$$\text{S} \qquad\qquad\qquad\qquad \text{S}$$
$$| \qquad\qquad\qquad\qquad |$$
$$\text{~~~CH—C}\text{=}\text{CH—CH}_2\text{—CH—C}\text{=}\text{CH—CH}_2\text{~~~}$$
$$\text{CH}_3 \qquad\qquad\qquad \text{CH}_3$$

Substances other then sulfur arc also components of rubber formulations. In particular, carbon black (Section 7.7) is often added as a reinforcing filler to make the rubber for tires stronger, abrasive, and easier to elongate, and to prevent it from breaking and tearing. A specified amount of previously used rubber (reclaimable rubber) is usually recycled by adding it to a new batch of rubber.

Solved Exercise 13.6

Why do foam rubber products catch fire easily within a hot clothes dryer or under a heating blanket, while vehicular rubber tires are comparatively difficult to ignite even upon exposure to intense heat?

Solution: Foam rubber products are produced by blowing and entrapping an inert gas within the complex structure of rubber until it hardens. This process causes an expansion of the rubber and an increase in its surface area. By contrast, an inert gas is not entrapped within the rubber used for the production of rubber tires. Following the principles in Section 5.5, an increase in the surface area of the reactants increases the reaction rate. Since the rubber in foam rubber products possesses a greater surface area than the rubber in rubber tires, foam rubber products ignite and burn at a comparatively faster rate than rubber tires.

Depending on the projected use of the rubber product, air or ammonium carbonate can be mixed into the rubber latex. When this mixture is subsequently heated, the ammonium carbonate decomposes to ammonia and carbon dioxide, which become entrapped within the complex rubber structure as it hardens. This product is called *foam rubber;* it is used in cushions and mattresses.

Natural rubber still accounts for approximately 35% of the demand for rubber in the United States. Today, however, the world no longer relies solely on the latex from rubber trees for its source of rubber. Chemists have synthesized elastomers having properties similar to those of natural rubber. One of the simplest raw materials used to produce a synthetic rubber is 1,3-butadiene. By using a Ziegler–Natta catalyst to initiate polymerization, 1,3-butadiene can be polymerized to produce *cis*-1,4-polybutadiene.

Other synthetic rubbers are manufactured using 1,3-butadiene as the raw material. A commercially important copolymer is *SBR* (styrene butadiene rubber), formerly known as *GRS* (Government rubber styrene) since it was first produced on behalf of the United States government during World War II. As the former name suggests, this is a copolymer of 1,3-butadiene and styrene, as follows:

$$CH_2=CH-CH=CH_2(g) \ + \quad CH=CH_2(g)$$

1,3-butadiene   styrene

$$\longrightarrow \ \text{\small ww}CH_2-CH=CH-CH_2CH_2-CH-\overset{\displaystyle CH_2\text{\small ww}}{\overset{\displaystyle |}{CH}}CHCH=CH_2(s)$$

(GRS or SBR)

SBR is vulcanized using elemental sulfur. Since *SBR* resists wear more than any other synthetic rubber, it is the raw material used to manufacture most tire treads.

Outside government circles, *SBR* became known historically as *Buna S* ("Bu" for butadiene, "na" for sodium, and "S" for styrene). A similar synthetic rubber was *Buna N*, a copolymer of butadiene and acrylonitrile ("N" for acrylonitrile). Buna N rubber has the unique feature of withstanding heat up to 350°F (177°C); other synthetic rubbers soften, melt, or burn at lower temperatures.

Another important synthetic rubber is *cis*-1,4-polychloroprene, commonly known as *neoprene*. This rubber is produced by polymerizing chloroprene, a process that is easily accomplished by heating. Chloroprene is also known as 2-chloro-1,3-butadiene. Research studies link chloroprene exposure with the incidence of cancer in humans; hence, chloroprene is classified as a known human carcinogen.

The polymerization of chloroprene occurs as follows:

$$CH_2=CH-\overset{\displaystyle |}{\underset{\displaystyle Cl}{C}}=CH_2(g) \ \longrightarrow \ \text{\small ww}\Big(\!CH_2-\overset{\displaystyle |}{\underset{\displaystyle Cl}{C}}=C-CH-CH_2\!\Big)_{\!n}\text{\small ww}(s)$$

chloroprene   neoprene

When vulcanized with zinc oxide, single strands of neoprene cross-link with oxygen atoms and replace the chlorine atoms. A segment of the vulcanized neoprene macromolecule possesses a structure such as the following:

$$\text{\small ~~~}CH_2-\overset{\displaystyle Cl}{\overset{\displaystyle |}{C}}=C-CH_2\text{\small ~~~}$$
$$\overset{\displaystyle |}{\underset{\displaystyle O}{}}$$
$$\text{\small ~~~}CH_2-\underset{\displaystyle Cl}{\underset{\displaystyle |}{C}}=C-CH_2\text{\small ~~~}$$

Neoprene rubber does not possess the resilience necessary for use in tires, but it is chemically resistant to petroleum products and ozone. This feature makes neoprene suitable for specialized uses such as roofing tile and flexible hose.

FIGURE 13.17 ◆ A fire within massive numbers of tires is sometimes activated by a lightning strike. As these tires burn, the smoke that evolves is intensely dense and black. To bring the fire under control, the use of special fire extinguishers is generally required. *Courtesy of Pyrocool Technologies, Inc., Monroe, Virginia.*

RESPONDING TO INCIDENTS INVOLVING THE BURNING OF RUBBER PRODUCTS

To understand the nature of the substances produced when the various synthetic rubbers burn, it is necessary to recollect the general features of their chemical formulations. During the combustion process, the constituents of these formulations combine with atmospheric oxygen in the following ways:

* As they burn, rubber products vulcanized with sulfur or sulfur-bearing compounds produce carbon monoxide, sulfur dioxide, and water vapor.
* The smoke associated with rubber fires is extraordinarily dense and black, as in the scene in Figure 13.17. Both features of these fires are associated with the presence of carbon black in the rubber formulation.
* As neoprene burns, carbon monoxide, water vapor, and hydrogen chloride are produced.

With virtual certainty, the atmosphere near burning rubber consists of a mixture of toxic gases. Their presence must be considered when determining a proper firefighting strategy. To prevent fatalities, the use of self-contained breathing apparatus is warranted.

■ ■

REVIEW EXERCISES

Nature of Polymerization

13.1 Write the chemical formula for macromolecular segments of each of the following polymers:
  (a) polyethylene having eight recurring units
  (b) polypropylene having five recurring units
  (c) poly(methyl methacrylate) having three recurring units

13.2 Addition and condensation polymerization reactions are often descriptively referred to as "chain growth" and "step growth" polymerization, respectively. Show that these terms appropriately describe the relevant polymerization phenomena.

13.3 Poly(vinylidene chloride) is produced by the addition polymerization of vinylidene chloride, or 1,1-dichloroethene. Write the stepwise equations that illustrate how this addition polymerization occurs.

13.4 The popular thermosetting resin called *Bakelite* is produced by the condensation polymerization of phenol and formaldehyde. In the first step of this reaction, water is eliminated when phenol and formaldehyde combine to produce two compounds, each of which has a $-CH_2OH$ group bonded to a carbon atom at the *ortho* and *para* positions to the hydroxy group in phenol, respectively.
  (a) Write the equation that shows how this condensation reaction occurs.
  (b) Why are Bakelite products much harder and more durable than most other products manufactured from synthetic polymers?

Autopolymerization

13.5 DOT prohibits shippers and carriers from accepting acrolein for transportation unless it has been "stabilized." DOT then denotes its proper shipping name as "acrolein, stabilized."
  (a) What specific property of acrolein must be stabilized?
  (b) Use Tables 6.1 and 6.4 to determine the basic description a carrier is required to enter on the waybill when acrolein is transported by railroad tankcar.

13.6 What role is rendered by hydroquinone, a substance often added by manufacturers to their products that autopolymerize?

13.7 When properly posted on the exterior wall of a building, how does the observation of an NFPA diamond convey to firefighters that an autopolymerizing substance is stored nearby?

Combustion of Polymers

13.8 Identify the combustion products that result when polymers having each of the following recurring units burn in air:

$$\text{(a)} \quad \sim\sim\sim CH_2 - \underset{\underset{\underset{\underset{CH_3}{|}}{O}}{\overset{\overset{\overset{H}{|}}{}}{\underset{C=O}{C}}}{C} \sim\sim\sim \qquad\qquad \text{(b)} \quad \sim\sim\sim NH - (CH_2)_5 - \overset{\overset{O}{\|}}{C} \sim\sim\sim$$

(c) ~~~$CH_2CH_2CH_2CH_2$—O—$\overset{\overset{O}{\|}}{C}$—⟨◯⟩—$\overset{\overset{O}{\|}}{C}$—O~~~

(d) ~~~CH_2—$\overset{\overset{Cl}{|}}{\underset{\underset{Cl}{|}}{C}}$~~~

(e) ~~~$CH_2CH_2CH_2$—S—S—CH_2CH_2~~~

(f) ~~~NH—CH—$\overset{\overset{O}{\|\!\!\!/}}{C}$
$\qquad\quad\;|\qquad\backslash$
$\qquad\quad CH_3\quad NH$—CH—$\overset{\overset{O}{\|\!\!\!/}}{C}$
$\qquad\qquad\qquad\;|\qquad\backslash$
$\qquad\qquad\quad CH_3$—CH$\quad$NH—CH—$\overset{\overset{O}{\|\!\!\!/}}{C}$~~~
$\qquad\qquad\qquad\qquad|\qquad\quad\;|$
$\qquad\qquad\qquad\quad CH_3\qquad CH_2$—

13.9 Which toxic gases are produced when a leather-covered couch smolders during a residential fire?

13.10 Cystine is a substance composed of macromolecules that are produced when two cysteine molecules oxidize by forming a sulfur–sulfur bridge (a disulfide bond) between them. It is abundant in animal hair, nails, horn, hoof, and feathers.
   (a) Write the chemical formula for cystine.
   (b) Why does the thermal decomposition of animal hair, nails, horn, hoof, and feathers produce hydrogen sulfide and ammonia?
   (c) Why does the burning of animal hair, nails, horn, hoof, and feathers produce sulfur dioxide, nitric oxide, and nitrogen dioxide?

Vegetable and Animal Fibers

13.11 What thermal decomposition products are produced when a woolen sweater is exposed to intense heat?

13.12 Short cotton fibers are called *cotton linters*. The DOT regulation at 49 CFR § 176.900 stipulates that baled cotton linters may be accepted for domestic transportation by watercraft only when each bale has been compressed to a minimum density of 32 lb/ft^3 (512 kg/m^3). The DOT regulation further specifies that a poorly compressed bale or a bale having damaged binding cannot be transported by watercraft. What is the most likely reason DOT requires cotton linters to be compressed before they can be accepted for domestic transportation on watercraft?

Vinyl Polymers

13.13 Which requires the application of a larger volume of water for effective total extinguishment: 1000 lb (454 kg) of burning polyethylene or the same mass of burning polystyrene?

13.14 Identify five products manufactured from poly(vinyl chloride) that are likely to be found in an average home.

13.15 What feature of polypropylene allows manufacturers to produce a more versatile range of products than can be produced from any other polymer?

13.16 When red-brown smoke evolves at the scene of a residential fire, why should the home's occupants be particularly wary about their well-being?

13.17 Write the chemical formula for each of the following nitrogenous organic compounds:

(a) methyl isocyanate   (b) nitromethane
(c) butenenitrile   (d) acetonitrile
(e) 2-aminopropane   (f) *p*-nitrochlorobenzene
(g) toluene 2,4-diisocyanate   (h) formamide
(i) 3-aminopropanoic acid   (j) hexamethylenediamine

13.18 Provide an acceptable name for each of the nitrogenous organic compounds having the following condensed formulas:

(a) [benzene ring]—NH_2

(b) [benzene ring]—CH_2—$\overset{\overset{\textstyle O}{\|}}{C}$—$NH_2$

(c) [benzene ring]—NO_2 (with CH_3 substituent)

(d) CH_3—CH—CH—$\overset{\overset{\textstyle O}{\|}}{C}$—$OH$ (with CH_3 and NH_2 substituents)

(e) CH_3—CH—CH_2CH_2—NH_2 (with CH_3 substituent)

13.19 *Dacron* fabrics are produced from poly(ethylene terephthalate) fibers. The recurring unit in the macromolecules of this polyester is the following:

$$\sim\sim\sim O-CH_2CH_2-O-\overset{\overset{\textstyle O}{\|}}{C}-[\text{benzene ring}]-\overset{\overset{\textstyle O}{\|}}{C}\sim\sim\sim$$

When Dacron fabrics burn, which combustion products are produced?

13.20 Poly(vinylidene chloride) decomposes at a fire scene when it has been heated to a temperature ranging from 437 to 527°F (225 to 275°C). The recurring unit in the macromolecules of poly(vinylidene chloride) is $\sim\sim\sim CH_2-CCl_2\sim\sim\sim$.

(a) Which toxic gases are most likely to be components of its decomposition products?

(b) In which ways could the presence of these gases at a fire scene adversely affect the health of emergency responders?

13.21 Why is the ash from a municipal solid waste incinerator usually a source of dioxin?

Polyurethane

13.22 Polyurethane foam is commonly used as the cushioning in automobile seats. Why is its use as cushioning in airplane seats avoided?

Heat- and Fire-Resistant Polymers

13.23 Why are Kevlar fibers incorporated into the body armor used by emergency-response personnel?

Rubber

13.24 Air monitoring provides the following contaminant concentrations at the approximate center of a fire scene at which acres of automobile tires are burning:

	Concentration, ppm
Carbon monoxide	1700
Carbon dioxide	7000
Sulfur dioxide	600

Ignoring synergistic effects between these gases, which individual concentrations are considered life-threatening to the firefighters at the scene?

CHAPTER 14

Chemistry of Some Explosive Materials

Chemical explosives are customarily components of the following:

- Ammunition for hunting and similar sporting activities
- Charges that are implanted during the mining of ores, drilling of oil wells, tunneling through mountains, clearing land, and loosening of underground rock formations
- Artillery and munitions that destroy cities, sink ships, and kill the enemy during wartime.

Less frequently, they are employed for fighting unusually large, uncontrolled fires in oil fields, prairies, forests, or segments of large metropolitan areas where the intent of their use is to create a firebreak.

Chemical explosives are also the active components of the weapons most commonly chosen by terrorists for mass destruction. In 1995, for example, the detonation of an explosive mixture of ammonium nitrate and fuel oil destroyed the Murrah Federal Building in Oklahoma City, Oklahoma, killed 168 people, and injured 518 (Figure 11.1). In 1998, terrorists downed Pan Am Flight 103 over Lockerbie, Scotland, by detonating a chemical explosive embedded within a suitcase. Two hundred fifty-nine (259) passengers and crew were killed along with 11 more people on the ground. In the same year, nearly simultaneous car bombings destroyed the American embassies in Kenya and Tanzania, killing 231 people including 12 Americans. Also in 1998, suicide attackers rammed an explosives-laden boat into the destroyer U.S.S. *Cole* in Yemen, killing 17 American sailors. International terrorists have also used chemical explosives to carry out acts of violence in Afghanistan, Columbia, Djerba, India, Indonesia, Iraq, Israel, Kenya, Lebanon, Morocco, Pakistan, Philippines, Russia, Saudi Arabia, Turkey, and elsewhere.

Individuals who use chemical explosives receive specialized training in the handling of explosives. Although such training requires more rigorous instruction than can be provided here, we shall learn in this chapter about the chemical composition of explosive materials, their different types, and the recommended procedures for responding to mishaps involving them.

Performance Goals for Section 14.1

- Distinguish between the combustion and the detonation of a substance.
- Distinguish between the nature of a chemical explosive and an explosive article, noting the examples in Tables 14.1 and 14.2.
- Describe the manner by which the brisance of a chemical explosive is determined.

◆ 14.1 GENERAL CHARACTERISTICS OF EXPLOSIVE MATERIALS

There are two general types of explosive materials:

◆ *Chemical explosives.* These are unique chemical substances whose transformation is activated by the application of friction, a mechanical impact, or heat. Their primary use is associated with the accomplishment of a specific act, such as demolition. In this latter regard, chemical explosives are distinguishable from substances like gasoline or flammable gases, which commercially serve other purposes, but appear to "explode" under certain unique circumstances. They are also distinguishable from nuclear explosives, which are associated with the occurrence of certain nuclear phenomena, such as nuclear fission (Section 15.6).

◆ *Explosive articles.* These are manufactured items that contain a chemical explosive as a component. Some examples of explosive articles are provided in Table 14.1. The chemical substance in an explosive article is intentionally initiated to detonate, as needed, by mechanical, electrical, chemical, or hydrostatic means.

TABLE 14.1 ◆ Some Explosive Articles[a]

Ammunition, illuminating	Cord, detonating
Ammunition, incendiary	Cord igniter
Ammunition, practice	Fireworks
Ammunition, proof	Flares
Ammunition, smoke	Flash powder
Ammunition, tear-producing	Fuse[b]
Ammunition, toxic	Fuze[b]
Black powder (gunpowder)	Igniters
Bombs, military	Lighters, fuse
Boosters	Mines
Cartridges, blank	Powder cake (powder paste)
Cartridges, flash	Powder, smokeless
Cartridges for weapons	Primers, cap type
Cartridges, oil well	Primers, tubular
Cartridges, power device	Projectiles
Cartridges, signal	Propellants, liquid
Cartridges, small arms	Propellants, solid
Charges, bursting	Rocket motors
Charges, demolition	Signals
Charges, depth	Sounding devices
Charges, expelling	Torpedoes
	Warheads

[a]49 CFR §173.59.
[b]"Fuse" refers to a cord-like igniting device, whereas "fuze" refers to a device used in ammunition that incorporates mechanical, electrical, chemical, or hydrostatic components to initiate a train by deflagration or detonation.

When explosive materials are mentioned in this chapter, the term is meant to imply both chemical and explosive articles.

A chemical explosive may be either an inorganic or organic compound. In either instance, it simultaneously represents both the oxidizing agent and reducing agent. Parts of its molecules are oxidized as other parts are reduced. The explosive undergoes a chemical transformation very suddenly by comparison to the rates of other chemical reactions, usually forming gases, vapors, and a large amount of energy. This phenomenon, called a *detonation,* is expressed by an equation similar to the following, where "A" is the formula of an explosive substance and "B" and "C" are the formulas of its decomposition products:

$$A(s,l) \longrightarrow B(g) + C(g)$$

A detonation is associated with the essentially instantaneous passage of an energy wave through the body of the chemical explosive at a supersonic rate (i.e., exceeding the speed of sound). The movement of this wave is called the *detonation velocity.* Most chemical explosives are powerful not because of the total energy emitted when they detonate, but because of the high rate at which the energy is released. On the average, the detonation velocity of chemical explosives is a staggering 4 mi/s (7 km/s).

An organic explosive is usually a relatively complex compound whose molecules contain a fuel and an oxidizer. Its molecules are typically composed of carbon, hydrogen, oxygen, and nitrogen atoms. When the explosive detonates, the associated transformation ordinarily produces carbon monoxide, carbon dioxide, nitrogen, oxygen, and/or water vapor as decomposition products, although carbon particulates (soot) and nitrogen oxides may also be produced. The products of these reactions absorb some of the energy that evolves, which heats them to profoundly high temperatures, typically around 5400°F (3000°C). This increase in temperature causes the substances to simultaneously expand so rapidly that the passage of the accompanying shock wave exerts an exceedingly high pressure upon the nearby surroundings.

Solved Exercise 14.1

Individuals who survive the shock effects affiliated with a typical explosion may still experience the risk of asphyxiation or poisoning from exposure to the gases produced during the detonation. Identify the most prevalent gases produced during a typical explosion and name those that are toxic.

Solution: Most organic explosives contain carbon, hydrogen, oxygen, and nitrogen atoms. When they detonate, the products of decomposition include carbon monoxide, carbon dioxide, nitrogen, oxygen, water, carbon particulates, nitric monoxide, and nitrogen dioxide. There are three toxic gases in this group of substances. They are carbon monoxide, nitric oxide, and nitrogen dioxide.

For illustration, consider the detonation of the explosive nitroglycerin. When it detonates, nitroglycerin transforms in less than one-millionth of a second into carbon dioxide, nitrogen, oxygen, and water vapor, as follows:

$$
\begin{array}{l}
\text{CH}_2-\text{O}-\text{NO}_2 \\
\quad | \\
4\text{CH}-\text{O}-\text{NO}_2 \qquad \longrightarrow \quad 12\text{CO}_2(g) + 6\text{N}_2(g) + \text{O}_2(g) + 10\text{H}_2\text{O}(g) \\
\quad | \\
\text{CH}_2-\text{O}-\text{NO}_2(l)
\end{array}
$$

     nitroglycerin   carbon dioxide   nitrogen   oxygen   water

The energy of detonation travels at the rate of 4.8 mi/s (7.8 km/s).

    Unlike a combustion reaction, a detonation does not require the interplay of atmospheric oxygen or other oxidizer. Instead, it is an oxidation–reduction reaction involving the rapid decomposition of a single substance that acts as both an oxidizing agent and a reducing agent.

    At the elevated temperature accompanying any detonation, the heated gases expand so rapidly that they could occupy approximately 10,000 times their initial volume. This prodigious expansion is routinely accompanied by the simultaneous production of shock waves, an extraordinarily loud sound, and a luminescence like that shown in Figure 14.1.

FIGURE 14.1 ◆ The detonation of a chemical explosive may be initiated in several ways, such as by heat or mechanical impact. This detonation was induced by exposing the material to pulsed, high-energy X rays. The X-ray machine is located in the domed building and is part of the Dynamic Testing Division at Los Alamos National Laboratory, Los Alamos, New Mexico. The machine is used to study explosions and the nature of explosive materials. *Courtesy of the U.S. Department of Energy, Washington, D.C.*

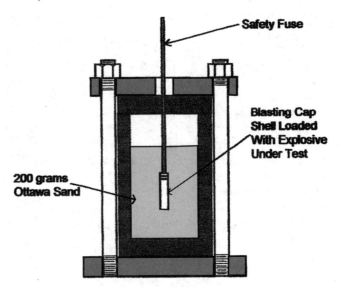

FIGURE 14.2 ◆ The type of equipment for conducting a sand test, which is used to determine the brisance of a chemical explosive. The sand test measures the quantity of Ottawa sand a chemical explosive crushes inside a heavy, thick-walled confining bomb. Ottawa sand is characterized by its inability to pass through a 30-mesh screen. After the chemical explosive is detonated, the sand is rescreened and the amount that now passes through a 30-mesh screen is weighed. The result is measured in grams of sand. Of the explosives noted in this book, PETN (Section 14.12) possesses the highest brisance, 62.7 g of sand.

The shock waves produced during the detonation of a chemical explosive are linked with its potential shattering power, called the *brisance*. The brisance is an important factor to consider when a chemical explosive is selected for a specific purpose. For example, when constructing a roadway, an engineer considers whether the brisance of an explosive is likely to clear a given amount of rock. The brisance of an explosive is measured as the weight of graded sand that is crushed when a given amount of the explosive is detonated in the device shown in Figure 14.2.

Some so-called chemical explosives are actually pyrotechnic mixtures of substances, at least one of which is an oxidizer such as ammonium nitrate (Section 11.8). These mixtures undergo oxidation–reduction reactions so vigorously that they *appear* to detonate like a single substance. They include the formulations used in blasting agents (Section 14.2) and gunpowder (Section 14.5). Mixtures of substances are also used as the active components of certain illuminating, incendiary, tear-producing, and smoke-producing devices (Solved Exercise 14.2). These mixtures may or may not contain ammonium nitrate. When offered for transportation, DOT regulates them as explosive materials, not oxidizers.

Solved Exercise 14.2

During warfare, the U.S. Army deploys smoke bombs to visually mask the movement of troops and vehicles during combat activities. A popular smoke bomb used during the Korean conflict was a pyrotechnic mixture containing aluminum dust, hexachloroethane, and zinc oxide. The military formulation for this mixture was 6.68% granulated aluminum, 46.66% zinc oxide, and 46.66% hexachloroethane by mass. When the components of this mixture were activated with a fuse, the products that appeared in the smoke were aluminum oxide, zinc chloride, and carbon (soot). The zinc chloride attracted atmospheric moisture to form a fog, the aluminum oxide deflected light rays, and the carbon colored the smoke cloud gray. Write a balanced chemical equation illustrating this chemical reaction.

   Solution: The reactants are aluminum, hexachloroethane, and zinc oxide, whose chemical formulas are Al, CCl_3—CCl_3, and ZnO, respectively. The products of the reaction are aluminum oxide, zinc chloride, and carbon, whose chemical formulas are Al_2O_3, $ZnCl_2$, and C, respectively. The unbalanced equation is now written.

$$Al(s) \ + \ CCl_3\text{—}CCl_3(s) \ + \ ZnO(s) \ \longrightarrow \ ZnCl_2(s) \ + \ Al_2O_3(s) \ + \ C(s)$$
$$\text{aluminum} \qquad \text{hexachloroethane} \qquad \text{zinc oxide} \qquad \text{zinc chloride} \qquad \text{aluminum oxide} \qquad \text{carbon}$$

This equation is then balanced, as follows:

$$2Al(s) + CCl_3\text{—}CCl_3(s) + 3ZnO(s) \longrightarrow 3ZnCl_2(s) + Al_2O_3(s) + 2C(s)$$

Solved Exercise 14.3

The chemical formula of trinitrotoluene (TNT) (Section 14.9) can be condensed to $C_7H_5(NO_2)_3$. Write the balanced equation that denotes the detonation of TNT.

   Solution: Before the equation representing the detonation of TNT can be written, we need to first identify the decomposition products that are produced. During any detonation, the hydrogen and oxygen atoms combine to form water, and the carbon and oxygen atoms combine to form either carbon monoxide or carbon dioxide, or both. If carbon and oxygen remain after production of carbon monoxide and/or carbon dioxide, they are released during the detonation as their respective elements. Since there are seven carbon atoms in each TNT molecule, some elemental carbon is inevitably produced during the detonation of TNT. The nitrogen atoms unite to form elemental nitrogen, or they react with oxygen atoms to produce either nitrogen monoxide or nitrogen dioxide, or both.

   After some consideration, it becomes apparent that the TNT detonation products are nitrogen, water vapor, carbon monoxide, and elemental carbon. The unbalanced equation is then written as follows:

$$C_7H_5(NO_2)_3(s) \ \longrightarrow \ CO(g) \ + H_2O(g) + N_2(g) + C(s)$$
$$\text{trinitrotoluene} \qquad\qquad \text{carbon monoxide} \qquad \text{water} \qquad \text{nitrogen} \qquad \text{carbon}$$

This equation is then balanced.

$$2C_7H_5(NO_2)_3(s) \ \longrightarrow \ 7CO(g) + 5H_2O(g) + 3N_2(g) + 7C(s)$$

◆ 14.2 CLASSIFICATION OF EXPLOSIVE MATERIALS AND BLASTING AGENTS

Individual chemical explosives are usually classified into the following two groups, depending on the rapidity and sensitivity with which they detonate:

- *High,* or *detonating explosives*
- *Low,* or *deflagrating explosives*

High and low explosives are distinguishable only in a broad sense by the differences in their detonation velocities and shock sensitivities. High explosives undergo chemical transformations quite rapidly, whereas low explosives transform hundreds of times more slowly and typically with less ease. The detonation velocity can be as great as 4 mi/s (6000 m/s) for a high explosive, but only 900 ft/s (270 m/s) for a low explosive.

The classification of explosives as high or low is not an entirely satisfactory one. Some high explosives can be detonated only when initiated by using the blasting caps in Figure 14.3 or similar activating devices. Some explosives can be categorized into either class, depending upon their state of subdivision. Smokeless powder (Section 14.8), for instance, has been typically considered a low explosive, but under appropriate conditions, it detonates like a high explosive.

Chemical explosives are also classified into either of the following two groups:

- *Primary explosives*
- *Secondary explosives*

Primary explosives are substances that are extremely sensitive to heat, mechanical shock, and friction; hence, they develop shock waves within an extremely short time

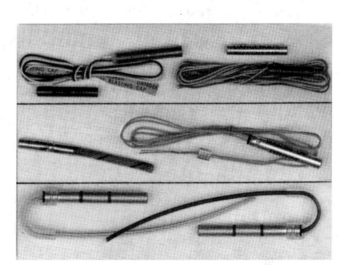

FIGURE 14.3 ◆ Various types of blasting caps used to detonate other chemical explosives, especially during mining and construction work.

period. Typical primary explosives are lead azide, mercury fulminate, and mercury styphnate (Section 14.14). By comparison, *secondary explosives* are relatively insensitive to heat, mechanical shock, and friction, and they typically require the use of a booster to bring about their detonation. Typical secondary explosives are cyclonite (Section 14.10), tetryl (Section 14.11), PETN (Section 14.12), and HMX (Section 14.13). Both primary and secondary explosives are dangerous hazardous materials since a considerable brisance is associated with their detonations.

Explosive materials that are classified as "low explosives" or "secondary explosives" should never be regarded as materials or articles possessing relatively lesser hazards compared to those of "high explosives" or "primary explosives," respectively. Low explosives can deflagrate like burning organic peroxides (that is, they can burn furiously and persistently); but low explosives *can* also detonate, causing considerable damage at the explosion site. Although a low explosive may first burn, the heat generated during its combustion could constitute the initiating mechanism that causes it to subsequently detonate.

From a chemical standpoint, the instability of explosives is often understood by examining the composition of their molecules. Some molecules possess one or more active groups of atoms called *explosophores*. The presence of multiple numbers of explosophores within the structural framework of a single substance is likely to cause it to decompose explosively when appropriately provoked. Many organic compounds having explosophores in their molecules are chemically unstable, but only certain ones are commercially suitable as chemical explosives.

The explosophore most commonly encountered in the molecular structures of the commercially available explosives is the *nitro* group ($-NO_2$). When multiple nitro groups are bonded within the structure of an organic compound, the substance is generally explosive to some degree. Examples include the commercial explosives trinitrotoluene (Section 14.9), cyclonite (Section 14.10), tetryl (Section 4.11), and HMX (Section 14.13). Large amounts of energy are released to the environment when the relatively weak chemical bonds in the molecules of the chemical explosive are transformed into the product molecules, especially nitrogen molecules.

BLASTING AGENTS

A *blasting agent* like that shown in Figure 14.4 is any single material or mixture that has been designated for blasting, has been subjected to prescribed tests, and has been shown to have little probability of initiating an explosion or of burning and then exploding. The term generally refers to ammonium nitrate mixtures sensitized with oil or wax.

DOT regulates the transportation of the following five types of blasting agents:

- *Type A* includes substances consisting of liquid organic nitrates, like nitroglycerin, or a mixture of ingredients with one or more of the following: nitrocellulose, ammonium nitrate, other inorganic nitrates, nitro-derivatives of organic compounds, or combustible materials, such as wood-meal and aluminum powder. Type A blasting agents exist in powdery, gelatinous, plastic, and elastic forms, which include dynamite, blasting gelatin, and gelatin dynamite (Section 14.7).
- *Type B* includes substances consisting of a mixture of ammonium nitrate or other inorganic nitrates with an explosive material such as trinitrotoluene, with or without other substances, such as wood-meal or aluminum powder, or a mixture of ammonium nitrate or other inorganic nitrates with other combustible substances that are not explosive ingredients. Type B blasting agents cannot contain nitroglycerin, similar liquid organic nitrates, or chlorates.

FIGURE 14.4 ◆ A blasting agent is a mixture of a fuel and an oxidizer (generally ammonium nitrate) that is intended exclusively for the purpose of blasting. None of the components of a blasting agent is an explosive material. A blasting agent is usually a Division 1.5 explosive material.

- *Type C* includes substances consisting of a mixture of either potassium chlorate or sodium chlorate, and potassium perchlorate, sodium perchlorate, or ammonium perchlorate, with a nitro derivative of an organic compound or combustible material, such as wood-meal or aluminum powder, or a hydrocarbon. Such explosives must not contain nitroglycerin or any similar liquid organic nitrate.
- *Type D* includes substances consisting of a mixture of organic nitrate compounds and combustible materials, such as hydrocarbons and aluminum powder. Such explosives must not contain nitroglycerin, any similar liquid organic nitrate, chlorate, or ammonium nitrate. The term generally includes plastic explosives (Section 14.10).
- *Type E* includes substances consisting of water as an essential ingredient and a high proportion of ammonium nitrate or other oxidizer, some or all of which is in solution. The other constituents include the nitro-derivatives of organic compounds, hydrocarbons, or aluminum powder. The term includes explosives that have been formulated as an emulsion, slurry, or watergel.

ARTILLERY AMMUNITION

The sensitivity of individual chemical explosives can be greatly appreciated by examining the role performed by each component of a round of artillery ammunition. The structure of a typical round is shown in Figure 14.5. Each of its six components performs a unique function, as follows:

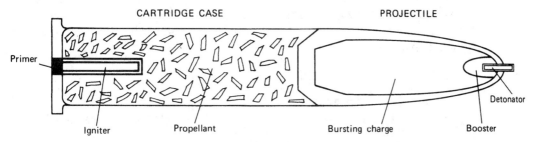

FIGURE 14.5 ◆ A typical round of artillery ammunition, illustrating its component parts. Each part plays a specific role in the proper functioning of the round. Chemical explosives are sometimes identified by the function that they perform in such a round (for example, primer or booster).

- *Primer.* In a round of artillery ammunition, the primer is generally a mixture of an oxidizer and some other arbitrary substance that readily burns. This mixture is usually confined within a cap or tube at one end of the round and is activated by a percussion action, such as a firing pin.
- *Igniter.* In a round of ammunition, the igniter is almost always black powder (Section 14.5). The oxidation–reduction reaction between the components of black powder is initiated by the heat released from burning the primer.
- *Propellant.* This is a material such as smokeless powder (Section 14.8) that deflagrates and produces a relatively large volume of gases and vapors. In a round of artillery ammunition, the quantity is intentionally limited to prevent a major explosion, but sufficiently sizable to generate an adequate volume of vapors and gases to propel the projectile forward. The explosive that performs the role of the propellant is physically separated from the other chemical explosives within the round by means of a thin wall.
- *Detonator.* This component of the round explodes from the mechanical shock it receives when it hits a target. In a round of artillery, this is present in a limited quantity.
- *Booster.* This component of the round is induced to explode by the heat generated from the explosion of the detonator. As with the detonator itself, the quantity of the booster is limited so as to generate a prescribed brisance. The function of the booster is to intensify the brisance resulting from the explosion of the detonator.
- *Bursting charge.* This component is typically a high explosive whose detonation is initiated by the shock resulting from the booster explosion. It is the main charge in a round of artillery ammunition.

The chain of events from initiation of the primer to detonation of the bursting charge is called an *explosive train*. For a round of ammunition to function effectively, each component must be activated in the proper order.

The components of all types of ammunition available today can be regarded as modifications of those found in a typical round of artillery ammunition. A rifle cartridge, for instance, contains only a primer and propellant, or a primer, igniter, and propellant; the projectile is generally a mass of metal such as lead or steel pellets. A bomb is also a modification of a round of artillery ammunition; generally, however, bombs lack a propellant since the detonator explodes from the force achieved by dropping them from high altitudes.

When explosives are used for nonmilitary purposes such as blasting, the presence of a propellant is ordinarily unnecessary. In these instances, a blasting cap holds a prescribed amount of a high explosive that is generally detonated by means of a booster.

The assembly containing the bursting charge and booster is connected to several hundred yards of electric wire and is activated from a safe distance by using a plunger-type magneto apparatus.

Performance Goal for Section 14.3

- Discuss the requirements OSHA imposes on an employer when explosives are stored in the workplace.

◆ 14.3 STORING EXPLOSIVE MATERIALS

Federal and state regulations and local ordinances require owners to store explosive materials within a specially constructed building called a *magazine* (Figure 14.6). OSHA regulates the manner by which a magazine is constructed and maintained, as

FIGURE 14.6 ◆ Explosives are stored in specially designed rooms or bunkers called *magazines*. The magazine noted in the foreground of this photograph is a bunker of masonry construction. All components are bullet, weather, and fire resistant. The entire roof is covered with sand, except structures that are required for interior ventilation. Inside the magazine, packages of explosives are stored flat with the top sides up and in stable configurations. The interior is ventilated to prevent dampness and heating of stored explosives. The land surrounding the magazine is kept clear of all combustible materials for a distance of at least 25 ft (7.6 m). *Courtesy of Los Alamos National Laboratory, Los Alamos, New Mexico.*

well as the actual manner by which explosive materials are stored in it. Some OSHA requirements are listed below:

◆ All explosive materials shall be stored in a magazine. Blasting caps, electric blasting caps, detonating primers, and primed cartridges shall not be stored in the same magazine with other explosive materials.

◆ The land surrounding a magazine shall slope away from the magazine and shall be kept clear of brush, dried grass, leaves, and other materials for a distance of at least 25 ft (7.6 m).

◆ A magazine shall be located away from inhabited buildings, passenger railways, public highways, and other magazines in conformity with prescribed distances established by the Institute of Makers of Explosives.*

◆ A magazine shall generally be bullet-, weather-, and fire-resistant and ventilated sufficiently to protect the integrity of the explosive material within the specific locality.

◆ The property upon which a magazine is located shall be posted with signs, as follows:

EXPLOSIVES—KEEP OFF

The implementation of these regulatory requirements constitutes measures that aim to prevent explosive materials from accidentally detonating or being directly involved in a fire.

Performance Goal for Section 14.4

◆ Discuss the special requirements DOT imposes on shippers and carriers who transport explosive materials.

◆ 14.4 TRANSPORTING EXPLOSIVE MATERIALS

As first noted in Section 6.7, DOT regulates the domestic transportation of chemical explosives and articles containing chemical explosives in six divisions, some examples of which are provided in Table 14.2. These divisions form a continuum of decreasing explosive hazard from 1.1 to 1.6.

Prior to submitting an explosive material for transportation, DOT requires the manufacturer to test the material by prescribed procedures, the results of which are then used to assign a division number. DOT requires explosive manufacturers to submit the test data on a given explosive material to the Bureau of Explosives, U.S. Department of Interior. In conjunction with DOT, bureau personnel evaluate these data and, if approved, assign an EX-number to it. When the explosive material is transported, DOT requires this number to be included in the basic description and marked on its packaging as EX-_____.

*The Institute of Makers of Explosives is the safety association of the commercial explosives industry in the United States and Canada. It promotes the safety and protection of employees, users, the public, and the environment throughout all aspects of the manufacture and use of explosive materials in industrial blasting and other essential operations.

Solved Exercise 14.4

The DOT regulation at 49 CFR §174.101(h) requires shippers and carriers of packages containing Division 1.1 and Division 1.2 explosives for detonators and detonating primers to securely block and brace them within the transport vehicle, so they are prevented from changing position, falling to the floor, or sliding into each other under normal transportation conditions. What is the most likely reason DOT regulates the loading of these explosive materials in this specific fashion?

Solution: Since the detonation of explosive materials can be initiated by friction, shock, and mechanical impact, DOT most likely requires packaged explosives to be loaded into a transport vehicle in the indicated fashion to reduce or eliminate the likelihood that any of these means will initiate their premature detonation. Additional measures of safety are afforded by preventing the packages from changing position, falling to the floor, or sliding into each other.

Solved Exercise 14.5

DOT prohibits the transportation of several explosive materials listed in Table 14.2 unless they have been wetted with a specified amount of water or alcohol or mixed with a specified amount of a plasticizer. What is the most likely reason DOT requires these explosive materials to be wetted or mixed with a plasticizer prior to acceptance for transportation?

Solution: As we first noted in Section 5.5, the rate of a chemical reaction is dependent upon the concentration of its reactants. The mixture of an explosive material with water, alcohol, or a plasticizer effectively dilutes the material, thereby reducing the likelihood that its component molecules can interact. Consequently, the explosive materials are sufficiently stabilized when wetted with water or alcohol, or when mixed with a plasticizer, so they can be transported without prematurely exploding.

Whenever explosives and explosive articles are transported, shippers and carriers are obligated to comply with DOT's marking, labeling, and placarding requirements. They are required to affix the relevant labels to packages containing these hazardous materials. DOT also requires the posting of EXPLOSIVE 1.1, EXPLOSIVE 1.2, and EXPLOSIVE 1.3 placards on the vehicle used to transport any quantity of the relevant explosive. DOT requires the posting of EXPLOSIVE 1.4, EXPLOSIVE 1.5, and EXPLOSIVE 1.6 placards only when the shipper or carrier transports an amount exceeding 1001 lb (454 kg).

DOT also requires shippers and carriers of explosive materials to evaluate the *compatibility group* of a given explosive substance or explosive article. There are 12 compatibility groups, each represented by one of the capital letters A, B, C, D, E, F, G, H, J, K, L, and S. The assignment of the compatibility group is based upon the

TABLE 14.2 ◆ Some Chemical Explosives and Explosive Articles Regulated by DOT for Transport

Proper Shipping Name	Label Codes
Ammonium nitrate–fuel oil mixture *containing only prilled ammonium nitrate and fuel oil*	1.5D
Ammonium nitrate, *with more than 0.2% combustible substances, including any organic substance calculated as carbon, to the exclusion of any other added substance*	1.1D
Ammunition, illuminating *with or without burster, expelling charge, or propelling charge*	1.2G
Ammunition, illuminating *with or without burster, expelling charge, or propelling charge*	1.3G
Ammunition, illuminating *with or without burster, expelling charge, or propelling charge*	1.4G
Ammunition, incendiary *liquid or gel, with burster, expelling charge, or propelling charge*	1.3J
Black powder, compressed *or* Gunpowder, compressed *or* Black powder, in pellets *or* Gunpowder, in pellets	1.1D
Black powder *or* Gunpowder, *granual or as a meal*	1.1D
Cyclotetramethylenetetranitramine, desensitized *or* Octogen, desensitized *or* HMX, desensitized	1.1D
Cyclotetramethylenetetranitramine, wetted *or* HMX, wetted *or* Octogen, wetted *with not less than 15% water by mass*	1.1D
Cyclotrimethylenetrinitramine, desensitized *or* Cyclonite, desensitized *or* Hexogen, desensitized *or* RDX, desensitized	1.1D
Cyclotrimethylenetrinitramine, wetted *or* Cyclonite, wetted *or* Hexogen, wetted *or* RDX, wetted *with not less than 15% water by mass*	1.1D
Explosive, blasting, type A	1.1D
Explosive, blasting, type B	1.1D
Explosive, blasting type B *or* Agent blasting, Type B	1.5D
Explosive, blasting, type C	1.1D
Explosive, blasting, type D	1.1D
Explosive, blasting, type E	1.1D
Explosive, blasting, type E *or* Agent, blasting, Type E	1.5D
Fireworks	1.1G
Fireworks	1.2G
Fireworks	1.3G
Fireworks	1.4G
Fireworks	1.4S
Lead azide, wetted *with not less than 20% water or mixture of alcohol and water, by mass*	1.1A
Lead styphnate, wetted *or* Lead trinitroresorcinate, wetted *with not less than 20% water, or mixture of alcohol and water, by mass*	1.1A
Mercury fulminate, *wetted with not less than 20% water, or mixture of alcohol and water, by mass*	1.1A
Nitrocellulose, dry *or wetted with less than 25% water (or alcohol), by mass*	1.1D
Nitrocellulose, plasticized *with not less than 18% plasticizing substance, by mass*	1.3C
Nitrocellulose, unmodified or *plasticized with less than 18% plasticizing substance, by mass*	1.1D
Nitroglycerin, desensitized *with not less than 40% non-volatile water-insoluble phlegmatizer, by mass*[a]	1.1D, 6.1

TABLE 14.2 ◆ *(continued)*

Proper Shipping Name	Label Codes
Nitroglycerin, solution in alcohol *with more than 1% nut not more than 10% nitroglycerin*	1.1D
Pentaerythrite tetranitrate *or* Pentaerythritol tetranitrate *or* PETN, *with not less than 7% wax by mass*	1.1D
Pentaerythrite tetranitrate, wetted *or* Pentaerythritol tetranitrate, wetted *or* PETN, wetted *with not less than 25% water, by mass or* Pentaerythrite tetranitrate, *or* Pentaerythritol tetranitrate, *or* PETN, desensitized *with not less than 15% phlegmatizer by mass*	1.1D
Trinitrotoluene and trinitrobenzene mixtures *or* TNT and trinitrobenzene mixtures *or* TNT and hexanitrostilbene mixtures[b]	1.1D
Trinitrotoluene mixtures containing trinitrobenzene and hexanitrostilbene	1.1D
Trinitrotoluene, *or* TNT, *dry or wetted with 30% water by mass*	1.1D
Trinitrotoluene, *or* TNT, *dry or wetted with not less than 30% water by mass*	4.1
Trinitrophenylmethylnitramine *or* Tetryl	1.1D

[a]A *phlegmatizer* is a substance capable of reducing the ability of an explosive to detonate.
[b]Stilbene is the common name for the substance having the chemical formula $C_6H_5—CH\!=\!CH—C_6H_5$. The proper name of this substance is *trans*-α,β-diphenylethene. Hexanitrostilbene is the name of the substance having the chemical formula $C_6H_2(NO_2)_3—CH\!=\!CH—C_6H_2(NO_2)_3$.

nature of the explosive substance or article as provided by its relevant description in Table 14.3. When classifying an explosive substance or article for transportation, shippers and carriers identify its compatibility group by noting the entries in column 3 of the Hazardous Materials Table published at 49 CFR §172.101.

DOT requires shippers and carriers to display the compatibility group on the EXPLOSIVE 1.1, 1.2, 1.3, 1.4, 1.5, or 1.6 labels that they affix to packaging containing the relevant explosive. DOT also requires the compatibility group to be displayed on the EXPLOSIVE placards posted on the transport vessel.

For certain listings of chemical explosives in the Hazardous Materials Table at 49 CFR §172.101, the word *forbidden* appears in column 3. This designation signifies that DOT prohibits the domestic transportation of the referenced commodity. Nonetheless, DOT may permit the same explosive to be transported when it has been suitably "desensitized" or reduced in its ability to detonate. This is typically accomplished by mixing the chemical explosive with an inert material.

When an explosive material is moved by public highway, DOT can designate the routing or impose curfews, time-of-travel restrictions, lane restrictions, or route-weight restrictions on the shipper or carrier. The identity of such routing designations must be made available to the public in the form of a map, list, and/or a road sign like the one shown in Figure 14.7. The highway DOT designates for the domestic transportation is called a *preferred route*, or *preferred highway*.

Finally, DOT requires special precautions to be implemented when explosive materials are loaded, unloaded, or handled onboard watercraft. In particular, a fire hose of sufficient length to reach every part of the loading area with an effective stream of water must be laid and connected to the water main, ready for use. In the event of fire, firefighting personnel may proceed immediately to prevent a premature cargo explosion.

TABLE 14.3 ◆ Compatibility Group Assignments[a]

Description of Substances or Article to be Classified	Compatibility Group	Classification Code
Primary explosive substance.	A	1.1A
Article containing a primary explosive substance and not containing two or more effective protective features. Some articles, such as detonators for blasting, detonator assemblies for blasting and primers, cap-type, are included, even though they do not contain primary explosives.	B	1.1B 1.2B 1.4B
Propellant explosive substance or other deflagrating explosive substance or article containing such explosive substance.	C	1.1C 1.2C 1.3C 1.4C
Secondary detonating explosive substance or black powder or article containing a secondary detonating explosive substance in each case without means of initiation and without a propelling charge, or article containing a primary explosive substance and containing two or more effective protective features.	D	1.1D 1.2D 1.4D 1.5D
Article containing a secondary detonating explosive substance, without means of initiation, with a propelling charge (other than one containing flammable liquid get or hypergolic liquid).[b]	E	1.1E 1.2E 1.4E
Article containing a secondary detonating explosive substance with its means of initiation, with a propelling charge (other than one containing flammable liquid get or hypergolic liquid) or without a propelling charge.	F	1.1F 1.2F 1.3F 1.4F
Pyrotechnic substance or article containing a pyrotechnic substance, or article containing both an explosive substance and an illuminating, incendiary, tear-producing, or smoke-producing substance (other than a water-activated article or one containing white phosphorus, phosphide or flammable liquid or gel or hypergolic liquid)	G	1.1G 1.2G 1.3G 1.4G
Article containing both an explosive substance and white phosphorus.	H	1.2H 1.3H
Article containing both an explosive substance and flammable liquid or gel.	J	1.1J 1.2J 1.3J
Article containing both an explosive substance and a toxic chemical agent.	K	1.2K 1.3K
Explosive substance or article-containing substance and presenting a special risk (e.g., due to water-activation or presence of hypergolic liquids, phosphides, or pyrophoric substances) needing isolation of each type.	L	1.1L 1.2L 1.3L
Articles containing only extremely insensitive detonating substances.	N	1.6N
Substance or article so packed or designed that any hazardous effects arising from accidental functioning are limited to the extent that they do not significantly hinder or prohibit firefighting or other emergency response efforts in the immediate vicinity of the package.	S	1.4S

[a]49 CFR §173.52
[b]A *hypergolic liquid* is a fluid that is capable of spontaneously igniting upon contact with the explosive substance.

FIGURE 14.7 ◆ DOT requires shippers and carriers to transport explosive materials only along certain designated routes and during assigned time periods. This road sign identifies a route that DOT has approved for the transportation of explosive materials on a military base.

```
▪ ▪ ▪ ▪ ▪ ▪ ▪ ▪ ▪ ▪ ▪ ▪ ▪ ▪ ▪ ▪ ▪ ▪ ▪ ▪ ▪ ▪ ▪ ▪ ▪ ▪ ▪ ▪ ▪ ▪ ▪ ▪
  Performance Goals for Section 14.5
        ◆ Identify the chemical composition of black powder and black gunpowder.
        ◆ Discuss the nature of the oxidation–reduction that occurs when they are activated.
▪ ▪ ▪ ▪ ▪ ▪ ▪ ▪ ▪ ▪ ▪ ▪ ▪ ▪ ▪ ▪ ▪ ▪ ▪ ▪ ▪ ▪ ▪ ▪ ▪ ▪ ▪ ▪ ▪ ▪ ▪ ▪
```

◆ 14.5 BLACK POWDER

The oldest known explosive material is *black powder*.* Although its chemical composition varies, black powder generally consists of an intimate mixture of charcoal, sulfur, and either potassium nitrate or sodium nitrate. It is a low explosive. Although black powder is often employed as a blasting agent, it is also a component of fireworks and certain forms of ammunition. A variation of black powder, called *black gunpowder,* is a mixture of 15 parts charcoal, 10 parts sulfur, and 75 parts potassium nitrate by mass.

When ignited, the components of black powder and black gunpowder undergo an oxidation–reduction reaction.

$$\underset{\text{sulfur}}{3S_8(s)} + \underset{\text{carbon}}{16C(s)} + \underset{\text{potassium nitrate}}{32KNO_3(s)} \longrightarrow$$

$$\underset{\text{potassium oxide}}{16K_2O(s)} + \underset{\text{carbon dioxide}}{16CO_2(g)} + \underset{\text{nitrogen}}{16N_2(g)} + \underset{\text{sulfur dioxide}}{24SO_2(g)}$$

No component of this mixture detonates in the true sense. Nevertheless, when the mixture is used as a blasting agent, the outcome can resemble the actual detonation of a chemical explosive. Since it is composed of such a reactive mixture of substances, black powder should always be stored, transported, handled, and used as if it actually could detonate.

*Black powder formulations were known to the ancient Chinese, who used them in warfare and pyrotechnic displays. Although the color of black powder is indeed black, the material is not named for its color, but for the direct translation of the German word "Schwarzpulver," named for Bernard Schwarz, who experimented with black powder formulations in the 14th century.

◆ **14.6 NITROGLYCERIN**

When alcohols react with nitric acid, the compounds called *nitrate esters* are produced. The relevant reactions may be denoted by the following general equation, where R is an arbitrary alkyl or aryl group:

$$R-OH(l) + HNO_3(l) \longrightarrow R-O-NO_2(l) + H_2O(l)$$

The best known nitrate ester is the chemical explosive properly named glyceryl trinitrate, but more commonly known as *nitroglycerin*.

$$
\begin{array}{l}
CH_2-O-NO_2 \\
| \\
CH-O-NO_2 \\
| \\
CH_2-O-NO_2
\end{array}
$$

It is prepared by slowly drizzling glycerol, a trihydroxy alcohol, into a cooled mixture of nitric acid and sulfuric acid, as follows:

$$
\begin{array}{l}
CH_2-OH \\
| \\
CH-OH \\
| \\
CH_2-OH(l)
\end{array}
\quad + 3HNO_3(l) \longrightarrow
\begin{array}{l}
CH_2-O-NO_2 \\
| \\
CH-O-NO_2 \\
| \\
CH-O-NO_2(l)
\end{array}
\quad + 3H_2O(l)
$$

glycerol   nitric acid   nitroglycerin   water

In the reaction, the sulfuric acid acts as a catalyst.

Pure nitroglycerin is a thick, oily liquid whose physical appearance otherwise resembles water. Other physical properties of this chemical explosive are noted in Table 14.4. When generally encountered, however, nitroglycerin is a viscous, pale yellow liquid, which is extremely sensitive to spontaneous decomposition. Even a slight jarring or dropping upon a hard surface may trigger a premature explosion.

Given this high degree of sensitivity, pure nitroglycerin is regarded as a high explosive. The liquid cannot be safely transported and is impractical for general use as a chemical explosive. In Section 14.1, we noted that when nitroglycerin detonates, it decomposes into carbon dioxide, water vapor, nitrogen, and oxygen. The resulting

TABLE 14.4 ◆ **Some Physical Properties of Nitroglycerin**

Specific gravity	1.60
Melting point	55°F (13.1°C)
Brisance (g of sand)	51.5
Detonation velocity	4.8 mi/s (7.8 km/s)
Sensitivity	Very high (almost a primary explosive)

brisance is approximately three times that of an equivalent amount of gunpowder and occurs roughly 25 times faster.

When exposed to atmospheric moisture, nitroglycerin hydrolyzes and produces a mixture of glycerol and nitric acid. When it has absorbed moisture, nitroglycerin becomes extremely sensitive to spontaneous decomposition. For this reason, the repeated exposure of nitroglycerin to the atmosphere is likely to produce a highly treacherous mixture, which could detonate with the slightest provocation. To prevent such an occurrence, the mixture can be cautiously neutralized with a solution of sodium sulfide in water, acetone, and denatured alcohol.

Aside from its potentially explosive nature, nitroglycerin is a toxic substance by ingestion, inhalation, and skin absorption. When absorbed into the blood system, nitroglycerin causes the small veins, capillaries, and coronary vessels to dilate. These effects are caused by nitric oxide, which is slowly released from the nitroglycerin. Under severe conditions, the nitric oxide causes the development of severe headaches, flushing of the face, and a drop in blood pressure. In severe instances, it causes death.

Notwithstanding the toxic nature of nitroglycerin, nontoxic doses are used medicinally for the treatment of heart and certain blood-circulation diseases. A compound like nitroglycerin that causes dilation of the blood vessels is called a *vasodilator*. Afflicted individuals put a tiny nitroglycerin tablet under the tongue when they experience a chest pain, or they consume a small amount of the medication called *spirits of nitroglycerin*, which consists of nitroglycerin dissolved in a solvent such as alcohol. The nitroglycerin slowly releases nitric oxide within the bloodstream. The nitric oxide relaxes nearby muscle cells and temporarily lowers blood pressure.

DOT prohibits the domestic transportation of nitroglycerin unless it has been desensitized against unwanted decomposition. Nitroglycerin is usually desensitized by dissolving it in a simple alcohol such as methanol, ethanol, or isopropanol. Since these alcohols are flammable, the commercially available forms of nitroglycerin other than those listed in Table 14.2 may be transported as flammable liquids.

When nitroglycerin is present in the workplace, OSHA requires employers to limit employee exposure by skin contact to a concentration of 0.2 mg/m^3, averaged over an 8-hour workday.

Performance Goals for Section 14.7
- Describe the differences in properties between nitroglycerin and dynamite that cause dynamite to be safer to handle and use.
- Identify the three commercial types of dynamite and describe their composition.

◆ 14.7 DYNAMITE

In 1867, Swedish engineer Alfred Nobel discovered that nitroglycerin could be absorbed into a porous material such as siliceous earth. The resulting mixture became known as *dynamite*. From his discovery and the subsequent manufacture of dynamite, Nobel acquired a fortune, which he used in part to establish a fund for the world-famous Nobel prizes.

Dynamite can be handled more safely than nitroglycerin, a property that allows it to be transported and used with less risk of spontaneous decomposition. In practice, dynamite requires a detonating cap to activate its detonation. Notwithstanding this fact, dynamite is a high explosive sensitive to heat, shock, and friction.

FIGURE 14.8 ◆ Cylindrical cartridges into which dynamite has been packed are generally referred to as *sticks* of dynamite. These sticks are commercially available in a number of sizes. *Courtesy of IRECO, Inc., Salt Lake City, Utah.*

Today, dynamite is produced by absorbing a mixture of nitroglycerin and diethyleneglycol dinitrate into wood pulp, sawdust, flour, starch, or similar carbonaceous materials. Diethyleneglycol dinitrate is also a chemical explosive, but its function in the production of dynamite is to depress the solution's freezing point. Calcium carbonate is often added to dynamite to neutralize the nitric acid produced by hydrolysis. Oxidizers are routinely added as a source of additional internal energy. This mixture of substances is then packed into cylindrical cartridges made of waxed paper, which vary in size from 7/8 to 8 in. (2 to 20 cm) in diameter and from 4 to 30 in. (10 to 76 cm) in length. The dynamite sticks in Figure 14.8 are cylinders, each of which is roughly 1.5 in. × 8 in. (3.8 cm × 20 cm) in size and weighing 0.5 lb (230 g).

Three commercial forms of dynamite are commonly encountered today: *ammonia dynamite, straight dynamite*, and *gelatin dynamite*. Although all consist of a mixture of nitroglycerin and diethyleneglycol dinitrate, ammonia dynamite and straight dynamite also contain ammonium nitrate and sodium nitrate, respectively. Gelatin dynamite contains about 1% nitrocellulose by mass, which serves to thicken the nitroglycerin. Additional information about these forms of dynamite is provided in Table 14.5.

The equation denoting the detonation of nitroglycerin was previously noted in Section 14.1. When their constituent nitroglycerin detonates, the three forms of dynamite produce more brisance than an equivalent amount of nitroglycerin, because the chemical action of ammonium nitrate, sodium nitrate, or nitrocellulose provides an additional source of energy.

TABLE 14.5 ◆ Physical Properties of Some Commercial Forms of Dynamite

	Straight Dynamite	*Ammonia Dynamite*	*Gelatin Dynamite*
Specific gravity	1.3	0.8–1.2	1.3–1.6
Chemical composition	20–60% nitroglycerin depending on grade, sodium nitrate, carbonaceous material, antacid, and moisture	20–60% nitroglycerin depending on grade, ammonium nitrate, carbonaceous materials, sulfur, antacid, and moisture	20–90% nitroglycerin depending on grade, sodium nitrate, and moisture, gelatinized in nitrocellulose
Detonation velocity	2.5–4 mi/s (4–6 km/s)	0.5–0.8 mi/s (0.8–1.2 km/s)	0.47–0.68 mi/s (0.75–1.1 km/s)
Sensitivity	High	High	High

FIGURE 14.9 ◆ The loading of a borehole with a water-gel cartridge. *Courtesy of E. I. Du Pont de Nemours & Co., Wilmington, Delaware.*

When they are involved in fires, small quantities of the dynamite may burn with a bluish flame without detonating. Nonetheless, the heat generated during these fires can also activate the detonation of the remaining nitroglycerin. It is for this reason that experts recommend not fighting fires involving dynamite.

From the 1920s through the 1930s, dynamite was the most popular explosive used for peacetime purposes. Accidental detonations involving dynamite occurred often. Since the 1930s, new chemical explosives that are far safer to transport, store, and use than dynamite have been developed. An example is the watergel cartridge in Figure 14.9. Today, even though these safer explosives are now commercially available, some explosives experts still prefer to use dynamite for unique demolition assignments.

Performance Goals for Section 14.8
- Describe the chemical nature of nitrocellulose.
- Compare the detonation and burning of nitrocellulose.

◆ 14.8 NITROCELLULOSE

Nonfire

Fire

As we first noted in Section 13.5, nitrocellulose is a polymer that is produced by reacting the cellulose in cotton with nitric acid. Since cellulose can be nitrated to different degrees, varying forms of nitrocellulose can be produced. One molecular structure of nitrocellulose is noted below:

In this structure, the cellulose is almost completely nitrated. Although the highly nitrated cellulose is used as a rocket propellant and chemical explosive, the lesser grades of nitrocellulose are utilized for other purposes (Section 13.5). In the United States, the explosive is now produced exclusively by the federal government.

Condensing the chemical formulas of cellulose and nitrocellulose to $[C_6H_7O_2(OH)_3]_n$ and $[C_5H_5(CH_2ONO_2)O_2(ONO_2)_2]_n$, respectively, the nitration of cellulose may be represented as follows, where n is approximately 5000:

$$[C_6H_7O_2(OH)_3]_n(s) + 3nHNO_3(l) \longrightarrow$$
cellulose   nitric acid

$$[C_5H_5(CH_2ONO_2)O_2(ONO_2)_2]_n(s) + 3nH_2O(l)$$
nitrocellulose   water

When the nitration produces a substance having a nitrogen content exceeding 13.2% by mass, the result is called *guncotton*.

Nitrocellulose is a white solid that resembles cotton in physical appearance. When intended for use as an explosive material, it can be blocked, gelled, flaked, granulated, or powdered. To desensitize it for domestic transportation and storage, nitrocellulose is usually wetted with either water or an aqueous solution of ethanol. Notwithstanding the presence of either desensitizing agent, every nitrocellulose-based explosive material is inherently unstable. The nitrocellulose slowly decomposes, producing traces of nitrogen oxides. For this reason, a *stabilizer* is also added to all nitrocellulose-based formulations. A typical stabilizer is diphenylamine, which functions by reacting with the nitrogen oxides to form innocuous compounds.

When nitrocellulose is activated, it usually deflagrates. Since its flashpoint is only 55°F (13°C), nitrocellulose burns readily, producing carbon dioxide, water vapor, and nitrogen, but not nitrogen dioxide. The deflagration of nitrocellulose is a furiously burning phenomenon that may be described as follows, where n is an integer:

$$2[C_5H_5(CH_2ONO_2)O_2(ONO_2)_2]_n(s) + 10nO_2(g) \longrightarrow$$
nitrocellulose   oxygen

$$12nCO_2(g) + 7nH_2O(g) + 3nN_2(g)$$
carbon dioxide   water   nitrogen

The burning of nitrocellulose occurs at a faster rate than virtually any other flammable solid. Tons of nitrocellulose can be consumed by fire within minutes.

Nitrocellulose detonates when it is confined or accumulated in large quantities. The detonation reaction may be expressed as follows:

$$2[C_5H_5(CH_2ONO_2)O_2(ONO_2)_2]_n(s) \longrightarrow$$
nitrocellulose

$$nCO_2(g) + 11nCO(g) + 7nH_2O(g) + 3nN_2(g)$$
carbon dioxide   carbon monoxide   water   nitrogen

Nitrocellulose is probably most commonly encountered as a component of small-arms ammunition, although it is also used frequently as the propellant in artillery ammunition. When intended for use as a chemical explosive in demolition work, it is almost always combined with a second explosive, such as nitroglycerin.

A formerly popular explosive article containing nitrocellulose is *smokeless powder*, a name most likely linked to the absence of visible smoke following its detonation. Using smokeless powder during warfare can provide a tactical advantage: the absence of visible smoke prevents the enemy from easily identifying the location from which the artillery was fired.

There are three varieties of smokeless powder available commercially, each referred to as a single-, double-, or triple-base formulation. Each consists of a number of ingredients including one, two, or three chemical explosives, respectively. A popular variation is the double-base consisting of nitrocellulose, nitroglycerin, and nonexplosive ingredients. This formulation is commonly encountered in illegally improvised devices such as pipe bombs.

◆ 14.9 TRINITROTOLUENE

2,4,6-Trinitrotoluene is a pale yellow solid, although its commercial grade is generally a yellow to dark brown. This chemical explosive is known more commonly by its acronym, *TNT*.

During World War II, TNT was used on a very large scale as a military explosive. Movie buffs are especially familiar with its use in depth charges, which were catapulted from destroyers to disable enemy submarines. During peacetime, it is still used to accomplish demolition during construction and mining projects.

The detonation of 500 tons (454 tonnes) of TNT is depicted in Figure 14.10. The equation denoting the detonation is provided in Solved Exercise 14.3. Some important physical properties of this high explosive are provided in Table 14.6.

TNT is prepared by nitrating toluene using a mixture of nitric acid and sulfuric acid.

| toluene | nitric acid | 2,4,6-trinitrotoluene | water |

In the reaction, sulfuric acid acts as a catalyst. TNT production occurs only at military arsenals.

Commercial explosives are often prepared by mixing TNT with other substances. A popular explosive containing TNT is *amatol,* a mixture consisting of 80% ammonium nitrate and 20% TNT by mass. It has been widely used as a blasting agent and military and industrial explosive. Two other commercial chemical explosives containing TNT are cyclonite (Composition B) (Section 14.10) and tetrytol (Section 14.11).

TNT is regarded as one of the most valuable commercial and military explosives since it possesses the following features:

FIGURE 14.10 ◆ The detonation of 500 tons (454 tonnes) of trinitrotoluene. The massiveness of this explosion may be noted by comparing its size to those of the objects in the foreground. *Courtesy of the U.S. Department of Energy, Las Vegas, Nevada.*

- TNT is safe to handle. Although it is a high explosive, TNT is unusually insensitive to heat, shock, and friction. When it is engulfed in fire, relatively small quantities often burn but do not detonate. TNT ordinarily detonates only when confined, or when relatively large amounts are intentionally activated.
- TNT does not react with atmospheric moisture.
- TNT is not susceptible to spontaneous decomposition, even after it has been kept in storage for years.
- TNT can be melted with steam with little fear of explosive decomposition. This feature is put to use when commercial explosive mixtures containing TNT are prepared. The molten TNT can be safely mixed with other explosives or oxidizers, cast into the shape of blocks, or poured into ammunition shells.

TNT is toxic by ingestion, inhalation, and skin absorption. When TNT is present in the workplace, OSHA requires employers to limit employee exposure by skin contact to a concentration of 1.5 mg/m^3, averaged over an 8-hour workday. NIOSH recommends limiting employee exposure to a concentration of 0.5 mg/m^3 over a 40-hour workweek.

TABLE 14.6 ◆ **Some Physical Properties of 2,4,6-Trinitrotoluene**

Specific gravity	1.65
Melting point	178°F (81°C)
Brisance (g of sand)	48
Detonation velocity	3.2–4.3 mi/s (5.1–6.9 km/s)
Sensitivity	Low

Exposure to TNT has also been linked in animals with the contracting of cancer. Epidemologists rank it as a probable human carcinogen.

Performance Goals for Section 14.10
- ◆ Describe the chemical nature of cyclonite.
- ◆ Describe the detonation of cyclonite.
- ◆ Identify the commercial compositions of cyclonite.

◆ 14.10 CYCLONITE

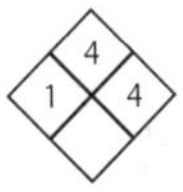

Cyclotrimethylenetrinitramine, or hexahydro-1,3,5-trinitro-1,3,5-triazine, is the active ingredient of the commercial and military chemical explosives known more commonly as *cyclonite, RDX,* and *hexogen.*

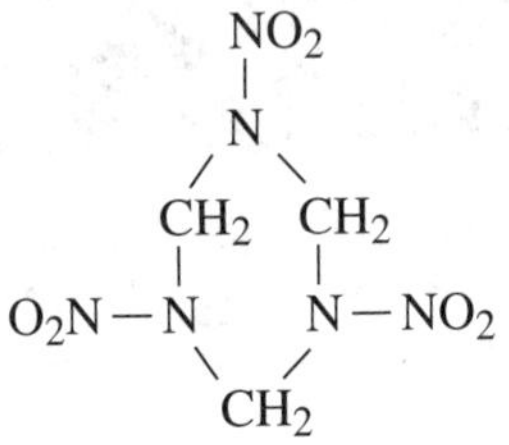

$$
\begin{array}{c}
NO_2 \\
| \\
N \\
/ \quad \backslash \\
CH_2 \quad CH_2 \\
| \qquad | \\
O_2N-N \qquad N-NO_2 \\
\backslash \quad / \\
CH_2
\end{array}
$$

RDX is the acronym for "Royal Deutsche Explosive" or "Royal Demolition Explosive." It is an important member of a class of explosives called *nitramines,* organic compounds having the following group of atoms:

$$
\begin{array}{c}
-N-NO_2 \\
|
\end{array}
$$

Other nitramines noted in this chapter include tetryl (Section 14.11) and cyclotetramethylenetetranitramine (Section 14.13).

Cyclonite is prepared from hexamethylenetetramine, nitric acid, ammonium nitrate, and acetic anhydride.

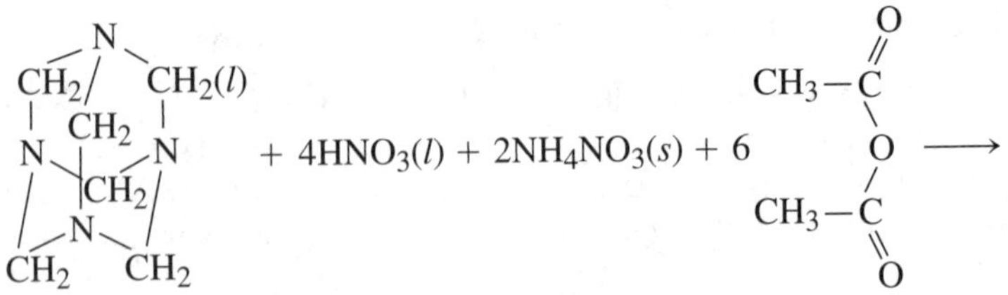

$$
\underset{\text{hexamethylenetetramine}}{
\begin{array}{c}
N \\
CH_2 / \backslash CH_2(l) \\
| \quad CH_2 \quad | \\
N \backslash \quad / N \\
| CH_2 | \\
/ N \backslash \\
CH_2 \quad CH_2
\end{array}}
+ 4HNO_3(l) + 2NH_4NO_3(s) + 6\;
\underset{\text{acetic anhydride}}{
\begin{array}{c}
O \\
// \\
CH_3-C \\
\backslash \\
O \\
/ \\
CH_3-C \\
\backslash\backslash \\
O
\end{array}} \longrightarrow
$$

hexamethylenetetramine   nitric acid   ammonium nitrate   acetic anhydride

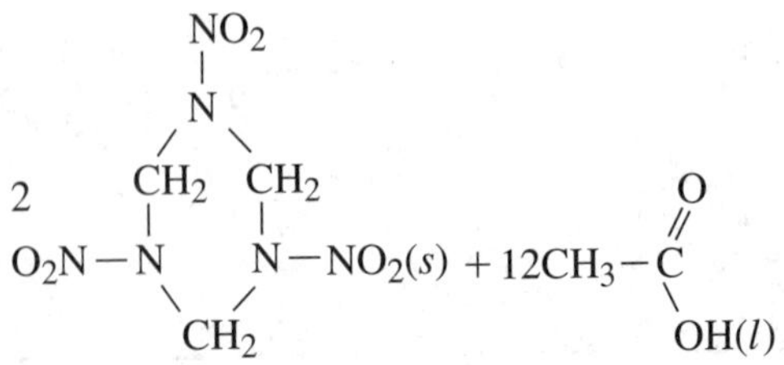

$$
2\;
\underset{\text{cyclonite}}{
\begin{array}{c}
NO_2 \\
| \\
N \\
/ \quad \backslash \\
CH_2 \quad CH_2 \\
| \qquad | \\
O_2N-N \qquad N-NO_2(s) \\
\backslash \quad / \\
CH_2
\end{array}}
+ 12\;
\underset{\text{acetic acid}}{
\begin{array}{c}
O \\
// \\
CH_3-C \\
\backslash \\
OH(l)
\end{array}}
$$

cyclonite   acetic acid

TABLE 14.7 ◆ Some Physical Properties of Cyclotrimethylenetrinitramine	
Specific gravity	1.82
Melting point	396°F (204°C)
Brisance (g of sand)	60.2
Detonation velocity	4.2–5.0 mi/s (6.8–8.0 km/s)
Sensitivity	High

It is a white crystalline solid. Other physical properties of cyclonite are listed in Table 14.7.

Upon condensing the chemical formula of cyclonite to $(CH_2)_3N_3(NO_2)_3$, the detonation of cyclonite is denoted as follows:

$$(CH_2)_3N_3(NO_2)_3(s) \longrightarrow \underset{\text{carbon monoxide}}{3CO(g)} + \underset{\text{nitrogen}}{3N_2(g)} + \underset{\text{water}}{3H_2O(g)}$$
$$\underset{\text{cyclonite}}{}$$

Cyclonite is approximately 1.5 times as powerful as TNT. As the pure substance, it is extremely sensitive to explosive decomposition. Mixed with beeswax, however, cyclonite is said to be "desensitized," since it acquires a stability even when exposed to high temperatures. The fact that cyclonite is thermally stable when desensitized makes it potentially useful when there is a need to use explosives during firefighting.

Mixed with beeswax in varying proportions, cyclonite was used widely through World War II as the bursting charge in aerial bombs, mines, and torpedoes. A mixture of cyclonite, TNT, and aluminum fines called *torpex* was also used. In warfare, torpex was used as the explosive material in mines, depth charges, and torpedo warheads.

The following formulations of cyclonite are still encountered:

* *Composition A* is a mixture of 91% cyclonite and 9% beeswax by mass. This chemical explosive is often selected by explosive experts because it detonates so rapidly. A charge of Composition A in a 1-ton bomb detonates in approximately 1/4 millisecond. This relatively high rate of detonation yields tremendous brisance.
* *Composition B* is a mixture of 60% cyclonite, 40% trinitrotoluene, and 1% beeswax by mass. This mixture has largely replaced Composition A in artillery shells.
* *Composition C-4* is a military explosive consisting of a mixture of 91% cyclonite, 5.3% di(2-ethylhexyl) sebacate,* 2.1% polyisobutylene, and 1.6% 20-weight motor oil.

*Sebacic acid is the common name of the substance whose proper name is 1,8-octane dicarboxylic acid. Its chemical formula is $HOOC-(CH_2)_8-COOH$. Di(2-ethylhexyl) sebacate is the ester of sebacic acid having the following chemical formula:

$$\begin{array}{c} O \\ \parallel \\ C-O-C_8H_{17} \\ | \\ (CH_2)_8 \\ | \\ C-O-C_8H_{17} \\ \parallel \\ O \end{array}$$

Cyclonite is sometimes mixed with a gummy binder and molded into a putty-like shape for use. This mixture is an example of a *plastic explosive*. When mixed with the binder, the chemical explosive retains its brisance but is less sensitive to heat, shock, and friction. The main disadvantage in the ethical use of a plastic explosive is that it becomes brittle in cold climates.

Although the most common use of a plastic explosive is connected with demolition work, international terrorists have also used plastic explosives as weapons of mass destruction. Cyclonite and PETN (Section 14.12) were identified as the active ingredients within the plastic explosive used to destroy Pan Am flight 103 in 1998. Residues of these chemical explosives were also identified following a terrorist action in India in 2003.

Cyclonite is toxic when ingested or inhaled. Exposure to high concentrations has been linked with the onset of seizures. When cyclonite is present in the workplace, OSHA requires workers to limit employee exposure to a concentration of 1.5 mg/m^3, averaged over an 8-hour workday. NIOSH recommends a short-term exposure limit of 1.5 mg/m^3 over a 40-hour workweek.

Exposure to cyclonite has also been linked in animals with cancer. Epidemologists rank cyclonite as a probable human carcinogen.

Performance Goals for Section 14.11
- ◆ Describe the chemical nature of tetryl.
- ◆ Describe the detonation of tetryl.

◆ 14.11 TETRYL

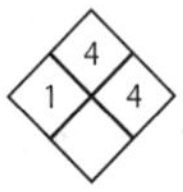

The active component of the commercial chemical explosive known as *tetryl* is *N*-methyl-*N*-2,4,6-tetranitroaniline.

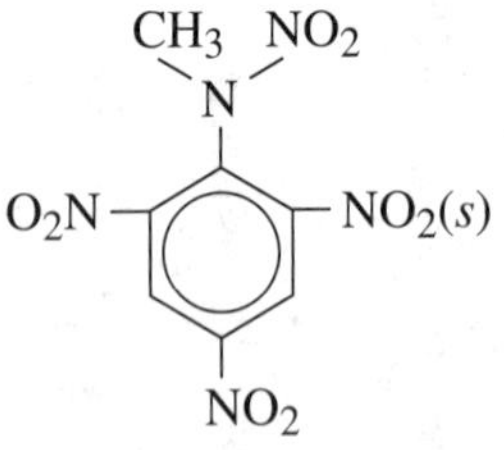

This substance is produced in several ways, one of which involves the nitration of 2,4-dinitrochlorobenzene in a step-wise process, as follows:

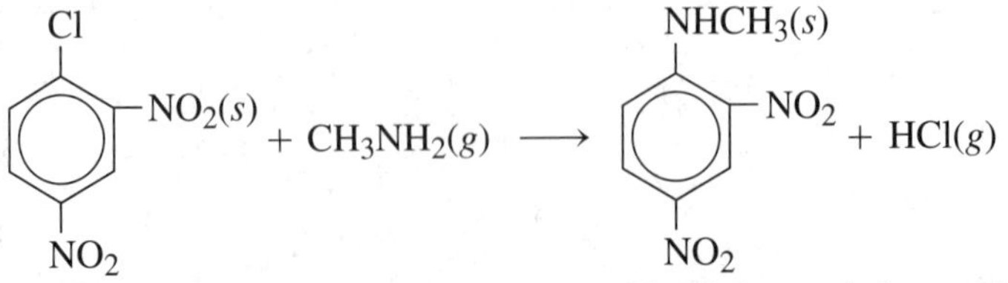

2,4-dinitrochlorobenzene   methylamine   *N*-methyl-   hydrogen chloride
2,4-dinitrobenzene

TABLE 14.8 ◆ Some Physical Properties of N-Methyl-N-2,4,6-Tetranitroaniline

Specific gravity (pressed)	1.30
Melting point	266°F (130°C)
Brisance (g of sand)	54.2
Detonation velocity	4 mi/s (7 km/s)
Sensitivity	High

Tetryl is a yellow solid; other physical properties are provided in Table 14.8.

Since World War II, tetryl has been the standard chemical explosive used by the military as the booster in artillery ammunition. Condensing its chemical formula to $(NO_2)_3C_6H_2N(NO_2)(CH_3)$, the detonation of tetryl is denoted as follows:

$$2(NO_2)_3C_6H_2N(NO_2)(CH_3)(s) \longrightarrow 3C(s) + 11CO(g) + 7N_2(g) + 5H_2O(g)$$
$$\text{tetryl} \qquad\qquad \text{carbon} \quad \text{carbon monoxide} \quad \text{nitrogen} \quad \text{water}$$

Tetryl is known for its exceptionally high brisance. When tetryl is mixed with molten trinitrotoluene and a small amount of graphite, the popular chemical explosive called *tetrytol* is produced. Tetrytol is sometimes selected by the military as the bursting charge in artillery ammunition.

Tetryl is toxic by inhalation, ingestion, and skin absorption. The afflicted experience a host of symptoms, including coughing, fatigue, headache, nosebleed, nausea, vomiting, and skin rashes. Research studies suggest that exposure to tetryl may also affect kidney, liver, and spleen function.

When tetryl is present in the workplace, OSHA requires employers to limit employee exposure to a concentration of 1.5 mg/m^3, averaged over an 8-hour workday. NIOSH recommends a maximum exposure limit of 1.5 mg/m^3 over a 40-hour workweek.

Performance Goals for Section 14.12
  ◆ Describe the chemical nature of PETN.
  ◆ Describe the detonation of PETN.

TABLE 14.9 ◆ Some Physical Properties of Pentaerythritol Tetranitrate	
Specific gravity	1.75
Melting point	282°F (139°C)
Brisance (g of sand)	62.7
Detonation velocity	4.92 mi/s (7.92 km/s)
Sensitivity	High

◆ 14.12 PETN

The active component of the commercial chemical explosive known as *PETN* (pronounced "pettin") is pentaerythritol tetranitrate.

$$O_2N-O-CH_2 \quad CH_2-O-NO_2$$
$$\diagdown \diagup$$
$$C$$
$$\diagup \diagdown$$
$$O_2N-O-CH_2 \quad CH_2-O-NO_2$$

The condensed formula of this substance is $C(CH_2-O-NO_2)_4$.

PETN is produced by the following two-step process:

◆ Preparation of pentaerythritol from formaldehyde, acetaldehyde, and sodium hydroxide.

$$4H-\overset{H}{\underset{\underset{O(aq)}{\parallel}}{C}} \quad + \quad CH_3-\overset{H}{\underset{\underset{O(aq)}{\parallel}}{C}} \quad + NaOH(aq) \longrightarrow \quad \overset{HO-CH_2 \; CH_2-OH}{\underset{HO-CH_2 \; CH_2-OH(s)}{C}} \quad + HCOONa(aq)$$

formaldehyde   acetaldehyde   sodium hydroxide   pentaerythritol   sodium formate

◆ Nitration of pentaerythritol.

$$\overset{HO-CH_2CH_2-OH}{\underset{HO-CH_2CH_2-OH(s)}{C}} \quad + 4HNO_3(l) \longrightarrow \quad \overset{O_2N-O-CH_2CH_2-O-NO_2}{\underset{O_2N-O-CH_2CH_2-O-NO_2(s)}{C}} \quad + 4H_2O(l)$$

pentaerythritol   nitric acid   pentaerythritol tetranitrate (PETN)   water

PETN is a white solid; some of its other physical properties are provided in Table 14.9.

PETN is sometimes used as the booster in artillery ammunition, but it is probably most commonly encountered as a form of "primacord," a detonation fuse consisting of a core of PETN wrapped in a fabric sheath. Its detonation is represented as follows:

$$C(CH_2-O-NO_2)_4(s) \longrightarrow 3CO_2(g) \quad + \quad 2CO(g) \quad + 2N_2(g) + 4H_2O(g)$$

PETN   carbon dioxide   carbon monoxide   nitrogen   water

Performance Goals for Section 14.13
◆ Describe the chemical nature of HMX.
◆ Describe the detonation of HMX.

TABLE 14.10 ◆ Some Physical Properties of Cyclotetramethylenetetranitramine	
Specific gravity	1.890
Melting point	527°F (275°C)
Brisance (g of sand)	60.4
Detonation velocity	5.66 mi/s (9.11 km/s)
Sensitivity	High

◆ **14.13 HMX**

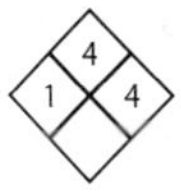

The secondary high explosive cyclotetramethylenetetranitramine is generally known by one of its commercial names: *HMX, Octogen,* and *Rowanex 2000.* The origin of HMX is most likely linked with either of the names "High-Molecular-Weight Explosive" or "High-Melting Explosive."

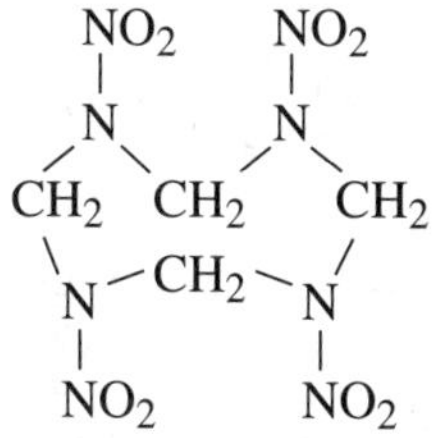

C-4

This molecular structure may be condensed to $(CH_2)_4(N—NO_2)_4$. Some physical properties of HMX are provided in Table 14.10.

When HMX detonates, the chemical change occurs as follows:

$$(CH_2)_4(N—NO_2)_4(s) \longrightarrow 4CO(g) + 4N_2(g) + 4H_2O(g)$$

HMX   carbon monoxide   nitrogen   water

HMX is regularly used in the military warfare actions of today, generally as shaped-charge warhead explosives and rocket propellants. The military also uses the explosive called *Octol,* a mixture of HMX and TNT.

Like cyclonite, HMX is produced from hexamethylenetetramine, nitric acid, ammonium nitrate, and acetic anhydride. HMX and cyclonite possess chemical similarities. Their molecules are eight- and six-membered-ring-shaped nitramines, respectively.

Studies conducted on rats, mice, and rabbits reveal that HMX may be harmful to the heptagenic (liver) and central nervous systems when ingested or absorbed through the skin.

Performance Goals for Section 14.14
- ◆ Describe the chemical nature of the common primary explosives.
- ◆ Describe their detonation.

◆ 14.14 PRIMARY EXPLOSIVES

Primary explosives are very sensitive to heat, shock, and friction. For this reason, they are high explosives and extremely dangerous to handle. In practice, primary explosives are used in very small quantities to initiate the detonation of a main explosive charge. When used for this purpose, they are called *initiators,* or *initiating explosives.*

Three primary explosives are frequently used as initiators in percussion caps, shells cartridges, detonators, and fuses. When activated, they produce the detonation wave that initiates the booster, or bursting charge. These primary explosives are mercuric fulminate, lead azide, and lead styphnate. As compounds of mercury and lead, they are highly toxic. For this reason, their use by explosives experts is declining. Their physical properties are provided in Table 14.11.

MERCURY FULMINATE

Mercury fulminate is synonymous with mercury(II) cyanate; its chemical formula is $Hg(CNO)_2$.

$$N\equiv C-O-Hg-O-C\equiv N$$

It is a white to gray solid produced by pouring a nitric acid solution of mercury(II) nitrate into ethyl alcohol, but the mechanism of the chemical reaction is not entirely understood.

When it detonates, mercury fulminate decomposes into mercury, carbon monoxide, and nitrogen, as follows:

$$Hg(NCO)_2(s) \longrightarrow Hg(g) + 2CO(g) + N_2(g)$$

It is used in the manufacture of caps and detonators for military, industrial, and sporting purposes.

LEAD AZIDE

Lead azide is unique among the commercial explosives in that it is the only one whose chemical composition does not include oxygen. Its chemical formula is $Pb(N_3)_2$.

$$N=N=N-Pb-N=N=N \qquad \text{or} \qquad \begin{matrix} N & & N \\ \| & & \| \\ N-Pb-N \\ / & & \backslash \\ N & & N \end{matrix}$$

TABLE 14.11 ◆ Physical Properties of Some Primary Explosives

	Mercury Fulminate	*Lead Azide*	*Lead Styphnate*
Specific gravity	4.42	—	3.1
Brisance (g of sand)	23.4	19	24
Detonation velocity	2.9 mi/s (4.7 km/s)	3.2 mi/s (5.1 km/s)	3.0 mi/s (4.8 km/s)
Sensitivity	High	High	High

It is a colorless solid prepared by reacting aqueous solutions of sodium azide and lead acetate, as follows:

$$2NaN_3(aq) + Pb(C_2H_3O_2)_2(aq) \longrightarrow Pb(N_3)_2(s) + 2NaC_2H_3O_2(aq)$$

When it detonates, lead azide decomposes into lead and nitrogen, as follows:

$$Pb(N_3)_2(s) \longrightarrow Pb(s) + 3N_2(g)$$

LEAD STYPHNATE

Lead styphnate and lead trinitroresorcinate are synonyms.

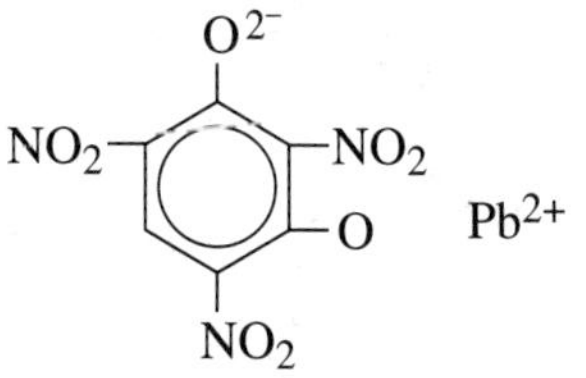

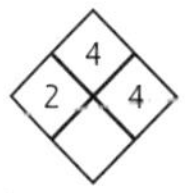

It is a yellow-orange solid made from styphnic acid, or trinitroresorcinol. Lead styphnate is produced when lead acetate reacts with styphnic acid, as follows:

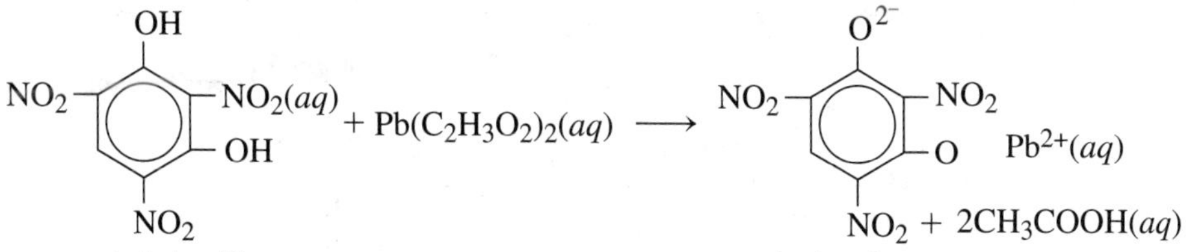

styphnic acid   lead styphnate

Condensing its formula to $Pb[C_6HO_2(NO_2)_3]$, the detonation of lead styphnate is denoted as follows:

$$2Pb[C_6HO_2(NO_2)_3](s) \longrightarrow$$
$$Pb(s) + 3N_2(g) + 5CO(g) + 5CO_2(g) + H_2O(g) + 2C(s)$$

Performance Goal for Section 14.15
- ◆ Describe the nature of the response actions to be implemented when explosive materials have been released into the environment.

◆ **14.15 RESPONDING TO INCIDENTS INVOLVING THE RELEASE OF EXPLOSIVE MATERIALS**

There are four potential types of emergency incidents affecting explosive materials: a transportation mishap; a magazine fire; disowned or abandoned ordnance; and the treatment or disposal of waste explosives. *A response action to each emergency incident should generally be undertaken only by competent individuals who have received special training in the handling of explosive materials.*

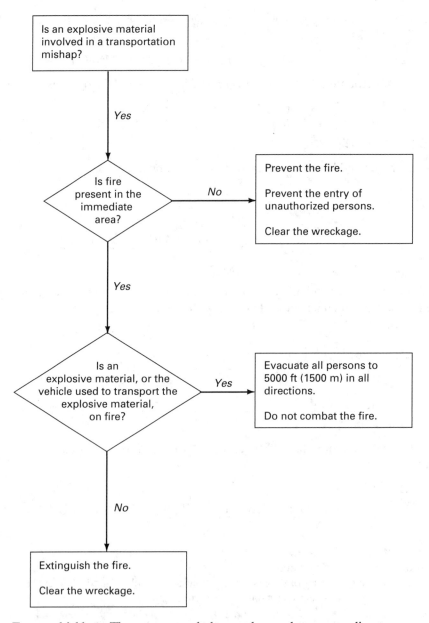

FIGURE 14.11 ◆ The recommended procedures when responding to a disaster involving a division 1.1 explosive material. *Adapted with permission of the American Society for Testing and Materials, from a figure in ASTM STP 825, A Guide to the Safe Handling of Hazardous Materials Accidents. Copyright © 1983, American Society for Testing and Materials.*

When explosive materials are involved in a transportation mishap, the first-on-the-scene responders are advised to implement the procedures in Figures 14.11, 14.12, and 14.13 when the emergency incident involves a Division 1.1 explosive, a Division 1.2 or 1.3 explosive, and a Division 1.4, 1.5, or 1.6 explosive, respectively. When identity of the division of the explosive cannot be established, all unnecessary personnel

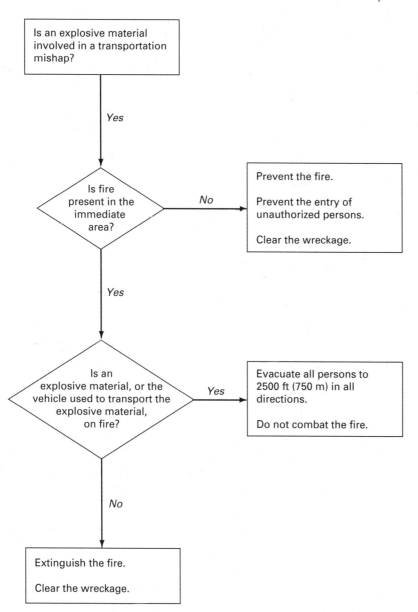

FIGURE 14.12 ◆ The recommended procedures when responding to a disaster involving a division 1.2 or 1.3 explosive material. *Adapted with permission of the American Society for Testing and Materials, from a figure in ASTM STP 825, A Guide to the Safe Handling of Hazardous Materials Accidents. Copyright © 1983, American Society for Testing and Materials.*

and other individuals should be evacuated to a distance of 0.5 mi (800 m). As a general rule, fires involving explosive materials should be fought *only* when the presence of a Division 1,4, 1.5, or 1.6 explosive can be conclusively established from the relevant shipping paper, labels, or placards. Personnel are advised to avoid fighting fires when the presence of a Division 1.1, 1.2, or 1.3 explosive has been conclusively verified.

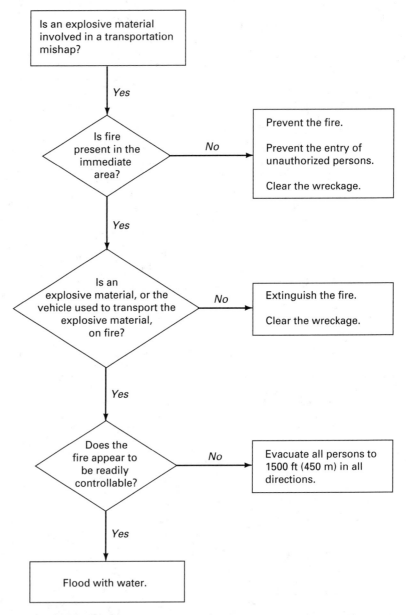

FIGURE 14.13 ◆ The recommended procedures when responding to a disaster involving a division 1.4, 1.5, or 1.6 explosive material. *Adapted with permission of the American Society for Testing and Materials, from a figure in ASTM STP 825, A Guide to the Safe Handling of Hazardous Materials Accidents. Copyright © 1983, American Society for Testing and Materials.*

When called to an emergency incident involving an explosive magazine, the Institute of Makers of Explosives advises first-on-the-scene responders to implement the procedures in Figure 14.14. Fires outside the magazine should be controlled to ensure that they do not reach the exterior perimeter of the magazine. Fires inside the magazine should not be fought.

FIGURE 14.14 ◆ An "Emergency Procedures" poster published by the Institute of Makers of Explosives. *Courtesy of Institute of Makers of Explosives, Washington, D.C.*

Solved Exercise 14.6

The first-on-the-scene-responders to a transportation mishap note the presence of an overturned motor van along a route specifically designated for the transportation of explosive materials. From a distance of approximately 300 ft (91 m), they also observe a small fire that appears to be spreading upward from a tire on the vehicle toward the overlying cargo van. Orange placards have been posted on the visible sides of the vehicle, although the DOT division of the explosive is not readily discernible. What immediate actions should these responders complete to protect public health, safety, and the environment?

Solution: DOT requires the posting of orange placards on a transport vehicle to warn that packages of explosive materials are stored onboard for transit. Given that the transport vehicle is on fire, the crew should promptly recognize that the highest degree of hazard is potentially associated with detonation of the explosive materials. Since the DOT subdivision of the explosive material cannot be discerned, it is prudent to assume that the subdivision is 1.1.

Using the information summarized in Figure 14.11, the responding crew should accomplish the following:

- Evacuate all persons to 5000 ft (1500 m) in all directions from the transportation incident. This includes stopping traffic in all directions.
- Do not attempt to extinguish or control the spread of the fire *until* the driver, shipper, or carrier can accurately determine that the subdivision of the explosive material on board is not 1.1 or 1.2.
- Without knowledge of the subdivision, acknowledge that the risk to the lives of the personnel is so great that combating the fire is unwarranted.

FIGURE 14.15 ◆ A total-containment vessel can be mounted on a transporter or trailer bed and driven to the location at which a suspect suitcase or package has been discovered. These vessels are marketed to withstand the detonation of 26 lb (12.5 kg) of Composition C–4. *Courtesy of Nabco, Inc., Canonsburg, Pennsylvania.*

When emergency responders are called to incidents involving material that has been disowned or abandoned, perhaps by activists or terrorists, it is prudent to suspect that this material may be unexploded ordnance. When responding to this type of incident, the suspect material is generally removed remotely into a total-containment device like that shown in Figure 14.15. Once the suspect material has been transferred, the device is driven to an isolated location and the chemical nature of the material is determined.

Disowned or abandoned waste explosive materials are RCRA-regulated hazardous wastes that exhibit the characteristic of reactivity. When a decision is made to treat or destroy them, EPA should be contacted for guidance relating to the manner of treatment or disposal that best protects public health and the environment.

∎∎∎

REVIEW EXERCISES

General Characteristics of Chemical Explosives

14.1 Why is a shock wave generated when a chemical explosive detonates?

14.2 Following the detonation of a chemical explosion, a plume often appears, much like the appearance of a smoke plume during a fire. Why is the detonation plume often red?

14.3 Why can an explosive charge effectively detonate when placed deep within a borehole, where the amount of atmospheric oxygen is limited or entirely absent?

14.4 The U.S. Bureau of Mines approves chemical explosives intended for use within coal mines. To achieve its approval, the explosives are subjected to certain standardized tests. What is the most likely reason the Bureau of Mines bears the responsibility for approving the use of these explosives?

14.5 The OSHA regulation at 29 CFR §1910.109(e)(2) stipulates that empty boxes, paper, and fiber packing materials that previously contained high explosives shall not be used again for any purpose; instead, they shall be destroyed by burning at an approved isolated location outdoors, and no person shall be nearer than 100 ft (31 m) after the burning has begun. What is the most likely reason OSHA requires the destruction of the packaging materials in this fashion?

14.6 Use equations to demonstrate that ammonium nitrate may detonate as a chemical explosive and it can also participate as an oxidizing agent in chemical reactions that proceed at explosive rates.

Transporting Chemical Explosives

14.7 Packages of the following explosive articles are labeled as indicated. Which of the two articles is likely to pose the greater detonation hazard during transportation mishaps?
  (a) EXPLOSIVE 1.4S or EXPLOSIVE 1.2G
  (b) EXPLOSIVE 1.1D or EXPLOSIVE 1.4S

14.8 The DOT regulation at 49 CFR §177.835(a) requires shippers and carriers to load and unload explosive materials from a motor vehicle only when the engine is not operating. What is the most likely reason DOT prohibits operation of the vehicle's engine during these processes?

14.9 The DOT regulation at 49 CFR §173.60 stipulates that nails, staples, and other closure devices made of metal with no protective covering cannot penetrate to the inside of the outer packaging of an explosive material unless the inner packaging adequately protects the explosive material against contact with the metal. What is the most likely reason DOT requires an explosive material to be protected in this specific fashion?

14.10 A carrier wishes to transport fourteen 1-lb boxes of trinitrotoluene in a sole-use truck.

(a) Use Table 6.1 to determine the basic description the carrier must use on the accompanying shipping paper if the material is wetted with less than 30% water by mass.

(b) Use Table 6.1 to determine the basic description the carrier must use on the accompanying shipping paper if the material is wetted with more than 30% water by mass.

(c) For the transport incidents described in (a) and (b), identify the labels, if any, DOT requires the carrier to affix to the boxes.

(d) For the transport incidents described in (a) and (b), identify the placards, if any, DOT requires the carrier to post on the transport vehicle.

Black Powder and Black Gunpowder

14.11 Black powder consists of a mixture of substances whose components, when ignited, undergo an oxidation–reduction reaction. What is the most likely reason DOT regulates the transportation of black powder as an explosive material rather than as an oxidizer?

14.12 When emergency responders encounter black gunpowder while combating a fire at a sporting goods store, how can they substantially reduce its explosive potential?

Nitroglycerin and Dynamite

14.13 Why is it essential to store nitroglycerin within a dry room?

14.14 Write the equation that shows the manner by which nitroglycerin undergoes hydrolysis.

14.15 When dynamite sticks are discolored, excessively soft, or crumbly, or when visible signs of exterior crystallization are evident, why is the best practice to handle them remotely?

14.16 Although nitroglycerin is a chemical explosive, low doses of this substance are often prescribed to cardiac patients. What is the general purpose of these prescriptions?

Nitrocellulose

14.17 Why do explosive experts strongly advise against storing nitrocellulose in bulk quantities?

14.18 Why does the deflagration of nitrocellulose produce a colorless plume, rather than a red plume?

Trinitrotoluene

14.19 What combination of properties is responsible for the selection of trinitrotoluene as a military explosive?

14.20 Why is 2,4,6-trinitrotoluene also named *symmetrical*-trinitrotoluene, or *sym*-trinitrotoluene?

14.21 The threshold limit values of trinitrotoluene and nitroglycerin are 0.5 and 0.05 mg/m^3, respectively. Which of these chemical explosives is more toxic?

Cyclonite, Tetryl, PETN, and HMX

14.22 Write an equation illustrating the detonation of each of the following chemical explosives:
  (a) cyclonite
  (b) tetryl
  (c) PETN
  (d) HMX

14.23 Why is cyclonite less susceptible to premature decomposition by physical impact when it is contained with hardened beeswax?

14.24 Military explosives containing cyclonite, tetryl, PETN or HMX are sometimes "aluminized," i.e., they are mixed with aluminum dust and an oxidizing agent. What role do these latter substances play during the explosive detonations?

Primary Explosives

14.25 Why is lead azide selected for use in virtually *all* blasting caps and other hot-wire initiated detonators?

14.26 Why can it be a dangerous practice to dispose of azide compounds by washing them into sinks?

14.27 Why are primary explosives ordinarily manufactured in very small quantities?

Responding to Emergencies Involving the Release of Explosives

14.28 Police suspect that a chemical explosive is a component of a package that has been abandoned within a heavily trafficked mall. What actions should be taken by the police and other emergency-response personnel to protect public health, safety, and the environment?

CHAPTER 15 ◆ Radioactive Materials

When we hear the term "radioactive," the occurrence of two fearsome incidents comes to mind: the accidental release of material from a nuclear power plant and the intentional deployment of nuclear weapons. Both devices contain radioactive materials. What is meant when we use the terms "radioactive" and "radioactivity"? We will discover in this chapter that radioactivity is a phenomenon associated with the occurrence of certain *nuclear* processes. The matter that displays this phenomenon is a radioactive material.

Throughout the history of civilization, the use of nuclear weapons was necessitated only twice—when the United States detonated nuclear bombs* in Japan to end World War II. Thereafter, especially through the years of the Cold War, it was the *threat* of nuclear weapon deployment that facilitated the balance of worldwide political power. There was not a target on planet Earth that could not be hit by the use of long-range, nuclear-armed ballistic missiles. Even today, strategically stockpiled throughout the world are massive numbers of nuclear warheads, which war-ridden countries could choose to employ at a moment's notice. The fear is that their flagrant use would bring about a holocaust capable of decimating individual civilizations and altering life on the entire planet.[†]

The energy of a nuclear bomb blast or the engagement of a nuclear warhead far surpasses the energy associated with the detonation of ordinary explosive materials. The explosion of a nuclear bomb, for instance, releases more energy than that involved in the detonation of a million tons of TNT. This amount of energy can immediately kill tens of thousands of people. It is for this reason that nuclear bombs and nuclear-armed ballistic missiles are regarded as weapons of mass destruction.

---

*These nuclear bombs were initially called atomic bombs. Use of the term "nuclear" is preferable to the use of "atomic," since the phenomenon responsible for the detonation of these bombs is a nuclear one.

[†]To encourage international nuclear disarmament and nonproliferation, 150 nations agreed in 1996 to the conditions of the Comprehensive Nuclear-Test-Ban Treaty. One condition of this treaty prohibits the testing of a nuclear device of any size within any environment. Inspectors for the International Atomic Energy Agency, Vienna, Austria, periodically monitor nuclear facilities worldwide to ensure that nuclear material has not been diverted to military uses.

Notwithstanding the agency's work, nonparticipating countries continue to test nuclear weapons. In 1995 and 1996, for instance, France conducted six nuclear tests in the Pacific Ocean, and in 1998, India and Pakistan separately conducted five tests involving nuclear weapons. Israel is widely perceived as possessing nuclear weapons, and North Korea has indicated that it possesses two nuclear bombs. These incidents signal all too clearly that the threat of mass killing and destruction from the deployment of nuclear weapons may again be realized.

In this chapter, we shall observe that exposure to radioactive materials can induce cancer, genetic mutations, and other adverse health effects. This means that while a significant population can be outrightly killed from the detonation of nuclear weapons, even the survivors remain at risk from long-term health-related problems.

Radioactive materials are not exclusively used for nuclear weapons. They are also used for peaceful purposes. Most notably, the nuclear power industry uses radioactive materials to generate electricity. Four hundred thirty-five (435) nuclear power plants produce 17.0% of the electrical energy used worldwide. One hundred three (103)—roughly one-fourth of the world's plants—operate within the United States alone. Most of them are nearing their originally projected operating lives of 40 years. Although they are subject to rigid regulatory controls and license requirements, radioactive materials could be released from their confinement vessels into the environment.

Among the radioactive material used at some nuclear power plants is plutonium fuel, a substance that ranks with the *botulinum* toxin as one of the world's most poisonous substances. The component of the fear about nuclear reactors is that the fuel could be inadvertently released into the environment through such means as mismanagement or the failure to follow established safety practices. In addition, radioactive waste products are generated during the operation of nuclear power plants, which could also be inadvertently released.

These latter scenarios are not merely hypothetical possibilities. There have been well-documented instances of airborne releases of radioactive materials from nuclear power plants in the United States, the countries of the former Soviet Union, and elsewhere. The worst disaster occurred in April 1986 at the Chernobyl nuclear power plant in Russia. A series of operator errors unleashed a power surge that triggered an explosion and partial meltdown of the fuel. Tons of radioactive fuel and by-products were discharged into the environment, the majority of which spread over Russia, Ukraine, Belarus, and parts of western Europe. This single incident turned bustling villages and towns into unpopulated areas and damaged the health of 3.5 million people.

In March 1979, the worst nuclear power accident in the United States occurred at the Three Mile Island plant in southeastern Pennsylvania. The coolant used to modulate the generated heat was inadvertently discharged to the environment. Operators shut down the plant and little radioactivity was released. No one died as an immediate result of the accident, but decades later, the cancer rates in the region surrounding the plant remain points of fierce debate.

In the United States, the growth of the nuclear power industry has spawned the fear of contracting cancer from exposure to the by-products, the so-called nuclear wastes. How can we safely dispose of them? In the United States, these wastes are now mainly stored at the plants where they were generated. Nuclear wastes generated elsewhere are now stored at temporary storage sites in 35 states. We store not only those nuclear wastes generated within the United States, but certain nuclear wastes generated outside the country as well. This latter practice has been implemented to keep nuclear fuel from reaching international terrorist organizations or rogue nations.

Acknowledging the magnitude of the disposal problem, Congress directed the U.S. Department of Energy (DOE) to locate and construct a national site at which high-level nuclear wastes could be permanently stored. The site in Figure 15.1 was selected by DOE and approved by Congress. It is a geologic repository within

FIGURE 15.1 ◆ The United States Department of Energy selected Yucca Mountain, NV, as a permanent repository for spent nuclear fuel and high-level radioactive waste. Plans call for the construction of over 100 miles (160 km) of tunnels within the mountain, which will hold 77,000 tons (70,000 tonnes) of nuclear fuel and radioactive waste. Trucks and rail will be used to transport the fuel and waste within shielded shipping containers. Upon arrival at Yucca Mountain, these material will be removed from their containers and placed in corrosion-resistant canisters for permanent disposal. The canisters will then be placed on supports within the tunnels. *Courtesy of United States Department of Energy, Yucca Mountain Project, Las Vegas, Nevada.*

Yucca Mountain in Nevada. The plan is for nuclear wastes to begin arriving at the repository for permanent entombment by 2010. Notwithstanding Congressional acceptance of the site selection, considerable apprehension continues to be voiced about the viability of safely disposing of nuclear wastes, not just in Nevada, but anywhere. After the Yucca Mountain repository is opened, it is slated to remain accessible for 300 years and to hold at least 77,000 tons (70,000 tonnes) of high-level nuclear waste.

To minimize the hazards associated with exposure to radioactive materials, regulatory bodies have adopted the responsibilities noted below:

- Under the auspices of the Atomic Energy Act of 1954, the U.S. Nuclear Regulatory Commission (NRC) licenses and oversees any facility that uses intensely radioactive materials. The NRC also regulates the construction and operation of the commercial nuclear power plants in the United States. In addition, it oversees the construction and operation of nuclear waste disposal sites.
- The Department of Energy (DOE) oversees the research and development of new and creative means for reducing our burgeoning supply of nuclear waste.
- EPA are responsible for establishing radiation exposure limits that are protective of public health. These limits apply to radiation arising in the environment, including natural radiation and the radiation from spent radioactive materials in storage.
- OSHA is responsible for establishing radiation exposure limits that are protective of employees using radioactive materials within the workplace.
- DOT is responsible for ensuring that radioactive materials are transported safely.

In combination, these various means serve to provide a degree of protection against the hazards associated with inadvertent exposure to radioactive materials. Nonetheless, when compared to the other classes of hazardous materials, exposure to radioactive materials can manifest the highest degree of hazard. What are the properties of radioactive materials that give rise to their particularly hazardous nature? What can be done to minimize the adverse effects caused by exposure to radiation? These questions shall be answered in this final chapter.

Performance Goals for Section 15.1
- Identify the major constituents of atomic nuclei.
- Distinguish between the composition of the hydrogen isotopes.
- Show that the nuclei of atoms of the same element may have different numbers of protons and neutrons.
- Describe the phenomenon of radioactivity.
- Describe the concept of a half-life.

◆ 15.1 FEATURES OF ATOMIC NUCLEI

In Section 4.4, we noted that there are two primary constituents of the atomic nucleus: protons and neutrons. The nuclei of all atoms of the same element possess the same number of protons, but they may differ by the number of neutrons they possess. These different nuclei of the same element are called the *isotopes* of that element.

The number of protons found in the nucleus of an atom is called the *atomic number*. The number of protons equals the number of electrons in a neutral atom. Thus, the number of protons or electrons can be readily identified for any atom by examining the periodic table. The total number of protons and neutrons possessed by a particular isotope is called its *mass number;* it corresponds to the atomic mass of an isotope, rounded off to the nearest whole number.

Hydrogen has an atomic number of 1; this means that every hydrogen atom possesses only one proton. Each hydrogen atom has one electron, but when a hydrogen atom is ionized, it is stripped of its electron and only its nucleus remains.

Hydrogen atoms exist in any of three isotopic forms having the following unique names and compositions:

- *Protonium.* This is simplest of the hydrogen isotopes, and in fact, the simplest of all atoms. When a protonium atom is ionized, only a proton remains.
- *Deuterium.* The nucleus of this second hydrogen isotope is composed of one proton and one neutron. When deuterium atoms are ionized, the remaining nuclei are composed of one proton and one neutron; these nuclei are called *deuterons*.
- *Tritium.* The nucleus of the third hydrogen isotope is composed of one proton and two neutrons. When tritium atoms are ionized, the remaining nuclei are composed of one proton and two neutrons; they are called *tritons*.

Each isotope is designated by a symbol $^A_Z X$, where Z and A are the atomic number and mass number, respectively; X is the element's chemical symbol. Using this format, the three hydrogen isotopes are designated by the symbols $^1_1 H$, $^2_1 H$, and $^3_1 H$, respectively. For each, the symbol Z equals 1, the number of protons. The number of neutrons is obtained by difference: for protonium, the number is 0; for deuterium, the number is 1; and for tritium, the number is 2.

Only the isotopes of hydrogen have unique names. The isotopes of other elements are named by identifying the name or symbol of the element and the mass number of the isotope at issue. Thus, nuclei designated as $^{12}_6 C$, $^{235}_{92} U$, and $^{40}_{19} K$ are named carbon-12, uranium-235, and potassium-40, or C-12, U-235, and K-40, respectively.

All elements have from three to approximately 25 isotopes. Many isotopes are stable; that is, they retain their structure and do not undergo spontaneous changes. On the other hand, many other nuclei are subject to spontaneous transformations or *transmutations*. The nuclei are said to disintegrate or decay; this related phenomenon is called *radioactivity*. The unstable nuclei at issue are said to be *radioactive* and are called *radioisotopes* or *radionuclides*. Two hydrogen isotopes, protonium and deuterium, are stable nuclear species, but tritium is a radioisotope.

Radioactivity is not generally affected by any physical or chemical change in a substance. Hence, when radioactive materials are subjected to changes in pressure, volume, temperature, or chemical nature, the spontaneous disintegration of the relevant radioisotope is not usually altered.

When a radioisotope undergoes a change, it usually emits a particle; less commonly, it absorbs an electron. Both processes are frequently accompanied by the simultaneous emission of energy. When the transformation occurs, the radioisotope has been converted into a new nucleus, which is either stable or radioactive itself. Frequently, radioisotopes undergo several transformations before they are converted into stable nuclei.

Each radioactive transformation is associated with a specific time period. The time during which an arbitrary number of nuclei are reduced to half its number is called the

Solved Exercise 15.1

Indium-111 is a radioisotope used in medicine as a tumoric diagnostic drug and for gastric and cardiac imaging. It decays by electron capture and possesses a half-life of 2.8 days.

(a) How many protons and neutrons are present in the indium-111 nucleus?

(b) Identify the product of its radioactive decay.

(c) What percentage of indium-111 remains in the bloodstream 8.4 days after administration of the drug?

Solution: By reference to either Figure 4.3 or the compilation on the inside back cover, the atomic number of indium is determined to be 49. The symbol for the indium-111 nucleus is $^{111}_{49}$In.

(a) Since the atomic number is the number of protons in a nucleus, there are 49 protons in the indium-111 nucleus. Since the atomic mass of $^{111}_{49}$In is 111, the total number of protons and neutrons in this nucleus is 111. The number of neutrons in $^{111}_{49}$In is $111 - 49$, or 62.

(b) Indium-111 decays by capturing an electron, a process represented as follows:

$$^{111}_{49}\text{In} + {}^{0}_{-1}e \longrightarrow {}^{111}_{48}\text{Cd}$$

The product of the decay is cadmium-111.

(c) For indium-111, a period of 8.4 days represents three half-lives. After the passage of these three half-lives, 12.5% of the administered drug remains in the bloodstream.

$$\text{Final percentage} = 100\% \times \tfrac{1}{2} \times \tfrac{1}{2} \times \tfrac{1}{2} = 12.5\%$$

half-life of that radioisotope. For instance, suppose a volume of tritium gas containing 500,000 molecules is set aside for 12.5 years. Since the tritium molecule is diatomic, there are 1 million atoms of tritium in this volume. After 12.5 years lapses, only 500,000 tritium atoms remain; and after another 12.5 years lapses, only 250,000 tritium atoms remain. Hence, 12.5 years is the half-life of tritium. This means that all supplies of tritium deplete naturally by over 5% each year.

Half-life periods vary appreciably. The half-life of uranium-238, a naturally occurring radioisotope, is 4.5 billion years, but the half-life of astatine-216, a radioisotope of an artificially produced halogen, is only 0.0003 second.

Each element has at least one radioisotope. The hundred or so elements collectively have nearly 1200 known radioisotopes. Table 15.1 shows that few radioisotopes remain in nature. If radioisotopes were present when Earth was initially formed, most of them disappeared long ago, since the age of the earth is 3.6 billion years. Some radioisotopes that are commercially available today are listed in Table 15.2. They are produced by artificial means in nuclear reactors or particle accelerators.

Solved Exercise 15.2

Cesium-137 is a radioisotope that is often used as a source of gamma radiation for the treatment of cancer. If a clinic purchases a source containing 7.60×10^{15} atoms of cesium-137 today, how many atoms will remain in 132 y?

Solution: The half-life of cesium-137 is listed in Table 15.2 as 33 y. The number of half-lives in 132 y is then determined by division:

$$\text{Number of half-lives} = 132 \text{ y}/33 \text{ y} = 4.00$$

This means that after the passage of 132 y, cesium-137 will have proceeded through four half-lives and one-sixteenth of the atoms will remain.

$$\tfrac{1}{2} \times \tfrac{1}{2} \times \tfrac{1}{2} \times \tfrac{1}{2} = (\tfrac{1}{2})^4 = \tfrac{1}{16}$$

Multiplying 7.60×10^{15} by $\tfrac{1}{16}$ gives 4.75×10^{14} atoms.

$$7.60 \times 10^{15} \text{ atoms} \times \tfrac{1}{16} = 4.75 \times 10^{14} \text{ atoms}$$

Thus, after the passage of 132 y, 4.75×10^{14} atoms of cesium-137 remain in the gamma radiation source.

TABLE 15.1 ◆ Naturally Occurring Radioisotopes

Radioisotope	Type of Disintegration	Half-Life in Years	Relative Isotopic Abundance
Tritium	β	12.26	0.00013
Carbon-14	β	5570	
Potassium-40	β, EC[a]	1.2×10^9	0.012
Rubidium-87	β	6.2×10^{10}	27.8
Indium-115	β	6×10^{14}	95.8
Lanthanum-138	β, EC	$\sim 2 \times 10^{11}$	0.089
Neodymium-144	α	$\sim 5 \times 10^{15}$	23.9
Samarium-147	α	1.3×10^{11}	15.1
Lutetium-176	β	4.6×10^{10}	2.60
Rhenium-187	β	$\sim 5 \times 10^{10}$	62.9
Platinum-190	α	$\sim 1 \times 10^{12}$	0.012
Radium-226	α	1622	
Thrium-232	α	1.4×10^{10}	100
Uranium-235	α	7.13×10^8	0.72
Uranium-238	α	4.5×10^9	99.28

[a]EC = electron capture.

TABLE 15.2 ◆ Some Commercially Available Radioisotopes

Radioisotope	Half-Life	Application
Tritium	12.26 y	Determination of total body water
Fluorine-18	1.8 h	Brain and bone imaging
Sodium-24	15.0 h	Detection of obstructions in the circulatory system
Phosphorus-32	14.3 d	Detection of skin cancer
Chromium-51	27.8 d	Determination of red blood cell volume and total blood volume
Cobalt-57	271.8 d	Instrument calibration; determination of the effectiveness of the body's uptake of vitamin B12
Iron-59	44.5 d	Measurement of the rate of formation and lifetime of red blood cells
Cobalt-60	5.2 y	Determination of the effectiveness of the body's uptake of vitamins; cancer therapeutic drug
Gallium-67	3.3 d	Tumoric diagnostic drug
Selenium-75	128 d	Pancreatic cancer diagnostic drug
Yttrium-90	2.7 d	Radioimmunotherapeutic drug
Technetium-99^m	5.9 h	Imaging of the brain, thyroid, liver, kidney, lung, and cardiovascular system
Palladium-103	17 d	Cancer therapeutic
Indium-111	2.8 d	Tumor diagnostic; gastric and cardiac imaging
Iodine-123	13 h	Brain, thyroid, and renal imaging
Iodine-125	59.4 d	Cancer therapeutic; brain, blood, metabolic function diagnostic drug
Iodine-131	8 d	Brain, pulmonary, and thyroid diagnostic
Cesium-137	33 y	Cancer therapeutic; food irradiation
Iridium-192	73.8 d	Cancer therapeutic
Thallium-201	3 d	Cardiac diagnostic
Radium-226	1590 y	Radiation therapy for cancer

Performance Goals for Section 15.2
- ◆ Describe each mode by which radioisotopes decay.
- ◆ Describe the differentiating features of alpha, beta, and gamma radiation.
- ◆ Write equations denoting the changes that occur when radioisotopes decay.

◆ 15.2 TYPES OF RADIATION AND MODES OF NUCLEAR DECAY

Radioisotopes undergo nuclear transformations by the six mechanisms illustrated in Figure 15.2: alpha particle emission, negatron emission, positron emission, electron capture, gamma ray emission, and spontaneous fission. Although some decay exclusively by only one means, most radioisotopes decay by several mechanisms. Only the very heavy radioisotopes decay by spontaneous fission. This mode of decay is noted further in Section 15.6.

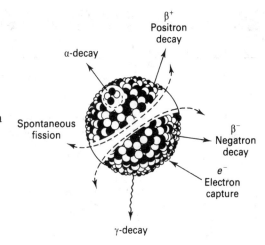

FIGURE 15.2 ◆ The modes by which a radioisotope may decay (open circles are protons, solid black circles are neutrons). Of these modes, the rarest is spontaneous fission. The most common modes are those associated with the production of beta and gamma radiation.

The transformations of radioisotopes are associated with the emission from the nucleus of three types of radiation: *alpha radiation* (α), *beta radiation* (β), and *gamma radiation* (γ). Since their passage through matter results in its ionization, alpha, beta, and gamma radiation are examples of ionizing radiation.

The energy associated with the emission of alpha, beta, and gamma radiation is typically cited in multiples of a unit called an *electron volt* (eV). One electron volt is the amount of energy acquired by an electron when it is accelerated by an electric potential of 1 volt. It is equivalent to 1.602×10^{-19} J. Nuclear radiation is typically associated with energies 1 million times greater than 1 electron volt. One million electron volts (MeV) is equivalent to 1.602×10^{-13} J.

ALPHA RADIATION

Many isotopes having atomic numbers greater than 83 disintegrate by emitting particles consisting of two protons and two neutrons. These particles are called *alpha particles*; they are the nuclei of doubly ionized helium atoms. Alpha particles are symbolized as $^4_2$He, but generally, we denote them by the Greek letter α. When the nuclei of many such atoms decay, many alpha particles are correspondingly emitted. This combination of alpha particles is called *alpha radiation*. It is associated with a relatively large amount of energy that ranges from 4 to 8 MeV, but since alpha particles are doubly ionized, this energy is readily dissipated by its passage through a few centimeters of air, or by absorption in a thin piece of matter. For instance, alpha radiation can be absorbed by the thickness of this page.

When a radioisotope emits an alpha particle, its atomic number decreases by two, and its mass number decreases by four. An example of a radioisotope that disintegrates by alpha particle emission is uranium-238, as follows:

$$^{238}_{92}\text{U} \longrightarrow {}^{234}_{90}\text{Th} + {}^4_2\text{He} \quad (\text{or, } \alpha)$$

The equation notes that the uranium-238 nucleus changes into the thorium-234 nucleus by emitting an alpha particle. The particle is written to the right of the arrow, designating that it has been emitted from the uranium-238 nucleus. When we write equations denoting nuclear phenomena, they are not balanced in the chemical sense. Instead, a nuclear equation is balanced when each of the following is fulfilled: The sums of the charges are the same on each side of the arrow. The sums of the mass numbers are also the same.

BETA RADIATION

The second mode of radioactive disintegration is associated with three different processes. The first is equivalent in result to the emission of an electron from the nucleus. When electrons are encountered in nuclear phenomena, they are called *negatrons* and designated as β^- or $_{-1}^{0}e$. When the nuclei of many such atoms decay, a substantial number of negatrons are correspondingly emitted. This is one form of *beta radiation.*

   When negatron emission occurs, the mass numbers of the associated nuclei remain unchanged, but the atomic number increases by one.* An example of a radioisotope that disintegrates by emitting a negatron is thorium-234. This is the nucleus produced when uranium-238 disintegrates. Upon emitting a negatron, thorium-234 becomes protoactinium-234, as follows:

$$_{90}^{234}\text{Th} \longrightarrow {}_{91}^{234}\text{Pa} + {}_{-1}^{0}e \quad (\text{or } \beta^-)$$

   The emission of a negatron from a nucleus raises a basic question: How can it be emitted *from* the nucleus when the electron is not a component *of* the nucleus? The process is apparently more involved than an equation is capable of representing. Nuclear scientists have determined that during negatron emission, each neutron within the unstable nucleus transforms into a proton and an electron. The proton then becomes part of the new nucleus and the electron is simultaneously emitted. This conversion of the neutron ($_{0}^{1}n$) into a proton and an electron is designated by the following equation:

$$_{0}^{1}n \longrightarrow {}_{1}^{1}\text{H} + {}_{-1}^{0}e$$

Negatrons possess a range of energies, but these energies are generally no greater than 4 MeV. They can usually be absorbed by a 1-cm-thick sheet of aluminum.

   A second process associated with the production of beta radiation involves the emission from the nucleus of a particle called the *positron*. This particle has one different feature than the electron: A positron is positively charged. It is symbolized as $_{+1}^{0}e$ or β^+.

   Radioisotopes emitting positrons retain their mass numbers but decrease in atomic number by 1. This nuclear event can be envisioned as the conversion of a proton into a neutron and positron, as follows:

$$_{1}^{1}\text{H} \longrightarrow {}_{0}^{1}n + {}_{+1}^{0}e$$

An example of a radioisotope that disintegrates by positron emission is sodium-22. This nucleus is spontaneously converted into neon-22. The event is denoted as follows:

$$_{11}^{22}\text{Na} \longrightarrow {}_{10}^{22}\text{Ne} + {}_{+1}^{0}e \quad (\text{or, } \beta^+)$$

   Positron emission represents a second form of beta radiation. Like negatrons, they possess a range of energies, but the energy is generally no greater than 3 MeV. Positrons can usually be absorbed by a 1-cm-thick sheet of aluminum.

   A third process associated with beta radiation involves the union of an unstable nucleus with an extranuclear electron. This phenomenon is called *electron capture*. A radioisotope that transforms by electron capture decreases its atomic

---

*Beta-decay processes are also associated with the production of neutral subatomic particles called neutrinos and antineutrinos. These particles are of no interest here.

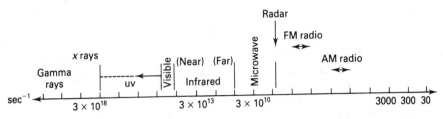

FIGURE 15.3 ◆ The components of the electromagnetic spectrum as a function of frequency. Gamma and X rays, shown to the far left, are forms of radiant energy associated with short wavelengths and high frequencies. Such forms are sufficiently energetic to ionize the matter through which they pass.

number by one, but retains its mass number. Each electron captured by the nucleus reacts with a proton, thereby forming a neutron, which then becomes part of the structure of the new nucleus. This nuclear event is represented by the following equation:

$$\,^1_1H + \,^0_{-1}e \longrightarrow \,^1_0n$$

An example of a radioisotope that disintegrates by electron capture is oxygen-15, as follows:

$$\,^{15}_8O + \,^0_{-1}e \longrightarrow \,^{15}_7N$$

GAMMA RADIATION

Nuclear transformations are frequently accompanied by the simultaneous emission of the third form of radiation: *gamma radiation*. This is a form of *electromagnetic radiation*. Like X radiation, infrared radiation, and ultraviolet radiation, gamma radiation has no mass or charge. In the portion of the electromagnetic spectrum shown in Figure 15.3, the various forms of radiant energy are characterized by their wavelengths. Ultraviolet, infrared, and radio waves have long wavelengths, while gamma radiation and X radiation have short wavelengths. The components of the electromagnetic spectrum with short wavelengths are very energetic, so much so that they ionize matter through which they pass. The components with long wavelengths are relatively nonenergetic and do not ionize matter.

The individual components of gamma radiation are called *gamma rays* or *photons*, represented as γ. Since they do not possess a charge, they are also extremely penetrating and can only be absorbed by dense forms of matter, such as thick blocks of lead.

When gamma rays are emitted from a radioisotope, no change occurs in either the atomic number or mass number. Instead, some fraction of the energy of excitation that causes the nucleus to be unstable is removed as the radioisotope undergoes internal conversion. Imagine a radioisotope that exists in only two energy states. The more energetic form, called the *excited state,* can emit one or more gamma rays from the nucleus. The phenomenon can be illustrated by the following equation, where the excited state is represented by an asterisk:

$$(\,^A_ZX)^* \longrightarrow \,^A_ZX + \gamma$$

In this process, the radioisotope gives up a fraction of its excitation energy to become a more stable form of the *same* radioisotope.

Although the excited states of nuclei generally possess extremely short half-lives, there are excited states that possess reasonably measurable half-lives. The excited state is then referred to as a metastable state of the radioisotope. The metastable state is designated by the use of an "*m*" adjacent to the isotope's mass number. For example, technetium-99^m is an excited state of technetium-99 that has a half-life of 6.0 h. It decays by gamma ray emission, as follows:

$$Tc^{99m} \longrightarrow Tc^{99} + \gamma$$

As noted in Table 15.2, the gamma ray emission from a source of technetium-99^m is specifically used in medicine for imaging various organs of the body.

The phenomenon of gamma radiation emission can also be represented by the following diagram, where each horizontal line designates a discrete energy state of the atomic nucleus:

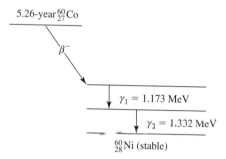

Cobalt-60 is another radioisotope that decays by gamma radiation. When it decays, cobalt-60 emits a negatron and first transformers into an excited state of nickel-60; in turn, this state emits two gamma rays having energies of 1.173 MeV and 1.332 MeV, as follows:

The gamma rays emitted from cobalt-60 are used during food-irradiation processes to disrupt the fast-growing cells of insects, molds, and microbes. Irradiation destroys the microorganisms that cause food spoilage, but it does not cause the food to become radioactive. A primary advantage of gamma-ray irradiation is its success at killing bacteria in raw meat; it is the only known method of eliminating the deadly bacterium *Escherichia coli* from raw meat. Gamma-ray irradiation also eliminates pests in spices and extends the shelf-life of fruits and vegetables (Figure 15.4) with little effect on their flavor, texture, or nutritive value. On the negative side, however, irradiation creates free radicals, the presence of which could negatively affect the quality of foods. For this reason, the widespread use of gamma-ray irradiation on foods remains a controversial practice.

Finally, in 2001, the U.S. Postal Service used gamma-ray irradiation to kill the spores on anthrax-laden mail addressed to two senators as well as mail addressed to government agencies.

FIGURE 15.4 ◆ After one month, the unirradiated orange on the right had shriveled even with refrigeration. The orange on the left was irradiated with gamma radiation and survived with refrigeration for over 2 months.

Performance Goals for Section 15.3
- ◆ Identify the units used to measure radiation.
- ◆ Convert a radiation measurement in one unit into its equivalent in another unit.

◆ 15.3 DETECTION AND MEASUREMENT OF RADIOACTIVITY

The most commonly encountered instrument for the detection of nuclear radiation is the Geiger counter, such as the type shown in Figure 15.5. Like all other radiation-detection instruments, the Geiger counter operates on the principle that nuclear radiation ionizes matter. Inside the case is an argon-filled tube containing two electrodes. The inner walls of the tube are coated with a material that is given a negative charge and a wire in the center of the tube that is given a positive charge. When radiation enters the tube, it strikes and ionizes the argon atoms, which briefly conduct a tiny electric current from the walls to the wire. The passage of this current is then amplified and converted to a clicking sound that is audibly perceived in the earphones worn by the operator. An increase in the number of clicking sounds means an increase in radiation intensity.

FIGURE 15.5 ◆ A portable Geiger counter, often used for surveying an area when monitoring for the presence of beta and gamma radiation. The operating voltage is supplied by batteries. The indicating meter is calibrated to record a radiation intensity unit, like milliroentgens per hour or counts per minute. For accurate interpretation, the instrument must be calibrated with radiation of the same type and energy whose intensity is to be measured. *Courtesy of Fisher Scientific Company, Pittsburgh, Pennsylvania.*

Radiation detectors are still based upon the design of the Geiger counter, but they have now been modernized to accommodate the needs of the user as in Figure 15.6.

As radiation-detection instruments were developed, a number of units of radiation measurement were simultaneously defined. The following units are in use today:

- *Roentgen* (R). This is the international unit of radiation quantity for gamma radiation, but the unit is also used to measure alpha and beta radiation. One roentgen is the amount of radiation that produces 2.1×10^9 units of charge in 1 mL of dry air at standard atmospheric pressure.
- *Rad* (D). This unit is the acronym for radiation-absorbed-dose. It is the amount of radiation absorbed per gram of body tissue. For most purposes, the roentgen and the rad are so close they are considered identical; thus, 1 R = 1 D.
- *Rem.* This unit is the acronym for roentgen-equivalent-man. It is the amount of body tissue damage caused by radiation relative to a dose of 1 roentgen of X rays. The relation of the rem to other dose units depends upon the biological effect under consideration. One rem is the amount of radiation that produces the same damage as 1 R of X rays or γ rays.
- *Gray* (Gy). The is the amount of radiation equivalent to the transfer of 1 joule of energy to 1 kilogram of living tissue. One gray is equal to 100 D.
- *Curie* (Ci). This unit measures the number of radioactive disintegrations occurring each second in a sample. One curie is the amount of radiation corresponding to 3.7×10^{10} disintegrations per second. Since the curie is an extremely large unit, radiation is more often measured in millicuries (mCi), microcuries (μCi), and picocuries (pCi).

$$1 \text{ mCi} = 3.7 \times 10^7 \text{ disintegrations/s}$$
$$1 \text{ μCi} = 3.7 \times 10^4 \text{ disintegrations/s}$$
$$1 \text{ pCi} = 3.7 \times 10^{-2} \text{ disintegrations/s}$$

FIGURE 15.6 ◆ A firefighter at Los Alamos National Laboratory notes a radiation reading on a Palm Pilot. *Courtesy of Los Alamos National Laboratory, Los Alamos, New Mexico; Leroy N. Sanchez, photographer.*

◆ *Becquerel* (Bq). This is the SI unit used to measure radioactive disintegrations per second. One Bq is equivalent to 1 disintegration per second, or

$$1 \text{ Ci} = 3.7 \times 10^{10} \text{ Bq}$$

A convenient unit for measuring radioactivity is the terabecquerel (Tbq), which equals a trillion becquerels.

$$1 \text{ TBq} = 10^{12} \text{ Bq}$$

Solved Exercise 15.3

To protect public health, EPA proposes maximum exposure limits from radiation sources within the environment. Presently, EPA proposes 15 mrem/year as the radiation limit for a whole-body dose from all exposure pathways. If an individual has been exposed to a whole-body dose of 200 μSv/year, has the person been exposed to radiation in excess of the recommended limit?

Solution: Since one sievert equals 100 rem, an annual dose of 200 μSv is equivalent to 20 mrem/year.

$$200 \text{ μSv/year} \times \text{Sv}/10^6 \text{ μSv} \times 100 \text{ rem/Sv} \times 10^3 \text{ mrem/rem} = 20 \text{ mrem/year}$$

This means that the individual has been exposed to a radiation dose in excess of EPA's recommended limit of 15 mrem/year.

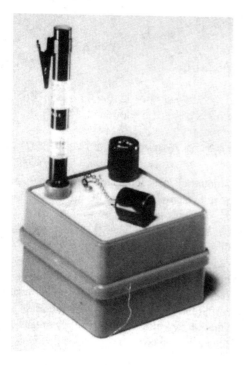

FIGURE 15.7 ◆ A pocket dosimeter on the left and three types of radiation-monitoring badges (film body, thermoluminescent body, and thermoluminescent ring) on the right. For individuals who work near radiation sources or operating X-ray equipment, these devices provide an estimate of the total amount of beta, gamma, and X radiation to which they have been exposed. *Courtesy of Lab Safety Supply Co., Janesville, Wisconsin [pocket dosimeter], and R. S. Landauer, Jr. and Co., Glenwood, Illinois [radiation-monitoring badges].*

The total amount of radiation emitted by a radioisotope is called its *activity*. The activity per unit mass of a radioactive material is called its *specific activity;* it is routinely measured in units such as curies per gram (Ci/g) or terabecquerels per gram (TBq/g).

The amount of radiation to which an individual has been exposed is often easily determined through the use of personal monitoring equipment, such as the film badges or pocket dosimeters shown in Figure 15.7. Film badges consist of photographic emulsions that are very sensitive to radiation. They are worn for a specified period, usually one each day, and then professionally developed. The radiation exposure is ascertained by comparing the developed film against those previously exposed to known amounts of radiation.

Pocket dosimeters are small ionization chambers, typically calibrated to read between 0 and 200 mR. After exposure to radiation, they are read through the use of an auxiliary reader instrument to establish the exposure.

Performance Goals for Section 15.4
- Identify the adverse health effects that result from exposure to different types of radiation.
- Identify the adverse health effects that result from exposure to the radioisotopes of iodine, strontium, and radium.

◆ 15.4 ADVERSE EFFECTS CAUSED BY EXPOSURE TO RADIATION

Each of us is constantly exposed to cosmic radiation and inescapable low-level ionizing radiation that is emitted from naturally occurring radioisotopes. For the average American, this background radiation is 300 mrem/y. We are exposed to ionizing radiation when X-ray patterns are obtained to trace defects in bones and teeth, and when radioactive pharmaceuticals are used to detect the presence of tumors and therapeutically kill diseased tissue. These exposures are associated with only minor epidemological risks, since the radiation intensity is relatively low and the exposure period is short. The largest single contribution to radiation exposure is associated with exposure to naturally occurring radon (Figure 15.8). We shall visit this issue in Section 15.10.

When human tissue is exposed to ionizing radiation, we are concerned with the quantity absorbed per unit of mass. This is called the *dose*. The biological consequences resulting from exposing the human organism to different short-term doses of radiation are provided in Table 15.3. The consequences of radiation exposure to specific body organs, especially the sexual and active blood-forming organs, are likely to cause chronic effects.

The SI unit of ionizing radiation exposure is the coulomb per kilogram (c/kg), which equals approximately 3876 R. The SI unit of dose-equivalent is the *sievert* (Sv), which equals 100 rem. The radiation dose-equivalent rate is called the *radiation level,* which is measured in millisieverts per hour (mSv/hr).

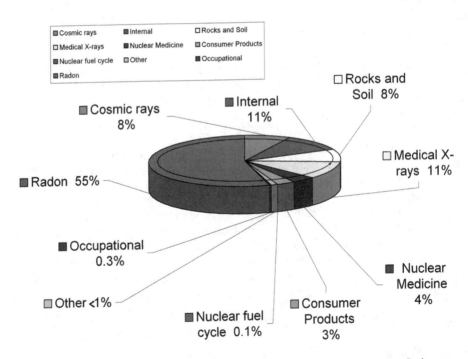

FIGURE 15.8 ◆ Sources of radiation exposure for the American population. Although most of the radiation exposure is associated with natural sources, there is concern that our exposure from medical sources, nuclear power plants, and radon is increasing. *Reprinted with permission of the National Council on Radiation Protection and Measurements, NCRP Report No. 93.*

TABLE 15.3 ◆ Biological Effects from Short-Term Radiation Doses on Humans

Dose (Rems)	Biological Effects
0–25	No detectable effect
25–100	Temporary decrease in white blood cell count
100–200	Nausea, vomiting, longer term decrease in white blood cell count
200–300	Vomiting, diarrhea, loss of appetite, listlessness; easy bruising, fatigue, hair loss, sterility
300–600	Vomiting, diarrhea, hemorrhaging, eventual death in some individuals
>600	Eventual death in most cases

To avoid the adverse effects associated with radiation exposure, it is essential to limit the radiation dose to the maximum extent possible. In the workplace, OSHA regulates employee exposure to radiation in the workplace, as follows (29 CFR §1910.96):

- No employer shall possess, use, or transfer sources of ionizing radiation in such a manner as to cause any individual to receive in any period of one calendar quarter from sources in the employer's possession or control a dose in excess of the limits provided in Table 15.4.
- An employer can permit an individual to receive a dose greater than those provided in Table 15.4 as long as:
  (a) The dose during any calendar quarter shall not exceed 3 rems.
  (b) The dose, when added to the accumulated occupational dose, shall not exceed $5(N - 18)$ rems, where N is the individual's age in years.
- No employer shall permit any employee who is under 18 years of age to receive in any period of one calendar quarter a dose in excess of 10% of the limits in Table 15.4.

Body tissue can be exposed to a certain amount of radiation without experiencing deleterious effects. Unless the radiation is unusually intense, no physical sensation ordinarily accompanies short periods of radiation exposure, such as during the taking of dental X-ray patterns. On the other hand, long-term exposure to radiation causes body tissue to become necrotic or ulcerating.

The biological impact of radiation exposure depends upon the type of radiation absorbed, its energy, the nature of specific radioisotopes, and the age of the individual. We shall note the impact of these factors individually.

TABLE 15.4 ◆ Permissible Radiation Doses[a]

	Rems per Calendar Year
Whole body; head and trunk; active blood-forming organs; lens of the eyes; or sexual organs	1.25
Hands and forearms; feet and ankles	18.75
Skin of the whole body	7.5

[a]29 CFR §1910.96(b).

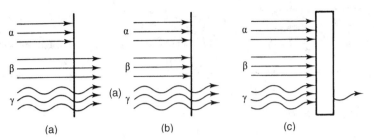

FIGURE 15.9 ◆ The relative penetrating power of alpha, beta, and gamma radiation: (a) sheet of paper, (b) sheet of aluminum, and (c) block of lead.

TYPES OF RADIATION AND THEIR ENERGIES

When evaluating the potential risk from radiation exposure, the specific nature of the radiation is relevant. Alpha, beta, and gamma radiation penetrate matter in the manner noted below and demonstrated in Figure 15.9:

- Alpha radiation is very energetic, but it is easily absorbed externally by the epidermis, the outer layer of the skin. Since the epidermis consists of dead skin cells, external exposure to alpha radiation could represent only a minor hazard. If it is ingested, however, the energetic alpha radiation can localize in a minute area of bone matter, where the subsequent biological damage can be severe.
- Beta radiation penetrates deeper into tissue than alpha radiation. A 3-MeV negatron, for instance, travels through 0.6 in. (15 mm) of tissue before being absorbed. Upon external exposure, beta radiation passes through the epidermis into the dermis, a layer of skin tissue containing living cells. This means that both external *and* internal exposure of the human organism to beta radiation can cause biological damage.
- Gamma and X radiation cause the most severe deleterious effects upon body tissues. Neither is easily absorbed. Gamma and X radiation pass through tissue and cause the ionization of the various substances they encounter. External and internal exposure to gamma and X radiation causes more severe biological damage than exposure to either alpha or beta radiation. As shown in Table 15.5, exposure to gamma and X radiation also affects the entire human organism.

EXPOSURE TO SPECIFIC RADIOISOTOPES

Aside from the general effects caused by radiation exposure, specific radioisotopes often cause damage in the human organism in relatively localized ways. Of special interest are the radioisotopes of strontium and radium that replace calcium in the structure of bones. Inside some bones are spaces occupied by marrow, a tissue that contains the cells capable of forming new blood cells. When radioisotopes of strontium or radium deposit within these spaces, they could radiate the marrow and cause leukemia. This is especially problematic since several radioisotopes of strontium and radium possess relatively long half-lives; for example, the half-lives of strontium-90 and radium-226 are 28.1 years and 1600 years, respectively.

Iodine radioisotopes can also cause localized damage within the body, specifically to the thyroid. The thyroid is a gland located near the base of the neck that manufactures and secretes the hormones responsible for regulating the rates of cellular metabolism

TABLE 15.5 ◆ X-Ray and Gamma-Ray Doses Required to Produce Various Somatic Effects

Dose (Rads)	Effect
0.3 weekly	No observable effect
60 (whole body)	Reduction of lymphocytes (white blood cells formed in lymphoid tissues as in the lymph nodes, spleen, thymus, and tonsils)
100 (whole body)	Nausea, vomiting, fatigue
200 (whole body)	Reduction of all blood elements
400 (whole body)	50% of an exposed group would probably die
500 (testicles)	Sterilization
1000 (skin)	Erythema (reddening of the skin)

and body growth. Two such hormones are thyroxine and triiodothyronine, whose chemical formulas are $HO-C_6H_2I_2-O-C_6H_2I_2-CH_2CH(NH_2)-COOH$ and $HO-C_6H_3I-O-C_6H_2I_2-CH_2CH(NH_2)-COOH$, respectively.

thyroxine   triiodothyronine

A supply of these hormones remains stored in the thyroid for use as the body demands. When the hormones are flooded with iodine radioisotopes, they serve as an irradiation source.

In the decade following the Chernobyl nuclear plant disaster, researchers identified 577 cases of thyroid cancer in youths 18 years of age and under, compared to just 59 cases in the previous 5 years. This increase was attributed to the presence in air and milk of iodine-131, which was among the radioisotopes discharged into the environment when the reactor failed. This iodine radioisotope concentrated in the children's thyroids, where the subsequent radiation caused the production of malignant tumors.

THE IMPACT OF AN INDIVIDUAL'S AGE

The chronic effect on the body's tissues depends upon the number and nature of their cells whose biological function has been altered. Since cell division occurs more frequently in children than it does in adults, exposure to radiation has a heavy impact on youngsters. Investigations following the Chernobyl nuclear disaster illustrated that more than one-third of the population who fell ill were children.

Performance Goal for Section 15.5

◆ Describe the action that occurs when radiation passes into or through matter.

◆ 15.5 EFFECTS OF IONIZING RADIATION UPON MATTER

When ionizing radiation passes through matter, its energy is dissipated by the ionization or excitation of the atoms or molecules of which the matter is composed. The primary action of the radiation is to ionize the substances through which it passes. This ionization can be represented for a molecule of an arbitrary substance A by the following equation:

$$A \rightsquigarrow A^+ + e^-$$

The wiggly arrow indicates that the relevant reaction was induced by ionizing radiation, and e^- symbolizes an electron. A^+ is the symbol of a *molecule-ion,* or *molecular ion,* a species with fewer electrons than protons.

When many such events occur, the number of ions and electrons correspondingly increase. Then, several secondary phenomena usually occur. In particular, the ions can combine with any of the electrons to form an excited state of A as follows:

$$A^+ + e^- \longrightarrow A^*$$

Here, A^* refers to a molecule of A that possesses excess energy. A^* is an unstable molecule. Possessing excessive energy, it can dissociate entirely into molecules of new substances as follows:

$$A^* \longrightarrow B + C$$

The ions that form from the initial action of radiation upon a substance can also react with neutral molecules. Such reactions produce new ions and free radicals. The direct exposure of a substance to radiation can also result in the production of free radicals. As we noted before, free radicals are highly reactive chemical species that can trigger a variety of chemical reactions.

Let's consider specifically what occurs when water is exposed to ionizing radiation. Water constitutes three-fourths of all the body's tissues; hence, considerable research efforts have been devoted to examining the nature of the chemical species produced when water is exposed to radiation. When foreign substances are formed in this water, the cellular biochemistry could be significantly altered, thereby causing damage or death to the affected cells.

When water is exposed to ionizing radiation, primary ionization occurs first, as follows:

$$H_2O(l) \rightsquigarrow (H_2O)^+(l) + e^-$$

Then, the water molecule-ion reacts with a neutral water molecule to form a hydroxyl radical:

$$(H_2O)^+(l) + H_2O(l) \longrightarrow (H_3O)^+(aq) + \cdot\ddot{O}-H(aq)$$

Exposure of water to radiation could also result in the production of hydrogen atoms and hydroxyl free radicals:

$$H_2O(l) \longrightarrow H\cdot(aq) + \cdot\ddot{O}-H(aq)$$

When this latter event has occurred many times, the concentration of hydroxyl radicals increases; eventually, two hydroxyl radicals unite to form hydrogen peroxide:

$$2 \cdot\ddot{O}-H(aq) \longrightarrow H_2O_2(l)$$

The production of these highly reactive chemical species within living cells causes serious disruptions in the mechanistic steps associated with metabolism and other biochemical processes. When radiation damages the cellular genetic material, the consequences are even more far-reaching. Damaging genetic material leads to mutations, cancer, and cell death. Even subtle changes profoundly affect the human organism by modifying an individual's ability to adapt and survive within its environment.

Performance Goals for Section 15.6
- Describe the phenomenon of nuclear fission.
- Differentiate between the impact of using fissile material in a nuclear reactor and that of using it in a nuclear bomb.
- Write equations denoting individual fission processes.
- Describe spontaneous fission.

◆ 15.6 NUCLEAR FISSION

Scientists have been able to synthetically produce numerous radioisotopes that are not found naturally. One method involves irradiating stable or long-lived isotopes in specially designed machines, like cyclotrons and bevatrons. Samples of such isotopes are generally irradiated with α particles, protons, deuterons, and other particles accelerated to relatively high velocities. These high-velocity particles possess increased energies. When they are used to bombard nuclei, *nuclear reactions* often occur. A nuclear reaction constitutes the artificial transmutation of one nucleus into another one. Many of the artificially produced radioisotopes used in medicine are prepared in this fashion.

Solved Exercise 15.4

Why do the workers at nuclear weapons facilities represent a group at a higher risk of contracting cancer compared to workers at nonnuclear facilities?

Solution: Workers at nuclear weapons facilities are exposed to radioactive materials more frequently than workers at nonnuclear facilities. These radioactive materials include the dozens of fission products produced when the nuclear fuel fissions, some examples of which are given in Table 15.5. Low concentrations of these fission products, especially the gaseous ones, leak through tiny openings in the reactor's structure into the surroundings. Since exposure to radioactive materials can induce various types of cancer, workers at nuclear weapons facilities are at a higher risk of contracting cancer compared to workers who are not occupationally exposed to radioactive materials. Among workers at nuclear weapons facilities, there are documented instances of excess cancers in the lung, brain, bladder, stomach, respiratory tract, larynx, trachea, as well as the body's blood-forming tissues (bone marrow and the lymph system).

A second method of production involves utilizing the nuclear reactions that occur within a nuclear reactor. A nuclear reactor is the heart of a nuclear power plant. It functions because of the special phenomenon called *nuclear fission*. As illustrated in Figure 15.10, fission is a process whereby a nucleus splits into two new nuclei, accompanied by the release of a relatively large amount of energy and several neutrons.

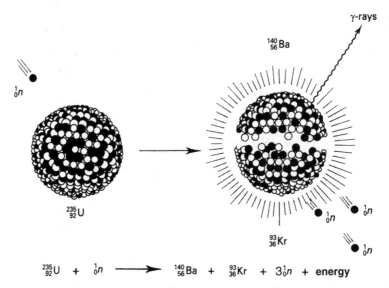

$$^{235}_{92}U + ^{1}_{0}n \longrightarrow ^{140}_{56}Ba + ^{93}_{36}Kr + 3^{1}_{0}n + \textbf{energy}$$

FIGURE 15.10 ◆ The fission of a uranium-235 nucleus induced by the low-velocity neutron entering from the upper left. In this instance, barium-140, krypton-93, and three neutrons are produced.

In the nuclear power industry, engineers harness this energy and convert it into electrical energy in the manner illustrated in Figure 15.11. The heat evolved during fission is absorbed by water, which passes through the reactor within pipes. The heated water is then piped from the reactor to a steam generator, where the water is vaporized and sent to a turbine. The steam spins the turbine blades, which generate electricity much like they do in a fossil-fuel-fired power plant.

The neutrons generated by fission can also be used to bombard other nuclei. This serves as a means of artificially producing radioisotopes that do not exist naturally. One such radioisotope is cobalt-60, whose half-life is 5.3 years. Cobalt-60 may be artificially produced by irradiating cobalt-59 with neutrons produced during fission.

$$^{59}_{27}Co + ^{1}_{0}n \longrightarrow ^{60}_{27}Co + \gamma$$

Cobalt-59 is the only naturally occurring, stable isotope of cobalt.

As shown in Section 15.2, when the cobalt-60 nucleus decays, it emits two gamma rays having energies of 1.173 MeV and 1.332 MeV. Radioactive material containing cobalt-60 serves as a potent source of gamma radiation. It has been used commercially to induce cross-linking in polyethylene, to vulcanize rubber, and as a radiation source in cancer therapy. To treat prostate cancer, for example, a therapist implants tiny pellets containing cobalt-60 in the prostate; the radiation emitted by the cobalt-60 kills cells, including cancerous cells.

The neutrons generated by fission can also be used to *induce* nuclear fission. This is the phenomenon that occurs when a nuclear bomb or nuclear warhead detonates. Only certain nuclei can be induced with neutrons to undergo fission; they are said to be *fissile*. Uranium-233, uranium-235, plutonium-238, plutonium-239, and plutonium-241 are fissile nuclei. Each is induced to undergo fission by exposing it to low-energy neutrons. For example, when uranium-235 captures a low-energy neutron, unstable uranium-236 forms. To achieve stability, each uranium-236 nucleus fissions into two other

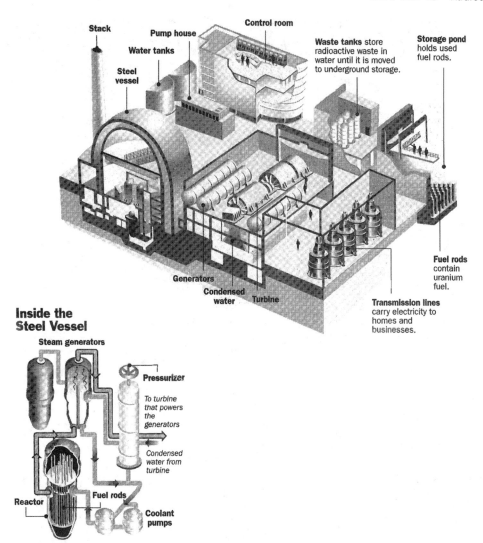

FIGURE 15.11 ◆ In a nuclear power plant, the energy generated by fission is used to produce electricity. A nuclear power plant does not generate air pollutants like a fossil-fuel-fired power plant does, but a nuclear power plant generates spent fuel and high-level radioactive waste, both of which require disposal.

nuclei, simultaneously releasing energy and several neutrons. One such event is represented as follows, where an asterisk is used to denote the unstable nucleus:

$$^{235}_{92}U + ^{1}_{0}n \longrightarrow [^{236}_{92}U]^{*} \longrightarrow ^{89}_{35}Br + ^{145}_{57}La + 2^{1}_{0}n + 192MeV$$

Dozens of individual fission events occur when the uranium-235 fuel in a reactor is consumed. The combination of these events produces numerous radioactive by-products, some of which are listed in Table 15.6. These radioisotopes are distributed among the elements from zinc ($Z = 30$) to terbium ($Z = 65$) and have mass numbers ranging from $A = 72$ to $A = 161$. In a reactor, this mixture of fission products forms its waste products; in a nuclear bomb, they form the components of atmospheric fallout. In both instances, these products are highly radiotoxic.

TABLE 15.6 ◆ Principal Radioactive Fission Products in Nuclear Waste

Radioisotope	Half-Life	Emitted Radiation
Krypton-85	4.4 hr	β, γ
Zirconium-95	65 d	β, γ
Strontium-89	54 d	β
Strontium-90	27 yr	β
Niobium-95	90 hr	β, γ
Technetium-99^m	5.9 hr	β
Ruthenium-103	39.8 d	β, γ
Rhodium-103	57 min	e^{-a}
Ruthenium-106	1 yr	β
Rhodium-106	30 s	β, γ
Tellurium-129	34 d	β, γ
Iodine-129	1.7×10^7 yr	β, γ
Iodine-131	8 d	β, γ
Xenon-133	2.3 d	β, γ
Xenon-135	9.2 h	β, γ
Cesium-137	33 yr	β, γ
Barium-140	12.8 d	β, γ
Lanthanum-140	40 hr	β, γ
Cerium-141	32.5 d	β, γ
Cerium-144	590 d	β, γ
Praseodymium-143	13.8 d	β, γ
Praseodymium-144	17 min	β
Promethium-147	2.26 yr	β

[a]e^- denotes internal conversion electrons. Internal conversion is a relatively rare process in which an unstable nucleus gives its excess energy directly to an orbital electron, instead of undergoing de-excitation by means of gamma-ray emission. The electron is subsequently ejected from the nucleus.

Solved Exercise 15.5

Identify the isotope produced when the fission phenomenon represented by the following equation occurs:

$$^{235}_{92}\text{U} + ^{1}_{0}n \longrightarrow ^{82}_{35}\text{Br} + \underline{\hspace{1cm}} + 2^{1}_{0}n$$

Solution: A nuclear equation is balanced when two conditions are fulfilled: the sums of the charges are the same on each side of the arrow, and the sums of the mass numbers are the same on each side of the arrow. On the left, the sum of the charges is $92 + 0 = 92$; on the right, the sum is $35 + x + 0 = 34 + x$, where x is the charge of the second isotope.

$$35 + x = 92$$
$$x = 57$$

By reference to either Table 4.3 or the compilation on the inside back cover, the element having an atomic number of 57 is lanthanum.

On the left, the sum of the mass numbers is $235 + 1 = 236$; on the right, the sum is $82 + 2 + y = 84 + y$, where y is the mass number of the second isotope.

$$236 = 84 + y$$
$$y = 152$$

Hence, the symbol of the isotope produced by this fission event is $^{152}_{57}$La.

Only 0.71% by mass of naturally occurring uranium is uranium-235. Uranium-238 is the more abundant isotope, but it is not fissile. The neutron-induced fission of uranium-235 is a particularly unusual phenomenon, since for each single neutron that is consumed by reaction, additional neutrons are produced. These newly formed neutrons can be utilized to induce the fission of other uranium-235 nuclei. The process thereby initiates the *chain reaction* in Figure 15.12. In a nuclear reactor, the chain reaction is controlled by workers and the energy is harnessed for the production of steam. By contrast, in the nuclear bomb or nuclear-armed missile, the chain reaction builds at an explosive rate and the energy is released to the environment with devastating effects.

Can the nuclear reactor at a power plant detonate like a nuclear bomb or nuclear-armed missile? Answering this question requires that we understand certain details about the fission phenomenon. For fission to occur, a certain minimum mass of fissile material must be present; this is called the *critical mass*. When the mass of the fissile material is less than the critical mass, most of the neutrons formed during fission events escape from the mass and do not induce the fission of other nuclei. The perpetuation of a chain reaction requires that at least as many neutrons be retained by the system as were consumed in the fission reaction from which they were produced. In this way, the neutrons can react in further fission events.

Solved Exercise 15.6

In 2003, the American Academy of Pediatrics recommended that households, schools, and child-care centers located near nuclear power plants keep potassium iodide tablets on hand. What is the most logical reason the academy recommended this action?

Solution: Table 15.5 indicates that iodine-129 and iodine-131 are among the radioactive fission products released to the environment during a nuclear power plant accident. When exposed to the radioactivity emitted during the accident, the thyroid gland preferentially absorbs the iodine radioisotopes, since the gland biochemically manufactures iodine-bearing hormones (Section 15.4). During the subsequent decay of these radioisotopes, the energy that evolves disrupts normal cellular processes and triggers the onset of thyroid cancer.

The American Academy of Pediatrics most likely recommended the availability of potassium iodide tablets for immediate use if a nuclear power plant accident occurs.* By consuming an acceptable dose of these tablets, the thyroid hormones become saturated with nonradioactive iodine, thereby blocking absorption of the iodine radioisotopes. This practice saves children from the major impact of an accidental or intentional radiation release, and lessens the risk that they will contract thyroid cancer from the exposure.

Individuals exposed to the radioactivity emitted during a nuclear power accident should also *immediately* evacuate the area upon learning of the incident. Increasing an individual's distance from the plant may be even more important than taking a dose of potassium iodide, since the latter reduces only the chance of developing thyroid cancer. Exposure to radiation causes not only thyroid cancer, but leukemia and other forms of cancer upon which potassium iodide has no impact.

*Although a prescription is unnecessary for the purchase of potassium iodide tablets, individuals should consult their doctor or pharmacist for the proper dosage for each family member.

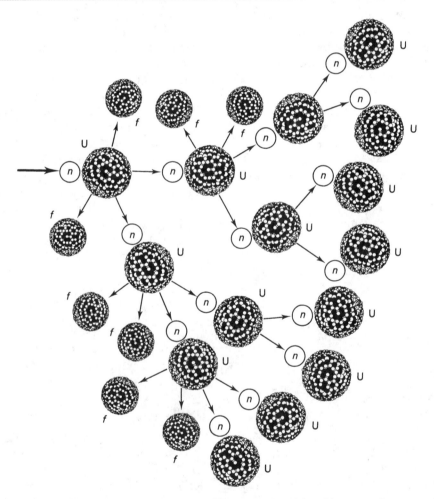

FIGURE 15.12 ◆ A nuclear chain reaction. The neutron (*n*) adjacent to the arrow is absorbed by a uranium-235 nucleus, which subsequently undergoes fission. Fission by-products (*f*) and additional neutrons are thus produced. These neutrons induce the fission of other uranium-235 nuclei. The chain reaction is self-perpetuating until either the fuel is depleted or the neutrons are absorbed by matter other than the fuel.

FIGURE 15.13 ◆ Top: "Little Boy," the nuclear bomb detonated over Hiroshima, Japan, on August 6, 1945. This bomb contained fissile uranium-235, was 28 in. (71 cm) in diameter, 120 in. (3 m) long, weighed about 9000 lb (4100 kg), and killed more than 200,000 people. Bottom: "Fat Man," the nuclear bomb detonated over Nagasaki, Japan, on August 9, 1945. This bomb contained fissile plutonium-239, was 60 in. (150 cm) in diameter, 128 in. (3.2 m) long, weighed about 10,000 lb (4500 kg), and killed more than 74,000 people. *Courtesy of the Los Alamos National Laboratory, Los Alamos, New Mexico.*

In the nuclear bombs shown in Figure 15.13, two pieces of fissionable material having a mass less than the critical mass were compressed into a relatively small volume by explosive charges. Numerous fission events subsequently occurred, releasing a cataclysmic amount of energy.

By contrast, an amount of fissionable material just slightly larger than a critical mass is used in the fuel rods in a nuclear reactor. There is far less fissionable material per unit volume in a reactor than there was in either nuclear bomb detonated over Japan. Furthermore, the number of neutrons available for future fission events can be restricted in a nuclear reactor by means of "control rods," long rods made of cadmium that are inserted into the reactor to effectively absorb extra neutrons and control their propagation. With the use of control rods, there is little fear that a reactor will explode. Even if the chain reaction in a nuclear reactor became uncontrollable, the reactor would most likely melt before it would detonate.

Notwithstanding its improbability, a criticality accident can occur whenever an amount of fissionable material that is equal to or greater than the critical mass is accumulated. Such an unintended event is rare, but not nonexistent. In 1999, a critical mass of uranium-235 was unintentionally accumulated at a manufacturing facility in Japan and caused a criticality accident.

SPONTANEOUS FISSION

As we first noted in Section 15.2, spontaneous fission is a mode of decay for the very heaviest radioisotopes. When a radioisotope disintegrates by spontaneous fission, it splits into two nuclei. An example of a radioisotope that decays by spontaneous fission is curium-250, whose half-life is 1.7×10^4 y. When it decays, curium-250 spontaneously fissions into two other nuclei. One such event is denoted as follows:

$$^{250}_{96}\text{Cm} \longrightarrow {}^{137}_{52}\text{Te} + {}^{113}_{44}\text{Ru}$$

■ **Performance Goal for Section 15.7**
  ◆ Describe the manner by which fissile material is produced in a breeder reactor.

◆ 15.7 THE BREEDER REACTOR

Nuclear reactors use fissile fuels to produce electricity, but simultaneously, they generate other fissile fuels. When the reactor uses ordinary uranium (99.3% uranium-238 and 0.71% uranium-235) as the fuel, the nuclear reactions denoted by the following equations occur:

$$^{238}_{98}\text{U} + {}^{1}_{0}n \longrightarrow {}^{239}_{92}\text{U} + \gamma$$
$$^{239}_{92}\text{U} \longrightarrow {}^{239}_{93}\text{Np} + \beta^-$$
$$^{239}_{93}\text{Np} \longrightarrow {}^{239}_{94}\text{Pu} + \beta^-$$

These equations illustrate that the fissionable plutonium-239 is produced during the routine operation of a nuclear power plant fueled by uranium.

The fissionable uranium-233 can also be produced within a nuclear reactor. In this instance, thorium-232 is blanketed about the reactor, so the neutrons that are not absorbed by the fuel *are* absorbed by the thorium. In effect, while the reactor is designed as a source of energy, it also produces the fissionable uranium-233. The relevant reactions are illustrated as follows:

$$^{232}_{90}\text{Th} + {}^{1}_{0}n \longrightarrow {}^{233}_{90}\text{Th} + \gamma$$
$$^{233}_{90}\text{Th} \longrightarrow {}^{233}_{91}\text{Pa} + \beta^-$$
$$^{233}_{91}\text{Pa} \longrightarrow {}^{233}_{92}\text{U} + \beta^-$$

A reactor that is operated to maximize the production of a fissionable radioisotope is called a *breeder reactor*. Uranium-238 and thorium-232 are respectively used to breed plutonium-239 and uranium-233.

There is international concern when rogue nations operate nuclear power plants as a guise for energy production when their actual interest is to maximize the production of fissionable material. The concern revolves around the potential use of this material to construct nuclear-armed ballistic missiles, which can be deployed against their enemies or sold to other nations as weapons of mass destruction. In such instances, delicate diplomatic efforts are required by the international community to ensure that the production of weapons-grade material is kept to a minimum.

Performance Goal for Section 15.8
- ◆ Discuss the special requirements DOT imposes on shippers and carriers who transport radioactive materials.

◆ 15.8 TRANSPORTING RADIOACTIVE MATERIALS

DOT regulates the transportation of radioactive materials, some examples of which are provided in Table 15.7. The DOT regulations require carriers and shippers to affix *RADIOACTIVE WHITE-I*, *RADIOACTIVE YELLOW-II*, or *RADIOACTIVE YELLOW-III* labels, as relevant, to packaging containing radioactive materials. DOT also requires shippers and carriers to post the RADIOACTIVE YELLOW placard on the bulk packaging, freight container, transport vehicle unit load device, or railcar used to transport any amount of a radioactive material. These labels and placards were previously illustrated in Figure 6.18.

Table 15.7 ◆ Some Radioactive Materials Regulated by DOT for Transport

Proper Shipping Name	*Label Code*
Radioactive material, excepted package—articles manufactured from natural or depleted uranium or natural thorium[a]	7
Radioactive material, fissile, n.o.s.	7
Radioactive material, low specific activity, *or* Radioactive material, LSA, n.o.s.	7
Radioactive material, n.o.s.	7
Radioactive material, special form	7
Thorium metal, pyrophoric	7, 4.2
Thorium nitrate	7, 5.1
Uranium hexafluoride, *fissile excepted or non-fissile*	7, 8
Uranium hexafluoride, fissile *(with more than 1% U-235)*	7, 8
Uranium metal, pyrophoric	7, 4.2
Uranyl nitrate hexahydrate solution	7, 8
Uranyl nitrate, solid	7, 5.1

[a]Natural uranium is the element with its naturally occurring distribution of uranium isotopes (approximately 0.711% by mass uranium-235 and the remainder essentially uranium-238). Depleted uranium is the element from which uranium-235 has been removed or reduced in concentration compared to the natural isotopic distribution. Natural thorium is the naturally occurring distribution of thorium isotopes (predominantly thorium-232).

PACKAGE MUST WITHSTAND NORMAL CONDITIONS
OF TRANSPORT ONLY WITHOUT LOSS OR DISPERSAL OF THE
RADIOACTIVE CONTROL CONTENTS.

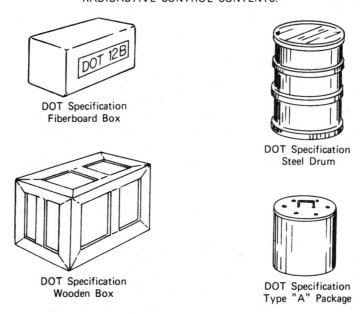

DOT Specification
Fiberboard Box

DOT Specification
Steel Drum

DOT Specification
Wooden Box

DOT Specification
Type "A" Package

FIGURE 15.14 ◆ Typical forms of type A packaging for containment of radioactive materials during transportation. These forms of packaging are designed to withstand the normal conditions of transport, as demonstrated by retention of the integrity of containment and shielding, to the extent required by DOT.

When shippers or carriers transport packaging that has been emptied of its class 7 contents, DOT requires them to affix EMPTY labels to the packaging. The EMPTY label is a white square with black lettering.

DOT regulates the transportation of all radioisotopes whose specific activities exceed 0.002 μCi/g, or 74 Bq/g. To minimize the radiation exposure to transportation personnel, DOT requires the radioisotopes to be containerized in either of two packaging types denoted as type A (Figure 15.14) and type B (Figure 15.15). Both packaging types are designed to retain the integrity of containment and shielding when they are subjected to normal conditions of transport. In addition, type B packaging is also designed to retain the integrity of containment and shielding when it is subjected to prescribed hypothetical accident test conditions.

When shippers and carriers intend to transport a radioisotope, they must first determine its specific activity. Most frequently, this is accomplished by using appropriate radiation-detection equipment. This specific activity is then compared to two published activity values for the radioisotope, A_1 and A_2:

◆ A_1 means the maximum activity of special form Class 7 radioactive material permitted in a Type A package.

PACKAGE MUST STAND BOTH NORMAL AND
ACCIDENT TEST CONDITIONS WITHOUT LOSS OF
CONTENTS.

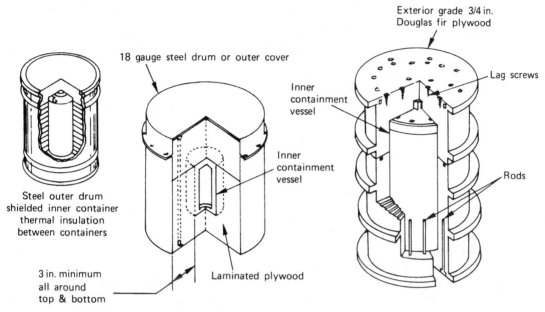

Exterior grade 3/4 in.
Douglas fir plywood

18 gauge steel drum or outer cover

Inner
containment
vessel

Lag screws

Inner
containment
vessel

Rods

Steel outer drum
shielded inner container
thermal insulation
between containers

3 in. minimum
all around
top & bottom

Laminated plywood

FIGURE 15.15 ◆ Typical forms of type B packaging for containment of radioactive materials during transportation. These forms of packaging are designed to withstand the damaging effects of a transportation accident, as demonstrated by the retention of the integrity of containment and shielding, to the extent required by DOT using specified test procedures.

- A_2 means the maximum activity of Class 7 radioactive material other than special form, low-specific-activity material (LSA material), or a surface-contaminated object (SCO) permitted in a Type A package. LSA material refers to radioactive material with limited specific activity that satisfies certain DOT descriptions and limits; SCO means a solid object that is not itself radioactive, but that has Class 7 radioactive material distributed on any of its surfaces.

The values of A_1 and A_2 are provided for some representative radioisotopes in Table 15.8. Based upon these values, the shipper selects either type A or type B packaging for the radioisotope.

DOT also requires the shipper and carrier to determine the relevant *transport index*. This is a dimensionless number used to designate the degree of control to be exercised during transportation. It is determined as follows:

- For nonfissile radioactive materials, the transport index is the number determined by multiplying the maximum radiation level in mSv/hr at 3.3 ft (1 m) from the external surface of the package by 100.
- For fissile radioactive materials, the transport index is the number determined by multiplying the maximum radiation level in mSv/hr at 3.3 ft (1 m) from any external surface of the package by 100; or to control criticality, the number obtained by dividing 50 by the allowable number of packages that can be transported together. Whichever number is larger is the transport index.

TABLE 15.8 ◆ Representative Activity Values[a]

Symbol of Radioisotope	A_1 [TBq]	A_1 [Ci]	A_2 [TBq]	A_2 [Ci]	Specific Activity	
					[TBq/g]	[Ci/g]
Bi-207	0.7	18.9	0.7	18.9	1.9	5.2×10^1
Co-60	0.4	10.8	0.4	10.8	4.2×10^1	1.1×10^3
Cs-137	2	54.1	0.5	13.5	3.2	8.7×10^1
K-40	0.6	16.2	0.6	16.2	2.4×10^{-7}	6.4×10^{-6}

[a]49 CFR §173.435

Then, using the information provided in Table 15.9, the proper label is selected using the measured radiation level at the surface of the package and the transport index. The space on the relevant RADIOACTIVE label must be inscribed with the name of the radioisotope, its activity, and on RADIOACTIVE-II and RADIOACTIVE-III labels, the transport index.

When intending to transport a nonfissile radioactive material, DOT requires shippers and carriers to include the following information in its basic description:

- The specific name of the radioisotope
- A description of the physical and chemical form of the material
- The activity of the radioisotope
- Identification of the required labels on the relevant packaging
- The transport index, when DOT requires the package to be labeled with either the RADIOACTIVE YELLOW-II or RADIOACTIVE YELLOW-III label

For example, suppose a shipper or carrier intends to transport $^{99}_{42}$Mo-tagged molybdenum dioxide having a specific activity of 200 TBq/g in a package that measures at the surface of the package a dose of 0.005 mSv/hr. The basic description of this substance is the following:

200   Radioactive material, n.o.s. (Molybdenum-99, as solid

TBq/g   molybdenum dioxide), 7, UN2982 (RADIOACTIVE WHITE-I)

TABLE 15.9 ◆ Selection of Label on Packages of Radioactive Materials[a]

Transport Index	Maximum Radiation Level at Any Point on the External Surface	Label
0	Less than or equal to 0.005 mSv/hr	WHITE-I
More than 0 but not more than 1	Greater than 0.005 mSv/hr but less than or equal to 0.5 mSv/hr	YELLOW-II
More than 1 but not more than 10	Greater than 0.05 mSv/hr but less than or equal to 2 mSv/hr	YELLOW-III
More than 10	Greater than 2 mSv/hr but less than or equal to 10 mSv/hr	YELLOW-III[b]

[a]49 CFR §172.403
[b]Must be transported under exclusive use provisions; see 49 CFR §173.44

DOT requires the proper shipping name of a fissile radioactive material to also include the words "Fissile Excepted" when the shipper or carrier has complied with designated requirements. For instance, when DOT requires the RADIOACTIVE YELLOW placard to be displayed on a motor vehicle, the carrier is further mandated to ensure that the motor vehicle is operated on the route that minimizes the public's risk of radiological exposure. This is accomplished by considering accident rates, transit time, population density and activities, and the time of day and time of week during which the transportation will occur. DOT then directs the carrier to operate the motor vehicle on the *preferred route* or *preferred highway* (Section 14.4). DOT permits the carrier to deviate from using the preferred route only under emergency conditions that would make continued use of the preferred route unsafe or, when necessary, to stop for rest, fuel, or vehicle repairs.

Performance Goal for Section 15.9

+ Describe the nature of the response actions to be taken when radioactive materials have been released into the environment.

◆ 15.9 RESPONDING TO INCIDENTS INVOLVING THE RELEASE OF RADIOACTIVE MATERIALS

Within the workplace, OSHA requires the conspicuous posting of a sign, like any of those illustrated in Figure 15.16 (a),(b), and (c), to warn individuals that a given area is any of the following:

+ *Radiation area.* This is an area in which the radiation exists at a level in which the body could receive in any 1 hour a dose in excess of 0.05 mSv, or in any 5 consecutive days a dose in excess of 1 mSv.
+ *High radiation area.* This is an area in which the radiation exists at such levels that a major portion of the body could receive in any 1 hour a dose in excess of 1 mSv.
+ *Airborne radioactivity area.* This is an area in which the airborne radiation is likely to exist at concentrations in excess of levels prescribed by the Nuclear Regulatory Commission.

These signs bear a magenta or purple three-bladed propeller on a yellow background, along with the word CAUTION and the words RADIATION AREA, HIGH RADIATION AREA, or AIRBORNE RADIOACTIVITY AREA, as relevant. OSHA also requires the conspicuous posting of a sign like that in Figure 15.16(d) within any area or room in which radioactive material is stored or used, as well as on the containers used to transport or store them. Although OSHA requires the posting of these signs to provide workers with an awareness that radioactive materials are present within the area, the signs also serve to inform emergency responders during an emergency that radioactive materials are present.

But, what should be done to protect them against unnecessary exposure to ionizing radiation while responding to an emergency? Three basic tenets can ensure that such radiation exposure will be reduced or eliminated: shielding, time, and distance.

When a radioactive material is encountered, it is generally containerized and stored so as to limit public exposure. The presence of a steel or concrete barrier between

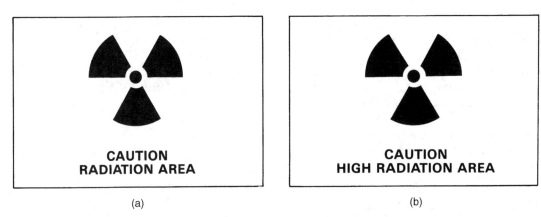

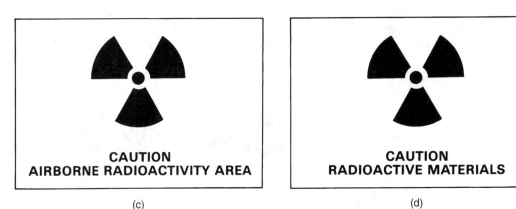

FIGURE 15.16 ◆ OSHA requires one or more of these caution signs to be conspicuously posted in areas where radioactive materials are stored or otherwise located. The three-bladed symbol is the conventional radiation caution sign. OSHA also requires that labels be affixed to certain radioactive material storage and transport containers; these labels resemble the caution sign illustrated in (a).

an individual and the radioactive material minimizes the radiation exposure dose. When it is essential to be exposed to ionizing radiation, the best method of protection is to limit the time of exposure and to maintain a position as far removed from the source as is practical. Limiting the time of exposure and staying far removed from the radiation, yet still being in a position that allows one to accomplish a given task, ensures that only the minimum radiation dose is received.

The intensity of radiation decreases as the inverse square of the distance from the source. This is the *inverse square law of radiation,* which can be arithmetically expressed as follows:

$$I = I_o/r^2$$

In this equation, I_o is the original intensity of a radiation source and I is the intensity at a distance r. Thus, if a radioactive material registers 1000 R on a Geiger counter held 1 ft (0.3 m) from the source, it registers only 250 R when held 2 ft (0.6 m) from the source. Doubling the distance between an individual and the source of radiation reduces the radiation exposure to one-fourth the original value.

Emergency-response crews also encounter radioactive materials at transportation mishaps. Most generally, the presence of a radioactive material in a shipment is verified by observing the following:

- The number 7 as a component of a basic description of a hazardous material on a shipping paper
- The word "RADIOACTIVE" upon a label affixed to packaging
- The word "RADIOACTIVE" upon a placard posted on the relevant transport vehicle

When a radioactive material has spilled, or when the packaging used for containment of a radioactive material is leaking or damaged, emergency responders should implement the procedures shown in Figure 15.17 to protect public health and the environment. For small and large spills, DOT recommends initial isolation distances of 100 ft (30 m) and 300 ft (95 m) in all directions, respectively. For small spills, DOT recommends protective-action distances of 0.1 mi (0.2 km) and 0.3 mi (0.5 km) during day and nighttime hours, respectively. For large spills, DOT recommends protective-action distances of 0.6 mi (1.0 km) and 1.9 mi (and 3.1 km) during day and nighttime hours, respectively.

Since the handling of radioactive materials requires special training, the regional office of the U.S. Department of Energy, U.S. Department of Defense, or the appropriate state or local radiological authorities should be notified of the mishap. When firefighters and other emergency-response personnel respond to such incidents, care should be exercised to avoid possible inhalation, ingestion, or contact with the radioactive materials. Loose radioactive material and its associated packaging should be segregated in an area pending disposal instructions from responsible radiological authorities. If a fire has occurred, it should be extinguished only from a suitable distance. All measures should be taken to prevent the spread of radioactivity by keeping the quantity of runoff water to a minimum.

Performance Goal for Section 15.10

- Identify the health concern that radon poses when it accumulates within buildings.

◆ 15.10 RESIDENTIAL RADON

Radioisotopes of radium continuously form from the decay of naturally occurring uranium and thorium radioisotopes. They are produced within areas where uranium and thorium are components of metal-bearing ores, granite, or black shale. For example, radium-228 and radium-224 are produced when the naturally occurring thorium-232 in a metal-bearing ore decays by the following series of steps:

$$\ce{^{232}_{90}Th} \longrightarrow \ce{^{228}_{88}Ra} \longrightarrow \ce{^{228}_{89}Ac} \longrightarrow \ce{^{228}_{90}Th} \longrightarrow \ce{^{224}_{88}Ra}$$

Although geological formations containing uranium and thorium are relatively uncommon, they do exist below Earth's upper stratum in specific areas of the United States and elsewhere.

Buried deep beneath Earth's surface, the presence of radium radioisotopes presents little problem. However, the decay of some radium radioisotopes produces

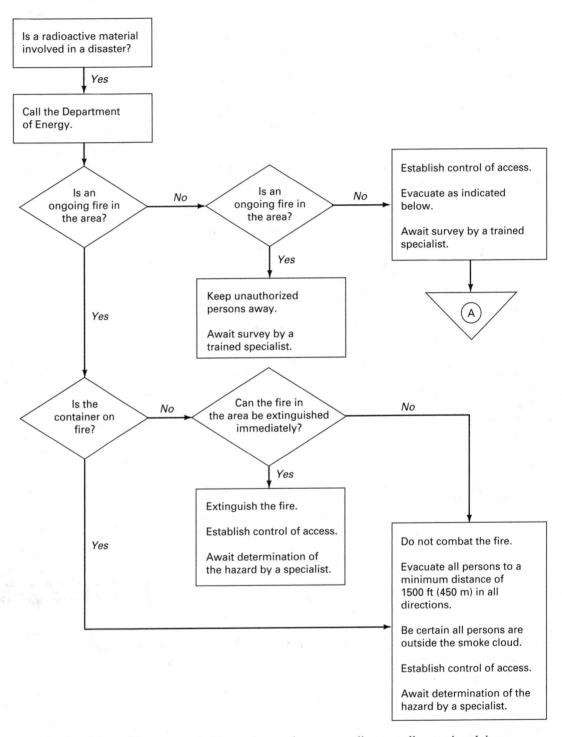

FIGURE 15.17 ◆ The recommended procedures when responding to a disaster involving a radioactive material. *Adapted with permission of the American Society for Testing and Materials, from a figure in ASTM STP 825, A Guide to the Safe Handling of Hazardous Materials Accidents. Copyright © 1983, American Society for Testing and Materials.*

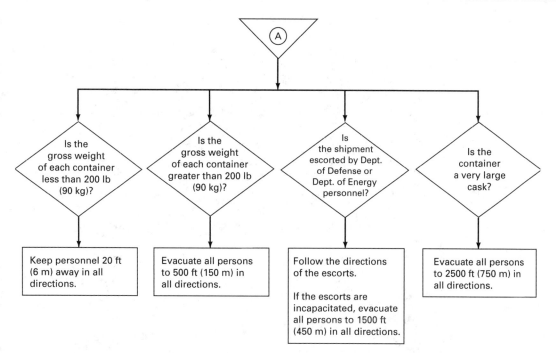

FIGURE 15.17 ◆ *(continued)*

the colorless, odorless noble gas called *radon*. This is illustrated by the decay of radium-224.

$$^{224}_{88}\text{Ra} \longrightarrow {}^{220}_{86}\text{Rn} + \alpha$$

Thus formed, the radon slowly percolates upward through the rock and soil.

Radon often enters the atmosphere, where it dissipates innocuously. The average concentration of the radon radioisotopes in the air is 0.2 pCi/L. On the other hand, radon can also dissolve in groundwater aquifers from which it can enter a home by means of the water supply and open sump pumps. In the basement of homes, radon can also sneak through mortar joints, the cracks in foundation floors, gaps between wall-to-wall joints, and pores in concrete blocks and other building materials.

The radon that enters basements and other components of a dwelling is called *residential radon.* It consists primarily of two radioisotopes. They are radon-220 and radon-222, which possess half-lives of 54.5 s and 3.82 d, respectively. Each decays by alpha emission. The transmutation products resulting from their decay are polonium radioisotopes, which also decay by alpha emission.

Inside a building, as in Figure 15.18, where there is little exchange between the inside and outside air, the concentration of the residential radon could become so elevated that it poses an inhalation health hazard. In the early 1980s, scientists demonstrated that exposure to residential radon caused lung cancer. The risk of contracting cancer from exposure to radon is proportional to the amount that enters a home and the length of time it remains within living areas. Most radon that is inhaled is immediately exhaled back to the atmosphere, but since the relevant half-lives are so short, some radon decays *before* it is exhaled. The solid polonium radioisotopes can then lodge within lung tissue and irradiate the nearby cells with alpha particles. It is the energy deposited in these cells that is believed to initiate carcinogenesis.

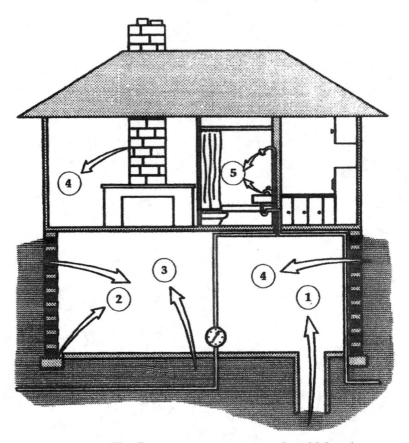

FIGURE 15.18 ◆ The five most common routes by which radon enters a home: (1) an open sump pump; (2) gaps between basement wall-to-wall joints; (3) foundation cracks in basement floor; (4) pores in concrete blocks and mortar joints; and (5) water supply.

Since the inhalation of radon has been shown to cause lung cancer in humans, radon is denoted as a known human carcinogen. Also denoted as known human carcinogens are the specific radon radioisotopes, radon-220 and radon-222. Exposure to radon is generally regarded as the *second* leading cause of lung cancer. The leading cause is exposure to tobacco smoke.

The risk of acquiring lung cancer from exposure to a radon-enriched atmosphere is exacerbated for those individuals who are smokers; that is, in combination, radon and tobacco smoke act synergistically (Section 10.17). Researchers have determined that smoking increases an individual's risk of acquiring cancer from radon exposure by as much as 15 times.

As an action guideline, EPA recommends a concentration of no more than 4 pCi/L of radon within a residence. In some parts of the United States, it is difficult to comply with this guideline. In the late 1990s, scientific researchers estimated that residential radon exposure causes 15,400 to 21,800 cases of lung cancer each year. Approximately one-third of them could have been avoided if the concentration of residential radon had been reduced to a level below EPA's action guideline. But, how can the amount of radon in a dwelling be reduced? Little improvement can be made to the structure of a home to prevent radon from accumulating. Nonetheless, EPA suggests increasing the air flow into and through the home by opening windows and using fans; when the homes have crawl spaces, the vents in these areas should be kept open year-round.

Far fewer individuals die from consumption of radon in household water, approximately 260 deaths annually. The current maximum allowable level is 11 Bq/L.

Performance Goals for Section 15.11

  ◆ Describe the composition of a radiological dirty bomb.
  ◆ Describe the nature of the response action to be implemented when a radiological dirty bomb has been activated.

◆ 15.11 THE RADIOLOGICAL DIRTY BOMB

A *dirty bomb* may be regarded as any conventional explosive device charged with a hazardous material that disperses into the environment as the explosive material detonates. In theory, the hazardous material could be a poisonous gas, flammable material, corrosive material, biological or chemical warfare agent, or radioactive material; however, it is by means of the widespread dissemination of a radioactive material that terrorists instill the highest degree of fear and panic in the affected populations by increasing their risk of contracting cancer from radiation exposure.

Initially following the detonation of a radiological dirty bomb, the immediate area of the incident would be contaminated with a radioisotope. Having been dispersed by explosive action, the material then diffuses within the air, where it is carried by air currents to far removed locations. In time, the radioactive material could disperse worldwide. It is for this reason that this type of dirty bomb is also called a *radiological dispersal device*.

The detonation of a radiological dirty bomb would not cause the immediate mass casualties or devastation associated with the detonation of nuclear weapons. Nonetheless, the level of radiation could cause regulatory agencies to cordon off sections of entire cities for long periods of time. The detonation of a radiological dirty bomb would also have the ultimate affect of increasing the background radiation to which we are all exposed. Depending on the nature of the radioactive material, the background radiation could remain elevated for hundreds or thousands of years. Human exposure to the increased level could dramatically increase the incidence of cancer.

Experts have concluded that the radioactive material most likely to be used by terrorists in a radiological dirty bomb is cesium chloride powder, which is used to irradiate food. As a powder, it would disperse readily when released into the environment by detonation. The cesium radioisotope used to irradiate food is cesium-137, whose half-life is 33 years. This means that if 1000 curies of cesium chloride were dispersed in the environment, 500 curies would still remain after the passage of 33 years, 250 curies after the passage of another 33 years, and so on.

What can an emergency-response team do to best serve the public when terrorists have detonated a radiological dirty bomb? To answer this question, remember that the three basic tenets associated with radiation protection are shielding, time, and distance. Translated into action, this means that the affected population should be removed as quickly as feasible from the area where the concentration of radioactive material is highest. These individuals can then seek out protective information from officials through local radio and television broadcasts.

■■■

REVIEW EXERCISES

Features of Atomic Nuclei

15.1 Identify the number of protons and neutrons contained in the following radioisotopes:
   (a) potassium-40
   (b) cerium-146
   (c) barium-135
   (d) lanthanum-133

15.2 Iridium-192 is a radioisotope used to treat cancer. Its half-life is 73.8 days, and it decays by negatron emission.
   (a) How many protons and neutrons are contained in iridium-192?
   (b) Write the equation denoting its transmutation.
   (c) What is the approximate percentage of iridium-192 that remains a year after its use for cancer treatment?

15.3 Selenium-75 is used in medicine to assist in diagnosing pancreatic cancer. Its half-life is 128 days, and it is a negatron emitter.
   (a) How many protons and neutrons are in a nucleus of selenium-75?
   (b) What is the product of its decay?
   (c) If the selenium-75 source contains 1.75×10^{20} atoms today, how many atoms will remain in 384 days?

15.4 Examples of radioisotopes that decay by electron capture include manganese-54, iron-55, and cobalt-57.
   (a) Write the equations denoting their transmutations.
   (b) What are the products of their decay?

15.5 What fraction of a radioisotope remains after five half-lives?

15.6 Potassium-42 is a radioisotope used by nutritionists to determine whether the body is effectively using potassium at the cellular level. The half-life of potassium-42 is 12.4 hours. Assuming that the cells assimilate all of the potassium chloride tagged with potassium-42 given to a patient, approximately what percentage of the dose will remain in 2.1 days?

15.7 A radiologist determines that exposure to a particular radioisotope, X, is considered safe at concentrations less than 1 mg/mile^2. The concentration of X at a nuclear explosion testing site is measured to be 10 mg/mile^2 immediately following the explosion of a nuclear bomb. If the half-life of X is 24 h, will the site be safe to enter in 4 d? If not, how many days must elapse before the concentration of X has been reduced to a level that entry to the site is considered safe?

Modes of Nuclear Decay

15.8 Identify the nucleus produced when each of the following nuclear transmutations occur:

(a) $^{212}_{86}\text{Rn} \longrightarrow$ _____ $+ \alpha$

(b) $^{222}_{88}\text{Ra} \longrightarrow$ _____ $+ \alpha$

(c) $^{27}_{12}\text{Mg} \longrightarrow$ _____ $+ \beta^-$

(d) $^{33}_{15}\text{P} \longrightarrow$ _____ $+ \beta^-$

(e) $^{18}_{9}\text{F} \longrightarrow$ _____ $+ \beta^+$

(f) $^{71}_{32}\text{Ge} + {}^{0}_{-1}e \longrightarrow$ _____

(g) $^{209}_{84}\text{Po} \longrightarrow$ _____ $+ \alpha$

(h) $^{197m}_{79}\text{Au} \longrightarrow$ _____ $+ \gamma$

15.9 Identify the means by which each of the following nuclei decays:

(a) $^{70}_{33}\text{As} \longrightarrow {}^{70}_{32}\text{Ge} +$ _____

(b) $^{86}_{37}\text{Rb} \longrightarrow {}^{86}_{38}\text{Sr} +$ _____

(c) $^{117}_{48}\text{Cd} \longrightarrow {}^{117}_{49}\text{In} +$ _____

(d) $^{215}_{86}\text{Rn} \longrightarrow {}^{211}_{84}\text{Po} +$ _____

(e) $^{234}_{92}\text{U} \longrightarrow {}^{230}_{90}\text{Th} +$ _____

(f) $(^{199}_{80}\text{Hg})^* \longrightarrow {}^{199}_{80}\text{Hg} +$ _____

Measurement of Radioactivity

15.10 The average annual dose from medical and dental X rays is approximately 1 rad. When expressed in millisieverts, what is the average annual X-ray dose that a person receives?

15.11 What is the specific activity in TBq/g of 10 g of $^{14}_{6}\text{C}$-tagged sodium bicarbonate whose activity is measured to be 1000 pCi?

15.12 To protect public health, EPA proposes 4 mrem/year as the maximum radiation limit an individual should receive from the radiation present in groundwater. If an individual consumes groundwater to receive a dose of 5 μSv/month, has the person been exposed to radiation in excess of EPA's recommended limit?

15.13 One type of smoke detector is an ionization device containing americium-241. This radioisotope ionizes the air molecules between a pair of electrodes, thereby permitting the passage of a small electrical current. When smoke particles enter this space, they adhere to the ionized molecules, thereby reducing the flow of current. This reduction in current activates an alarm. Ion detectors are especially useful for recognizing fast-moving fires, such as wastepaper basket and kitchen grease fires. If the specific activity of a source of americium-241 for use in manufacturing smoke detectors is measured as 3.4 Ci/g, what is its specific activity in TBq/g?

15.14 The DOT regulation at 49 CFR §172.403 requires shippers and carriers to affix RADIOACTIVE YELLOW labels to packaging when the maximum radiation level at any point on the surface is greater than 2 mSv/h but less than or equal to 10 mSv/h. Are shippers and carriers required to affix RADIOACTIVE YELLOW labels to a package if the maximum radiation level on the surface is measured to be 13 mrem/min?

Adverse Effects from Radiation Exposure

15.15 The OSHA regulation at 29 CFR §1910.96(b)(3) prohibits an employer from exposing individuals under 18 years of age to a radiation dose greater than

750 mrem per calendar quarter. What is the most likely reason OSHA selected 18 as the maximum age to which radiation exposure should be severely limited?

15.16 What biological effect is likely to be experienced by emergency responders, each of whom receives a short-term radiation dose of 400 mSv from exposure to cesium-137?

15.17 Table salt is often sold in an "iodized" form, meaning that the sodium chloride contains approximately 0.023% potassium iodide. What benefit is associated with the normal consumption of iodized table salt?

15.18 Why is the ingestion of lead radioisotopes especially hazardous to one's health?

15.19 The DOT regulation at 49 CFR §177.842 stipulates that shippers and carriers cannot place packages of radioactive materials bearing the RADIOACTIVE YELLOW-II and RADIOACTIVE YELLOW-III labels within a transport vehicle, storage location, or any other place closer than certain prescribed distances to any area that can be occupied by a passenger, employee, or animal. What is the most likely reason DOT requires the placement of these packages in this fashion?

Nuclear Fission

15.20 Identify the second isotope produced when the fission phenomenon represented by each of the following equations occurs:

(a) $^{239}_{94}Pu + ^{1}_{0}n \longrightarrow ^{91}_{36}Kr + \underline{\hspace{1cm}} + ^{1}_{0}n$

(b) $^{235}_{92}U + ^{1}_{0}n \longrightarrow ^{142}_{56}Ba + \underline{\hspace{1cm}} + 3^{1}_{0}n$

(c) $^{235}_{92}U + ^{1}_{0}n \longrightarrow ^{74}_{33}As + \underline{\hspace{1cm}} + 2^{1}_{0}n$

15.21 Atmospheric radioisotopic monitoring is used to establish conclusively that nuclear testing has occurred. Use Table 15.5 to determine the radioisotopes whose presence should be monitored?

15.22 What distinguishes weapons-grade uranium and plutonium from other grades or forms of these elements?

15.23 The DOT regulation at 49 CFR §173.417 requires shippers and carriers to limit packages of fissile radioactive material to certain prescribed amounts per transport vehicle intended for domestic shipment. What is the most likely reason DOT limits the amount of fissile material contained within a transport vehicle?

15.24 In 2003, Russia announced that it had sold a plutonium-239/plutonium-240 mixture to Iran, ostensibly for use in developing its nuclear power program. Why does the sale of nuclear fuel to Iran constitute cause for concern to officials of the American government?

Radon

15.25 Write the equations illustrating the decay of radon-220 and radon-222.

15.26 During the winter months, when the exchange of indoor and outdoor air is limited, 20 pCi of radon was found on the average in each liter of air in a home. Does this concentration exceed EPA's guideline for protection against acquiring cancer from exposure to radon?

15.27 Why is occupational exposure to carcinogens likely to be elevated compared to the norm for underground hard-rock miners?

Glossary

absolute pressure. The pressure exerted on any material; for confined gases, the sum of the gauge pressure and the atmospheric pressure

absolute zero. The lowest temperature attainable by any substance, equal to $-459.67°F$ ($-273.15°C$)

acetylene. The common name of the simplest alkyne; the hydrocarbon having the chemical formula C_2H_2

acid. A compound that forms hydrated hydrogen ions when dissolved in water

acidic anhydride. A nonmetallic oxide that chemically combines with water to produce an acid

activity series. A listing of the metals in order of their decreasing chemical activity

acrolein. A pungent smelling, intensely irritating lacrimator (tear-producing substance) formed during wood fires

activation energy. The minimum energy that must be supplied to a group of reactants before a chemical reaction can be initiated

activity series. An arrangement of the metals illustrating their decreasing ability of generating hydrogen by chemical reaction with water and acids

acute health effect. Impairment, injury, or disease that occurs abruptly and severely following a brief exposure to a substance

addition polymer. A polymer resulting from the addition of a molecule, one by one, to a growing chain

adsorption. A physical phenomenon characterized by the adherence or occlusion of atoms, ions, or molecules of a gas, vapor, or liquid to the surface of another substance

aerosol. A fine mist or spray containing minute particles

alcohol. An organic compound whose molecules possess a hydroxyl group ($-OH$)

alcohol, absolute. Dehydrated alcohol

alcohol, denatured. Ethanol to which a poisonous or noxious substance has been added to make it unfit for consumption

alcoholism. The disease associated with addiction to ethanol

alcohol-resistant aqueous film-forming foam. A foam prepared by mixing a commercially available concentrate with water, intended for use as a fire extinguisher on flammable liquid fires (abbreviated AR-AFFF)

aldehyde. An organic compound whose molecules are composed of an alkyl or aryl group, a carbonyl group, and one hydrogen atom

aliphatic hydrocarbon. Any compound containing only carbon and hydrogen atoms and whose molecules are not composed of benzene or benzene-like structures

alkali metal family. The elements comprising Group IA on the periodic table: lithium, sodium, potassium, rubidium, cesium, and francium

alkaline. Basic, as opposed to acidic; any aqueous solution or other material whose pH is greater than 7

alkaline earth metal family. The elements comprising Group IIA on the periodic table: beryllium, magnesium, calcium, strontium, barium, and radium

alkane. Any hydrocarbon whose molecules are composed only of carbon–carbon single bonds ($C—C$)

alkene. Any hydrocarbon whose molecules contain a carbon–carbon double bond ($C=C$)

alkylation. The preparation of a branched-chain alkane by the combination of an alkane and an alkene

alkyl group. Any group of atoms (e.g., methyl, ethyl, *n*-propyl, isopropyl) derived by removing one hydrogen atom from the molecular formula of an alkane

alkyne. Any hydrocarbon whose molecules are composed of at least one carbon–carbon triple bond ($C{\equiv}C$)

allotrope. Any variety of the same element possessing different physical and chemical properties

alloy. A solid mixture composed of two or more metals

altitude sickness. The inability of the human organism to extract sufficient oxygen for survival from the air at higher altitudes, where the atmospheric pressure is reduced

alpha particle. A particle (α or ^4_2He) emitted from certain radioisotopes and having the properties of a doubly ionized helium atom

alpha radiation. The radiation emitted from certain radioisotopes and composed of alpha particles

aluminum alkyl compound. Any of the group of substances whose molecules are composed of an aluminum atom covalently bonded to three carbon atoms, each of which is a component of an alkyl group

amalgam. An alloy of mercury with one or more metals

amine. An alkylated analog of ammonia; any organic compound derived by the substitution of one or more of the hydrogen atoms in a hydrocarbon molecule with the $-NH_2$ group of atoms

anemia. The condition represented by a below-normal reduction in the body's quantity of hemoglobin

anesthesia. The entire or partial loss of feeling or sensation caused by the inhalation of a substance or mixture of substances, with little or no adverse systemic action, although overexposure causes impaired judgment, dizziness, drowsiness, headache, unconsciousness, and even death

antiknock agent. A substance added to a petroleum fuel to reduce or eliminate the noises associated with its combustion

aqueous. Containing water as a component, usually the primary component

aqueous ammonia (see **household ammonia**) **aqueous film-forming foam.** A foam prepared by mixing a commercially available concentrate with water, intended for use as an extinguisher of fires involving flammable, water-insoluble liquids (abbreviated AFFF)

aromatic hydrocarbon. A compound whose molecules are composed solely of carbon and hydrogen atoms, at least some of which are structurally similar to benzene

aryl group. The group of atoms that remain after a hydrogen atom is removed from benzene or a benzene-like molecule (e.g., phenyl and benzyl)

asbestos. The generic name of a class of naturally occurring fibrous silicates

asbestosis. A scarring of lung tissue by asbestos fibers

asphalt. The residue remaining from the fractional distillation of crude petroleum

asphyxiant. A gas or vapor that, when inhaled, causes unconsciousness or death by suffocation

atactic polymers. A stereochemical type of polymer whose macromolecules possess randomly positioned side chains along a carbon–carbon chain

atmosphere. The layer of gases that surrounds Earth

atmospheric pressure. The force exerted on matter by the mass of the overlying air

atmospheric tank. A tank designed to store a flammable liquid at pressures ranging from atmospheric pressure through 0.5 psi_g (3 kPa)

atom. The smallest particle of an element that can be identified with that element

atomic nucleus. The region at the center of the atom composed of protons and neutrons

atomic number. The number of protons possessed by an atomic nucleus; the number of electrons possessed by an atom

atomic orbital. Any of the regions in space about an atomic nucleus in which electrons are most likely to be found

atomic weight. The mass of an atom compared to the mass of the carbon-12 isotope, which is assigned a mass of exactly 12

autoignition temperature. The minimum temperature of a liquid necessary to initiate the self-sustained combustion of its vapor in the absence of an ignition source

autopolymerization. Spontaneous polymerization

available chlorine. A means of comparing the effectiveness of commercial bleaching agents to elemental chlorine

Avogadro number. The number of atoms, molecules, or other units contained within a mole of a specified substance, 6.02×10^{23}

barometer. An instrument used for determining the atmospheric pressure, usually expressed as the height of a column of mercury

base. A compound whose units produce hydroxide ions, OH^-, when dissolved in water

base unit. Any of the fundamental units of measurement that describe length, mass, and other quantities

basic anhydride. A metallic oxide that chemically combines with water to produce a base

becquerel. The SI unit of radioactivity equivalent to one disintegration per second

benzene. The hydrocarbon having the chemical formula C_6H_6 with all six carbon atoms bonded one to the other in a hexagonal ring and with each carbon atom also bonded to a hydrogen atom

benzyl group. The group of atoms designated as $C_6H_5CH_2$—

beta radiation. Negatrons or positrons

binary compound. A compound composed of only two elements

biological warfare agent. Any etiological agent that inflicts disease or causes mass casualties when disseminated throughout a population

biphenyl. The hydrocarbon having the chemical formula C_6H_5—C_6H_5

black powder. A mixture of charcoal and sulfur with potassium nitrate or sodium nitrate in prescribed amounts

blasting agent. A material designated for blasting that has undergone prescribed tests and has been shown to have little probability of initiating an explosion, or of burning and then exploding

bleaching agent. A substance capable of removing color, in whole or in part, from a colored material

BLEVE. The phenomenon in which the rapid buildup of internal pressure within a container or tank is relieved by explosion; the acronym for a "boiling liquid evaporating vapor explosion"

blood alcohol concentration. The amount of alcohol in a specified volume of an individual's blood, typically measured in milligrams per deciliter (mg/dL) (abbreviated BAC)

blood agents. Chemical warfare agents that cause seizures, respiratory failure, and cardiac arrest by interfering with the blood's ability to absorb atmospheric oxygen

blood pressure. The constant pressure on the walls of the arteries

boiling point. The temperature at which the vapor pressure of a substance equals the average atmospheric pressure

boil-over. The phenomenon associated with the production of steam during a crude petroleum fire, which drives the burning petroleum up and over the walls of its confining tank

bottled gas (see **liquefied petroleum gas**).

Boyle's law. The observation that at constant temperature, the volume of a confined gas is inversely proportional to its absolute pressure

breeder reactor. A nuclear reactor that produces fissionable uranium-233 or plutonium-239

brisance. The shattering power of an explosive material

British thermal unit. The amount of heat required to raise the temperature of 1 pound of water 1 degree Fahrenheit (abbreviated Btu)

butane. The alkane having the chemical formula C_4H_{10}

calorie. The amount of heat required to raise the temperature of 1 gram of water 1 degree Celsius

cancer. A disease associated with any of various types of malignant tumors or other abnormal growths that develop in the body

carbon black. An intensely black form of carbon whose particles are submicron in diameter, used commercially in a number of products

carbonization. The process of converting an organic compound into carbon or a carbon-containing residue, typically conducted under intense temperature and pressure

carbonyl group. The $C{=}O$ functional group, as in a ketone or aldehyde

carboxyhemoglobin. The substance that forms when carbon monoxide reacts with the blood's hemoglobin

carboxyl group. The $-C{<}^{O}_{OH}$ functional group, as in a carboxylic acid

carboxylic acid. An organic compound containing the carboxyl group

carcinogen. A substance capable of causing or producing cancer in an organism

carcinogenesis. The process by which normal tissue becomes cancerous

cardiovascular. Relating to or involving the heart and blood vessels

cargo tank. Any tank permanently attached to or forming a part of any motor vehicle or any bulk liquid or compressed gas packaging not permanently attached to a motor vehicle that, by reason of its size, construction, or attachment to the motor vehicle, is loaded or unloaded without being removed from the motor vehicle

carrier. A person engaged in the transportation of passengers or property by land or water, as a common, contract, or private carrier, or civil aircraft

catalyst. A substance that increases or decreases the rate of a chemical reaction without itself being permanently altered

ceiling limit. The concentration of a substance to which an individual should not be exposed, even for an instant, to prevent the onset of ill effects or death

cellulose. The substance that forms the cell walls of all plants; the major constituent of wood and cotton

Celsius scale. The temperature scale on which water freezes at 0°C and boils at 100°C at 1 atm

central nervous system. The brain and spinal cord

chain reaction. (1) A multistep reaction in which a reactive intermediate formed in one step reacts during a subsequent step to generate the species needed in the first step; (2) A self-sustaining nuclear reaction that, once started, steadily provides the energy and matter necessary to continue the reaction

chalcogen family. The elements comprising Group VIA on the periodic table: oxygen, sulfur, selenium, tellurium, and polonium

charge. A characteristic of matter resulting from a deficiency or excess of electrons

Charles's law. The observation that at constant pressure, the volume of a confined gas is directly proportional to its absolute temperature

chemical bond. The force by which the atoms of one element become attached to, or associated with, other atoms in a compound

chemical change. Any change that results in an alteration of the chemical identity of a substance; a chemical reaction

chemical energy. The energy stored in the chemical bonds of a substance; the energy absorbed or released during a chemical reaction

chemical foam. A fire-extinguishing foam such as that produced from the reaction of aqueous sodium bicarbonate and aluminum sulfate

chemical formula. A method of expressing the number of atoms or ions of specific types in a molecule or unit of matter through the use of chemical symbols for the elements and numbers or proportions of each kind of atom

chemical property. Any of the types of behavior a substance exhibits when it undergoes a chemical change

chemical symbol. A notation using one or two letters to represent an element

chemical warfare agent. Any nerve gas, vesicant, or other substance that inflicts harm or causes mass casualties when disseminated

chemistry. The natural science concerned with the properties, composition, and reactions of substances

CHEMTREC. A national communications center that provides advice for emergency responders and others at the scene of a transportation emergency involving the release of a hazardous material

CHLOREP. The chlorine emergency plan devised by the Chlorine Institute to handle emergencies that involve the release of chlorine

chlorination. (1) A chemical reaction that involves the addition of chlorine to a substance;
(2) The treatment of contaminated drinking water to kill the microorganisms that cause diseases such as dysentery, cholera, typhoid, and hepatitis

chlorofluorocarbon. Any fluorinated derivative of chloromethane or chloroethane (abbreviated CFC)

chromium compound, hexavalent. Any compound containing chromium in the +6 oxidation state, the absorption of which has been shown to cause certain adverse health ailments including cancer

chronic health effect. Any long-lasting disorder, injury, or disease caused by exposure to a substance that occurs over a few days, weeks, or longer periods

cirrhosis. A chronic disease of the liver characterized by inflammation, pain, and jaundice (a yellowishness of the skin)

classes of fire. The four types of fire, each of which is differentiated by the nature of its fuel

Clean Air Act. The federal statue that empowers EPA to establish national ambient air quality standards for specified air contaminants and national emission standards for airborne substances that cause or are suspected to cause serious health problems

Clean Water Act (see **Federal Water Pollution Prevention and Control Act**)

coal distillation. Separation of some components of coal accomplished by heating the coal to vaporize its constituents, some of which may be condensed

coal tar. The black, viscous liquid produced by heating coal in the absence of air

coefficient of volume expansion. The change in the volume of a liquid per degree change in temperature

coma. A state of unconsciousness from which the victim cannot be aroused

combination reaction. The type of chemical reaction involving two substances that results in the formation of a single product

combined gas law. The observation that the volume of a confined gas is directly proportional to its absolute temperature and inversely proportional to its absolute pressure

combustible. Easily ignitable and free-burning

combustible liquid. (1) Any liquid having a flashpoint at or above 100°F (37.8°C) (OSHA); (2) Any liquid that does not meet the definition of a DOT hazard class other than 3 and possesses a flashpoint above 141°F (60.5°C) and below 200°F (93°C) (DOT)

combustible metal. Metals including the alkali metals, alkaline earth metals, aluminum, titanium, and zinc that can readily ignite under certain conditions and whose fires are denoted as class D fires

combustion. The rapid oxidation of a material in the presence of oxygen or air

combustion, spontaneous. Rapid oxidation initiated by the accumulation of heat in a material undergoing slow oxidation

commerce. Trade, traffic, transportation, importation, exportation, and similar activities engaged in by shippers and carriers

compatibility group letter. A designated alphabetical letter used by DOT to categorize different types of explosive materials for purposes of stowage and segregation

compound. A substance composed of two or more elements in chemical combination

Comprehensive Environmental Response, Compensation, and Liability Act. The federal statute that empowers EPA to identify and clean up sites at which hazardous substances were released into the environment (abbreviated CERCLA)

compressed gas. A gas or mixture of gases confined within a container and having an absolute pressure exceeding 40 psi_a (276 kPa) at 70°F (21°C), *or* regardless of the pressure at 70°F (21°C), an absolute pressure exceeding 104 psi_a (717 kPa) at 130°F (54°C), *or* any liquid flammable material having a vapor pressure exceeding 40 psi_a (276 kPa) at 100°F (37.8°C)

concentration. The relative amount of a minority constituent of a mixture or solution to the majority constituent, expressed in such units as grams per liter, pounds per cubic foot, parts per million, and percentage by mass and volume

condensation. (1) The conversion of a gas or vapor into a liquid; (2) Polymerization associated with the production of a polymer and a small molecule such as water or ammonia

condensation polymer. A polymer produced by a chemical reaction along with a small molecule such as water or ammonia

conduction. The mechanism by which heat is transferred to the parts of a stationary material or from one material to another with which it is in contact

contaminant. A foreign substance that causes a degradation in the quality of the containing medium

convection. The mechanism by which heat is transferred by the movement of the heated material itself from spot to spot

conversion factor. A fraction equal to 1 in which the magnitude of one unit is related in the numerator to the magnitude of another unit of the same type in the denominator

copolymer. A polymer made of repetitive unlike subunits

corrosion. Deterioration or destruction brought about by chemical action

corrosive material. Any liquid or solid that causes visible destruction or irreversible alterations in human skin tissue at the site of contact, or a liquid that has a severe corrosion rate on steel or aluminum, each measured in accordance with prescribed DOT testing procedures

corrosivity. The RCRA characteristic of a liquid waste possessing either of the following properties: It is an aqueous solution and has a pH less than or equal to 2 or equal to or greater than 12.5; *or* it is a liquid and corrodes steel at a rate greater than 0.250 in. (6.35 mm) per year at a test temperature of 130°F (54°C) using a specified test method (see 40 CFR §261.22)

cotton. A naturally occurring fiber of vegetable origin

Coulomb's law. The observation that unlike charges repel and like charges attract

covalent bond. A shared pair of electrons between two atoms

covalent compound. A substance whose constituent atoms are mainly bonded together by covalent bonds

cracking. An industrial process, typically conducted at a petroleum refinery, during which heat is used to break the molecules of hydrocarbons into smaller ones

cresol. Any of the methylated derivatives of phenol

criteria air pollutants. The six most commonly encountered air pollutants in the United States for which EPA has developed emission standards based on specific health criteria; carbon monoxide, nitrogen dioxide, ozone, lead, sulfur dioxide, and particulate matter

critical mass. The minimum mass of fissionable material that must be present before low-energy, neutron-induced fission becomes self-sustaining

critical pressure. The minimum pressure that causes a gas to liquefy at its critical temperature

critical temperature. The temperature above which the vapor of a liquid cannot be condensed at any pressure

cross-linking. Chemical bonds in linear and nonlinear directions, typically noted within the macromolecules of certain polymers

cryogenic liquid (cryogen). Any substance that has been cooled to less than approximately *90 K*

cyanosis. The adverse affliction resulting from exposure to cyanides and certain other substances, exhibited physically by the presence of a bluish tinge in the fingernail beds, lips, ear lobes, conjunctiva, mucous membranes, and tongue, as well as the occurrence of drowsiness, headache, nausea, and vomiting

cycloalkane. A hydrocarbon whose molecules are composed of a cyclic ring of carbon–carbon single bonds

cyclonite. The common name of a commercially available form of the high explosive trinitrotrimethylenetriamine

cylinder. A vessel having a circular cross-section and designed to safely store and transport compressed gases and liquefied compressed gases at pressures greater than 40 psi$_a$

dangerous-when-wet material. A material that, by interacting with water, becomes spontaneously flammable or evolves a flammable or toxic gas or vapor at a rate greater than 28 in.^{3}/lb (1 L/kg) per hour

decomposition reaction. The type of chemical reaction involving the breakup of one substance into two or more elementary substances

deflagration. The chemical process by which a substance burns intensely instead of detonating

defoliant. A herbicide that causes the leaves to drop from plants prematurely

denatured alcohol. Ethanol that has been made unfit for consumption as a beverage by the addition of a toxic or noxious substance

degree of hazard. A relative ranking of the ability of a hazardous material to cause harm due to its ignitability, corrosiveness, chemical reactivity, or other characteristics and to cause injury, disease, or death to humans upon exposure

dehydration. The removal of water from a compound

dehydrogenation. A chemical process typically conducted at high temperature and pressure and during which molecular hydrogen is removed from a compound

denaturant. A substance added to ethanol to make it unpalatable

density. The property of a substance that measures its compactness; the mass of a substance divided by the volume it occupies

depolymerization. The thermal decomposition process during which a polymer breaks down into simpler substances

detonation. The chemical transformation of certain chemical substances accompanied by the essentially instantaneous passage of a wave at a supersonic rate, the production of gases and vapors, and the evolution of energy

deuterium. The isotope of hydrogen consisting of one proton and one neutron

diastolic. The bottom number of a blood pressure reading, typically less than 85 mm Hg; the pressure exerted within an individual's arteries as the heart is relaxed between beats

dibenzofuran. A compound consisting of two benzene rings bonded through carbon–carbon and carbon–oxygen–carbon bridges

dibenzo-*p*-dioxin. An organic compound consisting of two benzene rings bonded through a pair of oxygen atoms

dielectric fluid. The fluid used in certain types of electrical equipment to increase their resistance to arcing and serve as a heat-transfer medium

diene. An organic compound whose molecules possess two carbon–carbon double bonds

diesel fuel. A petroleum fraction commonly used as the fuel for operating diesel engines

dimer. A compound produced by the combination of two like units

dioxin. The common name for 2,3,7,8-tetrachlorodibenzo-*p*-dioxin (abbreviated TCDD)

dirty bomb. A clandestine explosive device charged with a radioactive or other hazardous material that is disseminated into the environment as the bomb detonates

dissociation. A chemical reaction in which one substance breaks up into two or more simpler substances

distillation. A physical process in which a substance or group of substances is separated from a mixture contained within a vessel by heating the mixture to a specified temperature or range of temperatures, thereby converting one or more constituents into a vapor that is subsequently condensed in a another vessel

dome. The circular fixture at the top of a railroad tankcar, where the valves and pressure-relief devices are located

dose. The quantity of a substance per body weight administered directly to and absorbed by an organism

dosimeter. An instrument used to detect and measure radiation exposure

double bond. A covalent bond composed of four electrons shared between two atoms

double replacement reaction. The type of chemical reaction in which two different compounds exchange their ions to form two new compounds

downwind distance. The distance downwind from the scene of an emergency scene involving the release of a toxic gas or the vapor of a toxic liquid

dry ice. Solid carbon dioxide

ductile. The physical property of metals associated with their ability to be stretched into wires

dust explosion. The combustion of dust particles suspended in the air within a confined space

dynamite. A detonating explosive containing nitroglycerin, similar organic nitrate esters, and one or more oxidizers, all of which are mixed into a stabilizing absorbent

ebonite. A form of hard rubber used primarily to make the casings for automobile storage batteries

edema. The accumulation of watery fluid in the cells, tissues, or cavities of the body

elastomer. Any of a number of natural or synthetic polymers that are capable of elongating or stretching under strain, but incapable of retaining the deformation when the strain is released

electrolyte. A substance that dissociates into ions when fused (melted) or is dissolved in solution, thereby becoming capable of conducting an electric current

electromagnetic radiation. The entire range of energy that travels as waves through space

electron. A fundamental particle that bears a charge of -1

electron capture. A spontaneous mode of radioisotopic decay in which the nucleus captures an extranuclear electron

element. A substance composed of only one kind of atom

Emergency Planning and Community Right-To-Know Act. The federal statute that empowers EPA to require a facility handling chemical substances to either submit copies of MSDSs for certain chemical substances present at the facility in amounts greater than specified threshold quantities *or* to provide a list of such substances along with their acute health hazards, chronic health hazards, fire hazards, sudden-release-of-pressure hazards, and reactivity hazards (abbreviated EPCRA)

Emergency Response Guidebook. A DOT publication that provides emergency actions for implementation at hazardous materials transportation mishaps

endocrine. Any of the ductless glands, such as the adrenals, thyroid, and pituitary, whose secretions pass directly into the circulatory system

endothermic. Referring to a process that absorbs heat from its surroundings

energy. The property of matter that enables it to do work

English units of measurement. Units based upon the yard, pound, and quart, and their fractions and multiples

environment. The water, ambient air, and land, including the surface and subsurface strata, and the interrelationship that exists among and between them and all living things

enzyme. A biological catalyst

epidemiology. The branch of medical science dealing with the nature and transmission of diseases

epidermis. The outer layer of the skin

equation. An expression used to symbolically represent a chemical or nuclear reaction

esophagus. The passage between the throat and the stomach

ester. An organic compound having molecules in which an alkyl or aryl group has replaced the hydrogen atom in the carboxylate group (—COOH)

esterification. The process of producing an ester by the reaction of an organic acid with an alcohol

ether. Any organic compound having molecules that consist of an oxygen atom bonded between two alkyl and/or aryl groups

etiologic agent. A viable microorganism or its toxin that causes or can cause disease in humans or animals

ethyl group. The group of atoms designated as CH_3CH_2—

evaporation. The process by which the surface particles of a liquid escape into the vapor state

exothermic. Referring to a process that emits heat to its surroundings

explosive article. Commercially available items that contain a chemical explosive as a component

explosive, high. A substance that detonates as the result of heat or mechanical impact

explosive, initiating. A substance used as a component of blasting caps, detonators, and primers

explosive, low. A substance that usually deflagrates rather than detonates, such as black powder

explosive material. A substance that, when subjected to sudden shock, pressure, or high temperature, almost instantaneously decomposes with the release of pressure, gas, and heat

explosive range (see **flammable range**).

factor-unit method. A procedure for changing a quantity expressed in one unit to a quantity of another unit by multiplying, dividing, and arithmetically canceling numbers and units

Fahrenheit scale. The temperature scale on which water freezes as 32°F and boils at 212°F at 1 atm

family of elements. The group of elements listed in any vertical column of the periodic table, all of whose members possess similar chemical properties

Federal Food, Drug, and Cosmetic Act. The federal statute that empowers the Food and Drug Administration to ensure that foods are safe to eat, drugs are safe and effective, and the labeling of foods and drugs is truthful and informative

Federal Hazardous Materials Transportation Law. The federal statute that empowers DOT to regulate certain activities of shippers and carriers who offer or accept hazardous materials for intrastate, interstate, or international transportation

Federal Hazardous Substances Act. The federal statute that empowers the Consumer Product Safety Commission to protect the public against unsuspecting exposure to hazardous substances contained within household products

Federal Insecticide, Fungicide, and Rodenticide Act. The federal statute that empowers EPA to regulate the manufacture, use, and disposal of pesticides for agricultural, forestry, household, and other activities (abbreviated FIFRA)

Federal Water Pollution Prevention and Control Act. The federal statute that empowers EPA to restore and maintain the chemical, physical, and biological integrity of the waters of the United States

feedstock. The raw material from which a new substance is manufactured

fiber. Any polymeric substance that can be separated into threads or thread-like structures

fire. A rapid, persistent chemical reaction, most commonly the combination of a combustible material with atmospheric oxygen, accompanied by the release of heat and light

fire point. The lowest temperature at which a flammable liquid evolves vapors at a sufficient rate to support continuous combustion

fire tetrahedron. A four-sided pyramid, each face of which denotes an essential fire component (fuel, oxygen, heat, and free radicals)

fire triangle. A triangle, each leg of which is used to designate three of the four components of a fire (fuel, oxygen, and heat)

fireworks. Pyrotechnic devices intentionally designed for entertainment

first-degree burn. A superficial injury of the skin caused by exposure to heat

fissile. The ability of some nuclei to undergo fission, producing neutrons and a relatively large amount of energy

flame. The visible manifestation of a fire that results when minute particulate matter composed principally of incompletely burned fuel is heated to incandescence

flammable. Capable of being easily ignited, of burning intensely, or dispersing fire at a rapid rate

Flammable Fabrics Act. The federal statute that empowers the Consumer Safety Commission to require ignition and rate-of-burn tests on the fabrics used in apparel and furniture

flammable gas. Any material that is a gas at 68°F (20°C) or less and 14.7 psi (101.3 kPa) of pressure *and* is ignitable at 14.7 psi (101.3 kPa) when in a mixture of 13% or less by volume with air; *or* any material whose vapor possesses a flammable range at 14.7 psi (101.3 kPa) with air of at least 12%, regardless of the lower limit

flammable liquid. A liquid having a flashpoint below 100°F (37.8°C), other than a liquid mixture having components with flashpoints of 100°F (37.8°C) or greater, the total of which makes up 99% or more of the volume

flammable range. The numerical difference between a flammable substance's lower and upper explosive limits in air

flammable solid. Any of the following types of materials: wetted explosives; thermally unstable materials that can undergo a strongly exothermic decomposition even without the participation of atmospheric oxygen; and readily combustible solids

flare. Any article containing a pyrotechnic substance designed to illuminate, identify, signal, or warn

flashback. The path traversed by a flame along a vapor trail after ignition by a spark or other ignition source

flashover. The phenomenon associated with the spread of fire from the burning area to other areas physically isolated from the initial fire source

flashpoint. The minimum temperature at which the vapor of a liquid or solid ignites when exposed to sparks, flames, or other ignition sources

flowers of sulfur. The finely divided powder of elemental sulfur

foam. A material consisting of a mixture of small bubbles of air, water, and stabilizing agents that is designed to extinguish a fire by blanketing the fuel, excluding air, or preventing the escape of volatile vapor into the atmosphere

formalin. Any solution of formaldehyde in water and/or methanol

formula weight. The sum of the atomic weights of all the atoms in a chemical formula

fractional distillation. A method of separating multiple components of a mixture based upon the different boiling points of its constituents, during which the vapors are collected at specified temperatures or within specified temperature ranges and subsequently condensed to liquids

free radical. A reactive species resembling a molecule but possessing an unpaired electron

freezing point. The temperature at which the liquid and solid states of a substance coexist at 1 atm (101.3 kPa)

freight container. A reusable container having a volume of at least 640 ft^3 (18 m^3), designed and constructed to permit it to be lifted with its contents intact and intended primarily for containment of packages during transportation

Freon. The trademark of the chlorofluorocarbons formerly used in refrigeration and air-conditioning equipment

friable asbestos. Asbestos that is crumbly, pulverized, or powdery

fuel oil. A liquid petroleum product derived in whole or part from crude oil, which is typically burned for the generation of heat or power

functional group. Any of the atoms or groups of atoms that identify a particular organic compound as an alcohol, aldehyde, ester, ether, ketone, acid, and so forth, and determine many of its characteristic chemical properties

fusible plug. A piece of low-melting metal inserted as a pressure-relief device in cylinders, cargo tanks, stationary tanks, and railroad tankcars that are used to store or transport a compressed gas or flammable liquid

galvanize. The process of coating a metal with a protective layer of elemental zinc

gamma radiation. Electromagnetic radiation emitted from some radioisotopes

gas. Matter that possesses neither a definite volume nor a definite shape

gauge pressure. The amount of pressure by which a confined gas exceeds atmospheric pressure

geometrical isomer (see **isomer, geometrical**).

glycol. An alcohol, each of whose molecules possesses two hydroxyl groups

gray. The unit of absorbed dose of radiation equivalent to 1 joule per kilogram of living tissue

grain alcohol. The common name of ethanol produced by the fermentation of a grain

gram. One one-thousandth of the mass of 1 kilogram

greenhouse effect. Global warming caused by the decrease in the rate of escape of energy from the Earth's surface into outer space compared to the rate at which the sun's radiant energy enters the atmosphere and is absorbed by Earth

greenhouse gases. The components of the atmosphere that absorb energy and thus affect the rate of energy escape from the Earth's surface into outer space

guncotton. A commercial form of nitrocellulose

half-life. The time period during which an arbitrary amount of a radioisotope is transformed into half that amount

halogenated hydrocarbon. Any hydrocarbon derivative in which one or more of its hydrogen atoms have been substituted with halogen atoms

halogenation. Any chemical reaction in which one or more halogen atoms are incorporated into a compound

halogen family. The elements comprising Group VIIA on the periodic table: fluorine, chlorine, bromine, iodine, and astatine

halon. Any commercial fire suppressant primarily composed of a substance possessing one or two carbon atoms and from four to six atoms of fluorine, chlorine, or bromine per molecule

hazard class. The category of hazard DOT assigns to a hazardous material, such as flammable gas, flammable liquid, corrosive material, or dangerous-when-wet material

hazard warning. Any words, pictures, symbols, or combination thereof appearing on a label or other appropriate form of warning that conveys the hazard of a substance

hazardous. The condition having the potential to damage a site or initiate an event that could result in injury, illness, or death

hazardous air pollutant. Any substance designated by EPA pursuant to enactment of the Clean Air Act and listed at 40 CFR §61.01

hazardous material. Any substance or material in any form which, because of is quantity, concentration, chemical, corrosive, flammable, reactive, toxic, infectious, or radioactive characteristics, either separately or in combination with any other substance or substances, constitutes a present or potential threat to human health, safety, welfare, or the environment, when improperly stored, treated, transported, disposed or used, or otherwise managed (Hazardous materials are listed by DOT at 49 CFR §172.101)

hazardous substance. (1) Any substance designated by EPA pursuant to enactment of the Federal Water Pollution Prevention and Control Act and listed at 40 CFR §116.4; (2) Any substance designated by EPA and DOT pursuant to enactment of the Comprehensive Environmental Response, Compensation, and Liability Act and listed at 40 CFR §302.4 and Appendix A, 49 CFR §172.101

hazardous waste. Any substance or material designated by EPA pursuant to the Resource Conservation and Recovery Act and listed at 40 CFR §§261.31, 261.32, 261.33(e), and 261.33(f) as well as any substance or material having a characteristic set forth at 40 CFR §§261.21, 261.22, 261.23, and 261.24

hazardous waste characteristic. Any of the following four properties exhibited by a hazardous waste: ignitability, corrosivity, reactivity, and toxicity

hazard zone. Any of the four levels of hazard assigned by DOT to poison gases and one of the two levels assigned to liquids that are poisonous by inhalation [see 49 CFR §§173.116(a)]

heat. The form of energy transferred from one body to another because of a temperature difference between them; energy arising from atomic or molecular motion

heat capacity. The amount of heat needed to raise either 1 gram of a substance 1 degree Celsius or 1 pound of the substance 1 degree Fahrenheit

heat of combustion. The heat evolved to the surroundings when a compound is burned to yield carbon dioxide and water vapor

heatstroke. The condition caused by exposure to excessive heat

hemoglobin. The component of blood that transports oxygen to the tissues of the body

hemotoxicant. A substance that decreases the function of the blood's hemoglobin and deprives the tissues of oxygen

hepatotoxicant. A substance that causes liver damage

herbicide. A substance designed to control plant life

heterocyclic compounds. Compounds whose molecules contain one or more ring atoms other than carbon

hexogen. A commercial form of the explosive cyclotrimethylenetrinitramine

high-test-hypochlorite. A commercial oxidizer capable of producing chlorine in acidic water (abbreviated HTH)

hormone. Any chemical substances produced by an endocrine gland and carried through the bloodstream to organs of the body, where it serves to regulate certain body functions

household ammonia. An aqueous solution of ammonia; ammonium hydroxide

hydration. The addition of water to a substance

hydrocarbon. Any compound whose molecules are composed solely of carbon and hydrogen atoms

hydrogenation. The addition of hydrogen to a substance

hydrotreatment. The processing of gasoline and other petroleum products with hydrogen to reduce or eliminate their nitrogenous and sulfurous content

hydrolysis. The chemical reaction between a substance and water

hydrometer. A device calibrated on its stem so the specific gravity of a liquid can be determined by reading a number at the surface of the liquid

hydroxyl group. The —OH function group, as in an alcohol

hydroxyl radical. The group of atoms represented as ·OH

hyperbaric oxygen therapy. The treatment of various afflictions including carbon monoxide poisoning, during which the victim breathes oxygen under increased pressure within a sealed steel chamber

hyperthermia. The excessive overheating of the body

ignitability. The RCRA characteristic of a waste that is any of the following: a liquid, a sample of which possesses a flashpoint equal to or less than 140°F (60°C) when determined through use of a Pensky–Martens closed cup tester; a solid capable of causing fire through friction, absorption of moisture, or spontaneous chemical changes and that burns vigorously and persistently; an oxidizer; or an ignitable compressed gas (see 40 CFR §261.21)

immediately-dangerous-to-life-and-health level. The atmospheric concentration of any substance that poses an immediate threat to life, causes an irreversible or delayed adverse health effect, or interferes with an individual's ability to escape during a 30-minute period (abbreviated IDHL)

immiscible. Incapable of mixing so as to form a single phase

incendiary agent. A substance primarily used during warfare to intentionally initiate fire

inflammation. The protective response of the body's tissues to irritation or injury, characterized by redness, heat, swelling, or pain and accompanied by the loss of function

ingestion. The taking in of a substance through the mouth and into the digestive system

inhalation. The taking in of a substance in the form of a gas, vapor, fume, mist, or dust into the respiratory system

inhibitor. A substance or mixture of substances used to arrest or retard the speed of corrosion, polymerization, or other chemical phenomenon

initial isolation distance. The crosswind distance from an emergency scene involving the release of a toxic gas or liquid

initial isolation zone. The area surrounding an emergency scene in which persons may be exposed to dangerous, life-threatening concentrations of a toxic gas or the vapor of a toxic liquid

initiator. An explosive material used to detonate the main charge

inorganic chemistry. The chemistry of those substances that do not contain carbon in their composition

insecticide. A substance designed to control insect life

insoluble. A substance's inability to dissolve in a given solvent

inverse square law of radiation. The observation that the intensity of radiation decreases as the inverse square of the distance from its source

ion. An atom (or group of atoms bound together) with a net electric charge, which is due to the loss or gain of electrons

ionic bonding. The electrostatic force of attraction between oppositely charged ions

ionic compound. Any compound composed of atoms that are mainly bonded together by ionic bonds

irradiation. The exposure of a material to gamma radiation

irritant. A substance capable of injuring the body's tissues by causing inflammation at the site of contact

isomer, geometrical. Either of two compounds that have the same chemical formula but whose molecules differ with respect to the *cis-* or *trans-* position of the atoms or groups of atoms about a carbon–carbon double bond

isomer, structural. Any of several compounds that have the same molecular formula but whose atoms are structurally arranged in different fashions

isooctane. The common name given to 2,2,4-trimethylpentane within the petroleum industry

isotactic polymer. A stereochemical polymer whose macromolecules possess side chains that are positioned on the same side of a carbon–carbon chain

isotope. Any of a group of nuclei having the same number of protons but different number of neutrons

joule. The SI unit of energy equal to that possessed by a 2-kilogram mass moving at a velocity of 1 meter per second

kelvin. The unit of temperature on the Kelvin scale

kerosene. An oily liquid obtained by the fractional distillation of crude petroleum and often used as a tractor fuel, jet fuel, and heating fuel

ketone. An organic compound each of whose molecules are composed of a carbonyl group bonded to alkyl and/or aryl groups

kindling point. The temperature at which the combustion of a solid can proceed without further addition of heat from an external source

knocking. The noises associated with the incomplete combustion of a petroleum fuel, especially in internal combustion engines

Kyoto protocol. The international policy aimed at reducing the worldwide impact of global warming through curtailment of greenhouse gas emissions into the atmosphere

label. The written, printed, or graphic hazard warnings required by DOT, FDA, OSHA, and other governmental agencies to be affixed on product containers

lacrimator. A substance that causes the eyes to involuntarily tear and close

latent health effect. An injury or disease that manifests itself only after the passage of a considerable time period following one or more initial exposures to a substance

latent heat of fusion. The amount of heat required to convert a unit mass of a solid substance into a liquid

latent heat of vaporization. The amount of heat involved during a phase change from a liquid to a gas, or vice versa

latex, natural. A white liquid obtained from some species of shrubs and trees and consisting in part of natural rubber suspended in water

law of conservation of mass and energy. The observation that the total amount of mass and energy in the universe is constant

LC$_{50}$. The concentration of a substance that, when administered to laboratory animals, kills half of them

LD$_{50}$. The dose of a substance that is lethal to 50% of the organisms tested during a specified time

length. The distance between two points

leukemia. Cancer of the blood-forming tissues and organs

Lewis structure. A means of displaying the bonding between the atoms of a molecule by using dashes to represent a shared pair of electrons

Lewis symbol. The symbol of an element together with a number of dots representing its bonding electrons

lifting power. The natural ability of a confined gas that is less dense than air to rise in the atmosphere, measured as the difference between the mass of a given volume of air and that of the same volume of the gas at the same temperature and pressure

ligroin. A fraction of crude petroleum; light naphtha

linen. The naturally occurring fiber of the flax plant

liquefied compressed gas. A gas that is at least partially liquid at 70°F (21°C) when confined in a vessel under pressure

liquefied petroleum gas. The compressed and liquefied mixture obtained as a by-product during the refining of petroleum and consisting primarily of propane, propylene, butane, isobutane, and butylenes

liquid. Matter that possesses a definite volume but lacks a definite shape

liter. The volume occupied by a cube measuring 10 cm to an edge

lower explosive limit. The concentration of a gas or vapor in air below which a flame will not propagate upon exposure to an ignition source

low-pressure tank. A tank designed to store flammable liquids at pressures above 0.5 psi$_g$ (3 kPa) but not more than 15 psi$_g$ (100 kPa)

macromolecule. The giant molecule of which polymers are composed, comprising an aggregation of hundreds or thousands of atoms and typically consisting of repeating chemical units linked together into chains and cross-linked into complex three-dimensional networks

macrophage. A large cell that engulfs and destroys invading microorganisms

magazine. A building, room, or vessel used exclusively for receiving, storing, and dispensing of explosive materials

malignant. Tending to become progressively worse or life-threatening

malleable. A physical property of metals characterized by an ability of being hammered into sheets

marine pollutant. Any substance denoted in Appendix B, 49 CFR §172.101

marking. Any descriptive name, instruction, precautionary remark, weight, or construction specification note or mark required by DOT to be applied or otherwise placed upon the outside surface of a container, tank, or other vessel used to transport a hazardous material

mass. The quantity of matter possessed by an object regardless of its location

mass number. The total number of protons and neutrons in a given nucleus; the atomic mass of a specific nucleus rounded off to the nearest whole number

Material Safety Data Sheet. A technical bulletin prepared by manufacturers and marketers to provide workers with detailed information about the properties of a commercial product that contains hazardous substances (abbreviated MSDS) [see 29 CFR §1910. 1200 (g)]

matter. Anything that possesses mass and occupies space

mechanism of a chemical reaction. The step-by-step pathway by which reactants are converted into products

melting point (see **freezing point**).

mesothelioma. A cancer of the membranes lining the chest and abdominal cavities resulting from exposure to asbestos fibers

meta- (*m-*). Characterized by the substitution of the hydrogen atoms on the first and third carbon atoms on the benzene ring

metabolism. The process in which absorbed foods are broken down to release their energy, or in which absorbed nutrients are used by the body

metalloid. Certain metals that exhibit the general physical properties of both metals and nonmetals

meter. The SI unit of length in the metric system of measurement

metal. Any element that conducts electricity well and possesses high physical strength, ductility, and malleability

methane. The simplest alkane; the hydrocarbon having the chemical formula CH_4; the primary constituent of natural gas

methemoglobin. The substance produced when nitrogen dioxide reacts with the blood's hemoglobin

methemoglobinemia. The affliction resulting from exposure to nitrogen dioxide and similar substances, caused by the presence of methemoglobin in the blood, and exhibited by cyanosis, dizziness, headache, diarrhea, and anemia

methyl group. The group of atoms designated as CH_3—

metric system of measurement. The decimal system of measurement based upon the meter, kilogram, and cubic meter

microbiocide. A chemical substance capable of destroying microorganisms

micron. One millionth of a meter

mineral. An inorganic solid substance formed within Earth

mineral acid. Any of the inorganic acids, chiefly hydrochloric acid, sulfuric acid, nitric acid, perchloric acid, phosphoric acid, and hydrofluoric acid

mineral oil transformer. An electrical transformer containing mineral oil as its dielectric fluid

miscible. Soluble in all proportions; capable of mixing without subsequent separation into distinct components

mole. An Avogadro number of particles; the amount of a substance whose mass equals its molecular weight or formula weight, as relevant

molecular formula. A chemical formula that indicates the number of atoms of each element in one molecule of a substance

molecular weight. The sum of the atomic weights of all the atoms in a molecular formula

molecule. The smallest neutral unit of some elements and compounds, composed of two or more atoms

monomer. One or more of the single substances that combine to produce a polymer

Montreal protocol. The international agreement to phase out the production, manufacture, and use of halons, chlorofluorocarbons, and other ozone-depleting substances

muriatic acid. Technical-grade hydrochloric acid

mutation. The permanent change in the quality of the genetic material within a cell

napalm. The mixture of aluminum compounds produced by reacting aluminum hydroxide with certain substances, including coconut oil

naphtha. A petroleum fraction, often further divided into light naphtha (ligroin) and heavy naphtha, and used as a feedstock for production of certain petrochemicals

naphthalene. The simplest polynuclear aromatic hydrocarbon, the molecular structure of which consists of two fused benzene rings

National Fire Protection Association. An international membership organization that promotes and improves fire protection and prevention and establishes safeguards against loss of life and property by fire (abbreviated NFPA)

natural gas. The flammable gas—primarily consisting of methane—formed in nature by the decomposition of plant life

necrosis. The pathologic death of one or more cells, or a portion of tissue or an organ, typically resulting from exposure to certain substances or to radiation

negatron. A particle (β^-) that has all the properties of an electron but is emitted from certain radioisotopes

neoprene. A synthetic elastomer composed of *cis*-1,4-polychloroprene

nerve gas. A highly volatile liquid, the vapor of which acts as a neurotoxicant when inhaled

nephrotoxicant. A substance that causes kidney damage

neurotoxicant. A substance that adversely affects the central nervous system

neutralization. A double replacement reaction between an acid and a metallic hydroxide, resulting in the production of a salt and water

neutron. A fundamental particle of which matter other than protonium is composed, bearing no charge

NFPA diamond. A diamond-shaped figure divided into four quadrants, each of which is generally color-coded for each of three hazards (blue for the health hazard, red for the fire hazard, and yellow for the chemical reactivity hazard) and in which is marked a number from 0 to 4 to designate the degree of the relevant hazard

NFPA system. The procedure of assigning one of five numbers, 0 through 4, to the relative degree of three hazards (health, flammability, and chemical reactivity) for a given hazardous substance

nitrocellulose. A low explosive produced by chemically treating cellulose with a mixture of nitric acid and sulfuric acid

noble gas family. The elements in Group VIIIA on the periodic table: helium, neon, argon, krypton, xenon, and radon

nonbonding electron. Any electron possessed by an atom that does not participate in bonding to other atoms

nonflammable. Incapable of burning under ordinary conditions

nonflammable gas. Any material or mixture that exerts in its packaging an absolute pressure of 41 psi (280 kPa) or greater at 68°F (20°C), *or* any material or mixture that does not meet the definition of a DOT division 2.1 or 2.3 material

nonmetal. Any element that does not conduct electricity well, possesses a low physical strength, and is neither ductile nor malleable

nucleus (see **atomic nucleus**).

Occupational Safety and Health Act. The federal statute that empowers the Occupational Safety and Health Administration (OSHA) to protect employees from occupational illness and injuries caused by exposure to hazardous substances in the workplace

octane number. A number used to represent the antiknock properties of a petroleum fuel, based upon comparison to the performance of a mixture of *n*-heptane and 2,2,4-trimethylpentane

octet rule. The tendency for certain nonmetallic atoms to achieve electronic stability by gaining or sharing a total of eight electrons

olefin. An alkene

oleum. Concentrated sulfuric acid containing additional dissolved sulfur trioxide; fuming sulfuric acid

olfactory. Pertaining to the sense of smell

olfactory fatigue. The temporary deadening of the sense of smell

organic chemistry. The chemistry of substances that contain carbon in their chemical composition

organic hydroperoxide. A derivative of hydrogen peroxide in which one hydrogen atom in the H_2O_2 molecule has been substituted with an alkyl or aryl group; an organic compound having the general chemical formula H—O—O—R, where R is an alkyl or aryl group

organic peroxide. A derivative of hydrogen peroxide in which both hydrogen atoms in the H_2O_2 molecule have been substituted with alkyl or aryl groups; an organic compound having the general chemical formula R—O—O—R′, where R and R′ are alkyl or aryl groups

ortho (*o*-). Characterized by the substitution of the hydrogen atoms on adjacent (the first and second) carbon atoms on the benzene ring

other regulated material. The DOT classification of a specific hazardous material that, while not conforming to the definition of any of the nine hazard classes, is nonetheless considered to possess a dangerous property that could pose a risk to transportation personnel and equipment during its transit (abbreviated ORM)

outage. The percentage by volume by which a container of liquid falls short of being filled to the brim

oxidation. (1) A chemical reaction during which a substance unites with oxygen or another oxidizing agent; (2) A chemical process in which the oxidation number of a species increases, sometimes accompanied by the loss in a number of its electrons

oxidation number. A number assigned to an atom or ion to reflect its capacity for combining with other atoms or ions

oxidation–reduction reaction. A chemical reaction between one or more oxidizing and reducing agents; a redox reaction

oxidizer. The substance reduced in an oxidation–reduction reaction

oxidizing agent. (1) An oxygen-rich substance capable of readily yielding some of its oxygen; (2) The substance reduced during an oxidation–reduction reaction

oxygenate. An additive that promotes the complete combustion of vehicular fuels, thereby generating lesser amounts of carbon monoxide and other pollutants

oxyhemoglobin. The compound that forms when oxygen combines with the blood's hemoglobin

ozone (see also **stratospheric ozone**). The allotrope of oxygen whose molecules are represented as O_3, that is, three oxygen atoms per molecule

ozone layer. The region of the atmosphere rich in ozone, approximately 10 to 19 miles (16 to 30 km) from Earth's surface

package. Any packaging plus its contents

packaging. Any receptacle (fiber drum, glass bottle, steel drum, steel cylinder, and so forth) and any other component or material necessary for the receptacle to perform its containment function in conformance with DOT's packing requirements

para- (*p*-). Characterized by the substitution of the hydrogen atoms on the first and fourth carbon atoms on the benzene ring

particulate. A tiny liquid or solid particle typically suspended in the air or atmospheric emissions, comprising the nature of dust, fumes, mist, smoke, and smog

pascal. The SI unit of pressure, equal to a force of 1 newton applied to an area of 1 square meter

passivity. The property of metals such as aluminum and chromium that describes their loss in normal chemical reactivity after treatment with strong oxidizing agents

PCB transformer. Any transformer whose dielectric fluid contains polychlorinated biphenyls

period of elements. A horizontal row on the periodic table

periodic law. The observation that the chemical properties of the elements are periodic functions of their atomic numbers

periodic table. A compilation of the known elements into periods and families by increasing atomic number, so that elements with similar chemical properties are in the same column (or family)

permissible exposure limit. The time-weighted average threshold limit value of substances listed by OSHA at 29 CFR §1910.1000 to which workers can be exposed continuously during an 8-hour work shift of a 40-hour work week without suffering ill effects (abbreviated PEL)

pesticide. Any substance or mixture of substances intended for preventing, destroying, repelling, or mitigating any pest, or any substance or mixture of substances intended for use as a plant regulator, defoliant, or drying agent

petrochemical. Any substance obtained directly or indirectly from natural gas, crude petroleum, or a petroleum fraction

petroleum, crude. The liquid mixture primarily consisting of hydrocarbons formed in nature by the decomposition of plant life

petroleum distillate. A fraction of crude petroleum obtained by the vaporization of its components within a specific temperature range followed by their condensation

petroleum refinery. A facility primarily engaged in producing gasoline, kerosene, fuel oils, lubricants, and other products through fractionation, alkylation, redistillation of unfinished derivatives, cracking, blending, and other processes

pH. A numerical scale from 0 to 14 used to quantify the acidity of alkalinity of a solution with neutrality indicated as 7

phenol. An organic compound, each of whose molecules possesses a hydroxyl group (—OH) bonded directly to the benzene ring

phase diagram. A plot of the temperature and pressure conditions at which a sole substance exists as a gas, liquid, and solid

phenolic resins. Any of a group of polymeric substances obtained from the condensation of phenol and formaldehyde or chemically similar substances

phenyl group. The group of atoms designated as C_6H_5—

phlegmatizer. A substance that is capable of reducing the ability of a chemical explosive to detonate

photodecomposition. The breakdown of a substance by light

photosynthesis. The chemical process by which plants use carbon dioxide and water to construct cellulosic materials with the aid of sunlight and chlorophyll

photon. A massless packet of electromagnetic energy

physical change. Any change that does not result in an alteration of the chemical identity of a substance, such as vaporization or sublimation

physical property. Any type of behavior that a substance exhibits when it undergoes a physical change, such as melting and boiling

pickling. The combination of chemical reactions associated with removing surface impurities from metals by dipping them into an acid bath

placard. The written, printed, or graphic matter DOT requires to be posted on the bulk packaging, freight container, transport vehicle, unit containment device, or railcar used to transport hazardous materials

plasticizer. A substance, such as a phthalate ester, that, when added in a prescribed amount to a polymeric formulation, facilitates processing and provides flexibility and toughness to the end product

plastic. Capable of being molded and shaped; any polymeric substance that has been molded and shaped

plastic explosive. The combination of a chemical explosive and a dough-like binding material

polyatomic ion. An ion having more than one constituent atom

polychlorinated biphenyl. Any of the chlorinated derivatives of biphenyl, the manufacture, sale, and distribution of which has been banned in the United States; the major constituent of certain dielectric fluids formerly contained within capacitors, transformers, and other electrical equipment (abbreviated PCB)

polymer. A high-molecular-weight substance produced by the linkage and cross-linkage of its multiple subunits (monomers)

polyester. Any polymer consisting of recurring ester groups

polymerization. The chemical reaction in which monomer molecules are linked and cross-linked into macromolecules

polynuclear aromatic hydrocarbon. An aromatic hydrocarbon whose molecules possess two or more fused benzene and other rings (abbreviated PAH)

polyp. A protruding growth from a membrane

portable tank. (1) A closed container that is not fixed in position and possesses a liquid capacity of 60 gallons (227 L) or more (OSHA) (see 29 CFR §1910.106); (2) Bulk packaging designed primarily to be loaded onto, or on, or temporarily attached to a transport vehicle or ship and equipped with skids, mountings, or accessories to facilitate handling of the tank by mechanical means (DOT) (see 49 CFR §171.8)

positron. A particle (β^+) emitted from some radioisotopes that is positively charged but otherwise possesses the characteristics of an electron

preferred route (preferred highway). A highway designated for the transportation of highway-route-controlled quantities of explosive or radioactive materials

pressure. A force applied to a unit area

pressure-relief valve. A device typically located on cylinders, cargo tanks, stationary tanks, and railroad tankcars used to store or transport a compressed gas or flammable liquid that opens and closes at preestablished pressures

pressure tank. A tank designed to store flammable liquids at pressures above 15 psi_g (100 kPa)

product. A substance produced as a result of a chemical change

propane. The alkane having the chemical formula C_3H_8

proper shipping name. The name assigned to a hazardous material as provided in Roman print in the Hazardous Materials Table at 49 CFR §172.101

protective-action distance. The distance downwind from an emergency scene involving the release of a toxic gas or the vapor of a toxic liquid, at which persons may experience incapacitation or the incurrence of serious adverse, irreversible health effects

protective-action zone. The area downwind from an emergency scene involving the release of a toxic gas or the vapor of a toxic liquid, in which persons may become incapacitated and unable to take protective action and/or incur serious adverse, irreversible health effects

protein foam. A film-forming foam prepared from natural protein materials and used to extinguish fires

proton. A fundamental particle of which all atoms are composed, bearing a charge of +1

psi. The unit of pressure expressed as pounds per square inch (psi_a means pounds per square inch absolute; psi_g means pounds per square inch gauge)

pulmonary. Pertaining to the lungs

pulmonary agent. A chemical warfare agent that damages the respiratory tract and cause pulmonary edema

pulmonary edema. The excessive accumulation of fluid within the lungs

pyrolysis. The decomposition of a substance by heat

pyrophoric material. A substance that ignites spontaneously in air at a temperature of 130°F (54.4°C) or below (OSHA)

radiation. (1) The transmission of energy in the form of electromagnetic waves from certain forms of matter; (2) The mechanism by which heat is transferred between two materials not in contact

radiant energy. Any energy, including heat, light, and X rays, that is transmitted by radiation as electromagnetic waves

radiation, ionizing. Highly energetic, penetrating electromagnetic radiation emitted by radioisotopes (alpha radiation, beta radiation and gamma radiation)

radiation, nonionizing. Low-energy electromagnetic radiation such as microwaves and radio waves

radioactivity. The nuclear property associated either with the spontaneous emission of alpha, beta, and/or gamma radiation, or the capture of an extranuclear electron

radioisotope. Any atomic nucleus that undergoes a spontaneously change by emitting a particle or by absorbing an extranuclear electron

rate of reaction. The speed at which a chemical transformation occurs; the amount of a product formed, or reactant consumed, per unit of time

reactivity. The RCRA characteristic of a waste that exhibits any of the following: readily undergoes a violent change without detonating; reacts violently with water; forms potentially explosive mixtures with water; generates toxic gases, vapors, or fumes in a quantity sufficient to present a danger to human health or the environment; is capable of detonation or an explosive reaction if subjected to a strong initiating source or if heated under confinement; is readily capable of detonation or explosive decomposition or reaction at standard temperature and pressure; or is a "forbidden explosive" as defined by DOT (see 40 CFR §261.23)

redox reaction (see **oxidation–reduction reaction**).

reducing agent. (1) A substance that takes oxygen from another substance; (2) The substance oxidized during an oxidation–reduction reaction

reduction. (1) A chemical reaction in which oxygen or another oxidizing agent is removed from a compound; (2) A chemical reaction in which the oxidation number of a species decreases, sometimes accompanied by the gain of electrons

reformulated gasoline. A mixture of gasoline and an oxygen-boosting additive

release. Any spilling, leaking, pumping, pouring, emitting, emptying, discharging, injecting, leaching, dumping, or disposing of a hazardous material or hazardous substance into the environment

reportable quantity. The quantity of a "hazardous substance" within the meaning of DOT and CERCLA, the release of which triggers mandatory notification to the National Response Center (see 40 CFR §302.4 and 49 CFR §172.101) (abbreviated RQ)

residential radon. The radon that accumulates in a dwelling from the natural decay of uranium and thorium radioisotopes

respiratory tract. The structures and organs involved in breathing, including the nose, larynx, trachea, bronchi, bronchiole, and lungs

Resource Conservation and Recovery Act. The federal statute that empowers EPA to regulate the treatment, storage, and disposal of hazardous wastes (abbreviated RCRA)

rodenticide. A substance designed to kill rodents or to prevent them from damaging food, crops, or other materials

rubber. Any polymer that is simultaneously elastic, airtight, water-resistant, and long-wearing

rubber, natural. The polymer produced from the latex that exudes from the South American rubber tree

rubber, synthetic. Any polymer produced from substances other than, or in addition to, the latex from the South American rubber tree and having similar properties to those of natural rubber

rubber, styrene–butadiene. The synthetic rubber produced by the polymerization of styrene and butadiene

safety can. An approved closed container of not more than 5 gallons (19 liters) capacity, having a flash-arresting screen, spring-closing lid and spout cover, and so designed that it will safely relieve internal pressure when subjected to fire exposure

salt. A compound in which the hydrogen ion from an acid has been substituted with a metallic ion

Superfund Amendments and Reauthorization Act (SARA). The federal statute that requires EPA, when selecting a remedial action, to consider the standards and requirements found in other environmental laws and regulations and to include citizen participation in the decision-making process

saturated. Any organic compound whose molecules contain only carbon–carbon single bonds

secondary containment. A safeguarding method (such as diking around a primary containment vessel) used to prevent the unplanned release of a hazardous material into the environment

second-degree burn. An injury of the skin to a limited depth caused by exposure to heat; a partial-thickness burn

self-heating material. A material likely to self-heat in contact with the air, even when an energy source is absent

service pressure. The pressure designated by DOT in units of psi_g at 70°F (21°C) for each specific type of authorized compressed gas cylinder

shipping paper. The shipping order, bill of lading, air bill, manifest, or similar shipping document required by DOT at 49 CFR §§172.202, 172.203, and 172.204

short-term exposure limit. The concentration to which workers can be exposed continuously for a short period of time without suffering irritation, chronic, or irreversible tissue damage, or narcosis of sufficient degree to increase the likelihood of accidental injury, impair self-rescue, or materially reduce work efficiency, and provided that the daily threshold limit value, time-weighted average is not exceeded (abbreviated STEL)

short-term health effect. An injury or disease of relatively short duration caused by exposure to a substance and from which recovery occurs rapidly

SI. The abbreviation for the International System of Units (in French, le Système International d'Unités), a modernized metric system

sievert. The SI unit of dose-equivalent, equal to 100 rem

single replacement reaction. The type of chemical reaction in which one element replaces another within a compound

silk. A naturally occurring protein produced by the action of silkworms

SI system of units. The scientific standard of measurement that employs a set of units describing length, mass, time and other attributes of matter

smelter. A facility that separates or refines a metal from its ore by means of a chemical change

smoke. The visible plume associated with combustion and consisting of carbon particulates, ash, water droplets, and other matter

smokeless powder. A form of nitrocellulose

soda ash. The common name of sodium carbonate

solid. Matter that possesses a definite volume and a definite shape

soluble. Capable of being dissolved, that is, able to pass into a particular solvent such as water or oil to form a solution

solution. A homogeneous mixture of two or more substances

solvent. A substance capable of dissolving another substance to form a solution; the component of a solution that is present in the greater amount

soot. The agglomeration of carbon particulates generated during the incomplete combustion of carbonaceous materials

specific gravity. The mass of a given volume of matter compared to the mass of an equal volume of water

specific heat. The ratio of the heat capacity of a substance to the heat capacity of water at the same temperature

sphygmomanometer. A device used to measure blood pressure

spirits of nitroglycerin. A medication containing a nontoxic dose of nitroglycerin that is often prescribed to individuals afflicted with heart and circulatory diseases

spontaneous combustion (see **combustion, spontaneous**).

spontaneously combustible material. A pyrophoric material or a self-heating material

stowage. The placement of materials onboard a vessel

stratospheric ozone (see also **ozone**). The naturally occurring layer of ozone in the upper atmosphere that shields Earth's surface from the sun's harmful ultraviolet rays

strong. A designation for acids and bases that are completely ionized

structural isomer (see **isomer, structural**).

sublimation. A physical change in which a substance passes from the solid directly to the gaseous state of matter, without first liquefying

substance. A homogeneous material having a constant, fixed chemical composition; any element or compound

Superfund Act (see **Comprehensive Environmental Response, Compensation, and Liability Act**).

syndiotactic polymers. A stereochemical type of polymer whose macromolecules possess alternately positioned side chains along a carbon–carbon chain

synergism. The aspect of two substances that interact to produce an effect that is greater than the sum of the substances's individual effects of the same type

synthesis gas or **syngas** (see **water gas**).

synthesis reaction (see **combination reaction**).

systemic toxicity. The adverse affect on the body as a whole as opposed to a localized effect

systolic. The top number of a blood pressure reading, typically less than 130 mm Hg; the pressure exerted within an individual's arteries as the heart is beating

tank farm. Property upon which multiple stationary storage tanks have been situated

tear gas (see **lacrimator**).

temperature. The condition of a body that determines the transfer of heat to or from other bodies by comparison to established values for water, which is used as the standard

tetryl. The common name of the chemical explosive 2,4,6-trinitrophenylmethylnitramine

textile. Any fabric that has been produced by weaving fibers

thermite reaction. The chemical reaction during which elemental iron is produced from its oxide by reduction with elemental aluminum

thermogenesis. The physiological activity of microorganisms that results in the slow liberation of heat within a material, and when absorbed by the material, could cause its spontaneous combustion

thermoplastic polymers. Those plastics that soften when heated, but return to their original condition upon cooling to ambient temperature

thermosetting polymers. Those plastics that cannot be remolded once they have solidified

third-degree burn. An injury of the subsurface fat, nerves, and muscular structure, caused by exposure to heat; full-thickness burn

threshold limit value. Any of the guideline standards established by The American Conference of Governmental Industrial Hygienists (ACGIH) for airborne concentrations to which an average worker can be exposed day after day without experiencing adverse health effects (abbreviated TLV)

time-weighted average. The average exposure concentration of a substance measured over a period of time, distinct from a concentration obtained during an instantaneous exposure

tonne. A unit of mass equivalent to 1000 kilograms

torpex. The explosive material consisting of a mixture of cyclonite, trinitrotoluene, and aluminum fines

toxic. The nature of a substance or material associated with its ability to cause disease, injury, or death in exposed humans and animals at relatively low concentrations

toxicant. A substance capable of inflicting harm systemically or at specific organ sites

toxicity. Poisonous; the RCRA characteristic of a waste, as determined by implementing a specified test procedure, in which a representative sample of the waste is found to contain one or more of certain constituents identified by EPA at concentrations equal to or greater than prescribed levels (see 40 CFR §261.24)

toxicology. The study of the effects of toxic or poisonous substances on living organisms

toxic pollutant. Any substance designated by EPA under the provisions of the Federal Water Pollution Prevention and Control Act and listed at 40 CFR §401.15 which, after discharge and upon exposure, ingestion, inhalation, or assimilation into any organism will, on the basis of information, cause death, disease, behavioral abnormalities, cancer, genetic mutations, physiological malfunctions (including malfunctions in reproduction) or physical deformations, in organisms or their offspring

toxic substance. Any substance or mixture designated by EPA under the provisions of the Toxic Substances Control Act

Toxic Substances Control Act. The federal statute that empowers EPA to regulate all aspects of the manufacture of toxic substances other than pesticides and drugs, including the obtaining of production and test data from industry on those toxic substances whose manufacture, processing, distribution, use, or disposal could present an unreasonable risk of injury or damage to the environment (abbreviated TSCA)

toxin. A poisonous substance produced by a living organism

transformer. A device designed to transfer electrical energy between electrical circuits of different voltage

transmutation. The transformation of a radioisotope to another isotope as the result of a nuclear reaction

transport vehicle. A motor vehicle or railcar used for the transportation of cargo by any mode

triene. An organic compound, each of whose molecules possesses three carbon–carbon double bonds

triple bond. A covalent bond composed of six shared electrons between two atoms

tritium. The radioisotope of hydrogen consisting of one proton and two neutrons

tumor. A mass of tissue that persists and grows independently of its surrounding structures and has no known physiological function

ullage (see **outage**).

unsaturated. An organic compound whose molecules contain one or more carbon–carbon double bonds or carbon–carbon triple bonds

unstable. The characteristic of some substances or materials to vigorously polymerize, decompose, condense, or become self-reactive when shocked or exposed to certain pressure or temperature conditions

upper explosive limit. The concentration of a gas or vapor in the air above which it cannot burn when exposed to an ignition source

valence electron. Any of an atom's electrons that participates in chemical bonding to other atoms

valve. A device that controls the flow or pressure of a gas within a device

vapor. The gaseous form of a substance that exists either as a solid or liquid at normal ambient conditions

vapor density. The mass of a vapor or gas compared to the mass of an equal volume of another gas or vapor, generally air

vaporization. The process during which the molecules of a liquid escape as a gas or vapor

vapor pressure. The pressure exerted within a confinement vessel by the vapor of a substance in equilibrium with its liquid; a measure of a substance's propensity to evaporate

vasodilator. A substance, such as nitroglycerin, exposure to which is capable of causing the blood vessels to widen

vesicant. A substance used in chemical warfare to blister the skin and body tissues of the enemy

vessel. (1) Any tank, cylinder or similar device used for containment; (2) Any watercraft or other artificial contrivance used, or capable of being used, as a means of transportation on water

vinyl compound. Any substance having the group of atoms $-CH=CH_2$

vinyl halide. A halogenated derivative of an alkene in which one or more of the hydrogen atoms bonded to one or both of the carbon atoms comprising the carbon–carbon double bond have been substituted with halogen atoms

vinyl polymer. Any polymer produced from one or more vinyl compounds

viscosity. The quality of resistance to flow

volatile. The quality of a solid or liquid when it passes into the vapor state at a given temperature

volatile organic compounds. Certain substances having a boiling point less than approximately 392°F (200°C), including the constituents of gasoline vapor (abbreviated VOCs)

volume. The capacity of matter confined within a tank or container

vulcanization. The heating of natural or synthetic rubbers with elemental sulfur or sulfur-bearing compounds to produce disulfide cross-linking within their macromolecules

water gas. The mixture of carbon monoxide and hydrogen produced by blowing steam through a bed of red-hot coke

water-reactive substance. Any substance that, by its chemical reaction with water, is likely to become spontaneously flammable or to give off flammable or toxic gases in dangerous quantities

weak. A designation for acids and bases that are slightly ionized

weight. The force of gravity acting on the mass of a particular object

wood alcohol. The common name for methanol produced by heating wood in the absence of air

wool. The naturally occurring fiber obtained from the coats of sheep, llamas, and several other animals

Ziegler–Natta catalyst. A mixture of titanium tetrachloride and an aluminum alkyl compound, used to initiate addition polymerization

Index

Hazardous Materials Index

TABLE OF ELEMENTS AND ATOMIC WEIGHTS

	Symbol	Atomic Number	Atomic Weight		Symbol	Atomic Number	Atomic Weight
Actinium	Ac	89	227*	Meitnerium	Mt	109	—
Aluminum	Al	13	26.9815	Mendelevium	Md	101	256*
Americium	Am	95	243*	Mercury	Hg	80	200.59
Antimony	Sb	51	121.75	Molybdenum	Mo	42	95.94
Argon	Ar	18	39.948	Neodymium	Nd	60	144.24
Arsenic	As	33	74.9216	Neon	Ne	10	20.183
Astatine	At	85	210*	Neptunium	Np	93	237*
Barium	Ba	56	137.34	Nickel	Ni	28	58.71
Berkelium	Bk	97	245*	Niobium	Nb	41	92.906
Beryllium	Be	4	9.0122	Nitrogen	N	7	14.0067
Bismuth	Bi	83	208.980	Nobelium	No	102	253*
Bohrium	Bh	107	262*	Osmium	Os	76	190.2
Boron	B	5	10.811	Oxygen	O	8	15.9999
Bromine	Br	35	79.909	Palladium	Pd	46	105.4
Cadmium	Cd	48	112.40	Phosphorus	P	15	30.9738
Calcium	Ca	20	40.48	Platinum	Pt	78	195.09
Californium	Cf	98	248*	Plutonium	Pu	94	242*
Carbon	C	6	12.01115	Polonium	Po	84	210*
Cerium	Ce	58	140.12	Potassium	K	19	39.102
Cesium	Cs	55	132.905	Praseodymium	Pr	59	140.907
Chlorine	Cl	17	35.453	Promethium	Pm	61	145*
Chromium	Cr	24	51.996	Protactinium	Pa	91	231*
Cobalt	Co	27	58.9332	Radium	Ra	88	226*
Copper	Cu	29	63.54	Radon	Rn	86	222*
Curium	Cm	96	245*	Rhenium	Re	75	186.2
Darmatadtium	Ds	110	—	Rhodium	Rh	45	102.905
Dubnium	Db	105	262.11	Rubidium	Rb	37	85.47
Dysprosium	Dy	66	162.50	Ruthenium	Ru	44	101.07
Einsteinium	Es	99	247*	Rutherfordium	Rf	104	261*
Erbium	Er	68	167.26	Samarium	Sm	62	150.35
Europium	Eu	63	151.96	Scandium	Sc	21	44.956
Fermium	Fm	100	254*	Seaborgium	Sg	106	263*
Fluorine	F	9	18.9984	Selenium	Se	34	78.96
Francium	Fr	87	223*	Silicon	Si	14	28.086
Gadolinium	Gd	64	157.25	Silver	Ag	47	107.870
Gallium	Ga	31	69.72	Sodium	Na	11	22.9898
Germanium	Ge	32	72.59	Strontium	Sr	38	87.62
Gold	Au	79	196.967	Sulfur	S	16	32.064
Hafnium	Hf	72	178.49	Tantalum	Ta	73	180.948
Hassium	Hs	108	—	Technetium	Tc	43	99*
Helium	He	2	4.0026	Tellurium	Te	52	127.60
Holmium	Ho	67	164.930	Terbium	Tb	65	158.924
Hydrogen	H	1	1.00797	Thallium	Tl	81	204.37
Indium	In	49	114.82	Thorium	Th	90	232.038
Iodine	I	53	126.9044	Thulium	Tm	69	168.934
Iridium	Ir	77	192.2	Tin	Sn	50	118.69
Iron	Fe	26	55.847	Titanium	Ti	22	47.90
Krypton	Kr	36	83.80	Tungsten	W	74	183.85
Lanthanum	La	57	138.91	Uranium	U	92	238.03
Lawrencium	Lw	103	259*	Vanadium	V	23	50.942
Lead	Pb	82	207.19	Xenon	Xe	54	131.30
Lithium	Li	3	6.939	Ytterbium	Yb	70	173.04
Lutetium	Lu	71	174.97	Yttrium	Y	39	88.905
Magnesium	Mg	12	24.312	Zinc	Zn	30	65.37
Manganese	Mn	25	54.9381	Zirconium	Zr	40	91.22

*Mass number of isotope of longest known half-life.